Progress
in
Liquid Physics

Progress in Liquid Physics

Edited by

CLIVE A. CROXTON

Department of Mathematics,
University of Newcastle, New South Wales,
Australia.

A Wiley–Interscience Publication

JOHN WILEY & SONS

Chichester · New York · Brisbane · Toronto

Library of Congress Cataloging in Publication Data:

Main entry under title:

Progress in liquid physics.

'A Wiley–Interscience publication.'
1. Liquids – Addresses, essays, lectures.
I. Croxton, Clive A.
QC145.26.P76 530.4′2 76-40166

ISBN 0 471 99445 6

Typeset in Great Britain by Preface Ltd., Salisbury and printed by J. W. Arrowsmith Ltd., Bristol

Contents

Contributors

P. Barnes	*Department of Crystallography, Birkbeck College, University of London, England.*
A. Ben-Naim	*Department of Physical Chemistry, The Hebrew University, Jerusalem, Israel.*
C. E. Campbell	*School of Physics and Astronomy, University of Minnesota, U.S.A.*
S. Chandrasekhar	*Raman Research Institute, Bangalore, India.*
G. H. A. Cole	*Physics Department, University of Hull, England.*
C. A. Croxton	*Department of Mathematics, University of Newcastle, New South Wales, Australia.*
K. F. Freed	*The James Franck Institute and the Department of Chemistry, University of Chicago, U.S.A.*
T. Gaskell	*Department of Physics, The University, Sheffield, England.*
J. D. Gunton	*Department of Physics, Temple University, Philadelphia, U.S.A.*
R. A. Howe	*Department of Physics, University of Leicester, England.*
K. Kawasaki	*Department of Physics, Faculty of Science, Kyushu University, Fukuoka, Japan*
N. V. Madhusudana	*Raman Research Institute, Bangalore, India.*
R. D. Mountain	*National Bureau of Standards, Washington, D.C., U.S.A.*
H. J. Raveché	*National Bureau of Standards, Washington, D.C., U.S.A.*
J. V. Sengers	*Institute for Molecular Physics, University of Maryland, U.S.A. and National Bureau of Standards, Washington, D.C., U.S.A.*
J. M. H. Levelt Sengers	*National Bureau of Standards, Washington, D.C., U.S.A.*

E. R. Smith *Department of Mathematics, University of Melbourne, Parkville Victoria, Australia.*

H. N. V. Temperley *Department of Applied Mathematics, University College of Wales, Swansea, Wales.*

Preface

The most recent major reviews have outlined the essentially sound statistical mechanical foundation of the physics of simple liquids as it stood in the mid-sixties. Since then, significant advances have been made into more difficult and controversial fields, in particular quantal and critical systems, irreversibility, transport phenomena, phase transitions, molecular and polymeric systems, liquid crystals and the role of triplet effects, to mention some of the topics covered in this collection.

Given the recent background to what is a rapidly growing field, it was felt that we should not only attempt to consolidate the existing corpus of understanding, but also focus attention on some of the major outstanding concepts and difficulties which remain unresolved, at the same time promoting a *lateral* development of the subject away from simple fluid systems toward new applications in adjacent fields.

To expose the liquids community to current developments and difficulties as fully as possible demands something other than an exclusively retrospective review – rather, a shift of emphasis towards the unresolved aspects. Some controversy is inevitable, and indeed healthy, as a stimulus to both theoreticians and experimentalists: nevertheless, it is in this sense that the present volume is intended as a contribution to the subject's progressive and continued development.

CLIVE A. CROXTON

University of Newcastle
New South Wales
January 1977

Chapter 1

Kinetic Theory of Liquids

G. H. A. COLE

Physics Department,
University of Hull, North Humberside, England

1.1 Introduction

The behaviour of a macroscopic fluid in response to forces of various kinds is described using the statements of the conservation of mass, momentum and energy. Dissipative effects are accounted for by introducing transport coefficients such as the coefficient of dynamical (or shear) viscosity, the coefficient of thermal conductivity, the coefficient of electrical conduction, and so on. For a continuum the description is in terms of the equation of density continuity, the Navier–Stokes equation for the flow of momentum, and the equation for the total energy of the fluid. This last equation can be expressed in several alternative forms including that involving the temperature. Quantities such as pressure and temperature are more properly defined for conditions of thermal equilibrium but a flowing fluid containing a gradient of temperature is obviously not in that uniform state.[1] Application to non-equilibrium conditions is made by assuming the immediate neighbourhood of every point of the fluid to be in thermodynamic equilibrium although the equilibrium conditions for one point will not be the same as those at another.[2] This assumption of local thermodynamic equilibrium is an important aspect of the assumptions underlying the theory. Any region of discontinuity within the fluid is treated differently in that the statements of conservation are applied separately on each side of the shock discontinuity: the width of the shock is related directly to the magnitudes of the transport coefficients associated with the continuum fluid. These various ideas are readily extended to quantum conditions although the formal mathematical development is different because of the introduction then of probability as a matter of principle.

This description of macroscopic conditions is entirely self-contained and does not require reference to the molecular composition of the fluid. Indeed, provided the various entities involved in the theory (such as density, pressure, dissipative coefficients and so on) are accepted as the basic concepts of the description it is irrelevant to the theory whether the fluid has a molecular structure or not.

The macroscopic properties of the fluid are, however, some form of average of the collective effects of the molecular constituents, and the recognition of this allows the structure of the fluid to be explored in a wider context. The theory for the calculation of the collective properties of a large number of molecules in a specified continuous interaction forms statistical mechanics. In its application to systems in thermal equilibrium, the theory has been developed into a very full and beautiful form by many contributors and can account for quantum as well as classical systems. Again the conservation statements of mass, momentum and energy are invoked in a general form and the theory is applicable, in principle at least, to any physical system. The theory for systems not in thermal equilibrium is far less complete, although theories of restricted validity (such as of the Brownian motion, or of the heat conduction in crystals) have been available for many years.

The theory of microscopic matter is related to that for macroscopic matter explicitly through the respective expressions for the fluxes of mass, momentum and energy. For equilibrium these involve the thermodynamic functions while for non-equilibrium they involve the transport coefficients. As a result expressions for these various functions are obtained in terms of the properties of the molecules, and have proved of the greatest importance in exploring the molecular structure of matter.[3,4]

The generality of these arguments is certainly an important achievement from a theoretical point of view but has a crucial disadvantage for the practical use of the theory. The mathematical analysis is too complicated to solve in a closed form for a real system of molecules, and the situation is not improved if appeal is made to numerical analysis even using the most advanced computational aids. To explore real systems it is necessary to introduce a simplification, and here there is a choice. One approach is to appeal to sufficient mathematical approximations to allow the exact theory to be reduced to a form amenable to calculations: the other is to replace the exact equations by mathematically simpler ones chosen to exemplify physical principles thought characteristic of the particular physical system of interest and yet which allow full mathematical solutions to be found. The two approaches are linked in that the second approach will allow physical arguments to be associated with the mathematical manipulations associated with the first one. This reduction of the complexity of the full theory is the aim of the kinetic theory; the more simple equations of strictly limited validity are the kinetic equations. In this article we are concerned with the recent efforts to construct a kinetic theory of liquids.

1.2 The nature of the simplification

As the first step in constructing a kinetic theory of liquids it is necessary to isolate those physical features which make a liquid different from solids and gases. It has been found useful to centre these arguments on the distribution of the constituent molecules in space. For a crystal each molecule vibrates about an equilibrium position located on a lattice which in principle covers all space and in practice can cover a very large volume without imperfection. The potential energy of interaction between the molecules has a greater effect than the kinetic energy of molecular vibration and diffusion, and this is the essential simplification which allows the properties of crystals to be calculated. The collective motion of the lattice can be expressed in terms of phonons as an expression of the simultaneous action of a large number of coupled particles. The long range order of the closely packed periodic lattice is one limiting case of the possible arrangement of molecules in space, and the other limit is exemplified by the very dilute gas. Here the molecules are in random motion almost all the time, and interact with each other only for the very short time that it takes for two of them to collide. There is no collective motion now and the kinetic energy of the motion is only slightly perturbed by the potential energy of the interaction between molecular pairs. Collisions between more than two molecules will be so rare an event as to be negligible unless the density of the gas is increased substantially. The disorder of the gas is described by specifying the motion of a representative single particle in space.

For a dense fluid and a liquid the interaction between one particle and the rest is continuous and not that of a simple well-defined collision. On the other hand the molecules are not held to a lattice pattern. Effects of the kinetic energy of motion and the potential energy of interaction are comparable and both must be accounted for in calculating properties to be compared with observation. The effect on a particular molecule cannot be expressed precisely because it must depend upon the motion of all the neighbouring molecules, but we do know that the interaction with neighbours will fluctuate in time due to the movement of the neighbours.[5]

If the force on a molecule at the time t is written $\mathbf{F}(t)$ and at time $(t+\delta t)$ is $\mathbf{F}(t+\delta t)$, then the difference $[\mathbf{F}(t+\delta t)-\mathbf{F}(t)]$ is a measure of the movement of the neighbours during δt. The average of this quantity over all the molecules of the liquid, written $\langle \mathbf{F}(t+\delta t)-\mathbf{F}(t)\rangle_{\mathrm{Av}}$, can be used to characterise conditions within the fluid. Alternatively, the correlation between the forces acting on the molecule at the two times can be defined by the average $\langle \mathbf{F}(t+\delta t)\mathbf{F}(t)\rangle_{\mathrm{Av}}$. An average over all the molecules of the liquid is equivalent to an average over all the momenta and positions (i.e. the phase) available to them. If the frequency with which different values of the phase occur is known, the average force can be specified quantitatively. It is in these terms that a simplified theory of the liquid state is sought at the present time.

1.3 Distribution functions

1.3.1 Phase and configuration distributions

The fluid is represented by N molecules contained in a volume V, and the continuous interaction between the molecules is represented by the total potential function Ψ which depends on the location of the N particles but not on the time. The probability that, at some time t, a molecule will be found in the space volume $d\mathbf{r}_1$ about $\mathbf{r}_1$ with momentum in the range $\mathbf{p}_1$ to $\mathbf{p}_1+d\mathbf{p}_1$, and simultaneously another molecule will be located at $(\mathbf{r}_2, d\mathbf{r}_2)$ and with momentum in the range $\mathbf{p}_2$ to $\mathbf{p}_2+d\mathbf{p}_2$, and so on for the N molecules, will be denoted by

$$f^{(N)}(\mathbf{p}_1, \mathbf{p}_2, \ldots, \mathbf{p}_N, \mathbf{r}_1, \mathbf{r}_2, \ldots, \mathbf{r}_N, t)\, d\mathbf{p}_1\, d\mathbf{p}_2, \ldots, d\mathbf{p}_N\, d\mathbf{r}_1\, d\mathbf{r}_2, \ldots, d\mathbf{r}_N. \tag{1.3.1}$$

For the convenience of notation we introduce the symbol $\Gamma^{(N)}$ to denote the phase point $(\mathbf{p}_1, \mathbf{p}_2 \ldots \mathbf{p}_N, \mathbf{r}_1, \mathbf{r}_2 \ldots \mathbf{r}_N)$ so that the probability (1.3.1) that the total system will be located with a given phase $\Gamma^{(N)}$ within the range $d\Gamma^{(N)}$ is written $f^{(N)}(\Gamma^{(N)},t)d\Gamma^{(N)}$.

Such a description could apply to any physical system, but we have already agreed that the spatial distribution of a few molecules will be of especial significance for a fluid. This description can be obtained from the probability (1.3.1) by removing information about the unwanted momenta and location of particles. Let us consider a small group of $n(<N)$ particles where the probability corresponding to equation (1.3.1) is $f^{(n)}(\Gamma^{(n)},t)d\Gamma^{(n)}$, extending the earlier notation in an obvious way. The reduced n-th order phase distribution function $f^{(n)}$ is related to the N-th order distribution function $f^{(N)}$ by

$$f^{(n)}(\Gamma^{(n)}, t) = \int \ldots \int f^{(N)}(\Gamma^{(N)}, t)\, d\Gamma^{(N-n)} \tag{1.3.2}$$

where

$$d\Gamma^{(N-n)} = d\mathbf{p}_{n+1}\, d\mathbf{p}_{n+2} \ldots d\mathbf{p}_N\, d\mathbf{r}_{n+1}\, d\mathbf{r}_{n+2} \ldots d\mathbf{r}_N \tag{1.3.3}$$

The phase distribution function $f^{(n)}$ is reduced to a configuration distribution function $n^{(n)}$ by integration over the appropriate momenta

$$n^{(N)}(\mathbf{r}^N, t) = \int \ldots \int f^{(n)}(\Gamma^{(n)}, t)\, d\mathbf{p}^n \tag{1.3.4}$$

where $\mathbf{r}^N$ is a vector in configuration space of $3N$ dimensions with components $(\mathbf{r}_1, \mathbf{r}_2, \mathbf{r}_3 \ldots \mathbf{r}_N)$ and $d\mathbf{p}^n$ is the vector in momentum space of $3n$ dimensions with components $(d\mathbf{p}_1, d\mathbf{p}_2\, d\mathbf{p}_3 \ldots d\mathbf{p}_n)$, all referring to the time t.

The simplest distribution functions, which are unfortunately also those containing the least information, are those referring to one, two, three or four molecules. For the phase, these are obtained from equation (1.3.2) by setting n respectively equal to 1, 2, 3, or 4. For the configurational distributions we have the formulae, for $n^{(1)}$ and $n^{(2)}$

$$n^{(1)}(\mathbf{r}_1, t) = \int \ldots \int n^{(N)}(\mathbf{r}^{(N)})\, d\mathbf{r}^{N-1} \tag{1.3.5a}$$

$$n^{(2)}(\mathbf{r}_1, \mathbf{r}_2, t) = \int \ldots \int n^{(N)}(\mathbf{r}^{N})\, d\mathbf{r}^{N-2} \tag{1.3.5b}$$

and

$$n^{(1)}(\mathbf{r}_1, t) = \frac{1}{V} \int n^{(2)}(\mathbf{r}_1, \mathbf{r}_2, t)\, d\mathbf{r}_2 \tag{1.3.6a}$$

$$n^{(2)}(\mathbf{r}_1, \mathbf{r}_2, t) = \frac{1}{V} \int n^{(3)}(\mathbf{r}_1, \mathbf{r}_2, \mathbf{r}_3, t)\, d\mathbf{r}_3 \tag{1.3.6b}$$

and so on.

The various formulae form an obvious hierarchy of equations and are complete in principle if $f^{(N)}$ is once specified, for a given time. A knowledge of $n^{(1)}$ above is sufficient for describing the behaviour of a very dilute gas, while $n^{(2)}$ is required to describe a dense gas or a liquid.

1.3.2 Time evolution

The change of $f^{(N)}(t)$ with time is given by the Liouville equation, which is an equation of continuity for the total probability. Explicitly, this equation is

$$\frac{\partial f^{(N)}}{\partial t} + \sum_{j=1}^{N} \frac{\mathbf{p}_j}{m} \cdot \frac{\partial f^{(N)}}{\partial \mathbf{r}_j} + \sum_{j=1}^{N} \mathbf{F}_j \cdot \frac{\partial f^{(N)}}{\partial \mathbf{p}_j} = 0. \tag{1.3.7}$$

The force $\mathbf{F}_j$ in this expression is the force on the j-th particle due to the neighbours. For simple dense fluids, and these alone will concern us in this chapter, this force is supposed derivable from a potential energy $\Psi(j)$ which is the sum of the interactions between the j-th molecule and every other molecule taken separately. Expressed another way, the total potential energy for the total system of N molecules will be supposed equal to the sum of the potential energy between each pair of molecules in the fluid. If $\Psi(i,j)$ is the potential of the actual force between the molecules i and j we assume that we can write

$$\Psi(1, 2, 3, \ldots, N) = \sum_{i=1}^{N} \sum_{j>i} \psi(i,j). \tag{1.3.7a}$$

This assumption of pair forces in the fluid is central to most of the arguments of kinetic theory developed so far, and will be supposed to hold in all that follows unless the contrary is stated.

Equation (1.3.7) can be written in the operator form

$$i\frac{\partial f^{(N)}}{\partial t} = -\mathscr{L} f^{(N)} \tag{1.3.8a}$$

where

$$\mathscr{L} = -i \sum_{j=1}^{N} \frac{\mathbf{p}_j}{m} \cdot \frac{\partial}{\partial \mathbf{r}_j} + \sum_{j=1}^{N} \mathbf{F}_j \cdot \frac{\partial}{\partial \mathbf{p}_j}. \tag{1.3.8b}$$

The operator $\mathscr{L}$ is Hermitian. The equation (1.3.8a) has the formal solution

$$f^{(N)}(t) = \exp\{-i\mathscr{L}t\}f^{(N)}(0) \tag{1.3.9}$$

and we shall have more to say about this in a later section.

Integration over the momentum converts equation (1.3.7) into a form involving $n^{(N)}$. Again, formally we have

$$i\frac{\partial n^{(N)}}{\partial t} = -\int\ldots\int \mathrm{d}\mathbf{r}^N \exp\{-i\mathscr{L}t\}f^{(N)}(0). \tag{1.3.10}$$

Further integration over the appropriate configuration variables leads to equations for the time evolution of the lower order configuration distributions.

1.3.3 Van Hove correlation function; the pair distribution

The motion of a molecule in the fluid can be described by introducing a generalized configuration distribution function, G, called the van Hove correlation function.[6] This is proportional to the probability of finding a molecule at $\mathbf{r}_2$ at time t_2 given that a molecule (which may be the same one but not necessarily so) was at $\mathbf{r}_1$ at time t_1. Explicitly, let $f(\mathbf{p}_2, \mathbf{r}_2, t_2; \mathbf{p}_1, \mathbf{r}_1, t_1)$ be the probability of finding a molecule with phase $(\mathbf{p}_2, \mathbf{r}_2)$ at time t_2 given that a molecule had the phase $(\mathbf{p}_1, \mathbf{r}_1)$ at time t_1: the corresponding configurational distribution G is found by averaging over the momenta:

$$G(\mathbf{r}, \tau) = \int\int f(\mathbf{p}_2, \mathbf{r}_2, t_2; \mathbf{p}_1, \mathbf{r}_1, t_1)\,\mathrm{d}\mathbf{p}_1\,\mathrm{d}\mathbf{p}_2 \tag{1.3.11}$$

where $\mathbf{r} = \mathbf{r}_2 - \mathbf{r}_1$ and $\tau = t_2 - t_1$. G, which has the dimensions of $1/V$, is normalized to become equal to the average particle density n for the total system: thus, $n = N/V$ and $G(\mathbf{r}, \tau) \to$ n as $\mathbf{r} \to \infty$. This same limit is also assumed to apply for an indefinitely large time interval where the two events become statistically independent.

The alternative limit, when $\tau \to 0$, refers to conditions at two points at essentially the same time. The integral (1.3.11) then reduces to the distribution function $n^{(2)}$ defined in equation (1.3.5b) because f is identically $f^{(2)}$ then. In conformity with the limit already imposed on G for an indefinitely large time interval, we shall introduce the static pair correlation function $g^{(2)}(\mathbf{r}_1 \cdot \mathbf{r}_2)$ by

$$n^{(2)}(\mathbf{r}) = ng^{(2)}(\mathbf{r}). \tag{1.3.12}$$

For a fluid in thermodynamic equilibrium, the distribution of molecules is isotropic if each molecule is sensibly spherical (simple fluid) and the intermolecular forces are of short range, and in this case the spatial distribution will not depend upon the angle. Consequently, only the relative separation distance $|\mathbf{r}| = r$ between two molecules is relevant and the pair distribution function reduces to the radial distribution $g(r)$. If one molecule is taken as origin and $\mathrm{d}n$ is the number of molecules contained in a shell of thickness $\mathrm{d}r$ and radius r about the origin then

$$\mathrm{d}n = 4\pi r^2 n(r)\mathrm{d}r$$

where $n(r)$ is the number of molecules per unit volume at the distance r.

The ratio $n(r)/n$ is the radial distribution so that

$$\mathrm{d}n = 4\pi r^2 n\, g(r)\mathrm{d}r \tag{1.3.13}$$

in terms of the mean number density n of molecules in the fluid. The short range order characteristic of the fluid is summarized by the conditions on $g(r)$:

$$\lim_{r \to R} g(r) = 1 \tag{1.3.14a}$$

where R is some distance large in comparison with the average separation of the molecules, and

$$\lim_{r \to 0} g(r) = 0 \tag{1.3.14b}$$

expressing the impenetrability of the molecules. The radial distribution has with increasing r an oscillatory form about the value unity due to the action of the intermolecular force, and the amplitude of the oscillation will be larger the more marked the motion of the molecule. This means that $g(r)$ will be dependent upon the temperature.

The radial distribution can be determined from X-ray experiments which involve momentum transfer between the X-ray photons and the atoms of matter. If $I(\kappa)$ is the intensity of the radiation of wavelength λ scattered through the half angle θ, with $\kappa = (4\pi/\lambda)\sin\theta$, and I_0 is the intensity of the incident radiation then the scattering function $S(\kappa)$ is defined by

$$S(\kappa) = \frac{I(\kappa)}{I_0} - 1. \tag{1.3.15}$$

If $I(\kappa)$ and I_0 are measured in an experiment as functions of κ, $S(\kappa)$ is then determined. The theory of scattering shows that $g(r)$ is related to the Fourier transform of $S(\kappa)$: explicitly

$$g(r) = 1 + \frac{1}{8\pi^3 n} \int \{S(\boldsymbol{\kappa}) - 1\} \exp(\mathrm{i}\boldsymbol{\kappa} \cdot \mathbf{r})\, \mathrm{d}\boldsymbol{\kappa} \tag{1.3.16}$$

Other information is also available when $S(\kappa)$ is known. Thus, for the isothermal compressibility K_T we have

$$K_T = \frac{S(0)}{nk_B T} \tag{1.3.17}$$

where k_B is the Boltzmann constant and T is the temperature.

The arrangement of the atoms in space can also be deduced from the scattering of thermal neutrons, although there is energy as well as momentum transfer with the atoms. The scattering function S is now dependent on the time. If $\boldsymbol{k} = \mathbf{k}_0 - \mathbf{k}$ is the wavenumber change of the neutrons during a collision and ω is the angular frequency $\omega = \hbar/2m(\mathrm{k}_0^2 - \mathrm{k}^2)$, where m is the mass of the neutron and $\hbar$ the reduced Planck's constant, the scattering function $S(\boldsymbol{k}, \omega)$ is defined by

$$S(\boldsymbol{k}, \omega) = \frac{1}{2\pi} \iint \{G(\mathbf{r}, \tau) - n\} \exp[\mathrm{i}(\boldsymbol{k} \cdot \mathbf{r} - \omega t)]\, \mathrm{d}\mathbf{r}\, \mathrm{d}\tau \tag{1.3.18}$$

where G is the van Hove correlation function. The scattering function can be found experimentally from time-of-flight measurements of the neutrons undergoing scattering. The Fourier inversion of equation (1.3.18) and a time average of the resulting formula reduces it to the form (1.3.16) involving the pair distribution.

The physical description of the scattering function $S(\boldsymbol{k}, \omega)$ can be usefully assigned three aspects distinguished by the magnitudes of the wavenumber and frequency. The first is when $\boldsymbol{k}$ and ω are both very small: if L is a characteristic length for the molecular interaction and τ is a time characterizing the interaction we mean the limit where $\boldsymbol{k}L \ll 1$ and $\omega\tau \ll 1$. This is the situation for ordinary hydrodynamics and can be called the hydrodynamic limit. The second aspect is the alternative limit of very high frequencies, where $\omega\tau \gg 1$ for all $\boldsymbol{k}$. In this case the response of the fluid to forces is essentially instantaneous, and the static average form is the appropriate fluid structure in this case. The behaviour of the simple fluid, where the pair force assumption (1.3.7a) is valid, can be calculated using the radial distribution $g(r)$ of equation (1.3.13). The third aspect lies between the other two, with wavelengths in the range 10^{-11} to 10^{-9} m and frequencies in the range 10^{11} to 10^{13} s^{-1}. This region is difficult to account for theoretically, but is precisely that covered experimentally by the scattering of thermal neutrons.

1.3.4 Dynamic correlations

Dynamical processes in a simple fluid are associated with thermally excited fluctuations of the local number of molecules. Because most of the measured properties of a fluid are proportional to the number of molecules which interact to form the average collective effect in question, a knowledge of the thermal fluctuation of the number of molecules in any region is central to the elucidation of the physical processes in the dense fluid. According to equation (1.3.18) the fluctuation in time of the number of molecules per unit volume can be represented by the time-dependent scattering function $S(k, \omega)$ and so is open to measurement in ways other than by the elementary counting of molecules.

The specification of the thermal fluctuation can be made in terms of the van Hove correlation function. More precisely, let δn be the correlation function between the two events: first, that the i-th molecule is located at the point $\mathbf{r}_i - \mathbf{r}_2$ at time $t = 0$, irrespective of whether or not there is a molecule at the location $\mathbf{r}_2$; and second, that the j-th molecule will be located at the point $\mathbf{r}_2$ at time $t = \tau$ irrespective of whether there is then a molecule located at the point $\mathbf{r}_i - \mathbf{r}_2$. Although we distinguish between the i-th and j-th molecules, we will allow the possibility of them in fact being the same molecule that has moved the distance $\mathbf{r}_i - \mathbf{r}_2$ during the time τ. We can express δn in the form of the correlation function

$$\delta(\mathbf{r} + \mathbf{r}_i(0) - \mathbf{r}_2) \cdot \delta(\mathbf{r}_2 - \mathbf{r}_j(\tau)). \tag{1.3.19}$$

We must extend this expression to cover all pairs of molecules of the fluid and all locations for the position vector $\mathbf{r}_2$. The result is the classical van Hove function $G(\mathbf{r}, \tau)$ in the form

$$G(\mathbf{r}, \tau) = \frac{1}{N} \sum_{i,j} \delta(\mathbf{r} + \mathbf{r}_i(0) - \mathbf{r}_j(\tau)) \tag{1.3.20}$$

where account has been taken of the commutativity of correlations in classical physics. The particle number density can be defined by the expression

$$n(\mathbf{r}, \tau) = \sum_j \delta(\mathbf{r} - \mathbf{r}_j(\tau)) \tag{1.3.21}$$

so that equation (1.3.20) is transformed to the form of an ensemble average

$$nG(\mathbf{r}, \tau) = \langle n(0, 0) n(\mathbf{r}, \tau) \rangle. \tag{1.3.22}$$

The expression relates G directly to the fluctuations in the number density and is normalized so that $G \to n$ as $t \to \infty$, the two molecules becoming statistically independent after a very large time interval.

There are two contributions to G according to equation (1.3.20) of different character, distinguished by whether or not $i = j$. In the first case, if we consider the movement of one particular molecule (so that $i = j$), we are concerned with the probability that this molecule will move from the initial location $\mathbf{r}_1$ to the location $\mathbf{r}_2$ during the time τ. This function will be written $G_s(\mathbf{r}, \tau)$, called the self-correlation function. The second contribution is the general one involving two molecules and called the distinct correlation $G_d(\mathbf{r}, \tau)$. The summation in equation (1.3.20) allows us to write

$$G(\mathbf{r}, \tau) = G_s(\mathbf{r}, \tau) + G_d(\mathbf{r}, \tau). \tag{1.3.23}$$

These functions have the simple limiting form as the time interval τ vanishes:

$$G_s(r, 0) = \delta(\mathbf{r}) \quad G_d(\mathbf{r}, 0) = ng(r) \tag{1.3.24}$$

and so relate the dynamic correlations to the static singlet and pair correlation functions.

The calculation of the thermal fluctuations in a liquid can be expressed in terms of the calculation of the structure factor S, and recent developments in the kinetic theory of liquids have involved the construction of an appropriate equation for S.[7] We shall refer to this work later.

I. EQUILIBRIUM PROPERTIES

The calculation of the thermodynamic functions for a dense fluid in equilibrium might be expected to present the simplest problems from the theoretical point of view. For now the static approximation is appropriate and the structure of a simple fluid, where the interaction forces between the molecules are pair forces, is adequately specified by the radial distribution function $g(r)$. The appropriate kinetic equations are those for the calculation of $g(r)$.

1.4 Total and direct static correlations

For thermal equilibrium the radial distribution is defined by equations (1.3.13) and (1.3.14) related to the lower order configurational distribution functions (1.3.5) by the equation (1.3.12) expressed in a form which does not depend on the angles. In what follows we shall suppose for simplicity that the fluid is composed of one species of molecule, and that each molecule is spherically symmetric and without internal degrees of freedom.

The deviation of $g(r)$ about the mean value unity is called the total correlation, $h(r)$: explicitly

$$h(r) = g(r) - 1. \tag{1.4.1}$$

The short range order in a simple liquid is described by the condition that $h(r)$ shall become zero for large values of the intermolecular distance r.

The total influence of one molecule, say particle 1, on a second molecule 2 in the system of interacting molecules is the recombination of two separate effects: namely the direct effect of molecule 1 on molecule 2, and the indirect effect of molecule 1 on 2 through the intermediate effect of molecule 1 on molecules 3, 4, . . . N with the consequent effect of these molecules on molecule 2. Thus in the indirect contribution, molecule 1 will affect a representative third molecule directly, and this effect will in turn be transmitted to the molecule 2. The full indirect contribution for the system is the sum of all the separate indirect contributions for molecules 3, 4, . . . N, averaged over all positions available to the molecules in the volume V.

This argument has been used to define a direct correlation function $c(r)$ between two molecules, distance r apart, and in the presence of the remaining molecules, by the expression

$$h(1, 2) = c(1, 2) + n \int h(2, 3)c(1, 3)\, d\mathbf{r}_3. \tag{1.4.2}$$

The concept of the direct correlation seems first to have been considered by Ornstein and Zernike in a rather different connection.[8] It is important because it has physical meaning for a pair of molecules only when they are in the presence of the remaining molecules of the fluid.

The direct correlation is related to the radial distribution through equations (1.4.1) and (1.4.2) although not in a simple way.

The total and direct correlation functions can be determined experimentally for a given fluid. According to equations (1.3.15), (1.3.16) and (1.4.1) the total correlation is related to the relative scattered intensity in an X-ray experiment through its Fourier transform: explicitly:

$$\tilde{h}(k) = \mathrm{i}(k) = \frac{I(k) - I(0)}{I_0} \tag{1.4.3}$$

because the static scattering function S is the Fourier transform of h, denoted by $\tilde{h}$. Taking the Fourier transform of equation (1.4.2) we obtain

$$\tilde{h}(k) = \tilde{c}(k) + n\tilde{c}(k)\tilde{h}(k).$$

It follows that

$$\tilde{h}(k) = \mathrm{i}(k) = \frac{\tilde{c}(k)}{1 - nc(k)} \tag{1.4.5}$$

so that

$$\tilde{c}(k) = \frac{\mathrm{i}(k)}{1 + n\mathrm{i}(k)} \tag{1.4.6}$$

Apparently the direct correlation is itself a function that can be determined experimentally through its Fourier transform. This relation has been used to determine $c(r)$ experimentally for a range of liquids.

1.5 Thermodynamic functions

The essential element in the kinetic theory of simple dense fluids and simple liquids in thermal equilibrium is the representation of the actual short range spatial order in the fluid by the radial distribution function. This simplification is useful only if the thermodynamic functions can be expressed uniquely in terms of $g(r)$ and the pair interaction potential $\psi(r)$ between a representative pair of molecules. This can in fact be done as we shall now see.

1.5.1 The free energies and the partition functions

The condition of thermal equilibrium is specified by a particular form of the N-th order distribution functions. For a classical fluid, if $H(\mathbf{p}^N, \mathbf{r}^N)$ is the Hamiltonian (energy) function for the system of molecules and T is the fluid temperature for a system of N molecules contained in a volume V, we set

$$f^{(N)}(\mathbf{p}^N, \mathbf{r}^N) = \frac{1}{\Xi} \exp \frac{\mu N - H(\mathbf{p}^N, \mathbf{r}^N)}{kT} \tag{1.5.1}$$

where Ξ is the grand partition function

$$\Xi = \sum_{N>0} \exp \frac{\mu N}{kT} Z(N) \tag{1.5.2a}$$

$$Z(N) = \frac{1}{h^{3N} N!} \int \ldots \int \exp \left\{ -\frac{H(\mathbf{p}^N, \mathbf{r}^N)}{kT} \right\} \mathrm{d}\mathbf{r}^N \, \mathrm{d}\mathbf{p}^N \tag{1.5.2b}$$

and μ is the chemical potential. $Z(N)$ is called the partition function and refers to a system with a fixed number of molecules: the grand canonical distribution (1.5.1) is more general in that it allows N to vary. The factor $N\cdot$ arises in equation (1.5.2b) because all the molecules are supposed identical (a one component fluid). In (1.5.2b), h is Planck's constant introduced for reasons of dimensions to ensure $Z(N)$ is dimensionless. The partition function is the normalizing function to ensure that

$$\int \dots \int f^{(N)}(\Gamma^{(N)})\, \mathrm{d}\Gamma^{(N)} = 1 \tag{1.5.3}$$

when N is constant. The grand canonical partition function is the normalizing function when the restriction of constant N is lifted.

The link with macroscopic thermodynamics is achieved by linking the Helmholtz free energy F to Z according to

$$F(V, T) = k_B T \ln Z(V, T). \tag{1.5.4}$$

Expressions for other thermodynamic variables follow from the relation

$$F = U - TS \tag{1.5.5}$$

where U is the internal energy, T the temperature and S the entropy.
Explicitly we obtain the expressions

$$p = -\frac{\partial F}{\partial V}\bigg|_{T,N} \qquad S = -\frac{\partial F}{\partial T}\bigg|_{V,N} \qquad \mu = \frac{\partial F}{\partial N}\bigg|_{T,V} \tag{1.5.6}$$

for the pressure p and chemical potential μ, as well as the entropy. The internal energy is given by

$$U = -T^2 \frac{\partial}{\partial T}\left(\frac{F}{T}\right). \tag{1.5.7}$$

Finally the Gibbs free energy $G(p, T)$ is

$$G = -V^2 \frac{\partial}{\partial V}\left(\frac{F}{V}\right). \tag{1.5.8}$$

Equivalent formulae are obtained using the grand canonical distribution (1.5.1) and the grand partition function (1.5.2a). In this case the equation of state is given by

$$\frac{pV}{k_B T} = n\Xi(N, V, T) \tag{1.5.9a}$$

$$U = -k_B \frac{\partial}{\partial(1/T)} \Xi(N, V, T) \tag{1.5.9b}$$

$$F = N\mu - kT \ln \Xi(N, V, T) \tag{1.5.9c}$$

$$G = N\mu. \tag{1.5.9d}$$

This general formalism allows the thermodynamic functions to be calculated once one or other of the partition functions is evaluated from equations (1.5.2). This is the central problem; the evaluation of Z, involving as it does an integration over the phase of the N particles of the liquid, is technically impossible – at least at the present time. Instead it is necessary to simplify the analysis by making an explicit appeal to the fluid structure, which means introducing the radial distribution. The situation is easier for a dilute gas where the partition function can be evaluated by expansion techniques using the feature that the effects due to the interaction potential are much smaller than those due to kinetic energy.

1.5.2 Formulae involving the radial distribution

The Hamiltonian function for the system of molecules is

$$H(\mathbf{p}^N, \mathbf{r}^N) = \sum_{j=1}^{N} \frac{\mathbf{p}_j^2}{2m} + \Psi(\mathbf{r}^N) \tag{1.5.10}$$

where Ψ is the total interaction potential. The insertion of equation (1.5.10) into the expression (1.5.2b) for Z shows that the integration over the momenta can proceed separately from those over the configuration. More than this, the momentum integrations are elementary. Explicitly we find

$$Z(N) = \frac{1}{N!}\left(\frac{2\pi mkT}{h^2}\right)^{3N/2} \int \ldots \int \exp\left(-\frac{\Psi}{k_B T}\right) \mathrm{d}\mathbf{r}^N \tag{1.5.11}$$

leaving the multiple integrations over the configuration of the molecules as the central difficulty.

For a simple fluid of spherical molecules the total potential energy is equated to the sum of the potential energies between the constituent molecular pairs according to equation (1.3.7a). In this case the full set of integrations in equation (1.5.11) can be replaced by an integral involving the radial distribution. The mathematical basis for this substitution starts with the definition (1.3.4) for the N-th order configuration distribution function $n^{(N)}$. Because pair forces are operative, appeal can further be made to the pair distribution function (1.3.5b). Insert equations (1.5.10) and (1.3.7a) into equation (1.5.1) and form $n^{(N)}$ from equation (1.3.4); form $n^{(2)}$ from equation (1.3.5b); this allows Z given by equation (1.5.2b) to be expressed in terms of the radial distribution according to the definition equation (1.3.12); the internal energy of the fluid follows from equations (1.5.4) and (1.5.7). The result of this rather lengthy path is the formula for the internal energy per unit of fluid volume $u = U/V$

$$u = \frac{3nk_B T}{2} + \frac{n^2}{2}\int_0^\infty g(r)\,\psi(r)\,4\pi r^2\,\mathrm{d}r. \tag{1.5.12}$$

An expression for the pressure is obtained in an analogous way, and this is

$$p = nk_B T - \frac{n^2}{6}\int_0^\infty g(r)\left(\frac{\partial\psi(r)}{\partial r}\,r\right)4\pi r^2\,\mathrm{d}r. \tag{1.5.13}$$

In each expression the first term refers to the motion of single molecules and is dominant in a low density fluid; the second term is dominant for a liquid and includes the molecular interaction through the radial distribution.

The use of the grand partition function Ξ allows the effects of changing the number of molecules in a fixed volume to be treated, and so allows fluctuations to be included. Such arguments lead to a relation to be obtained between the dependence of the pressure on the number of molecules present (the isothermal compressibility of the fluid) and the fluid structure described by the radial distribution

$$k_B T \left.\frac{\partial n}{\partial p}\right|_T = 1 + 4\pi n \int_0^\infty h(r)\,r^2\,\mathrm{d}r \tag{1.5.14}$$

where $h(r)$ is the total correlation. Using equation (1.4.2) this equation can be inverted to the form

$$\frac{1}{k_B T}\left.\frac{\partial p}{\partial n}\right|_{T,N} = 1 - 4\pi n \int_0^\infty c(r) r^2 \, \mathrm{d}r \tag{1.5.15}$$

where now the direct correlation function appears in the formula.

Other functions can be expressed in a similar way; thus for the chemical potential

$$\mu = \frac{3}{2} k_B T \ln\left(\frac{h^2 n^{2/3}}{2\pi mkT}\right) + n \int_0^1 \mathrm{d}\xi \int_0^\infty \psi(r) g(r, \xi) 4\pi r^2 \, \mathrm{d}r \tag{1.5.16}$$

where ξ is a coupling parameter between the molecules which is zero for hypothetical uncoupled molecules but is unity for the full interaction for the real liquid. We have seen enough to realize that the representation of the liquid structure by the radial distribution reduces the complexity of the calculations and offers the hope of providing a theory of fluid thermodynamic equilibrium of practical interest.

1.6 Calculating the radial distribution

The calculation of the pair distribution has proved a difficult task and there is no entirely satisfactory procedure even now. The hierarchy of equations set out in Section 1.3.1 are not helpful directly because we know only the N-th order distribution $n^{(N)}$ explicitly in terms of the intermolecular potential energy and cannot perform the $3(N-3)$ integrations necessary to reduce this function to the radial distribution.

1.6.1 The mean potential: Born–Green equation

The mean force $\langle \mathbf{F}_1 \rangle$ acting on a representative molecule (say molecule 1) in the liquid is assumed conservative and equal to the negative gradient of a mean potential which includes the interaction between the molecule and its neighbours. For the pair interaction described by equation (1.3.7a), the mean potential for a pair of molecules is defined as

$$\Phi(1, 2) = -k_B T \ln g(r_{12}) \tag{1.6.1}$$

$g(r)$ being the radial distribution for the total fluid. The mean force on molecule 1 of the pair is

$$\langle F \rangle = k_B T \frac{\partial}{\partial \mathbf{r}_1} \ln g = \frac{k_B T}{g(r)} \frac{\partial g(r)}{\partial r}. \tag{1.6.2}$$

However, the mean force can be related alternatively to the actual force on the molecule and involving the pair interaction potential. There are two contributions: the first is the direct interaction between molecule 1 and molecule 2; the second is an indirect effect on molecule 2 by way of a direct effect on a third molecule. The direct interaction with molecule 2 is expressed by the negative gradient of $\psi(1,2)$ at the location of particle 1, i.e. by $-\partial\psi(1,2)/\partial\mathbf{r}_1$. The indirect interaction involves the actual force at r due to a third molecule, $-\dagger\psi(1,3)/\partial\mathbf{r}_1$, and the effect this will have on the spatial arrangement of molecules 1 and 2; the combined effects of all the molecules of the fluid (other than 1 and 2) which can act as the third molecule gives the full indirect contribution to the force. Explicitly this indirect contribution to the mean force will be

$$-n \int_0^\infty \frac{\partial\psi(1,3)}{\partial\mathbf{r}_1} \frac{g^{(3)}(1, 2, 3)}{g(1, 2)} \mathrm{d}\mathbf{r}_3$$

where $g^{(3)}$ is the normalized triplet distribution such that $n^{(3)}(1, 2, 3) = n^3 g^{(3)}(1, 2, 3)$. Combining these contributions and equating the resulting expression of the mean force to the right hand side of equation (1.6.2) provides an equation for $g(r)$; multiplication throughout by the function $g(1, 2)$ provides the final form

$$\frac{\partial g(1,2)}{\partial r_{12}} + \frac{1}{k_B T}\frac{\partial \psi(1,2)}{\partial r_{12}} g(1,2) = -\frac{n}{k_B T}\int_0^\infty \frac{\partial \psi(1,3)}{\partial r_{13}} g^{(3)}(1,2,3) 4\pi r_{13}^2 \, dr_{13}. \quad (1.6.3)$$

This equation was first derived (though by a different method to that used here) by Born and Green, and equivalent equations were derived independently by Yvon and by Kirkwood.[9] The equation is exact and relates the distribution of pairs of molecules in the liquid to the potential of the actual force between the pair and to the distribution of triplet arrangements of molecules. The Born–Green equation (1.6.3) is more directly derived by the direct differentiation of the expressions (1.3.10), or equivalently from the Liouville equation (1.3.7) applied to fluid equilibrium. Groupings of more than two molecules can be treated in the same way, and equations of the type (1.6.3) derived successively for the functions $g^{(3)}$, $g^{(4)}$, and so on.[10]

While the Born–Green equation describes the liquid short range order through small changes of the relative separation of a representative pair of molecules, the second derivative of $g(r)$ with respect to r is a measure of the deviation of $g(r)$ from its local mean value for a specified separation distance (which is different from the mean value of $g(r)$ for infinite separation). The second order equation which replaces equation (1.6.3) involves $g^{(4)}$ as well as $g^{(3)}$, and so describes the local liquid spatial ordering more widely than equation (1.6.3) which involves only $g^{(3)}$. The second order equation will not be written down here but represents one generalization of the equation (1.6.3)

1.6.2 Closure procedures

There is no practical possibility of using the Born–Green equation, either alone or as a member of the heirarchy of equations for the lower order distributions without introducing a simplifying closure procedure of some kind. The method now widely used is to appeal to a superposition approximation, as was first done by Kirkwood.[11] The idea is to use a physical argument to relate $g^{(n+1)}$ to $g^{(n)}$ independently of the exact equations; for equation (1.6.3) this amounts to expressing $g^{(3)}$ in terms of $g^{(2)}$. We approach this by way of the potential of the mean force.

Let the potential of the mean force for a group of three molecules in interaction be $\Phi(1, 2, 3)$: this can be expressed in terms of the potentials of the mean forces for the pair groups of three molecules as

$$\Phi(1,2,3) = \Phi(1,2) + \Phi(1,3) + \Phi(2,3) + W_3(1,2,3) \quad (1.6.4)$$

where $W_3(1, 2, 3)$ is the potential for the mean collective interaction between the three particles. By invoking the relation (1.6.1), and its generalization for three molecules as a group we can rearrange equation (1.6.4) into the form

$$g^{(3)}(1,2,3) = g^{(2)}(1,2) g^{(2)}(1,3) g^{(2)}(2,3) S_3(1,2,3) \quad (1.6.5)$$

where $S_3 = -k_B T \ln W_3$. The expression (1.6.5) is exact if S_3 is properly assigned but it is an approximation insofar as S_3 departs from its correct form. The simplest assumption, also chosen by Kirkwood, is to suppose the collective term W_3 in (1.6.4) to vanish, so that $S_3 = 1$; this is the Kirkwood superposition approximation. The arguments leading to equation (1.6.4) are readily extended to involve four molecules rather than three, and then provides a superposition relation for the quartet:

$$g^{(4)}(1,2,3,4)=g^{(3)}(1,2,3)g^{(3)}(1,2,4)g^{(3)}(1,3,4)g^{(3)}(2,3,4)S_4(1,2,3,4) \quad (1.6.6)$$

which again is exact if S_4 is known exactly. Fisher[12] has suggested the use of equation (1.6.6) under the condition $S_4 = 1$ as a superposition approximation. A hierarchy of such relations can be built up, but they are not directly useful if the functions S remain unknown.

Arguments designed to provide data for the functions S_j have been attempted, but with varying degrees of success. One recent approach has been based on the mathematical techniques of functional differentiation,[13] which is an extension of ordinary differentiation to involve the effect on one function of varying another function upon which it depends. Such an approach provides a systematic expansion of S_j in powers of the number density but the numerical details of the improvement of data for $g^{(2)}$ to which they lead is at present unknown. As an example, we quote the expression for $g^{(3)}$ derived in this way as far as the linear term in n, an expression which is exact to this order:

$$S_3(1,2,3) = 1 + n \int d\mathbf{r}_4 f(1,4)y(1,4)X(2,3,4)$$

$$X(2,3,4) = \frac{g^{(3)}(2,3,4)}{g(2,3)} - g(2,4) - g(3,4) + 1. \quad (1.6.5a)$$

It has not yet proved possible to relate the arguments of functional differentiation to physical criteria, and so to give a physical basis for the procedure.

1.6.3 Approximate equations

Because information about $g(r)$ cannot be deduced directly from exact equations it is necessary to appeal to approximate theories.

The first one can be obtained from the exact Born and Green equation using the superposition approximation (1.6.5) with $S_3 = 1$. Introduce the function $y(r)$ by

$$y(r) = g(r)\exp\left(\frac{\psi(r)}{kT}\right). \quad (1.6.7)$$

Using equation (1.6.1) we have alternatively

$$y(r) = \exp\frac{(\psi(r) - \Phi(r))}{kT} \quad (1.6.8)$$

so that $y(r)$ is a measure of the difference between the actual and mean forces on a molecule at a specific location. The equation (1.6.3) with (1.6.5) can then be rearranged into the form

$$\ln y(1,2) = n\int_0^\infty c(2,3)h(1,3)\,d\mathbf{r}_{13} \quad (1.6.9)$$

where h is given by (1.4.1) and $c(r)$ by

$$c(r) = \frac{1}{k_B T}\int_r^\infty g(s)\frac{d\psi(s)}{ds}\,ds. \quad (1.6.9a)$$

Modify this expression by introducing the potential of the mean force by adding and subtracting the same integral

$$\frac{1}{k_B T}\int_r^\infty ds\, g(s)\frac{d\Phi}{ds}\,ds. \quad (1.6.10)$$

The result is

$$c(r) = h(r) + \frac{1}{k_B T} \int_r^{\infty} g(s) \frac{d}{ds} [\psi(s) - \Phi(s)] \, ds. \tag{1.6.11}$$

The integral here can be performed if $g(s)$ is given its asymptotic value unity. Then, using the subscript H for this approximation, we have

$$c_H(r) = h(r) - \ln y(r) \tag{1.6.12}$$

which, referring to equation (1.4.2), is seen to be the direct correlation function. Inserting this expression into (1.4.2) gives the equation

$$\ln y(1,2) = n \int_0^{\infty} [h(1,3) - \ln y(1,3)] h(2,3) \, d\mathbf{r}_3 \tag{1.6.13}$$

which is called the hypernetted chain equation.[14] Although we have derived it here as a consequence of approximating an integrated form of the Born and Green equation (1.6.9), equation (1.6.13) is to be regarded as an equation for g in its own right for a simple liquid, based on the particular assumption (1.6.13) for the direct correlation. This particular form for $c(r)$ was originally derived on the basis of an analysis of the interactions between molecules using mathematical diagram techniques and an approximation according to which certain sets of diagrams are omitted from the analysis, for mathematical reasons – they could not be evaluated explicitly or related to other diagrams.[15]

Modified arguments lead to modified forms of $c(r)$. One important form of $c(r)$ is

$$c_P(r) = f(r) y(r) \tag{1.6.14}$$

where $f(r)$ is the so-called Mayer function

$$f(r) = \exp\left(-\frac{\psi(r)}{kT}\right) - 1. \tag{1.6.15}$$

It is seen that c_P will have a range comparable to that of the intermolecular force itself, which for a simple liquid is short-ranged. The insertion of equation (1.6.14) into equation (1.4.2) gives the Percus–Yevick equation[16]

$$y(1,2) = 1 - n \int_0^{\infty} \left[\exp \frac{\psi(1,3)}{k_B T} - 1\right] h(2,3) g(1,3) \, d\mathbf{r}_3. \tag{1.6.16}$$

This equation also can be related to approximations in the evaluation of diagram terms although the initial derivation was not directly related to such terms explicitly. Comparison of equation (1.6.14) with equation (1.6.12) shows that

$$c_H - c_P = y(r) - 1 - \ln y(r)$$

and the difference between the two approximate theories is expressed in terms of the differences between the actual and mean potentials in the two cases. This difference disappears if $y(r)$ is close to unity over all separation distances. To see this, write

$$y(r) = 1 - z(r)$$

where $z(r) \ll 1$ everywhere. Then $\ln y(r) \sim z(r)$ and $c_H = c_P$. In this way the Percus–Yevick equation can be regarded as an approximation to the hypernetted chain equation which is itself an approximation to the Born–Green equation. Like equation (1.6.13), equation (1.6.16) has a stronger deductive base for a dilute gas than for a liquid; for a liquid these equations must be assumed valid, and the validity checked by comparing the data for the radial distribution derived from them with experimental data. The equations set down so far, that is the Born–Green equation including the superposition approximation, the Percus–Yevick equation and the hypernetted-chain equation, have been derived independently using the mathematical techniques of functional differentiation, and this approach allows them to be generalized.[17] In this sense the equations are first order theories of liquid structure. The generalized (higher order) theories are mathematically most complicated and have not been used in numerical work for liquid densities. But quite separately from the mathematical complexity, the physical basis of this procedure is still obscure, which must limit the confidence with which it can be used.

1.7 Comments on some numerical results

We have not the space available here to attempt any detailed discussion of the wide range of numerical studies of the equilibrium theory carried out by a large number of workers over the last two decades. But we must indicate the general scope of the work and gather together in broad terms the type of calculations that has been made.

1.7.1 Fluid structure

The radial distribution is obtained by solving the equations of the last section for a chosen expression for ψ. One aim of such studies is to link particular details of $g(r)$ with specific details of ψ; another aim can be to obtain the best data for ψ for a chosen real fluid by way of the details of $g(r)$ deduced experimentally.[18] The hard-sphere, square-mound, square-well, Lennard–Jones and other potential forms having convenient analytical properties are also used, often in conjunction with simulation studies. A dipole interaction has been included by way of the Stockmayer equation.

For low fluid densities $g(r)$ is expressible as an expansion in powers of n of the form

$$y(r) = \sum_{j=0}^{\infty} a_j(r)n^j \tag{1.7.1}$$

where the a_j are coefficients to be determined from the particular equations for $g(r)$. Such an expansion is not valid for liquid densities and the radial distribution must be found then in a form applicable to all densities.

In a general way it is found that the measured features of $g(r)$ result from the repulsive features of the potential function ψ at small separation distances. The attractive properties of the potential modify the general features to a small extent, which will nevertheless have an important influence on the calculated data for thermodynamic functions. The best form of comparison between calculated and measured data is probably that between the Fourier transform of the calculated $g(r)$ and the experimental scattering data itself: there are a number of difficulties in determining the radial distribution experimentally and consequent ambiguities are made worse by making a Fourier transform to that data.

The arguments can be inverted to yield information about ψ for an experimentally determined function g by using the equations of Section 1.6 but with $\psi(r)$ treated as the unknown function. A range of liquids has been studied in this way, including liquid metals.[18] The actual potential found in this way for a simple liquid such as liquid argon is markedly different from that for a liquid metal, such as sodium. The curve for argon found this way applies with only minor changes over the full range of density, from vapour to solid, but the oscillatory form for a liquid metal depends upon the state. There is evidence to suggest that for short-range force potentials such as for argon, the Percus–Yevick approximation (1.6.16) is the most reliable whereas for liquid metals the Born–Green equation and the superposition approximation lead to the most reliable data. Second order theories may be different when eventually they are used. Of course, other methods are also available for constructing accurate ψ functions for rare gas atoms, and these again can act as a datum for testing data calculated by the methods described in this chapter.

1.7.2 Thermodynamic functions

Information about $g(r)$ is used in the formulae of Section 1.5 to calculate thermodynamic functions. For low densities the expansion (1.7.1) is valid and leads to expressions for the thermodynamic functions in the form of expansions in powers of n. For the pressure, equation (1.5.13), or equivalently equation (1.5.14), provide the virial expansion, with the virial coefficients related explicitly to the intermolecular pair potential ψ. The Born–Green, hyper-netted chain, and Percus–Yevick first order theories provide exact values for the second and third virial coefficients, but the higher ones are not given correctly. The incorrectness is shown by equations (1.5.13) and (1.5.14) leading to different data for the pressure and the degree of this inconsistency is a measure of the approximation introduced into the theory by the particular equation for $g(r)$. Second order theories provide exact information for the fourth virial coefficient as well as the second and third: this is also the case for the extended superposition expression (1.6.5) with the Born–Green equation (1.6.3). Higher order theories are able to add progressively higher order virial coefficients to the list of exact expressions. Most information about the virial expansion is known for a hard-core molecule gas, and for such a gas of hard spheres the first seven virial coefficients are known. An empirical interpolation formulation allows others to be inferred up to the density of condensation, associated with the density of dynamical close packing. Rather less information is available for other intermolecular potential forms, but some of this information is useful for engineering applications.

Numerical studies for liquid densities are available, although on a considerably more restricted scale than for gases. Generally speaking data of the correct order of magnitude result from the theory, and agreement with corresponding experimental values is probably better than 30%. The way in which such an inaccuracy is to be apportioned between the approximate nature of the theory and the approximate nature of the function ψ used in the calculations is not yet known. Much remains to be done, particularly in constructing more accurate structural theories than those of Section 1.6, and in deducing a more accurate pair potential with which to represent the interaction between selected molecules. The effects of approximations (such as the superposition approximation) estimated for gas densities are not necessarily applicable for liquid densities, because the physical circumstances of the two cases are different. The exact theory, of course, will remain exact in both cases.

1.7.3 Alternative developments

The methods of molecular dynamics have been used to calculate pV/T as a function of n for a range of densities and for different force potentials. The motion of individual molecules of a prescribed group is studied by solving the correct equations of motion numerically, and thermodynamic data inferred by performing the phase average required in Section 1.5. Because the form of ψ (however idealized) is prescribed, the method is invaluable in providing data to act as a datum (of the same nature as genuine experimental data) against which the various approximate theories can be compared and contrasted. We shall have need to refer to certain results of molecular dynamics calculations later.[20]

The representation of liquid structures by mechanical models has been investigated by several authors over a number of years. Originally the motion of mechanically agitated gelatine spheres suspended in a liquid of essentially the same density was observed and the static structure of the liquid inferred, but more recently the bonded structure of various liquids, including water, has been studied on the basis of simple models often including plasticine. Such studies must surely increase our physical understanding of liquids and can be used to extend our thinking to include amorphous materials of rheological interest.

II. NON-EQUILIBRIUM PROPERTIES

The equilibrium macroscopic state of a physical fluid is fully specified by the free energy as a function of volume and temperature, although other parameters are necessary for the specification of a non-equilibrium state even if it is held steady in time. The non-uniformities within such a system (generally involving the gradients of temperature, or concentration of mass, or velocity, or electrical potential, and so on) must be maintained by the performance of work and expenditure of energy outside the system and communicated to it through its boundary surface. If these outside links are severed the fluid is observed to move irreversibly to a stable equilibrium state, the various gradients diminshing until they eventually vanish not to reappear. The description of the move to equilibrium introduces a range of problems not met in the treatment of the equilibrium condition itself, and the arguments of Sections 1.4 to 1.7 are not applicable. The analysis of the earlier sections is, however, available for extension and refinement.

1.8 Irreversibility

It is assumed that the same molecular properties control the movement to equilibrium as control the equilibrium condition itself. This presents a problem of interpretation because the equations of motion for individual molecules are reversible (the Hamiltonian (energy) function is time independent and is a quadratic function of the momentum) whereas the collective properties of the molecules are irreversible as is witnessed by the move to equilibrium observed in macroscopic matter. This apparent paradox gave much trouble in its resolution but the solution of the problem has given insight into the nature of irreversible processes.

1.8.1 Recurrence of initial phases

Poincaré[21] showed long ago that a conservative dynamical system composed on many particles, where the energy is independent of the time and the force on each particle, depends only on the relative configuration of the particles and has the property that any given initial phase must recur to any arbitrarily specified degree of accuracy an indefinitely large number of times. The trajectories representing the system of interacting particles has a quasi-periodic structure so that any initial phase will reappear if we are willing, or able, to wait for it. The time taken for the initial phase to reappear is the Poincaré period τ_p, and the cycle is the Poincaré cycle. The initial phase determines the later phase because the phase trajectories do not cross; and the phase volume is constant.

The magnitude of τ_p was estimated by Boltzmann[22] for an abnormal state of dilute gas as 10^{143} years, which is enormously in excess of the age of the Universe. This particular initial state will never recur in any practical circumstances and the system therefore shows irreversible behaviour in practice even though the underlying laws of mechanics are strictly reversible. Later authors have treated other systems, more idealized perhaps, but the same conclusion has always been obtained. In particular, the further the system is from equilibrium initially the longer it takes for it to achieve the initial phase again. These ideas of irreversibility are closely related to those of fluctuation phenomena and the recognition of this link has proved of very great importance.

The general study of the molecular aspects of macroscopic fluctuations has developed from the initial work of Von Smoluchowski on colloidal statistics, the so-called theory of probability after-effects.[23,24] The analysis, therefore, is concerned with the fine details of the Brownian motion. A small volume V_1 containing colloidal particles is studied and the number of particles in the volume is counted for each of a succession of times separated by the same time interval τ. During each interval τ each colloidal particle undergoes a Brownian motion, at the end of which some particles initially in V_1 will have moved out and some initially outside V_1 will have moved in. Among the many questions that can be asked are: what is the mean life τ_L of a particular state, and how long must we wait before some chosen initial number $N(\tau_0)$ (the initial state) will recur (time of recurrence τ_R)?

These questions can be answered by a direct appeal to probability arguments, and answers were given by Smoluchowski. The arguments can be extended to include continuous observations and can be applied to atomic systems. The results are still interesting. If N is the number of particles observed in V_1 on a particular occasion and $\nu = \langle N \rangle$ is the average number in V_1 averaged over a set of observations, τ_L is proportional to $(N + \nu)^{-1}$ whereas τ_R involves N through an exponential factor. This difference of behaviour means that while τ_L decreases only slowly as N increases, τ_R increases dramatically with increasing N. Calculated numbers are rather astonishing, and as an example we quote data given by Smoluchowski for τ_R for oxygen gas with $\nu = 3 \times 10^{19}$ and $T = 3 \times 10^2$ K, the initial state for the small volume V_1 being chosen with the molecular concentration differing from the mean by 1%. The variable is the radius R of the sphere. For $R = 10^{-5}$ cm, $\tau_R \sim 10^{-11}$ s; $R = 2.5 \times 10^{-5}$ cm, $\tau_R \sim 1$ s; $R = 3 \times 10^{-5}$ cm, $\tau_R \sim 10^6$ s; $R = 5 \times 10^{-5}$ cm, $\tau_R \sim 10^{68}$ s; $R = 1$ cm, $\tau \sim 10^{10}$ s. Thus for $R = 3 \times 10^{-5}$ cm, τ_R is rather less than a year while for $R = 5 \times 10^{-5}$ cm, τ_R is orders of magnitude greater than the age of the Universe.

Although these arguments refer to a gas, they have a wider application and can be generalized to apply to condensed states of matter. In this way the reversibility of the dynamics of molecular motion is seen to be fully compatible with the observed irreversible behaviour provided the recurrence time is taken into account. In particular, the irreversible behaviour is linked directly to fluctuation phenomena well known in macroscopic statistical mechanics where the mean square amplitudes of fluctuation of the internal energy or the number of particles in a small volume are linked directly to thermodynamic functions (the specific heat at constant volume and the isothermal bulk modulus for these two cases).

1.8.2 Entropy

Thermodynamically, the equilibrium state is one of maximum entropy for a given energy; for a state not in equilibrium the entropy increases towards a maximum value. Maximum entropy corresponds to a minimum value of the Helmholtz free energy. For equilibrium the entropy S is conventionally defined microscopically as

$$S = -k_B \int \dots \int f_0^{(N)} \log f_0^{(N)} \, d\mathbf{p}^N \, d\mathbf{r}^N \equiv -k_B \langle \log f_0^{(N)} \mid f_0^{(N)} \rangle \qquad (1.8.1)$$

where $f_0^{(N)}$ is the canonical distribution (1.5.1) and the bracket is a convenient alternative way of writing the integration involved. The generalization of this definition to include non-equilibrium is not a simple thing to do. The replacement of $f_0^{(N)}$ for equilibrium by $f^{(N)}$ for non-equilibrium was made by Boltzmann and the resulting integral $\langle \log f^{(N)} | f^{(N)} \rangle$ called the H-function. Unfortunately, the entropy so defined does not increase to a maximum value and cannot be identified with the physical entropy. The distribution of phase points can be regarded as a fluid in the phase space of N-particles (the γ-space), and the equations of motion show this fluid to be incompressible. The phase points are distributed non-uniformly between two neighbouring energy surfaces (microcanonical ensemble) and the time evolution of the system does not change this distribution in phase because any small phase element cannot gain or lose phase points.

With the *number* of points fixed in this way, the only variable able to lead to a uniform distribution of points in phase is the shape of contiguous volume elements. The approach to uniformity must be pictured as a mixing process; the initial gross mosaic of phase volumes representing the non-uniform state is transformed in time into a fine mosaic structure. This fine structure can have all the appearances of a uniform distribution provided we agree to set a limit to the fineness in phase with which we examine the details of γ-space.[25] The appearance of a certain coarseness in the final phase specification is in agreement with the statistical form of the description and is important in calculating τ_p for particular cases.

1.8.3 Coarse-graining and time smoothing

The specification of the phase involving a minimum volume was given by Ehrenfest in terms of a coarse-grained probability phase distribution $\bar{\bar{f}}^{(N)}$ according to

$$\bar{\bar{f}}^{(N)} \Gamma^{(N)} \delta \mathbf{p}^N \delta \mathbf{r}^N = \int_{\delta \mathbf{p}\, \delta \mathbf{r}} f^{(N)} \Gamma^{(N)} \mathrm{d}\Gamma^{(N)}. \tag{1.8.2}$$

Here $(\delta \mathbf{p}\, \delta \mathbf{r})_N$ is a small though finite phase volume which forms the integration limit; it is now realized that it is related to the quantum of action, although this was not clear to Ehrenfest when he developed these ideas. The coarse-grained distribution defined in this way can be used to define an H-function which provides a broadly adequate account of the increase in entropy associated with a move towards equilibrium.

The arguments were taken one step further by Kirkwood,[26] who pointed out that coarse-graining in time is also necessary to account for the fact that any observation, to be reliable, must involve an interval of time on the microscopic scale of sufficient extent to allow many molecular interactions to occur, although it must be short enough on the macroscopic scale to refer to a small macroscopic time interval.

For these purposes we introduce the time-smoothed distribution $\bar{f}^{(N)}$ at the time t according to

$$\bar{f}^{(N)}(\Gamma^{(N)}, t) = \frac{1}{\tau} \int_0^\tau f^{(N)}(\Gamma^{(N)}, t + s)\, \mathrm{d}s \tag{1.8.3}$$

where τ is a time microscopically large but macroscopically small. It turns out in practice that τ need not be specified precisely.

1.8.4 Statistical nature of particle interactions

The specific features of a theory of non-equilibrium are introduced by the way in which statistical account is taken of the interaction between the constituent particles of the model, and this was referred to in Section 1.2.

Although the idea of the binary encounter allows the behaviour of a dilute gas to be understood, and formed the basis of the kinetic theory approach of Boltzmann, Chapman, Enskog, and later authors, the description of even a simple liquid requires a more detailed interaction model.

Kirkwood appealed to the auto-correlation of the force on a representative particle as a means of introducing the extended molecular interactions characteristic of a liquid. The continual extended interaction between a chosen particle 1 and the remaining particles involves an asymmetry of motion because, whereas the effect of neighbours on particle 1 causes the particle to undergo an appropriate sympathetic motion, the effect of particle 1 on the neighbours collectively will be virtually insignificant. It is, then, plausible to suppose the motion of particle 1 to have a random form in a nearly steady field due to the neighbours. More than this: the motion of the particle 1 can be supposed essentially uncorrelated with that after a small time interval τ_1. On this basis the particle motion is described as a stochastic process with associated correlation time τ_1, this condition being expressed by the mathematical statement

$$\langle \mathbf{F}_1(t) \cdot \mathbf{F}_1(t+s) \rangle_{N-1} = 0 \quad \text{for } s > \tau \tag{1.8.4}$$

for the force on particle 1, where the angular brackets describe an average over the phase of the remaining $(N-1)$ particles of the system of N particles. The assumption (1.8.4) has been found to have the ability to introduce the element of irreversibility into the treatment of a liquid system in a way analogous to that of the hypothesis of molecular chaos in the theory of the dilute gas.

1.9 Molecular fluxes and fluid transport

The macroscopic fluid properties are related to the collective microscopic properties, and the equations of macroscopic hydrodynamics are related to the microscopic fluxes of mass, momentum and energy within the fluid. There are two separate though related approaches to the calculation of these fluxes, depending upon whether the approach to equilibrium is supposed a unique event or whether it is part of a general fluctuation. We consider these two approaches separately.

1.9.1 General expressions for the flux

Consider the macroscopic observable $Q(t)$ regarded as a collective flux of the microscopic variable $Q(\Gamma^{(N)})$ which is a function of the total phase but not the time. We assert the validity of the expression

$$Q(t) \equiv \langle Q \mid f^{(N)} \rangle = \int \ldots \int Q(\Gamma^{(N)}) f^{(N)}(\Gamma^{(N)}, t) \, \mathrm{d}\Gamma^{(N)} \tag{1.9.1}$$

for the calculation of the observable quantity Q. Here $f^{(N)}$ is the appropriate distribution function, appropriately phase or time smoothed as the theory may require. For equilibrium, $f^{(N)}$ will have the canonical form and Q will be a thermodynamic function.

For non-equilibrium, equations of motion are obtained by considering the time change of Q so that we are concerned with

$$\frac{\partial}{\partial t} \langle Q \mid f^{(N)} \rangle = \int \ldots \int Q(\Gamma^{(N)}) \frac{\partial f^{(N)}}{\partial t} \, \mathrm{d}\Gamma^{(N)}. \tag{1.9.2}$$

The integration here is to be performed by using the Liouville equation (1.3.7) so we have, using equation (1.3.9), the operator form[27]

$$\frac{\partial}{\partial t}\langle Q \mid f^{(N)} \rangle = \int \ldots \int Q(\Gamma^{(N)})(-i\mathscr{L})\exp\{-i\mathscr{L}t\}\,\mathrm{d}\Gamma^{(N)}. \tag{1.9.3}$$

where $\mathscr{L}$ is the Liouville operator (1.3.8b). This general equation of conservation of Q is applied to a liquid by making appropriate choices for Q. For mass conservation we write

$$Q = m \sum_{j=1}^{N} \delta(\mathbf{x} - \mathbf{r}_j) \tag{1.9.4a}$$

where $\mathbf{x}$ is the macroscopic location; for the momentum we write

$$Q = \sum_{j=1}^{N} p_j \delta(\mathbf{x} - \mathbf{r}_j) \tag{1.9.4b}$$

and for the energy we have

$$Q = \sum_{j=1}^{N} \frac{p_j^2}{2m} \delta(\mathbf{x} - \mathbf{r}_j) + \frac{1}{2} \sum_{j \neq k}^{N} \psi(j,k)\delta(\mathbf{x} - \mathbf{r}_j) \tag{1.9.4c}$$

where $\psi(j, k)$ is the potential of the actual force between a pair of molecules. When the function $f^{(N)}$ is given on the basis of a particular kinetic, these expressions for Q allow the general equation (1.9.3) to be arranged into the usual form of the equations of continuum hydrodynamics. The problem that remains is the solution of the appropriate form of the Liouville equation (1.3.7).

The expressions for the momentum and energy fluxes derived in this way define the macroscopic fluxes only to an additive term of vanishing divergence. This gives no difficulty in practice because it is the divergence of the fluxes that is involved in macroscopic theories of fluid flows. The theory provides expressions for the transport coefficients, such as viscosity and thermal conductivity, in terms of molecular properties but we will not write down the general expressions for these at this stage.

1.9.2 Response to external forces

The maintenance of a non-equilibrium condition requires the expenditure of external work and the liquid will show a characteristic response to external systems when mechanical forces are applied instantaneously to the system. The distribution of phase within the system is then supposed to be the sum of a distribution without the forces, $f_0^{(N)}$ say, plus a *known* contribution to the Hamiltonian arising from the forces. This total distribution will be associated with a current flow: if a force of type A, F_{A} say, gives rise to a flux of type B, written J_{B}, a linear response to the force of frequency ω can be expressed in the form[28]

$$J_{\mathrm{B}}(\omega) = L_{\mathrm{AB}}(\omega)F_{\mathrm{A}}(\omega) \tag{1.9.5}$$

where L_{AB} is the transport coefficient for the flux B in response to the force A. For a periodic driving force $F_{\mathrm{A}}(\omega) = F_{\mathrm{A}} \exp(-\mathrm{i}\omega t)$, Mori obtained the formula

$$L_{\mathrm{AB}}(\omega) = \frac{1}{V k_{\mathrm{B}} T} \int_0^{\infty} \langle J_{\mathrm{B}}(t) J_{\mathrm{A}}(s - t) \rangle \exp(-\mathrm{i}\omega s)\,\mathrm{d}s \tag{1.9.6}$$

as the classical form of a more general quantum formulae. The upper limit of integration is allowed to approach infinity on the understanding that the time t is greater than the correlation time for the fluxes. The static coefficient $L_{\mathrm{AB}}(0)$ is obtained from equation (1.9.6) by setting $\omega = 0$.

The flux can be expressed as the time derivative of some quantity a (say), so that writing $J = \mathrm{d}a/\mathrm{d}t = \dot{a}$ we obtain a general expression for the static coefficient L_{ij} in the form

$$L_{ij} = c\int_0^t \langle \dot{a}_i(s)\dot{a}_j(0)\rangle \,\mathrm{d}s \tag{1.9.7}$$

where c is a constant. This expression is compatible with the linear regression law

$$\dot{a}_i = \sum_j L_{ij}X_j \tag{1.9.8}$$

which is related to the celebrated Onsager expressions of irreversible thermodynamics.[29]

These arguments express the transport coefficients as averages over the equilibrium ensemble of the system and have been used to obtain expressions for the viscosities, thermal conductivity and electrical conductivity. Numerical predictions are not easy to make and the theory has not yet been subjected to a rigorous comparison with experiment.

1.9.3 Perturbation of local thermodynamic equilibrium

Suppose an isolated fluid volume is prepared in a non-equilibrium state by the action of an external reservoir. The imposed non-uniformities may involve gradients of the fluid velocity, of the temperature, or the electrical potential, or other quantities. If the reservoir is withdrawn we can study the subsequent movement of the isolated system to equilibrium.

The study is based on the specification of local thermodynamic equilibrium for which collective parameters of the thermodynamic type apply locally, related by the conventional formulae of thermodynamics. The parameters appropriate to one location are different from those applying nearby, although the difference is small for two locations not too far apart. Each local equilibrium will have the canonical form for equilibrium although the Hamiltonian and the temperature will depend on the location and the time. The time evolution of the distribution will be dictated by the Liouville equation. Again the fluxes follow from the ensemble everage (1.9.1) once the Liouville equation has been solved, and may be used to obtain expressions for the transport coefficients. As an example, if T_{ij} is the i-th component of the momentum flux in the j-direction, the dynamical (shear) viscosity η and the dilatational viscosity ζ are expressed in the form of the auto-correlations

$$\eta = \frac{2}{Vk_{\mathrm{B}}T}\int_0^{\tau} \langle T_{ij}(t)T_{ij}(0)\rangle \,\mathrm{d}t$$

$$\zeta - \frac{2}{3}\eta = \frac{1}{Vk_{\mathrm{B}}T}\int_0^{\tau} \langle T_{jj}(t)T_{jj}(0)\rangle \,\mathrm{d}t. \tag{1.9.9a}$$

The thermal conductivity λ is correspondingly

$$\lambda = -\frac{1}{Vk_{\mathrm{B}}T}\int_0^{\tau} \langle \mathbf{Q}(t)\cdot\mathbf{Q}(0)\rangle \,\mathrm{d}t \tag{1.9.9b}$$

where $\mathbf{Q}$ is the energy flux vector. The essential ingredient in calculations is the solution of the Liouville equation or some acceptable approximation to this. This has proved difficult so far even within a linearized theory.

1.10 Stochastic theory: Brownian motion

The importance of the auto-correlation function in the description of liquid properties suggests the development of specific kinetic models based explicitly on stochastic arguments. An example of such a model of irreversible behaviour is the semi-macroscopic theory of the Brownian motion, where the central feature is the stochastic force acting on the Brownian particle.[24]

1.10.1 Equation of Langevin

It is supposed that a time interval exists during which the semi-macroscopic particle suffers a very large number of collisions with the molecules of the suspending fluid, but without the macroscopic phase of the particle changing in consequence by more than infinitesimal amounts. Although the particle is accelerated by collisions with individual neighbouring molecules it is continually retarded by interaction with them collectively through the macroscopic viscosity. In constructing the equation of motion for the Brownian particle, then, the acceleration is represented by a fluctuating acceleration $\mathbf{A}(t)$, which is supposed independent of location, and the deceleration is represented as a macroscopic friction $-\beta\mathbf{u}(t)$, where $\mathbf{u}(t)$ is the velocity of the Brownian particle at the time t and β is a macroscopic dynamical friction constant. For a spherical particle of radius R and mass m, β is given by the Stokes formula $\beta m = 6\pi R\eta$ where η is the shear viscosity corresponding to the temperature T of the equilibrium fluid. Finally, the whole system may be subjected to an acceleration $\mathbf{K}$ of external origin (such as gravity), but in general this acceleration could be dependent on location. The equation for the resultant instantaneous acceleration $\mathbf{a}(t)$ of the Brownian particle was expressed by Langevin in the form

$$\mathbf{a}(t) = -\beta\mathbf{u}(t) + \mathbf{A}(t) + \mathbf{K}(\mathbf{r}). \tag{1.10.1}$$

This expression can be regarded as a special form of the regression law (1.9.8). The energy transport in the mechanism has a cyclic character if the external force is neglected: random energy from the molecules of the fluid passes to the Brownian particle by collisions and the resulting kinetic energy of the particle is transferred back to random molecular motion by the action of the viscous force.

Equation (1.10.1) is integrated to provide expressions for the mean and mean square particle displacements during the time interval τ respectively $\langle \Delta\mathbf{r} \rangle$ and $\langle \Delta\mathbf{r} \cdot \Delta\mathbf{r} \rangle$ where the angular brackets denote an average over the possible velocities for the particle in conformity with the equations of motion. Explicitly we obtain the formulae

$$\langle \Delta\mathbf{r} \rangle = -(\beta\mathbf{u} - \mathbf{K})\tau \tag{1.10.2a}$$

$$\langle \Delta\mathbf{r} \cdot \Delta\mathbf{r} \rangle = \frac{6k_\mathrm{B}T}{m\beta}\left(\tau - \frac{1}{\beta}\,[1 - \exp(-\beta\tau)]\right). \tag{1.10.2b}$$

For small time intervals these formulae give $\langle \Delta\mathbf{r} \rangle \propto \tau$ and $\langle \Delta\mathbf{r} \cdot \Delta\mathbf{r} \rangle \propto \tau$, but for large values of τ we find instead

$$\langle \Delta\mathbf{r} \cdot \Delta\mathbf{r} \rangle = \frac{6k_\mathrm{B}T}{\beta m}\,\tau \tag{1.10.3}$$

and this long time behaviour is a characteristic feature of an irreversible process. Analogous formulae apply for the momentum increment during τ: in particular

$$\langle \Delta\mathbf{p} \cdot \Delta\mathbf{p} \rangle = 2mk_\mathrm{B}T\beta\tau. \tag{1.10.4}$$

If we interpret macroscopic diffusion as the random flights of a very large number of particles, then the constant coefficient in equation (1.10.3) is the diffusion coefficient

$$D = \frac{k_B T}{\beta m}. \tag{1.10.5}$$

The associated mobility coefficient is $(\beta m)^{-1}$.

1.10.2 Transition probabilities

Let $f(\mathbf{p}, \mathbf{r}, t)\,\mathrm{d}\mathbf{p}\mathrm{d}\mathbf{r}$ be the probability of finding the Brownian particle with momentum $\mathbf{p}$ and position $\mathbf{r}$ at time t within the elementary range $\mathrm{d}\mathbf{p}\mathrm{d}\mathbf{r}$. Suppose $W(\Delta\mathbf{p}, \Delta\mathbf{r})$ to be the probability that the particle will undergo the phase increment during the time interval τ. Continuity of f is expressed by the integral equation

$$f(\mathbf{p} + \Delta\mathbf{p}, \mathbf{r} + \Delta\mathbf{r}, t + \tau) = \int\int f(\mathbf{p}, \mathbf{r}, t)\, W(\Delta\mathbf{p}, \Delta\mathbf{r})\,\mathrm{d}(\Delta\mathbf{p})\,\mathrm{d}(\Delta\mathbf{r}). \tag{1.10.6}$$

If the increments $\Delta\mathbf{p}$ and $\Delta\mathbf{r}$ are sufficiently small the functions f and W can be expanded in a Taylor series to yield the integral equation

$$\begin{aligned} &\frac{\partial f}{\partial t} + \mathbf{p} \cdot \frac{\partial f}{\partial \mathbf{r}} = -\frac{\partial}{\partial \mathbf{p}} \cdot \left[\frac{\langle \Delta\mathbf{p} \rangle}{\tau} f + \frac{1}{2} \frac{\partial}{\partial \mathbf{p}} \frac{\langle \Delta\mathbf{p} \cdot \Delta\mathbf{p} \rangle}{\tau} f \right] - \mathbf{K} \cdot \frac{\partial f}{\partial \mathbf{p}} \\ &\langle \Delta\mathbf{p} \rangle = \int \Delta\mathbf{p} W(\Delta\mathbf{p})\,\mathrm{d}(\Delta\mathbf{p}) \\ &\langle \Delta\mathbf{p} \cdot \Delta\mathbf{p} \rangle = (\Delta\mathbf{p} \cdot \Delta\mathbf{p})\, W(\Delta\mathbf{p})\,\mathrm{d}(\Delta\mathbf{p}). \end{aligned} \tag{1.10.7}$$

This equation is often called the Fokker–Planck equation, and is the generalization of the Liouville equation to include a stochastic force. Application is made to the Brownian motion by making an appropriate specification of the moments $\langle \Delta\mathbf{p} \rangle$ and $\langle \Delta\mathbf{p} \cdot \Delta\mathbf{p} \rangle$.

Integrating $f(\mathbf{p}, \mathbf{r}, t)$ over the momentum provides a configuration distribution $n(\mathbf{r}, t)$ according to

$$n(\mathbf{r}, t) = \int f(\mathbf{p}, \mathbf{r}, t)\,\mathrm{d}\mathbf{p} \tag{1.10.8}$$

and we can construct an expression for the continuity in n from

$$n(\mathbf{r} + \Delta\mathbf{r}, t + \tau) = \int n(\mathbf{r}, t)\,\omega(\Delta\mathbf{r})\,\mathrm{d}(\Delta\mathbf{r}) \tag{1.10.9}$$

where $\omega(\Delta\mathbf{r})$ is the configurational transition probability. Performing a Taylor expansion as before provides the equation

$$\frac{\partial n}{\partial t} = \frac{\partial}{\partial \mathbf{r}} \cdot \left[\frac{\langle \Delta\mathbf{r} \cdot \Delta\mathbf{r} \rangle}{6\tau} \frac{\partial n}{\partial \mathbf{r}} - \frac{\langle \Delta\mathbf{r} \rangle}{\tau} \right] \tag{1.10.10}$$

which is often named after von Smoluchowski.

If the relaxation time for the system is small enough for the momentum distribution to reach an equilibrium form before the configuration distribution, the phase equation (1.10.7) then reduces to the form (1.10.10). These equations apply to the Brownian motion if the expressions for the moments given in Section 1.10.1 are inserted into them. The condition that equation (1.10.7) shall reduce to equation (1.10.10) is that β shall be sufficiently large.

1.10.3 Correlation functions

The Langevin equation can be integrated and arranged into a form specifically showing correlations of the particle velocity and of the fluctuating force. Expressions for the diffusion coefficient D and Stokes' friction constant derived in this way are

$$D = \frac{1}{3}\int_0^\infty \langle \mathbf{u}(t+s) \cdot \mathbf{u}(t) \rangle \, \mathrm{d}s \tag{1.10.11a}$$

$$\beta = \frac{m}{6k_{\mathrm{B}}T}\int_0^\infty \langle \mathbf{A}(t+s) \cdot \mathbf{A}(t) \rangle \, \mathrm{d}s \tag{1.10.11b}$$

for large times.

The representation of diffusion as the random motion of many particles allows formulae such as (1.10.11) to be evaluated on the bases of computer experiments (molecular dynamics), and the behaviour of the various correlation functions studied for a range of time intervals, not necessarily large.

1.11 Stochastic motion in liquids

The arguments of the last section give a full physical insight into one theory of irreversibility which, on the basis of the arguments of Section 1.8, can be expected to have some relevance in the treatment of simple liquids. The continual interaction between the many molecules of the liquid can be expected to lead to a stochastic motion of broadly the Brownian type, but there is one significant difference: whereas the Brownian particle is substantially more massive than the molecules that give it apparent life, all the molecules in a single component liquid will have the same mass and size. This will mean that the molecule under discussion will have a smaller inertia than the surroundings and the arguments of Section 1.8.4 will apply. This is the basis of the theory first explored by Kirkwood and subsequently developed by a number of authors.

1.11.1 Phase equation

Irreversibility will be introduced through a friction constant analogous to (1.10.1) and (1.10.11b) by taking the time average of the force correlation function (1.8.4). Thus

$$\beta^{(n)} = \frac{1}{3mk_{\mathrm{B}}T}\int_{t+\tau}^{\tau} \langle \mathbf{F}(t) \cdot \mathbf{F}(t+s) \rangle_{N-n} \, \mathrm{d}s, \tag{1.11.1}$$

where m is the mass of the particle. Provided the correlation function vanishes after a time τ_{R}, say, the integral will achieve a constant value for $\tau > \tau_{\mathrm{R}}$ and is to be interpreted as a friction constant of the Stokes type. Kirkwood showed the validity of this interpretation by determining the way in which the full Liouville equation can be reduced to the Fokker–Planck equation (1.10.7) or the Smoluchowski equation (1.10.10) for a single molecule or for a pair of molecules.

The theory is concerned with a time-averaged coarse grained distribution in phase (in the sense of Section 1.8.3) and the Liouville equation is integrated over the phase of $(N-1)$ or $(N-2)$ unwanted molecules. The phase of two molecules can be expressed as the relative phase of one compared with that of the other treated as the standard so the phase equation for a pair of molecules is expressible in terms of a singlet distribution. The result of such a reduction is the construction of a singlet (or relative singlet) equation of the form

$$\frac{\partial \bar{f}^{(1)}}{\partial t} + \frac{\mathbf{p}_1}{m} \cdot \frac{\partial \bar{f}^{(1)}}{\partial \mathbf{r}_1} = \frac{\Delta \bar{f}^{(1)}}{\tau} \tag{1.11.2}$$

where $\Delta\bar{f}^{(1)}$ contains the contributions due to interactions between the molecules. The form of this term is determined by the nature of the interaction. If we suppose that the molecular movement in a liquid is characterized by the momentum increment during the time interval τ being small in comparison with both the initial and final momenta, then $\Delta\bar{f}^{(1)}$ takes the form

$$\frac{\Delta\bar{f}^{(1)}}{\tau} = \beta^{(1)} \frac{\partial}{\partial \mathbf{p}_1} \cdot \left(\left(\frac{\mathbf{p}_1}{m} - \mathbf{u} \right) \bar{f}^{(1)} + k_B T \frac{\partial \bar{f}^{(1)}}{\partial \mathbf{p}_1} \right) \tag{1.11.3}$$

and equation (1.11.2) the form of the Fokker–Planck equation (1.10.7). $\beta^{(1)}$ is the friction constant which results when n is set equal to unity in equation (1.11.1).

Kirkwood[30] traced the form of the Langevin equation associated with equation (1.11.3), showing the validity of the interpretation of $\beta^{(1)}$ as a friction constant in this way.

1.11.2 *Friction constant*

Various attempts have been made to calculate the friction constant $\beta^{(1)}$ from the defining expression (1.11.1). The methods involve approximations and the last word has not been said on this very difficult subject.

Kirkwood attempted the first estimate of $\beta^{(1)}$. Writing

$$\langle \mathbf{F}_1(t) \cdot \mathbf{F}_1(t+s) \rangle = \langle \mathbf{F}_1(t) \cdot \mathbf{F}_1(t) \rangle \phi(s). \tag{1.11.4}$$

Assuming $\phi(s)$ will decay in the same way as the average momentum then $\phi = m/\beta^{(1)}$. Assuming further that the environment of the molecule is in local equilibrium we can express the mean square force in terms of the static pair equilibrium distribution according to:

$$\frac{\langle F_1^2 \rangle}{k_B T} = \frac{4\pi N}{V} \int_0^\infty g(r)(\nabla^2 \psi) r^2 \, dr$$

Consequently

$$(\beta^{(1)})^2 = \frac{4\pi m}{3} \int_0^\infty g(r)(\nabla^2 \psi) r^2 \, dr \tag{1.11.5}$$

which is an expression derived by Rice and Kirkwood. Alternative expressions have been pioneered, but the numerical magnitudes are all generally the same. Generally the estimates fall in the range 10^{-10} to $10^{-11}\,\mathrm{g\,s^{-1}}$. The precise form will depend on the details of the potential of the pair interaction ψ.

1.11.3 *Configuration equation*

The expressions (1.11.4) and (1.11.5) refer to configuration space, and if we suppose that the correlation time τ is sufficiently large for $\beta^{(1)}\tau \gg 1$, the Fokker–Planck equation (1.10.7) is replaced by the Smoluchowski equation (1.10.10). The detailed transition from phase to configuration spaces is based on equation (1.3.10). The result is the equation of continuity in singlet space

$$\frac{\partial n^{(1)}}{\partial t} = -\frac{\partial \mathbf{j}}{\partial \mathbf{r}} \tag{1.11.6}$$

where the mass current $\mathbf{j}$ is, retaining only terms of order $(1/\beta)$

$$\mathbf{j} = \frac{\partial}{\partial \mathbf{r}} \left(\frac{k_B T}{m} n^{(1)} \right) - \frac{\langle F \rangle}{\beta m} n^{(1)} + O(1/\beta^2) \tag{1.11.7}$$

and $\langle \mathbf{F} \rangle$ is the mean force on the molecule. This is written in terms of the potential of the mean force by

$$\langle \mathbf{F} \rangle = -\frac{\partial \Phi}{\partial \mathbf{r}} = \frac{\partial}{\partial \mathbf{r}} [k_B T \ln g(r)]. \tag{1.11.8}$$

The insertion of equations (1.11.7) and (1.11.8) into equation (1.11.6) provides the final equation for $n^{(1)}$. The boundary conditions to be applied depend upon the particular non-uniformity being studied.[31]

Terms of higher power in $(1/\beta)$ than the first can be included in an expansion procedure to account for cases where the friction constant is not high enough for the simple transition described in section 1.10.2 to be sufficient description. Rather lengthy algebraic manipulation leads to a fourth order equation to replace the second order equation (1.10.10); it has proved too complicated to use in calculations so far but it has importance in connection with the boundary conditions to be applied.

1.11.4 Equations for pair distributions

If the deviation from equilibrium is not too great the differences between the non-equilibrium and equilibrium space arrangements of the particles will be small.[32] Let $\bar{g}^{(2)}$ be the relative non-equilibrium pair distribution. For laminar viscous flow over a flat surface we suppose $\dot{\boldsymbol{\epsilon}}$ to be the rate of shear and take $\bar{g}^{(2)}$ to be related to the equilibrium distribution $g^{(2)}$ by

$$\bar{g}^{(2)}(\mathbf{r}) = g^{(2)}(r)[1 + \omega(\mathbf{r})] \tag{1.11.9a}$$

where

$$\omega(\mathbf{r}) = \frac{\beta}{k_B T}\left(\frac{r_i \dot{\epsilon}_{ij} r_j}{r^2} - \frac{1}{3}\frac{\partial u_1}{\partial x_1}\delta_{ij}\right)\omega_2(r) P_2(\cos\theta) + \frac{\beta}{6k_B T}\frac{\partial u_1}{\partial x_1}\omega_0(r) \tag{1.11.9b}$$

and ω_2 and ω_0 are two functions to be obtained by inserting the expressions (1.11.9) into the equations (1.11.6)–(1.11.8).

The function ω_2 refers to pure shear while ω_0 to pure dilatation. The two forms of the Smoluchowski equation for these functions are

$$\frac{d^2\omega_2}{dr^2} + \left(\frac{2}{r} - \frac{1}{k_B T}\frac{d\Phi}{dr}\right)\frac{d\omega_2}{dr} - \frac{6\omega_2}{r} = -\frac{r}{k_B T}\frac{d\Phi}{dr} \tag{1.11.10a}$$

$$\frac{d^2\omega_0}{dr^2} + \left(\frac{2}{r} - \frac{1}{k_B T}\frac{d\Phi}{dr}\right)\frac{d\omega_0}{dr} = -\frac{r}{k_B T}\frac{d\Phi}{dr}. \tag{1.11.10b}$$

These equations are solved subject to the boundary conditions proposed by Kirkwood[33]

$$\left.\begin{aligned} r^2 g(r)\frac{d\omega_0}{dr} &= 0 \quad (r \to \infty) \\ \omega_0(r) &= 0 \quad (r \to 0) \end{aligned}\right. \tag{1.11.11}$$

and

$$\left.\begin{aligned} r^3\omega_2(r) &= 0 \quad (r \to \infty) \\ \frac{d\omega_2}{dr} &= 0 \quad (r \to \infty). \end{aligned}\right. \tag{1.11.12}$$

For a linear temperature gradient,

$$T = T_0(1 + \boldsymbol{\alpha} \cdot \mathbf{r}) \tag{1.11.13}$$

and the appropriate Smoluchowski equation is

$$\frac{d^2\omega_1}{dr^2} + \left(\frac{2}{r} - \frac{1}{k_B T}\frac{d\Phi}{dr}\right)\frac{d\omega_1}{dr} - \frac{2\omega_1}{r} = \frac{\alpha}{(k_B T)^2}\frac{d\Phi}{dr} \tag{1.11.14}$$

with the boundary conditions

$$\frac{\omega_1}{r} = \frac{d\omega_1}{dr} = -|\boldsymbol{\alpha}| \quad (r \to \infty). \tag{1.11.15}$$

1.11.5 Transport coefficients

The insertion of the expression (1.11.9a) for the non-equilibrium pair distribution into the equations of Section 1.9.1 for the flux of momentum gives the two equations for the dynamical (shear) viscosity η and dilatational (bulk) viscosity ζ

$$\eta = \frac{nmk_B T}{2\beta} + \frac{\pi\beta}{15k_B T} n^2 \int_0^\infty \frac{\partial\psi(r)}{\partial r}\omega_2(r)g(r)r^3\,dr \tag{1.11.16a}$$

$$\zeta = \frac{\pi\beta}{ak_B T} n^2 \int_0^\infty \frac{\partial\psi(r)}{\partial r}\omega_0(r)r^3 g(r)\,dr. \tag{1.11.16b}$$

For the thermal conductivity λ_T we have

$$\lambda_T = \frac{n^2\pi k_B T}{\beta}\left[\frac{1}{3}\int_0^\infty r\frac{\partial\psi(r)}{\partial r} - \psi(r)g(r)\frac{\partial\omega_1}{\partial r}r^3\,dr + \int_0^\infty \left(\psi(r) - \frac{r}{3}\frac{\partial\psi}{\partial r}\right)\omega_1 r^2\,dr\right]. \tag{1.11.17}$$

These various expressions are to be applicable to a monatomic insulating liquid and require, for their evaluation, a knowledge of the static radial distribution function and the corresponding pair distribution for the liquid. It is seen that the evaluation of the transport coefficients according to these kinetic theory arguments requires that related problems of the equilibrium structure have been solved.

1.11.6 More general interaction

The spatial distribution of fluid molecules is determined primarily by the strong repulsive features of the intermolecular force and the volume available to each molecule is restricted by the presence of neighbours. The representation of the molecular motion as a simple Brownian movement does not account for the restrictive effects of neighbours, and the analysis developed so far must be modified on this account. The stochastic movement for which $\Delta\mathbf{p}$ is small will be broken after a certain time by a binary encounter with a neighbouring molecule characterized by the condition that $\Delta\mathbf{p}$ is large. These binary encounters will constrict the Brownian type movement and will make a contribution to the overall microscopic flux of mass, momentum and energy.

The effect can be treated separately from the Brownian motion and is included in the theory through equation (1.11.2) by making a particular specification of the right hand side. This has the form of the collision term of Boltzmann, well known in the kinetic theory of the dilute gas. Taking account of the conservation of mass, momentum, and energy during a binary encounter between the stochastic molecule 1 and a restraining neighbour 2 we have

$$d\mathbf{r}_1\, d\mathbf{p}_1 \frac{\Delta f^{(1)}}{\tau} = \frac{N}{\tau}\int\int [f^{(2)}(t-\tau) - f^{(2)}(t)]\, d\mathbf{p}_2\, d\mathbf{r}_2 \equiv J d\mathbf{r}_1\, d\mathbf{p}_1. \tag{1.11.18}$$

The integral is rearranged into the conventional form by introducing the hypothesis of molecular chaos which is an expression for the statistical independence of the two molecules: explicitly

$$f^{(2)}(\mathbf{p}_1, \mathbf{p}_2, \mathbf{r}_1, \mathbf{r}_2) = f^{(1)}(\mathbf{p}_1, \mathbf{r}_1, t) f^{(1)}(\mathbf{p}_2, \mathbf{r}_2, t). \tag{1.11.19}$$

Provided the frequency of binary encounters is sufficiently low to allow the Brownian type motion to develop between collisions, the equation of motion can be taken to be equation (1.11.2) with the right hand side the sum of a Fokker–Planck term (1.11.3), denoted by M, and the Boltzmann term J given by the combination (1.11.18) and (1.11.19). This is the kinetic equation developed some time ago by Rice and Allnatt.[34]

The molecular interaction will have the form of a strong repulsion at small separation distances and a weak interaction beyond this. These two contributions can be kept separate in the theory and each will make a characteristic contribution to the transport coefficients and to the total friction constant.

For the friction constant, a hard and soft contribution can be recognized referring respectively to a strong repulsion and to the remaining interaction. For the hard core contribution β_H it is found that, for a hard sphere repulsion of radius σ

$$\beta_H = \tfrac{8}{3} n\sigma^2 g(\sigma)(\pi m k_B T)^{1/2} \tag{1.11.20}$$

and the soft contribution β_S is given by

$$\beta_S = \frac{k_B T}{Dm} - \beta_H. \tag{1.11.21}$$

Because β_H can be calculated and D measured, β_S can be inferred. We see that β_S and β_H will be expected to respond differently to change of temperature.

1.11.7 Brief comments on numerical results

Generally speaking, the calculated data for viscosity and thermal conductivity agree with the measured data for a monatomic liquid surprisingly well. The calculations depend on a knowledge of the equilibrium pair distribution function for a specific intermolecular pair interaction potential, and this is not known very well from the present point of view. There is also uncertainty in connection with the friction constant. Neverthless, data of the correct magnitude are found for molecular variables adapted to liquid argon, and for other simple liquids. Taking the radial distribution from X-ray data, and assigning ψ a Lennard–Jones (12–6) form but with a hard core, it was found by Zwanzig, Kirkwood, Stripp, and Oppenheim the $\eta = 0.84 \times 10^{-3}$ P and $\zeta = 0.42 \times 10^{-3}$ P on the basis of the simple Brownian type interaction mechanism. Apparently η and ζ are of the same magnitude; the calculation of the value for ζ is a definite theoretical advance since macroscopic hydrodynamics is unable to assign a magnitude to this coefficient. The thermal conductivity was found to be $\lambda = 2.4 \times 10^{-4}$ cg^{-1} sK^{-1}. While $\eta \propto \beta$, $\lambda \propto 1/\beta$, the product $\eta\lambda$ is independent of the friction constant so that the calculation of the product $\eta\lambda$ provides a test of the theory independent of β. Theory predicts the value $\eta\lambda = 2.02 \times 10^{-7}$ whereas experiment yields the value about 4×10^{-7}. The order of magnitude is correct but the numerical factor is wrong by a factor 2. Zwanzig *et al.*[35] found that the calculated value of λ is critically dependent on the distance scale and that by replacing $g(r)$ by $g(1.026r)$ the calculated value of d is doubled – this provides a rough numerical agreement with experiment.

The Rice–Allnatt theory, involving the equation (1.11.18), has also been used to calculate η and λ. Here a new contribution enters the calculation arising from the Boltzmann-type interaction. The formulae are complicated and the reader is referred elsewhere for details. The phase equation (1.11.3) is reduced to a configuration equation and it is this latter equation which is solved for the spatial pair distribution. The friction constant is deduced by equations (1.11.5) and (1.11.21), using experimentally-inferred diffusion data. The calculated data agree with experimental data probably to better than 50% in spite of the uncertainties of g and ψ already referred to. The temperature dependence is also of the correct magnitude and sign. The Rice–Allnatt theory is probably the best as yet available for the overall calculation of transport coefficients. Rice and his colleagues have also attempted to satisfy, at least partially, the present need for more extensive experimental data for the transport coefficients.

1.12 More general approaches

Developments over the last decade or so have concentrated on two approaches; one is the formal solution of the Liouville equation and the other is the evaluation of the correlation functions, particularly involving the scattering function S introduced in equation (1.3.18). The latter evaluation provides a direct link with experiment, since X-ray and neutron scattering data can be arranged to yield S directly. We will consider these two aspects separately.

1.12.1 Formal development of the Liouville equation

The Liouville equation (1.3.7) has the formal solution (1.3.9) involving the exponential factor which is defined by the expansion

$$\exp\{-i\mathscr{L}t\} = 1 + \sum_{j=1}^{\infty} (-1)^j (i\mathscr{L}t)^j (j!)^{-1}. \tag{1.12.1}$$

The exponential operator is called the time propagation operator, or sometimes the propagator, and its form will depend on the form of the Liouville operator $\mathscr{L}$. If $\hat{f}^{(N)}$ is the Laplace transform of the distribution function $f^{(N)}$ then

$$\hat{f}^{(N)}(k) = \int_0^\infty f^{(N)}(t)\exp(-kt)\,dt \tag{1.12.2}$$

and the Laplace transform of the Liouville equation (1.3.7) is

$$ik\hat{f}^{(N)} - if^{(N)}(0) = \mathscr{L}\hat{f}^{(N)}(k) \tag{1.12.3}$$

remembering that $\mathscr{L}$ is independent of the time. Rearrangement of this last expression gives

$$\hat{f}^{(N)}(k) = -i(\mathscr{L} - ik)^{-1} f^{(N)}(0) \tag{1.12.4}$$

which is the Laplace transform of the propagator equation (1.3.9). Here the operator $(\mathscr{L} - ik)^{-1}$ is the inverse of the operator $(\mathscr{L} - ik)$. The integral inverse of $\hat{f}^{(N)}(k)$ will provide an expression for the distribution $f^{(N)}(t)$ itself, viz.

$$f^{(N)}(t) = -\frac{1}{2\pi i}\int_{\infty+ic}^{-\infty+ic} (\mathscr{L} - z)^{-1}\exp(-izt)f^{(N)}(0)\,dz \tag{1.12.5}$$

where $z = ik$ is the complex variable. Because $\mathscr{L}$ is Hermitian it has real eigenvalues only so that $(\mathscr{L} - z)^{-1}f^{(N)}(0)$ will have singularities restricted to the real axis. The integral (1.12.5) is an example of the Bromwich integral well known in the theory of the complex variable. It represents a formal solution of the Liouville equation, but the evaluation of the integral presents formidable difficulties looked at from the point of view of a theory of liquids because it is merely another way of looking at the problem of solving the Liouville equation for N interacting particles contained in the volume V.

The only present technique for the evaluation of (1.12.5) is to use a perturbation expansion based on the characteristics of the interaction between the molecules. This formalism, which is closely related to that met in connection with the Schrödinger equation of quantum physics, has been developed particularly by Prigogine and his school.[36]

The operator $\mathscr{L}$ is assigned eigenfunctions ϕ_k with eigenvalues λ_k so that

$$\mathscr{L}\phi_k = \lambda_k \phi_k. \tag{1.12.6}$$

The linearity of the Liouville equation allows $f^{(N)}$ to be expanded in the form

$$f^{(N)}(t) = \sum_k a_k(t)\phi_k(\mathbf{p}^N, \mathbf{r}^N) \tag{1.12.7}$$

where (from the Liouville equation (1.3.7) and (1.12.6), the coefficients a_k are given by

$$a_k(t) = C_k \exp(-i\lambda_k t). \tag{1.12.8}$$

The C_k are numerical coefficients determined by $f^{(N)}(0)$.

The eigenfunctions ϕ_k are given a physical association by choosing one physical system where the solution of equation (1.3.7) is known exactly. This can be taken to be the case of N non-interacting particles in a cubic box of side L. The Liouville operator in this case is $\mathscr{L}_0$ where

$$\mathscr{L}_0 = -i \sum_j \frac{\mathbf{p}_j}{m} \cdot \frac{\partial}{\partial \mathbf{r}_j} \tag{1.12.9}$$

and it follows easily that

$$\phi_k = \left(\frac{1}{\mathscr{L}}\right)^{3/2} \exp\{i(\mathbf{k}_j \cdot \mathbf{r}_j)\} \tag{1.12.10a}$$

where

$$\mathbf{k}_j = \frac{2\pi}{\mathscr{L}} \mathbf{n}_j \tag{1.12.10b}$$

is the wave vector corresponding to the vector $\mathbf{n}_j$ with integer components. If the particle interactions are not too strong it is plausible to use (1.12.10a) in the expansion (1.12.7) and so obtain

$$f^{(N)}(\mathbf{p}^N, \mathbf{r}^N, t) = \left(\frac{2\pi}{\mathscr{L}}\right)^{3N} \sum_{\{\mathbf{k}\}} \rho_{\{\mathbf{k}\}}(\mathbf{p}^N, t) \exp\left\{ i \sum_j \left[\mathbf{k} \cdot \left(\mathbf{r}_j - \frac{\mathbf{p}_j}{m} t \right) \right] \right\} \tag{1.12.11}$$

with

$$\rho_{\{\mathbf{k}\}} = \left(\frac{\mathscr{L}}{4\pi^2}\right)^{3N/2} a_{\{\mathbf{k}\}}. \tag{1.12.11a}$$

This particular treatment of the Liouville equation has effectively replaced the system of interacting particles by a system of coupled wave fields. The effect of the coupling (i.e. particle interactions) is described by the specification of the coefficients ρ in (1.12.11). These are obtained explicitly by inserting (1.12.11) into (1.3.7) or equivalently by evaluation of the integral (1.12.5).

The integral (1.12.5) is expanded in an analogous way. We divide the Liouville operator into a purely kinetic and a purely interaction component according to

$$\mathscr{L} = \mathscr{L}_0 + \lambda\delta\mathscr{L} \tag{1.12.12}$$

where λ is a parameter which measures magnitude of the interaction forces. From equation (1.12.12) we can write

$$(\mathscr{L} - z)^{-1} - (\mathscr{L}_0 - z)^{-1} = (\mathscr{L}_0 - z)^{-1}\,[(\mathscr{L}_0 - z) - (\mathscr{L} - z)](\mathscr{L} - z)^{-1}.$$

That is,

$$(\mathscr{L} - z)^{-1} = (\mathscr{L}_0 - z)^{-1} - (\mathscr{L}_0 - z)^{-1}(\lambda\delta\mathscr{L})(\mathscr{L} - z)^{-1}. \tag{1.12.13}$$

The rearrangement of this expression to give $(\mathscr{L} - z)^{-1}$ in terms of $\mathscr{L}_0$ requires approximation. If $\lambda = 0$,

$$(\mathscr{L} - z)^{-1} = (\mathscr{L}_0 - z)^{-1}$$

which is the approximation of non-interacting particles. The next approximation is the insertion of this into the right hand side of (1.12.13); continued iteration in this way gives

$$(\mathscr{L} - z)^{-1} = \sum_{j=0}^{n} (-\lambda)^n (\mathscr{L}_0 - z)^{-1}\,[\delta\mathscr{L}(\mathscr{L}_0 - z)^{-1}]^n. \tag{1.12.14}$$

where $\delta\mathscr{L}$ is defined by

$$\delta\mathscr{L} = \mathrm{i}\lambda \sum_j \sum_k [\nabla_{r_k}\psi(j,k)] \cdot [\nabla_{\mathrm{p}_j} - \nabla_{\mathrm{p}_k}]$$

The interaction terms can now be included piecemeal in a systematic way, and the single integration necessary for the solution of the original Liouville equation is introduced as a large number of integrations over restricted particle interactions. Each component of interaction is summed to all powers of the density. The complete interaction is accounted for by evaluating all the component series separately, and in practice this has presented the most severe difficulties. The application of the analysis to liquid densities is restricted by purely mathematical difficulties – but these are quite severe. Progress has involved the use of a modification of the Feynman notation previously developed in quantum physics, but only two of the series have been evaluated directly so far. The detailed discussion of the various terms is too lengthy and complicated to attempt here.

The complete interaction series is specified in terms of the three parameters λ (of equation (1.12.12)), $n = N/V$ and the time t. When only the hard core contribution to the pair potential is included, the theory provides the Boltzmann equation for the simple dilute gas. On the other hand, the Fokker–Planck equation is obtained when the remaining weakly repulsive and attractive features are included in terms of order $(\lambda^2 nt)^s$, where the integer $s > 0$, in the limit of times very long in comparison with those for a single interaction between a pair of particles. For short times, of the order $(\lambda^2 n)^s$, the Vlasov equation results, well known in connection with a gaseous plasma. Other terms have been explored, but the application of this pragmatic approach to liquid densities is not based on clear-cut physical arguments.

The theory has been applied by several authors[37] to the calculation of viscosity and thermal conductivity, and the work is continuing. In parallel, the first steps have been taken to link the general theory to the restricted theories already available, and the whole field is moving from a fragmented discussion with a number of theories to a unified theory for simple dense fluids. The extension to complicated liquids has hardly been developed so far.

1.12.2 *Master equations*

The complexity of the theory outlined so far makes it highly desirable to have available more simple models of matter which, although they are inadequate as a basis for the calculation of data to be associated with observed properties of real matter, nevertheless show specific general features of practical interest. Of especial interest now are equations showing irreversible features, ofter referred to as master equations. They can be derived from the hierarchy of equations (1.12.5) by neglecting interaction contributions on mathematical rather than physical grounds. Thus an elementary master equation follows by retaining only diagrams of order $(\lambda^2 t)^n$, referring to the time evolution of the momentum distribution in a weak coupling aproximation. The most general deduction is based on quantum arguments, but we refer the reader elsewhere.

1.12.3 *Projection operators*

Master equations, in the weak coupling approximation, can be derived by introducing a time-dependent linear projection operator as was shown by Zwanzig.[38] The importance of the technique lies in that it allows formal functions of great complexity to be rearranged into a form where they can be evaluated for at least a limited range of physical situations.

A projection operator $\mathscr{P}$ is introduced which divides the full distribution $f^{(N)}(t)$ into a component $f_1(t)$ which is directly relevant to a particular fluid non-uniformity of interest and a remainder $f_2(t)$ which is irrelevant. This projection of relevant information from the full specification allows the analysis involving $f^{(N)}$ to be rearranged into an alternative form involving f_1 explicitly but f_2 only implicitly through initial conditions. Without specifying $\mathscr{P}$ for the moment other than requiring that it be linear and time-independent, assume that f_1 and f_2 are defined by

$$f_1(t) = \mathscr{P} f^{(N)}(t) \tag{1.12.15}$$

$$f_2(t) = (1 - \mathscr{P}) f^{(N)}(t) = f^{(N)}(t) - f_1(t).$$

We see that f_1 and f_2 are not singlet and pair distributions by these definitions, although application to liquid densities may require them so to be. Using the Liouville equation (1.3.8) the functions f_1 and f_2 are easily found to satisfy the two coupled equations

$$\mathrm{i}\frac{\partial f_1}{\partial t} = \mathscr{P}\mathscr{L}(f_1 + f_2) \tag{1.12.16a}$$

$$\mathrm{i}\frac{\partial f_2}{\partial t} = (1 - \mathscr{P})\mathscr{L}(f_1 + f_2) \tag{1.12.16b}$$

and equation (1.12.16b) is integrated into the form

$$f_2(t) = \{\exp[-\mathrm{i}t(1 - \mathscr{P})\mathscr{L}]\} f_2(0) - \mathrm{i}\int_0^t \{\exp[-\mathrm{i}s(1 - \mathscr{P})\mathscr{L}]\}(1 - \mathscr{P})\mathscr{L} f_1(t+s)\,\mathrm{d}s. \tag{1.12.17}$$

In this way $f_2(t)$ is represented by initial information and integrated information about f_1. The insertion of equation (1.12.17) into (1.12.16a) gives the final equation for f_1

$$\mathrm{i}\frac{\partial f_1(t)}{\partial t} = \mathscr{P}\mathscr{L}\{\exp[-\mathrm{i}t(1-\mathscr{P})\mathscr{L}]\}f_2(0) + \mathscr{P}\mathscr{L}f_1(t)$$

$$-\mathrm{i}\int_0^t \mathscr{P}\mathscr{L}\{\exp[-\mathrm{i}s(1-\mathscr{P})\mathscr{L}]\}(1-\mathscr{P})\mathscr{L}f_1(t-s)\mathrm{d}s. \tag{1.12.18}$$

The time change of f_1 at any instant depends on the past history of the system through the integral term: such a system is non-Markovian. By contrast, the Fokker–Planck and Smoluchowski equations describe Markovian systems.

The equation (1.12.18), due to Zwanzig, is exact and can be written in the alternative form

$$\frac{\partial f_1}{\partial t} = \mathscr{D}_1(t)f_2(0) + \mathscr{D}_2(t)f_2(t) + \int_0^t \mathscr{K}(s)f_1(t-s)\mathrm{d}s \tag{1.12.19}$$

where the operators $\mathscr{D}_1$ $\mathscr{D}_2$ and $\mathscr{K}$ are identifiable from (1.12.18). The important feature is the term involving the integral, which introduces the memory effect. If t is large in comparison with the time of interaction the correlation is broken and the memory effect disappears. The evolution of f_1 then depends on the present state of the system and not on past configurations. Thermodynamic equilibrium is one such condition of this type but is not the only one. Once details of any initial non-uniformity are lost and the uniformity of equilibrium is reached, the system cannot 'remember' anything else and so will not move away from equilibrium. The term $\mathscr{D}_2f_2$ in equation (1.12.20) has a form which depends on the operator $\mathscr{P}$, but is found to vanish in many cases of interest.

Application to particular physical problems is made by defining an appropriate projection operator for the problem, and there is room for skill in doing this.[39] A fully systematic physical criterion for doing the selection is still awaited. As one example, due to Zwanzig, suppose the projection is that onto the particle momentum space so that

$$\mathscr{P} = \frac{1}{V^N}\int \mathrm{d}\mathbf{r}^N. \tag{1.12.21}$$

If the fluid is supposed homogeneous initially, so that $f_1(0) = f_2(0) = 0$ then (1.12.19) becomes

$$\frac{\partial f_1(\mathbf{p}^N, t)}{\partial t} = \lambda^2 \int_0^\infty \mathscr{K}(s) f_1(\mathbf{p}^N, s)\mathrm{d}s \tag{1.12.22a}$$

and the kernel is

$$\mathscr{K}(s) = -\mathscr{P}\delta\mathscr{L}\exp(-\mathrm{i}s\mathscr{L}_0)\delta\mathscr{L} \tag{1.12.22b}$$

using the notation of Section 1.12.1. Equations (1.12.22) are in fact the Prigogine–Brout equations derived independently from the interaction representation by retaining only the terms of order $(\lambda^2 t)^n$. The method can be used to derive equations of both the Boltzmann and Fokker–Planck types, and generalizations of them. There is a close link with the Prigogine theory and the projection operator approach promises many future possibilities.

1.12.4 Generalized Langevin equation

The ideas summarized in the Langevin equation (1.10.1) have been seen to be central to the development of the kinetic theory of non-equilibrium. The equation itself requires generalization if it is to form the basis of a full kinetic theory of liquids. This generalization is the subject of much current work and is still very much under development. Its usefulness for liquid densities is not entirely clear; because its mathematical formalism is complicated we will not give a detailed treatment here but rather give a brief review of its main features.

Kirkwood showed how the Langevin equation (1.10.1) can be obtained from the Liouville equation under appropriate assumptions, and the generalization also follows from the Liouville equation. In particular the equation can involve any arbitrary dynamical variable $Q(t)$ for which the Liouville equation is

$$Q(t) = \exp\{\mathrm{i}\mathscr{L}t\}Q(O) \tag{1.12.23}$$

for mechanical motion, and non-stochastic forces.

Collective oscillations of Q are accounted for by the term

$$\mathrm{i}\Omega Q = \langle \dot{Q}(0) \cdot Q(0) \rangle Q^{-2} \tag{1.12.24}$$

referring to the initial time. Again the friction (damping) term ($-\beta\mathbf{u}$ of equation (1.10.1)) is expressible in terms of the random force $\mathbf{F}$, and the empirical term $\beta\mathbf{u}$ in equation (1.10.1) is written as the integral

$$\int_0^t \mathrm{d}\tau \langle \mathbf{F}(\tau) \cdot \mathbf{F}(0) \rangle Q(t-\tau) Q^{-2} \tag{1.12.25}$$

where the brackets in (1.12.24) and (1.12.25) are an average over the equilibrium ensemble. The integral in equation (1.12.25) introduces a 'memory' component into the analysis, accounting for the correlation of the random force on the particle over the time interval t. The generalization of the original Langevin equation becomes

$$\frac{\partial Q(t)}{\partial t} + \frac{\langle Q(0) \cdot Q(0) \rangle}{\langle Q(0) \cdot Q(0) \rangle} Q(t) + \int_0^T \mathrm{d}t \frac{\langle F(\tau) \cdot F(0) \rangle}{\langle Q(0) \cdot Q(0) \rangle} Q(\tau - t) = \mathbf{A}(t) \tag{1.12.26}$$

for the fluctuating variable $Q(t)$. In the long-time limit this expression approaches the simpler form equation (1.10.1): this could be regarded as a hydrodynamic limit. For intermediate times the random variations of Q could be rapid. If this were so, the new variable could be introduced into equation (1.12.26) by averaging over a constrained equilibrium ensemble, and in this case the fluctuating force will also be represented by an average. Such an average might be expected to vanish for small departures from equilibrium but otherwise it could introduce nonlinear terms into the equation of motion.

The analysis up to this point can be exact, but the application to physical systems has involved the introduction of approximations of one kind or another, and particularly so for liquids. Two choices of the dynamical variable Q have been studied in some detail, particularly by Akcasu and Daniels.[40] In one the dynamical variables are the microscopic densities in configuration space (such as the mass, momentum and energy flows). The result is an exact hydrodynamic description of correlation functions providing transport parameters which depend on the frequency and wavenumber. The second chooses the dynamical variables to be the microscopic phase density function which provides an exact kinetic equation for the density correlation function, which is essentially the van Hove function (1.3.20). Application of both these equations involves approximation, especially in connection with the damping term. If the damping term is neglected altogether, the kinetic equation takes on a Vlasov-type form which has been used in the interpretation of neutron scattering experiments in liquids (see chapter 13 by R. A. Howe). Alternatively, a perturbation expansion of the damping kernel based on the strength of the coupling term can be attempted, and the result here is an equation of the Fokker–Planck or Boltzmann form, depending on the precise nature of the interparticle force potential.

In general terms, these approaches are proving particularly useful in the hydrodynamic regime, and especially for the ranges of wavenumber and frequency encountered in light and neutron scattering. Future experimental work can be expected to support this part of the theory, and our knowledge within the hydrodynamic approximation will develop at least for simple fluids.

1.13 Conclusion

The kinetic theory of dense fluids has been the subject of considerable work over the last thirty years and a wide range of approaches have been made which are closely inter-related. At the same time the rather restricted experimental information given by X-ray diffraction and by the study of simple hydrodynamic flows has been substantially extended by more recent studies involving light scattering or the diffraction of neutrons. The natural method of description of the molecular behaviour of the liquid is that involving the time or space correlation of various dynamical variables, and the description is then represented both by the recognition of variables that are correlated, and those which are not.

There is no lack of exact kinetic equations to allow the calculation of dynamical variables or the correlations between them. They are all, one way or another, alternative forms of the Liouville equation (1.3.8) which is an expression of the conservation of probability for the total number of particles in the fluid. The difficulty is the mathematical one of solving the equations, even within an approximation of physical interest. It is proving very difficult to pass beyond the Boltzmann type and the Fokker–Planck type approximations, being particular forms of the actual molecular interactions. It could be that some alternative representation of this interaction will be needed to make significant further progress; or it may be that a different physical picture is necessary. Only time will tell.

References

There is a wide range of standard books to act as references for the general statistical description of matter. Among them we can quote the following:

R. C. Tolman (1938). *The Principles of Statistical Mechanics*, London, Oxford University Press. This is still an important work for the study of equilibria.

J. E. Meyer and M. G. Meyer (1940). *Statistical Mechanics*, New York, Wiley: this is a very useful book, not, perhaps, known well enough these days.

R. H. Fowler (1936). *Statistical Mechanics*, London, Cambridge University Press. Remains an important standard reference.

J. W. Gibbs (1948). *Collected Works*, New Haven, Yale University Press. This is a very stimulating collection of the work of one of the founders of the ensemble approach to statistical physics.

L. D. Landau and E. M. Lifshitz (1958). *Statistical Physics*, London, Pergamon Press.

T. L. Hill (1956). *Statistical Mechanics*, New York, McGraw-Hill, is concerned particularly with application to the equilibrium states of dense fluids and liquids.

For non-equilibrium conditions a recent very full account of the modern statistical theories is given by:

D.N. Zubarev (1974). *Non-Equilibrium Statistical Mechanics*, 1974, Consultants Bureau, New York.

For dilute gases, a standard detailed account is contained in:

S. Chapman and T. G. Cowling (1939). *The Mathematical Theory of Non-Uniform Gases*, London, Cambridge University Press.

Other accounts of particular use for dense fluids are:

R. K. Eisenschitz (1958). *Statistical Theory of Irreversible Processes*, London, Oxford University Press.

N. N. Bogoliubov (1946). *Dynamical Theory in Statistical Physics*. (English translation: 'Problems of a Dynamical Theory in Statistical Physics', in *Studies in Statistical Mechanics Vol. I* (Ed. de Boer and Uhlenbeck), Amsterdam, North-Holland.)

I. Z. Fisher (1964). *Statistical Theory of Liquids*, Chicago University Press.

G. H. A. Cole (1967). *The Statistical Theory of Classical Simple Dense Fluids*, Oxford, Pergamon Press.

C. A. Croxton (1974). *Liquid State Physics – A Statistical Mechanical Introduction*, London, Cambridge University Press.

S. A. Rice and P. Gray (1965). *The Statistical Mechanics of Simple Liquids*, Interscience.

P. Egelstaff (1967). *An Introduction to the Liquid State*, New York, Academic Press.

I. Prigogine (1962). *Non-Equilibrium Statistical Mechanics*, New York, Interscience.

1. L. Landau and E. M. Lifschitz (1958). *Fluid Mechanics*, Oxford, Pergamon Press.
2. I. Gyarmati (1970). *Non-Equilibrium Thermodynamics*, Berlin, Springer-Verlag.
3. D. N. Zubarev (1974). *Non-Equilibrium Statistical Mechanics*, Consultants Bureau, New York.
4. G. H. A. Cole (1967). *The Statistical Theory of Classical Simple Dense Fluids*, Oxford, Pergamon Press.
   (1968). *Rep. Prog. Phys.*, **31**, 419, London, Institute of Physics.
   (1970). *Rep. Prog. Phys.*, **33**, 737, London, Institute of Physics.
5. J. R. Copley and S. W. Lovesey (1975). *Rep. Prog. Phys.*, **38**, 461, London, Institute of Physics.
6. L. Van Hove (1954). *Phys. Rev.*, **95**, 249.
7. C. A. Croxton (1974). *Liquid State Physics – A Statistical Mechanical Introduction*, Cambridge University Press, London.
   D. Forster and P. C. Martin (1970). *Phys. Rev.*, **A2**, 1575.
   H. R. Leribaux and N. K. Pope (1971). *Phys. Rev.*, **A3**, 1752.
   G. F. Mazenko (1971). *Phys. Rev.*, **A3**, 2121; (1972). *Phys. Rev.*, **A5**, 2545.
   P. C. Martin, E. D. Siggia and H. A. Rose (1973). *Phys. Rev.*, **A8**, 423.
   M. Rao (1974). *Phys. Rev.*, **A9**, 2220.
   M. S. Jhon and D. Forster (1975). *Phys. Rev.*, **A12**, 254.
8. L. S. Ornstein and F. Zernike (1914). *Proc. Acad. Sci. Amst.*, **17**, 793.
9. M. Born and H. S. Green (1949). *A General Kinetic Theory of Liquids*, London, Cambridge University Press.
   J. Yvon (1935). *La Théorie Statistique des Fluides et L'Equation d'Etat, Act. Sci. Ind.* 203, Paris, Herman.
   J. G. Kirkwood (1935). *J. Chem. Phys.*, **3**, 300.
   J. G. Kirkwood and E. M. Boggs (1942). *J. Chem. Phys.*, **10**, 394.
10. G. H. A. Cole (1958). *J. Chem. Phys.*, **28**, 912.
11. J. G. Kirkwood (1935). *J. Chem. Phys.*, **3**, 300.
12. I. Z. Fisher (1962). *Uspekhi Fiz. Nauk*, **76**, 499 (English translation: *Soviet Physics, Uspekhi*, **5**, 239); (1964). *Statistical Theory of Liquids*, Chicago University Press.
13. G. H. A. Cole and A. Moreton (1967). *Mol. Phys.*, **13**, 501.
14. G. S. Rushbrooke and H. I. Scoins (1953). *Proc. Roy. Soc. (London)*, **A216**, 203.
15. G. Stell (1963). *Physica*, **29**, 517.
16. J. K. Percus and G. J. Yevick (1958). *Phys. Rev.*, **110**, 1.
    J. K. Percus (1962). *Phys. Rev. Letters*, **8**, 462.
17. L. Verlet (1964). *Physica*, **30**, 95.
18. An extensive discussion has recently been given by Croxton (1974) and other analyses have been given by Rice and Gray (1965), Cole (1967), and others.
19. M. D. Johnson, P. Hutchinson and N. H. March (1964). *Proc. Roy. Soc. (London)*, **A282**, 283.
    M. D. Johnson and N. H. March (1963). *Phys. Lett.*, **3**, 313.
    N. H. March (1967). *Liquid Metals*, Oxford, Pergamon Press.
20. For a good general survey see Croxton (1974) where various references are also given.
21. H. Poincaré (1894). *Rev. Gen. des Sciences*, **516**.
22. L. Boltzmann (1896). *Ann. Physik*, **57**, 773; (1897). *ibid*, **60**, 392.
23. M. von Smoluchowski (1916). *Phys. Zeits*, **17**, 557, 585.

24. A very good summary is given by: S. Chandrasekhar (1943). *Rev. Mod. Phys.*, **15**, 1, where there is a full historical account with references.
25. P. Ehrenfest and T. Ehrenfest (1911). *Encyclop. der Math. Wissenschaften,* **4**.
26. J. G. Kirkwood (1946). *J. Chem. Phys.*, **14**, 180; (1947). *ibid*, **15**, 72.
27. R. Eisenschitz (1955). *Phys. Rev.*, **99**, 1059.
28. G. H. A. Cole (1970). *Rep. Prog. Phys.*, **33**, 737.
    H. Mori (1958). *Phys. Rev.*, **112**, 1829.
    R. Kubo (1952). *J. Phys. Soc. Japan,* **12**, 570, 1203.
29. L. Onsager (1931). *Phys. Rev.*, **37**, 405; **38**, 2265.
30. S. A. Rice and J. G. Kirkwood (1959). *J. Chem. Phys.*, **31**, 901.
31. R. K. Eisenschitz and A. Suddaby (1954). *Proc. 2nd Int. Congress. Rheol.*, London, Butterworth.
    A. Suddaby and J. R. N. Miles (1961). *Proc. Phys. Soc. (London)*, **77**, 1170.
32. The Smoluchowski equation need not be derived from the Fokker–Planck equation as has been stressed by Gray (1964).
    P. Gray (1964). *J. Mol. Phys.*, **7**, 235.
33. J. G. Kirkwood, F. P. Buff and M. S. Green (1949). *J. Chem. Phys.*, **17**, 988.
34. S. A. Rice and A. R. Allnatt (1961). *J. Chem. Phys.*, **34**, 2144.
    A. R. Allnatt and S. A. Rice (1961). *J. Chem. Phys.*, **34**, 2156.
35. R. W. Zwanzig, J. G. Kirkwood, K. P. Stripp, and I. Oppenheim (1953). *J. Chem. Phys.*, 21, 2050.
36. I. Prigogine (1962). *Non-Equilibrium Statistical Mechanics*, New York, Interscience.
37. P. M. Allen and G. H. A. Cole (1968). *Mol. Phys.*, **14**, 413; **15**, 549, 557.
    I. Prigogine, G. Nicolis and J. Misguich (1965). *J. Chem. Phys.*, **43**, 4516.
    P. M. Allen (1970). *Mol. Phys.*, **18**, 349.
    J. Misguich (1969). *J. Phys., Paris,* **30**, 221.
    J. Misguich, G. Nicolis, J. A. Palyvos and H. T. Davis (1968). *J. Chem. Phys.*, **48**, 951.
    J. A. Palyvos, H. T. Davis, J. Misguich and G. Nicolis (1968). *J. Chem. Phys.*, **49**, 4088.
    M. J. Foster and G. H. A. Cole (1971). *Mol. Phys.*, **20**, 417; **21**, 385.
    W. E. Hagston (1974). *Mol. Phys.*, **28**, 1473.
38. R. Zwanzig (1961). *Phys. Rev.*, **124**, 983.
39. R. Brout and I. Prigogine (1956). *Physica,* 22, 621.
40. A. Z. Akcasu and J. J. Duderstadt (1969). *Phys. Rev.*, **188**, 479.

Chapter 2

Atomic Processes at the Liquid Surface

CLIVE A. CROXTON

Department of Mathematics, University of Newcastle,
New South Wales 2308, Australia

2.1 Introduction

Formal statistical mechanical expressions for the principal thermodynamic functions of the free liquid surface have been available for some time for classical systems, and recent extensions to quantal, molecular, multi-component and metallic systems have also been made. Current interest, however, centres not so much on the numerical evaluation of these statistical thermodynamic expressions, but rather upon the determination of the surface modification of the one- and two-particle distribution functions. The surface excess free energy per unit area or surface tension γ, and the surface excess energy per unit area u_s, are both expressed in terms of integrals over the modified distributions, and as such are relatively insensitive to the detailed structural modification of the liquid surface.

The form of the transition profile, or single-particle density distribution $\rho_{(1)}(z)$ along a z-axis directed normally across a planar interface from liquid to vapour is subject to the usual constraints governing a stable two-phase co-existence – constancy of the chemical potential and normal pressure across the transition zone:

$$\begin{aligned} \mu_{\text{liquid}} &= \mu(z) = \mu_{\text{vapour}} \\ P_{\perp\,\text{liquid}} &= P_{\perp}(z) = P_{\perp\,\text{vapour}}\,. \end{aligned} \tag{2.1.1}$$

$P_{\perp}$ represents the normal component of the pressure tensor, and its continuity across the interface ensures the mechanical stability of the free surface, whilst constancy of the chemical potential, or local Gibbs free energy per particle, ensures thermodynamic stability of the surface; this follows from minimizing the Helmholtz free energy of the entire system (liquid + gas + transition region) at constant N, V, T. Partial expressions apply in the case of multi-component systems. Both the chemical potential and normal pressure are functionals of the two lowest order distributions $\rho_{(1)}(z)$ and $\rho_{(2)}(z_1, \mathbf{r})$, and appeal to the constraints (2.1.1) should yield the equilibrium single-particle distribution $\rho_{(1)}(z)$. In the absence of an exact and explicit knowledge of the anisotropic pair distribution $\rho_{(2)}(z_1, \mathbf{r})$ to which $\rho_{(1)}(z)$ is hierarchically related, approximation is inevitable, and so far appeal to one or other of the constraints (2.1.1) has not yielded a transition profile which is demonstrably both mechanically and thermodynamically stable. Moreover, the transition profiles so obtained are highly controversial in as far as both monotonic and oscillatory profiles have been reported, each yielding numerically satisfactory estimates of the principal thermodynamic functions of the liquid surface. Experimental evidence as to which profile is correct is largely indirect, circumstantial and subject to considerable experimental uncertainty. Computer simulations of the liquid surface have not helped – both monotonic and oscillatory transitions have been reported!

In this chapter an assessment of the theoretical analyses will be made in the context of the available experimental evidence, including the computer simulations, and in particular we shall focus attention on those aspects of the problem which appear to underlie the controversy, yet which remain as yet unsatisfactorily resolved.

2.2 Basic approaches

The theoretical analyses of the structure of the liquid–vapour transition zone in terms of the single-particle distribution $\rho_{(1)}(z)$ may be broadly classified into the thermodynamic, the quasi-thermodynamic and the statistical mechanical. The first approach has limiting validity only in the vicinity of the critical point and has been reviewed elsewhere[99]. The latter two approaches are strictly molecular theories and appeal to the constraint on chemical potential and mechanical stability, respectively, and both analyses are appropriate to the entire fluid range. Both oscillatory and monotonic transition profiles have been obtained on the basis of the statistical mechanical treatments, whilst those based on the constancy of the chemical potential appear inevitably monotonic. Of course, as $T \to T_c$ all these approaches should yield structureless monotonic profiles: the controversy arises in the vicinity of the triple point.

We have to consider whether the thermodynamic functions appropriate to homogeneous assemblies can be extended to the specification of local point functions for the chemical potential or free energy density. Hill[100] has adapted the expression for the chemical potential appropriate to a homogeneous system to the interfacial region by writing

$$\mu(z_1) = \text{constant} = kT \ln \rho_{(1)}(z_1)\Lambda^3 + \int_0^1 \int \Phi(r_{12})\rho_{(1)}(z_2)g(z_1, \mathbf{r}_{12}, \xi)\mathrm{d}\mathbf{r}_{12}\,\mathrm{d}\xi \qquad (2.2.1)$$

$$\Lambda = h/(2\pi mkT)^{1/2}$$

where ξ is the Kirkwood coupling parameter. Hill then *defines* the right-hand side of the above equation to be the local or point chemical potential $\mu(z_1)$, the expression reducing correctly to μ_l or μ_g in either bulk phase. The integral represents the work necessary to 'charge up' a new molecule at z_1. These point functions, however, are functionals of the density profile $\rho_{(1)}(z)$ and do *not* depend merely upon the local density, but are coupled through the long range part of $\Phi(r_{12})$ to other regions of the transition zone. Whether we can expect *a priori* that the defined chemical potential point function is related to its environment in the same way as for uniform point functions is not clear, and it would seem that the possible intervention of functional dependence upon the density gradient has been suppressed. Certainly in the development of Hill's expression for the point function there is the implicit assumption that at each point homogeneous interrelations between local chemical potential and density obtain, and to this extent equation (2.2.1) is a mean field approximation. A functional expansion in the density about Hill's local value should be made, and terms in the functional derivatives should supplement equation (2.2.1). It may well be that this suppression of density gradient dependence is responsible for the invariably monotonic profiles obtained on the basis of this quasi-thermodynamic approach, particularly in the vicinity of the triple point when the density varies by a factor $\sim 10^3$ over two or three molecular diameters.

Both this and the statistical mechanical approach involve the hierarchically-related anisotropic two-particle distribution $\rho_{(2)}(z_1,\mathbf{r})$, and in the absence of an accurate and explicit expression for this function, some closure device has to be adopted. It appears that in the specification of $\rho_{(2)}(z_1,\mathbf{r})$ oscillations may inadvertently be induced in the single-particle distribution – both oscillatory and monotonic profiles have been reported on the basis of statistical mechanical analyses – and it seems that further attention has to be devoted to the closure devices which, so far, have been adopted on a largely *ad hoc* basis.

Despite the difficulties of direct experimental investigation and theoretical resolution of the controversy, the implications of a structured transition zone – one developing stable density oscillations – are considerable for a variety of physical, chemical, biological and industrial processes. The possiblity arises, for example, that $d\gamma/dT$ may actually show *positive* slopes just beyond the triple point, and such surface tension characteristics have been reported for a variety of liquid metal and liquid crystal systems. It must be acknowledged, however, that it is possible to contrive some temperature-dependent migration process of impurity species at the liquid surface to account for these 'anomalous' results. A related phenomenon is that of surface faceting of certain liquid metal droplets upon solidification: both phenomena bear interpretation in terms of a structured rather than a monotonic transition zone. The consequences for atomic and electronic transport processes at the liquid surface are considerable, in particular for the surface optical and electronic properties of liquid metal systems.

In the case of molecular fluids and liquid crystals, the development of *orientational* structure may occur and, again, the surface thermodynamics and optical properties are likely to be substantially modified with respect to a 'classical' description of the surface structure.

In the case of quantal fluids, further novel features may develop, notably quantum interference effects at the free surface and anisotropic zero point effects. These will be discussed later in this chaper.

2.3 The single-particle distribution $\rho_{(1)}(z)$

The dimensions and structure of the transition zone depend sensitively upon the temperature along the liquid–vapour equilibrium line. In the vicinity of the critical point the difference between the liquid and vapour densities suggests that the transition zone is spatially extended, and the expedient of subdividing the interfacial region into elemental strata and considering their mutual thermodynamic equilibrium is justified. Indeed, it is possible to develop a complete thermodynamic theory of the critical transition region on this basis.[1] Clearly, far from the critical point the underlying assumption that the density does not vary significantly over a correlation length within the liquid is inappropriate and an approach at a molecular rather than a thermodynamic level is enforced.

The quasi-thermodynamic approach

Stability of the interface requires the constraints (2.1.1) to be simultaneously satisfied. However, in none of the profiles so far proposed on a variety of theoretical analyses has this been explicitly shown – indeed, in some cases it is clear that the constraints cannot be simultaneously satisfied.

The constraint of constancy of the chemical potential across the transition zone involves the specification of thermodynamic point functions, in particular the point chemical potential $\mu(z)$. It should be emphasized that by a point function we do not mean a function whose value is completely determined by properties such as density and temperature at that point alone. Instead, we mean a function that is well-defined at that point, and will generally be a function of the state of the system in the neighbourhood of the point in question, the range of the neighbourhood being determined, of course, by the range of the intermolecular potential. Quite how the point chemical potential is to be defined, however, is another matter.

Taking Kirkwood's equation for the bulk chemical potential

$$\mu = kT \ln \rho_L \Lambda^3 + \rho_L \int_0^1 \int \Phi(r_{12}) g_{(2)}(r_{12}, \xi) d\mathbf{r}_{12}\, d\xi \tag{2.3.1}$$

where $\Lambda = h/(2\pi mkT)^{1/2}$ and ξ is Kirkwood's coupling parameter, Hill[100] eliminates the second r.h.s. term in Kirkwood's integral equation (2.3.2) for the single distribution function

$$kT \ln \rho_{(1)}(z_1) = kT \ln \rho_L + \rho_L \int_0^1 \int \Phi(r_{12}) g^L_{(2)}(r_{12}, \xi) d\mathbf{r}_{12}\, d\xi$$

$$- \int_0^1 \int \Phi(r_{12}) \rho_{(1)}(z_2) g_{(2)}(z_1, \mathbf{r}_{12}, \xi)\, d\mathbf{r}_{12}\, d\xi \tag{2.3.2}$$

to yield

$$\mu = \text{constant} = kT \ln \rho_{(1)}(z_1)\Lambda^3 + \int_0^1 \int \Phi(r_{12}) \rho_{(1)}(z_2) g_{(2)}(z_1, \mathbf{r}_{12}, \xi) d\mathbf{r}_{12}\, d\xi \tag{2.3.3}$$

where the r.h.s. of (2.3.3) is *defined* as the local or point chemical potential. The implicit assumption is that at each point z, the homogeneous relation (2.3.1) obtains. Whether or not this defined point function represents the actual point chemical potential is open to question, despite the fact that it reduces to the correct form in either bulk homogeneous phase. Moreover, whether constancy of this point function as defined in (2.3.3) does lead to the specification of the correct transition profile is now uncertain.

As we observed in Section 2.2, the point function is coupled through the long range part of the intermolecular potential to the inhomogeneous regions of the transition zone, and a functional expansion in the density should be made:

$$\mu = \bar{\mu} + \int_1 \left[\frac{\delta \mu}{\delta \rho_{(1)}(\mathbf{r}_1)} \right]_0 \Delta \rho_{(1)}(\mathbf{r}_1) d\mathbf{r}_1$$

$$+ \frac{1}{2} \int_2 \int_1 \left[\frac{\delta^2 \mu}{\delta \rho_{(1)}(\mathbf{r}_1) \delta \rho_{(1)}(\mathbf{r}_2)} \right]_0 \Delta \rho(\mathbf{r}_1) \Delta \rho(\mathbf{r}_2) d\mathbf{r}_1\, d\mathbf{r}_2 + \ldots \tag{2.3.4}$$

where $\Delta \rho_{(1)}(\mathbf{r}_1)$ represents a variation of the whole profile from the equilibrium distribution and $[\ ]_0$ represents derivatives taken with respect to the equilibrium distribution. $\bar{\mu}$ is the homogeneous quantity (2.3.1). Eliminating the second integral in (2.3.2) with the aid of equations (2.3.4) and (2.3.1) yields a much more complicated expression than (2.3.3) for the point chemical potential, but one which will be functionally dependent upon the density derivatives.

An early approach by Hill[3] based on equation (2.3.3), and subsequently refined by Plesner and Platz,[4] expresses the constancy of the chemical potential across the transition zone of a system of hard spheres interacting through a long-range van der Waals tail. From a knowledge of the equation of state, and the interaction energy $\Psi(z)$, it is straightforward to establish the constancy of the chemical potential. For example, using the Reiss–Frisch–Lebowitz[5] hard sphere equation of state, Plesner and Platz obtain

$$\ln \frac{\eta(z)}{1-\eta(z)} + \frac{7\eta(z)}{1-\eta(z)} + \frac{15\eta^2(z)}{2[1-\eta(z)]^2} + \frac{3\eta^3(z)}{[1-\eta(z)]^3} + \Psi(z) = \text{constant} \tag{2.3.5}$$

in terms of the reduced density $\eta(z) = \pi\sigma^3 \rho_{(1)}(z)/6$. It should be noted that there is, of course, no dependence upon the density gradient. The specification of $\Psi(z)$ implicitly involves a knowledge of the distribution of neighbouring particles, and these authors set

$$\begin{aligned} g_{(2)}(r) &= 1.00 \qquad r \geqslant \sigma \\ &= 0 \qquad r < \sigma. \end{aligned} \tag{2.3.6}$$

This assumption suppresses from the outset any possibility of a structured transition zone. Moreover, the pair interaction and pair distribution are statistically inconsistent, not to mention the incorrect assumption of isotropy of the pair distribution in the vicinity of the liquid surface.

Buff and Stillinger have made a similar analysis of the distribution of electrolyte against a metallic surface.[101]

Such a development as that outlined above is more appropriate to low-density, high-temperature systems in which the elemental strata into which the transition zone is subdivided constitute thermodynamic entities. In this case it is meaningful to discuss them in a quasi-thermodynamic rather than a statistical mechanical sense, and, in particular, represent systems over which the local density remains sensibly constant. For low-temperature systems possessing a relatively sharp transition zone the elemental subdivision would have to be so fine ($\ll \sigma$) that it is no longer meaningful to regard them as thermodynamic subsystems, and the propriety of discussion in terms of thermodynamically related point functions becomes questionable.

Nevertheless, the single-particle distribution $\rho_{(1)}(z)$ determined from a relation such as (2.3.5) is related to a local pressure appropriate to the stratum $P(z)$ via an assumed equation of state. The surface tension γ may then be determined from the pressure tensor relation[6]

$$\gamma = \int_{-\infty}^{\infty} [P - P_{\perp}(z)]\,\mathrm{d}z$$

where P is the system pressure. Again, whether the pressure $P(z)$ may legitimately be related to an equation of state under these circumstances is debatable. None of the equations of state is dependable at liquid densities, and the low-temperature, high-density region – the region with which we are primarily concerned – is probably quite unreliable. The condition for mechanical stability of the free surface, that the normal pressure $P_{\perp}(z)$ should be constant across the transition is, of course, *not* satisfied. The pressure within each elemental stratum is isotropic, and this has occurred simply because macroscopic thermodynamics were used to relate the local pressure to the local density. Such an approach may be justified in the vicinity of the critical point[1]. Then the transition zone is sufficiently delocalized that the density is sensibly constant within each semi-macroscopic stratum, $P(z)$ is virtually isotropic, the density profile is almost certainly structureless, and the surface tension is virtually zero.

An extension of Hill's analysis[3] has been made by Toxvaerd[7] by applying a perturbation expansion[8,9] of the two-particle distribution at the liquid surface. In the perturbation treatments the structure is assumed to be determined essentially by the repulsive core of the two-particle interaction, in which case realistic interactions are resolved into a short-range repulsive component $\Phi_0(r)$ and a weaker long-range component $\Phi_1(r)$. Toxvaerd adapts the approach to an inhomogeneous system, enabling him to express the constancy of the chemical potential across the transition zone for a system of square well particles as follows:[7]

$$\text{constant} = \frac{\mu_0(z)}{kT} + 2\pi \int_{-\infty}^{\infty} \rho_{(1)}(z + z')\,\mathrm{d}z' \int_{|z'|}^{\infty} \frac{\Phi_1(r)}{kT} \times \left(g^{0}_{(2)}[r \mid \rho_{(1)}(z)] + \frac{1}{2}\frac{\Phi_1(r)}{kT} g^{1}_{(2)}[r \mid \rho_{(1)}(z)] + \ldots \right) r\,\mathrm{d}r \tag{2.3.7}$$

where $\mu_0(z)$ is the chemical potential of a uniform fluid of the local density $\rho_{(1)}(z)$ interacting through the hard core component potential $\Phi_0(r)$. As before, the analysis is open to the objection that the dependence upon density gradient is suppressed, being based on Hill's expression (2.3.3). Accordingly, we anticipate a systematic and progressive error in the single-particle distribution as the triple point is approached. Equation (2.3.7) does, however, take some explicit account of the local structure through the isotropic pair distributions $g^{0}_{(2)}(r)$, $g^{1}_{(2)}(r)$, appropriate to the local density, but nevertheless remains a 'mean field' theory. Moreover, the important angular dependence of $g_{(2)}(z_1,\mathbf{r})$ is suppressed, though this is a very common approximation.

As we mentioned above, a structureless profile is almost inevitable on the basis of a theory which neglects density gradient dependence. This is not to say, however, that a structured profile *should* develop, but rather that an adequate theory should in principle be *capable* of describing a structured transition zone which the present quasi-thermodynamic treatments appear incapable of doing.

Toxvaerd[7] iteratively solves (2.3.7) for a square-well fluid at the reduced temperature $T^*(=kT/\epsilon) = 1$. Verlet and Weiss have given a parametrized expression for $g^{0}_{(2)}(r/\rho)$ on the basis of computer simulations, whilst $g^{1}_{(2)}(r/\rho)$ is tabulated by Smith, Henderson and Barker.[10] A structureless transition profile is obtained (Figure 2.3.1). Toxvaerd goes on to calculate the surface tension of the square-well fluid,[7] but such a calculation is again open to questions of thermodynamic propriety and mechanical stability as discussed above.

Whether such an approach is capable *in principle* of resolving the issue regarding density oscillations is not clear. Mechanical and thermodynamic stability of the solid–vapour interface is, of course, similarly characterized by the constraints (2.1.1), and but for the specification of the pair distribution, Toxvaerd's analysis should bear extension to the solid–vapour transition when oscillations undoubtedly develop. Neglect of the functional dependence of the chemical potential upon the density gradient would then become fully apparent and provide a quantitative assessment of its role in the stabilization of the transition zone.

Toxvaerd also obtains a perturbation expansion of the (point function) Helmholtz free energy[7,11] in the vicinity of the density transition:

$$a(z) = a^0(z) + \frac{1}{2}\int \beta\Phi_1(r_{12})\rho_{(1)}(z + z_{12})g^{0}_{(2)}[r_{12} \mid \rho_{(1)}(z)]\,\mathrm{d}\mathbf{r}_{12} - \frac{1}{4}\int [\beta\Phi_1(r_{12})]^2 \rho_{(1)}(z + z_{12})g^{0}_{(2)}[r_{12} \mid \rho_{(1)}(z)]\,\frac{\partial\rho(z + z_{12})}{\rho_0\,\partial\beta}\,\mathrm{d}\mathbf{r}_{12} + \ldots \tag{2.3.8}$$

The density distribution $\rho_{(1)}(z)$ is determined which minimizes the interfacial free energy (and

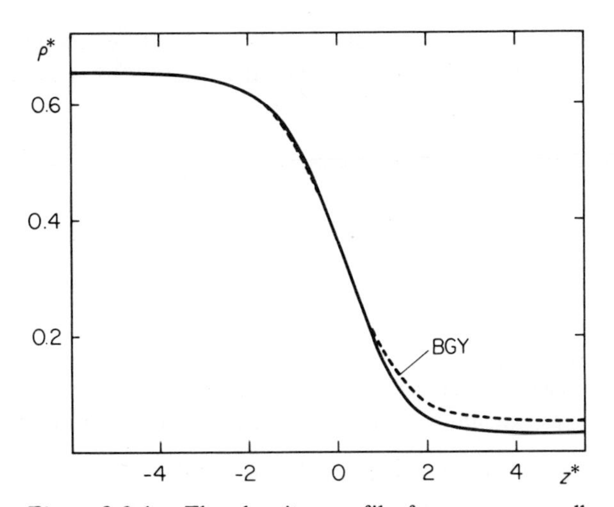

Figure 2.3.1 The density profile for a square-well fluid ($\Phi(r) = -\epsilon$, $\sigma \leqslant r < 1.5\sigma$) at $T^* = 1.00$. The full line represents the profile obtained from the criterion of constancy of the chemical potential across the transition zone. This is graphically indistinguishable from the profile obtained by minimizing the free energy using the trial function (2.3.9). The broken curve represents the BGY profile. (From Toxvaerd,[7] reproduced by permission of the American Institute of Physics)

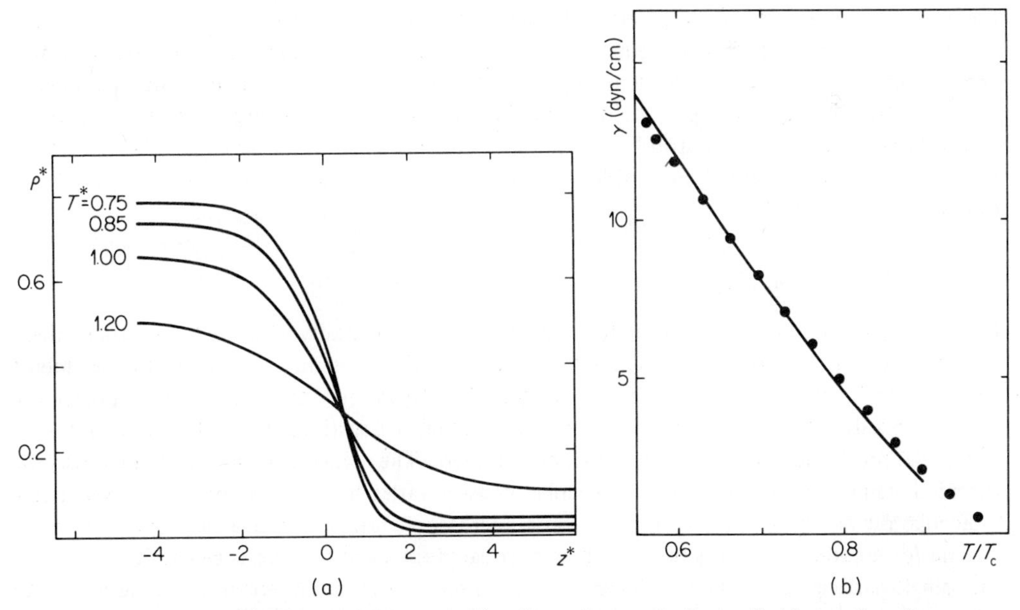

Figure 2.3.2 (a) The transition profiles for a Lennard–Jones fluid as a function of reduced temperature ($T^* = kT/\epsilon$) on the basis of free energy minimization of the tanh trial function (2.3.9).[7] (b) The surface tension of a Lennard–Jones fluid based on the tanh profiles. The agreement with the experimental points is seen to be good, although the $\gamma(T)$ curve is relatively insensitive to the detailed form of the transition profile. (From Toxvaerd,[7] reproduced by permission of the American Institute of Physics)

at the same time, of course, permits a calculation of the surface tension or excess surface free energy per unit area). Toxvaerd makes a variational determination for a trial profile of tanh form[7]

$$\rho_{(1)}(z) = \{\rho_L \exp[a(b-z)] + \rho_V\}/\{\exp[a(b-z) + 1\} \qquad (2.3.9)$$

where

$$a = a_1 \quad (z < b) \qquad b = \left(\frac{1}{a_1} - \frac{1}{a_2}\right) \ln 2.$$
$$\;\; = a_2 \quad (z > b)$$

The two variable parameters of this profile, a_1 and a_2, are adjusted so as to variationally minimize the interfacial free energy though, as Toxvaerd points out, the trial function (2.3.9) does not have the correct asymptotic form. An asymptotically correct trial function

$$\rho_{(1)}(z) = (\rho_L + \rho_V)/2 - (\rho_L - \rho_V)/\pi \tan^{-1} [a(z-b)]^3$$

with

$$b = \frac{1}{\sqrt{3}}\left(\frac{1}{a_1} - \frac{1}{a_2}\right)$$

increases γ by 5–10%. Toxvaerd's results are shown in Figures 2.3.1 and 2.3.2 for the square-well and Lennard–Jones systems, respectively. The possibility of a structured profile is, of course, explicitly excluded. Nevertheless, agreement between the chemical potential and the free energy profiles is striking, although perhaps not surprising since the qualitative features of the profile, in particular its range, are related more or less directly to the attractive branch of the pair interaction and the temperature through the Boltzmann factor.

Toxvaerd[7] observes that agreement between the square-well chemical potential and free energy surface tensions is excellent. In the case of the Lennard–Jones fluid, parametric adjustment of the tanh profile to yield the minimum superficial free energy leads to the profiles and $\gamma(T)$ characteristics shown in Figure 2.3.2. Unfortunately, few conclusions may be drawn from the $\gamma(T)$ results and trial profiles regarding the central issue – the single particle distribution.

The statistical mechanical approach

In the vicinity of the critical point the transition zone is spatially extended, and under these circumstances it is legitimate to subdivide the interphasal region into thermodynamic elements within which the density varies inappreciably. However, near the triple point the quasi-thermodynamic point function approach becomes questionable, and a more rigorous statistical mechanical formulation must be adopted. The dangers in applying macroscopic thermodynamics in regions where molecular density varies by a factor of $\sim 10^3$ over a few molecular diameters enforces an approach at a microscopic rather than macroscopic level.

The formulation of the majority of the statistical mechanical theories is generally in terms of the single-particle Born–Green–Yvon (BGY) integro-differential equation, generalized to inhomogeneous systems. There is the usual hierarchical relationship between the one- and two-particle distributions: in the present case the single-particle function $\rho_{(1)}(z_1)$ is related through the BGY equation to the anisotropic pair distribution $\rho_{(2)}(z_1, \mathbf{r}_{12})$. This latter function may be expressed in terms of an anisotropic radial distribution:

$$\rho_{(2)}(z_1, \mathbf{r}_{12} \mid \rho) = \rho_{(1)}(z_1)\rho_{(1)}(z_2)g_{(2)}(z_1, \mathbf{r}_{12} \mid \rho). \qquad (2.3.10)$$

Generalized to the inhomogeneity at the liquid surface, the BGY equation for a classical system of structureless particles may be written:

$$kT\nabla_1\rho_{(1)}(z_1) + \int_2 \nabla_1\Phi(r_{12})\rho_{(2)}(z_1, \mathbf{r}_{12})\mathrm{d}2 = 0. \tag{2.3.11}$$

Introducing a single particle potential of mean force $\Psi(z_1)$ related to a surface constraining field as $-\nabla\Psi(z_1)$,

$$\rho_{(1)}(z_1) = \rho_\mathrm{L} \exp\left(\frac{-\Psi(z_1)}{kT}\right) \tag{2.3.12}$$

it follows directly that (2.3.11) expresses the mechanical stability of the transition zone, and is exact to within the assumption of pairwise additivity of the total potential. Equivalently, solutions to the single-particle BGY equation (2.3.11) are characterized by constancy of the normal component of the pressure tensor across the interface.

The statistical mechanical description of the liquid surface may now proceed in formal terms to within the specification of the one- and two-particle distribution functions at the surface. Formal expressions for the principal thermodynamic functions of the liquid surface are to be found in the literature.[25] Fowler,[26] as an initial approximation, and Kirkwood and Buff,[25] for the purposes of numerical evaluation, resort to the expedient of shrinking the transition zone to a surface of density discontinuity coincident with the Gibbs dividing surface. In this step model of the liquid surface the formulations of Fowler and of Kirkwood and Buff become identical, both assuming that the liquid remains homogeneous right up to the liquid surface. In this case the single-particle distribution assumes the form of a step function, whilst the two-particle distribution remains identical to that of the bulk-anisotropy arising as a consequence of the truncation of the symmetric pair distribution by the Gibbs surface. Since Kirkwood and Buff's initial estimates for argon,[25] Shoemaker *et al.*[27] have utilized more recent scattering determinations of the isotropic pair distribution, and Freeman and McDonald[28] have evaluated the statistical mechanical expressions for the surface thermodynamic functions at a step surface by a Monte Carlo technique: these results, together with those of Shoemaker *et al.*, represent the best possible estimates of the surface tension and energy on the basis of a step model of the liquid surface.

Instructive though these calculations are, it is clear that the assumption of a step discontinuity in the single-particle distribution is unjustified, even at the triple point when spatial delocalization of the surface is at a minimum. There is a progressive discrepancy between the step model and experiment with increasing temperature, as we should expect. And since the surface tension represents the excess surface free energy per unit area, we understand that the inequalities $\gamma_{\mathrm{step}} > \gamma_{\mathrm{expt}}, u_{\mathrm{s\ step}} < u_{\mathrm{s\ expt}}$ generally hold.

Attempts have been made to incorporate more realistic, though nevertheless *ad hoc*, model transition profiles. Exponential,[29] linear,[30,31] cubic[31] and tanh[7] profiles have been proposed, but these afford little physical insight into the atomic processes operating at the liquid surface. Certainly the principal thermodynamic functions, arising as integrals over the distributions, are relatively insensitive to their detailed form. It is precisely for this reason that theoretical and experimental emphasis has shifted to the determination of $\rho_{(1)}(z)$.

In the absence of any knowledge of the anisotropic pair distribution $\rho_{(2)}(z_1, \mathbf{r}_{12})$ in (2.3.11), a variety of closures[11–15] has been proposed which permit a solution of the BGY equation. Both oscillatory and monotonic transition profiles have been obtained on the basis of the BGY equation, and assessment of the various treatments is made largely in terms of the closure procedure. We have few experimental or theoretical guidelines regarding the specification of the anisotropic distribution, and although detailed examination of the two-particle distributions developed in the machine simulations would help immensely, no reports appear in the literature.[16] Nevertheless, the usual hierarchical relation between adjacent orders of distribution holds, and we may test any prescription against the condition[17]

$$-1 = \int_2 \rho_{(1)}(2)[g_{(2)}(1,2 \mid \rho) - 1]\, d2 \tag{2.3.13}$$

which all closure approximations should satisfy. None of them does, however, and no attempts have been made to ascertain the consequences involved in assuming an approximate closure despite its bearing on the central issue – the detailed form of the single-particle distribution. Failure to satisfy (2.3.13) does mean, of course, that the one- and two-particle distributions are inconsistent. How serious this is remains unclear: the two- and three-particle distributions in the superposition approximation for bulk homogeneous liquid systems are similarly inconsistent, yet still yield qualitatively acceptable results. The surprisingly satisfactory nature of the Kirkwood superposition approximation for the triplet distribution has been demonstrated explicitly by computer simulation,[18] and is understood in terms of extensive self-cancellation in a diagrammatic density expansion of the three-particle distribution.[19] A similar conclusion regarding the widely used analogue of the superposition approximation[20] (c.f. (2.3.10))

$$\rho_{(2)}(z_1, \mathbf{r}_{12}) = \rho_{(1)}(z_1)\rho_{(1)}(z_2)g^{L}_{(2)}(r_{12}) \tag{2.3.14}$$

(where $g^{L}_{(2)}(r)$ is the bulk liquid isotropic distribution) would be most welcome.

The specification of the surface constraining field, or the closure approximation which amounts to its specification, is a central issue in the statistical mechanical analyses, since interfacial density oscillations in the single-particle distribution develop as a response to the collective surface field or boundary condition. The strongly oscillatory profiles reported for the softly-coupled Lennard–Jones fluids[14,15] undoubtedly develop in response to a two-particle closure which implies an over-constrained surface. At least, if such pronounced density oscillations do develop at the surface of Lennard–Jones fluids, then the result is wholly inconsistent with the body of available thermodynamic data – there is, of course, no direct structural information available.

Before considering the approximate closure in any detail, we first discuss the interrelation between the assumed closure prescription and the development of the single-particle distribution in terms of the BGY equation generalized to an inhomogeneous system. Insertion of the exact closure (2.3.10) in (2.3.11), and integration subject to the boundary condition $\rho_{(1)}(-\infty) = \rho_L$, yields

$$\begin{aligned}\rho_{(1)}(z_1) &= \rho_L \exp\left(-\frac{1}{kT}\int_{-\infty}^{z_1}\int_2 \nabla_1\Phi(r_{12})\rho_{(1)}(z_1)g_{(2)}(z_1, r_{12} \mid \rho)\, d2\, dz\right)\\ &= \rho_L \exp\left(-\frac{\Psi(z_1)}{kT}\right)\end{aligned} \tag{2.3.15}$$

(c.f. 2.3.12). Apart from the temperature-dependence of the two-particle distribution, it is clear that for classical systems the profile $\rho_{(1)}(z_1)$ is related to the constraining potential through a Boltzmann factor, and that it will exhibit spatial delocalization with increasing temperature.

The iterative solution of (2.3.15) will generally involve the adoption of an approximate closure for $\rho_{(2)}(z_1, \mathbf{r}_{12})$ or $g_{(2)}(z_1, \mathbf{r}_{12})$, and an initial guess at the transition profile – most usually a step function. As the centre of integration z_1 moves towards the surface, a particle located at this point experiences an anisotropic distribution of force $-\int_2 \nabla_1\Phi(r_{12}) \times \rho_{(1)}(z_1)g_{(2)}(z_1, \mathbf{r}_{12} \mid \rho)\, d2$ directed into the bulk fluid: clearly the specification of the closure specifies the surface constraining field, and we speculate that an 'over-constrained' closure device induces oscillations in the transition profile, whilst 'under-constraint' results in a monotonic density transition. Moreover, the kernel of the integral appearing in (2.3.15) is extremely sensitive to the details of the pair potential and two-particle distribution and, as Borstnik and Azman[102] have recently observed, the solution of the BGY equation can strongly depend upon the details of the input quantities in the case of realistic systems.

In view of the complexity of the mathematical formulation of the problems discussed above, it is interesting to consider the effect of an extreme constraint such as that imposed by an ideal infinitely high potential wall for which a full and rigorous solution is possible.[21,22] In the case of a square-well fluid, for example, Fisher and Bokut[21] find a strongly oscillatory transition profile in the vicinity of the boundary; such effects have been simulated by Bernal[23] for hard spheres against both rigid and soft constraining boundaries, and more recently by machine simulation.[119]

In the case of liquid metals, the intervention of quantum interference effects of the conduction electronic distribution at the surface may modify the distribution to ionic centres, and it would appear that for certain systems at least, the electronic processes might well amplify any tendency towards the development of stable density oscillations in the ionic profile. We shall discuss the important case of the liquid metal surface in Section 2.6.

2.4 Closures and solutions

In the absence of an accurate and explicit expression for the two-particle distribution $\rho_{(2)}(z_1, \mathbf{r})$ arising in the generalized BGY equation, a variety of approximate closures have been proposed, all of which are hierarchically inconsistent with the single-particle distribution, and none of which satisfy the condition (2.3.13). As discussed in Section 2.3, the inconsistency may prove not to be as serious as *a priori* considerations might suggest, although this is a purely speculative observation, and is a point worthy of detailed attention.

Toxvaerd,[11] for example, solves the BGY equation iteratively for the square-well and Lennard–Jones systems having formed a linear interpolation between the isotropic bulk liquid and vapour distributions:

$$g_{(2)}(z_1, \mathbf{r}_{12} \mid \rho) = \alpha g_{(2)}[r_{12} \mid \rho_{(1)}(z_1)] + (1 - \alpha) g_{(2)}[r_{12} \mid \rho_{(1)}(z_2)] \tag{2.4.1}$$

where the weighting factor α is a simple numeric which Toxvaerd supposes to be proportional to the 'effective local density'

$$\rho_{\text{eff}} = \gamma \rho_{(1)}(z_1) + (1 - \gamma) \rho_{(1)}(z_2)$$

$$\alpha = \frac{\rho_{\text{eff}} - \rho_{\text{V}}}{\rho_{\text{L}} - \rho_{\text{V}}}. \tag{2.4.2}$$

Clearly difficulties would arise in the specification of ρ_{eff} should the transition profile not be monotonic: in fact equations (2.4.1, 2.4.2) ensure that $\rho_{(1)}(z)$ is a monotonic function. Indeed, on the basis of Toxvaerd's distributions $g_{(2)}(r \mid \rho_{\text{eff}})$, but using the more realistic estimate of the effective local density

$$\rho_{\text{eff}} = \frac{1}{z} \int_{z_1}^{z_1 + z} \rho(z')\,\mathrm{d}z'$$

which, incidentally, takes account of any structural features of the density profile, *oscillatory* profiles are obtained at reduced temperatures $\leqslant T^* = 1.00$[124]. We therefore conclude that the density-smearing implied in (2.4.1, 2.4.2) is responsible for the suppression of structure in the

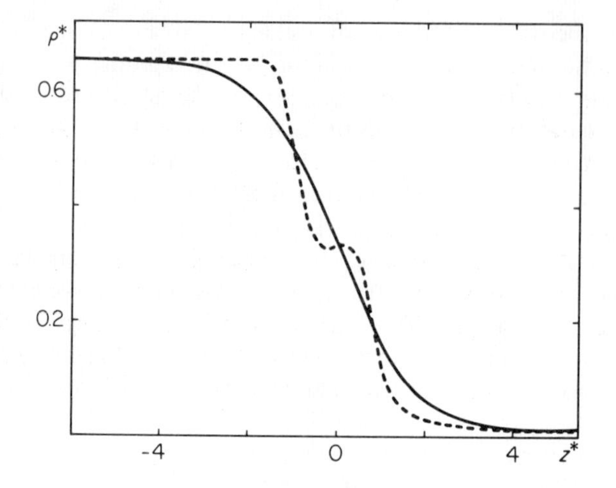

Figure 2.4.1 Successive iterations in the BGY transition profile for a square-well fluid. The zeroth or initial profile is a step function which, after the second iteration, develops an oscillation and subsequently relaxes to the monotonic profile after 40 iterations. Parameters are the same as for Figure 2.3.1

transition profile. As Borstnik and Azman[102] have recently pointed out, Toxvaerd's closure (2.4.1, 2.4.2) is unphysical in as far as it is symmetric with respect to particle exchange only when $\gamma = 0.5$, whilst the author[11] reports convergence of the BGY equation only when $\gamma = 0.7$–0.8, and application to multicomponent systems is clearly impossible since the distributions are not invariant with respect to particle exchange. Beyond this, the closure is difficult to assess in terms of its implications for the atomic processes at the liquid surface and its consistency with the single particle distribution, though the asymmetry of the two-particle distribution with respect to particle exchange casts considerable doubt upon the conclusions drawn by the author.[11] Nevertheless, this closure evidently implies a soft surface field, since in both the square-well[11] (at the reduced temperature $T^* = kT/\epsilon = 1.0$) and the Lennard–Jones systems[24] ($T^* = 0.75$, 0.85, 1.00, 1.20) Toxvaerd obtains structureless monotonic profiles which are graphically indistinguishable from those obtained by minimization of the interfacial free energy for a parametrically adjusted tanh profile (2.3.7) (Figure 2.3.2). There is no reason to anticipate significantly different transition profiles for these systems, as observed in Section 2.3 – particularly since Toxvaerd's determinations are for reduced temperatures substantially above the triple point. However, the 'nested' approximation (equations 2.4.1, 2.4.2) obscures the physical implications for the surface field. Toxvaerd takes the agreement between the BGY and quasi-thermodynamic profiles as evidence of the essential correctness of the latter approach. The two treatments are essentially distinct, however, appealing only to one or other of the stability conditions (2.1.1): moreover, the profiles may be legitimately compared only as $T \rightarrow T_c$. Toxvaerd makes the interesting observation that in the course of iterative solution of the BGY equation using the closure (2.4.1, 2.4.2), the profile initially developed an oscillation which subsequently died out (Figure 2.4.1) as the distribution relaxed from a step function. This may well be attributable to the density-smearing process implicit in (2.4.2).

Toxvaerd's closure would adopt a particularly simple form for a system of square-well particles against an ideal wall for which ρ_V and $g_{(2)}(r|\rho_V)$ are identically zero – a problem solved analytically by Fisher and Bokut.[21] In this case stable density oscillations in the vicinity

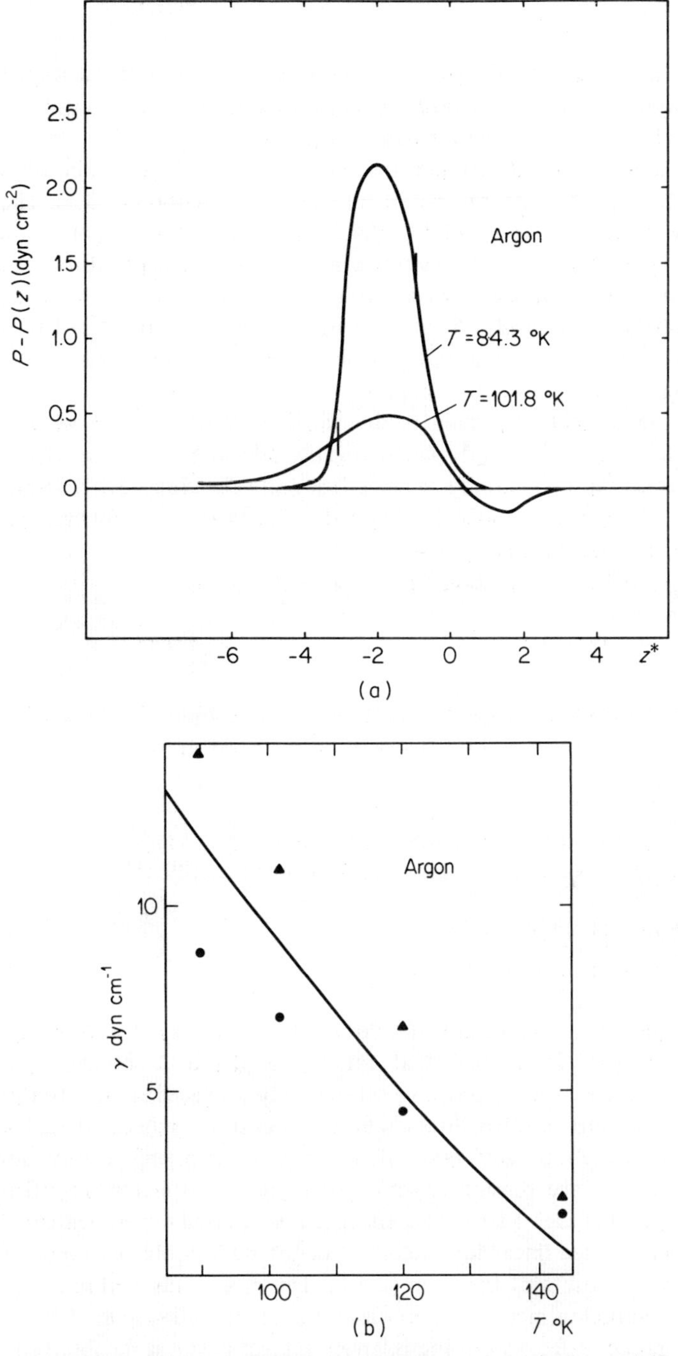

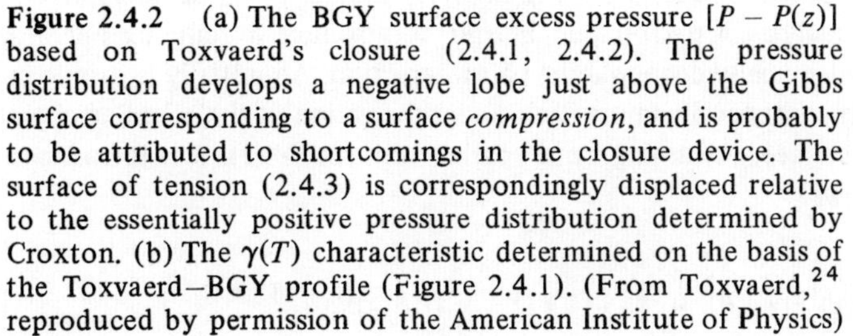

Figure 2.4.2 (a) The BGY surface excess pressure $[P - P(z)]$ based on Toxvaerd's closure (2.4.1, 2.4.2). The pressure distribution develops a negative lobe just above the Gibbs surface corresponding to a surface *compression*, and is probably to be attributed to shortcomings in the closure device. The surface of tension (2.4.3) is correspondingly displaced relative to the essentially positive pressure distribution determined by Croxton. (b) The $\gamma(T)$ characteristic determined on the basis of the Toxvaerd–BGY profile (Figure 2.4.1). (From Toxvaerd,[24] reproduced by permission of the American Institute of Physics)

of the boundary are known for sure to occur, but it has never been demonstrated explicitly that oscillations *do* develop with the Toxvaerd closure, which, of course, they should – although it is difficult to see how they could on the basis of a ρ_{eff} model.

Toxvaerd[24] goes on to determine the z-dependence of the reduced pressure tensor (Figure 2.4.2*a*) required for the calculation of the surface tension (2.3.4). The BGY pressure tensor using the Toxvaerd closure (2.4.1, 2.4.2) shows a negative region in the vapour phase which does not appear in any other quasi-thermodynamic or statistical mechanical treatment. Such a region has the significance of a surface *compression* in the vapour phase which is difficult to understand physically. Such behaviour can only be attributed to shortcomings in the closure device, and objections have already appeared in the literature.[102] To be more specific is difficult on account of the complex form of the closure.

Relevant to the present discussion is the work of Berry *et al.*[29] where an exponential transition profile is assumed and parametrically adjusted to bring the theoretical and experimental surface tensions into agreement. Using the closure (2.3.14) a negative lobe in the z-dependence of the pressure tensor is obtained – but in such a contrived model it is again difficult to draw any useful conclusion.

The *surface of tension* is located on a plane $z_{\mathrm{ST}} = 0$ such that

$$0 = \int_{-\infty}^{\infty} z\,[P - P(z)]\,\mathrm{d}z \tag{2.4.3}$$

and is indicated on the excess tangential pressure curves by a line (Figure 2.4.2). The distance δ between the surface of tension and the Gibbs dividing surface z_Γ is

$$\delta = z_{\mathrm{ST}} - z_\Gamma \tag{2.4.4}$$

and is of importance in that it determines the curvature dependence of the surface tension. According to the rigorous thermodynamic Gibbs–Tolman equation:[32]

$$\frac{\mathrm{d}(\ln \gamma)}{\mathrm{d}(\ln r)} = \frac{2(\delta/r)[1 + \delta/r + \frac{1}{3}(\delta^2/r^2)]}{1 + 2\delta/r[1 + \delta/r + \frac{1}{3}(\delta^2/r^2)]} \tag{2.4.5}$$

where r is the radius of the spherical surface of tension, and γ is the surface tension referred to the surface of tension. It is clear that for Toxvaerd's BGY pressure tensor shown in Figure 2.4.4*a* the surface of tension will be substantially displaced relative to those analyses for which $[P - P(z)]$ remains positive throughout the transition. Indeed, the quantity δ is some three times greater than in any other determination, implying a very strong curvature dependence for the surface tension (2.4.5). There are unfortunately insufficient nucleation data[33] to decide whether such a strong dependence is supported experimentally.

Jouanin[30] has also tried linear interpolations of the bulk liquid and vapour distributions to form an approximate closure of the form (2.4.1). This closure inserted in the generalized BGY equation yields structureless monotonic profiles for a square-well system of the same qualitative form as those proposed by Toxvaerd, and is largely subject to the same observations.

In a recent analysis, Croxton and Ferrier[12] introduce a coupling operator $\Phi^*(z)$ whose effect is to anisotropically decouple the pair interaction. An initially isotropic bulk distribution is allowed to self-consistently adjust itself under the action of the coupling operator, effectively creating a free surface. Assuming some spatial form for $\Phi^*(z)$, it is introduced into the BGY equation as follows:

$$\rho_{\mathrm{L}} kT \nabla_1 g_{(1)}(z_1) + \rho_{\mathrm{L}}^2 \int_2 \nabla_1 [\Phi(r_{12})\Phi^*(z_2)] g_{(2)}^{\mathrm{L}}(r_{12})\mathrm{d}2 = 0 \tag{2.4.6}$$

where ρ_L and $g^L_{(2)}(r)$ represent the isotropic bulk liquid number density and radial distribution, respectively. Anisotropy in the integrand (2.4.6) has been introduced through the coupling operator, whilst we retain the bulk pair distribution $g^L_{(2)}(r)$ throughout. The assumption of this distribution will be least satisfactory in the vapour phase, though structural coupling of the liquid to the low density vapour at the triple point will be of negligible importance. $\Phi^*(+\infty)$ is chosen so as to yield the correct thermodynamic properties of the vapour phase (whose structure is still assumed to be $g^L_{(2)}(r)$), whilst $\Phi^*(-\infty) = 1.00$, of course. Otherwise, no interpolated closure between liquid and vapour is assumed.

Equation (2.4.6) may be integrated subject to the boundary condition $\rho_{(1)}(-\infty) = \rho_L$, whereupon

$$\rho_{(1)}(z_1) = \rho_L \exp\left(-\frac{\rho_L}{2kT}\int_{-\infty}^{z_1}\int_2 (\Phi\nabla_1\Phi^* + \Phi^*\nabla_1\Phi)g^L_{(2)}(r)\,\mathrm{d}2\,\mathrm{d}z\right) \tag{2.4.7}$$

This expression is of Boltzmann form, and comparison with (2.3.15) shows that $\int_2 \mathrm{d}2$ represents the mean normal force acting on a particle located at z_1. The term $\Phi\nabla_1\Phi^*$ represents the constraining effect of the operator, or the anisotropic decoupling at the liquid surface, whilst $\Phi^*\nabla_1\Phi$ represents the modification of the usual correlative forces developed amonst neighbouring particles in an otherwise isotropic assembly. Of course, it remains to specify the spatial form of the operator $\Phi^*(z)$. Certainly $\Phi^*(z)$ will assume constant values in each bulk phase, as mentioned above. In the transition region, however, there are a number of possibilities: Croxton and Ferrier[12] adopt an analytic form related to that of the attractive component of the Lennard–Jones interaction. The resulting single-particle distribution is shown in Figure 2.4.3, and but for a slight 'shoulder', density oscillations appear not to develop. Also shown are the components of the integrand (2.4.7): the role of the surface constraining field in the initiation of density oscillations is isolated explicitly in this approach. In this, as in all the statistical mechanical analyses, there is no guarantee that the condition on the chemical potential (2.1.1) is satisfied. Mechanical stability, on the other hand, is assured to within the approximate expression of the closure. Croxton and Ferrier go on to determine the surface tension and surface energy for argon at the triple point, and the results are shown in Table 2.4.1, together with other analyses. Numerical coincidence between the theoretical and experimental values does not provide a sensitive criterion of the acceptability of the one- and two-particle distributions at the surface; as we pointed out earlier, the principal thermodynamic functions are expressed as integrals over the distributions and correspondingly are relatively insensitive to their detailed features.

Nazarian[36] emphasizes the importance of the statistical mechanical approach, and obtains strongly oscillatory transition profiles for liquid argon at the triple point using two forms of closure, both of which are symmetric with respect to particle exchange:

$$\begin{array}{ll} g^L_{(2)}(r_{12}) & \text{if } z_1 + z_2 \leqslant 0 \\ g^V_{(2)}(r_{12}) & \text{if } z_1 + z_2 > 0 \end{array} \tag{2.4.8}$$

and a linear variation

$$g_{(2)}(z_1, z_2, r_{12}) = g^L_{(2)}(r_{12}) + \left[\left(\frac{z_2}{z_{12}}\right)A(z_2) - \left(\frac{z_1}{z_{12}}\right)A(z_1)\right][g^V_{(2)}(r_{12}) - g^L_{(2)}(r_{12})] \tag{2.4.9}$$

where $A(z)$ is the unit step function. The transition profiles are seen to be of extreme oscillatory form. Whilst the absolute value of the surface tension is largely independent of the surface structure, its temperature dependence is not, and such a profile as this is most unlikely

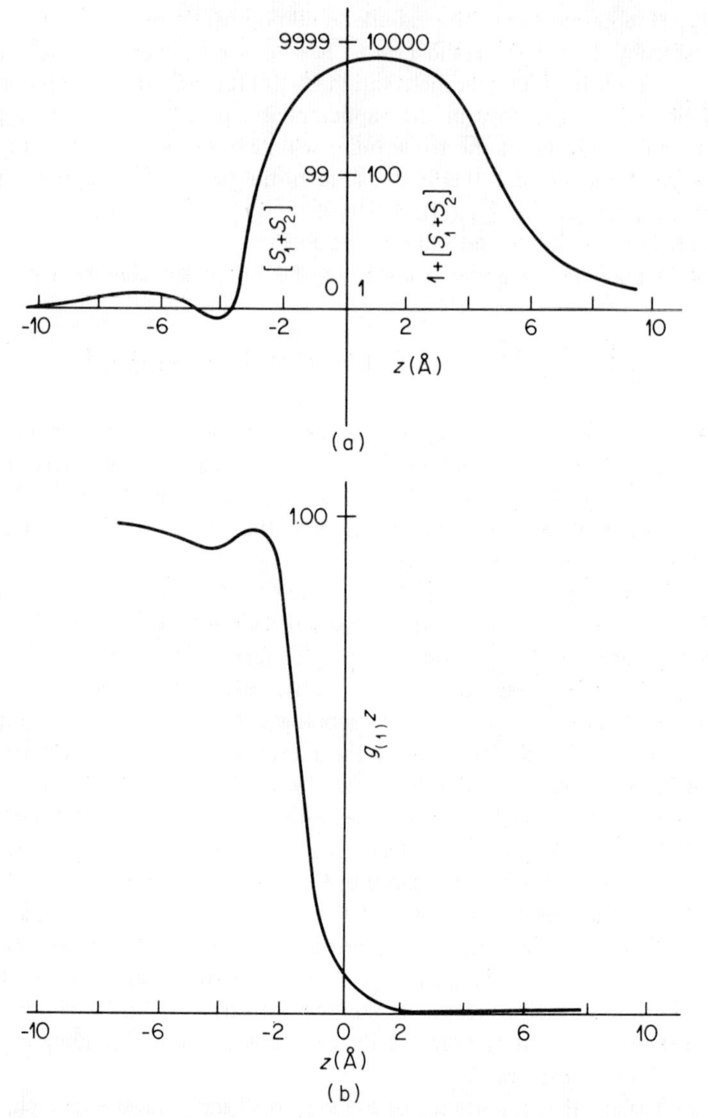

Figure 2.4.3 The single-particle distribution of Croxton and Ferrier for argon at 84 °K ($\sigma = 3.405$ °A, $\epsilon/k = 119.8$ °K). Also shown are the components of the integrand appearing in equation (2.4.7)

Table 2.4.1 Comparison of results for surface properties of liquid argon at the triple point (84.3°K)

Argon	H[3]	H†	PP[4]	SPC[27]	KB††[25]	FM[28]	CF[34]	Expt[35]
γ (dyn cm^{-1})	6.91	21.6	16.55	15.6	16.84	13.7	13.48	13.45
u_s (erg cm^{-2})	19.43	60.59	50.55	27.08	44.3	27.6	35.35	35.01
δ (Å)	2.67	2.65	2.01				3.84	

† Hill's original calculation corrected by PP.
†† Estimated from their 90°K values by $\gamma = \gamma_0(1 - T/T_c)^{1.28}$, $u_s = \gamma - T(\partial\gamma/\partial T)$.

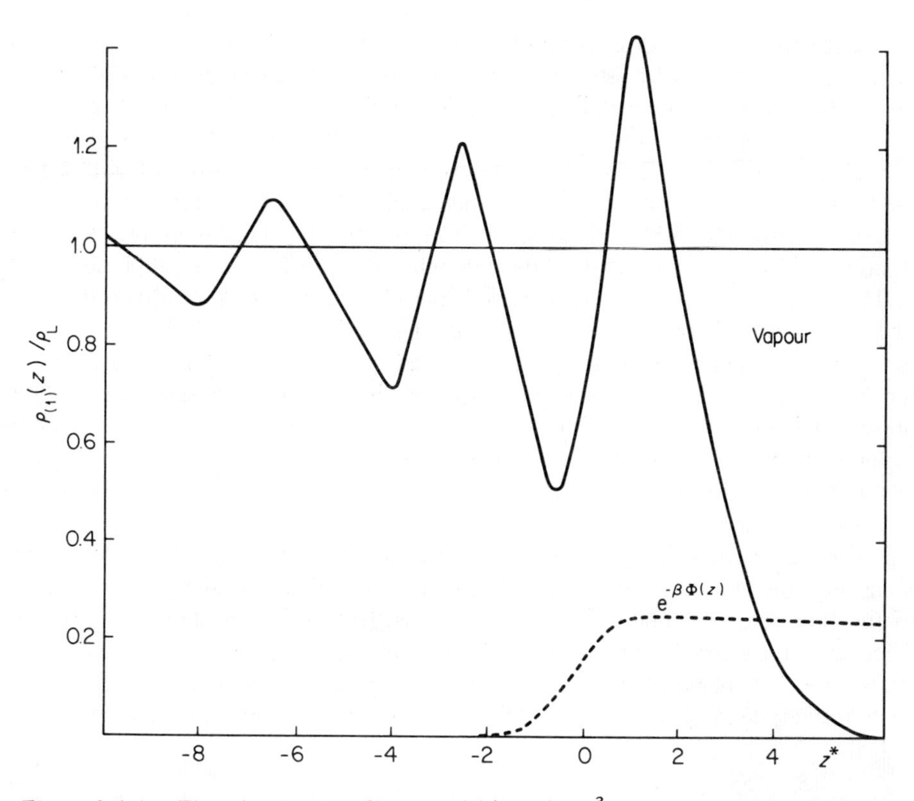

Figure 2.4.4 The density profile $\rho_{(1)}(z)/\rho_L$ ($\rho_L\sigma^3 = 0.85$) for a system of particles with a hard sphere core plus an attractive interaction $-\epsilon(\sigma/r)^6$, for $\epsilon/kT = 1.5$. The broken curve represents the Boltzmann exclusion function

for liquid argon and is certainly inconsistent with what must be regarded as a reliable body of experimental evidence. Nazarian's oscillations undoubtedly arise from a closure which implies too hard a constraining field at the surface, and this is clearly apparent from (2.4.8). As we observed earlier, the kernel of the BGY integral is highly sensitive to the choice of input functions, and Borstnik and Azman[102] find it impossible to obtain convergence using Nazarian's algorithms in conjuction with a Lennard–Jones pair potential ($\epsilon/k = 120\,^\circ$K, $\sigma = 3.4 \times 10^{-8}$ cm) and Verlet's[112] molecular dynamic pair distribution. These authors[102] did obtain convergence with these input quantities, however, using a modified Nazarian closure, and again obtained a strongly oscillatory profile.

Nevertheless, the development of stable density oscillations as a reponse to a strong surface field is clearly feasible and underlines the necessity for the correct specification of the closure.

A recent determination by Perram and White,[37] based on an idea of Helfand, Frisch and Lebowitz,[38] considers a bulk two-component system AB of collision diameters σ_A, σ_B and number densities ρ_A and ρ_B, respectively. Taking the limits $\sigma_B \to \infty$, $\rho_B \to 0$, one has effectively created an interface of type A particles interacting through a pair potential $\Phi_{AA}(r_{AA})$ whilst interacting with the type B 'particle' through $\Phi_{AB}(z)$. For the single-particle distribution about B we then have

$$\rho_{(1)}(z) = \rho_L \lim_{\substack{\sigma_B \to \infty \\ \rho_B \to 0}} g_{(2)}^{AB}(z) \tag{2.4.10}$$

If now we regard particle B as a vacuum bubble, the net force on particles A into the bulk of the fluid may be simulated by assigning a repulsive (i.e. surface constraining) field between the vacuum bubble and the surrounding particles. Working in the PY approximation for an assembly of hard spheres with an $-\epsilon(\sigma/r)^6$ tail, the repulsive potential centred on the bubble and the distribution $g_{(2)}^{AB}(z)$ are self-consistently determined and shown in Figure 2.4.4. The role of the exclusion potential within the bubble is analogous to the determination of $\Phi^*(z)$ in the Croxton–Ferrier approach, and, again, shows explicitly how oscillations may be induced in the transition profile. Perram and White determine the profile at the (unphysical) density $\rho\sigma^3 = 0.85$ with $\epsilon/kT = 1.5$, but at a *curved* interface of radius 20 Å which, it should be pointed out, does not represent a determination of $g_{(1)}(z)$.

As Croxton[39] has observed, however, the excess pressure or exclusion field associated with a vacuum bubble of only 20 Å radius necessary for stability is enormously large and undoubtedly initiates oscillations in the surrounding distribution. As it stands, the theory is more applicable to cavitation studies. Unfortunately, no estimate of δ (equation 2.4.4) is given: a knowledge of δ in the Gibbs–Tolman equation would enable us to determine the curvature-dependence of δ on the basis of this approach.

Experimental determinations of the surface tensions and surface energies of the liquid inert gases are probably the most reliable of all the thermodynamic measurements of the surface excess functions. Experimental difficulties are substantially greater in the case of liquid metal measurements, for example, where surface activity in the form of oxidation and associated problems arise. We conclude therefore that the development of low entropy surface states in the form of stable density oscillations is unlikely in the case of the liquid inert gases since

$$\frac{\partial \gamma}{\partial T} = -(S_\sigma - S_\beta) \tag{2.4.11}$$

(where S_σ, S_β are the surface and bulk entropies per unit area) is invariably negative for these systems (i.e., $S_\sigma > S_\beta$), and the liquid inert gas transtition profile is relatively structureless at and beyond the triple point.

Recent optical scattering experiments in the vicinity of the critical point appear to confirm that the density transition profile is a monotonic decreasing function from the liquid to the vapour phase. It has to be realized, however, that such optical measurements provide an assessment only of the *dielectric* profile, and whilst for critical systems this may be reasonably assumed proportional to the local density, at lower temperatures a more subtle interpretation is required. Light incident at the Brewster angle should be linearly polarized upon reflection for a discontinuously sharp surface: a measurement of the ellipticity of the scattered radiation therefore provides an assessment of the dielectric profile.[103] Drude's[104] analysis incorrectly assumes that the dielectric tensor remains isotropic throughout the transition zone: Buff and Lovett[105] have more recently reformulated the analysis in such a way that it is valid at low temperatures.

The optical reflectivity of normally incident light has been measured over a range of wavelengths,[106] but, again, as for all these optical experiments, one sees only the long wavelength Fourier components of the density profile, the more detailed structure which may be present being completely washed out. However, a recent investigation of the reflection spectrum of liquid mercury in the range 4500–10,000 Å shows the normal reflectivity to be very close to that predicted from a Drude dispersion relation based on the valence electron density and the d.c. conductivity of the liquid. Ellipsometric measurements, on the other hand, yield optical constants for liquid mercury considerably in excess of those predicted on the basis of the Drude theory. Bloch and Rice[107] show that such an apparent inconsistency can be resolved provided the *conductivity profile* passes through a maximum as the transition zone is traversed. The implications for the microstructure of the liquid mercury surface remain obscure, but the possibility of a structured transition zone at certain liquid metal surfaces appears sustained. Indeed, the speculation has been made[108] that there may well be a close connection between the anomalous Drude behaviour and the anomalous surface tension characteristics observed by a number of authors.

Finally, the surface tension, but not the transition profile, has been determined by Zollweg, Hawkins and Benedek[109] from the spectrum of light inelastically scattered from thermal excitations in the transition zone.

More direct investigations of the transition profile by means of reflection electron diffraction techniques, for example, are extremely difficult, and no conclusive investigations have been reported as yet. Machine simulations, however, have been made and appear to offer the opportunity of a detailed assessment of the microstructure of the transition zone. These investigations will be discussed in Section 2.10.

Finally, we should mention an alternative statistical mechanical expression for the surface tension in terms of the anisotropic direct correlation $c(z_1, \mathbf{r}_{12})$, obtained by Lovett, *et al.*[113] Since the analysis is directed toward the determination of the surface tension rather than the transition profile we shall not consider the analysis in any detail here: the approach has been reviewed by Toxvaerd.[2] Of course, the anisotropic function $c(z_1, \mathbf{r}_{12})$ is wholly unknown: so far c has been approximated[113] by its *isotropic* low density asymptotic form – the Mayer f-function.

2.5 Molecular fluids and liquid crystals

A formal extension of the Kirkwood–Buff[25] analysis to rigid molecular systems is readily made,[110] and Gubbins and Gray[40] have recently obtained the following rigorous, generalized expression for the surface tension as a sum of two components:

$$\gamma = \gamma_R + \gamma_\theta \tag{2.5.1}$$

where

$$\gamma_R = -\frac{\Omega^2}{2}\int_{-\infty}^{\infty}\int P_2^0(\cos\theta_{12})r_{12}\left\langle \rho_{(2)}(z_1, \mathbf{r}_{12}, \omega_1, \omega_2)\frac{\partial\Phi(\mathbf{r}_{12}, \omega_1, \omega_2)}{\partial r_{12}}\right\rangle_{\omega_1\omega_2} \mathrm{d}\mathbf{r}_{12}\,\mathrm{d}z_1 \tag{2.5.2}$$

$$\gamma_\theta = \frac{3\Omega^2}{4}\int_{-\infty}^{\infty}\int \sin\theta_{12}\cos\theta_{12}\left\langle \rho_{(2)}(z_1, \mathbf{r}_{12}, \omega_1, \omega_2)\frac{\partial\Phi(\mathbf{r}_{12}, \omega_1, \omega_2)}{\partial r_{12}}\right\rangle_{\omega_1\omega_2} \mathrm{d}\mathbf{r}_{12}\,\mathrm{d}z_1 \tag{2.5.3}$$

$\langle \ldots \rangle_{\omega_1\omega_2}$ represents an unweighted average over the molecular orientations ω_1, ω_2. The minimum number of Euler coordinates needed to specify ω depends on the symmetry of the molecule – it is two ($\omega = \theta\phi$) if the molecules are linear, and three ($\omega = \phi\theta\chi$) if they are nonlinear and rigid. $\Omega = \int \mathrm{d}\omega$ and is 4π for linear and $8\pi^2$ for nonlinear molecules. In addition to the introduction of the γ_θ term for orientationally-dependent interactions, the angular pair distribution $\rho_{(2)}(z_1, \mathbf{r}_{12}, \omega_1, \omega_2)$ appears, and is defined so that

$$\rho_{(2)}(z_1, \mathbf{r}_{12}, \omega_1, \omega_2)\mathrm{d}\mathbf{r}_1\,\mathrm{d}\omega_1\,\mathrm{d}\mathbf{r}_2\,\mathrm{d}\omega_2 \tag{2.5.4}$$

represents the probability of finding a molecule in each of the volume elements $\mathrm{d}\mathbf{r}_1$, $\mathrm{d}\mathbf{r}_2$ at $\mathbf{r}_1$, $\mathbf{r}_2$ and in the orientational elements $\mathrm{d}\omega_1$, $\mathrm{d}\omega_2$ at ω_1 and ω_2. This distribution may be expressed in terms of the correlation function as for centrosymmetric systems:

$$\rho_{(2)}(z_1, \mathbf{r}_{12}, \omega_1, \omega_2) = \rho_{(1)}(z_1, \omega_1)\rho_{(1)}(z_2, \omega_2)g_{(2)}(z_1, \mathbf{r}_{12}, \omega_1, \omega_2) \tag{2.5.5}$$

where $\rho_{(1)}(z_1, \omega_1)$ is the angular single-particle transition profile, and represents the probability of finding a molecule at z_1 with orientation ω_1.

In the absence of an accurate and explicit knowledge of $\rho_{(1)}(z_1, \omega_1)$ and $g_{(2)}(z_1, \mathbf{r}_{12}, \omega_1, \omega_2)$, Gubbins and Gray[40] reduce (2.5.1) to what is the analogue of the Fowler step-model of the liquid surface in which surface modification of the distribution $\rho_{(2)}(z_1, \mathbf{r}_{12}, \omega_1, \omega_2)$ is neglected. The component γ_θ in (2.5.1) disappears in this case.

Here, however, we are more concerned with the specification of the distribution $\rho_{(1)}(z_1, \omega_1)$. There is no difficulty in writing a formal expression for the distribution, using the exact closure (2.5.5)

$$\rho_{(1)}(z_1, \omega_1) = \rho_{\mathrm{L}} \exp\left(-\frac{1}{kT}\int_{-\infty}^{z_1}\int_0^{\omega_1}\int_2\int_{\hat{2}} \nabla\Phi(r_{12}, \omega_1, \omega_2) \times \rho_{(1)}(z_2, \omega_2)g_{(2)}(z_1, \mathbf{r}_{12}, \omega_1, \omega_2)\,\mathrm{d}2\,\mathrm{d}\hat{2}\,\mathrm{d}z_1\right) \tag{2.5.6}$$

where $\int_2\int_{\hat{2}}\mathrm{d}2\mathrm{d}\hat{2}$ represents integration over all relative positions and orientations of molecule 2. Equation (2.5.6) may be expressed in Boltzmann form (c.f. 2.3.15):

$$\rho_{(1)}(z_1, \omega_1) = \rho_{\mathrm{L}} \exp\left\{\frac{-\Psi(z_1, \omega_1)}{kT}\right\} \tag{2.5.7}$$

where $\Psi(z_1, \omega_1)$ is now the potential of mean force and torque, and expresses the existence of a mean torque acting on an arbitrarily-oriented linear molecule due to its neighbours, in addition to the usual central mean force component. Croxton and Osborn[41] set

$$\rho_{(1)}(z_1, \omega_1) = \rho_{(1)}(z_1)_{\omega_1}\,\rho_{(1)}(\omega_1)_{z_1} \tag{2.5.8}$$

where $\rho_{(1)}(z_1)_{\omega_1}$, $\rho_{(1)}(\omega_1)_{z_1}$ represent the distributions determined subject to the subscripted variables held constant. Equation (2.5.6) then adopts the form

$$\rho_{(1)}(z_1, \omega_1) = \rho_{\mathrm{L}} \exp\left\{-\frac{1}{kT}\int_2\int_{\hat{2}}\left[\int_{-\infty}^{z_1} \nabla_1\Phi(1, 2, \hat{1}, \hat{2})\rho_{(1)}(2, \hat{2})g_{(2)}(1, 2, \hat{1}, \hat{2})\,\mathrm{d}1 + \int_0^{\omega_1} \nabla_{\Omega_1}\Phi(1, 2, \hat{1}, \hat{2})\rho_{(1)}(2, \hat{2})g_{(2)}(1, 2, \hat{1}, \hat{2})\,\mathrm{d}\hat{1}\right]\mathrm{d}\hat{2}\,\mathrm{d}2\right\} \tag{2.5.9}$$

where ∇_{Ω_1} represents the angular gradient operator.

The first term in the square bracket (2.5.9) represents the mean force acting on the centre of gravity of a linear molecule located at z_1, and expresses the development of the usual surface constraining field. The second term expresses the torque field which develops in a region of density gradient. There is evidently a tendency to molecular orientation at the liquid surface, and the development of orientated states will depend upon the density anisotropy or density gradient along the z-axis. At higher temperatures spatial delocalization and thermal disorientation at the surface make surface oriented states unlikely. At lower temperatures, however, the development of oriented surface states would have important thermodynamic consequences for strongly anisotropic molecular systems, in particular the liquid crystals.[42] Thus the possibility arises of orientationally ordered states at the surface of molecular systems, in addition to the development of density oscillations in the form of a surface smectic zone.

Of course, there remains the problem of closure in (2.5.9), or more precisely, the specification of $g_{(2)}(1,2,\hat{1},\hat{2})$. There is every reason to anticipate precisely the same difficulties as for centrosymmetric systems, and the development or otherwise of surface oriented states will undoubtedly prove as controversial as the development of density oscillations.

In the absence of any knowledge of the distribution function $\rho_{(2)}$ in the vicinity of the molecular surface, we may assume the bulk liquid distribution $\rho^{L}_{(2)}$

$$\rho_{(2)}(1,2,\hat{1},\hat{2}) = \rho_{(1)}(1,\hat{1})\rho_{(1)}(\hat{2},\hat{2})g^{L}_{(2)}(1,2,\hat{1},\hat{2}) \tag{2.5.10}$$

which is analogous to Green's approximation[20] (2.3.14) for centrosymmetric systems. A number of determinations of the distribution $g^{L}_{(2)}(1,2,\hat{1},\hat{2})$ have been made,[43] and these permit an iterative solution of (2.5.9) to proceed.

Molecular theories of surface tension in nematic systems are difficult because of the anisotropic pair potential developed between the molecules. Very recently, however, Parsons,[44] working in what is essentially a Kirkwood–Buff model of the nematic liquid surface, obtains expressions for the surface tensions $\gamma_{\parallel}$ and $\gamma_{\perp}$ corresponding to parallel and perpendicular orientation of the surface molecules. As in the Kirkwood–Buff theory, the density of the bulk phase is taken as being constant up to the Gibbs dividing surface and zero in the vapour phase:

$$\begin{aligned} \rho_{(1)}(z) &= \rho_{L} \quad z<0 \\ \rho_{(1)}(z) &= 0 \quad z>0 \end{aligned} \tag{2.5.11}$$

and

$$\rho_{(2)}(\mathbf{r}_1,\mathbf{r}_2,\boldsymbol{\omega}_1,\boldsymbol{\omega}_2) = \rho_{(1)}(z)\rho_{(1)}(z)g_{(2)}(\mathbf{r}_1,\mathbf{r}_2,\boldsymbol{\omega}_1,\boldsymbol{\omega}_2)$$

Moreover, it is assumed that

$$g_{(2)}(\mathbf{r}_1,\mathbf{r}_2,\boldsymbol{\omega}_1,\boldsymbol{\omega}_2) = g_{(2)}(r_{12})f(\omega_1)f(\omega_2)$$

where the two angular distributions $f(\omega)$ and the radial distribution of the molecules are assumed mutually independent: short range radial and angular correlations are neglected. Any z-dependence of the distributions $f(\omega)$ is ignored, and this amounts to the Zwetkoff order parameter S remaining constant up to the surface. Croxton and Chandrasekhar[42] have discussed the $S(z)$ profile for a more realistic delocalized surface, but further progress along these lines is impeded by our lack of knowledge of $\rho_{(1)}(z)$.

Parsons finally obtains, for van der Waals interactions between the molecules,

$$\gamma_{\parallel} = \gamma_0 \left[1 - \frac{4}{9}S + \frac{8}{27}S^2\right] \tag{2.5.12a}$$

and

$$\gamma_{\perp} = \gamma_0 \left[1 + \frac{2S}{3} + \frac{S^2}{6}\right] \tag{2.5.12b}$$

where γ_0 is a constant. Clearly $\gamma_{\parallel} < \gamma_{\text{v}}$ for all values of the order parameter S, and so it is concluded that the total free energy is minimized when the orientation is parallel to the free surface. The discontinuous drop in S at the nematic–isotropic phase transition implies a discontinuous variation in the surface tension:

$$\frac{\Delta\gamma}{\gamma} = -\frac{4}{9}\Delta S + \frac{8}{27}(\Delta S)^2 \tag{2.5.13}$$

where ΔS is the discontinuity in the order parameter across T_{NI}. By way of illustration, taking

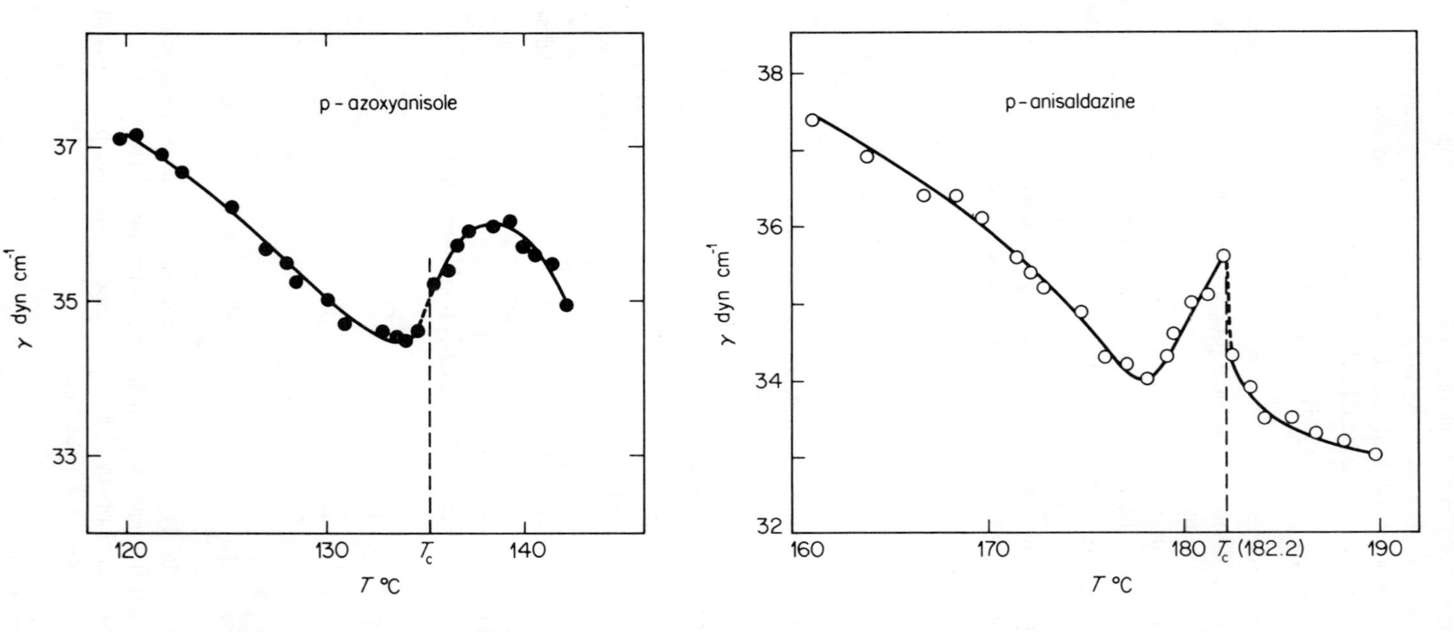

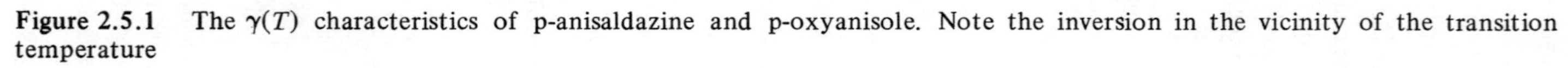

Figure 2.5.1 The $\gamma(T)$ characteristics of p-anisaldazine and p-oxyanisole. Note the inversion in the vicinity of the transition temperature

$S(T \lesssim T_{NI}) \sim \frac{1}{3}$, $S(T \gtrsim T_{NI}) = 0$ we obtain a negative drop in the surface tension of approximately 10%. There will, of course, be a density discontinuity across the nematic–isotropic phase transition, but this is subordinate (~0.6%) to the dependence upon the order parameter.

Unfortunately the theory is inconsistent in as far as the preferred orientation of the molecules parallel to the surface implies that there is a surface orientational excess order, which is contrary to the initial assumption. Clearly some self-consistent prescription for $S(z)$ remains to be determined: nevertheless, this appears to be the first molecular theory of surface tension in nematic liquid crystals, and certainly provides a qualitative explanation for the experimental $\gamma(T)$ curves[82] (Fig. 2.5.1).

There have been suggestions that this variation in surface tension could be attributed to changes in molecular orientation in the vicinity of the transition temperature.[84] Certainly this could account for the features of the $\gamma(T)$ characteristic; however, there is evidence[83] that in p-azoxyanisole the molecules are almost parallel to the free surface, and that the orientation is practically independent of temperature in the nematic range. The surface orientation in anisaldazine is not known.

The numerical evaluation of (2.5.6) for the nonspherical molecular interfacial density profile is impeded by our lack of knowledge concerning the function $g_{(2)}(z_1, r_{12}, \boldsymbol{\omega}_1, \boldsymbol{\omega}_2)$. Some progress has been made, however, for *slightly* anisotropic molecular systems for which a perturbation about a reference isotropic interaction can be made, the resulting integrals being evaluated over isotropic distributions. Thus, Haile, Gubbins and Gray[122] linearize (2.5.6) and expand in terms of the harmonic components of a Pople-type anisotropic interaction[121]: they obtained for axially symmetric molecular systems (note a slight change in nomenclature between reference 122 and the present discussion)

$$\rho_{(1)}(z_1, \theta_1) = \frac{\overset{0}{\rho}_{(1)}(z_1)}{\Omega} + \frac{cP_2(\cos\theta_1)}{\Omega kT}\,\overset{0}{\rho}_{(1)}(z_1)\int_2 P_2(\cos\theta_{12})\overset{0}{\rho}_{(1)}(z_2)\overset{0}{g}_{(2)}(z_1, z_2, r_{12})r_{12}^{*-n}\,\mathrm{d}2 \tag{2.5.14}$$

since for axially symmetric systems the orientation dependence upon ω_1 reduces to a function of the polar Euler angle θ_1 alone. $\Omega = \int \mathrm{d}\omega$, $r^* = r/\sigma$, $c = -8\delta\epsilon$ and $n = 12$ for anisotropic overlap, whilst for dispersion $c = 4\kappa\epsilon$, $n = 6$. κ is the dimensionless anisotropy of the polarizability, δ is a dimensionless overlap parameter, and ϵ and σ are isotropic potential parameters. For the two reference distributions $\rho^0_{(1)}(z_1)$, $g^0_{(2)}(z_1, z_2, r_{12})$, Haile *et al.* use Toxvaerd's model[11] (equation 2.4.1, *et seq.*).

The results are shown in Figure 2.5.2 using parameters which are roughly appropriate to liquid nitrogen. Haile *et al.* observe that the anisotropic dispersion forces tend to orient linear molecules parallel to the interfacial plane below $z_1^* \sim 0.44$, reaching a maximum at $z_1^* \sim -0.5$. For $z_1^* > 0.44$ there appears a tendency for a perpendicular alignment on the vapour side of the dividing surface. Certainly this is difficult to understand physically, and Haile, Gubbins, and Gray comment that such an alignment is quite unexpected and are unable to conclude whether the effect is real or an artifact of the use of Toxvaerd's model for $g^0_{(2)}(z_1, z_2, r_{12})$. For overlap interactions of rodlike molecules, the molecules appear to stand perpendicular to the interfacial plane on the liquid side of the interface.

In the present author's opinion the effect may be almost certainly ascribed to the use of inconsistent one- and two-particle distributions and the use of an interpolated pair distribution of the form (2.4.1) which is too heavily weighted toward the bulk liquid distribution for $z_1^* > 0.44$. Qualitative considerations suggest that the use of these distributions implies the development of an anomalous torque field above the dividing surface (analogous to the

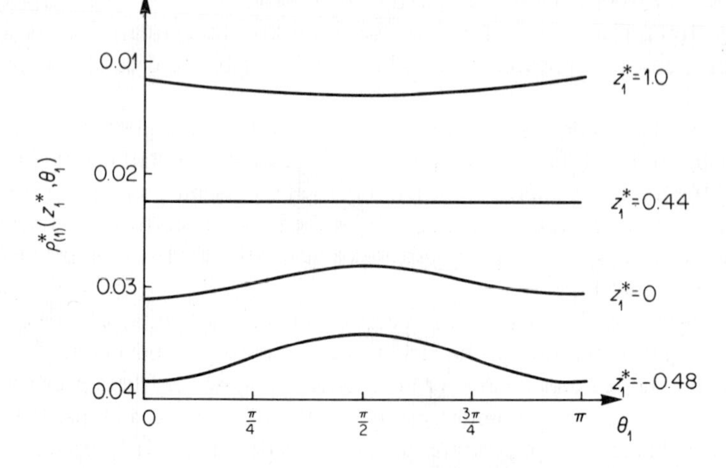

Figure 2.5.2 Density-orientation profile calculated on the basis of (2.5.13) by Haile, Gubbins and Gray[122] for a system of axially symmetric molecules at $T^* = kT/\epsilon = 0.85$. $\rho^*_{(1)} = \sigma^2 \rho_{(1)}$. (From Haile, Gubbins and Gray,[122] reproduced by permission of the American Institute of Physics)

development of anomalous constraining fields at the surface of symmetrical fluids), and as such accounts for the anomalous orientations.

Nevertheless, the model does exhibit the important qualitative feature of molecular alignment at the surface of molecular fluids, and presumably orientational contributions to the surface excess entropy would result in a somewhat less negative slope to the $\gamma(T)$ characteristic than that associated with the leading isotropic term in (2.5.14). However, these perturbative extensions remain inappropriate to strongly anisotropic systems, and unfortunately offer no prospect of a quantitative description of the $\gamma(T)$ characteristics for liquid crystal systems in the vicinity of the order–disorder transition.

For anisotropic overlap interactions ($\delta < 0$) corresponding to 'platelike' molecules, Haile *et al.* obtain similar results to those obtained for dispersion.

2.6 Liquid metals

Despite extensive theoretical and technological interest in the surface tension and energy of liquid metals, the statistical mechanical description remains largely undeveloped. In addition to the determination of the ionic distribution at the liquid surface, the electronic profile, to which the ionic profile is self-consistently related, has also to be established. Moreover, quantum interference effects in the electronic distribution and the density dependence of the Friedel oscillations have also to be incorporated. In general the electronic and ionic distributions $\bar{\rho}_{(1)}(z)$, $\overset{+}{\rho}_{(1)}(z)$ cannot be expected to coincide, and we therefore anticipate the development of electrostatic double-layer contributions to the surface energy. Finally, collective surface plasmon contributions to the surface energy will have to be incorporated. Until these problems have been extensively investigated both theoretically and experimentally there is little to be learned about such surface processes from the magnitude of γ alone. Its temperature variation may be more informative, however, and we shall consider this in some detail later in this section.

Liquid metal surface tensions are generally large in comparison with those of the liquid inert gases ranging from ~70 dyn cm^{-1} for Cs to ~2700 dy cm^{-1} for Re at their melting points.[45] In the past, attention has focused mainly on *either* the electronic or the ionic contributions to the surface tension and energy. In fact, of course, we need to account simultaneously and self-consistently for both processes.

The purely electronic analyses have concentrated on specifying how the electrons 'leak' out of a solid ionic jellium at $T = 0$ when surface excess entropy contributions do not arise and $\gamma \equiv u_s$. The initial electronic work[46] has been improved recently by Smith[47] and by Lang and Kohn.[48] Schmidt and Lucas[49] have recently emphasized the role of collective plasma modes at the liquid metal surface, and estimate that about 90% of the surface energy of most simple metals can be assigned to the shift in zero point energy of plasma oscillations in the process of creation of a free surface.

Lang and Kohn[48] use the Hohenberg–Kohn–Sham theory of an inhomogeneous high density electron gas to calculate the self-consistent electronic profile $\rho^{-}_{(1)}(z)$ against a positive jellium background terminating in a step function (Figure 2.6.1). Quantum interference effects at a sharp and *infinite* surface potential boundary[50] result in Friedel oscillations in the electron density of the form[48,50,51]

$$\rho^{-}_{(1)}(z) = \rho^{-}_{L}\left(1 + \frac{3\cos(2k_F z)}{4k_F^2 z^2} - \frac{3\sin(2k_F z)}{8k_F^2 z^3}\right) \tag{2.6.1}$$

where k_F is the Fermi wavenumber and ρ^{-}_{L} the average electron density in the bulk liquid. Equation (2.6.1) closely resembles that of Lang and Kohn[48], and the oscillations persist even when the ions are included explicitly.[52] As we see from Figure 2.6.1, an electrostatic double layer develops, and in addition to this Lang and Kohn include contributions to u_s arising from ion–electron and ion–ion interactions. The last two terms are positive, and for the purposes of calculation the jellium has to be replaced by an ionic lattice, or rather pseudo-potentials located on lattice sites. These authors take no account of the transverse collective plasmon modes –

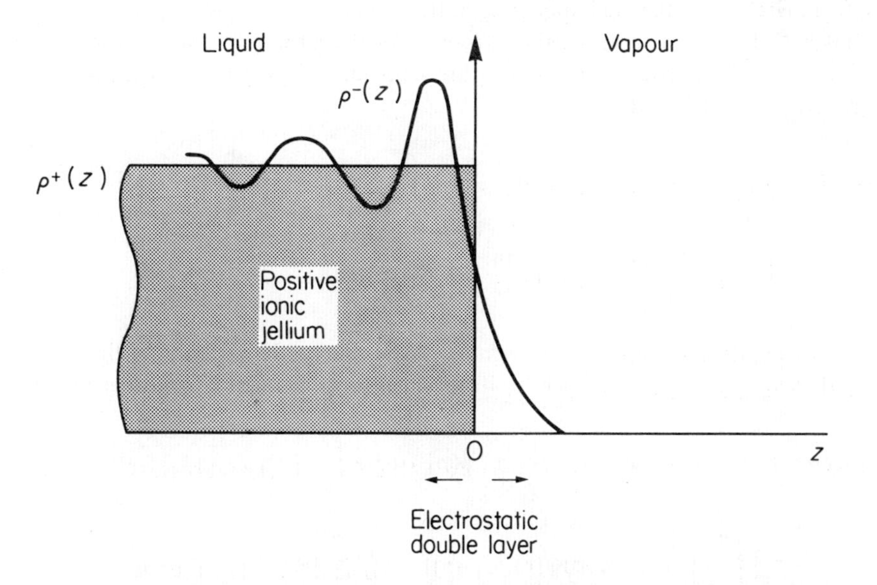

Figure 2.6.1 Lang and Kohn's model of the liquid metal surface. $\rho^{-}(z)$ represents the development of the electronic density profile against a positive jellium $\rho^{+}(z)$ terminating in a step function at the surface. Since the electronic and ionic profiles do not coincide, an electrostatic double layer will develop

nevertheless, their results are in reasonable agreement with a surface tension extrapolated to absolute zero.

Essentially ionic computations of the thermodynamic parameters of the liquid metal surface have been given by Johnson *et al.*,[53] and more recently by Waseda and Suzuki.[54] These authors work in what is the Fowler step model of the liquid surface, adopting pair potentials determined from experimental bulk liquid structure factors. This approach takes no account of the electronic processes mentioned above, and although the surface tensions are in reasonable agreement with experimental determinations, the results must, unfortunately, be considered largely fortuitous.

Considering a pseudo-ion model of a liquid metal, the interaction energy of such a system with a surface will be[55]

$$\Phi_N = \sum_i \phi[\rho^-_{(1)}(z_i)] + \tfrac{1}{2} \sum_{i>j} [\Phi(r_{ij} \mid \rho^-_{(1)}(z_i)) + \Phi(r_{ij} \mid \rho^-_{(1)}(z_j))] + \text{electrostatic terms} \tag{2.6.2}$$

where $\rho^-_{(1)}(z)$ is the single-particle electronic distribution. The first term represents the self-energy of a pseudo-atom; that is, the energy of a pseudo-ion i surrounded by its neutralizing cloud of electrons of average local density $\rho^-_{(1)}(z_i)$. The effective pairwise interactions between pseudo-ions located at $\mathbf{r}_i$ and $\mathbf{r}_j$ where $r_{ij} = |\mathbf{r}_i - \mathbf{r}_j|$ depends upon the local electronic densities. The pair interaction term in (2.6.2) is written so as to preserve symmetry between i and j. In the case of a bulk liquid metal a perturbation analysis readily yields $\phi(\rho^-_\mathrm{L})$ and $\Phi(r_{12} \mid \rho^-_\mathrm{L})$: in the absence of any knowledge of these functions in an inhomogeneous electron gas, we can only assume the same analytic density dependence of the self and pair terms as in the bulk. This *ansatz* implies that the self-energy and pair interaction of a surface pseudo-atom at z_i are equal to the corresponding quantities for a bulk metal of average electron density $\rho^-_{(1)}(z_i)$, and that this is the only effect of the surface on the interaction. This is a reasonable assumption for the liquid metal phase, but such a nearly free electron model is, of course, quite inappropriate, though fortunately of negligible importance in the vapour.

On this basis Evans[55] has obtained expressions for the surface tension and energy of a liquid metal. Assuming $Z^{-1}\rho^-_{(1)}(z_i)$ and $\rho^+_{(1)}(z_i)$ are coincident (Z = valence), then no electrostatic terms will arise in (2.6.2) and we obtain:

$$\gamma(T) = -\int_{-\infty}^{\infty} \rho^+_{(1)}(z) \frac{\mathrm{d}\phi}{\mathrm{d}z}(\rho^-_{(1)}(z)) z \,\mathrm{d}z - \tfrac{1}{2}\int_{-\infty}^{\infty}\int \rho^+_{(2)}(z, \mathbf{r}) \frac{\partial \Phi}{\partial z}(r \mid \rho^-_{(1)}(z)) z \,\mathrm{d}\mathbf{r}\,\mathrm{d}z$$

$$+\int_{-\infty}^{\infty}\int \frac{x^2 - z^2}{r} \rho^+_{(2)}(z, \mathbf{r}) \frac{\partial \Phi}{\partial r}(r \mid \rho^-_{(1)}(z)) \mathrm{d}\mathbf{r}\,\mathrm{d}z. \tag{2.6.3}$$

For an insulating fluid the self-energy term $\mathrm{d}\phi/\mathrm{d}z$ and the spatial variation of the pair potential $\partial\Phi/\partial z$ both vanish, and (2.6.3) reduces to its inert gas form. Similarly, for the surface energy

$$u_\mathrm{s}(T) = \int_{-\infty}^{0} \{\rho^+_{(1)}(z)\phi(\rho^-_{(1)}(z)) - \rho^+_\mathrm{L}\phi(\rho^-_\mathrm{L})\} \,\mathrm{d}z + \int_{0}^{\infty} \{\rho^+_{(1)}(z)\phi(\rho^-_{(1)}(z)) - \rho^+_\mathrm{V}\phi(\rho^-_\mathrm{L})\} \,\mathrm{d}z$$

$$+\frac{1}{2}\int_{-\infty}^{0}\int \{\rho_{(2)}(\mathbf{r}, z)\Phi(\mathbf{r} \mid \rho^-_{(1)}(z)) - \rho^2_\mathrm{L} g^\mathrm{L}_{(2)}(r)\Phi(r \mid \rho_\mathrm{L})\} \,\mathrm{d}\mathbf{r}\,\mathrm{d}z$$

$$+\frac{1}{2}\int_{0}^{\infty}\int \{\rho_{(2)}(\mathbf{r}, z)\Phi(\mathbf{r} \mid \rho^-_{(1)}(z)) - \rho^2_\mathrm{V} g^\mathrm{V}_{(2)}(r)\Phi(r \mid \rho_\mathrm{V})\} \,\mathrm{d}\mathbf{r}\,\mathrm{d}z. \tag{2.6.4}$$

The appearance of terms depending upon the self-energy $\phi(\rho^-)$ in (2.6.4) is a result of the explicit dependence of the interaction potential on $\rho^-(z)$. If $\phi(\rho^-)$ were constant, independent of z, the first two terms in (2.6.4) would sum to zero provided the origin of coordinates is located on the Gibbs dividing surface. Equation (2.6.4) then reduces to its insulating fluid form since then Φ is independent of $\rho^-(z)$.

At this stage Evans adopts an exponential electronic profile $\rho^-_{(1)}(z)$ at the liquid metal surface, in conjuction with two limiting forms for the ionic distribution – a step function, and an exponential ionic profile $\rho^+_{(1)}(z) = Z^{-1}\rho^-_{(1)}(z)$ which ensures electrostatic neutrality throughout the transition zone and avoids the development of double layer contributions. Whether the profiles do coincide in this way awaits self-consistent determinations of $\rho^+_{(1)}(z)$ and $\rho^+_{(1)}(z)$ (see equation 2.6.7 below): consequently, electrostatic terms may be important (Frenkel[56] once ascribed the entire surface tension and energy to this effect and obtained $\gamma = 472$ dyn cm^{-1} for mercury, which is in excellent agreement with modern determinations!). Evans concludes that the main contributions to γ and u_s come from the change in self-energy of the pseudo-atoms arising from the electron-gas exchange energy and band structure term dependence upon $\rho^-_{(1)}(z)$. To this extent the large surface tensions and surface energies which characterize the liquid metals are accounted for. However, as we see from (2.6.3), an oscillatory rather than monotonic electronic profile $\rho^-_{(1)}(z)$ and hence local self-energy $\phi(\rho^-_{(1)}(z))$ may substantially modify this conclusion.

Our interest centres more on the specification of the profiles $\rho^-_{(1)}(z), \rho^+_{(1)}(z)$, however. The thermodynamic functions of the liquid surface, γ and u_s, are expressed as integrals over the ionic and electronic distributions and, as we observed earlier, evaluation and comparison of (2.6.3, 2.6.4) with experiment will tell us little about the transition profiles.[85] Nevertheless, there is reason to believe that the density varies by a factor $\sim 10^3$ over an ionic diameter or so. A number of authors[90,91] have observed that the product of the surface tension and the bulk isothermal compressibility $\gamma\chi_T$ is virtually independent of the nature of the liquid at the triple point, and represents a characteristic length proportional to the surface thickness. A comparison of liquid metals with nonmetallic systems is significant because inter-ionic forces in metals are density-dependent. Thus, had the density gradient at the interface not been sharp, it might have been expected to show up as a difference in the value of $\gamma\chi_T$ for metals and insulators, and no such difference is found. The relation appears to hold for systems of quite dissimilar binding,[91] including molten salts, organic liquids and aqueous solutions in addition to liquid metals and the liquid inert gases – indeed, $\gamma\chi_T$ for ^{4}He and ^{3}He extrapolated to 0 °K agrees with the classical liquids.

There is no formal difficulty in determining the ionic profile $\rho^+_{(1)}(z)$ *given the electronic distribution* $\rho^-_{(1)}(z)$:

$$
\begin{aligned}
-kT\frac{\partial\rho^+_{(1)}(z_1)}{\partial z_1} &= \rho^+_{(1)}(z_1)\frac{d\phi}{dz_1}(\rho^-_{(1)}(z_1)) + \frac{1}{2}\int \rho^+_{(2)}(\mathbf{r}_{12}, z_1)\frac{\partial\Phi}{\partial z_1}(r_{12} \mid \rho^-_{(1)}(z_1))\,d\mathbf{r}_{12} \\
&\quad - \frac{1}{2}\int \rho^+_{(2)}(\mathbf{r}_{12}, z_1)\frac{\partial}{\partial r_{12}}\{\Phi(r_{12} \mid \rho^-_{(1)}(z_1)) \\
&\quad + \Phi(r_{12} \mid \rho^-_{(1)}(z_1 + z_{12}))\}\frac{z_{12}}{r_{12}}\,d\mathbf{r}_{12}.
\end{aligned}
\tag{2.6.5}
$$

This integro-differential equation expresses the mechanical stability of the ionic component of the liquid metal surface. The first term on the right represents the force on the ionic component arising from the asymmetric screening by the conduction electrons. In addition to the structural inhomogeneity at the liquid surface, asymmetry in the pair potential itself accounts for the second force term acting on the ionic distribution. The final term is the conventional contribution arising from the configurational anisotropy of the ionic configuration. Equation (2.6.5) does, of course, reduce to rare-gas form in the absence of the terms $d\phi/dz$ and $\partial\Phi/\partial z$ (c.f. 2.3.11).

Since we have, as usual,

$$\rho^{+}_{(2)}(\mathbf{r}_{12}, z_1) = \rho^{+}_{(1)}(z_1)\rho^{+}_{(1)}(z_2)g^{+}_{(2)}(\mathbf{r}_{12}, z_1) \tag{2.6.6}$$

integration of (2.6.5) subject to the boundary condition $\rho^{+}_{(1)}(-\infty) = \rho^{+}_{\mathrm{L}}$ yields

$$\begin{aligned}\rho^{+}_{(1)}(z_1) = \rho^{+}_{\mathrm{L}} \exp\Bigg[-\frac{1}{kT}\int_{-\infty}^{z_1}\Bigg(\frac{d\phi}{dz_1}\rho^{-}_{(1)}(z_1)\\ &+\frac{1}{2}\int \rho^{+}_{(1)}(z_2)g^{+}_{(2)}(\mathbf{r}_{12}, z_1)\frac{\partial\Phi}{\partial z_1}(r_{12}\,|\,\rho^{-}_{(1)}(z_1))\Bigg)d\mathbf{r}_{12}\\ &-\frac{1}{2}\int \rho^{+}_{(1)}(z_2)g^{+}_{(2)}(\mathbf{r}_{12}, z_1)\frac{\partial}{\partial r_{12}}\Big(\Phi(r_{12}\,|\,\rho^{-}_{(1)}(z_1))\\ &+\Phi(r_{12}\,|\,\rho^{-}_{(1)}(z_1+z_{12})\Big)\frac{z_{12}}{r_{12}}\,d\mathbf{r}_{12}\Bigg]dz_1.\end{aligned} \tag{2.6.7}$$

If we assume the same analytic dependence of ρ and Φ upon ρ^{-} as in the bulk liquid metal, and if we adopt some form for $\rho^{-}_{(1)}(z)$ which incorporates quantum interference effects (e.g. 2.6.1), then a consistent ionic profile may be determined by iterative solution of (2.6.7) in conjunction with the Lang–Kohn[48] prescription for $\rho^{-}_{(1)}(z)$, given $\rho^{+}_{(1)}(z)$. No self-consistent profiles $\rho^{-}_{(1)}(z)$, $\rho^{+}_{(1)}(z)$ have been reported in the literature: determinations of the above kind are, however, in progress.[57]

The role of surface plasmon contributions is entirely neglected in Evans' treatment. Certainly it is true that plasmon contributions do arise, but their status is not yet sufficiently understood to be usefully incorporated into the extant theories of the liquid metal surface: indeed their contribution is the centre of further controversy.[58]

This pseudo-atom approach can, in principle, be extended to the liquid transition metals,[55] but the failure of the perturbative description of the strong ion–electron interaction means that, unlike the simple metals, we have no way of establishing the pairwise or self-energy terms. Smith,[47] however, in a description of the work function and related surface properties in liquid transition metal systems, regarded the d electrons as being 'free' with some success. The resulting very high electronic density offers an explanation of the very large surface tensions and energies which characterize the transition metal series. Such qualitative remarks must be regarded as purely speculative at this stage, however. Kumaradivel and Evans[85] have recently calculated the surface energy of a number of liquid metals on the basis of the above pseudo-atom approach, and report good agreement with experiment for the alkali metals, but over-estimate the magnitude of the surface energy for several of the polyvalent systems. Unfortunately, implications of the numerical coincidence with experiment fails to resolve the problem regarding the surface structure.

The temperature variation $\partial\gamma/\partial T$ of the surface tension is more sensitive to the structure of transition zone than γ itself. For a single component system it is readily shown that[2]

$$\frac{\partial\gamma}{\partial T} = -S_{\mathrm{s}} \tag{2.6.8}$$

where S_{s} is the surface excess entropy per unit area. Substitution of equations (2.6.3, 2.6.4) in the Gibbs–Helmholtz equation and integration by parts yields[55]

$$T\frac{\partial\gamma}{\partial T} = \int_{-\infty}^{\infty} \phi(\rho_{(1)}^{-}(z)) \frac{\partial\rho_{(1)}^{+}(z)}{\partial z} z\,dz + \text{pairwise contributions} \tag{2.6.9}$$

from which we see the great sensitivity of the temperature coefficient $\partial\gamma/\partial T$ to the development of density oscillations in the ionic profile $\rho_{(1)}^{+}(z)$. In addition to the initiation of oscillations by the surface constraining field, quantum interference effects on the electronic profile and Friedel oscillations in the ionic screening may combine to produce a strongly oscillatory density transition at temperatures just above the triple point.[59]

These low entropy surface states correspond to a positive temperature coefficient $\partial\gamma/\partial T$ (2.4.11) as we see from (2.6.8) and (2.6.9). In Table 2.6.1[60] figures are given for a number of liquid metals for which the slopes $\partial\gamma/\partial T$ should be fairly reliable.[61-63] Certain of the liquid metals – Cu, Zn and Cd, in particular – show positive slopes with subsequent inversion over a temperature range ~100° above the melting point. The crudest explanation is that the liquid surface of these metals is virtually crystalline,[60] and since the entropy of melting of Cn, Zn and Cd is about 1.2 K per atom, we would need to regard the top three layers as crystalline to account for the data in Table 2.6.1. At higher temperatures, of course, the liquid surface will thermally delocalize and systems exhibiting positive $\gamma(T)$ slopes will ultimately invert to yield the familiar monotonic decreasing function of temperature, going to zero as $(T_c - T)^{\mu}$: the exponent is as yet unknown for liquid metals. Such a $\gamma(T)$ curve exhibiting an inversion is certainly contrary to the majority of experimental evidence, and probably all theoretical models.

The important series of *equilibrium* $\gamma(T)$ measurements by White[62,63] has been reviewed elsewhere,[2] and we shall restrict ourselves to the observation that only when precautions are taken to ensure that the liquid surface is in equilibrium with its vapour are inversions in the $\gamma(T)$ characteristic observed. It is possible that effects of surface impurity migration with temperature could account for the inversion, but there seems no reason to doubt White's observations. Evaporation rates of only ~100 atomic layers per second appear to substantially modify the $\gamma(T)$ and $\partial\gamma/\partial T$ characteristics. This is at first sight a somewhat surprising result since the surface would reach equilibrium in a time ~10^{-11} sec. However, a net flux of atoms

Table 2.6.1 Surface entropies of liquid metals (after Faber[60]). (These data are for temperatures close to T_m, except in the case of Hg)

Metal	$\partial\gamma/\partial T$ (dyn cm^{-1} $°C^{-1}$)	Surface entropy per atom (units of K)
Li	−0.14	0.80
Na	−0.1	0.85
K	−0.06	0.78
Rb	−0.06	0.94
Cs	−0.05	0.78
Ag	−0.13	0.65
Al	−0.135	0.68
In	−0.096	0.61
Sn	−0.083	0.55
Cu	+0.75	−2.9
Zn	+0.5	−2.2
	+0.5	−2.9
Hg (25°C)	−0.20	+1.2

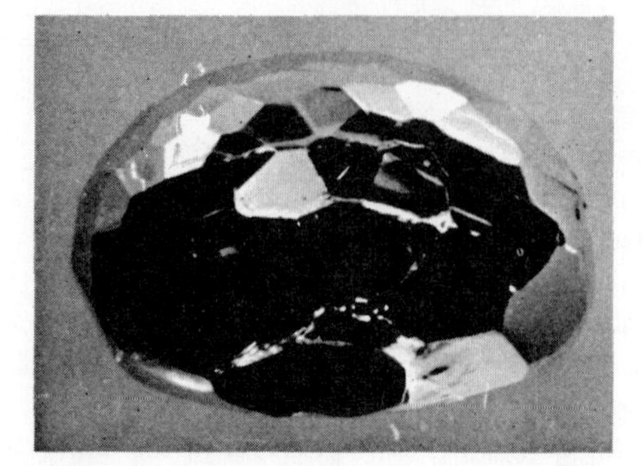

Figure 2.6.2 Faceting at the surface of a solidified droplet of zinc. Sectioning reveals that the grains initiated at surface sites, quite independently from those at the base

across the surface implies that the single-particle transition profile cannot be of equilibrium form, even though it may be time-independent.

One other striking qualitative observation of White[64] concerns the form of the sessile drops on solidification. It was found that the liquid metal sessile droplets used in the $\gamma(T)$-investigation developed surface faceting to an extent and perfection directly related to the system's tendency to show $\gamma(T)$ inversion. Thus Zn and Cd solidified into beautifully faceted droplets (Figure 2.6.2), unlike those showing classical behaviour which solidify into perfectly smooth drops. It is difficult to escape the conclusion that the development of stable density oscillations at the liquid surface serves as a nucleation centre in the solidification process. X-ray investigations show that the facets are basal planes – planes of maximum packing density. Indeed, on sectioning the zinc droplets, White[64] found that the configuration of grains is consistent with solidification starting from several locations at the base and growing upwards, having begun independently at several locations at the surface. The shrinkage cavity in more than thirty drops sectioned, though usually located just below the surface, in no case broke out into the surface.

2.7 Multi-component transition profiles

There has been very little attention devoted to the statistical mechanical description of the transition profile in classical, non-metallic multi-component systems. The partial profiles would, of course, be expected to show differing thermal dependences, and to display features of surface migration of whichever species minimize the excess Helmholtz free energy.

Suppose the bulk fluid is comprised of species $\alpha, \ldots, \omega$, present in number densities $\rho_{L\alpha}, \ldots, \rho_{L\omega}$ such that $\rho_L = \Sigma_{\lambda=\alpha}^{\omega} \rho_{L\lambda}$. It is then straightforward to write down the single-particle BGY equation for the λ-th partial profile:[123]

$$kT\nabla_1 \rho_{(1;\lambda)}(z_1) = -\sum_{\mu=\alpha}^{\omega} \int_2 \nabla_1 \Phi_{\lambda\mu}(12)\rho_{(2;\lambda\mu)}(\mathbf{12})\,\mathrm{d}2. \tag{2.7.1}$$

$\rho_{(1;\lambda)}(z_1)$ represents the single-particle distribution of species λ, whilst $\Phi_{\lambda\mu}(12)$ and $\rho_{(2;\lambda\mu)}(\mathbf{12})$ represent the pair potential and the *anisotropic* two-particle density distribution developed between particles of species λ and μ, respectively. For $\rho_{(2;\lambda\mu)}(\mathbf{12})$ we may write the exact closure analogous to (2.3.10):

$$\rho_{(2;\lambda\mu)}(\mathbf{12}) = \rho_{(1;\lambda)}(\mathbf{1})\rho_{(1;\mu)}(\mathbf{2})g_{(2;\lambda\mu)}(\mathbf{12}) \tag{2.7.2}$$

and in the absence of any explicit knowledge of $g_{(2;\lambda\mu)}(\mathbf{12})$ we may, following Green[20] (c.f. 2.3.14), assume

$$\rho_{(2;\lambda\mu)}(\mathbf{12}) = \rho_{(1;\lambda)}(\mathbf{1})\rho_{(1;\mu)}(\mathbf{2})g^{\mathrm{L}}_{(2;\lambda\mu)}(\mathbf{12}) \tag{2.7.3}$$

that is, adopt the *isotropic* bulk liquid radial distribution $g^{L}_{(2;\lambda\mu)}(12)$ of the species μ about λ. We should emphasize that the adoption of $g^{L}_{(2;\lambda\mu)}(12)$ in (2.7.3) neglects the functional dependence of the pair distribution upon the local number densities $\rho_{(1;\lambda)}(\mathbf{1}), \rho_{(1;\mu)}(\mathbf{2})$, and to that extent is an approximation additional to that of Green. Under these circumstances equation (2.7.1) reduces to

$$kT\frac{\nabla_1 \rho_{(1;\lambda)}(z_1)}{\rho_{(1;\lambda)}(z_1)} = -\sum_{\mu=\alpha}^{\omega}\int_{\mathbf{2}} \nabla_1 \Phi_{\lambda\mu}(12)\rho_{(1;\mu)}(\mathbf{2})g^{L}_{(2;\lambda\mu)}(12)\,\mathrm{d}\mathbf{2}. \tag{2.7.4}$$

Integrating subject to the boundary condition $\rho_{(1;\lambda)}(-\infty) = \rho_{L\lambda}$, we have

$$\rho_{(1;\lambda)}(\mathbf{1}) = \rho_{L\lambda}\exp\left(-\frac{1}{kT}\int_{-\infty}^{z_1}\sum_{\mu=\alpha}^{\omega}\int_{\mathbf{2}} \nabla_1 \Phi_{\lambda\mu}(12)\rho_{(1;\mu)}(\mathbf{2})g^{L}_{(2;\lambda\mu)}(12)\,\mathrm{d}\mathbf{2}\,\mathrm{d}z_1\right) \tag{2.7.5}$$

where the distributions $g^{L}_{(2;\lambda\mu)}(12)$ are generally available from scattering experiments. The specific difficulty which arises in the solution of equations (2.7.4) or (2.7.5) is the cross-dependence of the profile $\rho_{(1;\lambda)}(\mathbf{1})$ upon all the other unknown partial profiles $\rho_{(1;\mu)}(\mathbf{2})$. A possible – if tedious – approach might be to solve the $\alpha, \ldots, \omega$ coupled equations by initially adopting partial transition profiles of step function form according to

$$\begin{aligned}\rho_{(1;\mu)}(\mathbf{1}) &= \rho_{L\mu} \quad && z \leqslant 0\\ &= 0 && z > 0\end{aligned} \tag{2.7.6}$$

and then iteratively determining the self-consistent partial distributions $\rho_{(1;\alpha)}(\mathbf{1}), \ldots, \rho_{(1;\omega)}(\mathbf{1})$ in a similar manner to that proposed for the ionic and electronic partial distributions at the surface of a liquid metal (Section 2.6). More formally, we may express (2.7.5) in matrix form: setting

$$\hat{\boldsymbol{\rho}} = \begin{bmatrix} -kT\ln\rho_{(1;\alpha)}(\mathbf{1})\\ \vdots \\ -kT\ln\rho_{(1;\omega)}(\mathbf{1})\end{bmatrix}$$

the square, symmetric matrix

$$\mathbf{K} = \begin{bmatrix} \nabla_1\Phi_{\alpha\alpha}(12)g^{L}_{(2;\alpha\alpha)}(12) & \ldots & \nabla_1\Phi_{\alpha\omega}(12)g^{L}_{(2;\alpha\omega)}(12)\\ \vdots & \ddots & \vdots \\ \nabla_1\Phi_{\omega\alpha}(12)g^{L}_{(2;\omega\alpha)}(12) & \ldots & \nabla_1\Phi_{\omega\omega}(12)g^{L}_{(2;\omega\omega)}(12)\end{bmatrix}$$

and

$$\boldsymbol{\rho} = \begin{bmatrix} \rho_{(1;\alpha)}(\mathbf{2})\\ \vdots \\ \rho_{(1;\omega)}(\mathbf{2})\end{bmatrix}$$

then (2.7.5) may be written

$$\hat{\boldsymbol{\rho}} = \int_{-\infty}^{z_1}\int_{\mathbf{2}} \mathbf{K}\boldsymbol{\rho}\,\mathrm{d}z\,\mathrm{d}z_1. \tag{2.7.7}$$

A useful discussion of the solution of simultaneous, nonlinear integral equations of the above form in the present context has been given by Plesner, Platz and Christiansen[120].

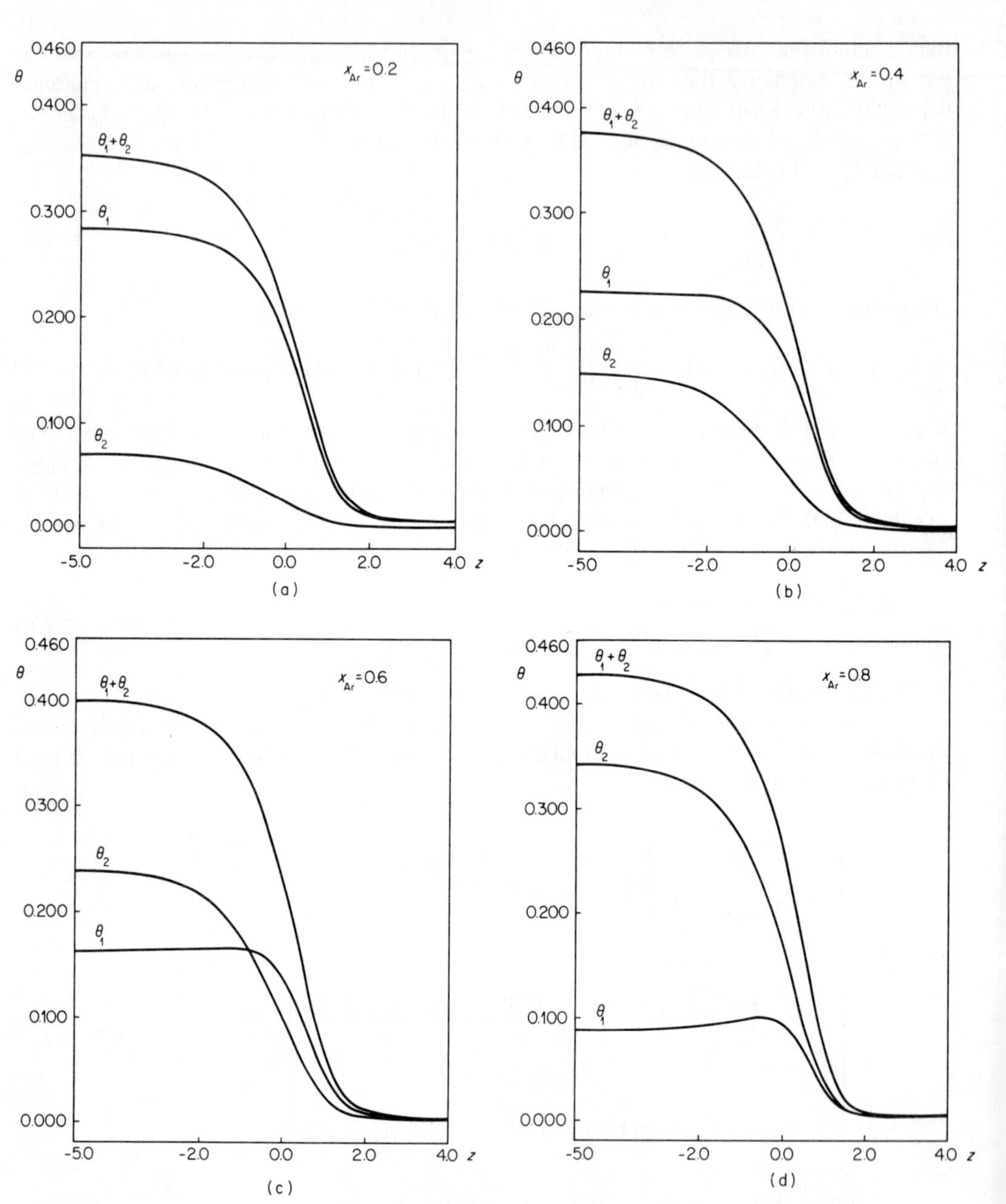

Figure 2.7.1 Density transition curves for Ar–N_2 mixtures. (a)X_{Ar} = 0.2; (b)X_{Ar} = 0.4; (c)X_{Ar} = 0.6; (d)X_{Ar} = 0.8. (From Plesner, Platz and Christiansen,[120] reproduced by permission of the American Institute of Physics)

The structural features of the partial profiles may be discussed in terms of surface partial constraining fields, just as for single-component systems. Indeed, the partial constraining field appropriate to the component λ is, from (2.7.5),

$$\sum_{\mu=\alpha}^{\omega} \int_2 \nabla_1 \Phi_{\lambda\mu}(12)\rho_{(1;\mu)}(2)g_{(2;\lambda\mu)}(12)\mathrm{d}2 \tag{2.7.8}$$

in the present closure approximation. Whether this surface field drives the component λ towards or away from the surface underlies the process of surface migration. If the surface were discontinuously sharp, then a surface migration of those components tending to reduce the surface excess free energy would be apparent at *all* molar fractions, and conversely for components tending to increase the surface free energy. In fact, of course, softer partial constraining fields prevail and the development or otherwise of surface partial densities greater than those in the bulk depends sensitively upon the detailed features of the functions appearing the constraint (2.7.8).

Whilst necessarily $\sum_{\mu=\alpha}^{\omega} \rho_{(1;\,\mu)}(-\infty) = \rho_L$ for reasons of stability, the partial profiles may show preferential surface migration of one species with respect to another, and in consequence we anticipate 'non-classical' thermal dependences of the surface excess functions, in particular the surface tension–temperature characteristic.

Plesner, Platz and Christiansen[120] have determined the partial profiles for a binary argon–nitrogen mixture at 84 °K for a range of mole fractions. The transition profiles were determined on the basis of the criterion of constancy of the chemical potential across the transition zone, and are open to precisely the same objections discussed in Section 2.3 for single-component systems. In addition, there are problems associated with the specification of the pair interaction and correlation functions for a multicomponent system. Nevertheless, the results of Plesner *et al*[120] provide the first qualitative assessment of the features of a binary transition zone. Their partial and total transition profiles are shown in Figure 2.7.1, and we see that for liquid mole fractions of argon $\geqslant 0.6$, the transition layer exhibits features of surface migration of N_2.

Unfortunately, the drastic approximations in the theory[120] restrict conclusions to purely qualitative observations. Although the calculated surface tensions as a function of mole fraction are in reasonable agreement with experiment, we have to remember that, as for single-component systems, their numerical coincidence is an insensitive criterion of the adequacy of the calculated partial profiles. We do observe, however, that the profiles bear comparison with the computer simulations of Chapela *et al.*,[119] to be discussed in Section 2.10.

2.8 The quantum liquid surface

The deviation of the quantum liquids from the classical estimates of the superficial thermodynamic functions is well known. The reduced quantal surface tension $\gamma^* = \gamma\sigma^2/\epsilon$, for example, is generally depressed below the corresponding state estimate for a classical Lennard–Jones system, and even neon shows a systematic departure from other inert gas liquids, particularly at low temperatures. In this section we shall therefore extend the scope of the discussion to include quantal modifications of the surface tension and energy, in addition to the determination of the single-particle distribution $\rho_{(1)}(z)$.

Atkins[65,88] has determined the limiting law of temperature dependence of superfluid ^{4}He in terms of quantized hydrodynamic surface capillarly waves. The surface part of the free energy per unit area may be identified with the surface tension, and Atkins obtains

$$\gamma = \gamma_0 - \frac{\hbar}{4\pi}\left(\frac{\rho}{\gamma_0}\right)^{2/3} \int_0^\infty \frac{\omega^{4/3}}{e^{\hbar\omega/kT}-1}\,d\omega \sim \gamma_0 - \frac{(kT)^{7/3}}{4\pi\hbar^{4/3}}\left(\frac{\rho}{\gamma_0}\right)^{2/3} \Gamma\left(\frac{7}{3}\right)\zeta\left(\frac{7}{3}\right)$$

$$= \gamma_0 - \frac{0.134(kT)^{7/3}\rho^{2/3}}{\hbar^{4/3}\gamma_0^{2/3}} \qquad (2.8.2)$$

where Γ and ζ are the gamma and zeta functions, respectively. γ_0 represents the residuum of free energy at absolute zero which may be identified with the surface tension at $T = 0$. The appropriate capillary wave dispersion relationship has been assumed in writing down (2.8.1), since at low temperatures only capillary and not Rayleigh waves are likely to be excited. It must be pointed out that this derivation applies only to superfluid ^{4}He. Capillary waves of this kind do not exist in the Fermi liquid ^{3}He since there the viscosity increases indefinitely as $T \to 0$.

Singh[66] has pointed out that we may equivalently regard the ^{4}He system as an ideal degenerate assembly of non-interacting phonons. This implies a different dispersion relation to that assumed in Atkins' analysis, and Singh obtains

$$\gamma = \gamma_0 - \frac{\pi m \zeta(2) k T^2}{2h^2}$$

$$= \gamma_0 - 7.5 \times 10^{-3} T^2. \tag{2.8.3}$$

Of course, in addition to the normal mode excitations we should include other elementary excitations in the form of turbulence and vortex flow. Equations (2.8.2, 2.8.3) are difficult to distinguish experimentally, though measurements of the surface tension[86,89] seem to suggest a temperature dependence which is larger than that predicted by the Atkins expression (2.8.2). Chan,[67] however, claims to have recently confirmed Atkins' limiting law of surface tension–temperature dependence in superfluid ^{4}He. Edwards, Eckardt and Gasparini[87] have subsequently improved Atkins' formulation by incorporating effects due to compressibility and phonon dispersion.

These models do not, of course, give any indication of the transition profile $\rho_{(1)}(z)$. Indeed, the implicit assumption is that the surface is discontinuously sharp. Beyond this, the more specific approaches may be resolved into two main classes – the phenomenological and the variational. Representative of the former category are the treatments of Regge[68] and of Padmore and Cole[69] in which local energy functionals of the single-particle density distribution $F[\rho_{(1)}(r)]$ are proposed, and the profile determined by the condition $\delta F/\delta \rho_{(1)} = 0$. Neither functional is based on microscopic considerations and since different functionals were adopted, markedly different profiles were obtained – in one case oscillatory[68] and in the other monotonic.[69]

The variational approaches may be subdivided as to whether a trial profile or trial wave function is adopted in the variational minimization of the surface energy.

The ground state properties of an infinite sample of ^{4}He have been quite successfully determined using the Jastrow wave function

$$\Psi(\mathbf{1}, \ldots, \mathbf{N}) = \exp\left[-\frac{1}{2}\sum_{i<j}^{N} u(r_{ij})\right]. \tag{2.8.4}$$

The special boundary condition on the wave function Ψ arising from the presence of the free surface is that $\Psi \to 0$ as $z \to +\infty$, and the form generally adopted is

$$\Psi(\mathbf{1}, \ldots, \mathbf{N}) = \exp\left[-\frac{1}{2}\sum_{i>j}^{N} u(r_{ij}) - \frac{1}{2}\sum_{i}^{N} t(z_i)\right]. \tag{2.8.5}$$

Far outside the drop $t(z_i) \to +\infty$, whilst in the bulk $t(z_i) = 0$, this governing the density. In their expression for the surface energy, Shih and Woo[70] and Chang and Cohen[71] eliminate the t-dependence through the relation

$$\nabla\rho_{(1)}(z) = \rho_{(1)}(z)\nabla t(z) + \int \rho_{(2)}(\mathbf{r}_1, \mathbf{r}_2)\nabla u(\mathbf{r}_1, \mathbf{r}_2)\,d\mathbf{r}_2 \tag{2.8.6}$$

and variationally minimize the surface energy through single-particle trial functions $\tilde{\rho}_{(1)}(z)$. As usual, some form of closure has to be specified for $\rho_{(2)}(\mathbf{r}_1, \mathbf{r}_2)$. The approximation adopted is to write

$$\rho_{(2)}(\mathbf{r}_1, \mathbf{r}_2) = \rho_{(1)}(z_1)\rho_{(1)}(z_2)g^{\mathrm{L}}_{(2)}(r \mid \rho_{\mathrm{eff}}(z_1, z_2)) \tag{2.8.7}$$

where $g^{\mathrm{L}}_{(2)}(\mathbf{r} \mid \rho_{\mathrm{eff}})$ is the *isotropic* radial distribution appropriate to a uniform system at local density ρ_{eff}. A variety of forms for ρ_{eff} were adopted:

$$\rho_{\mathrm{eff}}(z_1, z_2) = [\rho_{(1)}(z_1)\rho_{(1)}(z_2)]^{1/2} \quad {}^{70} \tag{2.8.8a}$$

$$\rho_{\mathrm{eff}}(z_1, z_2) = \tfrac{1}{2}[\rho_{(1)}(z_1) + \rho_{(1)}(z_2)] \quad {}^{71} \tag{2.8.8b}$$

$$\rho_{\mathrm{eff}}(z_1, z_2) = \rho_{(1)}[\tfrac{1}{2}(z_1 + z_2)] \quad {}^{71}. \tag{2.8.8c}$$

Chang and Cohen[71] report that (2.8.8c) yields an inferior result to (a) and (b), both of which give essentially the same result.

The $g^{\mathrm{L}}_{(2)}(r \mid \rho_{\mathrm{eff}}(z_1,z_2))$ are determined from the BGY,[70] HNC[71] and PY[71] equations.

Liu, Kalos and Chester[72] have tested the internal consistency of such an approach by adopting a trial distribution $\rho_{(1)}(z)$ and using (2.8.6) to determine $t(z)$ which, they point out, shows pronounced curvature accounting for an over-estimate of the surface energy. The wave function (2.8.5) thus determined may be substituted in

$$\rho_{(1)}(z) = \frac{N \int \Psi^2(1, \ldots, N)\,\mathrm{d}2 \ldots \mathrm{d}\mathbf{N}}{\langle \Psi \mid \Psi \rangle} \tag{2.8.9}$$

and the distributions $\rho_{(1)}(z)$ compared. Discrepancy in the profiles may be attributed to the approximation introduced through equations (2.8.7) and (2.8.8). Liu *et al.*[72] report an increase of 20% in the surface energy of the output distribution, together with a pronounced peak where none existed in the input function. Moreover, the specification of a monotonic trial profile does, of course, eliminate the possible development of structural features at the surface. Again, the assumption of locally isotropic distributions in regions of pronounced inhomogeneity is open to objection precisely as in the case of classical systems.

Liu *et al.* adopt a wave function Ψ of the form (2.8.5), but discuss the surface structure of a thin ^{4}He film, in which case $t(z_i) \to \infty$ as $|z_i| \to \infty$. Writing

$$\exp\left[-\frac{1}{2}\sum_{i=1}^{N} t(z_i)\right] = \prod_{i=1}^{N} h(z_i) \tag{2.8.10}$$

two trial functions were adopted:

$$h(z) = [1 + \exp k(|z| - z_0)]^{-1} \tag{2.8.11a}$$

and

$$h(z) = \begin{cases} 1 & |z| < z_0 \\ 1 + \exp k(|z| - z_0)^q & |z| > z_0. \end{cases} \tag{2.8.11b}$$

The variational parameters k and q control the width of the transition zone, whilst z_0 determines its location.

Using the trial functions (2.8.11), the surface energy may be variationally determined with respect to the parameters k and q, and the transition profile thereby determined through (2.8.9). Liu *et al.* conclude that $\rho_{(1)}(z)$ shows a weak oscillatory structure. Moreover, these authors believe that an exact solution would reveal an *enhancement* of the oscillations, basing their conclusions on analogous calculations for ^{4}He in a channel.

In another variational determination of a trial wave function, Croxton[73] takes a symmetrical correlated ground-state Bijl–Dingle–Jastrow wave function which modifies in the vicinity of the ^{4}He surface:

$$\Psi(1,\ldots,N) = \prod_{i>j=1}^{N} \exp[\tfrac{1}{2}u(r_{ij})\Gamma(z_i,\theta,\phi)]$$

$$= \prod_{i>j=1}^{N} \exp\left[\tfrac{1}{2}u(r_{ij}) \sum_{l=0}^{\infty} \sum_{m=0}^{l} A_{lm}(z_i)P_l^m(\cos\theta)\Phi(\pm im\phi)\right]. \quad (2.8.12)$$

The coefficients of the harmonic expansion in the vicinity of the free surface specify the spatial dependence of the wave function. In the bulk liquid, (2.8.12) reduces to the isotropic Jastrow function:

$$\prod_{i>j=1}^{N} \exp[\tfrac{1}{2}u(r_{ij})\Gamma(z_i,\theta,\phi] \to \prod_{i>j=1}^{N} \exp[\tfrac{1}{2}u(r_{ij})] \quad (2.8.13)$$

as $z_i \to -\infty$.

In the present case the function $u(r_{ij})$ was determined from the Abe relation.[74,75]

The angular dependence of the trial function was restricted partly on symmetry grounds and partly on grounds of computational expediency: obviously practical interest is in a trial function containing a relatively small number of components beyond which the difficulties involved in solving the secular determinant become prohibitive. The trial function was, in fact, restricted to the first two zonal harmonics ($l = 0,1$; $m = 0$) – the coefficients $A_{00}(z)$, $A_{10}(z)$ controlling the z-dependence of the wave function. The expectation value $\langle \mathscr{H} \rangle$ is then variationally minimized with respect to $A_{00}(z)$ and $A_{10}(z)$.

A novel feature is the dependence of the kinetic contribution to $\langle \mathscr{H} \rangle$ upon the distribution $\Psi(1,\ldots,N)$. This kinetic anisotropy at the free surface is peculiar to quantal systems, and represents the effect of a kinetic residuum present even at absolute zero. It is shown[73], that this represents a quantum kinetic temperature tensor $\boldsymbol{\tau}(\rho,T)$ in regions of atomic anisotropy. Moreover, $|\tau| > T$, the thermodynamic temperature, which serves to lower the surface tension below its classical estimate. Mazo and Kirkwood,[76] in their discussion of homogeneous quantal fluids, observed that the calculation of all quantum effects, dynamical and statistical, may be transferred to the determination of τ. Indeed, these authors go so far as to suggest that a quantum fluid at temperature T behaves like a classical fluid at temperature τ, at the same density. In this sense, then, we can understand the quantal depression of the surface tension below its classical or corresponding state value.

Recently interest has been aroused by experiments of Esel'son and co-workers[77] in which it was found that the surface tension of superfluid ^{4}He was lowered by the presence of ^{3}He as an impurity by an amount greater than that expected from the usual rule of additivity. The results were interpreted phenomenologically by Andreev[78] in terms of the development of bound ^{3}He surface states at the ^{4}He surface – an extra binding energy at the surface which causes the ^{3}He atoms to accumulate at the superfluid ^{4}He surface. Alternatively and equivalently, he proposed that there exist surface states of ^{3}He with a minimum energy below that in the bulk. Clearly, any analytic description will depend essentially upon the detailed nature of the ^{4}He surface.[70,79] Indeed, the theoretical justification for Andreev's proposal has prompted several[70,71] of the recent attempts to specify the ^{4}He density profile.

Shih and Woo[70] have made a detailed determination of the surface states of ^{3}He in dilute ^{3}He–^{4}He solutions on the basis of their ^{4}He calculations. They replace the N-th ^{4}He atom by a ^{3}He atom which changes the ^{4}He Hamiltonian (H_4) as follows

$$H = H_4 + \left(1 - \frac{M_4}{M_3}\right)\frac{\hbar^2}{2M_4}\nabla_N^2. \tag{2.8.14}$$

Adopting a trial wave function of the form

$$\Psi(1, \ldots, N) = \Psi_4(1, \ldots, N-1)\exp[\tfrac{1}{2}\phi(z_N)] \tag{2.8.15}$$

they variationally minimized the expectation energy $\langle \Psi | H | \Psi \rangle / \langle \Psi | \Psi \rangle$ for a variety of trial functions $\phi(z)$. Each resulted in a symmetrical distribution for the ^{3}He atom centred on the Gibbs surface. The lowest energy corresponded to a Gaussian distribution $\exp[-(z/\lambda)^2]$ of width $\lambda = 2.5$ Å. Shih and Woo found that the extra surface binding energy ϵ_0 driving the surface adsorption of ^{3}He atoms to be 1.6 °K, in good agreement with surface tension measurements by Zinov'eva and Boldarev[80] who determined $\epsilon_0 = 1.7 \pm 0.2$ °K. Guo *et al.*[81] have more recently determined ϵ_0 to be 1.95 ± 0.1 °K, although at a lower temperature and in the fully degenerate region where the ^{3}He system behaves like a dilute two-dimensional Fermi gas.

Shih and Woo go on to regard the ^{3}He particle as moving in a single-particle well, in which case the oscillator frequency was determined as $\hbar\omega \sim 2.6$ °K. In other words, from the binding energy ϵ_0 we conclude that the surface states of ^{3}He in ^{3}He–^{4}He solutions consist of a single bound state: this conclusion is in agreement with that of Saam.[79]

2.9 Dilute polymer solutions

A statistical theory of the conformational modification of polymer chains in the vicinity of a solvent boundary is almost entirely undeveloped, although surface adsorption phenomena have been discussed in statistical mechanical terms by a number of authors.[92] However, these authors either neglect the excluded volume effect (i.e. intrachain interactions) or treat it in a relatively crude way.

Bellemans[93] has recently discussed the influence of the excluded volume effect on the basis of some exact numerical results for short polymer chains on regular lattices. Each site on the lattice can be occupied either by a theta-solvent molecule or a polymer element, and the solution is assumed to be athermal and sufficiently dilute so that all polymer chains behave independently. Bellemans[93] identifies two processes which will modify the surface thermodynamics:

(i) an entropy effect related to the decrease in the available number of internal configurations near the boundary of the lattice (liquid–gas interface),

(ii) an energetic effect related to the energy change ϵ resulting from the replacement of a solvent molecule by a polymer element in the surface layer.

Bellemans shows that the modification of the surface tension relative to a pure solvent is given by

$$\Delta\gamma = \frac{CkT}{a}\left(\frac{l_n}{2c_n} + \frac{\sum_{\nu=1}^{n} g_n(\nu)[1 - \exp(\nu\epsilon/kT)]}{2c_n}\right) + \ldots \tag{2.9.1a}$$

$$= \frac{CkT}{a}\left(\tfrac{1}{2}\langle d_n\rangle + \sum_{\nu=1}^{n}\frac{g_n(\nu)}{2c_n}[1 - \exp(\nu\epsilon/kT)]\right) + \ldots \tag{2.9.1b}$$

where

V = total number of lattice sites (i.e., volume)
A = total number of superficial sites (i.e., surface area)

Vc_n = total number of configurations allowed to a chain of n elements when boundary effects are neglected (i.e., assuming periodic boundary conditions and $V^{1/3} > n$)

$\frac{1}{2}Al_n$ = number of configurations lost on account of the boundary

$\frac{1}{2}Ag_n(\nu)$ = number of configurations allowed to the chain when ν elements occupy superficial sites

C = volume fraction of polymer molecules

a = area per superficial site

$\langle d_n \rangle$ = average molecular span of polymer chain.

At present, analyses are made largely in terms of the asymptotic form for d_n which appears in expression (2.9.1b) for the modification $\Delta\gamma$ of the solvent surface tension. When no account is taken of excluded volume effects, i.e., the polymer chain is allowed to cross itself on the lattice, we find for a three-dimensional simple cubic lattice

$$\langle d_n \rangle = 2\left(\frac{2}{3\pi}\right)^{1/2} n^{1/2} = 0.92132 n^{1/2} \tag{2.9.2}$$

whilst for self-avoiding walks with $n \leqslant 13$, Bellemans concludes that the asymptotic mean span is

$$\langle d_n \rangle = (0.6006 \pm 0.0004) n^{2/3} \tag{2.9.3}$$

which shows a markedly different n-dependence; this stronger n-dependence in the entropy term is understood in terms of the greater spreading of the chain when excluded volume effects are incorporated.

A closely related quantity is the mean square end-to-end span of a walk $\langle \rho_n^2 \rangle$ which also measures the mean extent of a walk. Careful analysis[94] of the data leads one to expect the asymptotic forms

$$\langle d_n \rangle \sim n^{\delta} \qquad \langle \rho_n^2 \rangle \sim n^{\gamma}$$

to be inter-related by[98] (where the index conventionally designated γ should not be confused with the surface tension)

$$\delta = \tfrac{1}{2}\gamma. \tag{2.9.4}$$

Computer-generated self-avoiding chains on a variety of lattices[94,95,111] show that $\gamma \sim 1.2$–1.3, i.e. $\delta \sim 0.6$–0.65.

Curro *et al.*[96] have recently expressed the partition function for a polymer chain in terms of a cluster expansion, and in their subsequent approximation neglect classes of diagram which results in a Percus–Yevick type of integral equation for the excluded volume problem. The equation is solved exactly using a hard-core potential for the special case of the hard-core diameter equal to the polymer segment length (the 'pearl-necklace' model). Numerical results for the diameter ranging from zero to the segment length leads to exponent values ranging from $\gamma = 1.0$–2.0 (i.e. $\delta = 0.5$–1.0, from (2.9.4)).

Croxton[97] has subsequently shown that neglect of certain classes of diagrams in the above analysis, notably the small watermelons and their elementary derivatives, results in an over-rigidity of the polymer chain in the PY analysis for the exponent γ. Croxton gives an expression for the partial correction of Curro's exponent, resulting in a suppression of the n-dependence in both $\langle \rho_n^2 \rangle$ and $\langle d_n \rangle$, and consequently in the leading term of the expression (2.9.1b) for $\Delta\gamma$.

Determinations of the surface adsorption isotherm for polymeric systems in monomeric solvents have depended upon a variety of assumptions[127] – invariably the solvent liquid surface is discontinuously sharp, and usually a quasi-crystalline model is adopted in which sites may be occupied by a solvent or an N-mer segment. Explicit effects of the surface in restricting the accessible macromolecular configurations have even been neglected, with some success. Here,[123] however, we establish the surface distribution of polymer segments in the vicinity of a *realistic* solvent surface density transition profile. We consider a freely jointed chain of N hard sphere segments of unit diameter, where $\Phi(r_{1\mathrm{p}})$ and $\Phi(r_{1\mathrm{s}})$ (p and s referring to polymer and solvent particles) represents the interaction potentials developed between the first and non-adjacent polymer segments and the first segment and solvent molecules respectively. It is then straightforward to show that the single particle BGY equation generalized to give the distribution $\rho^N_{(1)}(1)$ of the first element in an N-segment polymer chain at a solvent liquid surface is given as

$$\rho^N_{(1)}(1) = \rho_{\mathrm{L}} \exp\left\{-\frac{1}{kT}\int_{-\infty}^{z_1}\left(\int_{\mathrm{p}}\sum_{p=3}^{N} \nabla_1 [\Phi_{pp}(1p) - \Phi_{ps}(1p)]\hat{g}^N_{(2)}(1p)\rho_{(1)}(\mathbf{p})\,\mathrm{dp} + \int_{\mathrm{s}} \nabla_1 \Phi_{ps}(1s) g_{(2)}(1s)\rho_{(1)}(\mathbf{s})\,\mathrm{ds}\right)\right\} \tag{2.9.5}$$

$\hat{g}^N_{(2)}(1p)$ represents the isotropic *internal* spatial distribution of segments 1 and p in the N-mer, whilst $\rho^N_{(1)}(\mathbf{p})$ is the appropriate single particle distribution for which (2.9.5) is to be solved self-consistently. ρ_{L} is the bulk number density of the polymer segments. $\hat{g}^N_{(2)}(1p)$ is, of course, quite distinct from $g^p_{(2)}(1p)$ – that is, from the spatial distribution of segments 1 and p in a p-mer: we shall establish $\hat{g}^N_{(2)}(1p)$ in terms of $g^p_{(2)}(1p)$ below. $g_{(2)}(1s)$ is simply the radial distribution of solvent molecules about a polymer segment. Here we adopt the approximate closure $\rho_{(2)}(ij) = \rho_{(1)}(\mathrm{i})\rho_{(1)}(\mathrm{j})g_{(2)}(ij)$ for the anisotropic pair distribution in the vicinity of the surface, in terms of the *bulk* distribution $g_{(2)}(ij)$.

The internal distribution $\hat{g}^N_{(2)}(1p)$ $(p<N)$ within the N-mer may be easily shown to be

$$\hat{g}^N_{(2)}(1p) = g^p_{(2)}(1p)\int g^N_{(2)}(1N)g^{N-p}_{(2)}(pN)\,\mathrm{dN} \tag{2.9.6}$$

working in what is essentially a superposition approximation. More generally, for entirely internal distributions, we may write

$$\hat{g}^N_{(2)}(pp') = g^{p-p'}_{(2)}(pp')\int \hat{g}^N_{(2)}(1p)\hat{g}^N_{(2)}(1p')\,\mathrm{d1} \tag{2.9.7}$$

where $\hat{g}^N_{(2)}(1p)$, $\hat{g}^N_{(2)}(1p')$ are defined in equation (2.9.6) Equation (2.9.7) gives the distribution of segments (p, p') within an N-mer where neither p or p' is an end particle. Determination of the distribution $g_{(2)}(1N)$ necessarily involves the determination of all lower-order distributions $g_{(2)}(13), g_{(2)}(14), \ldots, g_{(2)}(1N)$, and so the *internal* distributions $\hat{g}^N_{(2)}(1p)$ in an N-mer may be simply established from (2.9.6). In fact, there is a progressive and systematic discrepancy in the $g^{\mathrm{PY}}_{(2)}(1N)$ distributions amounting to a molecular over-rigidity, and Croxton[97] has determined a partial correction in terms of the subset of small watermelon diagrams neglected in the PY approximation.

$$ {}^{(k)}g^{PY}_{(2)}(1p) = \frac{(2\pi)^{2-p}}{r_{1p}} \sum_{l=0}^{k-1} (-1)^l \binom{N-2}{l} \frac{(r_{1p} - l - 1)^{N-3}}{(N-3)!} \tag{2.9.8} $$

$$ k \leqslant r_{1l} \leqslant k+1, \quad k \geqslant 1, $$

where ${}^{(k)}g^{PY}_{(2)}(1p)$ is piecewise analytic over the interval r_{1p}.

2.10 Machine simulations

The difficulty of direct experimental investigation of the density transition profile at the liquid surface has been emphasized in the preceding sections. In the case of homogeneous systems, machine simulation yielded information representative of the structure and thermodynamics of an infinite assembly of particles, and it would seem that an extension of these techniques to the inhomogeneous surface region would finally resolve the form of the transition profile. Unfortunately, simulation of the interfacial zone proves substantially more difficult than for homogeneous systems and, in consequence, both oscillatory and monotonic profiles for liquid argon have been reported in the literature.

The difficulties specific to the simulation of the liquid surface prove to be just those which are possibly responsible for the initiation of density oscillations in the transition zone.

Simulation of an inhomogeneous system necessarily involves a greater number of particles than its homogeneous counterpart, if detailed information regarding the inhomogeneity is to be obtained. Moreover, persistance of long-lived fluctuations and long-ranged correlations require $\sim 10^7$ configurations for a Monte Carlo ensemble, or time averaging over $\sim 10^{-9}$ sec., for a molecular dynamics simulation for an assembly of $\sim 10^3$ particles if adequate phase sampling is to be achieved. Indeed, it has been suggested that the oscillatory transition profiles reported for simulated Lennard–Jones fluids can be attributed to the extremely slow convergence of the density profiles, and that much longer runs would have yielded monotonic single-particle distributions.

The dimensions of the fundamental cell may also modify the final result since the periodic boundary conditions suppress thermal fluctuations of greater wavelength than the fundamental dimensions. The height of the cell, z, imposes no great restriction since it must in any case extend from the bulk liquid into the bulk vapour (Figure 2.10.1). The lower boundary must not, however, initiate any density oscillations which might extend into the transition zone region. The x and y dimensions, on the other hand, in addition to suppressing much shorter wavelength fluctuations, also impose an artificial planarity on the surface[119] since the condition for periodic replication in the xy plane effectively constrains the transition region, and may act as a kind of surface constraining field responsible for the initiation of density oscillations. An increase in the xy dimensions of the fundamental cell naturally leads to a substantial increase in the computing time: fortunately, however, it appears that xy dimensions as small as 5σ do not initiate spurious oscillations in the profile, although they may be responsible for inhibiting the relaxation of density fluctuations.

The first reported simulation of a free liquid argon surface was that of Croxton and Ferrier[114] in their *two-dimensional* molecular dynamics simulation for 200 LJ argon atoms at an equivalent temperature of 84.4 °K. The cell dimensions ($x = z = 100$ Å ($\sim 30\sigma$)) eliminated any possibility of surface constraint due to replication. The bottom of the cell was coupled to a non-static model bulk fluid whose effect was to prevent the initiation of density oscillations at

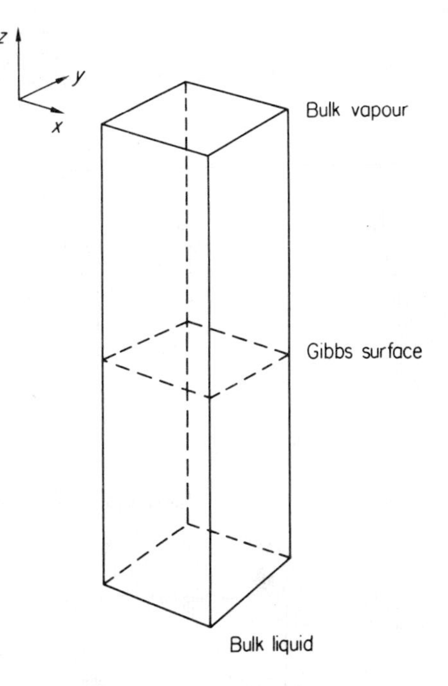

Figure 2.10.1 The fundamental cell in machine simulations of the liquid surface. Replication is generally in the xy-plane, whilst z must extend from bulk liquid to bulk vapour, the Gibbs surface being located well away from the upper or lower boundaries

the lower boundary. Using a time step of 10^{-14} sec. the simulation was followed for 10^{-11} sec., which is short with respect to the lifetime of fluctuations in the assembly. It was concluded that the transition zone was essentially oscillatory, and should not differ qualitatively for a three-dimensional assembly. There was a slight shift in the phase of the oscillations in the duration of the simulation, and a longer computation might have shown that the oscillations were indeed transient. Subsequent numerical simulations[115–117] of Lennard–Jones fluids have shown the development of a pronounced layer structure, but all can be criticized on the grounds of inadequate sampling of phase space: the layering is almost certainly an artifact of the extremely slow convergence of the density profile – much slower than for bulk fluid systems.

Abraham, Schreiber and Barker[118] formed a 256-particle liquid slab from four replications of a cube of bulk Monte Carlo LJ liquid, as shown in Figure 2.10.2(a). The cube contained 64 LJ atoms at 84 °K, the side of the periodic cube being 14.82 Å (4.4σ). Approximately 1.2×10^6 configurations (18,750 configurations/atom) were generated in the basic cube before the slab was assembled. Further simulations (1.7×10^6 configurations) allowed the free surfaces to relax to equilibrium.

Abraham *et al.* then show four density profiles for a LJ fluid at 84 °K obtained from averaging over 0.8, 2.6, 4.2 and 6.2 million configurations of a 256-atom system. The pronounced oscillations in the xy and z profiles relax monotonically, and the authors conclude it would have taken $\sim 10^7$ configurations to achieve the smooth profile. The persistance of spurious oscillations over long Monte Carlo chains is remarkable and emphasizes the necessity of adequate phase sampling.

A Monte Carlo simulation by Liu[116] (89 °K and $\sim 5 \times 10^4$ configurations/atom) for a Lennard–Jones fluid yields a profile which the author interprets as showing 'a striking layered structure'. In fact, the oscillations are indistinguishable from the noise, and the consensus of opinion is that Liu's results substantiate the conclusion that the profile is indeed monotonic. Both these results have been supported by the very recent Monte Carlo and molecular dynamics

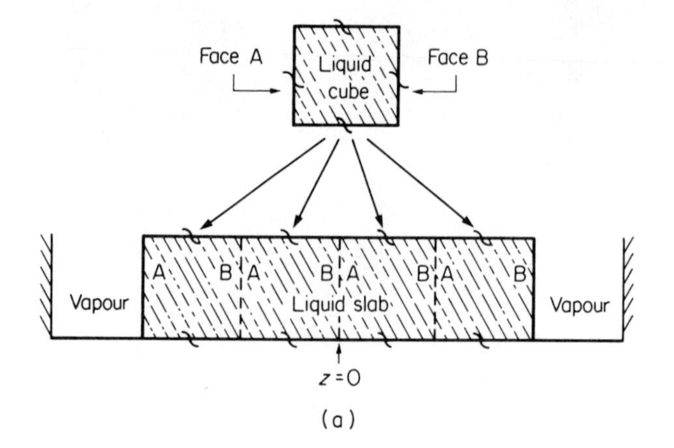

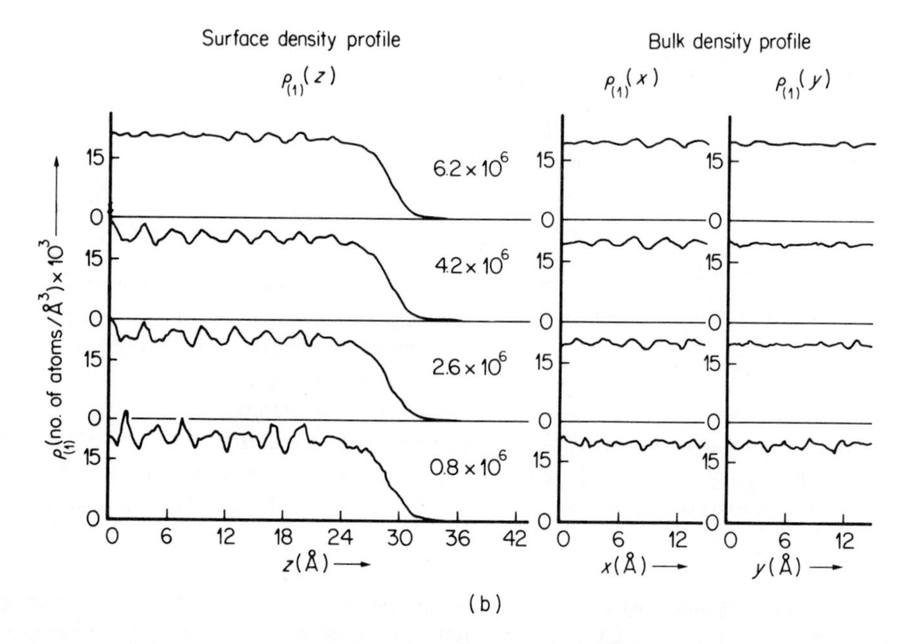

Figure 2.10.2 (a) Formation of liquid slab from component bulk liquid cubes for the Monte Carlo calculation of Abraham *et al.*[118] The symbol ~ implies periodic replication about that boundary. (b) Surface density profile $\rho_{(1)}(z)$ and bulk density profiles $\rho_{(1)}(x)$ and $\rho_{(1)}(y)$ for a Lennard–Jones fluid at 84 °K obtained from Monte Carlo averaging over 0.8, 2.6, 4.2 and 6.2 x 10^6 configurations of a 256-atom system. (From Abraham, Schreiber and Barker,[118] reproduced by permission of the American Institute of Physics)

results of Chapela, Saville and Rowlinson[119] over the reduced temperaturee range $T^* = kT/\epsilon = 0.701$, 0.708, 0.759, 0.918, 1.127. Both methods lead to similar results, though the molecular dynamics approach appears to sample phase space more efficiently. Their basic cell is of the form shown in Figure 2.10.1, with xy dimensions of 5σ and a height $z = 25\sigma$. Replication is in the x and y coordinates, whilst the planar upper and lower boundaries are composed of a molecularly homogeneous substance whose density is that of the bulk gas for $z > 25\sigma$, or the bulk liquid for $z < 0$. All the profiles show strong oscillations in the vicinity of the lower

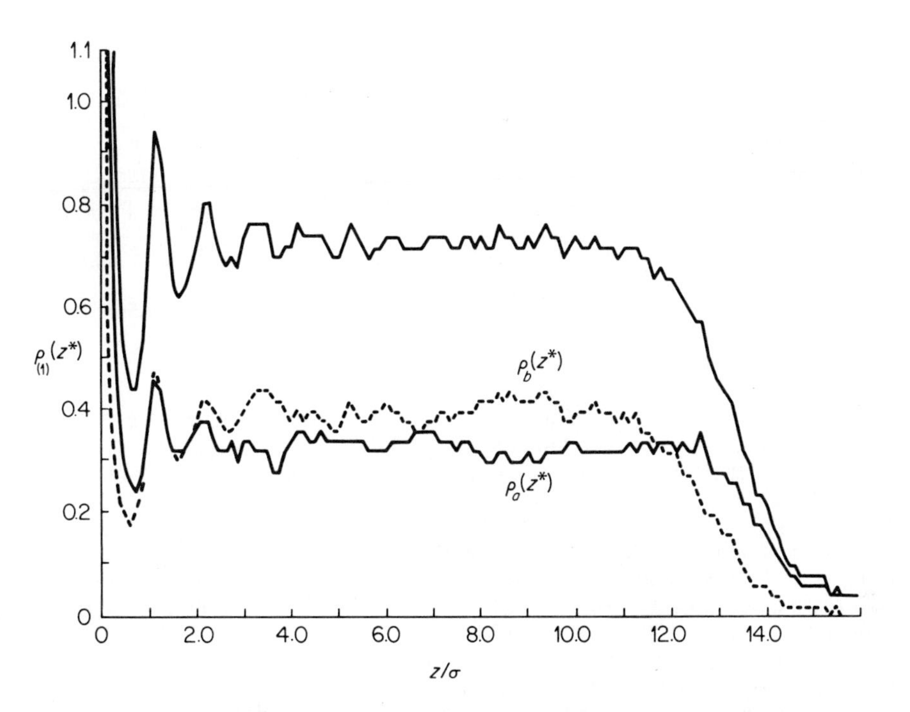

Figure 2.10.3 Density profile $\rho(z)$ for an equimolar binary total mixture at equivalent temperatures $T_a^* = 0.918$, $T_b^* = 0.701$. The component profiles are shown by the broken lines

boundary, as would be expected. In every case, however, the oscillations disappear beyond $z > 5\sigma$ and a monotonic, if noisy, transition profile is obtained. The Monte Carlo runs extended over $>10^6$ configurations, or $>10^4$ time steps (10^{-14} sec.) in the case of the molecular dynamics simulations. The necessity for extended runs is shown clearly by the presence of quite pronounced oscillations in $\rho_{(1)}(z)$ at intermediate stages of the simulation, and concurs with the Monte Carlo calculations of Abraham *et al.* Parsonage[125] makes the interesting point that the position of the Gibbs surface on the z-axis needs only to fluctuate by $\sim\sigma/2$ in these simulations for the oscillations to be smeared out, and suggests that the positions of the oscillations should be measured with respect to the Gibbs surface rather than the cell boundary. Fluctuation in the position of the interface would be slow since it would require the concerted effort of a large number of particles but could well explain the transient nature of the oscillations which seem to be frequently observed.

The only simulation of the free surface of an equimolar binary mixture appears to be that of Chapela *et al.*,[119] obtained by Monte Carlo simulation (15×10^6 configurations) at $T_a^* = kT/\epsilon_a = 0.918$, $T_b^* =,0.701$, and by molecular dynamics simulation (8.68×10^4 time steps) at $T_a^* = 0.934 \pm 0.002$, $T_b^* = 0.713 \pm 0.002$.

The profiles are shown in Figure 2.10.3, from which it is seen that the lighter component is adsorbed at the surface as a monolayer in qualitative agreement with the theoretical prediction of Plesner, Platz and Christiansen.[120]

Acknowledgements

It is a great pleasure to thank Søren Toxvaerd and John Rowlinson for their generous provision of manuscripts and results prior to publication.

References

1. J. D van der Waals (1894). *Z. Physik. Chem.* **13**, 657.
   L. D. Landau and E. M. Lifshits (1944). *Mekhanika Sploshnykh Sred'*, izd. 1-oe Gostekhizdat (*The Mechanics of Complex Media*), 1st ed., State Technical Press.
   J. W. Cahn and J. E. Hilliard (1958). *J. Chem. Phys.* **28**, 258. See also J. W. Cahn (1959). *J. Chem. Phys.* **30**, 1121.
   E. W. Hart (1959). *Phys. Rev.* **113**, 412 *Phys. Rev.* **114**, 27 (1959); *J. Chem. Phys.* **39** 3075 (1963); *J. Chem. Phys.* **34**, 1471 (1961).
   B. Widom (1964). *J. Chem. Phys.* **41**, 1633 *J. Chem. Phys.* **43**, 3892 (1965).
   B. Widom and O. K. Rice (1955). *J. Chem. Phys.* **23**, 1250.
   S. Fisk and B. Widom (1969). *J. Chem. Phys.* **50**, 3219.
2. C. A. Croxton (1973). *Advances in Physics,* **22**, 385.
   C. A. Croxton (1974). *Liquid State Physics – A Statistical Mechanical Introduction*, Cambridge University Press, London. Chap. 4.
   S. Toxvaerd (1976). In K. Singer (Ed.), *Statistical Mechanics,* The Chemical Society, London. **2**, 256.
3. T. L. Hill (1952). *J. Chem. Phys.*, **20**, 141.
4. I. W. Plesner and O. Platz (1968). *J. Chem. Phys.*, **48**, 5361.
5. H. Reiss, H. L. Frisch and J. L. Lebowitz (1961). *J. Chem. Phys.*, **34**, 1037.
6. See, for example, C. A. Croxton (1975). *Introduction to Liquid State Physics,* John Wiley, London and New York. p. 173.
7. S. Toxvaerd (1971). *J. Chem. Phys.*, **55**, 3116.
8. J. A. Barker and D. Henderson (1967). *J. Chem. Phys.*, **47**, 2856.
9. D. Chandler and J. D. Weeks (1970). *Phys. Rev. Lett.*, **25**, 149.
   J. D. Weeks, D. Chandler and H. C. Andersen (1971) *J. Chem. Phys.*, **54**, 5237.
10. W. R. Smith, D. Henderson and J. A. Barker (1970). *J. Chem. Phys.*, **53**, 508.
11. S. Toxvaerd (1972). *J. Chem. Phys.*, **57**, 4092.
    S. Toxvaerd (1972). *Prog. Surface Sci.*, **3**, 189.
12. C. A. Croxton and R. P. Ferrier (1971). *J. Phys. C.*, **4**, 1909.
    C. A. Croxton and R. P. Ferrier (1971). *Phil. Mag.*, **24**, 493.
13. G. M. Nazarian (1972). *J. Chem. Phys.*, **56**, 1408.
14. C. Jouanin (1969). *C.r. Lebd. Séanc. Acad. Sci.*, Paris, B, **268**, 1597.
15. J. W. Perram and L. White (February 1975). Private communication.
16. C. A. Croxton (1969). *Ph.D. Thesis*, University of Cambridge (unpublished).
17. C. A. Croxton (1974). *Liquid State Physics – A Statistical Mechanical Introduction*, Cambridge University Press, London. p. 158.
18. A. Rahman (1964). *Phys. Rev. Lett.*, **12**, 575.
    P. A. Egelstaff, D. I. Page and C. R. T. Heard (1971). *J. Phys. C.*, **4**, 1453.
19. S. A. Rice and J. P. Lekner (1965). *J. Chem. Phys.*, **42**, 3559.
20. H. S. Green (1960). *Handb. Phys.*, **10**, 79.
21. I. Z. Fisher and B. V. Bokut (1957). *Zh. Fiz. Khim.*, **31**, 200.
    I. Z. Fisher (1964). *Statistical Theory of Liquids,* University of Chicago Press, Chicago. p. 170.
22. B. U. Felderhof (1970). *Phys. Rev.* A., **1**, 1185.
23. J. D. Bernal (1964). *Proc. Roy. Soc.*, **A280**, 299.
24. S. Toxvaerd (1973). *Mol. Phys.*, **26**, 91.
25. J. G. Kirkwood and F. P. Buff (1949). *J. Chem. Phys.*, **17**, 338.
26. R. H. Fowler (1937). *Proc. Roy. Soc.*, **A159**, 229.
27. P. D. Shoemaker, G. W. Paul and L. E. Marc de Chazal (1970). *J. Chem. Phys.*, **52**, 491.
28. K. S. C. Freeman and I. R. McDonald (1973). *Mol. Phys.*, **26**, 529.
29. M. V. Berry, R. F. Durrans and R. Evans (1972). *J. Phys.* A., **5**, 166.
30. C. Jouanin (1969). *C.r. Lebd. Séanc. Acad. Sci.*,, Paris, B, **268**, 1597.
31. D. Fitts (1969). *Physica,* **42**, 205.
32. R. Tolman (1948). *J. Chem. Phys.*, **16**, 758.
33. A. C. Zettlemoyer (1969). *Nucleation*, Dekker, New York.
34. C. A. Croxton and R. P. Ferrier (1971). *J. Phys.* C., **4**, 2433.
    C. A. Croxton and R. P. Ferrier (1971). *Phys. Lett.* **36A**, 183.

35. F. B. Sprow and J. M. Prausnitz (1966). *Trans. Faraday Soc.*, **62**, 1097.
36. G. M. Nazarian (1972). *J. Chem. Phys.*, **56**, 1408.
37. J. W. Perram and L. White (January 1975). Private communication.
    J. W. Perram and L. White, Faraday General Discussion, *Physical Adsorption in Condensed Phases,* April 1975.
38. E. Helfand, H. L. Frisch and J. L. Lebowitz (1961). *J. Chem. Phys.*, **34**, 1037.
39. C. A. Croxton, Faraday General Discussion, *Physical Adsorption in Condensed Phases,* April 1975. p. 53.
40. K. E. Gubbins and C. G. Gray (1975). *Mol. Phys.*, in press.
41. C. A. Croxton and T. R. Osborn (1976). *Phys. Lett.*, **55A**, 415.
42. C. A. Croxton and S. Chandrasekhar (1973). *Proc. 2nd Intl. Conf. Liquid Crystals,* Bangalore, (Prāmana, Suppl. No. 1, 1975). p. 237.
43. S. Sung and D. Chandler (1972). *J. Chem. Phys.*, **57**, 1930.
    J. W. Perram and R. L. White (1974). *Mol. Phys.*, **28**, 527.
44. J. D. Parsons (1976). *Journal de Physique,* **37**, 1187.
45. V. K. Semenchenko, *Surface Phenomena in Metals and Alloys,* Pergamon, Oxford.
46. J. Frenkel (1928). *Z. Phys.*, **51**, 232.
    J. Bardeen (1936). *Phys. Rev.*, **49**, 653.
    K. Huang and G. Wyllie (1949). *Proc. Phys. Soc.*, **A62**, 180.
    H. B. Huntingdon (1951). *Phys. Rev.*, **31** 1035.
    R. Stratton (1953). *Phil. Mag.*, **44**, 1236.
47. J. R. Smith (1969). *Phys. Rev.*, **181**, 522.
48. N. D. Lang and W. Kohn (1970). *Phys. Rev.*, **B1**, 4555.
49. J. Schmidt and A. A. Lucas (1972). *Sol. St. Comm.*, **11**, 415, 419
50. J. Bardeen (1936). *Phys. Rev.*, **49**, 653.
    W. J. Suriatecki (1961). *Proc. Phys. Soc.*, **A64**, 226.
    P. P. Ewald and H. Juretschke (1953). In R. Gomer and C. S. Smith (Eds.), *Structure and Properties of Solid Surfaces,* University of Chicago Press, Chicago. p. 82.
    R. D. Berg and L. Wilets (1955). *Proc. Phys. Soc.*, **A68**, 229.
51. D. M. Newns (1970). *Phys. Rev.*, **B1**, 3304.
52. J. A. Appelbaum and D. R. Hamann (1972). *Phys. Rev.*, **B6**, 2166.
53. M. D. Johnson, P. Hutchinson and N. H. March (1964). *Proc. Roy. Soc.*, **A282**, 283.
54. Y. Waseda and K. Suzuki (1972). *Phys. Stat. Solidi,* (b), **49**, 643 *Phys. Stat. Solidi,* **57**, 351 (1973).
55. R. Evans (1974). *J. Phys. C.*, **7**, 2808.
56. J. Frenkel (1917). *Phil. Mag.*, **33**, 297.
57. C. A. Croxton, to be published.
58. W. Kohn (1973). *Sol. St. Comm.*, **13**, 323.
    R. A. Craig (1973). *Sol. St. Comm.*, **13**, 1517.
    See also Reference (49) and references contained therein.
59. C. A. Croxton (1972). *Proc. 2nd Intl. Conf. Liquid Metals,* Tokyo.
60. T. E. Faber (1972). *An Introduction to the Theory of Liquid Metals,* Cambridge University Press, London.
61. J. R. Wilson (1965). *Metall. Rev.*, **10**, 381.
    D. Gerner and H. Mayer (1968). *Z. Phys.*, **210**, 391.
62. D. W. G. White (1966). *Trans. Metall. Soc. A.I.M.E.*, **236**, 796.
63. D. W. G. White, *Metals, Materials and Metallurgical Reviews,* July 1968. p. 73.
64. D. W. G. White (1971). *J. Inst. Metals,* **99**, 287.
65. See L. D. Landau and E. M. Lifshits (1969). *Statistical Physics,* Pergamon Press, Oxford. p. 457.
66. A. D. Singh (1962). *Phys. Rev.*, **125**, 802.
67. S. L. Chan (1972). *Can. J. Phys.* **50**, 1139.
68. T. Regge (1972). *J. Low Temp. Phys.*, **9**, 123.
69. T. C. Padmore and M. W. Cole (1974). *Phys. Rev.*, **A9**, 802.
70. Y. M. Shih and C. W. Woo (1973). *Phys. Rev. Lett.*, **30**, 478.
71. C. C. Chang and M. Cohen (1973). *Phys. Rev.*, **A8**, 1930.
72. K. S. Liu, M. H. Kalos and G. V. Chester (June 1974). Private communication.
73. C. A. Croxton (1973). *J. Phys.* C, **6**, 411.
    C. A. Croxton (1972). *Phys. Lett.* A, **41**, 413.

74. R. Abe (1958). *Prog. Theoret. Phys. (Kyoto)*, **19**, 57; *Prog. Theoret. Phys. (Kyoto)*, **19**, 407 (1958).
75. C. A. Croxton (1974). *Liquid State Physics – A Statistical Mechanical Introduction* Cambridge University Press, London. pp. 184–189.
76. R. M. Mazo and J. G. Kirkwood (1958). *J. Chem. Phys.*, **28**, 644.
77. B. N. Esel'son and N. G. Bereznyak (1954). *Dokl. Akad. Nauk, SSSR*, **98**, 564.
    B. N. Esel'son, V. G. Ivantsov and A. D. Shvets (1963). *Zh. Eksp. i Teor. Fiz.*, **44**, 483. [*Sov. Phys. JETP*, **17**, 330 (1963).]
78. A. F. Andreev (1966). *Zh. Eksp. i Teor. Fiz.* **50**, 1415. [*Sov. Phys. JETP*, **23**, 939 (1966).]
79. J. Lekner (1970). *Phil. Mag.*, **22**, 669.
    W. F. Saam (1971). *Phys. Rev.*, **A4**, 1278.
80. N. K. Zinov'eva and S. T. Boldarev (1969). *Zh. Eksp. i Teor. Fiz.* **56**, 1089. [*Sov. Phys. JETP*, **29**, 585 (1969).]
81. H. M. Guo, D. O. Edwards, R. E. Sarwinski and J. T. Tough (1971). *Phys. Rev. Lett.*, **27**, 1259.
82. S. Krishnaswamy and R. Shashidar (1973). *Proc. 2nd Intl. Conf. Liquid Crystals*, Bangalore, (Prāmana Suppl. No. 1, 1975.) p. 247.
83. D. Langevin and M. A. Bouchiat (1972). *J. de Physique*, **33**, C1–77.
84. F. Jahnig (1973). *Proc. 2nd Intl. Conf. Liquid Crystals*, Bangalore. (Prāmana Suppl. No. 1, 1975.) p. 246 .
    F. M. Leslie (1973). *Proc. 2nd Intl. Conf. Liquid Crystals*, Bangalore. (Prāmana Suppl. No. 1, 1975.) p. 252.
85. R. Kumaradival and R. Evans (1975). *J. Phys.* C, **8**, 793.
86. K. R. Atkins and Y. Narahara (1965). *Phys. Rev.*, **138**, A437.
    J. R. Eckardt. *Ph.D. Dissertation*, Ohio State University (unpublished).
    F. M. Gasparini, J. R. Eckardt, D. O. Edwards and S. Y. Shen (1973). *J. Low Temp. Phys.*, **13**, 437.
87. D. O. Edwards, J. R. Eckardt and F. M. Gasparini (1974). *Phys. Rev.*, **9**, A2070.
88. K. R. Atkins (1953). *Can. J. Phys.*, **31**, 1165.
89. S. T. Boldarev and V. P. Peshkov (1973). *Physica*, **69**, 141.
90. P. A. Egelstaff and B. Widom (1970). *J. Chem. Phys.*, **53**, 2667.
91. R. D. Present (1974). *J. Chem. Phys.*, **61**, 4267.
92. R. Simha, H. L. Frisch and F. R. Eirich (1953). *J. Phys. Chem.*, **57**, 584.
    R. J. Rubin (1965). *J. Chem. Phys.*, **43**, 2392.
    A. Silberberg, (1968). *J. Chem. Phys.*, **48**, 2835.
93. A. Bellemans (1972). *J. Polymer Sci.*, **C39**, 305.
94. A. Bellemans (1974). *Physica*, **73**, 368; *Physica*, **74**, 441 (1974).
95. F. Wall and J. Erpenbeck (1959). *J. Chem. Phys.*, **30**, 637.
    L. Gallacher and S. Windwer (1966). *J. Chem. Phys.*, **44**, 1139.
    C. Domb, J. Gillis and G. Wilmers (1965). *Proc. Phys. Soc. (London)*, **85**, 625.
    A. Bellemans (1973). *J. Chem. Phys.*, **58**, 823.
    J. L. Martin and M. G. Watts (1971). *J. Phys.* A, **4**, 457.
    D. S. McKenzie (1973). *J. Phys.* A, **6**, 338.
    A. Bellemans (1973). *Physica*, **68**, 209.
96. J. G. Curro, P. J. Blatz and C. J. Pings (1969). *J. Chem. Phys.*, **50**, 2199.
97. C. A. Croxton (1975). *Phys. Lett.*, **51A**, 31.
98. P. J. Gans (1965). *J. Chem. Phys.*, **42**, 4159.
99. S. Toxvaerd (1976). In K. Singer (Ed.), *Statistical Mechanics*, The Chemical Society, London. **2**, 256.
100. T. L. Hill (1959). *J. Chem. Phys.*, **30** 1521.
101. F. P. Buff and F. H. Stillinger (1963). *J. Chem. Phys.*, **39**, 1911.
    I. W. Plesner and I. Michaeli (1974). *J. Chem. Phys.*, **60**, 3016.
102. B. Borstnik and A. Azman (1975). *Mol. Phys.*, **29**, 1165.
103. F. P. Buff and R. A. Lovett (1968). In H. L. Frisch and Z. W. Salsburg (Eds.), *Simple Dense Fluids* Academic Press, New York. p. 17.
104. P. Drude (1959). *Theory of Optics*, Dover, New York. pp. 287–95.
105. F. P. Buff and R. A. Lovett (1966). *1966 Saline Water Conversion Report*, U.S. Government Printing Office, Washington D.C. p. 26.

106. E. S. Wu and W. W. Webb (1973). *Phys. Rev.* A, **8**, 2065.
107. A. A. Bloch and S. A. Rice (1969). *Phys. Rev.*, **185**, 933.
108. C. A. Croxton (1974). *Liquid State Physics – A Statistical Mechanical Introduction*, Cambridge University Press, London. p. 157.
109. J. Zollweg, G. Hawkins and B. Benedek (1971). *Phys. Rev. Lett.*, **27**, 1182.
110. H. T. Davis (1975). *J. Chem. Phys.*, **62**, 3412.
111. S. Toxvaerd (November 1975). Private communication. (To be published in *J. Chem. Phys.* 1976.)
112. L. Verlet (1968). *Phys. Rev.*, **165**, 201.
113. R. Lovett, P. W. de Haven, J. J. Vieceli, Jr., and F. P. Buff (1973). *J. Chem. Phys.*, **58**, 1880.
114. C. A. Croxton and R. P. Ferrier (1971). *J. Phys.* C, **4**, 2433.
115. J. K. Lee, J. A. Barker and G. M. Pound (1974). *J. Chem. Phys.*, **60**, 1976.
116. K. S. Liu (1974). *J. Chem. Phys.*, **60**, 4226.
117. A. C. L. Opitz (1974). *Phys. Lett.*, **47**, 439.
118. F. F. Abraham, D. E. Schreiber and J. A. Barker (1975). *J. Chem. Phys.*, **62**, 1958.
119. G. A. Chapela, G. Saville and J. S. Rowlinson (1975). *Chem. Soc. Faraday Disc. No. 59*; Private communication, December 1975, to be published.
120. I. W. Plesner, O. Platz and S. E. Christansen (1968). *J. Chem. Phys.*, **48**, 5364.
121. M. S. Anath, K. E. Gubbins and C. G. Gray (1975). *Mol. Phys.*, **28**, 1005.
P. A. Egelstaff, C. G. Gray and K. E. Gubbins (1975). In A. D. Buckingham (Ed.), *Molecular Structure and Properties, International Review of Science*, Physical Chemistry, Ser. 2, Vol. 2. Butterworth, London.
122. J. M. Haile, K. E. Gubbins and C. G. Gray (1976). *J. Chem. Phys.*, **64**, 1852.
123. C. A. Croxton. *Phys. Lett.*, (in press).
124. T. R. Osborn and C. A. Croxton. *Mol. Phys.* (to be published).
125. N. G. Parsonage (1975). *Faraday Disc. Chem. Soc.*, **59**, 51.
126. C. A. Croxton. *Phys. Lett.* (in press).
127. I. Prigogine and J. Maréchal (1952). *J. Colloid Sci.*, **7**, 122.
R. Defay, I. Prigogine, A. Bellemans and D. H. Everett (1966). *Surface Tension and Adsorption*, Wiley, New York. Ch. 13.
A. Silberberg (1968). *J. Chem. Phys.*, **48** 2835.
K. S. Siow and D. Patterson (1973). *J. Phys. Chem.*, **77**, 356.

Chapter 3

Boundary Conditions for Numerical Simulation of Fluids with Dipolar Interactions

E. R. SMITH

Department of Mathematics, University of Melbourne
Parkville, Victoria 3052, Australia

3.1 Introduction

In this chapter we discuss a problem which arises in calculating non-dynamic properties of classical fluids. Thus we may ignore contributions to the Hamiltonian from kinetic energy. We consider Hamiltonians which may be written as a sum of pairwise potentials

$$H_N = \frac{1}{2} \sum_{\substack{i=1 \\ i \neq j}}^{N} \sum_{j=1}^{N} \phi_{ij}. \tag{3.1.1}$$

We are concerned with numerical simulation of real systems and, in particular, some of the problems which arise in the simulation of water and ionic aqueous solutions. Thus a typical Hamiltonian should include many-body forces. A. Ben-Naim, in Chapter 10 of this book (e.g. equation 10.2.1) shows how important many-body forces can be in modelling water. However, he and others[1,2,3] have pointed out that considerable progress may be made by using an effective pairwise potential. Further, in this chapter we are concerned with difficulties which arise when the two-body part of the Hamiltonian includes dipole–dipole interactions. Thus the Hamiltonian (3.1.1) is sufficiently general for this discussion.

The thermodynamic properties of a fluid may be calculated using many different ensembles. To focus our discussion we mention two commonly used ones: the canonical and grand-canonical ensembles. In the canonical ensemble we attempt to calculate the bulk free energy density

$$f = \lim_{V\to\infty} -\frac{1}{V} kT \log Z_N(V) \tag{3.1.2}$$

where $Z_N(V)$ is the canonical partition function for N particles in a volume V and the limit is taken with the density $\rho = N/V$ kept constant. In the grand-canonical ensemble we attempt to calculate the pressure as a function of fugacity z:

$$P(z) = \lim_{V\to\infty} \frac{kT}{V} \log \Xi(z, V) \tag{3.1.3}$$

where $\Xi(z, V)$ is the grand-canonical partition function for a fluid of volume V at fugacity z. Equations (3.1.2) and (3.1.3) both assume that the limits involved exist, and generally it is assumed that the results are independent of the conditions obtaining at the boundary of the container of volume V. While the analysis of these assumptions can sometimes be a fairly esoteric mathematical pursuit of doubtful physical use, we shall see that it is crucial in the interpretation of simulation data for systems with dipolar interactions.

If the pair potential function ϕ_{ij} falls off more rapidly than $r^{-3-\varepsilon}$ (where r is the distance between the interacting particles and $\varepsilon > 0$) then Ruelle[4] and Fisher[5] have shown that the limiting free energy density and pressure both exist and are independent of the surface properties of the container. Thus, for a finite container it is often assumed that we can write

$$Z_N(V) \sim \exp\{-Vf/kT + V^{2/3}g + \ldots\} \tag{3.1.4a}$$

and

$$\Xi(z, V) \sim \exp\{VP(z)/kT + V^{2/3}\pi(z) + \ldots\} \tag{3.1.4b}$$

where

$$g = \lim_{V\to\infty} V^{-2/3}[\log Z_N(V) + Vf/kT] \tag{3.1.5a}$$

and

$$\pi(z) = \lim_{V\to\infty} V^{-2/3}[\log \Xi(z, V) - VP(z)/kT]. \tag{3.1.5b}$$

The terms $V^{2/3}g$ (in $Z_N(V)$) and $V^{2/3}\pi(z)$ (in $\Xi(z, V)$) are surface contributions to the free energy and pressure. On physical grounds, we expect the limits in equations (3.1.5a, b) to exist, but it should be said that there is no general proof. The term proportional to $V^{2/3}$ in the exponents in these two equations are surface contributions, proportional to the surface area of the container. Of course, if we choose a container of sufficiently bizarre shape that its surface area to volume ratio is not $O(V^{-1/3})$ then these expansions may be invalid. In this chapter we shall contrive to study only pedestrian containers. The surface terms will also depend upon the conditions imposed at the surface of the container. Thus, if we impose conditions labelled $\{A\}$ at the surface of the container, then equations (3.1.4) and (3.1.5) might be rewritten

$$Z_N(V, \{A\}) \sim \exp\left(-\frac{1}{kT} Vf + V^{2/3}g_{\{A\}} + \ldots\right) \tag{3.1.6a}$$

and

$$\Xi(z, V, \{A\}) \sim \exp\left(\frac{VP(z)}{kT} + V^{2/3}\pi_{\{A\}}(z) + \ldots\right) \tag{3.1.6b}$$

to show the dependence of the surface terms on the boundary conditions.

In carrying out either molecular dynamics or Monte-Carlo simulations of a system with a short ranged potential (i.e. a potential which falls off faster than $1/r^{4+\epsilon}$)[6,7], it is customary[8,9,10,11] to use periodic boundary conditions. For small volumes, the partition functions in equations (3.1.6) will contain sizeable surface effects. Thus the probability distributions on which they are based and any quantities derived from the probability distributions, such as correlation functions, will also contain sizeable surface effects. The idea of using periodic boundary conditions is to suppress the surface effects in the simulation so that a smaller system can be used to get some given accuracy in estimating bulk properties than is necessary with ordinary boundary conditions. It is possible[6] to prove that for fluids with potentials which decay faster than $1/r^4$, the limits in equations (3.1.2) and (3.1.3) exist and are the same for the corresponding non-periodic system. Further, the surface effects are smaller than in the non-periodic system. This analytic progress in showing the suppression of surface effects lends weight to opinions formed by studying the success of simulation methods using quite small volumes.[10,11,12,13]

For fluids with electrostatic interactions, the situation is not so simple. The problem of the existence of thermodynamic limiting properties is much more complex than for shorter-ranged potentials. For charge–charge interactions, for example, it is necessary to invoke a quantum mechanical description of the fluid to prove the existence of bulk thermodynamic quantities[14,15,16], unless some highly repulsive core is added to the interactions.[17] Either way, it is necessary to have the interior of the system electrically neutral: the inclusion of a charge imbalance proportional to the surface area of the container of the system gives rise to surface charge densities which in turn create an effective external field on the bulk system[15,16]. Such an electric field is determined by surface or boundary conditions, and naturally changes the bulk values of thermodynamic quantities.

For fluids with dipole–dipole interactions, the existence and shape dependence questions for thermodynamic properties are not as well understood as we might like. Griffiths[18] has proved the existence and shape independence of the free energy for fluids with no external field. The problem of including an external field is not a trivial one, but it is important if we are to understand things like the dielectric response of a dipolar fluid. We have a fairly good idea[19] of how to study the effect of an external field on a macroscopic piece of matter, by the use of Maxwell's equations, constitutive relations and appropriate boundary conditions at the surface of the piece of matter. However, how best to include the effect of boundary conditions into the microscopic statistical mechanics of a fluid is not entirely clear. A start has been made by Penrose and Smith[20] who write the Hamiltonian for a system of N particles in a volume V as

$$H_N = \frac{1}{8\pi}\int_V d^3\mathbf{x}\mathbf{E}^2(\mathbf{x}) \qquad (3.1.7)$$

where the field $\mathbf{E}(\mathbf{x})$ is determined from Maxwell's equations (in the non-relativistic limit, $c \to \infty$) with boundary conditions given by the positions of the particles and the conditions imposed on the boundary of the volume V (i.e. the external field values). The Hamiltonian (3.1.7) can be shown[20] to be a sum of ordinary dipole–dipole interactions and interactions of the dipoles with an external field. Unfortunately, it is necessary to be fairly particular about the boundary conditions and it is not known how the thermodynamic properties from this approach are related to those in more conventional boundary conditions[21]. It is interesting to note that the bulk free energy calculated by this method is shape independent, but this shape independence would appear to be due to the formation of many almost-independent domains of orientation of the dipoles in the infinite sample.

For the finite pieces of matter encountered in the real world (and particularly in simulation studies) the size of the sample can make it impossible for large numbers of independent domains to form, so that the bulk free energy measured in an experiment or calculated by a simulation may be shape dependent. Thus for fluids with dipole–dipole interactions, the representations in equations (3.1.6) may not be quite correct. It appears probable that we should write, instead,

$$Z_N(V, \{A\}) \sim \exp\left(-\frac{V}{kT} f\{A\} + V^{2/3} g\{A\} + \ldots\right) \tag{3.1.8}$$

with the bulk free energy density term also depending on the boundary conditions used. We shall show, in this chapter, that this representation is correct. It is an important point, since if we use some set of boundary conditions in a simulation in an attempt to reduce the size of terms like $V^{2/3} g\{A\}$, we *must* understand what effects these boundary conditions have on the bulk terms.

In the next section we discuss several boundary conditions which have been proposed by various workers as appropriate for use in simulations of dipolar fluids or for studying the effects of external fields on dipolar fluids. These boundary conditions have generally been proposed in an attempt to reduce either surface contributions to the results, or the effects which other boundary conditions have upon the bulk behaviour of the fluid. The boundary conditions are generally proposed without an attempt to quantify the effects that they have on bulk behaviour. They may generally be replaced by an effective dipole–dipole potential. In Section 3.3 we use this effective potential to suggest a method which will quantify the changes in bulk behaviour caused by changing boundary conditions. We conclude with a short discussion section.

3.2 Boundary conditions

(A) 'Free' boundary conditions

In these boundary conditions we assume the fluid particles are confined to a cubic box of side V. No attempt is made to suppress surface effects. These will be rather inferior boundary conditions as a result, and are not used in simulation studies. If the fluid particles are assumed to have a highly repulsive core, then the repulsion of the particles from the surface of the container will induce a lot of structure in the fluid[22,23,24], and we may expect to see large surface contributions to the results.

(B) Minimum image boundary conditions

The object of these boundary conditions is to suppress surface effects due to structuring of the sample by the effect of the walls of the container on the repulsive cores in the particles. The cubic container (of volume $V = L^3$ and thus of side L) containing some configuration of the particles is surrounded by a cubic array of copies of the container and its configuration of particles. The potential energy of a particular particle is calculated by adding up its energies of interaction with the closest periodic copy of each of the other particles. (This closest periodic copy may, of course, be the original particle in the original container.) A particle interacts only with those particles of the periodic array which are inside a cube of side L centred on the particle. For hard sphere interactions, these boundary conditions are the same as periodic boundary conditions. They suppress that part of the surface terms which come from interference of the repulsive cores of the particles with the container wall. However, they do not make any attempt to handle the effects of the finite size of the sample on the contribution to the properties of the sample from dipole–dipole and other long range electrostatic interactions.

(C) Periodic boundary conditions

The cubic container of side L is again surrounded by a cubic array of copies of the container. A particle in the original container not only interacts with another particle in the original container, but also with every periodic copy of that other particle. If in free boundary conditions the potential function is $\phi(\mathbf{r})$, in periodic boundary conditions the effective potential becomes

$$\psi(\mathbf{r}) = \sum_{l=-\infty}^{\infty} \sum_{m=-\infty}^{\infty} \sum_{n=-\infty}^{\infty} \phi((Ll, Lm, Ln) + \mathbf{r}). \tag{3.2.1}$$

For some electrostatic interactions (including dipole–dipole) the sum in (3.2.1) is divergent, and it is necessary to resort to some analytic subterfuge to get a convergent effective potential to use in the Hamiltonian. The method described here is chosen because it is simpler than other methods in dealing with the dipole–dipole interactions. We start with a charge–charge interaction, but with $1/r^{2s}$ in place of $1/r$:

$$\psi(\mathbf{r}_{ij}; s) = \frac{q_i q_j}{4\pi\epsilon_0 L} \sum_{l=-\infty}^{\infty} \sum_{m=-\infty}^{\infty} \sum_{n=-\infty}^{\infty} [(l + x_{ij})^2 + (m + y_{ij})^2 + (n + z_{ij})^2]^{-s} \tag{3.2.2}$$

where ϵ_0 is the dielectric permittivity of vacuum, q_i is the charge on the *i-th* particle and we have written $\mathbf{r}_{ij} = L\boldsymbol{\rho}_{ij} = L(x_{ij}, y_{ij}, z_{ij})$. We want to study the function

$$\xi(\boldsymbol{\rho}; s) = \sum_{l=-\infty}^{\infty} \sum_{m=-\infty}^{\infty} \sum_{n=-\infty}^{\infty} [(l + x)^2 + (m + y)^2 + (n + z)^2]^{-s}. \tag{3.2.3}$$

This function is an analytic function of s for $\mathrm{Re}(s) > 3/2$. For charge–charge interactions, we want to use $\xi(\boldsymbol{\rho}; 1/2)$ and so we must find a way of making an analytical continuation of $\xi(\boldsymbol{\rho}; s)$ from the region $\mathrm{Re}(s) > 3/2$ to $s = 1/2$. The Hamiltonian for the system may be written

$$H(s) = \frac{1}{8\pi\epsilon_0 L} \sum_{i=1}^{N} \sum_{j=1}^{N} q_i q_j \xi\left(\frac{\mathbf{r}_{ij}}{L}; s\right) \tag{3.2.4}$$

where we define $\xi(0; s)$ as the sum in (3.2.3) with $\boldsymbol{\rho} = 0$ but excluding the term (l, m, n) = (0, 0, 0). Thus we include in the Hamiltonian the interaction of a charge with its own periodic lattice of images. It may be shown[25,26,27] that we may write

$$\xi(\boldsymbol{\rho}; s) = \frac{\pi^s}{\Gamma(s)(s - 3/2)} + \xi^*(\boldsymbol{\rho}; s) \tag{3.2.5}$$

where $\Gamma(s)$ is the gamma function and $\xi^*(\boldsymbol{\rho}; s)$ is a function analytic in s in the whole half s-plane $\mathrm{Re}(s) > 0$ and given by

$$\xi^*(\boldsymbol{\rho}; s) = \frac{1}{\Gamma(s)} \int_{\pi}^{\infty} \mathrm{d}t\, t^{s-1} \mathrm{e}^{-\rho^2 t} \vartheta_3\left(ixt \left| \frac{it}{\pi}\right.\right) \theta_3\left(iyt \left| \frac{it}{\pi}\right.\right) \theta_3\left(izt \left| \frac{it}{\pi}\right.\right)$$

$$+ \frac{\pi^{3/2}}{\Gamma(s)} \int_0^{\pi} \mathrm{d}t\, t^{s-5/2} \left[\theta_3\left(x\pi \left| \frac{i\pi}{t}\right.\right) \theta_3\left(y\pi \left| \frac{i\pi}{t}\right.\right) \theta_3\left(z\pi \left| \frac{i\pi}{t}\right.\right) - 1\right] \tag{3.2.6}$$

where θ_3 is a Jacobi theta function. In the Hamiltonian (3.2.4) we can insert $\xi^*(0; s) = 0$ since the $\boldsymbol{\rho} = 0$ terms correspond to a constant energy which we may ignore. In spite of the complexity of (3.2.6), numerical evaluation of $\xi^*(\boldsymbol{\rho}; s)$ may proceed very rapidly indeed[25,27]. If we insert (3.2.5) into (3.2.4) we have a Hamiltonian analytic in the region Re(s) > 0 except

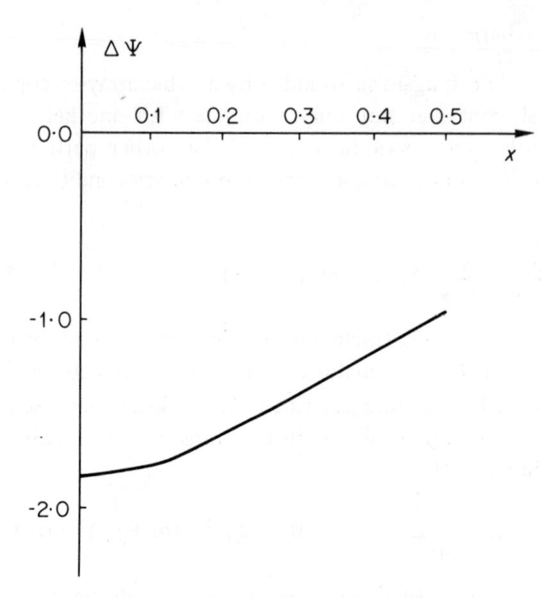

Figure 3.2.1 Plot of charge–charge effective potential change between periodic and minimum image boundary conditions for $\mathbf{r}_{ij} = (x, x, x)$ and a lattice distance $L = 1$ for $0 \leqslant x \leqslant 0.5$

for a simple pole of residue

$$\frac{1}{8\pi\epsilon_0 L} \frac{\pi^s}{\Gamma(s)} \sum_{i=1}^{N} \sum_{j=1}^{N} q_i q_j$$

at $s = 3/2$. But the charge neutrality of the system gives

$$\sum_{i=1}^{N} q_i = 0$$

so that

$$\sum_{i=1}^{N} \sum_{j=1}^{N} q_i q_j = \left(\sum_{i=1}^{N} q_i \right)^2 = 0$$

and the pole vanishes from the Hamiltonian. For charge–charge interactions, the Hamiltonian then becomes, by analytic continuation to $s = 1/2$,

$$H(\tfrac{1}{2}) = \frac{1}{8\pi\epsilon_0 L} \sum_{\substack{i=1 \\ i \neq j}}^{N} \sum_{j=1}^{N} q_i q_j \xi^* \left(\frac{\mathbf{r}_{ij}}{L}; \frac{1}{2} \right). \tag{3.2.7}$$

It is clearly of interest to see how much the effective potential in the Hamiltonian (3.2.7) differs from the $1/|\mathbf{r}|$ potential. In Figure 3.2.1 we plot $\xi^*(\boldsymbol{\rho}; 1/2) - 1/|\boldsymbol{\rho}|$ as a function of x for $\boldsymbol{\rho} = (x, x, x)$. It can be seen that the change in potential induced by the periodic boundary conditions can be quite large.

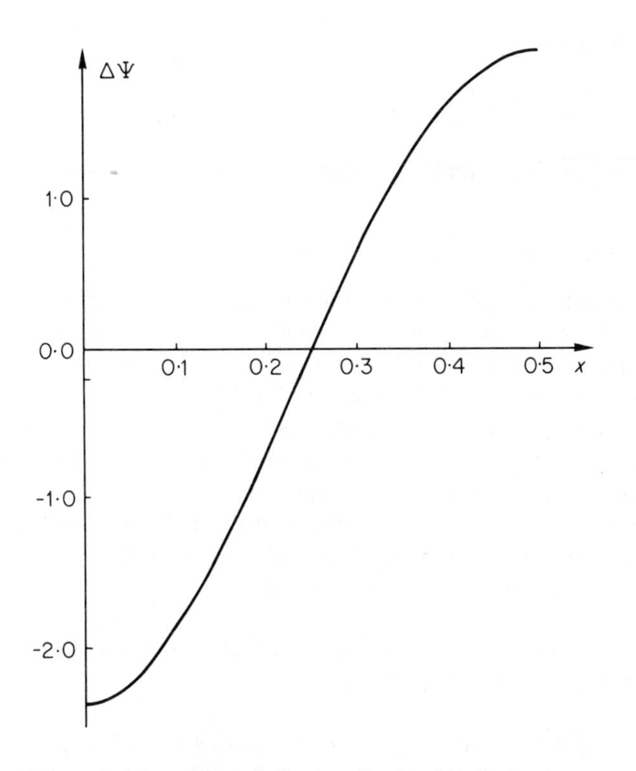

Figure 3.2.2 Plot of dipole–dipole effective potential change between periodic and minimum image boundary conditions for $\boldsymbol{\mu}_i = \boldsymbol{\mu}_j = (1, 0, 0)$, $\mathbf{r}_{ij} = (x, x, x)$ and a lattice distance $L = 1$ for $0 \leqslant x \leqslant 0.5$

In general, in periodic boundary conditions, it is possible to replace the $1/|\mathbf{r}|$ potential of ordinary electrostatics by the potential

$$\frac{1}{L}\xi^*\left(\frac{\mathbf{r}}{L}; \frac{1}{2}\right).$$

This replacement is not exact for dipoles, as the pole at $s = 3/2$ in $\xi(\boldsymbol{\rho}; s)$ has some effect also. The effective dipole–dipole potential in periodic boundary conditions for dipoles $\boldsymbol{\mu}_i, \boldsymbol{\mu}_j$ separated by $\mathbf{r}_{ij}$ may be written

$$\Psi_{\mathrm{p}}(\boldsymbol{\mu}_i, \boldsymbol{\mu}_j; \mathbf{r}_{ij}) = \frac{1}{8\pi\epsilon_0 L^3}\boldsymbol{\mu}_i \cdot \left[\frac{4\pi}{3}I - \nabla\nabla\xi^*(\boldsymbol{\rho}; \tfrac{1}{2})\Big|_{\boldsymbol{\rho} = (\mathbf{r}_{ij}/L)}\right] \cdot \boldsymbol{\mu}_j \tag{3.2.8}$$

where I is the 3 x 3 unit matrix. In Figure 3.2.2 we plot

$$\Delta\Psi(\boldsymbol{\mu}_i, \boldsymbol{\mu}_j; \mathbf{r}_{ij}) = \Psi_{\mathrm{p}}(\boldsymbol{\mu}_i, \boldsymbol{\mu}_j; \mathbf{r}_{ij}) - \Psi_{\mathrm{f}}(\boldsymbol{\mu}_i, \boldsymbol{\mu}_j; \mathbf{r}_{ij}) \tag{3.2.9}$$

where Ψ_{f} is the dipole–dipole potential in free boundary conditions:

$$\Psi_{\mathrm{f}}(\boldsymbol{\mu}_i, \boldsymbol{\mu}_j; \mathbf{r}_{ij}) = -\boldsymbol{\mu}_i \cdot \nabla\nabla(1/|\mathbf{r}_{ij}|) \cdot \boldsymbol{\mu}_j \tag{3.2.10}$$

for the case $\boldsymbol{\mu}_i = \boldsymbol{\mu}_j = (1, 0, 0)$, $L = 1$ and $\mathbf{r}_{ij} = (x, x, x)$ as a function of x. A crucial point which follows from equations (3.2.8, 3.2.9 and 3.2.10) is that the difference $\Delta\Psi(\boldsymbol{\mu}_i, \boldsymbol{\mu}; \mathbf{r}_{ij})$

may be written in the form

$$\Delta\Psi(\boldsymbol{\mu}_i, \boldsymbol{\mu}_j; \mathbf{r}_{ij}) = \boldsymbol{\mu}_i \cdot \frac{1}{L^3}\zeta\left(\frac{\mathbf{r}}{L}\right) \cdot \boldsymbol{\mu}_j \qquad (3.2.11)$$

and the 3 x 3 matrix $\zeta(\boldsymbol{\rho})$ has bounded components as $\boldsymbol{\rho} \to 0$.

(D) Boundary conditions via Maxwell's equations

The object of these boundary conditions is not to suppress surface effects but to provide some way of including the effect of an external field expressed via Maxwell's equations into a statistical mechanical calculation. We do this with the method of Penrose and Smith[20]. The method is to construct solutions of the equations for the electrostatic potential by the method of images. We may specify either the normal or transverse component of the electric field at the surface. If we set the normal or transverse component to be that of a fixed external field, then the method of images gives a potential energy equal to that when the fixed external field is zero plus the energy of the dipoles in the fixed field. Thus the important problem to solve is that of finding an effective potential between dipoles in a cube with zero transverse or normal components of electric field on the surface of the cube.

(i) Zero normal component. A dipole $\boldsymbol{\mu}_i = (\mu_i^X, \mu_i^Y, \mu_i^Z)$ at $\mathbf{r}_i = (x_i, y_i, z_i)$ sets up eight periodic lattices of images of period $2L$. We have[21]

$$\boldsymbol{\mu}_i^1 = (\mu_i^X, \mu_i^Y, \mu_i^Z) \quad \text{at} \quad \mathbf{r}_i^1 = (2lL + x_i, 2mL + y_i, 2nL + z_i),$$

$$\boldsymbol{\mu}_i^2 = (-\mu_i^X, \mu_i^Y, \mu_i^Z) \quad \text{at} \quad \mathbf{r}_i^2 = (2lL - x_i, 2mL + y_i, 2nL + z_i),$$

$$\boldsymbol{\mu}_i^3 = (\mu_i^X, -\mu_i^Y, \mu_i^Z) \quad \text{at} \quad \mathbf{r}_i^3 = (2lL + x_i, 2mL - y_i, 2nL + z_i),$$

$$\boldsymbol{\mu}_i^4 = (\mu_i^X, \mu_i^Y, -\mu_i^Z) \quad \text{at} \quad \mathbf{r}_i^4 = (2lL + x_i, 2mL + y_i, 2nL - z_i),$$

$$\boldsymbol{\mu}_i^5 = (-\mu_i^X, -\mu_i^Y, \mu_i^Z) \quad \text{at} \quad \mathbf{r}_i^5 = (2lL - x_i, 2mL - y_i, 2nL + z_i),$$

$$\boldsymbol{\mu}_i^6 = (-\mu_i^X, \mu_i^Y, -\mu_i^Z) \quad \text{at} \quad \mathbf{r}_i^6 = (2lL - x_i, 2mL + y_i, 2nL - z_i),$$

$$\boldsymbol{\mu}_i^7 = (\mu_i^X, -\mu_i^Y, -\mu_i^Z) \quad \text{at} \quad \mathbf{r}_i^7 = (2lL + x_i, 2mL - y_i, 2nL - z_i)$$

and

$$\boldsymbol{\mu}_i^8 = (-\mu_i^X, -\mu_i^Y, -\mu_i^Z) \quad \text{at} \quad \mathbf{r}_i^8 = (2lL - x_i, 2mL - y_i, 2nL - z_i) \qquad (3.2.12)$$

Thus a given dipole will interact with eight periodic lattices of images of another dipole. The effective dipole–dipole potential may then be written as

$$\Psi_{E_n=0}(\boldsymbol{\mu}_i, \boldsymbol{\mu}_j; \mathbf{r}_i, \mathbf{r}_j) = -\sum_{k=1}^{8} \frac{1}{64\pi\epsilon_0 L^3} \boldsymbol{\mu}_i \cdot [\nabla\nabla\, \xi^*(\boldsymbol{\rho}; \tfrac{1}{2})|_{\boldsymbol{\rho}=(\mathbf{r}_j^k - \mathbf{r}_i)/2L}] \cdot \boldsymbol{\mu}_j^k. \qquad (3.2.13)$$

The difference between $\Psi_{E_n=0}$ and Ψ_f has a property similar to that of $\Delta\Psi_p$ (equation 3.2.11), but because $\Psi_{E_n=0}$ depends on both $\mathbf{r}_i$ and $\mathbf{r}_j$, and not just on $\mathbf{r}_{ij}$, we write the difference $\Delta\Psi$ in the form

$$\Delta\Psi(\boldsymbol{\mu}_i, \boldsymbol{\mu}_j; \mathbf{r}_i, \mathbf{r}_j) = \boldsymbol{\mu}_i \cdot \frac{1}{L^3} \zeta_{E_n=0}\left(\frac{\mathbf{r}_i}{L}, \frac{\mathbf{r}_j}{L}\right) \cdot \boldsymbol{\mu}_j \qquad (3.2.14)$$

where, again, $\zeta_{E_n=0}$ is a 3 x 3 matrix with bounded components as $(\mathbf{r}_i - \mathbf{r}_j)/L \to 0$.

(ii) Zero transverse component. A set of eight periodic lattices of images is set up in this case also, giving rise to an effective potential similar in structure to (3.2.13) and whose difference from the free dipole–dipole potential also has the property given in equation (3.2.14).

(E) Barker–Watts reaction field boundary conditions[1,28]

The energy of a given dipole in these boundary conditions is calculated in the following way. First, set up a periodic array of copies of the cubic container of side L. Then construct a sphere of radius $1/2L$ about the dipole. Include in the potential energy of the dipole the free dipole–dipole interaction energy with all the other dipoles within the sphere. The region outside the sphere is imagined as filled with a dielectric of field permittivity, usually chosen to be close to the dielectric permittivity expected for the fluid. The dipoles inside the sphere polarize the dielectric, and the polarization sets up a reaction field on the original dipole. The interaction energy of the dipole with the 'uniform approximation' to this field (i.e. the first term in the multi-pole expansion) is included in the energy of the dipole.

The object of these boundary conditions is first to suppress boundary effects caused by short range interactions, in the same way as minimum image or periodic boundary conditions do. Also, the boundary effects due to long range interactions are expected to be suppressed, but without the cost of long range correlations which periodic boundary conditions imply. One criticism which may be levelled is that this method requires that, in calculating the dielectric permittivity, it is necessary to assume a value close to the one expected for the fluid being studied. The effective potential resulting from these boundary conditions is the free dipole–dipole potential plus a potential of the form (3.2.14).[29]

It is possible to use the full expansion of the reaction field[29,30] instead of just the first term. In this case it may be easier to use a cube, instead of a sphere. The reaction field is then an absolutely convergent sum (so the convergence problems of periodic boundary conditions are absent) of image potentials. The absolute convergence of the sum is assured by the fact that the magnitude of the image dipoles decreases rapidly with distance. Friedman[29] has suggested that the fine details of the reaction field may be important; and the sums which arise with a cube surrounding the dipole appear to be more easily evaluated than the sums for the spherical case, so that numerical evaluation of the exact reaction field potential energy may be much faster for a cube than for a sphere. The effective potential for the cubic case also has the form (3.2.14).

(F) Friedman's boundary conditions[29,30]

Friedman has suggested[29] that the Barker–Watts boundary conditions still have too much periodicity in them and hence induce too much long range correlation in the fluid sample. He suggests that the only way to avoid these correlations may be to forget about trying to suppress effects due to short range forces. Friedman suggests carrying out a simulation in a spherical cavity inside an infinite dielectric, chosen to have permittivity equal to that expected to be calculated from the simulation. This choice of permittivity should suppress sample size effects from the electrostatic interactions. In these boundary conditions, the resulting effective potential is again a sum of the free dipole–dipole interaction potential and a reaction field potential which again has the form given in equation (3.2.14). An alternative would be to use a cubic cavity in the dielectric rather than a spherical one. This again has the advantage that the sum of image potentials which give the exact reaction field may be more speedily evaluated than in the spherical case. The resulting effective potential is once again a sum of the free dipole–dipole potential and a potential with the form given in equation (3.2.14).

The various boundary conditions proposed in this section have all had, as their aim, the handling of some or all of the problems of finding bulk properties of dipolar fluids discussed in Section 3.1. None of them seem to be entirely successful and so it seems that at present there are no 'correct' boundary conditions to use. What would appear to be necessary, then, is some way of relating the results of a simulation using one set of boundary conditions to the results of a simulation with different boundary conditions.

In the next section we suggest a method for doing this.

3.3 Effects of different boundary conditions

For all the boundary conditions discussed in Section 3.2 we found that their result was an effective potential of the form, for boundary conditions $\{X\}$,

$$\Psi_{\{X\}}(\mu_i,\mu_j;\mathbf{r}_i,\mathbf{r}_j)=\frac{1}{8\pi\epsilon_0}\left(\frac{\mu_i\cdot\mu_j}{|\mathbf{r}_{ij}|^3}-\frac{3(\mu_i\cdot\mathbf{r}_{ij})(\mu_j\cdot\mathbf{r}_{ij})}{|\mathbf{r}_{ij}|^5}\right)+\mu_i\cdot\frac{1}{L^3}\zeta_{\{X\}}\left(\frac{\mathbf{r}_i}{L},\frac{\mathbf{r}_j}{L}\right)\cdot\mu_j \tag{3.3.1}$$

where $\zeta_{\{X\}}(\rho,\sigma)$ is a 3×3 matrix whose elements are bounded as $|\rho-\sigma|\to 0$. We are interested in relating the results of one simulation to those of another in different boundary conditions. To do this we relate the effective dipole–dipole potentials for each pair of boundary conditions. From equation (3.3.1) we see that we can write

$$\begin{aligned}\Psi_{\{X\}}(\mu_i,\mu_j;\mathbf{r}_i,\mathbf{r}_j)&=\Psi_{\{X\}}+\{\Psi_{\{Y\}}-\Psi_{\{X\}}\}\\ &\equiv\Psi_{\{X\}}(\mu_i,\mu_j;\mathbf{r}_i,\mathbf{r}_j)+\mu_i\cdot\frac{1}{L^3}\Theta_{\{X,Y\}}\left(\frac{\mathbf{r}_i}{L},\frac{\mathbf{r}_j}{L}\right)\cdot\mu_j,\end{aligned} \tag{3.3.2}$$

with $\Theta_{\{X,Y\}}(\rho,\sigma)$ a 3×3 matrix whose elements are bounded as $|\rho-\sigma|\to 0$. Equation (3.3.2) only covers the dipole–dipole part of the interaction. If we write down the Hamiltonian for the sample of N fluid particles in boundary conditions $\{Y\}$ we have

$$H_{\{Y\}}(N)=\sum_{i=1}^{N}\sum_{\substack{j=1\\ i\neq j}}^{N}[q(\mu_i,\mu_j;\mathbf{r}_i,\mathbf{r}_j)+\Psi_{\{Y\}}(\mu_i,\mu_j;\mathbf{r}_i,\mathbf{r}_j)] \tag{3.3.3}$$

where the function q carries a factor 1/2 to avoid double counting and includes all the short ranged interactions (and any electrostatic interactions shorter ranged than the dipole–dipole interaction). If we define $p_{\{X\}}(\mu_i,\mu_j;\mathbf{r}_i,\mathbf{r}_j)=q(\mu_i,\mu_j;\mathbf{r}_i,\mathbf{r}_j)+\Psi_{\{X\}}(\mu_i,\mu_j;\mathbf{r}_i,\mathbf{r}_j)$ then we can write (3.3.3) in the form

$$\begin{aligned}H_{\{Y\}}(N)&=\sum_{i=1}^{N}\sum_{\substack{j=1\\ i\neq j}}^{N}p_{\{Y\}}(\mu_i,\mu_j;\mathbf{r}_i,\mathbf{r}_j)\\ &=\sum_{i=1}^{N}\sum_{\substack{j=1\\ i\neq j}}^{N}\left\{p_{\{X\}}(\mu_i,\mu_j;\mathbf{r}_i,\mathbf{r}_j)+\mu_i\cdot\frac{1}{L^3}\Theta_{\{X,Y\}}\left(\frac{\mathbf{r}_i}{L},\frac{\mathbf{r}_j}{L}\right)\cdot\mu_j\right\}.\end{aligned} \tag{3.3.4}$$

As suggested by the plot for one case in Figure 3.2.2, the matrix function $\Theta_{\{X,Y\}}(\rho,\sigma)$ has bounded components for all ρ, σ belonging to the unit cube (the division of $\mathbf{r}$ by L ensures that since the original cube is of side L, the vectors which concern us all belong to the unit cube $0 \leqslant x \leqslant 1$, $0 \leqslant y \leqslant 1$, $0 \leqslant z \leqslant 1$ which we call Γ from now on). Thus, as the size of the sample cube of fluid becomes large (L increases) the potential $\boldsymbol{\mu}_i \cdot (1/L^3)\Theta\{X,Y\}(\boldsymbol{\rho}, \boldsymbol{\sigma}) \cdot \boldsymbol{\mu}_j$ becomes small. This suggests that we connect the properties of the fluid in boundary conditions {Y} to those in boundary conditions {X} by using the {X} case as a reference state and calculating the properties for the {Y} case by some adaptation of Barker–Henderson perturbation theory,[31] with $\boldsymbol{\mu}_i \cdot (1/L^3)\Theta\{X,Y\}(\boldsymbol{\rho}, \boldsymbol{\sigma}) \cdot \boldsymbol{\mu}_j$ as the (hopefully) small perturbation potential. However, if we want to analyse the problem as usefully as possible, we would like to develop an expansion for the change in properties which clearly identifies bulk terms (and perhaps surface terms) in the difference in properties.

Lebowitz, Stell and Baer[32] and Lebowitz and Penrose[33] have studied the properties of fluid systems with a pairwise interaction potential of the form

$$w(r) = q(r) + \gamma^3 \phi(\gamma \mathbf{r}) \tag{3.3.5}$$

where γ is to be very small (and eventually tend to zero in the work of Lebowitz and Penrose[33]). Lebowitz, Stell and Baer have studied such systems for the case when

(i) $q(r)$ is exponentially decreasing for large r,
(ii) $\int d^3\mathbf{x}\phi(\mathbf{x}) = \alpha < \infty$ exists and
(iii) $\phi(\mathbf{y})$ is bounded for all $\mathbf{y}$.

They are able to develop power series expansions in γ for the correlation functions and for the thermodynamic properties of the system. The condition (i) is not crucial to the analysis but merely makes some of the proofs very much simpler than without it. We should note that the first correction term in these expansions (at least for the Helmholtz free energy, the internal energy and the pressure) is proportional to γ°. When the limit $\gamma \to 0$ is taken in the correct way, changes to the bulk pressure occur which give rise to the van der Waals–Maxwell liquid–gas phase transition[33].

With a little imagination, it can be seen that the pair potential $p\{Y\}$ in equation (3.3.4) has the same structure as the potential $w(r)$ (see equation 3.3.5) considered by Lebowitz, Stell and Baer. We have merely introduced the extra particle coordinates μ_i (not crucial to the way the expansion works) and set $\gamma = 1/L$. This value of γ does lead to some difficulties since the idea of Lebowitz, Stell and Baer is to take the thermodynamic limit first and then let γ become small: we want to do both at once. However, the difficulties are not insurmountable and expressions for the free energy in the limit $L \to \infty$ can be developed.[26,34,35] The $1/L$ expansion (equivalent to the γ expansion of Lebowitz, Stell and Baer) is a little more complicated, but the first terms for the thermodymic properties may be calculated without too much difficulty. They all tend to become messy, except the expression for the internal energy density which may be written down fairly easily. We find

$$U\{Y\} = U\{X\} + \int_\Gamma d^3\boldsymbol{\rho} \int_\Gamma d^3\boldsymbol{\sigma} \int_\Omega d\boldsymbol{\mu}_1 \int_\Omega d\boldsymbol{\mu}_2 \, [\boldsymbol{\mu}_1 \cdot \Theta\{X,Y\}(\boldsymbol{\rho},\boldsymbol{\sigma}) \cdot \boldsymbol{\mu}_2] \, n^{(2)}_{\{X\}}(L\boldsymbol{\rho}, \boldsymbol{\mu}_1; L\boldsymbol{\sigma}, \boldsymbol{\mu}_2)$$
$$+ O(1/L) \tag{3.3.6}$$

where $n^{(2)}_{\{X\}}$ is an appropriate normalized two-particle correlation function in boundary conditions {X} and Ω is the unit sphere. The factor $1/L^3$ from calculating the internal energy and the factor $1/L^3$ in the difference potential have been used to change the natural variables of integration ($\mathbf{r}_1 = L\boldsymbol{\rho}$; $\mathbf{r}_2 = L\boldsymbol{\sigma}$) to variables $\boldsymbol{\rho}$, $\boldsymbol{\sigma}$ which range over the unit cube Γ. The terms in (3.3.6) which are $O(1/L)$ are surface contributions to the internal energy. Equation (3.3.6) shows that the largest changes to the internal energy due to change in boundary conditions are changes in the bulk values of the thermodynamic properties. This occurs also with the pressure and with the Helmholtz free energy.[21,26,34,35] Thus our suggested form for the partition functions (equation 3.1.8) is seen to be correct. The connection with the theory of Lebowitz, Stell and Baer provides us with a framework within which to compare data from simulations done in different boundary conditions.

3.4 Discussion

In the Introduction we pointed out that the boundary conditions used in a numerical simulation in statistical mechanics are almost always chosen to suppress surface effects, and the choice made is usually that of periodic boundary conditions. This choice leads to complicated sums for the effective potential and hence to longer computing time for given simulations. There has been a tendency to regard a simulation using the full periodic boundary conditions as inherently superior to a simulation using minimum image conditions[10,11,36] and that the use of periodic boundary conditions solves many of the finite size problems occurring in simulations. In Chapter 9 of this book, Barnes states: 'bulk conditions can be approximated by using a relatively small number of molecules (50–100) by the use of periodic boundaries.' A small number of workers[1,28,29] have seen that periodic boundary conditions imitate a very large sample of highly correlated fluid, and, in an attempt to reduce the effects of these correlations, have introduced their own boundary conditions. However, each piece of work seems to use the author's own boundary conditions fairly uncritically and without any real attempt to quantify the differences caused by changing the boundary conditions. To quote a fairly isolated critic,[37] 'some concerted attention to these matters seems due.'

In Section 3.3 we showed that some of the thermodynamic quantities have their bulk values changed by changing the boundary conditions. To get an estimate of how large the corrections to bulk properties can be, we consider equation (3.3.6) for the case $\{Y\}$ are periodic boundary conditions and $\{X\}$ are minimum image boundary conditions so that $\Theta\{X, Y\}$ is given by equations (3.2.9, 3.2.10 and 3.2.11). Of course, we do not know $n^{(2)}_{\{X\}}(\mathbf{r}_1, \boldsymbol{\mu}_1; \mathbf{r}_2, \boldsymbol{\mu}_2)$, but assume

$$n^{(2)}_{\{X\}} = \rho^2 \delta(\boldsymbol{\mu}_1 - \mathbf{m})\delta(\boldsymbol{\mu}_2 - \mathbf{m}) \tag{3.4.1}$$

i.e. that the sample is uniformly distributed with density ρ and uniform polarization density $\mathbf{m}$. The correction term in (3.3.6) is then

$$U_{\{Y\}} - U_{\{X\}} = \frac{4\pi}{3}\rho^2 \mathbf{m}^2. \tag{3.4.2}$$

As we pointed out in the Introduction, these changes are expected to vanish in the thermodynamic limit because of the formation of many domains of polarization which, while being large domains, are small compared with the size of the whole sample. The finite size of the samples usually considered in numerical simulations preclude the formation of such domains.

The size of the correction estimate in equation (3.4.2) shows that the suspicion of Barker and Watts[1,28] and Friedman[29] that periodic boundary conditions can introduce unwanted effects is correct. However, it seems likely that their boundary conditions, too, will introduce unwanted effects into the bulk properties of the fluid being simulated. For water, perhaps the most interesting quantity is the dielectric permittivity. Unfortunately we have not yet developed the $1/L$ expansion for this quantity. We have, however, shown that there is a general framework of theory which is suited to estimating the effects of a change in the boundary conditions imposed on a finite sample simulation of a dipolar fluid. We hope that the ability to make quantitative comparisons between the results of simulations in different boundary conditions will eventually enable us to calculate correct bulk properties for dipolar fluids from simulation data. An advantage of the corrections to the bulk properties is that their calculation requires only the appropriate effective potential and the two-particle correlation function. This information is usually included in published reports of simulations. We intend to carry out detailed comparisons of the effect of different boundary conditions on simulation results for the hard sphere dipole fluid[37].

An important question is whether a similar set of corrections can be developed for charge–charge interactions. There the effective potential change looks like $(1/L)\,\phi(\mathbf{r}/L)$ and hence the Lebowitz, Stell and Baer expansion scheme may not be used so easily. It would appear that the shielding effects characteristic of ionic systems must play a part in determining the effects of a change in effective potential. We do not yet know how to include these shielding effects.

I thank G. Stell and C. Hoskins for some valuable conversations, and the Australian Research Grants Commission for financial support of the numerical work.

References

1. Barker, J. A. and Watts, R. O. (1973). *Mol. Phys.*, **26**, 789.
2. Rowlinson, J. S. (1951). *Trans. Faraday Soc.*, **47**, 120.
3. Barnes, P. Chapter 9 of this book and references cited therein.
4. Ruelle, D. (1963). *Helv. Phys. Acta.*, **36**, 183.
5. Fisher, M. E. (1964). *Arch. Rat. Mech. Anal.*, **17**, 377.
6. Fisher, M. E. and Lebowitz, J. L. (1970). *Comm. Math. Phys.*, **19**, 251.
7. For $1/r^4$ potentials, convergence to the bulk thermodynamic limit seems poor (H. De Witt, private communication).
8. Metropolis, N., Rosenbluth, A. W., Rosenbluth, M. N., Teller, A. H. and Teller, E. (1953). *J. Chem. Phys.*, **21**, 1087.
9. Wood, W. W. and Jacobsen, J. D. (1957). *J. Chem. Phys.*, **27**, 1207.
   Alder, B. J. and Wainwright, T. E. (1957). *J. Chem. Phys.*, **27**, 1208.
10. Hansen, J. P. (1973). *Phys. Rev.* A., **8**, 3096.
    Pollock, E. L. and Hansen, J. P. (1973) *Phys. Rev. A.*, **8**, 3110.
    Hansen, J. P., McDonald, I. R. and Pollock, E. L. (1975). *Phys. Rev.* A., **11**, 1025.
    Hansen, J. P. and McDonald, I. R. (1975). *Phys. Rev.* A., **11**, 2111.
    Vieillefosse, P. and Hansen, J. P. (1975). *Phys. Rev.* A., **12**, 1106.
11. Woodcock, L. V. and Singer, K. (1970). *Trans. Faraday Soc.*, **67**, 12.
12. Hansen, J. P. and Verlet, L. (1969). *Phys. Rev.*, **184**, 151.
13. Watts, R. O., Barker, J. A. and Henderson, D. J. (1970). *Bull. Amer. Phys. Soc.*, **15**, 772.
14. Dyson, F. J. and Lenard, A. (1967). *J. Math. Phys.*, **8**, 423.
15. Lebowitz, J. L. and Lieb, E. (1969). *Phys. Rev. Lett.*, **22**, 631.
16. Lebowitz, J. L. and Lieb, E. (1972). *Adv. Math.*, **9**, 316.
17. Fisher, M. E. and Ruelle, D. (1966). *J. Math. Phys.*, **7**, 260.
18. Griffiths, R. B. (1969). *Phys. Rev.*, **172**, 655.
19. E.g. Landau, L. D. and Lifshitz, E. M. (1960). *Electrodynamics of continuous media* Pergamon, London.
20. Penrose, O. and Smith, E. R. (1972). *Commun. Math. Phys.*, **26**, 53.
21. Smith, E. R. and Perram, J. W. (1975). *J. Phys.* A., **8**, 1130.
22. Perram, J. W. and Smith, E. R. In preparation.
23. Henderson, D. J., Abraham, F. and Barker, J. A. *J. Chem. Phys.*, (in press).
24. Henderson, D. J. and Barker, J. A. (1975). *J. Chem. Phys.*, **62**, 2716.
25. Smith, E. R. and Perram, J. W. (1975). *Mol. Phys.*, **30**, 31.
26. Smith, E. R. and Perram, J. W. (1975). *J. Aust. Math. Soc.*, **19**B, 116.
27. Hoskins, C. S. and Smith, E. R. *Chemical Physics* (in press).
28. Barker, J. A. and Watts, R. O. (1969) *Chem Phys. Letts.*, **3**, 144.
29. Friedman, H. L., (1975). *Mol. Phys.*, **29**, 1533.
30. Fröhlich, H. (1959). *Theory of Dielectrics,* Clarendon Press, Oxford.
31. Barker, J. A. and Henderson, D. J. (1967). *J. Chem. Phys.*, **47**, 4714.

32. Lebowitz, J. L., Stell, G. and Baer, S. (1965). *J. Math. Phys.*, **6**, 1282.
33. Lebowitz, J. L. and Penrose, O. (1966). *J. Math. Phys.*, **7**, 98.
34. Smith, E. R. and Perram, J. W. (1974). *Phys. Lett.*, **50A**, 294.
35. Smith, E. R. and Perram, J. W. (1975). *Phys. Lett.*, **53A**, 121.
36. Jansoone, V. M. (1974). *Chemical Physics*, **3**, 79.
37. Valleau, J. P. and Whittington, S. G. 'Conventional Monte Carlo techniques in statistical mechanics'. To appear in B. Berne (Ed.) *Theoretical Chemistry*. Plenum Publishing Corp., New York.

Chapter 4

Critical Phenomena in Classical Fluids

JAN V. SENGERS

Institute for Physical Science and Technology,
University of Maryland, Maryland 20742
and
National Bureau of Standards, Washington, D.C. 20234

and

J. M. H. LEVELT SENGERS

National Bureau of Standards, Washington, D.C. 20234

4.1 Introduction

This chapter will be concerned with the nature of the thermodynamic behaviour of fluids in the vicinity of a critical point. We shall restrict ourselves to classical fluids, which means that we shall not consider the superfluid transition in liquid helium. However, the gas–liquid critical regions of ^{3}He, ^{4}He and other light fluids do fall within the domain of this chapter. We shall begin with a discussion of the nature of a critical point from thermodynamic considerations, and shall then conclude this Introduction with an outline of the subjects covered in this chapter on critical behaviour of fluids.

According to Gibbs' phase rule, a one-component fluid has in general two thermodynamic degrees of freedom. However, when two phases of a one-component fluid coexist, only one degree of freedom is left. Since all intensive properties or 'fields' are equal in the two coexisting phases, the condition of coexistence defines a curve in the space of the two independent field variables. On this coexistence curve, extensive properties or 'densities', such as entropy and number density, are generally not the same in the two coexisting phases. The critical point, however, is an endpoint on a coexistence curve at which not only all fields but also all densities have become equal in the two fluid phases. The two phases can no longer be distinguished from each other and the coexistence curve terminates.

When two phases coexist in a two-component fluid, the system still has two thermodynamic degrees of freedom. The condition of coexistence then determines a two-dimensional surface in the three-dimensional space of independent field variables. On this coexistence surface the density variables such as number density, entropy and concentration generally differ in the two coexisting phases. A coexistence surface may, however, end in a critical line at which not only all fields but also all densities of the coexisting phases are equal, i.e. the coexisting phases have become identical.

Critical points are focal points of exceptional thermodynamic behaviour, not only when they are approached from within the two-phase region but also when approached from the one-phase region. They are points of incipient instability. Thermodynamic stability requires that the determinant of the matrix of second derivatives of the energy $U(S, V, \ldots)$ as a function of its characteristic extensive variables, entropy S, volume V, etc., be positive definite. For a one-component fluid this matrix has the form

$$\text{Stiffness matrix} \begin{pmatrix} \dfrac{\partial^2 U}{\partial S^2} & \dfrac{\partial^2 U}{\partial S \partial V} \\[2ex] \dfrac{\partial^2 U}{\partial S \partial V} & \dfrac{\partial^2 U}{\partial V^2} \end{pmatrix} = \begin{pmatrix} \dfrac{T}{C_V} & -\dfrac{T\alpha_P}{K_T C_V} \\[2ex] -\dfrac{T\alpha_P}{K_T C_V} & \dfrac{C_P}{V K_T C_V} \end{pmatrix} \tag{4.1.1}$$

where C_P and C_V are the heat capacities at constant pressure and constant volume, respectively,

$\alpha_P = V^{-1}(\partial V/\partial T)_P$ the isobaric thermal expansion coefficient and $K_T = -V^{-1}(\partial V/\partial P)_T$ the isothermal compressibility. This matrix was called the stiffness matrix by Tisza (1961). Tisza chose this wording to indicate an analogy with mechanics, where 'stiffness' denotes the increase of stress when the system is strained. The analogy of strain in thermodynamics is a change in volume or entropy, and the increase of stress is the corresponding change in pressure $P = -(\partial U/\partial V)_S$ or in temperature $T = (\partial U/\partial S)_V$. The determinant of the stiffness matrix is given by

$$D(S, V) = \frac{T}{VK_T C_V}. \tag{4.1.2}$$

At the critical point the compressibility K_T diverges strongly, that is, at least as fast as the inverse temperature difference with the critical point. Thus the determinant $D(S, V)$ approaches zero at the critical point, indicating marginal stability.

Tisza also introduced the so-called compliance matrix whose elements are the second derivatives of $-G(T, -P)$. where G is the Gibbs free energy. (Note that G reduces to the chemical potential when taken per mole or per particle.)

$$\text{Compliance matrix} \begin{pmatrix} -\dfrac{\partial^2 G}{\partial T^2} & \dfrac{\partial^2 G}{\partial T \partial P} \\ \dfrac{\partial^2 G}{\partial T \partial P} & -\dfrac{\partial^2 G}{\partial P^2} \end{pmatrix} = \begin{pmatrix} \dfrac{C_P}{T} & V\alpha_P \\ V\alpha_P & VK_T \end{pmatrix} \tag{4.1.3}$$

This choice of name again indicates an analogy with mechanics; the compliance is the strain of a mechanical system due to an increase in stress. The compliance matrix is the inverse of the stiffness matrix. Since the determinant of the stiffness matrix approaches zero, *all* elements of the compliance matrix diverge at the critical point. Thus the strong divergence of K_T implies similar divergences in α_P and C_P. It was discovered in the 1960s that the constant-volume heat capacity C_V of fluids diverges weakly at the critical point, that is, roughly as the logarithm of the difference between the actual temperature and that of the critical point. Thus the determinant of the compliance matrix,

$$D(T, P) = VK_T C_V/T \tag{4.1.4}$$

diverges 'strongly times weakly' in the language of Griffiths and Wheeler (1970). In multicomponent fluids, the stiffness matrix contains the second derivatives of the energy with respect to all independent density variables. The determinant of the corresponding compliance matrix, however, diverges 'strongly times weakly', just as that of a one-component fluid. The description of the thermodynamic anomalies in a fluid mixture, therefore, bears a close analogy to that in one-component fluids.

Closely associated with the marginal stability and diverging compressibility is the presence of large thermal fluctuations in the vicinity of the critical point (Klein and Tisza 1949). These critical fluctuations in turn lead to anomalous behaviour of various dynamical properties of fluids near critical points. The average size of a critical fluctuation is indicated by the correlation length. Near a critical point, this correlation length becomes much longer than the range of molecular interaction. The fluid behaves as a collection of 'droplets' or aggregates of molecules of macroscopically fluctuating density, in which the individual molecular interactions are of less importance. Consequently, critical behaviour shows considerable similarity in a large variety of fluids. This similarity is expressed by the term 'universality' which we shall define precisely. The presence of a new length scale, the correlation length, gives thermodynamic properties in the critical region a character of homogeneity. The homogeneity property makes it possible to reduce the number of independent variables by proper choice of scale.

Scaling is the main theme of this chapter. In Section 4.2 the concepts of homogeneity and scaling are introduced as properties of the classical or mean-field critical-region equations of state. In Section 4.3 the scaling laws and the hypothesis of universality are formulated for the thermodynamic behaviour of one-component fluids and compared with the experimental results. In Section 4.4 the description is generalized to include fluid mixtures. In Section 4.5 the postulate of homogeneity is formulated for the correlation function and universality of the correlation function, both with three and with two scale factors is explored.

The literature on critical point phenomena is expanding rapidly. A bibliography covering the literature on equilibrium critical phenomena over the period 1950 through 1967 was published by the National Bureau of Standards (Michaels *et al.* 1970). A subsequent bibliography edited by Stanley (1973) includes articles on critical phenomena to mid-1972. In this chapter we shall restrict ourselves to reviewing the current status of the description of equilibrium critical phenomena in fluids. For a discussion of the critical behaviour of dynamical properties of fluids the reader is referred to reviews that have appeared elsewhere (Anisimov 1975, Sengers 1971, 1972, 1973, Swinney and Henry 1973) as well as to the chapter of Kawasaki and Gunton in this volume (Chapter 5).

4.2 Concepts for describing critical phenomena in fluids

4.2.1 Critical exponents for thermodynamic properties

Let A be the Helmholtz free energy, S the entropy, μ the chemical potential per particle and ρ the number density. We also find it convenient to introduce a symmetrized isothermal compressibility χ_T as

$$\chi_T \equiv (\partial\rho/\partial\mu)_T = \rho^2 K_T. \tag{4.2.1}$$

The thermodynamic properties are made dimensionless by expressing them in units of appropriate combinations of the critical temperature T_c, the critical density ρ_c and the critical pressure P_c. Specifically, we define

$$\begin{aligned} &T^* = T/T_c \quad \rho^* = \rho/\rho_c \quad P^* = P/P_c \\ &A^* = A/VP_c \quad \mu^* = \mu\rho_c/P_c \quad \chi_T^* = \chi_T P_c/\rho_c^2 \\ &S^* = ST_c/VP_c \quad C_V^* = C_V T_c/VP_c. \end{aligned} \tag{4.2.2}$$

Note that the reduced extensive thermodynamic properties A^*, S^* and C_V^* are taken per unit volume. In addition we define the differences

$$\begin{aligned} &\Delta T^* = (T - T_c)/T_c \\ &\Delta\rho^* = (\rho - \rho_c)/\rho_c \\ &\Delta\mu^* = [\mu(\rho, T) - \mu(\rho_c, T)]\rho_c/P_c \end{aligned} \tag{4.2.3}$$

where $\mu(\rho_c, T)$ is the chemical potential on the critical isochore at temperature T.

In the description of the anomalous critical behaviour of a physical property it is assumed that sufficiently close to the critical point the property varies as a simple power of the temperature difference or the density difference from the critical point. The exponent of the power law will depend on the property chosen, the path along which the critical point is approached and the way the distance from the critical point is measured. The most commonly used power laws for thermodynamic properties are summarized in Table 4.2.1. The paths along which these power laws are defined, namely the critical isochore $\Delta\rho^* = 0$, the critical isotherm $\Delta T^* = 0$ and the coexistence curve $\Delta\rho^* = \Delta\rho_{cxc}$ are schematically indicated in Figure 4.2.1 together with the corresponding exponents.

Table 4.2.1 Power laws for thermodynamic properties

Path	Power law
$T \leqslant T_c, \rho = \rho_{cxc}$	$\Delta\rho^*_{cxc} = \pm B \mid \Delta T^* \mid^{\beta}$
$T = T_c$	$\Delta\mu^* = D(\Delta\rho^*) \mid \Delta\rho^* \mid^{\delta - 1}$
$T \geqslant T_c, \rho = \rho_c$	$\chi^*_T = \Gamma \mid \Delta T^* \mid^{-\gamma}$
$T \leqslant T_c, \rho = \rho_{cxc}$, one-phase	$\chi^*_T = \Gamma' \mid \Delta T^* \mid^{-\gamma'}$
$T \geqslant T_c, \rho = \rho_c$	$C^*_V/T^* = \dfrac{A^+}{\alpha}\{\mid \Delta T^* \mid^{-\alpha} - 1\}$
$T \leqslant T_c, \rho = \rho_{cxc}$, one-phase	$C^*_V/T^* = \dfrac{A^-_{I}}{\alpha'}\{\mid \Delta T^* \mid^{-\alpha'} - 1\}$
$T \leqslant T_{c'}, \rho = \rho_{c'}$, two-phase	$C^*_V/T^* = \dfrac{A^-_{II}}{\alpha''}\{\mid \Delta T^* \mid^{-\alpha''} - 1\}$
$T \leqslant T_{c'}$, two-phase	$d^2 P^*_{cxc}/dT^{*2} \propto \mid \Delta T^* \mid^{-\theta_P}$
$T \leqslant T_{c'}$, two-phase	$d^2 \mu^*_{cxc}/dT^{*2} \propto \mid \Delta T^* \mid^{-\theta_\mu}$

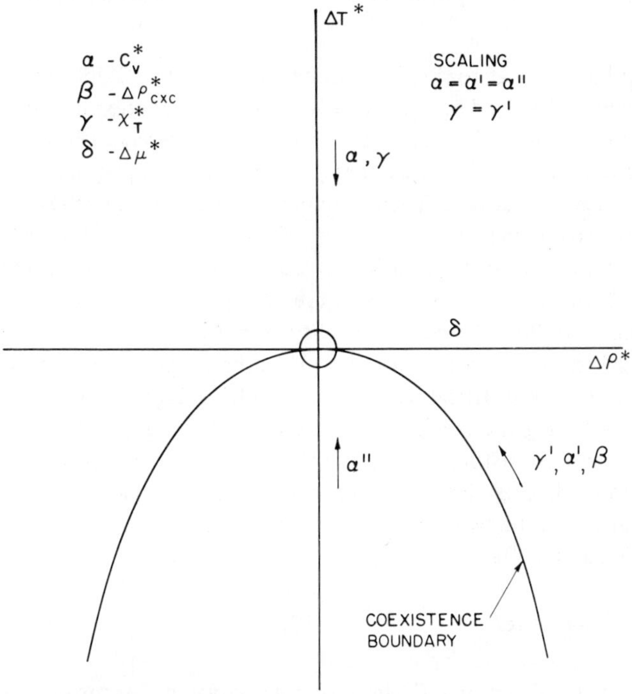

Figure 4.2.1 Special paths in the $\Delta\rho^* - \Delta T^*$ plane and power law exponents defined along them. (From Levelt Sengers *et al.* 1976, reproduced by permission of the American Institute of Physics)

The critical exponents introduced in Table 4.2.1 are not all independent of each other. The laws of thermodynamics impose several rigorous inequalities between combinations of the 'thermodynamic' exponents α, β, γ, δ, θ_P, θ_μ (Griffiths 1965a,b, 1972, Rowlinson 1969). In particular

$$
\begin{array}{lll}
\text{(Griffiths 1965a, Rushbrooke 1965)} & 2 - \alpha'' \leqslant \beta(\delta + 1) & \\
\text{(Rushbrooke 1963, Fisher 1964)} & \alpha'' + 2\beta + \gamma' \geqslant 2 & \\
\text{(Liberman 1966)} & \beta(\delta - 1) \leqslant \gamma' & (4.2.4) \\
\text{(Griffiths 1965b)} & \theta_P \leqslant \alpha'' + \beta & \\
\text{(Mermin and Rehr 1971)} & \theta_\mu \leqslant \alpha'' + \beta. &
\end{array}
$$

Furthermore, the scaling hypothesis for thermodynamic properties, to be introduced in Section 4.3.2, leads to the exponent relations

$$
\begin{aligned}
\alpha &= \alpha' = \alpha'' = \theta_P \\
\gamma &= \gamma' \\
2 - \alpha &= \beta(\delta + 1) \\
\gamma &= \beta(\delta - 1)
\end{aligned}
\tag{4.2.5}
$$

so that only two thermodynamic exponents can be chosen independently.

4.2.2 A model of a critical-point phase transition: I. The classical equation with third-degree isotherm

The first successful attempt at formulating an equation of state exhibiting a critical point was that of Van der Waals (1873). The place of his work in the context of critical phenomena was recently commemorated at the Van der Waals centennial conference (Prins 1974). The equation of Van der Waals is presently viewed as one of a large class of equations of state that are called classical or mean-field equations. Their common feature is that they assume an analytic dependence of the Helmholtz free energy or of the pressure on volume and temperature, while the critical point is characterized by the conditions $(\partial P/\partial V)_T = 0$, $(\partial^2 P/\partial V^2)_T = 0$ and $(\partial^3 P/\partial V^3)_T \neq 0$. Above the critical temperature, the pressure is a monotonically decreasing function of volume, so that the compressibility is positive and finite. At the critical point there is a horizontal inflection point on the P–V isotherm, at which point the compressibility diverges. Isotherms below the critical temperature exhibit a 'loop', on part of which the pressure rises with volume, violating the condition for mechanical stability. The system then splits into two mechanically stable coexisting phases. The Maxwell equal area construction, which replaces the 'loop' by a straight line parallel to the volume axis, ensures that the temperatures, pressures and chemical potentials of the two coexisting phases are equal.

Van der Waals' equation reads

$$\left(P + \frac{a}{\tilde{V}^2}\right)(\tilde{V} - b) = RT \tag{4.2.6}$$

where $\tilde{V}$ is the molar volume and R the molar gas constant. According to this equation the critical behaviour of the P–V isotherm comes about through competition between the hard core repulsion of the molecules, represented by the excluded volume term b, and the longer-range attraction between the molecules, represented by a pressure term a/V^2 or an internal energy term a/V. Van der Waals' approximation for the attractive interaction is rigorous in the limit of weak long-range intermolecular forces (Kac *et al.* 1963, Uhlenbeck *et al.* 1963, Hemmer *et al.* 1964, Van Kampen 1964, Lebowitz 1974). In real fluids, however, the attractive forces are usually not long range, as was realized by Van der Waals. This is the reason why the mean-field theories fail to represent the observed thermodynamic behaviour near the critical point.

The horizontal inflection point for the Van der Waals equation is located at P_c, $\tilde{V}_c$, T_c with

$$P_c = \frac{a}{27b^2} \qquad \tilde{V}_c = 3b \qquad RT_c = \frac{8a}{27b}.$$

If pressure, volume and temperature are measured in units of P_c, $\tilde{V}_c$ and T_c the *reduced* equation of state

$$\left(P^* + \frac{3}{V^{*2}}\right)\left(V^* - \frac{1}{3}\right) = \frac{8}{3}T^* \qquad (4.2.7)$$

is obtained in which there is no explicit appearance of parameters characteristic of a particular substance. Van der Waals fluids have identical reduced equations of state and are said to obey the law of corresponding states.

As mentioned earlier, Van der Waals' equation is an example of a classical or mean-field equation of state which is analytic in volume and temperature at the critical point. Most equations of state used in engineering applications are of this nature. All classical equations lead to a specific characteristic pattern of the critical anomalies. This pattern can be explored by studying the Taylor expansion of the classical equation around the critical point. Early investigations of this type were carried out by Van der Waals (1893, 1894) and by Van Laar (1912). Recent studies were made by Baehr (1963a,b), Barieau (1966, 1968), Levelt Sengers (1970) and Mulholland (1973).

Before developing the Taylor expansion a choice of variables has to be made. The relation between the variations of the field variables pressure P, temperature T and chemical potential μ is given by the Gibbs–Duhem equation

$$S\,\mathrm{d}T - V\mathrm{d}P + N\mathrm{d}\mu = 0 \qquad (4.2.8)$$

where N is the total number of molecules, if the chemical potential μ is taken per particle. In order to scale the conjugate extensive variables they are usually divided by N, or more specifically, by the number of moles N/N_0, where N_0 is Avogadro's number. This procedure yields the relation

$$\mathrm{d}\tilde{G} = -\tilde{S}\,\mathrm{d}T + \tilde{V}\mathrm{d}P \qquad (4.2.9)$$

where $\tilde{G}$, $\tilde{S}$ and $\tilde{V}$ are the molar values of the Gibbs free energy, entropy and volume, respectively. By a Legendre transformation one obtains the relation

$$\mathrm{d}\tilde{A} = -\tilde{S}\mathrm{d}T - P\mathrm{d}\tilde{V} \qquad (4.2.10)$$

for the Helmholtz free energy per mole $\tilde{A}(\tilde{V}, T)$ as a function of *volume* and temperature. Differentiation of $\tilde{A}$ with respect to $\tilde{V}$ then yields an equation of state $P(\tilde{V}, T)$ for the pressure as a function of $\tilde{V}$ and T. However, alternate but equivalent thermodynamic relations are obtained when the extensive properties in (4.2.8) are scaled by the volume V so that instead of (4.2.9)

$$\mathrm{d}P = s\,\mathrm{d}T + \rho\mathrm{d}\mu \qquad (4.2.11)$$

where $s = S/V$ is the entropy density and $\rho = N/V$ the number density. A Legendre transformation then yields, instead of (4.2.10), a relation

$$d(A/V) = -s dT + \mu d\rho \tag{4.2.12}$$

for the Helmholtz free energy density A/V as a function of *density* and temperature. Differentiation of the Helmholtz free energy relation then leads to an equation of state $\mu(\rho, T)$ for the chemical potential as a function of density and temperature. From the point of view of thermodynamics the two methods of description are completely equivalent. In practice, a description in terms of $P(\tilde{V}, T)$ is often preferred because of the more direct accessibility of the variables to measurement. However, for real fluids near the critical point a description in terms of $\mu(\rho, T)$ deserves preference for reasons of increased symmetry to be discussed in Section 4.3.1. For this reason we shall formulate here the Taylor expansion of the Helmholtz free energy density in terms of ρ, rather than the more conventional expansion of the Helmholtz free energy per mole $\tilde{A}$ in terms of $\tilde{V}$. However, if desired, the interested reader can easily work out the analogous expansion for $\tilde{A}(\tilde{V}, T)$.

Using the reduced variables defined in (4.2.2), the fundamental postulate of the classical theory can be formulated as

$$\begin{aligned}
A^* &= \sum_{m=0}^{\infty} \sum_{n=0}^{\infty} \frac{1}{m!n!} A_{mn} (\Delta\rho^*)^m (\Delta T^*)^n \\
\mu^* &= \sum_{m=0}^{\infty} \sum_{n=0}^{\infty} \frac{1}{m!n!} \mu_{mn} (\Delta\rho^*)^m (\Delta T^*)^n \\
P^* &= \sum_{m=0}^{\infty} \sum_{n=0}^{\infty} \frac{1}{m!n!} P_{mn} (\Delta\rho^*)^m (\Delta T^*)^n .
\end{aligned} \tag{4.2.13}$$

Since $\mu^* = (\partial A^*/\partial \rho^*)_T$ and $P^* = \mu^* \rho^* - A^*$, the expansion coefficients A_{mn}, μ_{mn} and P_{mn} are interrelated by

$$\begin{aligned}
\mu_{m,n} &= A_{m+1,n} && (m \geqslant 0) \\
P_{m,n} &= A_{m+1,n} + (m-1) A_{m,n} && (m \geqslant 0) \\
P_{m,n} &= \mu_{m,n} + (m-1) \mu_{m-1,n} && (m \geqslant 1).
\end{aligned} \tag{4.2.14}$$

The conditions of criticality imply that

$$A_{20} = A_{30} = 0 \quad \mu_{10} = \mu_{20} = 0 \quad P_{10} = P_{20} = 0. \tag{4.2.15}$$

We can arrange the expansion coefficients into matrices of the form

A_{mn}	1	ΔT^*	$(\Delta T^*)^2$	μ_{mn}	1	ΔT^*	$(\Delta T^*)^2$	P_{mn}	1	ΔT^*	$(\Delta T^*)^2$
1	A_{00}	A_{01}	A_{02}		μ_{00}	μ_{01}	μ_{02}		P_{00}	P_{01}	P_{02}
$\Delta\rho^*$	A_{01}	A_{11}	A_{12}		0	μ_{11}	μ_{12}		0	P_{11}	P_{12}
$(\Delta\rho^*)^2$	0	A_{21}	A_{22}		0	μ_{21}	μ_{22}		0	P_{21}	P_{22}
$(\Delta\rho^*)^3$	0	A_{31}	A_{32}		μ_{30}	μ_{31}	μ_{32}		P_{30}	P_{31}	P_{32}
$(\Delta\rho^*)^4$	A_{40}	A_{41}	A_{42}		μ_{40}	μ_{41}	μ_{42}		P_{40}	P_{41}	P_{42}

(4.2.16)

with

$$\begin{aligned}
&\mu_{00} = A_{10} \quad && P_{00} = A_{10} - A_{00} = P_c^* = 1 \\
&\mu_{30} = A_{40} && P_{30} = A_{40} = \mu_{30} \\
&\mu_{40} = A_{50} && P_{40} = A_{50} + 3A_{40} = \mu_{40} + 3\mu_{30} \\
&\mu_{01} = A_{11} && P_{01} = A_{11} - A_{01} \\
&\mu_{11} = A_{21} && P_{11} = A_{21} = \mu_{11} \\
&\mu_{21} = A_{31} && P_{21} = A_{31} + A_{21} = \mu_{21} + \mu_{11}
\end{aligned} \tag{4.2.17}$$

We assume here that $P_{30} = \mu_{30} = A_{40} \neq 0$. This is the case of 'three-point contact' between the critical isobar and the critical isotherm in the terminology of Baehr, since both the first derivative $(\partial P/\partial \rho)_T$ and the second derivative $(\partial^2 P/\partial \rho^2)_T$ vanish at the critical point, but not the third derivative $(\partial^3 P/\partial \rho^3)_T$. It is also possible to construct a classical equation of state in which $P_{30} = \mu_{30} = A_{40} = 0$. Thermodynamic stability then requires that $P_{40} = \mu_{40} = A_{50} = 0$ as well. If then $P_{50} = \mu_{50} = A_{60} \neq 0$ we obtain the case of 'five-point contact' to be discussed in Section 4.2.3. In the case of 'three-point contact' mechanical stability requires $P_{30} = \mu_{30} = A_{40}$ to be positive. For the compressibility to be positive at temperatures above T_c the coefficient P_{11} must be positive as well. Thermal stability requires A_{02} to be negative.

We first obtain the values of the critical exponents δ, γ and α. The asymptotic shape of the critical isotherm ($\Delta T^* = 0$) follows immediately from the form of the matrix (4.2.16).

$$P^* - P_c^* = \mu^* - \mu_c^* = P_{30}(\Delta\rho^*)^3/6 + \ldots \tag{4.2.18}$$

so that $\delta = 3$ and $D = P_{30}/6$. For the compressibility χ_T^* along the critical isochore ($\Delta\rho^* = 0$, $\Delta T^* \geqslant 0$) we find

$$\chi_T^{*-1} = (\partial\mu^*/\partial\rho^*)_T = P_{11}\Delta T^* + \ldots \tag{4.2.19}$$

so that $\gamma = 1$ and $\Gamma = P_{11}^{-1}$. The specific heat C_V^* along the critical isochore ($\Delta\rho^* = 0$, $T^* \geqslant 0$) is given by

$$C_V^*/T^* = -(\partial^2 A^*/\partial T^{*2})_\rho = -A_{02} + \ldots . \tag{4.2.20}$$

The specific heat C_V^* in the one-phase region remains finite and $\alpha = 0$.

As a next step we construct the two-phase region. For this purpose we consider the lowest-order terms in the expansions of the pressure and the chemical potential

$$\begin{aligned}
P^* &= P_c^* + P_{01}(\Delta T^*) + P_{11}(\Delta\rho^*)(\Delta T^*) + P_{30}(\Delta\rho^*)^3/6 + \ldots \\
\mu^* &= \mu_c^* + \mu_{01}(\Delta T^*) + \mu_{11}(\Delta\rho^*)(\Delta T^*) + \mu_{30}(\Delta\rho^*)^3/6 + \ldots .
\end{aligned} \tag{4.2.21}$$

The term linear in ΔT^* changes sign at T_c. Its negative slope below T_c causes the appearance of the 'Van der Waals loop.' The system then splits into a vapour and a liquid phase such that

$$P_{\text{vap}} = P_{\text{liq}} \qquad \mu_{\text{vap}} = \mu_{\text{liq}} \qquad T_{\text{vap}} = T_{\text{liq}} \tag{4.2.22}$$

while $\Delta\rho_{\text{vap}}^* \neq \Delta\rho_{\text{liq}}^*$. To lowest order we find for the coexisting phases

$$P_{11}(\Delta\rho_{\text{vap}}^*)(\Delta T^*) + P_{30}(\Delta\rho_{\text{vap}}^*)^3/6 = P_{11}(\Delta\rho_{\text{liq}}^*)(\Delta T^*) + P_{30}(\Delta\rho_{\text{liq}}^*)^3/6. \tag{4.2.23}$$

Since $\mu_{11} = P_{11}$ and $\mu_{30} = P_{30}$, the condition $\mu_{\text{vap}} = \mu_{\text{liq}}$ does not give any new information to this order. The condition (4.2.23) may be rearranged to read

$$\frac{(\Delta\rho^*_{\text{liq}} + \Delta\rho^*_{\text{vap}})^2 - (\Delta\rho^*_{\text{liq}})(\Delta\rho^*_{\text{vap}})}{\Delta T^*} = -\frac{6P_{11}}{P_{30}}. \tag{4.2.24}$$

Since the right-hand side of (4.2.24) is independent of temperature, we conclude that to lowest order $\Delta\rho^*_{\text{liq}}$ and $\Delta\rho^*_{\text{vap}}$ must vary as $|\Delta T^*|^{1/2}$. Thus the top of the coexistence curve is quadratic, that is $\beta = 1/2$.

Algebraic consistency leads us then to expect that the coexisting densities can be expanded in powers of $|\Delta T^*|^{1/2}$ so that (Van Laar 1912)

$$\Delta\rho^*_{\text{liq, vap}} = B_1^{\pm}|\Delta T^*|^{1/2} + B_2^{\pm}|\Delta T^*| + B_3^{\pm}|\Delta T^*|^{3/2} + \ldots \tag{4.2.25}$$

where the + sign refers to the liquid and the − sign to the vapour branch of the coexistence curve. When this equation is substituted into the expansions (4.2.21) for pressure and chemical potential, the coefficients of (4.2.25) are then determined by the condition that in each order the pressures and chemical potentials of the two branches are to be the same. We thus find (Levelt Sengers 1970)

$$B = B_1^+ = -B_1^- = (6\mu_{11}/\mu_{30})^{1/2} = (6P_{11}/P_{30})^{1/2} \tag{4.2.26a}$$

$$B_2 = B_2^+ = B_2^- = \frac{\mu_{21}}{\mu_{30}} - \frac{3\mu_{11}\mu_{40}}{5\mu_{30}{}^2} = \frac{P_{21}}{P_{30}} + \frac{4P_{11}}{5P_{30}} - \frac{3P_{11}P_{40}}{5P_{30}{}^2} \tag{4.2.26b}$$

and

$$\Delta\rho_{\text{liq, vap}} = \pm B_1|\Delta T^*|^{1/2} + B_2|\Delta T^*| + \ldots. \tag{4.2.27}$$

We conclude that the coexistence curve has a quadratic top which is symmetric. For the sum of the two coexisting densities we find

$$\frac{(\Delta\rho^*_{\text{liq}} + \Delta\rho^*_{\text{vap}})}{2} = B_2|\Delta T^*| + \ldots. \tag{4.2.28}$$

Asymptotically, this sum is linear in the temperature. For the classical equation this 'law of the rectilinear diameter' is a direct consequence of the assumed analyticity of the equation of state. It therefore not only applies in terms of the density, but it would also apply in terms of volume.

For the compressibility in the one-phase region along the coexistence curve we find

$$\chi_T^{*-1} = \mu_{11}\Delta T^* + ½\mu_{30}(\Delta\rho^*_{\text{liq, vap}})^2 + \ldots = 2P_{11}|\Delta T^*| + \ldots \tag{4.2.29}$$

so that $\gamma' = 1$ and $\Gamma' = (2P_{11})^{-1}$. The specific heat in the one-phase region along the coexistence curve reduces asymptotically again to (4.2.20), so that $\alpha' = 0$.

Finally, we consider the pressure $P^*_{\text{cxc}} = P^*_{\text{liq}} = P^*_{\text{vap}}$ and the chemical potential $\mu^*_{\text{cxc}} = \mu^*_{\text{liq}} = \mu^*_{\text{vap}}$ in the two-phase region obtained when the expansion (4.2.27) is substituted into (4.2.21)

$$P^*_{\text{cxc}} = P_{00} + P_{01}\Delta T^* + \frac{1}{2}\left[P_{02} + \frac{P_{11}^2}{P_{30}}\left(\frac{16}{5} - 2\frac{P_{21}}{P_{11}} + \frac{3P_{40}}{5P_{30}}\right)\right](\Delta T^*)^2 + \ldots$$

$$\mu^*_{\text{cxc}} = \mu_{00} + \mu_{01}\Delta T^* + \frac{1}{2}\left[\mu_{02} + \frac{\mu_{11}^2}{\mu_{30}}\left(-2\frac{\mu_{21}}{\mu_{11}} + \frac{3\mu_{40}}{5\mu_{30}}\right)\right](\Delta T^*)^2 + \ldots. \tag{4.2.30}$$

We note that the coefficient of the term proportional to $|\Delta T^*|^{3/2}$ vanishes. An algebraic proof that only integer powers of ΔT^* contribute to the vapour pressure $P^*_{\text{cxc}}(T^*)$ and the saturation chemical potential $\mu^*_{\text{cxc}}(T)$ was recently given by Mulholland (1973).

For the pressure and chemical potential along the critical isochore ($\Delta\rho^* = 0$) above the critical temperature we find

$$\begin{aligned} P^*(\Delta\rho^* = 0, \Delta T^* \geqslant 0) &= P_{00} + P_{01}\Delta T^* + \tfrac{1}{2}P_{02}(\Delta T^*)^2 + \ldots \\ \mu^*(\Delta\rho^* = 0, \Delta T^* \geqslant 0) &= \mu_{00} + \mu_{01}\Delta T^* + \tfrac{1}{2}\mu_{02}(\Delta T^*)^2 + \ldots . \end{aligned} \tag{4.2.31}$$

On comparing this with (4.2.30) we thus conclude

$$\frac{\mathrm{d}P^*_{\text{cxc}}}{\mathrm{d}T^*} = \left(\frac{\partial P^*}{\partial T^*}\right)_{\rho_c} \qquad \frac{\mathrm{d}\mu^*_{\text{cxc}}}{\mathrm{d}T^*} = \left(\frac{\partial \mu^*}{\partial T^*}\right)_{\rho_c} . \tag{4.2.32}$$

Table 4.2.2 Properties of the classical equation with 3-point contact

Taylor expansions (one-phase)	
$\mu^* = \sum_{m=0}^{\infty}\sum_{n=0}^{\infty} \frac{\mu_{mn}}{m!n!}(\Delta\rho^*)^m(\Delta T^*)^n \quad \mu_{10} = \mu_{20} = 0 \quad \mu_{30} > 0$	
$P^* = \sum_{m=0}^{\infty}\sum_{n=0}^{\infty} \frac{P_{mn}}{m!n!}(\Delta\rho^*)^m(\Delta T^*)^n \quad P_{10} = P_{20} = 0 \quad P_{30} > 0$	
$A^* = \sum_{m=0}^{\infty}\sum_{n=0}^{\infty} \frac{A_{m,n}}{m!n!}(\Delta\rho^*)^m(\Delta T^*)^n \quad A_{20} = A_{30} = 0 \quad A_{40} > 0$	
(For relations between coefficients, see (4.2.12))	
Power laws	
$\alpha = \alpha' = \alpha'' = 0$	
$\beta = \tfrac{1}{2}$	$B = (3!\mu_{11}/\mu_{30})^{1/2}$
$\gamma = \gamma' = 1$	$\Gamma = P_{11}^{-1} \quad \Gamma' = \tfrac{1}{2}P_{11}^{-1}$
$\delta = 3$	$D = P_{30}/3!$
$\theta_P = \theta_\mu = 0$	
Coexistence curve	
$\Delta\rho^*_{\text{cxc}} = \pm B\|\Delta T^*\|^{1/2} + B_2\|\Delta T^*\| \pm B_3\|\Delta T^*\|^{3/2} + \ldots$	$B_2 = \frac{\mu_{21}}{\mu_{30}} - \frac{3\mu_{11}\mu_{40}}{5\mu_{30}{}^2}$
Vapour pressure	
$P^*_{\text{cxc}} = 1 + P_{01}\Delta T^* + P_2\|\Delta T^*\|^2 + \ldots$	$P_2 = \tfrac{1}{2}P_{02} + \frac{8P_{11}^2}{5P_{30}} - \frac{P_{11}P_{21}}{P_{30}} + \frac{3P_{11}^2P_{40}}{10P_{30}^2}$
Saturation chemical potential	
$\mu^*_{\text{cxc}} = \mu_{00} + \mu_{01}\Delta T^* + \mu_2(\Delta T^*)^2 + \ldots$	$\mu_2 = \tfrac{1}{2}\mu_{02} - \frac{\mu_{11}\mu_{21}}{\mu_{30}} + \frac{3\mu_{11}^2\mu_{40}}{10\mu_{30}^2}$
Specific heat ($\rho = \rho_c'$, $T \leqslant T_c$)	
$C_V^*/T^* = C_0 + C_1\Delta T^* + \ldots$	$C_0 = -A_{02} + \frac{3\mu_{11}^2}{\mu_{30}}$

Table 4.2.3 Coefficients of power laws for van der Waals' equation $P^* = 8\rho^* T^*(3 - \rho^*) - 3\rho^{*2}$

Compressibility	$\Gamma = 1/6, \Gamma' = 1/12$
Critical isotherm	$D = 3/2$
Coexistence curve	$B = 2, B_2 = 2/5$
Vapour pressure	$P_{01} = 4, P_2 = 24/5$
Saturation chemical potential	$\mu_2 = -6/5$
Jump in C_V at $\rho = \rho_c$	$(C_V^*/T^*)_{II} - (C_V^*/T^*)_I = 12$

The vapour pressure below the critical temperature and the pressure along the critical isochore above the critical temperature have the same limiting slope, as has already been demonstrated by Van der Waals (1900). On the other hand, the second derivatives $(\partial^2 P^*/\partial T^{*2})_{\rho_c}$ and $(\partial^2 \mu^*/\partial T^{*2})_{\rho_c}$ change discontinuously at the critical point

$$\frac{d^2 P^*_{cxc}}{dT^{*2}} - \left(\frac{\partial^2 P^*}{\partial T^{*2}}\right)_{\rho_c} = \frac{P_{11}^2}{P_{30}}\left(\frac{16}{5} - 2\frac{P_{21}}{P_{11}} + \frac{3P_{40}}{5P_{30}}\right) \tag{4.2.33a}$$

$$\frac{d^2 \mu^*_{cxc}}{dT^{*2}} - \left(\frac{\partial^2 \mu^*}{\partial T^{*2}}\right)_{\rho_c} = \frac{\mu_{11}^2}{\mu_{30}}\left(-2\frac{\mu_{21}}{\mu_{11}} + \frac{3\mu_{40}}{5\mu_{30}}\right). \tag{4.2.33b}$$

Using a thermodynamic relation of Yang and Yang (1964)

$$\frac{C_V^*}{T^*} = \left(\frac{\partial^2 P^*}{\partial T^{*2}}\right)_\rho - \rho^*\left(\frac{\partial^2 \mu^*}{\partial T^{*2}}\right)_\rho \tag{4.2.34}$$

we find that the specific heat, on crossing the phase boundary from the one-phase region at the critical density, increases discontinuously by an amount

$$(C_V^*/T^*)_{II} - (C_V^*/T^*)_I = 3P_{11}^2/P_{30}. \tag{4.2.35}$$

The principal results for the classical equation with three-point contact are summarized in Table 4.2.2. In Table 4.2.3 we present values that the coefficients in the power law expansion for the various properties assume in the case of Van der Waals' equation.

4.2.3 A model of a critical-point phase transition: II. The classical equation with fifth-degree isotherm

At the turn of the century it became evident that equations of the Van der Waals' type did not predict correctly the asymptotic shape of the coexistence curve and the critical isotherm of real fluids. Verschaffelt (1900) found that the coexistence curve of isopentane is approximately cubic while the critical isotherm is of a degree slightly larger than 4. Therefore, Verschaffelt (1901, 1904, 1923) and Wohl (1914) attempted to formulate equations of state with $\beta = 1/3$ and $\delta = 4$. However, the non-analytic character of these equations was very troubling to engineers. A historical study of the early determinations of the values for the critical exponents was recently made by one of the present authors (Levelt Sengers 1976). Because of the reluctance to accept non-analytic equations of state, attempts were made, first by Van Laar (1912) and subsequently by Plank (1936), to formulate an analytic equation that has a flatter coexistence curve and critical isotherm than equations of the Van der Waals' type. This goal was achieved by setting not two but four derivatives of P with respect to V equal to zero at the critical point. The most extensive study of this case was made by Baehr (1963a,b) who revealed some of its surprising features.

The expansion of A^*, μ^* and P^* have the same form as (4.2.13) and the relations between the coefficients are again given by (4.2.14). However, the conditions for criticality are now

$$\begin{aligned} &A_{20} = A_{30} = A_{40} = A_{50} = 0 \\ &\mu_{10} = \mu_{20} = \mu_{30} = \mu_{40} = 0 \\ &P_{10} = P_{20} = P_{30} = P_{40} = 0 \end{aligned} \tag{4.2.36}$$

while $P_{50} = \mu_{50} = A_{60} > 0$ and $P_{11} = \mu_{11} > 0$. The equation for the critical isotherm is

$$P^* - P_c^* = \mu^* - \mu_c^* = \tfrac{1}{120} P_{50}(\Delta\rho^*)^5 + \ldots \tag{4.2.37}$$

so that $\delta = 5$ and $D = P_{50}/120$. The behaviour of the compressibility along the critical isochore is the same as in the case of three-point contact so that $\gamma = 1$ and $\Gamma = P_{11}{}^{-1}$. The coexistence curve to lowest order is now determined by the condition

$$\frac{P_{50}}{5!}(\Delta\rho_{\text{liq}}^*)^5 + P_{11}(\Delta\rho_{\text{liq}}^*)(\Delta T^*) = \frac{P_{50}}{5!}(\Delta\rho_{\text{vap}}^*)^5 + P_{11}(\Delta\rho_{\text{vap}}^*)(\Delta T^*). \tag{4.2.38}$$

Following the same reasoning as before, one finds

$$\Delta\rho_{\text{liq,vap}}^* = \pm B\,|\,\Delta T^*\,|^{1/4} + B_2\,|\,\Delta T^*\,|^{1/2} \pm B_3\,|\,\Delta T^*\,|^{3/4} + \ldots \tag{4.2.39}$$

with

$$B = \left(\frac{5!P_{11}}{P_{50}}\right)^{1/4} \tag{4.2.40a}$$

$$B_2 = \frac{B^2}{4}\left(\frac{\mu_{21}}{3\mu_{11}} - \frac{\mu_{60}}{7\mu_{50}}\right) = \frac{B^2}{4}\left(\frac{P_{21}}{3P_{11}} - \frac{P_{60}}{7P_{50}} + \frac{8}{21}\right). \tag{5.2.40b}$$

Thus $\beta = 1/4$, while the sum of the coexistence densities becomes

$$\tfrac{1}{2}(\Delta\rho_{\text{liq}}^* + \Delta\rho_{\text{vap}}^*) = B_2\,|\,\Delta T^*\,|^{1/2} + \ldots. \tag{4.2.41}$$

The law of the rectilinear diameter is no longer valid for this classical equation, neither in terms of density nor in terms of volume. For the compressibility along the coexistence curve one finds $\gamma' = 1$ and $\Gamma' = (4P_{11})^{-1}$. The temperature expansions for the vapour pressure and the saturation chemical potential become

$$\begin{aligned} P_{\text{cxc}}^* &= P_{00} + P_{01}\Delta T^* + B^2\left(\frac{P_{11}P_{60}}{42P_{50}} - \frac{P_{21}}{6} + \frac{8P_{11}}{21}\right)|\,\Delta T^*\,|^{3/2} + \ldots \\ \mu_{\text{cxc}}^* &= \mu_{00} + \mu_{01}\Delta T^* + B^2\left(\frac{\mu_{11}\mu_{60}}{42\mu_{50}} - \frac{\mu_{21}}{6}\right)|\,\Delta T^*\,|^{3/2} + \ldots. \end{aligned} \tag{4.2.42}$$

Thus the slopes of the pressure and the chemical potential along the critical isochore and the saturation curve are still continuous at the critical point. However, in contrast to the case of three-point contact, the second derivatives $\mathrm{d}^2P_{\text{cxc}}^*/\mathrm{d}T^{*2}$ and $\mathrm{d}^2\mu^*/\mathrm{d}T^{*2}$ and hence C_V^*/T^* in the two-phase region now diverge as $|\,\Delta T^*\,|^{-1/2}$, so that $\theta_P = \theta_\mu = \alpha'' = 1/2$. Of course $\alpha = \alpha' = 0$, since the Helmholtz free energy is analytic in the one-phase region. The properties of the classical equation with five-point contact are summarized in Table 4.2.4.

Table 4.2.4 Properties of the classical equation with 5-point contact

Taylor expansions (one-phase)

$$\mu^* = \sum_{m=0}^{\infty}\sum_{n=0}^{\infty}\frac{\mu_{mn}}{m!n!}(\Delta\rho^*)^m(\Delta T^*)^n \qquad \mu_{10}=\mu_{20}=\mu_{30}=\mu_{40}=0$$

$$P^* = \sum_{m=0}^{\infty}\sum_{n=0}^{\infty}\frac{P_{mn}}{m!n!}(\Delta\rho^*)^m(\Delta T^*)^n \qquad P_{10}=P_{20}=P_{30}=P_{40}=0$$

$$A^* = \sum_{m=0}^{\infty}\sum_{n=0}^{\infty}\frac{A_{m,n}}{m!n!}(\Delta\rho^*)^m(\Delta T^*)^n \qquad A_{20}=A_{30}=A_{40}=A_{50}=0$$

(For relations between coefficients see (4.2.14))

Power laws

$\alpha = \alpha' = 0 \quad \alpha'' = \frac{1}{2}$

$\beta = \frac{1}{4}$ $\qquad B = (5!\mu_{11}/\mu_{50})^{1/4}$

$\gamma = \gamma' = 1$ $\qquad \Gamma = P_{11}^{-1} \quad \Gamma' = \frac{1}{4}P_{11}^{-1}$

$\delta = 5$ $\qquad D = P_{50}/5!$

$\theta_P = \theta_\mu = \frac{1}{2}$

Coexistence curve

$$\Delta\rho^*_{\mathrm{cxc}} = \pm B|\Delta T^*|^{1/4} + B_2|\Delta T^*|^{1/2} \pm B_3|\Delta T^*|^{3/2} + \ldots \qquad B_2 = \frac{B^2}{4}\left(\frac{\mu_{21}}{3\mu_{11}} - \frac{\mu_{60}}{7\mu_{50}}\right)$$

Vapour pressure

$$P^*_{\mathrm{cxc}} = 1 + P_{01}\Delta T^* + P_{3/2}|\Delta T^*|^{3/2} + \ldots \qquad P_{3/2} = B^2\left(\frac{P_{11}P_{60}}{42P_{50}} - \frac{P_{21}}{6} + \frac{8P_{11}}{21}\right)$$

Saturation chemical potential

$$\mu^*_{\mathrm{cxc}} = \mu_{00} + \mu_{01}\Delta T^* + \mu_{3/2}|\Delta T^*|^{3/2} + \ldots \qquad \mu_{3/2} = B^2\left(\frac{\mu_{11}\mu_{60}}{42\mu_{50}} - \frac{\mu_{21}}{6}\right)$$

Specific heat ($\rho = \rho_c, T \leqslant T_c$)

$$C_V^*/T^* = \frac{A_{\mathrm{II}}^-}{\alpha''}|\Delta T^*|^{-1/2} \qquad A_{\mathrm{II}}^- = \mu_{11}B^2/8$$

4.2.4 Generalized homogeneous functions

The modern description of the thermodynamic behaviour of a system near a critical point is based on the assumption of homogeneity of the basic thermodynamic functions To introduce the concept of homogeneous functions we consider here functions of two variables only; generalization to functions of more variables is obvious.

A function $f(u, v)$ of two variables u and v is called a generalized homogeneous function if it satisfies the relation (Stanley 1971, 1972)

$$f(\lambda^{a_u}u, \lambda^{a_v}v) = \lambda f(u, v) \tag{4.2.43}$$

for two fixed exponents a_u and a_v and for all values of the parameter λ. It is noted that a relation of the type $f(\lambda^{a_u}u, \lambda^{a_v}v) = \lambda^P f(u, v)$ can always be reduced to the form (4.2.43) by a proper redefinition of the parameter λ.

When a function has the property of homogeneity one can always deduce a scaling law, i.e. the dependence on the two variables can be reduced to the dependence on one new variable by an appropriate change of scale. For this purpose we take $\lambda^{a_u} = u^{-1}$ so that

$$\frac{f(u, v)}{u^{1/a_u}} = f\left(1, \frac{v}{u^{a_v/a_u}}\right) \tag{4.2.44a}$$

where, for simplicity, we consider only positive values of the variables u and v. Hence, the function $f(u, v)$ after scaling with the factor u^{1/a_u} becomes a function of the single variable $v/u^{a_v/a_u}$. Another possible choice is $\lambda^{a_v} = v^{-1}$ so that

$$\frac{f(u, v)}{v^{1/a_v}} = f\left(\frac{u}{v^{a_u/a_v}}, 1\right). \tag{4.2.44b}$$

From (4.2.44) we note that a generalized homogeneous function satisfies a simple power law along any line $u/v^{a_u/a_v} = B$

$$f(u, v) = f(B, 1)v^{1/a_v} \tag{4.2.45}$$

where $f(B, 1)$ is a constant coefficient. In particular, along the special lines $u = 0$ and $v = 0$ the function behaves as

$$f(0, v) = f(0, 1)v^{1/a_v} \qquad f(u, 0) = f(1, 0)u^{1/a_u}. \tag{4.2.46}$$

4.2.5 *Homogeneity and scaling property of classical equations*

In this section we consider to what extent in the vicinity of the critical point the classical equation of state becomes a generalized homogeneous function of the independent thermodynamic variables and satisfies a scaling law.

As mentioned in Section 4.2.1, thermodynamics imposes a number of inequality relationships between the critical exponents. Of the five exponent relations given in (4.2.4) we find that three are obeyed with the equal sign for both classical equations, namely $2 - \alpha'' = \beta(\delta + 1)$, $\beta(\delta - 1) = \gamma'$ and $\alpha'' + 2\beta + \gamma' = 2$. The other two relations, $\theta_P \leqslant \alpha'' + \beta$ and $\theta_\mu \leqslant \alpha'' + \beta$, are obeyed as inequalities, since for the case of three-point contact $\theta_P = \theta_\mu = \alpha'' = 0$ and for the case of five-point contact $\theta_P = \theta_\mu = \alpha'' = 1/2$. We further note that in the case of three-point contact $\gamma = \gamma' = 1$, $\alpha = \alpha' = \alpha'' = 0$, while in the case of five-point contact $\gamma = \gamma' = 1$, $\alpha = \alpha' = 0$, $\alpha'' = 1/2$.

To formulate the scaling property of the classical equation we develop our arguments again in terms of the equation of state for $\mu(\rho, T)$. We do stress, however, that for the classical equations the same arguments would hold for any choice of variables, including $P(\tilde{V}, T)$ and $P(\rho, T)$. The Taylor expansions of the classical equations read to lowest order

$$\mu^* - \mu_c^* = \mu_{01}\Delta T^* + \frac{1}{3!}\mu_{30}(\Delta\rho^*)^3 + \mu_{11}(\Delta\rho^*)(\Delta T^*)$$

(three-point contact)

$$\mu^* - \mu_c^* = \mu_{01}\Delta T^* + \frac{1}{5!}\mu_{50}(\Delta\rho^*)^5 + \mu_{11}(\Delta\rho^*)(\Delta T^*). \tag{4.2.47}$$

(five-point contact)

It is convenient to define

$$\Delta\mu^* = \mu^* - \mu_c^* - \mu_{01}\Delta T^*. \tag{4.2.48}$$

Introducing the critical exponents $\delta = 3$, $\beta = 1/2$ (three-point contact) and $\delta = 5$, $\beta = 1/4$ (five-point contact), both equations (4.2.47) can be rearranged as

$$\Delta\mu^* = D(\Delta\rho^*)^\delta \left(1 + \frac{1}{x_0} \frac{\Delta T^*}{(\Delta\rho^*)^{1/\beta}}\right) \tag{4.2.49}$$

where D and $x_0 = B^{-1/\beta}$ are readily expressible in terms of μ_{mn}. Hence to lowest order $\Delta\mu^*$ does indeed satisfy the homogeneity property (4.2.43) $\Delta\mu^*(\lambda^{a_\rho}\Delta\rho^*, \lambda^{a_T}\Delta T^*) = \lambda\Delta\mu^*(\Delta\rho^*, \Delta T^*)$ with $a_\rho = 1/\delta$ and $a_T = 1/\beta\delta$. If $\Delta\mu^*$ is 'scaled' by $(\Delta\rho^*)^\delta$

$$\frac{\Delta\mu^*}{(\Delta\rho^*)^\delta} = D(1 + x/x_0) \tag{4.2.50}$$

the result is a function of only one independent scaling variable $x = \Delta T^*/(\Delta\rho^*)^{1/\beta}$.

The homogeneity and scaling properties of the nonanalytic equations of state will be discussed extensively in Section 4.3. Here it is appropriate to point out the limitations of scaling for the classical equations. The quantity $\Delta\mu^*$ defined in (4.2.48) is equal to $\mu^*(\rho, T) - \mu^*(\rho_c, T)$ above T_c to linear order in ΔT^*. Since the chemical potential $\mu^*(\rho_c, T)$ along the critical isochore and the saturation chemical potential μ^*_{cxc} have continuous slopes, $\Delta\mu^*$ is also equal to $\mu^*(\rho, T) - \mu^*_{cxc}$ below T_c to lowest order in ΔT^*. Thus the scaled equation (4.2.50) is valid in the one-phase region both above and below T_c to linear order in ΔT^*. However, the argument cannot be extended to include higher order terms in ΔT^*. If, for instance, in the case of three-point contact the term $\frac{1}{2}\mu_{02}(\Delta T^*)^2$ is added to (4.2.47) and then included in the definition of $\Delta\mu^*$, the quantity $\Delta\mu^* \neq \mu^*(\rho, T) - \mu^*_{cxc}$ below T_c, since the second derivative of $\mu^*(\rho_c, T)$ is discontinuous at the critical point. Alternatively, if all terms contributing on the $(\Delta T^*)^2$ level to μ^*_{cxc} are included in (4.2.47), the feature of homogeneity is lost. Thus, although the classical equations of state obey a scaling law to lowest order, the second derivative, and consequently the specific heat, in general does not scale even in lowest order.

4.2.6 A model of a critical-point phase transition: III. Ising model and lattice gas

Lenz and Ising formulated a model for the ferromagnetic phase transition in solids in which the assumption of long-range forces implicit in the classical or mean-field theories was dropped (Ising 1925). In this Ising model, magnetic spins are placed on a regular array, one spin at each site, which may either point 'upwards' or 'downwards'. In the simplest form of the model, the interactions between spins are limited to nearest neighbours only. These interactions may be ferromagnetic, when the lowest energy state is that of parallel spins on neighbouring sites, or antiferromagnetic, when it is energetically favourable for neighbouring spins to line up in an antiparallel configuration. Lenz and Ising were only able to solve the one-dimensional version of the model. However, in one dimension the system remains paramagnetic at all finite temperatures and does not exhibit a phase transition to a ferromagnetic state. The reason is that only one pair of opposite spins is enough to destroy the macroscopic magnetic moment.

In the 2- and 3-dimensional versions of the model a phase transition does occur (Peierls 1936, Griffiths 1972). Moreover, the 2-dimensional Ising model in zero magnetic field can be solved exactly as first achieved by Onsager (1944), who showed that the specific heat at constant field $H = 0$ diverges logarithmically at the critical point. This result was in direct contradiction with the assumption of analyticity of the Helmholtz free energy at the critical point in the classical theory.

The importance of the Ising model is enhanced by the fact that it can serve as a model for a variety of phase transitions (Green and Hurst 1964). We have already seen that by changing the sign of the nearest-neighbour interaction both the ferromagnetic and the antiferromagnetic phase transitions can be described. If spins pointing upwards are replaced by particles A, while spins pointing downwards are replaced by particles B, we obtain a model for phase transitions in binary solids. If the energetically favourable configuration is that of unlike nearest neighbours, the order–disorder phase transition in binary alloys can be modelled. If the configuration with like nearest neighbours is more favourable, demixing in the solid phase can be described. To the extent that a liquid can be approximated by a lattice model, we obtain also a description of phase separation in partially miscible liquids. Finally, the Ising model can be used as a model for the gas–liquid transition. An early attempt in this direction was made by Cernuschi and Eyring (1939) who devised a so-called hole theory of liquids. The rigorous translation of the Ising model into a model for a gas on a lattice was formulated by Lee and Yang (1952).

In the ferromagnetic Ising model, a spin variable σ_i is assigned to each lattice site i, where σ_i can assume the value +1 or −1 depending on whether the spin points 'up' or 'down'. The energy E of a microstate $\{\sigma_i\}$ is given by

$$E\{\sigma_i\} = -J \sum_{\langle i,j \rangle} \sigma_i \sigma_j - H \sum_i \sigma_i \tag{4.2.51}$$

where J is an interaction constant and H the magnetic field and where the sum is to be taken over all pairs of nearest neighbours $\langle ij \rangle$. The partition function z_s as a function of T, H and the total number of spins or sites N_s is then given by

$$Z_s(T, H, N_s) = \sum_{\{\sigma_i\}} \exp\left(\frac{J}{k_B T} \sum_{\langle ij \rangle} \sigma_i \sigma_j + \frac{H}{k_B T} \sum_i \sigma_i\right) \tag{4.2.52}$$

where k_B is Boltzmann's constant. The (Gibbs) free energy $F(T, H)$ is then obtained as

$$F = -k_B T \ln Z_s. \tag{4.2.53}$$

To obtain a lattice gas each site is identified with the centre of a cell with volume v_0 such that

$$V = v_0 N_s. \tag{4.2.54}$$

To each cell an occupancy variable τ_i is assigned such that $\tau_i = +1$ when the cell is occupied by a molecule and $\tau_i = 0$ when the cell is empty. Multiple occupancy is forbidden so as to account for the finite size of the molecules. To each pair of molecules in adjacent cells an attractive energy $-\epsilon$ is assigned independent of the positions of the molecules within the cells. The energy of a microstate is then

$$E\{\tau_i\} = -\epsilon \sum_{\langle ij \rangle} \tau_i \tau_j \tag{4.2.55}$$

and the partition function $Z(T, V, N)$ of the lattice gas becomes

$$Z(T, V, N) = \left(\frac{v_0}{\Lambda^3}\right)^N \sum_{\{\tau_i\}}{}' \exp\left(\frac{\epsilon}{k_B T} \sum_{\langle ij \rangle} \tau_i \tau_j\right) \tag{4.2.56}$$

with $\Lambda = (h^2/2\pi m k_B T)^{1/2}$ where h is Planck's constant and m the molecular mass. The accent on the summation sign indicates that the summation is restricted to microstates with $\Sigma \tau_i = N$. This awkward constraint is removed by considering the grand partition function

$$\Xi(T, V, \mu) = \sum_{N=0}^{\infty} \exp(N\mu/k_B T) = \sum_{\{\tau_i\}} \exp\left(\frac{\epsilon}{k_B T} \sum_{\langle ij \rangle} \tau_i \tau_j + \frac{\mu - k_B T \ln(\Lambda^3/v_0)}{k_B T} \sum_i \tau_i\right). \tag{4.2.57}$$

The relation between the grand partition function Ξ of the lattice gas and the partition function Z_s of the Ising model is obtained by noting that $\tau_i = \frac{1}{2}(\sigma_i + 1)$ so that $\Sigma_i\tau_i = \frac{1}{2}(\Sigma_i\sigma_i + N_s)$ and $\Sigma_{\langle ij\rangle}\tau_i\tau_j = \frac{1}{4}\Sigma_{\langle ij\rangle}\sigma_i\sigma_j + q\Sigma_i\sigma_i + \frac{1}{2}qN_s$, where q is the coordination number of the lattice. Hence, the grand partition function of the lattice gas can be written in the form

$$\Xi(T, V, \mu) = \exp[(H - qJ/2)N_s/k_BT]\, Z_s(T, H, N_s) \tag{4.2.58}$$

with

$$\tfrac{1}{4}\epsilon = J \tag{4.2.59}$$

$$\Delta\mu \equiv \mu + \frac{\epsilon q}{2} - k_BT \ln \frac{\Lambda^3}{v_0} = 2H. \tag{4.2.60}$$

Since $PV = Pv_0N_s = k_BT \ln \Xi$, it also follows that

$$v_0P = (H - \tfrac{1}{2}qJ) - F/N_s. \tag{4.2.61}$$

Furthermore, since $\rho = \langle \Sigma_i\tau_i \rangle / V$ and $M/M_0 = \langle \Sigma_i\sigma_i \rangle / N_s$, we find $2v_0\rho = 1 + M/M_0$, where M_0 is the saturation magnetization of the ferromagnetic Ising model. At the critical point of the Ising model $M = 0$ so that for the lattice gas $\rho_c = 1/2v_0$ and

$$\Delta\rho^* = M/M_0 = M^*. \tag{4.2.62}$$

On the critical isochore $M = 0$, the field $H = 0$ above and below T_c. Thus it follows from (4.2.60) that the chemical potential on the critical isochore of the lattice gas is given by

$$\mu(\rho_c, T) = k_BT \ln \frac{\Lambda^3}{v_0} - \frac{\epsilon q}{2}, \tag{4.2.63}$$

and the quantity $\Delta\mu$ introduced in (4.2.60) may be identified with $\mu(\rho, T) - \mu(\rho_c, T)$. The properties of the lattice gas follow immediately from those of the Ising model via the relations given above (Fisher 1966, 1967a).

The results obtained for the Ising model have been discussed in many places in the literature. Appropriate references may be found in the review papers of Fisher (1967a) and Domb (1974). The values of the thermodynamic critical exponents for the 2-dimensional Ising model are $\alpha = \alpha'' = 0$, $\beta = 1/8$, $\gamma = \gamma' = 7/4$, $\delta = 15$; of these exponents α β and γ are known exactly (Fisher 1967a), while δ is known numerically to within 0.5% (Gaunt and Sykes 1972).

The 3-dimensional Ising model has not been solved exactly, but many numerical results have been obtained from analyses of series expansions (Domb 1974). The current exponent values deduced from series expansions are (Gaunt and Sykes 1972, 1973, Gaunt and Guttman 1974, Meijer and Ferrell 1975, Camp *et al.* 1976):

$$\alpha = 0.125 \pm 0.020, \quad \beta = 0.312 \pm 0.005,$$
$$\gamma = 1.250 \pm 0.003, \quad \delta = 5.00 \pm 0.05, \tag{4.2.64}$$

independent of the lattice structure and the magnitude of the spin. It is also believed that the values of the critical exponents do not change if the interactions extend beyond nearest neighbours as long as their range remains finite; it must be remarked, however, that the numerical evidence is not conclusive (Dalton and Wood 1969; Farrell and Meijer 1972). Anticipating that the critical exponents may turn out to be rational numbers, one frequently adopts as estimated values for the critical exponents of the 3-dimensional Ising model

$$\alpha = \alpha'' = \frac{1}{8}, \quad \beta = \frac{5}{16}, \quad \gamma = \gamma' = \frac{5}{4}, \quad \delta = 5. \tag{4.2.65}$$

The theoretical and numerical evidence gathered for the Ising model (as well as other lattice models) appears to be in accord with the postulate that the free energy is a generalized homogeneous function of ΔT^* and $H^* = H/k_B T$. From this postulate it follows that the equation of state can be written in the form (Griffiths 1967, Fisher 1967b):

$$H^* = M^* \mid M^* \mid^{\delta - 1} h_s(\Delta T^*/\mid M^* \mid^{1/\beta}). \tag{4.2.66}$$

An equation of this form for the 3-dimensional Ising model was originally proposed by Domb and Hunter (1965).

The equation of state for the lattice gas follows from (4.2.66) using (4.2.60) and (4.2.62). The properties of this equation of state will be discussed in Section 4.3.4. We also note from (4.2.63) that the chemical potential of the lattice gas is a regular function of temperature along the critical isochore, in contrast to its behaviour in the classical theory. From the Yang–Yang relation (4.2.34) it then follows that the anomaly in the specific heat C_V is equal to the anomaly in d^2P/dT^2 so that for the lattice gas $\alpha = \alpha'' = \theta_p$, while $\theta_\mu = 0$.

The regular behaviour of $\mu(\rho_c, T)$ is a consequence of the particle–hole symmetry of the lattice gas. Since the Ising hamiltonian is invariant under reversal of the field H, while the magnetization M changes sign upon reversal of the field, we note that $\Delta\mu$ along an isotherm is antisymmetric with respect to the critical isochore

$$\mu\Delta^*(-\Delta\rho^*, \Delta T^*) = -\Delta\mu^*(\Delta\rho^*, \Delta T^*). \tag{4.2.67}$$

As a consequence the derivative $\chi_T = \rho^2 K_T = (\partial\rho/\partial\mu)_T$ of the lattice gas is a symmetric function of $\Delta\rho^*$. The pressure P, however, given by (4.2.61), contains a symmetric and an antisymmetric term and therefore does not exhibit any special symmetry. In constrast to the classical equations, the equation of state for the chemical potential of the lattice gas has a symmetry which is absent in the equation of state for the pressure.

4.2.7 Renormalization group theory

On approaching a critical point a system exhibits large fluctuations in the order parameter (the magnetization near the Curie point of a spin system or the density near the gas–liquid critical point of a fluid). The range of these fluctuations is characterized by a correlation length ξ, a precise definition of which will be given in Section 4.5.1. For the systems under consideration this correlation length becomes much larger than the range of the intermolecular forces. Kadanoff (1966) gave a plausibility argument that the long-range nature of the critical fluctuations causes the singular part of the free energy to become a generalized homogeneous function of its variables. He also introduced the idea of universality: the critical singularities do not depend on those parameters in the hamiltonian that characterize the microscopic nature of the system on a length scale of the order of the intermolecular distances, but depend only on some gross features of the system such as the dimensionality of the system and the number of components of the order parameter (Kadanoff 1966, 1971, 1976). These ideas have been given a firm theoretical basis by the renormalization group theory of critical phenomena, formulated by Wilson (1971) and further developed by many investigators.

In order to elucidate the method of the renormalization group theory, let us consider a generalization of the Ising model called Ising-like spin systems (Niemeijer and Van Leeuwen 1976). An Ising-like spin system is a lattice system in which the interactions are not restricted to nearest neighbours only and in which the spin variable σ_i may assume an arbitrary number of values. In accordance with (4.2.53) the reduced free energy per spin $F^* = -F/N_s k_B T$ may be written in the form

$$F^* = \ln \sum_{\{\sigma_i\}} \exp \mathscr{H}, \tag{4.2.68}$$

where $\mathscr{H}$ is a generalized hamiltonian which depends on a number of parameters such as the temperature T', the field H, the number of states that the spin variable σ_i may assume, the interaction constants between nearest neighbours, next nearest neighbours, etc. These parameters may be formally indicated by the set of variables $\{K\} = K_1, K_2, \ldots$.

For a spin system with N_s lattice sites one may attempt to calculate the free energy from (4.2.68) by first summing only over groups that are within cells of length l, where l is measured in terms of the lattice constant. The result may then be interpreted in terms of the properties of a new spin system for which the lattice sites correspond to the cells of the original lattice. The number N_s' of this new spin system is related to the number N_s of the original lattice by $N_s'/N_s = l^{-d}$, where d is the dimensionality of the system. The reduced free energy $F^{*\prime}$ per spin of the new spin system may be written as

$$F^{*\prime} = F\{K'\} = \ln \sum_{\{\sigma_i'\}} \exp \mathscr{H}\{K'\}, \tag{4.2.69}$$

where the summation is to be conducted over all possible values of the spin variables σ_i' of the new spin system and where the hamiltonian $\mathscr{H}\{K'\}$ will be determined by a new set of values K_α' of the parameters of the hamiltonian. The transformation

$$\{K'\} = \mathscr{R}_l\{K\} \tag{4.2.70}$$

is called a renormalization group transformation. It satisfies the semigroup property $\mathscr{R}_{l_1}\ \mathscr{R}_{l_2} = \mathscr{R}_{l_1+l_2}$. Apart from an integration constant which is an analytic function of the parameters K_α', the free energy $F^*\{K'\}$ of the new spin system is related to the free energy $F^*\{K\}$ of the original system by

$$F^*\{K\} = l^{-d} F^*\{K'\}. \tag{4.2.71}$$

Upon iteration of the renormalization transformation one traverses a trajectory in the space of the parameters of the hamiltonian. This means that one is studying the system with respect to a length scale which becomes larger and larger. After having summed over the short range contributions the properties of the spin systems thus generated will vary little until one reaches a length scale comparable with the correlation length ξ. At the critical point, however, the correlation length is infinite. Hence, the crucial observation of the renormalization group theory is that for a system at a critical point the procedure always leads to a fixed point $\{K^*\}$ of the transformation at which

$$\{K^*\} = \mathscr{R}_l\{K^*\}. \tag{4.2.72}$$

When the system is not at a critical point but close to it, the correlation length ξ is large but not infinite. Upon iteration of the renormalization transformation the trajectory in the space of the parameters of the hamiltonian will approach the fixed point as long as the length scale remains smaller than ξ and will move away from the fixed point when the length scale becomes larger than ξ. In the vicinity of the fixed point the transformation may be presented by a linear approximation

$$K_\alpha' - K_\alpha^* = \sum_\beta \left(\frac{\partial K_\alpha'}{\partial K_\beta}\right)_{K^*} (K_\beta - K_\beta^*). \tag{4.2.73}$$

Let us denote the eigenvalues of the matrix $(\partial K_\alpha'/\partial K_\beta)_{K^*}$ (which are assumed to be real and positive) by l^{y_α} and the corresponding eigenfunctions, usually referred to as scaling fields, by u_α (Wegner 1972). Replacing the parameters K_α by the scaling fields $u_{\alpha'}$, the transformation (4.2.73) in the vicinity of the fixed point reads

$$u_{\alpha'} = l^y u_\alpha. \tag{4.2.74}$$

Hence, if we consider the free energy F as a function of the scaling fields $u_{\alpha'}$, it follows from (4.2.71) and (4.2.72) that in the vicinity of the critical point the free energy satisfies the relation

$$F^*(u_1, u_2, \ldots) = l^{-d} F^*(\lambda^{y_1} u_1, \lambda^{y_2} u_2, \ldots). \tag{4.2.75}$$

Thus the free energy becomes a generalized homogeneous function of the type defined in (4.2.43). The exponents y_α can be calculated by determining the eigenvalues of the linearized renormalization group transformation defined in (4.2.73). The relation (4.2.75) applies to that part of the free energy which we shall identify as the singular part F^*_{sing}. In addition there is a regular contribution due to the summations over the short-range interactions.

The renormalization procedure is not restricted to lattice systems. In fact, most calculations based on the renormalization group theory have been performed for the so-called Landau–Ginzburg–Wilson model (Wilson 1971, Wilson and Kogut 1974). This model is a generalization of the Ising-like systems in which the spin variable σ_i is no longer associated with discrete lattice sites but is replaced by a spin function $\sigma(\vec{x})$ which is a continuous function of the position $\vec{x}$. The renormalization transformation is then obtained by integrating over those Fourier components of the spin function $\sigma(\vec{x})$ that correspond to wavelengths smaller than the correlation length ξ.

The scaling fields with $l^{y_\alpha} > 1$ (i.e. $y_\alpha > 0$) are called relevant and the scaling fields with $l^{y_\alpha} < 1$ (i.e. $y_\alpha < 0$) are called irrelevant. In some special cases one may also encounter a marginal field with $l^{y_\alpha} = 1$. Upon iteration of the transformation (4.2.75) we conclude that the singular part of the free energy in the vicinity of the critical point will assume the same form independent of the starting values of the irrelevant parameters. For Ising-like systems of a given dimensionality $d < 4$, there are two relevant scaling fields which asymptotically may be identified with the temperature ΔT^* and the field H^*. Thus near the critical point the free energy satisfies the relation

$$F^*_{\text{sing}}(\Delta T^*, H^*) = l^{-d} F^*_{\text{sing}}(l^{y_1} \Delta T^*, l^{y_2} H^*). \tag{4.2.76}$$

Choosing l such that $l^{y_2} = H^{*-1}$, we conclude that the singular part of the free energy satisfies a scaling law of the form

$$F^*_{\text{sing}}(\Delta T^*, H^*) = H^{*d/y_2} F^*_{\text{sing}} \left(\frac{\Delta T^*}{H^{*y_1/y_2}}, 1 \right). \tag{4.2.77}$$

The renormalization group theory confirms that the singular behaviour of the free energy, and thus of all thermodynamic properties, depends on only two critical exponents y_1 and y_2. Since $M^* = (\partial F^*/\partial H^*)_T$, one readily verifies that a scaled equation of state of the form postulated in (4.2.66) is recovered if the exponent β is identified with $(d - y_2)/y_1$ and the exponent δ with $y_2/(d - y_2)$. The renormalization group theory shows how these critical exponents can be calculated from the eigenvalues of the linearized renormalization group transformation. The scaling function of the free energy and other related properties can be obtained by studying their dependence on the parameters of the hamiltonian in the vicinity of the fixed point. For further details the reader is referred to the literature (Wilson 1971, Wilson and Kogut 1974, Ma 1973, 1976, Di Castro *et al.* 1974, Schroer 1974, Fisher, 1974, Van Leeuwen 1975, Niemeijer and Van Leeuwen 1976, Wallace 1976).

An important consequence of the renormalization group theory is that all systems whose hamiltonians differ with respect to the irrelevant parameters only, will have the same critical exponents and the same scaling functions. Such systems are said to belong to the same universality class. It is widely assumed that universality classes for homogeneous, isotropic systems with short-range forces may be assigned according to the spatial dimensionality and the number of components of the order parameter. Although the renormalization procedure has not yet been carried out for systems with hamiltonians resembling fluids, it is nevertheless expected that fluids near the gas–liquid critical point and binary liquids near the critical mixing point should belong to the same universality class as the Ising model and the Landau–Ginzburg–Wilson model (Hubbard and Schofield 1972). This hypothesis will be further discussed in Sections 4.3.5 and 4.3.8.

Using the method of the renormalization group theory estimates have been obtained for the critical exponents of this universality class (Kadanoff *et al.* 1976, Baker *et al.* 1976, Golner and Riedel 1975, 1976). The most accurate estimates appear those reported by Baker *et al.* for the Landau–Ginzburg–Wilson model. In particular they found for the critical exponents β and γ

$$\beta = 0.320 \pm 0.015, \gamma = 1.241 \pm 0.002. \tag{4.2.78}$$

There exists a small unresolved discrepancy between the exponent values obtained for the Landau–Ginzburg–Wilson model on the basis of the renormalization group theory and the values (4.2.64) obtained for the Ising model from series expansions. We shall return to this point in Section 4.5.5.

4.2.8 *Gravity effects*

Experimentation near a critical point is difficult because of the strong divergence in the thermal expansion coefficient and the compressibility and the slow rate of decay towards equilibrium. For a discussion of the type of thermodynamic information available and the experimental difficulties encountered the reader is referred to a review published elsewhere (Levelt Sengers 1975). Here we restrict the discussion to one major feature encountered in all critical region experimentation with fluids, namely the effect of the earth's gravitational field.

As noted by Gouy (1892), near the gas–liquid critical point the compressibility becomes so large that gravity will induce an appreciable density gradient. We may assume that at each level h in the cell the local chemical potential $\mu(\rho_h, T)$ equals the chemical potential of a system with uniform density ρ_h at temperature T in the absence of gravity. Since in the presence of gravity the total chemical potential is the sum of the local chemical potential $\mu(\rho_h, T)$ and the gravitational potential mgh, we have

$$\mu(\rho_h, T) - \mu(\rho_{h_0}, T) = -mg(h - h_0) \tag{4.2.79}$$

where m is the molecular mass if the chemical potential is taken per particle. In terms of dimensionless quantities

$$\mu^*(\rho_h, T) - \mu^*(\rho_{h_0}, T) = -\frac{h - h_0}{h_c} \tag{4.2.80}$$

with

$$h_c = P_c/mg\rho_c. \tag{4.2.81}$$

In Table 4.2.5 we present values of the parameter h_c^{-1} for a number of fluids. Note that the gravity effect on the chemical potential will be proportional to h_c^{-1}. Thus at a given $\Delta\rho^*$ and

Table 4.2.5 The parameter $h_c^{-1} = m\rho_c g/P_c$ for a number of fluids

Fluid	$h_c^{-1} \times 10^3$ (m^{-1})	Fluid	$h_c^{-1} \times 10^3$ (m^{-1})
^3He	3.48	CO	0.84
^4He	3.00	CO_2	0.62
Ne	1.72	NH_3	0.20
Ar	1.08	N_2O	0.61
Kr	1.62	SF_6	1.90
Xe	1.86	CH_4	0.35
H_2	0.24	C_2H_4	0.42
N_2	0.91	C_2H_6	0.42
O_2	0.85	C_3H_8	0.51
H_2O	0.14	$CClF_3$	1.50
D_2O	0.16		

ΔT^* the gravity effects in helium will be much larger than in steam. Since h_c is the order of 10^5 cm, gravity will contribute on the 10^{-5} level to the chemical potential in cells with a height of about 1 cm. Since $(\partial\rho^*/\partial\mu^*)_T$ diverges, this small variation in μ^* will cause an appreciable density gradient. As an example we show in Figure 4.2.2 the predicted size of such gravitationally induced density profiles in xenon at a number of temperatures. In binary liquids near the critical mixing point gravity will induce a concentration gradient (Yvon 1937, Voronel and Giterman 1965, Mistura 1971).

Gravity effects impose serious limitations upon the information that can be obtained from conventional *PVT* experiments in the vicinity of the critical point. The experiments become unreliable when the average or bulk density as measured begins to deviate from the local density prevailing at the level where the pressure is measured. The magnitude of the range where the data are obscured by these effects depends on a variety of factors such as the height

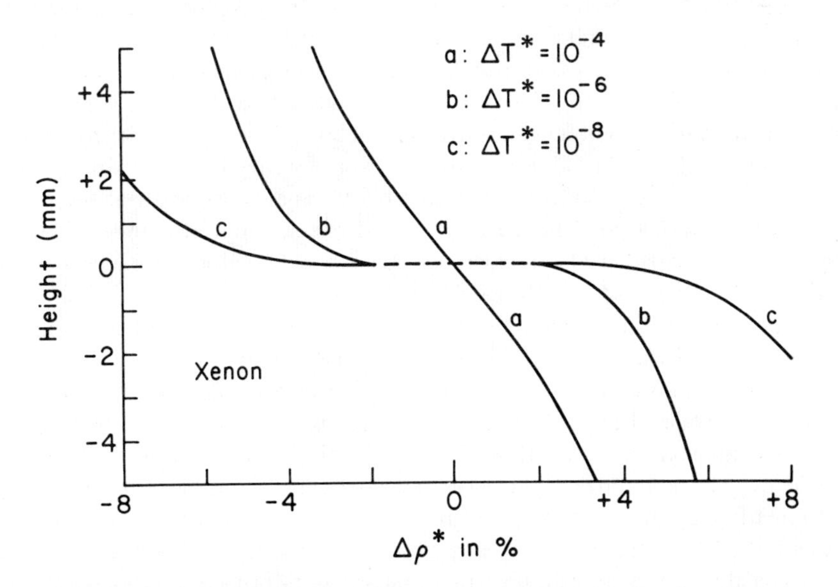

Figure 4.2.2 Calculated density profiles near the critical point of xenon in the earth's gravitational field. At the broken part of the curves the density varies so rapidly that the assumption of local thermodynamic equilibrium is no longer valid

of the vessel, the value of h_c and the precision of the temperature control. Generally no accurate information is obtained from conventional *PVT* experiments in the temperature range $|\Delta T^*| < 5 \times 10^{-4}$.

However, gravity effects can be used to advantage if the density is measured as a function of height. In accordance with (4.2.80) such measurements yield ρ as a function of μ over a total pressure span of a few hundred Pascals, i.e. a few millibars. Thus a resolution of the equation of state can be obtained that is several orders beyond what is attainable with conventional techniques.

Basically three different methods have been used to measure the local density as a function of height: floats, capacitance and refractive index as a function of height. As early as the beginning of the century Teichner (1904) used floats to indicate the existence of a density profile in CCl_4 near the critical point.

Since the 1950s optical techniques have become increasingly popular. There exist principally two complementary optical methods, namely that in which the refractive index is measured as a function of height and that in which the refractive index gradient is measured as a function of height. The Lorentz–Lorenz relation is used to relate refractive index to density. The result of the first method is a density profile and that of the second method a compressibility profile. Capacitance measurements provide a fast and accurate method of determining fluid densities. Here the Clausius–Mosotti relationship is used to convert dielectric constants to densities. In a density profile experiment, a stack of capacitors is used to indicate the density at various levels.

For density of compressibility profile experiments to become quantitative, great care has to be taken to eliminate thermal gradients. Due to the strong divergence of the thermal expansion coefficients, even small temperature gradients yield appreciable contributions to the density profile. In view of the more awkward arrangements due to the need for windows, optical experiments have been more prone to thermal gradient errors than capacitance experiments. Moreover, as emphasized by Verschaffelt (1905), uncontrolled impurities may affect the observed density profile appreciably. Thus the earlier work (Lorentzen 1953, Palmer 1954, Schmidt and Traube 1962, Schmidt 1966), though pioneering, yielded conflicting results and only recently have profile studies become more quantitative. Straub (1967) improved the method of Schmidt and Traube for measuring the refractive index profile. A number of Russian investigators have used both the float method for measuring the density profile and the Schlieren method for measuring the refractive index gradient (Naumenko *et al.* 1967, Artyukhovskaya *et al.* 1971). Density profiles by measuring the dielectric constant as a function of height have been obtained by Weber (1970) and Thoen and Garland (1974). The existence of concentration gradients near the critical mixing point of a binary liquid has also been confirmed experimentally (Lorentzen and Hansen 1966, Blagoi *et al.* 1970, Greer *et al* 1975, Giglio and Vendramini 1975, Maisano *et al.* 1976).

A very important experimental development was initiated by Wilcox and co-workers (Wilcox and Balzarini 1968, Estler *et al.* 1975) who perfected an optical interferometric method originally proposed by Gouy (1880) and who were able to reach a temperature stability to 20 microdegrees. A parallel beam of coherent light impinging on the cell will be deflected downward. Maximum deflection will occur at the level of maximum $\mathrm{d}n/\mathrm{d}z$, where n is the refractive index. Above and below this level, rays that experience the same deflection can be brought to interference after passage through a lens. The resulting interference pattern can be studied as a function of temperature; the location and order of the various maxima, counting from the maximally deflected beam upwards, contains all information necessary to obtain the equation of state. When combined with superior temperature control, this method has great potential. In practice, there are limitations to the number of decades in ΔT^* or $\Delta\rho^*$ where the method can be applied. For the method to work away from the critical point it is necessary that the level of maximum dn/dz stay somewhere near the centre of the cell. Since the locus of maximum dn/dz as a function of temperature probably does not coincide with a curve of precisely constant density, even the most careful filling of the cell cannot prevent the locus from moving out of the cell, limiting ΔT^* to values smaller than 10^{-4}. When the critical point is approached another cut-off is caused by gravity. When a band of strong refractive index gradients develops in the cell, the deflections of the rays become so large that they pass through fluid layers of widely varying density. This effect is proportional to the square of the thickness of the cell and provides a lower limit to all optical experiments. In practice, with cells a few mm thick, a range of $\Delta T^* < 10^{-5}$ is excluded, leaving a range of about one decade in temperature suitable for this technique (Moldover *et al.* 1976).

When the density profile is measured using the capacitance method other limitations appear. First, dielectric constant experiments, unlike optical experiments, do not permit continuous sampling of the density as a function of height. Furthermore, dielectric constant experiments generally do not permit direct observation of the temperature of meniscus disappearance. Finally, the results can no longer be easily interpreted when the density variation between the plates of one capacitor exceeds the desired experimental accuracy.

The limitations imposed by gravity are very severe in calorimetric experiments. Since in calorimeters the bulk phase is heated, redistribution of matter in the cell yields an additional contribution to the specific heat which becomes appreciable at $\Delta T^* < 10^{-4}$ (Schmidt 1971, Hohenberg and Barmatz 1972). Since the specific heat anomaly is weak, background contributions are large everywhere except very close to the critical point. Thus the cut-off imposed by gravity restricts the precision with which the critical exponent α can be determined in earth-bound experiments.

In laser light scattering, measurements in the gravity-affected range can be carried out reliably if the intensity of the scattered light is measured as a function of the height in the cell (Alekhin *et al.* 1969a,b, Chalyi and Alekhin 1971, Golik *et al.* 1969, Krupskii and Shimanskii 1972, White and Maccabee 1975). Gravity imposes limitations when the compressibility varies over the height of the scattering volume (Dobbs and Schmidt 1973). However, the most severe limitations in the case of light scattering are not due to gravity but to multiple scattering and attenuation of the light beam.

Finally, there exists an intrinsic limit as to how closely the critical point can be approached in earth-bound experiments. Near the gas–liquid critical point a fluid exhibits large fluctuations in the density. The spatial extent of these fluctuations is characterized by a correlation length to be defined in Section 4.5.1. If the system were in true homogeneous thermodynamic equilibrium the compressibility and the correlation length would diverge at the critical point. However, the presence of a gravitational field prevents the fluctuations from growing indefinitely. At each level the system can be expected to be in local thermodynamic equilibrium, when the local fluid properties do not vary appreciably over the distance of one correlation length. While the correlation length increases on approaching the critical point, the gravity induced density profile becomes more and more pronounced. Finally, a situation is reached where the fluid properties such as the compressibility and the correlation length vary non-negligibly over a height of the order of the correlation length. Then the assumption of local equilibrium ceases to be valid and the fluid properties themselves are modified by the gravitational field. These phenomena will be encountered at $|\Delta T^*| < 10^{-6}$; present-day experimental techniques are on the verge of entering this range.

In Table 4.2.6 we present some estimates of the range where no accurate information is obtained in experiments conducted near the critical point of a fluid like xenon in the earth's gravitational field. For further details the reader is referred to a report by Moldover *et al.* (1976).

Table 4.2.6 Limitations imposed by gravity on critical-region experiments†

Experiment	Property measured	Characteristic length	Nature of limitation	Excluded range	
				$\rho = \rho_c$ ΔT^*	$T = T_c$ $\Delta\rho^*$
PVT	Density	Cell 1 cm high	Density gradient	6×10^{-4}	8×10^{-2}
Float densimeter	Density	Height of float 2.5 mm	Density gradient	2×10^{-4}	5×10^{-2}
Capacitance	Density	Spacing between plates 0.2 mm	Density gradient	2×10^{-5}	2×10^{-2}
Calorimetry	Heat capacity	Cell 1 cm high	Redistribution of matter	3.5×10^{-4}	6×10^{-2}
Refractive index gradient	Compressibility	Path in cell 3 mm	Curved path of beam in cell	1.5×10^{-5}	2×10^{-2}
Light scattering	Compressibility correlation length	Path in cell 3 mm	Turbidity	5×10^{-5}	3×10^{-2}
All	Any	Correlation length	Non-local effects	1.5×10^{-6}	1.5×10^{-2}

†The estimates refer to xenon in the earth's gravitational field when the indicated property is to be measured within 1%. For details of obtaining these estimates see Moldover *et al.* (1976) and Hohenberg and Barmatz (1972).

4.3 Scaling laws for thermodynamic properties of one-component fluids

4.3.1 Choice of variables

In this section we describe the choice of variables we have made in scaling the thermodynamic properties of fluids, and the reasons for that choice. A critique of our choice of variables is postponed until Section 4.3.7.

As mentioned earlier, thermodynamics itself does not specify uniquely which set of variables is to be preferred in describing the critical behaviour of fluids. The regularities found for the classical equations of state, e.g. the symmetry of the top of the coexistence curve, are present for any choice of variables, such as $\rho(T)$ or $V(T)$ or even $\rho(P)$, in which we care to describe the coexistence curve. On the other hand, in the lattice gas one system of variables is to be preferred distinctly. The lattice gas has perfect antisymmetry with respect to the critical isochore when μ is considered as a function of ρ, while this symmetry property is lacking in any other set of variables. The analyticity of μ on the critical isochore gives the possibility of scaling to a higher order in ΔT^* than is possible for the classical equations. Thus if the assumption of scaling is to be tested on numerical data for the lattice gas, it is clearly advisable to consider $\mu(\rho, T)$ rather than $P(V, T)$. The first choice of variables guarantees a much larger range of asymptotic validity of scaling and the possibility of scaling the specific heat C_V.

For real fluids we do not know *a priori* which variables are to be preferred in describing critical behaviour, and the choice we have made is based on empirical considerations of symmetry (Tisza and Chase 1965, Vicentini-Missoni *et al.* 1969a, Levelt Sengers 1970). The shape of the coexistence curve can be studied by plotting either the coexisting volumes or the coexisting densities as a function of temperature as shown for argon in Figure 4.3.1. In terms of

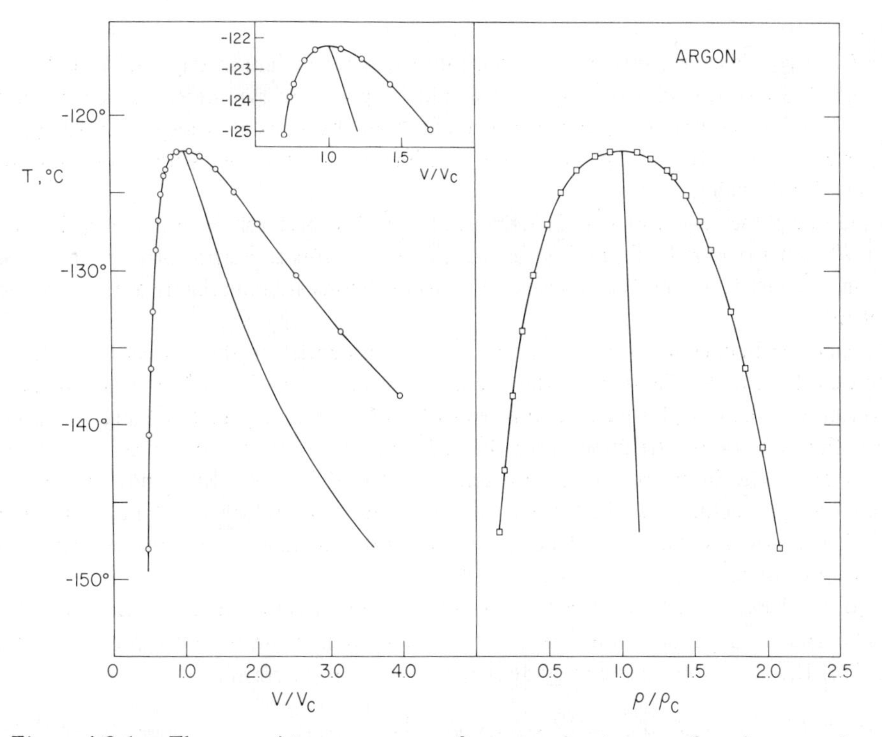

Figure 4.3.1 The coexistence curve of argon in terms of volume and temperature and in terms of density and temperature

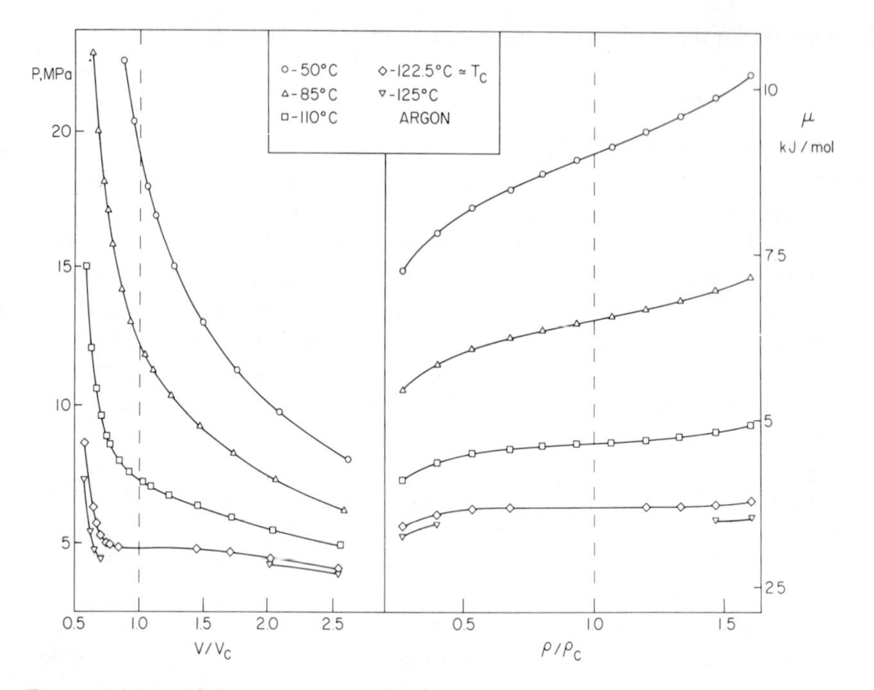

Figure 4.3.2 $P(V)$ isotherms and $\mu(\rho)$ isotherms of argon in the critical region. In contrast to the $P(V)$ isotherms, the $\mu(\rho)$ isotherms are nearly antisymmetric with respect to ρ_c

ρ and T the coexistence curve shows much more symmetry than in terms of V and T. In the lattice gas the coexistence curve $\rho_{cxc}(T)$ would be perfectly symmetric. In a real fluid, like argon, the lack of complete symmetry is evident from the locus of average coexisting densities which is approximately a straight line, the 'rectilinear diameter', but which does not coincide with the critical isochore $\rho = \rho_c$.

In the one-phase region of real fluids we notice a clear preference for a description in terms of $\mu(\rho, T)$ rather than $P(V, T)$. In Figure 4.3.2 $P(V)$-isotherms and $\mu(\rho)$-isotherms are shown for argon; in contrast to the $P(V)$-isotherms, the $\mu(\rho)$-isotherms are remarkably antisymmetric with respect to ρ_c.

The analytic behaviour of pressure and chemical potential on the critical isochore appears also to be different. In Figure 4.3.3 the two-phase specific heat $C_V{}^*/T^*$ of steam (Amirkhanov and Kerimov, 1963) is plotted as a function of $\Delta\rho^*$. According to the Yang–Yang relation (4.2.34) the slope of the tie lines equals $\mathrm{d}^2\mu/\mathrm{d}T^2$ and it does not vary much in a temperature range of 10%. The intercepts of the tie lines equal $\mathrm{d}^2P/\mathrm{d}T^2$ and they show an appreciable increase on approaching T_c. In the lattice gas $\mathrm{d}^2\mu/\mathrm{d}T^2$ would be constant, while $\mathrm{d}^2P/\mathrm{d}T^2$ would diverge weakly like C_V. Hence, the asymptotic symmetry features of real fluids are reminiscent of those of the lattice gas.

Based on these arguments we choose as independent variables the lattice-gas variables $\Delta\rho^*$ and ΔT^* and as dependent variable for the equation of state we take $\Delta\mu^* = \mu^*(\rho^*, T^*) - \mu^*(\rho_c^*, T^*)$. The Helmholtz free energy density is written in the form

$$A^* = A_0^*(T^*) + \rho^*\mu^*(\rho_c^*, T^*) + A_{\text{sing}}^*(\Delta\rho^*, \Delta T^*) \tag{4.3.1}$$

such that $(\partial A_{\text{sing}}/\partial\Delta\rho^*)_T = \Delta\mu^*$. $A_0^*(T^*)$ is an analytic background term depending on

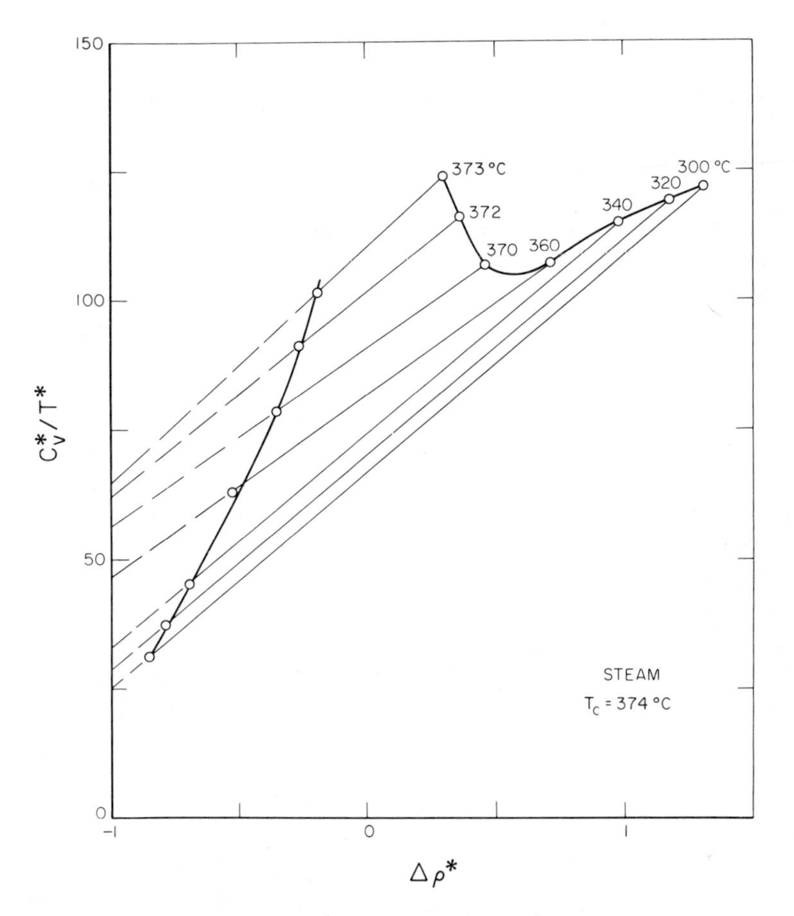

Figure 4.3.3 The two-phase specific heat C_V^*/T^* of steam as a function of $\Delta\rho^*$. The slope of the tie lines equals $d^2\mu^*/dT^{*2}$ and the intercept equals d^2P^*/dT^{*2}

temperature, while A^*_{sing} is the singular part representing the anomalous thermodynamic behaviour near the critical point. It is assumed that A^*_{sing} is symmetric and $\Delta\mu^*$ antisymmetric in $\Delta\rho^*$

$$\begin{aligned} A^*_{sing}(-\Delta\rho^*, T^*) &= A^*_{sing}(\Delta\rho^*, \Delta T^*) \\ \Delta\mu^*(-\Delta\rho^*, T^*) &= -\Delta\mu^*(\Delta\rho^*, \Delta T^*). \end{aligned} \tag{4.3.2}$$

The range over which these symmetry relations hold depends on the nature of the fluid and the precision of the experimental data: that is, the more precise the data, the sooner they will reveal departures from symmetry. For many available equations of state data a rough guideline for the range of symmetry is $\Delta\rho^* = \pm 0.25$. The symmetry relations (4.3.2) ignore the difference between the critical isochore and the rectilinear diameter when applied below the critical temperature. The coexistence dome of most fluids is so wide that a state with $\Delta\rho^* = \pm 0.25$ is reached within $|\Delta T^*| = 3 \times 10^{-3}$, in which range the departure from the rectilinear diameter from ρ_c is very small. Modifications in the choice of variables that may be needed to incorporate departures from perfect symmetry around ρ_c will be discussed in Section 4.3.7.

4.3.2 Homogeneity postulate

It has been observed experimentally that the singular parts of various thermodynamic properties follow a power law when the critical point is approached along the critical isochore $\Delta\rho^* = 0$, the critical isotherm $\Delta T^* = 0$ and along the coexistence curve $\Delta\rho^*/|\Delta T^*|^\beta = \pm B$ (Heller 1967). Hence, in analogy with (4.2.45) and (4.2.46), one is led to assume that the singular parts of these thermodynamic functions are generalized homogeneous functions of $\Delta\rho^*$ and ΔT^*. This homogeneity postulate was first formulated by Widom (1965a) for the singular part of the chemical potential. Here we adopt the formulation of Griffiths (1967) assuming that the singular part A^*_{sing} of the Helmholtz free energy density in the one-phase region is a generalized homogeneous function of its characteristic variables $\Delta\rho^*$ and ΔT^*

$$A^*_{\text{sing}}(\lambda^{a_\rho}\Delta\rho^*, \lambda^{a_T}\Delta T^*) = \lambda A^*_{\text{sing}}(\Delta\rho^*, \Delta T^*). \tag{4.3.3}$$

This hypothesis of homogeneity finds support in the known properties of the lattice gas, confirmed by the renormalization group theory. However, since the lattice gas is a highly artificial model of a fluid, the assumption of homogeneity and the choice of variables for real fluids remain empirical postulates which may have to be modified.

The homogeneity property (4.3.3) for A^*_{sing} implies that the chemical potential difference $\Delta\mu^*$, the isothermal compressibility $\chi^*_T = (\partial\mu^*/\partial\rho^*)_T$ and the singular contribution to the entropy $S^*_{\text{sing}} = -(\partial A^*_{\text{sing}}/\partial T^*)_\rho$ and specific heat $C^*_{V,\text{sing}}/T^* = -(\partial^2 A^*_{\text{sing}}/\partial T^{*2})_\rho$ are also generalized homogeneous functions of $\Delta\rho^*$ and ΔT^*. After differentiating (4.3.3) and redefining the parameter λ appropriately one obtains (Hankey and Stanley 1972, Levelt Sengers 1975)

$$\Delta\mu^*(\lambda^{a_\rho/(1-a_\rho)}\Delta\rho^*, \lambda^{a_T/(1-a_\rho)}\Delta T^*) = \lambda\Delta\mu^*(\Delta\rho^*, \Delta T^*) \tag{4.3.4a}$$

$$\chi_T^{*-1}(\lambda^{a_\rho/(1-2a_\rho)}\Delta\rho^*, \lambda^{a_T/(1-2a_\rho)}\Delta T^*) = \lambda\chi_T^{*-1}(\Delta\rho^*, \Delta T^*) \tag{4.3.4b}$$

$$S^*_{\text{sing}}(\lambda^{a_\rho/(1-a_T)}\Delta\rho^*, \lambda^{a_T/(1-a_T)}\Delta T^*) = \lambda S^*_{\text{sing}}(\Delta\rho^*, \Delta T^*) \tag{4.3.4c}$$

$$\frac{C^*_{V,\text{sing}}}{T^*}(\lambda^{a_\rho/(1-2a_T)}\Delta\rho^*, \lambda^{a_T/(1-2a_T)}\Delta T^*) = \lambda\frac{C^*_{V,\text{sing}}}{T^*}(\Delta\rho^*, \Delta T^*). \tag{4.3.4d}$$

Along the critical isochore $\Delta\rho^* = 0$ and along the critical isotherm $\Delta T^* = 0$ these thermodynamic properties will vary according to power laws analogous to (4.2.46). In particular

$$\chi^{*-1}(0, \Delta T^*) = \chi^{*-1}(0, 1)(\Delta T^*)^{(1-2a_\rho)/a_T} \quad (\Delta T^* \geqslant 0)$$

$$\frac{C^*_{V,\text{sing}}}{T^*}(0, T^*) = \frac{C^*_{V,\text{sing}}}{T^*}(0, 1)(\Delta T^*)^{(1-2a_T)/a_T} \quad (\Delta T^* \geqslant 0)$$

$$\Delta\mu^*(\Delta\rho^*, 0) = \pm\Delta\mu^*(1, 0)\,|\Delta\rho^*|^{(1-a_\rho)/a_\rho}.$$

For the two coexisting phases below the critical temperature $\Delta\mu^* = 0$, while $\Delta\rho^*_{\text{cxc}} \neq 0$. From the homogeneity assumption (4.3.4a) it follows that at coexistence

$$\Delta\mu^*(|\Delta T^*|^{-a_\rho/a_T}\Delta\rho^*, -1) = 0,$$

so that

$$|\Delta T^*|^{-a_\rho/a_T}\Delta\rho^*_{\text{cxc}} = \pm B,$$

where B is constant. The power law behaviour of the compressibility and the specific heat along the coexistence curve $\Delta\rho^*/|\Delta T^*|^{a_\rho/a_T} = \pm B$ follows from (4.3.4b) and (4.3.4d) in analogy with (4.2.45)

$$\chi^{*-1}(\Delta\rho^*_{\text{cxc}}, \Delta T^*) = \chi^{*-1}(B, -1)\,|\,\Delta T^*\,|^{(1-2a_\rho)/a_T}$$

$$\frac{C^*_{V,\text{sing}}}{T^*}(\Delta\rho^*_{\text{cxc}}, \Delta T^*) = \frac{C^*_{V,\text{sing}}}{T^*}(B, -1)\,|\,\Delta T^*\,|^{(1-2a_T)/a_T}.$$

The specific heat anomaly in the two-phase region is obtained by observing that the singular part A^*_s of the free energy inside the two-phase region is independent of the density and equal to its value at the phase boundary: $A^*_{\text{sing,II}} = A^*_{\text{sing}}(\Delta\rho^*_{\text{cxc}}, \Delta T^*) = A^*_{\text{sing}}(B, -1)\,|\,\Delta T^*\,|^{1/a_T}$. Differentiation yields for the specific heat anomaly in the two-phase region

$$\left(\frac{C^*_{V,\text{sing}}}{T^*}\right)_{\text{II}} = -\frac{1}{a_T}\left(\frac{1}{a_T} - 1\right) A^*_{\text{sing}}(B, -1)\,|\,\Delta T^*\,|^{(1-2a_T)/a_T}.$$

We thus have recovered the thermodynamic power laws introduced in Table 4.2.1 with

$$\begin{aligned} &\alpha = \alpha' = \alpha'' = (2a_T - 1)/a_T \qquad \beta = a_\rho/a_T \\ &\gamma = \gamma' = (1 - 2a_\rho)/a_T \qquad \delta = (1 - a_\rho)/a_\rho \end{aligned} \tag{4.3.5}$$

so that

$$a_\rho = \frac{1}{\delta + 1} = \frac{\beta}{2 - \alpha} \qquad a_T = \frac{1}{\beta(\delta + 1)} = \frac{1}{2 - \alpha}. \tag{4.3.6}$$

The thermodynamic critical exponents satisfy the relations (4.2.5) and all power laws are determined by two independent exponents only.

4.3.3 *Thermodynamic scaling laws*

The homogeneity postulate (4.3.3) implies in analogy with (4.2.44) that the symmetric part of the Helmholtz free energy density satisfies a scaling law of the form

$$\frac{A^*_{\text{sing}}(\Delta\rho^*, \Delta T^*)}{|\,\Delta\rho^*\,|^{\delta+1}} = A^*_{\text{sing}}(1, x) \tag{4.3.7}$$

where the scaling variable x is defined as

$$x = \Delta T^*/|\,\Delta\rho^*\,|^{1/\beta}. \tag{4.3.8}$$

We find it convenient to write $A^*_{\text{sing}}(1, x) = Da(x/x_0)$, where D is the amplitude of the power law $\Delta\mu^* = D(\Delta\rho^*)\,|\,\Delta\rho^*\,|^{\delta-1}$ for the critical isotherm and where x_0 is related to the amplitude B of the power law $\Delta\rho^*_{\text{cxc}} = \pm B\,|\,\Delta T^*\,|^{\beta}$ for the coexistence curve by

$$x_0 = B^{-1/\beta}. \tag{4.3.9}$$

Thus the Helmholtz free energy density (4.3.1) can be written in scaled form as

$$A^* = A^*_0(T^*) + \rho^*\mu^*(\rho^*_c, T^*) + |\,\Delta\rho^*\,|^{\delta+1}Da(x/x_0). \tag{4.3.10}$$

For the chemical potential difference $\Delta\mu^* = (\partial A^*_{\text{sing}}/\partial\Delta\rho^*)_T$ we obtain the scaling law

$$\Delta\mu^* = \Delta\rho^*\,|\,\Delta\rho^*\,|^{\delta-1}Dh(x/x_0) \tag{4.3.11}$$

where the functions $a(x/x_0)$ and $h(x/x_0)$ are related by the differential equation (Griffiths 1967, Levelt Sengers *et al.* 1976)

$$\beta h(w) = -wa'(w) + \beta(\delta + 1)a(w) \tag{4.3.12}$$

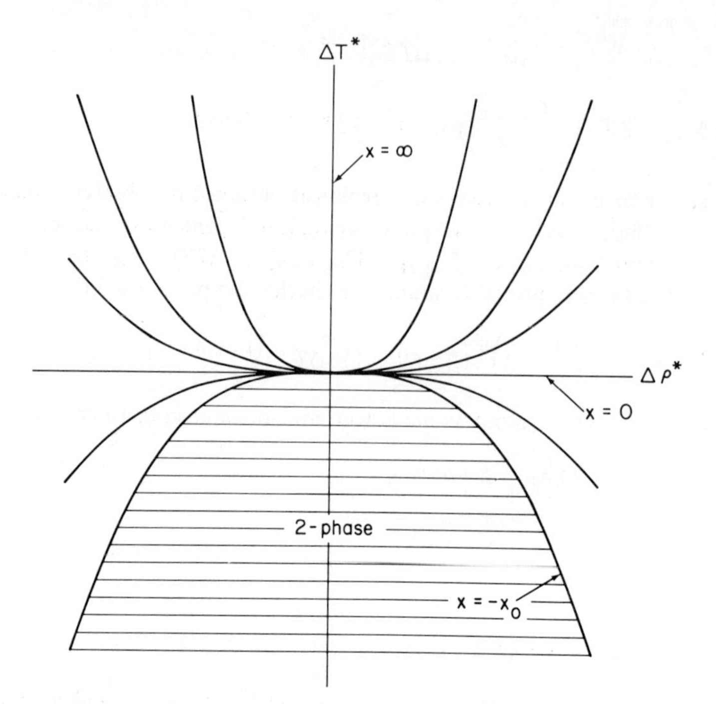

Figure 4.3.4 Curves of constant x in the $\Delta\rho^* - \Delta T^*$ plane

Table 4.3.1 Scaled expressions for thermodynamic functions

Chemical potential

$$\mu^* = \mu^*(\rho_c, T^*) + \Delta\rho^*|\Delta\rho^*|^{\delta-1} D h\left(\frac{x}{x_0}\right)$$

Compressibility

$$\chi_T^{*-1} = (\rho^{*2} K_T^*)^{-1} = |\Delta\rho^*|^{\delta-1} D\left[\delta h\left(\frac{x}{x_0}\right) - \frac{x}{\beta x_0} h'\left(\frac{x}{x_0}\right)\right]$$

Helmholtz free energy

$$A^* = A_0^*(T^*) + \rho^*\mu^*(\rho_c^*, T^*) + |\Delta\rho^*|^{\delta+1} D a\left(\frac{x}{x_0}\right)$$

Pressure

$$P^* = -A_0^*(T^*) + D\left\{\Delta\rho^*|\Delta\rho^*|^{\delta-1} h\left(\frac{x}{x_0}\right) + |\Delta\rho^*|^{\delta+1}\left[h\left(\frac{x}{x_0}\right) - a\left(\frac{x}{x_0}\right)\right]\right\}$$

Entropy

$$-S^* = \frac{dA_0^*(T^*)}{dT^*} + \rho^* \frac{d\mu^*(\rho_c^*, T^*)}{dT^*} + |\Delta\rho^*|^{(1-\alpha)/\beta} \frac{D}{x_0} a'\left(\frac{x}{x_0}\right)$$

Heat capacity

$$-\frac{C_V^*}{T^*} = \frac{d^2 A_0^*(T^*)}{dT^{*2}} + \rho^* \frac{d^2\mu^*(\rho_c^*, T^*)}{dT^{*2}} + |\Delta\rho^*|^{-\alpha/\beta} \frac{D}{x_0^2} a''\left(\frac{x}{x_0}\right)$$

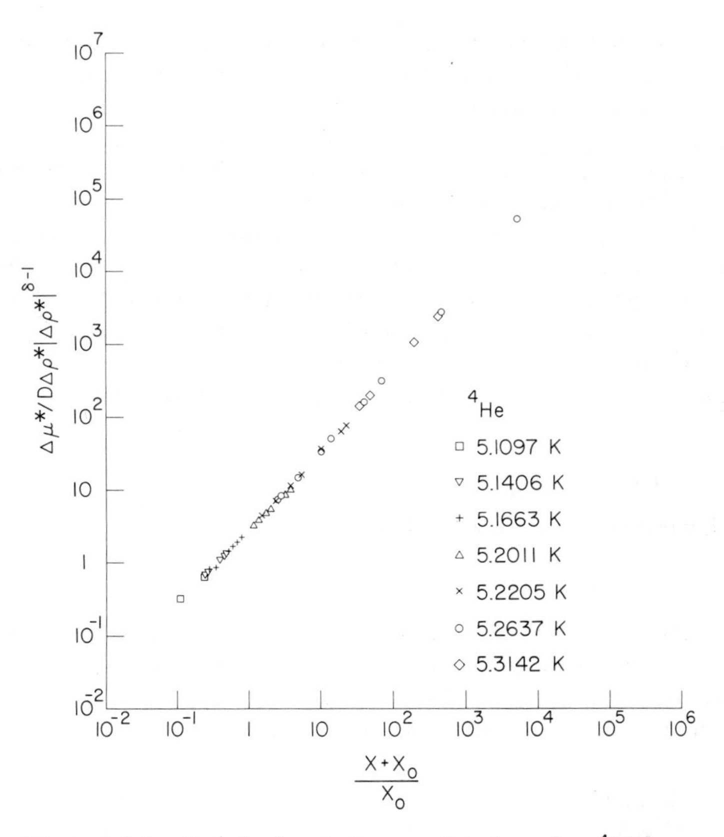

Figure 4.3.5 Scaled chemical potential data for ^{4}He as a function of $(x + x_0)/x_0$. Reduction parameters are taken from Table 4.3.4

with $a'(w) = da/dw$. Curves of constant x in the $\Delta\rho^* - \Delta T^*$ plane are schematically indicated in Figure 4.3.4. The curve $x = -x_0$ is the coexistence curve, the curve $x = 0$ is the critical isotherm and the curve $x = \infty$ is the critical isochore. The function $h(x/x_0)$ is normalized such that at the critical isotherm $h(0) = 1$. From (4.3.10) and (4.3.11) one can deduce scaled expressions for all thermodynamic properties in terms of $a(x/x_0)$ and $h(x/x_0)$; some of these expressions are given in Table 4.3.1.

When the experimental values of the ratios $\Delta\mu^*/(\Delta\rho^*) \mid \Delta\rho^* \mid^{\delta - 1}$ are plotted as a function of the scaling variable x/x_0, then the scaling law (4.3.11) predicts that different isotherms should all collapse onto one single curve. An example of such a scaled plot is presented in Figure 4.3.5 for ^{4}He as deduced from the data of Roach (1968) assuming $\beta = 0.355$ and $\gamma = 1.19$ (Levelt Sengers *et al.* 1976).

4.3.4 Scaled equations of state

The scaling laws are a conjecture regarding the asymptotic behaviour of the thermodynamic properties in the vicinity of the critical point. When the scaling laws are applied in a finite range of densities and temperatures the results may be affected by the presence of less singular correction terms not taken into account. As a consequence the actual range of the validity of the scaling laws will depend on the precision of the experimental data. The more precise data, the sooner deviations from asymptotic scaling behaviour will be detected. We shall first discuss a number of results deduced from an analysis of equation of state data in the range $5 \times 10^{-4} \leqslant |\Delta T^*| \leqslant 3 \times 10^{-2}$ not significantly affected by gravity. In Section 4.3.6 we shall then comment on some recent density gradient profile measurements obtained at temperatures $|\Delta T^*| < 10^{-4}$ and indicate how they affect our understanding of the thermodynamic behaviour of real fluids near the critical point.

In order to make a quantitative analysis of the experimental thermodynamic data one needs an expression for the scaling functions $h(w)$ or $a(w)$, where $w = x/x_0$. In the absence of an *a priori* theoretical expression for fluids, one has formulated approximate expressions of an empirical nature. The choices for the function $h(w)$ are restricted by a number of conditions formulated by Griffiths (1967). Some of these conditions arise from the requirements of thermodynamic stability. Mechanical stability requires the compressibility to be positive, so that

$$\beta\delta h(w) \geqslant wh'(w). \tag{4.3.13}$$

Thermal stability requires the specific heat to be positive, so that

$$a''(w) \leqslant 0. \tag{4.3.14}$$

Additional conditions are imposed on $h(w)$ by the assumption that $\mu(\rho, T)$ is an analytic function throughout the one-phase region with the exception of the critical point and perhaps the phase boundary. Thus $h(w)$ should be analytic in the range $-1 < w < \infty$; specifically for small values of w, $h(w)$ should have an expansion of the form

$$h(w) = 1 + \sum_{n=1}^{\infty} h_n w^n. \tag{4.3.15}$$

Analyticity of μ at large w implies that $h(w)$ can be expanded around $w = \infty$ as

$$h(w) = \sum_{n=1}^{\infty} \eta_n w^{\beta(\delta+1-2n)}. \tag{4.3.16}$$

It has not been possible to formulate a closed form for $h(w)$ that satisfies all these conditions. Vicentini-Missoni, Levelt Sengers and Green (1969a,b) proposed an approximate expression to which we refer as the MLSG equation (NBS equation in the original literature)

$$h(w) = (1+w)\left[\frac{1+E_2(1+w)^{2\beta}}{1+E_2}\right]^{(\gamma-1)/2\beta} \tag{4.3.17}$$

where E_2 is an additional adjustable parameter. In this approximation the equation of state (4.3.11) reads

$$\Delta\mu^* = \Delta\rho^* |\Delta\rho^*|^{\delta-1} E_1 \left(1+\frac{x}{x_0}\right)\left[1+E_2\left(1+\frac{x}{x_0}\right)^{2\beta}\right]^{(\gamma-1)/2\beta} \tag{4.3.18}$$

with

$$E_1 = \frac{D}{(1+E_2)^{(\gamma-1)/2\beta}}. \tag{4.3.19}$$

The MLSG equation has been used to represent experimental equation of state data for a variety of fluids (Vicentini-Missoni *et al.* 1969a,b, Wallace and Meyer 1970, Levelt Sengers *et al.* 1976, Gulari and Pings 1973). As an example of the quality of the fit we show in Figure 4.3.6 a

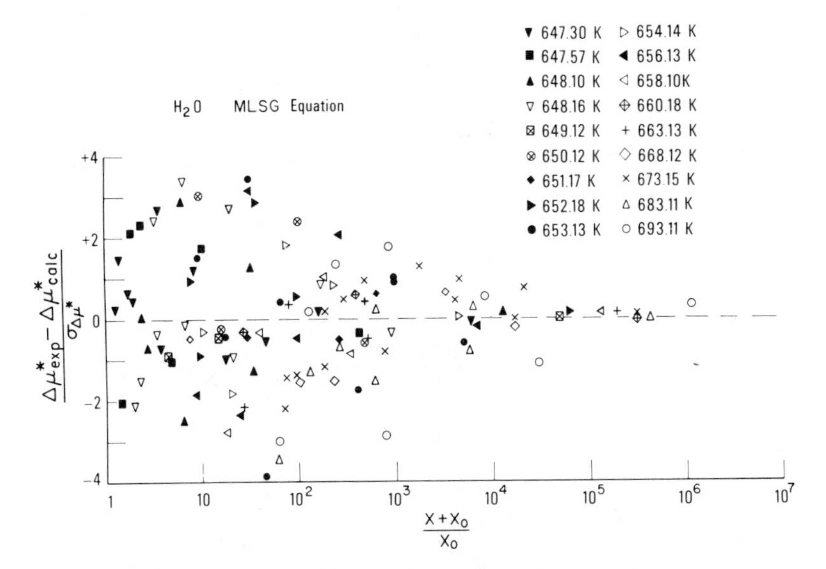

Figure 4.3.6 Plot of normalized deviations $(\Delta\mu^*_{\text{exp}} - \Delta\mu^*_{\text{calc}})/\sigma_{\Delta\mu^*}$ as a function of $(x + x_0)/x_0$ for H_2O. $\sigma_{\Delta\mu^*}$ is the estimated experimental uncertainty; $\Delta\mu^*_{\text{calc}}$ is calculated from the MLSG equation with parameters $\rho_c = 322.2$ kg/m^3, $T_c = 647.05$ K, $\beta = 0.350$, $\delta = 4.50$, $x_0 = 0.100$, $E_1 = 1.215$, $E_2 = 0.372$ (From Levelt Sengers *et al.* 1976, reproduced by permission of the American Institute of Physics)

deviation plot when the chemical potential data deduced from the *PVT* data of Rivkin *et al* (1962, 1963, 1964, 1966) for steam are fitted to the MLSG equation (Levelt Sengers *et al.* 1976).

The MLSG equation has the advantage that it expresses the equation of state in terms of the primary variables $\Delta\rho^*$ and ΔT^*. However, it has two disadvantages. Firstly, while the equation does satisfy the analyticity requirement for small values of w, it reproduces correctly only the first two terms of the series expansion (4.3.16) for large values of w. Secondly, the equation cannot be integrated analytically to yield a closed form for the function $a(w)$ (Schmidt 1971, Lentini and Vicentini-Missoni 1973). Hence, the equation does not provide a satisfactory basis for developing a fundamental equation and it can only be fitted to experimental *PVT* data after they have been integrated numerically to yield chemical potential data.

The two problems can be overcome by using parametric equations introduced by Schofield (1969) and Josephson (1969). This formulation entails a transformation from the physical variables ΔT^* and $\Delta\rho^*$ into two parametric variables, r and θ. The variable r is meant, in some sense, to describe how closely the critical point is approached and the variable θ a 'location on a contour of constant r.' All anomalies are then incorporated by the power laws in the r-dependence, while the θ-dependence is kept analytic.

The manner in which the thermodynamic variables are expressed in terms of r and θ is not unique (Fisher 1971). The constraints that the scaling laws are preserved are met by the following choice

$$\begin{aligned} \Delta T^* &= rT(\theta) \\ \Delta\rho^* &= r^{\beta}M(\theta) \\ \Delta\mu^* &= r^{\beta\delta}H(\theta). \end{aligned} \tag{4.3.20}$$

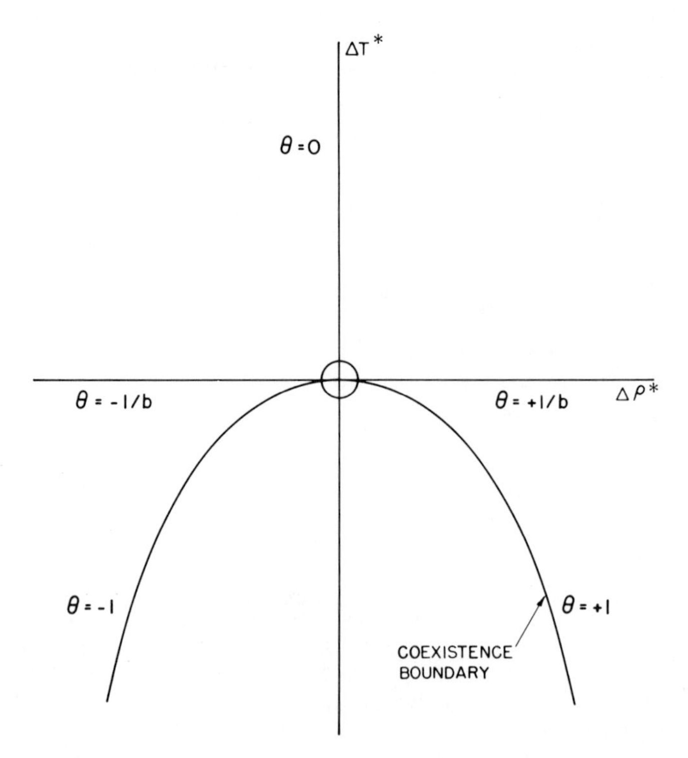

Figure 4.3.7 The variable θ in the parametric equations of state. (From Levelt Sengers *et al.* 1976, reproduced by permission of the American Institute of Physics)

On constructing the ratios $\Delta\mu^*/(\Delta\rho^*) \mid \Delta\rho^* \mid^{\delta - 1}$ and $x = \Delta T^*/ \mid \Delta\rho^* \mid^{1/\beta}$ one sees immediately that both ratios depend on θ alone, so that the scaling law (4.3.11) is implied by the parametric representations (4.3.20). The parameter θ can be chosen to span the range -1 to $+1$, such that it equals zero on the critical isochore and ± 1 on the coexistence boundary, as indicated schematically in Figure 4.3.7. Choices compatible with the assumed lowest order symmetry are those for which $T(\theta)$ is a symmetric function of θ and $M(\theta)$ and $H(\theta)$ are antisymmetric functions of θ. The two most popular choices compatible with these requirements are either the *linear* model (Schofield 1969) for which

$$\begin{aligned} T(\theta) &= 1 - b_1^2\theta^2 \\ M(\theta) &= k_1\theta \\ H(\theta) &= a_1\theta(1 - \theta^2) \end{aligned} \tag{4.3.21}$$

or the *cubic model* (Ho and Litster 1970) for which

$$\begin{aligned} T(\theta) &= 1 - b_2^2\theta^2 \\ M(\theta) &= k_2\theta(1 + c\theta) \\ H(\theta) &= a_2\theta(1 - \theta^2) \end{aligned} \tag{4.3.22}$$

The linear model contains three constants k_1, a_1, b_1 and the cubic model four constants k_2, a_2 b_2, c which have to be determined. Slightly more eleborate parametric equations have beei

Table 4.3.2 Amplitudes of power laws

MLSG equation

$$B = x_0^{-\beta}$$

$$D = E_1(1 + E_2)^{(\gamma - 1)/2\beta}$$

$$\Gamma = x_0^{\gamma}/E_1 E_2^{(\gamma - 1)/2\beta}$$

$$\Gamma' = \beta x_0^{\gamma}/E_1$$

Parametric equations†

$$B = k(1 + c)/(b^2 - 1)^{\beta}$$

$$D = a(b^2 - 1)b^{3(\delta - 1)}/k^{\delta}(b^2 + c)^{\delta}$$

$$\Gamma = k/a$$

$$\Gamma' = (b^2 - 1)^{\gamma - 1}\{1 - b^2(1 - 2\beta) - c[b^2(3 - 2\beta) - 3]\}k/2a$$

$$A^{+} = -ak(2 - \alpha)(1 - \alpha)\alpha f_0$$

$$A_{\mathrm{I}}^{-} = ak\alpha(b^2 - 1)^{\alpha}\left(\frac{(1 - \alpha)(1 + 3c)(s_0 + s_2 + s_4) - 2\beta(1 + c)(s_2 + 2s_4)}{1 - b^2(1 - 2\beta) - c\{b^2(3 - 2\beta) - 3\}}\right)$$

$$A_{\mathrm{II}}^{-} = -ak(2 - \alpha)(1 - \alpha)\alpha(f_0 + f_2 + f_4 + f_6)/(b^2 - 1)^{2 - \alpha}$$

†For definition of $f_i (i = 0, 2, 4, 6)$ and $s_j (j = 0, 2, 4)$ see Table 4.3.3.

considered by Kierstead (1973) and Estler *et al*(1975). The relationships between the constants of the scaled equations presented here and the amplitudes of the thermodynamic power laws introduced in Table 4.2.1 are given in Table 4.3.2.

Both the linear model and the cubic model can be integrated analytically to yield the Helmholtz free energy density and, hence, the other thermodynamic functions (Hohenberg and Barmatz 1972). The parametric representations for various thermodynamic properties are given in Table 4.3.3. For reasons to be discussed in the subsequent section, one often uses restricted versions of these parametric equations in which the constant b_1 of the linear model or the constants b_2 and c of the cubic model are fixed by the conditions presented at the bottom of Table 4.3.3.

Various authors have analyzed experimental data in terms of the linear model (Anisimov *et al.* 1974, Ho and Litster 1969, Hohenberg and Barmatz 1972, Huang and Ho 1973, Levelt Sengers *et al.* 1974, 1976, Murphy *et al.* 1973, 1975, Thoen and Garland 1974, White and Maccabee 1975). It turns out that the linear model, as well as the cubic model (Huang and Ho 1973, Murphy 1975), yields a satisfactory representation of the experimental equation of state data of a quality comparable to that obtained with the MLSG equation (Levelt Sengers *et al.* 1976). As an example we show in Figure 4.3.8 a deviation plot when the chemical potential data for steam are fitted to the linear model; the corresponding deviation plot when fitted to the MLSG equation was given in Figure 4.3.6. Unlike the MLSG equation, it is also possible to fit the linear model and the cubic model directly to the original pressure data (Murphy *et al.* 1973, 1975).

A stringent test on the validity of any equation of state is the extent to which it is consistent with experimental specific heat data as well. However, efforts to describe experimental *PVT* data and C_V data simultaneously in terms of one scaled equation of state like the linear model, cubic model, or the MLSG equation have had only limited success (Lentini and Vicentini-Missoni 1973, Huang and Ho 1973, Barmatz *et al.* 1975, White and Maccabee 1975).

Table 4.3.3 Parametric representation of thermodynamic functions (Linear model, $c = 0$; cubic model, $c \neq 0$)

Variables

$$\Delta T^* = r(1 - b^2\theta^2)$$

$$\Delta\rho^* = r^\beta k\theta(1 + c\theta^2)$$

Chemical potential

$$\mu^* = \mu^*(\rho_c^*, T^*) + r^{\beta\delta} a\theta(1 - \theta^2)$$

Compressibility

$$\left(\frac{\partial\rho^*}{\partial\mu^*}\right)_T = \chi_T^* = r^{-\gamma}\frac{k}{a}\frac{1 - (b^2 - 2\beta b^2 - 3c)\theta^2 - (3 - 2\beta)b^2 c\theta^4}{1 - (b^2 - 2\beta\delta b^2 + 3)\theta^2 - (2\beta\delta - 3)b^2\theta^4}$$

Helmholtz free energy

$$A^* = A_0^*(T^*) + \rho^*\mu^*(\rho_c^*, T^*) + r^{2-\alpha}akf(\theta)$$

where

$$f(\theta) = f_0 + f_2\theta^2 + f_4\theta^4 + f_6\theta^6$$

$$f_0 = -\frac{\beta(\delta - 3) - b^2\alpha\gamma}{2b^4(2 - \alpha)(1 - \alpha)\alpha} + \frac{c[b^2(1 + \alpha)(3\gamma + 2\beta) - 6\gamma]}{2b^6(2 - \alpha)(1 - \alpha)\alpha(1 + \alpha)}$$

$$f_2 = +\frac{\beta(\delta - 3) - b^2\alpha(1 - 2\beta)}{2b^2(1 - \alpha)\alpha} - \frac{c[b^2(1 + \alpha)(3\gamma + 2\beta) - 6\gamma]}{2b^4(1 - \alpha)\alpha(1 + \alpha)}$$

$$f_4 = -\frac{1 - 2\beta}{2\alpha} + \frac{c[b^2(1 + \alpha)(3 - 2\beta) - 3\gamma]}{2b^2\alpha(1 + \alpha)}$$

$$f_6 = -\frac{(3 - 2\beta)}{2(1 + \alpha)}c$$

Pressure

$$P^* = -A_0^*(T^*) + r^{\beta\delta} a\theta(1 - \theta^2) + r^{2-\alpha}ak[\theta^2(1 - \theta^2)(1 + c\theta^2) - f(\theta)'$$

Entropy

$$S^* = -\frac{\mathrm{d}A_0^*(T^*)}{\mathrm{d}T^*} - \rho^*\frac{\mathrm{d}\mu^*(\rho_c^*, T^*)}{\mathrm{d}T^*} + r^{1-\alpha}aks(\theta)$$

where

$$s(\theta) = s_0 + s_2\theta^2 + s_4\theta^4$$

$$s_0 = -(2 - \alpha)f_0 \qquad s_2 = -(2 - \alpha)b^2(1 - 2\beta)f_0 - \gamma f_2 \qquad s_4 = -3\gamma c/2b^2(1 + \alpha)$$

Pressure coefficient

$$\left(\frac{\partial P^*}{\partial T^*}\right)_T = -\frac{\mathrm{d}A_0^*(T^*)}{\mathrm{d}T^*} + r^{1-\alpha}aks(\theta) + (1 + \Delta\rho^*)r^{\beta\delta - 1}a\beta\theta\left(\frac{\delta(1 - \theta^2)(1 + 3c\theta^2) - (1 - 3\theta^2)(1 + c\theta^2)}{1 - (b^2 - 2\beta b^2 - 3c)\theta^2 - (3 - 2\beta)b^2c\theta^4}\right)$$

Heat capacity

$$\frac{C_V^*}{T^*} = -\frac{\mathrm{d}^2A_0^*(T^*)}{\mathrm{d}T^{*2}} - \rho^*\frac{\mathrm{d}^2\mu^*(\rho_c^*, T^*)}{\mathrm{d}T^{*2}} + r^{-\alpha}ak\left(\frac{(1 - \alpha)(1 + 3c\theta^2)s(\theta) - 2\beta(1 + c\theta^2)(s_2\theta^2 + 2s_4\theta^4)}{1 - (b^2 - 2\beta b^2 - 3c)\theta^2 - (3 - 2\beta)b^2c\theta^4}\right)$$

Table 4.3.3 (*Continued*)

Conditions for restricted linear model
$b^2 = \dfrac{\delta - 3}{(\delta - 1)(1 - 2\beta)} \qquad c = 0$
Conditions for restricted cubic model
$b^2 = \dfrac{3}{3 - 2\beta} \qquad c = \dfrac{2\beta\delta - 3}{3 - 2\beta}$

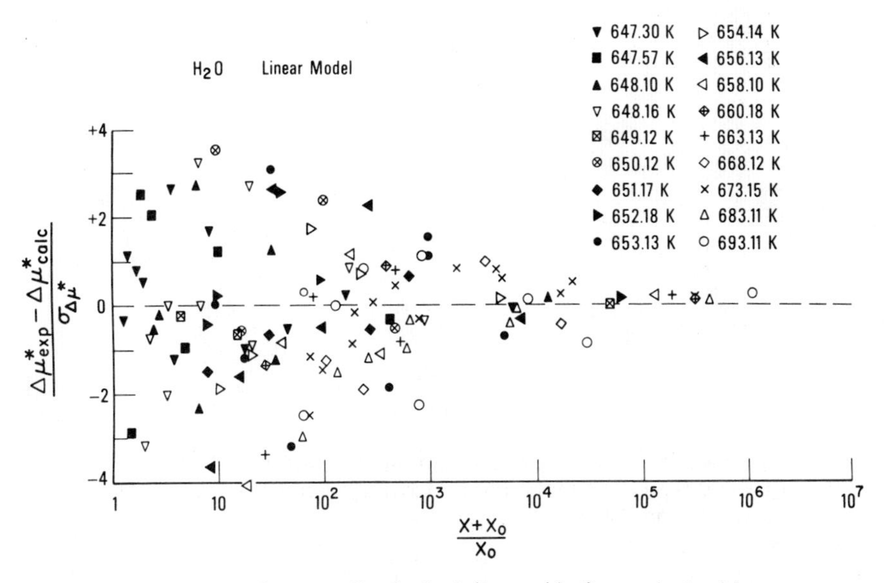

Figure 4.3.8 Plot of normalized deviations $(\Delta\mu^*_{\mathrm{exp}} - \Delta\mu^*_{\mathrm{calc}})/\sigma_{\Delta\mu^*}$ as a function of $(x + x_0)/x_0$ for H_2O. $\sigma_{\Delta\mu^*}$ is the estimated experimental uncertainty; $\Delta\mu^*_{\mathrm{calc}}$ is calculated from the linear model with parameters $\rho_c = 322.2$ kg/m^3, $T_c = 647.05$ K, $\beta = 0.350$, $\delta = 4.50$, $k_1 = 1.664$, $b_1^2 = 1.4286$, $a_1 = 24.47$. (From Levelt Sengers *et al.* 1976, reproduced by permission of the American Institute of Physics)

4.3.5 Universality of critical behaviour

The theoretical studies of various model systems have led to the formulation of the hypothesis of universality. According to this hypothesis systems having the same basic symmetries are expected to have identical critical exponents and scaling functions, and are said to belong to the same universality class (Jasnow and Wortis 1968, Watson 1969, Griffiths 1970, Kadanoff 1971, 1976, Betts *et al.* 1971, Milosevic and Stanley 1972, Aharony 1976). As discussed in Section 4.2.7 the hypothesis is supported by the renormalization group theory of critical phenomena.

Specifically, the universality hypothesis implies that systems belonging to the same universality class should asymptotically obey the same scaled equation of state apart from only two adjustable constants. That is, the critical exponents and the function $h(x/x_0)$ in (4.3.11) should be the same for all these systems leaving only two substance-dependent parameters, namely D and x_0. Hence, the hypothesis predicts that scaling plots of $\Delta\mu^*/(\Delta\rho^*)\,|\,\Delta\rho^*\,|^{\delta-1}$ versus x/x_0

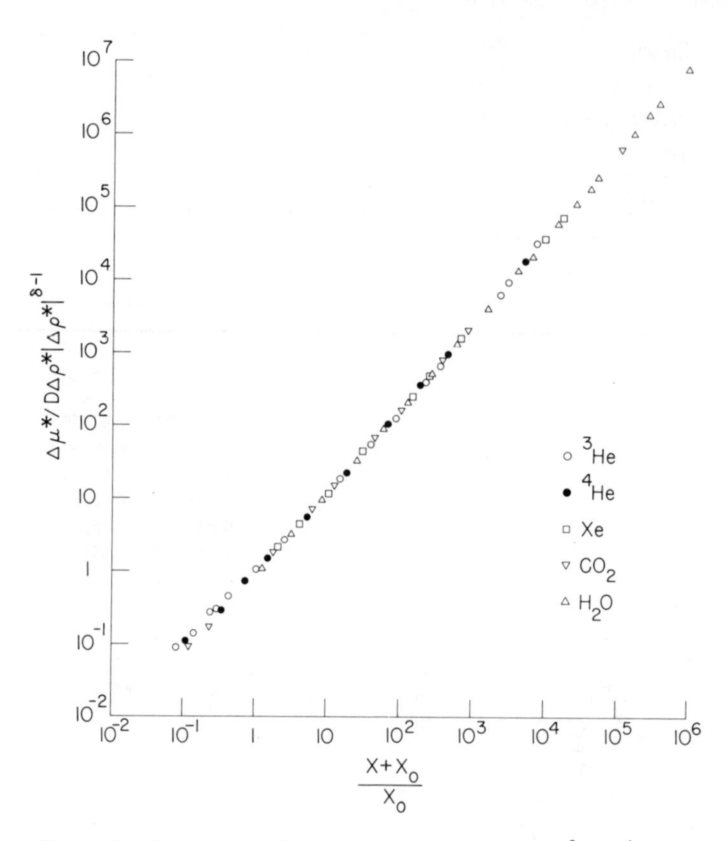

Figure 4.3.9 Scaled chemical potential data for ^3He, ^4He, Xe, CO_2 and H_2O as a function of $(x + x_0)/x_0$. Reduction parameters are taken from Table 4.3.4

for different fluids, such as the one shown in Figure 4.3.5 for ^4He, can be made to coincide with proper choice of the adjustable scale factors D and x_0. In Figure 4.3.9 we show that the scaling plots of five different gases, namely ^3He, ^4He, Xe, CO_2 and H_2O, can indeed be brought into one single universal curve, thus supporting the hypothesis of universality for these five gases (Levelt Sengers 1974).

In order to make a quantitative study of the validity of this hypothesis, it is necessary to use an explicit expression for the scaled equation of state. Using least-squares techniques, a goodness-of-fit criterion is then readily established in the form of the standard deviation; varying precision is taken care of by weighting, the deviations of the individual points are compared with their estimated standard error to check whether systematic errors are present, and a number of parameters such as T_c, ρ_c, the critical exponents and the adjustable constants D and x_0 are varied to optimize the fit (Levelt Sengers *et al.* 1976). The specific scaled equations of state described in the preceding section are only suitable candidates for a universal equation of state if the number of adjustable parameters is restricted to two. Thus when in the MLSG equation (4.3.18) E_1 and x_0 are treated as adjustable parameters, not only the critical exponents but also the constant E_2 must be independent of the fluid under consideration. In practice, E_2 turns out to be small, and indeed roughly the same for a number of fluids.

Likewise, the principle of universality allows us to treat k_1 and a_1 in the linear model (4.3.21) as adjustable parameters, but the parameter b_1 must be kept at a fixed value. A popular choice for the parameter b_1 in the linear model is that recommended by Schofield *et al.* (1969)

$$b_1^2 = \frac{\delta - 3}{(\delta - 1)(1 - 2\beta)}. \tag{4.3.23}$$

With this choice for b_1^2 the representation of the singular part of the specific heat in terms of the linear model parameters reduces to

$$\frac{C_{V,\text{sing}}^*}{T^*} = \frac{A^+}{\alpha} r^{-\alpha} = a_1 k_1 \frac{(1 - 2\beta)\gamma(\gamma - 1)(\delta - 1)}{2\alpha(\delta - 3)} r^{-\alpha} \tag{4.3.24}$$

so that a contour of constant r may be interpreted as a contour of constant anomalous specific heat. We refer to the linear model with the special choice (4.3.23) for b_1^2 as the *restricted linear model*. If this special choice for b_1^2 is substituted in

$$E_2^{-1} = \left(\frac{b_1^{(\delta - 3)/(\gamma - 1)}}{b_1^2 - 1}\right)^{2\beta} - 1 \tag{4.3.25}$$

one obtains a corresponding restricted version of the MLSG equation (Levelt Sengers *et al.* 1976).

With the cubic model, the compressibility χ_T can be made independent of θ by the choice (Ho and Litster 1970)

$$b_2^2 = \frac{3}{3 - 2\beta} \qquad c = \frac{2\beta\delta - 3}{3 - 2\beta}. \tag{4.3.26}$$

For this *restricted cubic model*

$$\chi_T^* = \frac{k_2}{a_2} r^{-\gamma} \tag{4.3.27}$$

and contours of constant r correspond to contours of constant compressibility χ_T^*.

It has been found that the cubic model does yield a satisfactory representation of the equation of state data for the 3-dimensional Ising model though with values of b_2^2 and c that differ from the choice (4.3.26) (Wallace and Zia 1974, Tarko and Fisher 1975, Wallace 1976).

To investigate the hypothesis of universality for fluids we made a statistical analysis of the equation of state data for ^{3}He (Wallace and Meyer 1970), ^{4}He (Roach 1968), Xe (Habgood and Schneider 1954), CO_2 (Michels *et al.* 1937), H_2O (Rivkin *et al.* 1962, 1963, 1964, 1966) and density profile data for O_2 (Weber 1970). It was found that the experimental data of all these fluids could be fitted to the same scaled equation of state for which we chose the restricted linear model. With the exception of O_2 the range of the fits corresponded approximately to

$$5 \times 10^{-4} \leqslant |\Delta T^*| \leqslant 3 \times 10^{-2} \qquad |\Delta\rho^*| \leqslant 0.25. \tag{4.3.28}$$

In this range the critical exponents of those fluids were compatible with the 'universal' values (Levelt Sengers and Sengers 1975)

$$\begin{aligned} \alpha &= 0.10 \pm 0.04 \qquad \beta = 0.355 \pm 0.007 \\ \gamma &= 1.19 \pm 0.03 \qquad \delta = 4.35 \pm 0.10. \end{aligned} \tag{4.3.29}$$

These exponent values are similar to the values found by other authors for ^{3}He (Wallace and Meyer 1970, Chase and Zimmerman 1973), ^{4}He (Roach 1968, Moldover 1969, Brown and Meyer 1972 Kierstead 1973, Tominaga 1974), Ar (Lin and Schmidt 1974a, Wu and Pings 1976), Kr (Gulari and Pings 1973), Xe (Edwards *et al.* 1968, Smith *et al.* 1971, Cornfeld and Carr 1972, Thoen and Garland 1974), CO_2 (Lipa *et al.* 1970, Levelt Sengers *et al.* 1971,

Table 4.3.4 Critical region parameters for a number of fluids assuming effective universal exponents

	Critical point parameters			MLSG equation		Restricted linear model		Restricted cubic model		Correlation length parameters
	P_c (MPa)	ρ_c (kg/m^3)	T_c (K)	x_0	E_1	k_1	a_1	k_2	a_2	ξ_0 (Å)
^3He	0.11678	41.45	3.3099	0.489	2.96	0.924	4.58	0.818	4.05	2.7
^4He	0.22742	69.6	5.1895	0.369	2.67	1.021	6.40	0.904	5.66	2.2
Ar	4.865	535	150.725	0.183	2.27	1.309	16.1	1.160	14.2	1.6
Kr	5.4931	908	209.286	0.183	2.27	1.309	16.1	1.160	14.2	1.7
Xe	5.8400	1110	289.734	0.183	2.27	1.309	16.1	1.160	14.2	1.9
p-H_2	1.285	31.39	32.935	0.260	2.34	1.156	9.6	1.024	8.5	1.9
N_2	3.398	313.9	126.24	0.164	2.17	1.361	18.2	1.206	16.1	1.6
O_2	5.043	436.2	154.580	0.183	2.21	1.309	15.6	1.160	13.9	1.6
H_2O	22.06	322.2	647.13	0.100	1.20	1.622	21.6	1.438	19.1	1.3
D_2O	21.66	357	643.89	0.100	1.20	1.622	21.6	1.438	19.1	1.3
CO_2	7.3753	467.8	304.127	0.141	2.01	1.436	21.3	1.273	18.9	1.6
NH_3	11.303	235	405.4	0.109	1.37	1.573	21.4	1.394	19.1	1.4
SF_6	3.7605	730	318.687	0.172	3.08	1.337	23.9	1.185	21.2	2.0
CH_4	4.595	162.7	190.555	0.164	2.03	1.361	17.0	1.206	15.1	1.7
C_2H_4	5.0390	215	282.344	0.168	2.17	1.350	17.5	1.197	15.5	1.9
C_2H_6	4.8718	206.5	305.33	0.147	2.03	1.416	20.2	1.255	17.9	1.8
C_3H_8	4.247	221	369.82	0.137	1.83	1.451	20.2	1.286	17.9	2.0

Notes: $\alpha = 0.100$, $\beta = 0.355$, $\gamma = 1.190$, $\delta = 4.352$. $E_2 = 0.287$. $b_1^2 = 1.3909$, $b_2^2 = 1.3100$, $c = 0.0393$. $\nu = 0.633$, $2 - \eta = 1.879$.

Murphy *et al.* 1973, White and Maccabee 1975), N_2O and $CClF_3$ (Levelt Sengers *et al.* 1971) and reviewed elsewhere (Levelt Sengers 1974).

If we represent the equation of state data in the range (4.3.28) by a scaled equation of state with the exponent values (4.3.29) and *assume* universality between fluids, we may then try to impose the same equation of state with the same exponents upon the data for other fluids and determine the resulting values for the two remaining adjustable parameters. Equation of state parameters thus obtained for a variety of fluids in terms of a universal equation of state (restricted MLSG equation, restricted linear model and restricted cubic model) are presented in Table 4.3.4. The parameters should be treated as informed estimates corresponding to the range (4.3.28). For accurate calculations one should return to the parameter values obtained in the original statistical fits to the data of the individual fluids (Levelt Sengers *et al.* 1976). The last column of Table 4.3.4 contains correlation length parameters to be discussed in Section 4.5.7.

The hypothesis of universality implies also universality for the amplitude ratios Γ/Γ', $\Gamma DB^{\delta-1}$ and A^+/A^-_{II} (cf. Table 4.3.2). The parameter values given in Table 4.3.4 yield $\Gamma/\Gamma' = 4.0$, $\Gamma DB^{\delta-1} = 1.5$ and $A^+/A^-_{II} = 0.47$ to be compared with the values $\Gamma/\Gamma' = 5.07$, $\Gamma DB^{\delta-1}$ 1.7 and $A^+/A^-_{II} = 0.51$ for the 3-dimensional Ising model (Aharony and Hohenberg 1976).

4.3.6 *Questions raised by experiments*

According to the hypothesis of universality, fluids are expected to belong to the same universality class as the Ising model and the Landau–Ginzburg–Wilson model. In the previous section we arrived at the conclusion that the experimental PVT data of six fluids could be represented in the temperature range $5 \times 10^{-4} \leqslant |\Delta T^*| \leqslant 3 \times 10^{-2}$ by a universal scaled equation of state with exponent values $\beta = 0.355 \pm 0.007$ and $\gamma = 1.19 \pm 0.03$. However, these exponent values exclude the values quoted in (4.2.64) and (4.2.78) for this universality class.

The picture of universality of fluid critical behaviour with the exponent values found in (4.3.29) has to be challenged as soon as, for at least one fluid, a different critical exponent value is found. Such experimental challenges do exist. One example is the exponent for the coexistence curve of SF_6 which has been measured in three independent experiments (Balzarini and Ohrn 1972, Rathjen and Straub 1973, Weiner *et al.* 1974). In each case, a value of β near 0.34 was found; in the last case, the value of β was 0.340 ± 0.001 in the range $2 \times 10^{-5} < |\Delta T^*| < 5 \times 10^{-2}$. In binary liquids, to be discussed in Section 4.4.5, a value of β near 0.34 is the rule rather than the exception (Stein and Allen 1974).

With respect to the exponent γ, the value $\gamma = 1.19 \pm 0.03$ in (4.3.28) is to be compared with a value $\gamma = 1.14$ recently deduced for ^3He by Behringer *et al.* (1976) from highly accurate *PVT* data, with a value $\gamma = 1.19 \pm 0.01$ determined for C_2H_4 by Hastings and Levelt Sengers (1976) from precise *PVT* data, and with γ values of about $\gamma = 1.22$ from light scattering data that are usually obtained at temperatures closer to T_c (Puglielli and Ford 1970, Lunacek and Cannell 1971, Cannell 1975).

The most serious challenge, however, has come from the study of gravity-induced density gradients by optical techniques developed by Wilcox and co-workers and discussed in Section 4.2.7. From an optical study of the coexistence curve of SF_6, Balzarini and Ohrn (1972) noted that the exponent β decreases when the temperature range is reduced. Very recently, Hocken and Moldover (1976) measured density gradient profiles in Xe, CO_2 and SF_6 in the range $-1.5 \times 10^{-5} < |\Delta T^*| < 5 \times 10^{-5}$. Analyzing the data in terms of a scaled equation of state, they obtained β values around 0.324 and γ values around 1.25, indicating that the critical exponents of fluids do approach Ising-like values sufficiently close to the critical point. Since these exponent values are only attained for $|\Delta T^*| < 10^{-4}$, it must be concluded that corrections to asymptotic scaling must be present in the entire range where conventional *PVT* experiments are accurate. Since these correction terms may vary from fluid to fluid, they may cause the deviations from universality of the apparent critical exponents determined in the conventional ranges.

4.3.7 Scaling fields and corrections to scaling

The treatment of the thermodynamic behaviour of fluids in the critical region outlined in the preceding sections has been challenged by theory in two respects. The first challenge pertains to the choice of scaling variables and the second one regards the corrections to be applied when the range of the asymptotic validity of the scaling laws is exceeded.

The choice of variables is best discussed in terms of intensive or 'field' variables, P, μ and T. It can be readily verified that the thermodynamic scaling laws formulated in Section 4.3.3 are equivalent to the statement in terms of field variables that the pressure has the form

$$P^* = P^*_{\text{sing}}(\Delta\mu^*, \Delta T^*) + \Delta\mu^* - A_0^*(T^*) \qquad (4.3.30)$$

with

$$P^*_{\text{sing}}(\Delta\mu^*, \Delta T^*) = |\Delta T^*|^{2-\alpha} f(\Delta\mu^*/|\Delta T^*|^{\beta\delta}). \qquad (4.3.31)$$

In the plane of the independent field variables μ^* and T^* one direction is singled out as special, namely the slope $c_1 = \mathrm{d}\mu^*_{\text{cxc}}/\mathrm{d}T^*$ of the coexistence curve at the critical point. To linear order in ΔT^* our first scaling field $\Delta\mu^*$ has the form $\Delta\mu^* = \mu^* - \mu_c^* - c_1\Delta T^*$, where $\mu_c^* = \mu^*(\rho_c, T_c)$. As the second scaling field in (4.3.31) we simply chose the temperature difference ΔT^*. There is, however, no physical reason why this second scaling field could also not be a function of $\mu^* - \mu_c^*$ and ΔT^*. In fact, the renormalization group theory of critical phenomena indicates that the singular part of the pressure, which is analogous to the free energy of a spin system discussed in Section 4.2.7, will in general have the form (Wegner 1972, Rehr and Mermin 1973):

$$P^*_{\text{sing}}(u_h, u_t) = |u_t|^{2-\alpha} f(u_h/|u_t|^{\beta\delta}), \qquad (4.3.32)$$

where the scaling fields u_h and u_t are analytic functions of μ^* and T^*

$$\begin{aligned} u_h &= (\mu^* - \mu_c^*) - c_1\Delta T^* + \ldots \\ u_t &= \Delta T^* - c_2(\mu^* - \mu_c^*) + \ldots . \end{aligned} \qquad (4.3.33)$$

The term $\Delta\mu^* - A_0^*(T^*)$ in (4.3.30) represents an analytic background term in terms of the field variables ΔT^* and $\Delta\mu^*$. Allowing for a more general analytic background one obtains a revised scaled equation of the form

$$P^* = |u_t|^{2-\alpha} f(u_h/|u_t|^{\beta\delta}) + \sum_i \sum_j P_{ij}(\Delta T^*)^i(\mu^* - \mu_c^*)^j \qquad (4.3.34)$$

with $P_{00} = P_c^* = 1$ and $P_{01} = \rho_c^* = 1$.

This equation reduces asymptotically to (4.3.30). However, the revised equation (4.3.34) with the scaling fields (4.3.33) leads to some additional features not contained in (4.3.30). In particular, the expression for the coexistence densities implied by (4.3.34) has, in addition to the leading term of the form $|\Delta T^*|^\beta$, a correction term with the entropy-like structure $|\Delta T^*|^{1-\alpha}$ (Widom and Rowlinson 1970, Rehr and Mermin 1973). This term drops out when the difference $\rho^*_{\text{liq}} - \rho^*_{\text{vap}}$ is formed, but it persists in the sum $\rho^*_{\text{liq}} + \rho^*_{\text{vap}}$. Hence, it is predicted that the coexistence curve diameter is not straight, but that it has a small 'hook' at the critical point. Thus far this hook has been observed in one experiment only (Weiner *et al.* 1974), but the theoretical arguments in favour of its existence in all fluids are rather strong. The reason that it has almost always escaped experimental detection is probably due to the fact that the $|\Delta T^*|^{1-\alpha}$ term is followed by a (ΔT^*) term, so that the effect of the former is only visible extremely close to the critical point (Zollweg and Mulholland 1972). The hook on the diameter implies that critical densities estimated by extrapolating the rectilinear diameter are overestimated, but the error is generally less than 0.5%. The effects of the different choices of scaling fields on the behaviour of fluids in the one-phase region have not yet been studied in any detail.

We now turn to the so-called corrections to scaling. Suppose that the scaling fields u_h and u_t have been defined properly and that the scaling hypothesis (4.3.32) is made. Then one can expect this hypothesis to be valid only in a limited range around the critical point. The renormalization group approach to the theory of critical phenomena also provides estimates of the form and the size of the correction terms beyond the asymptotic range, as pointed out by Wegner (1972). They are obtained by retaining in (4.2.75) not only the relevant scaling fields $u_1 = u_h$ and $u_2 = u_t$, but also irrelevant scaling fields $u_3, u_4, \ldots$. In analogy to the free energy of a spin system the singular part of the pressure then assumes the form (Wegner 1972, Ley-Koo and Green 1976)

$$P^*_{\text{sing}}(u_h, u_t, u_3, u_4, \ldots) = |u_t|^{2-\alpha} f(u_h\,|u_t|^{-\beta\delta}, u_3|u_t|^{\Delta_1}, u_4\,|u_t|^{\Delta_2}, \ldots) \qquad (4.3.35)$$

Unlike the relevant scaling fields u_t and u_h, the irrelevant scaling fields do not approach zero at the critical point. They scale with different powers of u_t, indicated by the so-called gap exponents Δ_1, Δ_2, ... These gap exponents are again expected to be universal. Current estimates of the first two gap exponents for Ising-like systems in three dimensions are $\Delta_1 = 0.50$ and $\Delta_2 = 1.5$ (Golner and Riedel 1975, Saul *et al.* 1975, Baker *et al.* 1976, Camp *et al.* 1976).

The irrelevant scaling fields do not contribute to the leading anomalies at the critical point, but they do lead to corrections to the asymptotic scaling laws. For instance, the expansions for the coexisting densities obtain the form (Ley-Koo and Green, 1976):

$$\frac{\Delta\rho^*_{\text{liq}} + \Delta\rho^*_{\text{vap}}}{2} = B_{M1}\,|\Delta T^*|^{1-\alpha} + P_{11}\Delta T^* + B_{C3}\,|\Delta T^*|^{1-\alpha+\Delta_1} + \ldots \qquad (4.3.36)$$

$$\frac{\Delta\rho^*_{\text{liq}} - \Delta\rho^*_{\text{vap}}}{2} = B\,|\Delta T^*|^{\beta} + B_{C1}\,|\Delta T^*|^{\beta+\Delta_1} + B_{C2}\,|\Delta T^*|^{\beta+2\Delta_1} + B_{M2}|\Delta T^*|^{\beta+1} + \ldots, \qquad (4.3.37)$$

where the coefficients with subscript M refer to terms originating from 'mixing' the μ and T variables in forming the scaling fields and those with subscript C to corrections to scaling.

4.3.8 Assessment of status of fluid critical behaviour

As discussed in Section 4.3.4 and 4.3.5, when equation of state data of fluids are considered in a range of ± 0.25 in $\Delta\rho^*$ and ± 0.03 in ΔT^*, the data can usually be represented in terms of a simple scaled equation of state with critical exponents that vary only slightly from fluid to fluid, but that are distinctly different from those of the Ising model. However, recent optical experiments at $|\Delta T^*| < 5 \times 10^{-5}$ have shown that in that range the fluid critical exponents are approaching those for the Ising model (Hocken and Moldover 1976).

The optical data can be reconciled with the data obtained in more conventional ranges if a Wegner expansion (4.3.35) with a sufficient number of terms is used. Since the gap exponent in this expansion is small and the coefficients are expected to be of order unity, the Wegner expansion will converge only slowly and a large number of terms will contribute in the range $10^{-4} < |\Delta T^*| < 10^{-1}$. Ley-Koo and Green (1976) recently fitted the difference of coexisting densities of SF_6. They found that a simple power law $B\,|\Delta T^*|^{\beta}$ with $\beta = 0.327 \pm 0.003$ was restricted to a range $|\Delta T^*| < 7 \times 10^{-4}$. A temperature range $|\Delta T^*| < 2 \times 10^{-2}$ already required three terms in the expansions (4.3.36) and (4.3.37). Had they imposed the value $\beta = 0.312$ suggested from series expansions for Ising-like systems, the asymptotic range would have been even smaller and more correction terms would have been required. A value for β slightly larger than 0.312 is thus supported by the optical experiments of Hocken and Moldover (1976), by an analysis of the coexistence curve data for SF_6 (Ley-Koo and Green 1976) and, as we shall see in Section 4.4.5, by data on coexisting phases in partially miscible binaries.

A value of β slightly larger than 0.312 is also supported by calculations on the basis of the renormalization group theory. As quoted in (4.2.78) the calculations of Baker *et al.* (1976) yield $\beta = 0.320 \pm 0.015$. Exponent values recently calculated by Golner and Riedel imply $\beta = 0.322$ with undetermined error. It is as yet not clear how the small discrepancy with the series expansion estimate for the exponent β will be resolved.

From a practical point of view, the representation of the thermodynamic data of fluids by a Wegner expansion in extended ranges around the critical point will be quite involved and unappealing. Hence, it is expected that the phenomenological expressions with effective exponents presented in Section 4.3.5 will remain useful for many practical purposes.

4.4 Critical phenomena in fluid mixtures

4.4.1 Phase transitions in fluid mixtures

In the one-phase region, a fluid mixture of n components possesses $n + 1$ thermodynamic degrees of freedom. For two phases to coexist, the number of degrees of freedom is reduced by one. The intensive or field variables are the same in the two coexisting phases. The thermodynamic states with two coexisting phases correspond to a n-dimensional surface in the $(n + 1)$-dimensional space of independent field variables. The condition of criticality is that the two coexisting phases become identical; this condition again reduces the number of available degrees of freedom by one. Thus the n-dimensional coexistence surface may terminate in a $(n - 1)$-dimensional critical surface.

As an example we show schematically in Figure 4.4.1 the geometry of the gas–liquid transition in a binary fluid in the 3-dimensional space of independent field variables. Since the chemical potentials μ_1 and μ_2 of the two components diverge to $-\infty$ at infinite dilution, it is often convenient to use instead the fugacities $f_1 = \exp(\mu_1/k_B T)$ and $f_2 = \exp(\mu_2/k_B T)$. As the three independent field variables we choose P, T and f_2. The coexistence surface is a 2-dimensional surface that terminates in a 1-dimensional critical line. The vapour–liquid critical points located on the critical line are called 'plait points', a term dating back to Van der Waals. In this particularly simple example the critical line connects the critical points of the two pure fluids. In many cases, the critical line is interrupted and more complicated phase behaviour results. The interested reader is referred to some excellent reviews on this subject (Kay 1968, Rowlinson 1969, Schneider 1970, 1972, Scott 1972, Streett 1974, Hicks and Young 1975).

In addition to the coexistence of a vapour and a liquid in a binary fluid, there exists also the possibility that two liquid phases of different composition coexist. This gives rise to another coexistence surface in the space of independent field variables. Such binary liquids are usually studied in the presence of a vapour phase; the resulting system has one degree of freedom and traces a curve on the coexistence surface. The possibility exists for the two liquid phases to

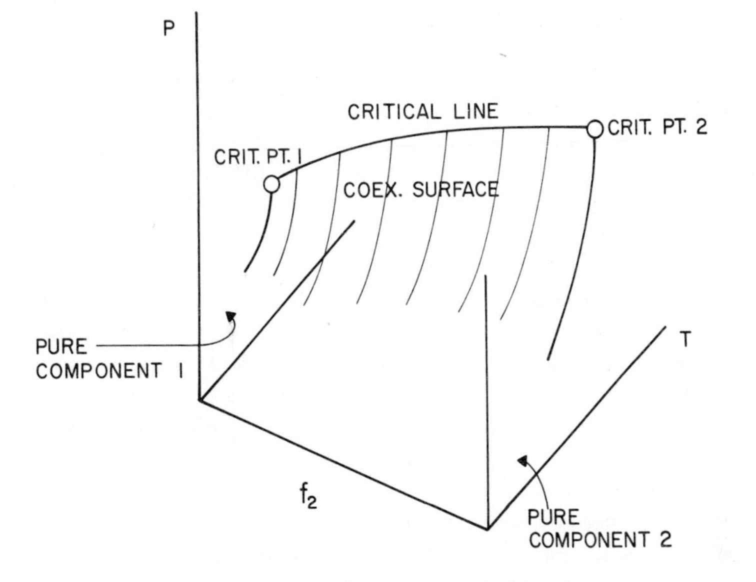

Figure 4.4.1 The gas–liquid coexistence surface and critical line of a binary fluid in the space of independent field variables pressure P, temperature T and fugacity f_2

become identical; such a critical point is called a consolute point. This point is part of a line of consolute points terminating the coexistence surface. In P–T space, the line of consolute points is usually very steep, since the pressure has only a small effect on the critical solution temperature. In general, the plait point critical curve and the consolute critical curve are separated, but examples exist where one curve goes continuously over into the other. In some cases where the critical points of the two pure components are not connected, the critical line, starting from the critical point of the least volatile component (i.e. the one with the higher T_c and P_c) moves up to higher temperatures and pressures. Since a phase separation now occurs at temperatures and pressures above the critical ones of both components, one speaks about gas–gas equilibria. The simplest systems in which gas–gas separation occurs are mixtures of light and heavy noble gases. They have been the subject of several recent studies (De Swaan Arons and Diepen 1966, Streett 1965, 1967, Trappeniers and Schouten 1974).

Although the variety of phase behaviour, even for systems of two components, may seem somewhat bewildering, the description of critical-region phase behaviour in multicomponent fluids is basically simple. Particular insight is provided by the theory of Griffiths and Wheeler (1970) who developed the geometry of the critical region in the space of field variables. If we exclude a special direction of any coordinate axis, the critical behaviour of a binary fluid at any type of critical point is a straightforward generalization of that in a one-component fluid, while no new features are introduced when three- or more-component fluids are considered.

Special orientations occur when critical lines pass through maxima or minima in temperature or pressure, or when critical azeotropy is encountered. Azeotropy occurs in binary fluids when two coexisting phases, which in general have different values of all extensive variables or densities, happen to have the same concentration. An azeotrope may trace out a curve on the coexistence surface; if it reaches the critical line one speaks of critical azeotropy. All these cases are considered separately in the theory of Griffiths and Wheeler. A brief account of this theory is given in the subsequent sections.

4.4.2 Introduction to theory of Griffiths and Wheeler

Griffiths and Wheeler (1970) developed a description of thermodynamic behaviour in the vicinity of critical surfaces in multicomponent systems that leads to a classification of the types of divergences to be expected in certain thermodynamic derivatives. The theory is most easily introduced by first reconsidering the description of the thermodynamic behaviour near the critical point of a one-component fluid in terms of field variables. As the independent field variables we take temperature and pressure and as dependent field variable the chemical potential. (In Sections 4.2, 4.3 and 4.5 the chemical potential is taken per particle. In this section we find it more convenient to take the chemical potential per mole. Molar values of extensive thermodynamic quantities are indicated by a tilde.) We have seen that the slope of the vapour pressure curve (coexistence curve in the P–T plane) has a finite limiting value at the critical point and the P- and T-axis do not have a special direction parallel to this coexistence curve. In addition we note that the second derivatives $-(\partial^2\mu/\partial T^2)_P = \tilde{C}_P/T$ and $-(\partial^2\mu/\partial P^2)_T = \tilde{V}K_T$ diverge strongly at the critical point as discussed in Section 4.1. *It is concluded that second derivatives of the dependent field variable with respect to directions that are oblique to the coexistence surface are strongly divergent.*

Next we consider a second derivative of μ taken along the coexistence surface, i.e. coexistence curve for a one-component fluid. For this purpose Griffiths and Wheeler introduce the thermodynamic identity

$$-\left(\frac{\partial^2\mu}{\partial T^2}\right)_Y = \frac{\tilde{C}_V}{T} + \tilde{V}K_T\left[\left(\frac{\partial P}{\partial T}\right)_Y - \left(\frac{\partial P}{\partial T}\right)_V\right]^2 \tag{4.4.1}$$

where Y is any direction in the P–T plane. It is seen that $\partial^2\mu/\partial T^2$ will diverge strongly, as K_T, for any direction Y that does not coincide with the direction of the critical isochore $V = V_c$. However, if the direction Y does coincide with that of the critical isochore at the critical point, then the term in brackets vanishes and $\partial^2\mu/\partial T^2$ will diverge only weakly, as C_V. Thus, a second derivative taken in this special direction is a derivative in which an extensive variable or 'density' (V) rather than a field (P) is kept constant.

It remains to be shown that the direction $V = V_c$ is asymptotically the same as that of the coexistence curve. We have seen that this is indeed the case for the classical equations and the lattice gas. In general, one makes use of the thermodynamic identity

$$\left(\frac{dV}{dT}\right)_{cxc} = \left[\left(\frac{dP}{dT}\right)_{cxc} - \left(\frac{\partial P}{\partial T}\right)_V\right]\left(\frac{\partial V}{\partial P}\right)_T \tag{4.4.2}$$

which relates the slope dV/dT and the compressibility $(\partial V/\partial P)_T$ on either side of the phase boundary. The slope dV/dT has a different sign on the two sides of the phase boundary, but the compressibility has the same sign. Thus the term in square brackets must vanish at the critical point and, in fact, it must go to zero as $|\Delta T^*|^{\gamma+\beta-1}$. Thus the continuity of the slope of the vapour pressure curve and the critical isochore at the critical point holds for real fluids as well. *It is concluded that second derivatives of the dependent field variable with respect to the independent ones, but taken along the coexistence surface, are in fact derivatives in which a density is held constant and that these derivatives diverge weakly.*

Finally, in multicomponent fluids a third direction exists, namely the one along the critical surface. One-component fluids do not provide us with any clues for the behaviour of derivatives along this additional direction. However, we have seen that in binary fluids critical lines are smooth curves in field space. We thus expect second derivatives taken in this direction to be nondivergent; as we shall see, they correspond to derivatives in which not one but two densities are held constant.

4.4.3 Classification of thermodynamic anomalies in fluid mixtures

In order to make the ideas about the geometry of critical-point phase transitions in mixtures more precise, Griffiths and Wheeler considered the properties of the compliance matrix introduced in Section 4.1. We have seen that second derivatives with respect to the field variables,

$$D(P) = -(\partial^2\mu/\partial P^2)_T \quad D(T) = -(\partial^2\mu/\partial T^2)_P \tag{4.4.3}$$

are strongly divergent if no special geometric relation exists between the field variables and the coexistence surface or the critical surface. Although each element of the compliance matrix diverges strongly, its determinant $D(T, P) = VK_TC_V/T$ is seen to diverge only as the product of a strongly (K_T) and a weakly (C_V) diverging quantity. The factor K_T corresponds to a second derivative taken at an angle with the coexistence curve. For an n-component mixture, the determinant of the full $(n + 1)$ by $(n + 1)$ compliance matrix is still no more divergent than strongly times weakly because, in the space of independent field variables, a base can be chosen with only one vector at an angle to the coexistence surface ('strong' direction) and only one vector in the coexistence surface, but not in the critical surface ('weak' direction). Thus the behaviour near a critical point in a multicomponent mixture is no more complicated than critical behaviour in a binary mixture.

Similar considerations can be applied to any subset A of m independent field variables with $m \leqslant n + 1$. If in this subset a base can be chosen such that one vector is oblique to the coexistence surface and one lies in the critical surface, then the determinant $D(A)$ of the matrix of corresponding second derivatives will be strongly times weakly divergent. If the base lacks a vector oblique to the coexistence surface, $D(A)$ is weakly divergent. If it contains a vector oblique to the coexistence surface but lacks one in the critical surface, then $D(A)$ is strongly divergent. If the base has no vectors other than in the critical surface, $D(A)$ is nondivergent.

As an example, let us consider a binary mixture with field variables T, P, μ_1, μ_2. As the dependent variable we consider $-\mu_2(T, -P, \Delta)$ as a function of T, $-P$ and $\Delta = \mu_1 - \mu_2$. In terms of these variables the Gibbs–Duhem equation assumes the form

$$\mathrm{d}(-\mu_2) = \tilde{S}\mathrm{d}T + \tilde{V}\mathrm{d}(-P) + X\mathrm{d}\Delta \tag{4.4.4}$$

where X is the mole fraction of component 1. The second derivatives

$$\tilde{C}_{P\Delta}/T = -\left(\frac{\partial^2\mu_2}{\partial T^2}\right)_{P\Delta} \quad \tilde{V}K_{T\Delta} = -\left(\frac{\partial^2\mu_2}{\partial P^2}\right)_{T\Delta} \quad \chi_{TP} = -\left(\frac{\partial^2\mu_2}{\partial \Delta^2}\right)_{TP} \tag{4.4.5}$$

are expected to diverge strongly, since the derivatives are taken in directions that are oblique to the coexistence surface. In the terminology developed above, these quantities are of the form $D(A)$ where A is a single vector not parallel to the coexistence surface.

However, with the exception of χ_{TP} which can be studied by light scattering, derivatives with two fields constant are usually not very accessible to the experimenter who tends to conduct measurements at constant concentration rather than at constant Δ. Hence, one is more interested in the behaviour of derivatives like K_{TX}, K_{SX}, C_{PX} and C_{VX}, i.e. derivatives in which one or more densities are held constant. Using Jacobians such derivatives can be expressed in terms of ratios of determinants of matrices of second derivatives, as pointed out by Griffiths and Wheeler (1970). A few examples will suffice to give the general idea:

$$\tilde{V}K_{TX} = -\left(\frac{\partial \tilde{V}}{\partial P}\right)_{TX} = -\frac{\partial(\tilde{V}X)}{\partial(P\Delta)} \cdot \frac{\partial(P\Delta)}{\partial(PX)} = -\frac{\partial(\tilde{V}X)}{\partial(P\Delta)} \Big/ \frac{\partial X}{\partial \Delta} = \frac{D(P\Delta)}{D(\Delta)} \tag{4.4.6a}$$

and, likewise,

$$\frac{\tilde{C}_{PX}}{T} = \frac{D(T\Delta)}{D(\Delta)}. \tag{4.4.6b}$$

A second derivative with one density constant, such as K_{TX} or C_{PX}, can be written as the quotient of a determinant, such as $D(P\Delta)$ or $D(T\Delta)$ which diverges strongly times weakly, and a determinant, such as $D(\Delta)$ which is strongly divergent. Hence we expect derivatives with one density held constant to be weakly divergent. A derivative at constant density is taken along a path asymptotically parallel to the coexistence surface.

Derivatives in which two densities are held constant can be handled in a similar manner. For example

$$\tilde{V}K_{SX} = -\left(\frac{\partial \tilde{V}}{\partial P}\right)_{SX} = -\frac{\partial(\tilde{V}\tilde{S}X)}{\partial(TP\Delta)} \cdot \frac{\partial(T\Delta)}{\partial(\tilde{S}X)} = \frac{D(TP\Delta)}{D(T\Delta)}, \tag{4.4.7a}$$

and, likewise,

$$\frac{\tilde{C}_{VX}}{T} = \frac{D(TP\Delta)}{D(P\Delta)}. \tag{4.4.7b}$$

Since both determinants diverge strongly times weakly, we expect the derivatives with two densities held constant to be nondivergent. These derivatives are taken along a path that lies asymptotically in the critical surface.

We should expect exceptions to these rules when the critical surface or the coexistence surface bears a special relation to any of the coordinate axes. For instance, critical lines often pass through pressure extrema as a function of Δ. At such an extremum the critical curve is parallel to the T–Δ plane. In this plane we can choose a base with one vector parallel to the critical line and one vector at an angle to the coexistence surface. As a consequence $D(T\Delta)$ no longer diverges strongly times weakly, but is only strongly divergent. Thus we conclude from (4.4.6b) and (4.4.7a) that C_{PX} is nondivergent and VK_{SK} weakly divergent, just the opposite of the general case. Similarly, if a critical line goes through an extremum in temperature, $D(P\Delta)$ is strongly divergent and we conclude that VK_{TX} is nondivergent, while C_{VX} is weakly divergent, again the reverse of the general case.

Along an azeotrope the volumes of the two coexisting phases, $\tilde{V} = (\partial\mu_2/\partial P)_{T\Delta}$, and the entropies, $\tilde{S} = -(\partial\mu_2/\partial T)_{P\Delta}$ are unequal, while the concentrations $X = -(\partial\mu_2/\partial\Delta)_{PT}$ are the same. By applying (4.4.4) to the two coexisting phases it can be readily shown that under these conditions

$$\left(\frac{\partial P}{\partial \Delta}\right)_T = \left(\frac{\partial T}{\partial \Delta}\right)_P = 0. \tag{4.4.8}$$

Thus the coexistence surface is parallel to the Δ-axis and $D(\Delta)$ is weakly divergent in the case of critical azeotropy. From (4.4.6) we conclude that K_{TX} and C_{PX} are strongly divergent as in a one-component fluid, while C_{VX} remains finite. Other cases of exceptional orientation of a coordinate axis with respect to the coexistence surface can be treated in a similar manner.

The thermodynamics of binary mixtures near the gas–liquid critical line was also considered by Saam (1970).

4.4.4 Thermodynamic behaviour near a critical line in binary fluids in terms of scaling laws

A procedure for describing the critical behaviour of binary fluids using scaling laws was recently developed by Leung and Griffiths (1973). They consider the case where the critical line

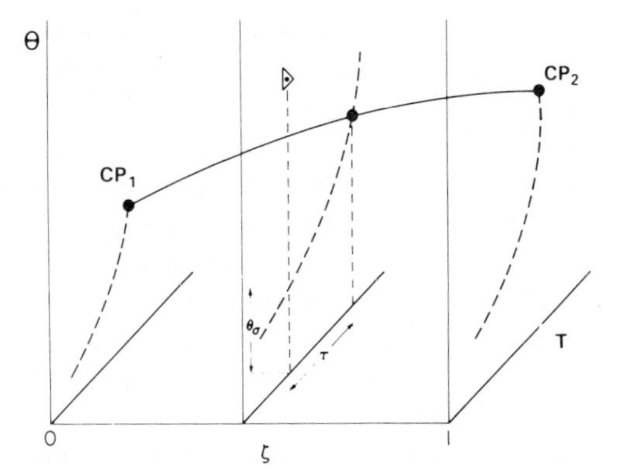

Figure 4.4.2 Independent field variables used in the Leung–Griffiths equation of state near the critical line of a binary gas–liquid system

connects the critical points of the two pure fluids as indicated in Figures 4.4.1 and 4.4.2. As the dependent field variable they take the pressure P; as the independent field variables they take the temperature T and two variables Θ and ζ which are related to the fugacities f_1 and f_2 by

$$\Theta = c_1 f_1 + c_2 f_2 \qquad \zeta = c_2 f_2/\Theta \tag{4.4.9}$$

where c_1 and c_2 are positive constants. The value $\zeta = 0$ corresponds to the pure fluid 1 and the value $\zeta = 1$ corresponds to the pure fluid 2. The critical line may be characterized by $T_c(\zeta)$. We now consider the behaviour of the mixture as a function of T and θ for a given value of ζ (see Figure 4.4.2). The intersection of the coexistence surface with a plane ζ = constant defines a coexistence curve that terminates in a point on the critical line but that can conceptually be extended into the one-phase region. For $\zeta = 0$ and $\zeta = 1$ this intersection reproduces the coexistence curves of the pure fluids 1 and 2, respectively. Leung and Griffiths now observe that the behaviour of the mixture in the plane ζ = constant is completely analogous to the behaviour of a one-component fluid. In this plane a coordinate system can be chosen with coordinates $\tau = T - T_c(\zeta)$ and $h = \ln(\Theta/\Theta_\sigma)$, where $\Theta_\sigma(\zeta, \tau)$ is the value of Θ on the coexistence surface for the given values of ζ and τ. The variables τ and h are closely related to the variables ΔT^* and $\Delta\mu^*$ in the scaling laws for one-component fluids. In the formulation of Leung and Griffiths it is assumed that the pressure in the plane ζ = constant can be written as the sum of a regular background term $P_r(\zeta, \tau, h)$ which is an analytic function of its variables and a singular term $P_{sing}(\zeta, \tau, h)$ which satisfies the same scaling law in terms of τ and h as the singular part of the pressure of one-component fluids does in terms of ΔT^* and $\Delta\mu^*$ (cf. (4.3.31)). In addition, P_{sing} contains a factor which is a smooth function of ζ so that one can interpolate between the scaled equations of state of the two pure components.

The approach of Leung and Griffiths is an application of the geometric considerations outlined in the previous section. They applied it with success to describe the experimental equation of state data obtained by Wallace and Meyer (1972) near the critical line of mixtures of ^{3}He and ^{4}He. Doiron *et al.* (1976) showed that the same equation also is in good agreement with the experimental data of the pressure coefficient $(\partial P/\partial T)_{VX}$ for these mixtures. However, the cusp-like behaviour of C_{VX} and $(\partial P/\partial T)_{VX}$, and the weak divergence of K_{TX}, predicted by the theory and incorporated in the equation of Leung and Griffiths, were not observed experimentally. Leung and Griffiths (1973) estimated that the cusp-like behaviour of C_{VX} and the divergence of K_{TX} would not be visible unless the critical line is approached to within

10^{-5} K. Outside this region, C_{VX} seems to diverge just as in pure fluids (Brown and Meyer 1972) and K_{TX} appears to rise to a finite maximum (Wallace and Meyer 1972).

The ^3He–^4He system has the simple feature that the critical line is a straight line connecting the critical points of the two pure components. Approximating the critical line by a quadratic curve D'Arrigo *et al.* (1975) applied the method of Leung and Griffiths to a mixture of CO_2 and C_2H_4 which exhibits critical azeotropy. They were able to describe the experimental dew and bubble curves satisfactorily. As discussed in Section 4.4.3 the theory of Griffiths and Wheeler predicts that K_{TX} will diverge weakly away from the critical azeotrope and strongly *at* the critical azeotrope. As in the case of the ^3He–^4He mixture it was found that these predicted anomalies in practice elude experimental observation, and it may be difficult to find systems in which they can be tested experimentally.

Moldover further generalized the method of Leung and Griffiths to systems in which the critical line is an arbitrary smooth curve connecting the critical points of the two pure components. He applied the method to the systems CO_2–C_2H_6, SF_6–C_3H_8 and C_3H_8–C_8H_{18} with satisfactory results (Moldover and Gallagher 1976).

4.4.5 Critical behaviour near the consolute point of binary liquids

The approach of Leung and Griffiths towards describing experimental data near a plait point is based on the continuity of critical phenomena in one- and two-component fluids. Thus, if universality of critical behaviour is valid for one-component fluids, it is expected to extend to critical behaviour in fluid mixtures as well.

The theory of Griffiths and Wheeler applies to all kinds of critical points in fluid mixtures, but the specific approach of Leung and Griffiths in its present form does not apply near the consolute point of partially miscible binary liquids. Nevertheless, a direct analogy can be made between critical phenomena near the gas–liquid critical point of a one-component fluid and critical phenomena near the critical mixing point of a binary liquid. When the pressure is held constant, the fundamental differential relation (4.4.4) for a binary mixture reduces to

$$d\mu_2 = -\tilde{S}dT - Xd\Delta. \tag{4.4.10}$$

Since the effect of pressure on critical mixing is small, and since experiments in binary liquids are usually conducted with the vapour phase present, most experimental data may indeed be interpreted as obtained at constant pressure. On comparing (4.4.10) with the corresponding relation $-dP = -sdT - \rho d\mu$, given in (4.2.11) for one-component fluids, we conclude that the behaviour of the fundamental equation $\mu_2(T, \Delta)$ for binary liquids at constant P near the critical mixing point will be analogous to that of $P(T, \mu)$ near the critical point of a one-component fluid. A number of thermodynamic properties exhibiting analogous critical behaviour are listed in Table 4.4.1. The concentration X is now the order parameter analogous to the magnetization of a ferromagnet near the Curie point and the density of a fluid near the gas–liquid critical point. It appears that the coexistence curve is considerably more symmetric when the concentration of either species is taken per unit of volume of the mixture rather than per unit of mole (Hildebrand *et al.* 1970). Therefore, the experimental data near the consolute critical point of a binary liquid are preferably analysed with the concentration measured in terms of volume fractions.

The critical exponents α, β, γ, δ now represent the power law behaviour of C_{PX} as a function of temperature, the composition difference between the two coexisting phases as a function of temperature, the osmotic susceptibility χ_T as a function of temperature and the critical isotherm of μ_2 as a function of concentration. A power law analysis of coexistence curve data for nine binary liquids mixtures was made by Stein and Allen (1974). A

Table 4.4.1 Analogy between critical behaviour of one-component fluids and critical mixing of binary liquids

	One-component	Binary liquid (P = constant)
Fundamental Equation	$P(T, \mu)$	$\mu_2(T, \Delta)$
'Density'	$\rho = (\partial P/\partial \mu)_T$	$X = -(\partial \mu_2/\partial \Delta)_{T,P}$
Coexistence curve	$\rho(T)$	$X(T)$
Equation of state	$\mu(\rho, T)$	$\Delta(X, T)$
Susceptibility (strongly divergent)	$\chi_T = (\partial^2 P/\partial \mu^2)_T$	$\chi_T = -(\partial^2 \mu_2/\partial \Delta^2)_{T,P}$
Specific heat (weakly divergent)	$C_V/V = T(\partial^2 P/\partial T^2)_\delta$	$C_{P,X} = -T(\partial^2 \mu_2/\partial T^2)_{X,P}$

comprehensive review of critical exponent values found experimentally for binary fluids has been presented by Scott (1977). When the coexistence curve data for various fluids were represented by a simple power law one most frequently deduced effective values for the exponent β between 0.33 and 0.35 (Stein and Allen 1974, Scott 1977). The experimental information concerning the exponents α and δ of binary liquids is rather limited. Morrison and Knobler (1976) recently reported $0.08 \leqslant \alpha \leqslant 0.14$ for the isobutyric acid–water system.

Recently, experimental information has become available which does support the hypothesis of universality for binary liquid mixtures. Greer (1976) obtained a set of very accurate coexisting density data for the isobutyric acid–water system using a magnetic-float densimeter. After converting the data to volume fractions she found that the coexistence curve could be represented by a simple power law $B\,|\Delta t^*|^\beta$ in the range $|\Delta T^*| < 6 \times 10^{-3}$ with $\beta = 0.328 \pm 0.004$. She also showed that an analysis of the coexistence curve data reported by Gopal *et al.* (1973, 1974) for carbondisulphide-nitromethane in terms of a 3- term Wegner expansion of the type presented in (4.3.37) yields a leading exponent $\beta = 0.316 \pm 0.008$. Earlier Balzarini (1974) had already noted that the exponent β of aniline-cyclohexane assumed the constant value $\beta = 0.328 \pm 0.007$ in a range $|\Delta T^*| < 10^{-2}$. The osmotic susceptibility $\chi_T = (\partial X/\partial \Delta)_{T,P}$ and, hence the exponent γ can be determined by measuring the intensity of scattered light (Fabelinski 1968). From a review of the light scattering data Chu (1972) suggested as the most probable value $\gamma = 1.23 \pm 0.02$. Recent light scattering measurements obtained by Chang *et al.* (1976) for 3-methylpentane-nitroethane at temperatures corresponding to $|\Delta T^*| < 3 \times 10^{-3}$ gave a value $\gamma = 1.240 \pm 0.014$. The close agreement between the exponent values currently found experimentally for β and γ of binary liquids with the values (4.2.78) calculated theoretically from the renormalization group theory does provide substantial evidence in support of the hypothesis of universality. The asymptotic power laws in binary liquids near the critical point appear to hold over a somewhat larger temperature range than in fluids near the gas–liquid critical point. Whether there exists a physical reason for this phenomenon or whether it is just the consequence of a fortuitous cancellation between the first few correction terms in the Wegner expansion is not clear yet.

Attempts have also been made to represent experimental chemical potential data of binary mixtures near the critical mixing point by a scaled equation of state analogous to (4.3.11) (Simon *et al.* 1972). However, progress has been limited due to a lack of sufficient experimental data off the critical isochore and coexistence curve.

4.5 Critical fluctuations

4.5.1 *Correlation function and power laws*

When a system approaches a critical point, its thermodynamic states are accompanied by large fluctuations in the order parameter. Thus a fluid near the gas–liquid critical point exhibits large density fluctuations, and a liquid mixture near the critical mixing point exhibits large concentration fluctuations. Here we shall develop an explicit description of these fluctuations in a fluid near the gas–liquid critical point. The corresponding equations for a binary liquid near the critical mixing point are obtained if the density is replaced with the concentration of either component.

The magnitude and spatial character of these fluctuations are described in terms of a correlation function defined as

$$\rho^2 G(|\mathbf{r}-\mathbf{r}'|) = \langle \{\rho(\mathbf{r})-\rho\}\{\rho(\mathbf{r}')-\rho\}\rangle = \langle \rho(\mathbf{r})\rho(\mathbf{r}')\rangle - \rho^2 \tag{4.5.1}$$

where the brackets $\langle\,\rangle$ indicate an equilibrium average over a grand canonical ensemble. $\rho(\mathbf{r})$ is the local number density at position $\mathbf{r}$ and $\rho = \langle \rho(\mathbf{r}) \rangle$ is the average equilibrium density which is independent of the position $\mathbf{r}$ (not considering the presence of external forces such as gravity). The correlation function $G(|\mathbf{r}-\mathbf{r}'|)$ measures the joint probability of finding molecules in volume elements $\mathrm{d}\mathbf{r}$ and $\mathrm{d}\mathbf{r}'$ minus the average number of pairs.† For an isotropic fluid this probability is only a function of the distance $|\mathbf{r}-\mathbf{r}'|$. The zeroth moment of the correlation function is related to the compressibility by the fluctuation theorem (De Boer 1949)

$$k_{\mathrm{B}}T\chi_T = \rho^2 \int \mathrm{d}\mathbf{r}\, G(r) \tag{4.5.2}$$

where $\chi_T = \rho^2 K_T$ is the symmetrized isothermal compressibility introduced in (4.2.1). The correlation function G is a function of $\Delta\rho^*$ and ΔT^* as well as of r. Whenever we want to indicate the dependence on the thermodynamic variables explicitly, we write $G(\Delta\rho^*, \Delta T^*; r)$ instead of $G(r)$.

A static structure factor may be defined as††

$$\chi(\Delta\rho^*, \Delta T^*; k) = \frac{\rho^2}{k_{\mathrm{B}}T}\int \mathrm{d}\mathbf{r}\, \exp(i\mathbf{k}\cdot\mathbf{r})\, G(\Delta\rho^*, \Delta T^*; r) \tag{4.5.3}$$

so that

$$\chi(\Delta\rho^*, \Delta T^*; 0) = \chi_T(\Delta\rho^*, \Delta T^*). \tag{4.5.4}$$

This structure factor as a function of the wavenumber k can be measured, since it is directly proportional to the intensity of scattered electromagnetic radiation as a function of scattering angle (Fabelinskii 1968, McIntyre and Sengers 1968). We find it also convenient to introduce a dimensionless structure factor χ^* defined as

$$\chi^* = \frac{P_{\mathrm{c}}}{\rho_{\mathrm{c}}^2}\chi. \tag{4.5.5}$$

†In the literature the correlation function $G(|\mathbf{r}-\mathbf{r}'|)$ is usually defined such that $\rho^2 G(|\mathbf{r}-\mathbf{r}'|) + \rho\delta(\mathbf{r}-\mathbf{r}') = \langle\rho(\mathbf{r})\rho(\mathbf{r}')\rangle - \rho^2$, where the δ-function accounts for the self correlation at $\mathbf{r}'=\mathbf{r}$ (Hirschfelder *et al.* 1954). Here we prefer to absorb this δ-function in the definition of $G(|\mathbf{r}-\mathbf{r}'|)$. Because of their long-range character the difference between the two correlation functions becomes irrelevant sufficiently close to the critical point.

††In the definition of the structure factor we have included a factor $k_{\mathrm{B}}T$ so that χ has the same dimension as the symmetrized compressibility χ_T. This factor may for all practical purposes be identified with $k_{\mathrm{B}}T_{\mathrm{c}}$ in the region where the scaling laws apply, but may have to be treated more carefully when corrections to the asymptotic scaling laws are considered.

The range of the correlation function and, hence, the spatial extent of the fluctuations is characterized by a correlation length ξ. It may be defined as

$$\xi^2 = \frac{1}{2d} \frac{\int \mathrm{d}\mathbf{r}\, r^2 G(r)}{\int \mathrm{d}\mathbf{r}\, G(r)} \tag{4.5.6}$$

where d is the dimensionality of the system or, in terms of the structure factor,

$$\xi^2 = -\lim_{k \to 0} \left(\frac{\partial \ln \chi(k)}{\partial k^2} \right)_{\Delta\rho^*,\, \Delta T^*} . \tag{4.5.7}$$

The correlation length $\xi = \xi(\Delta\rho^*, \Delta T^*)$ is a function of density and temperature which diverges at the critical point. In particular, along the critical isochore $\Delta\rho^* = 0$ and along the coexistence curve $\Delta\rho^* = \Delta\rho_{\mathrm{cxc}}$ it behaves as

$$\xi(0, \Delta T^*) = \xi_0 (\Delta T^*)^{-\nu} \qquad (\Delta T^* \geqslant 0) \tag{4.5.8a}$$

$$\xi(\Delta\rho^*_{\mathrm{cxc}}, \Delta T^*) = \xi_0' \,|\, \Delta T^* \,|^{-\nu'} \qquad (\Delta T^* \leqslant 0) \tag{4.5.8b}$$

which defines the critical exponents ν and ν'.

Another exponent η is introduced to specify the nature of the dependence of the correlation function on the distance r. It is defined such that at the critical point $\Delta\rho^* = 0$, $\Delta T^* = 0$

$$G(0, 0; r) \propto \frac{1}{r^{d-2+\eta}} \tag{4.5.9}$$

or, equivalently

$$\chi(0, 0; k) \propto k^{\eta - 2} . \tag{4.5.10}$$

The exponent η is zero in the classical theory of Ornstein and Zernike (Fisher 1964). However, the exponent η is known to be different from zero for the lattice models with short range interactions that have been solved theoretically. Series expansion estimates for the correlation function exponents of the 3-dimensional Ising model are (Moore *et al.* 1969, Camp *et al.* 1976)

$$\nu = 0.638 \pm {0.002 \atop 0.008}, \qquad \eta = 0.041 \pm {0.006 \atop 0.003} \tag{4.5.11}$$

while the calculations of Baker *et al.* (1976) for the Landau–Ginzburg–Wilson model yield the values

$$\nu = 0.627 \pm 0.01, \qquad \eta = 0.021 \pm 0.02. \tag{4.5.12}$$

Using some heuristic arguments Thompson (1976) proposed the formula

$$\nu = \frac{d+2}{4(d-1)} \qquad (d \leqslant 4) \tag{4.5.13}$$

which implies $\nu = 5/8$ for dimensionality $d = 3$.

4.5.2 Homogeneity postulate for the structure factor

In formulating the homogeneity postulate for the thermodynamic properties in Section 4.3.2 we started from the observation that the chemical potential difference $\Delta\mu$ is an odd and the compressibility $\chi_T = \rho^2 K_T$ is an even function of the density $\Delta\rho^*$. We then expect from (4.5.3), (4.5.4) and (4.5.7) that also the structure factor $\chi(\Delta\rho^*, \Delta T^*; k)$ and the correlation length $\xi(\Delta\rho^*, \Delta T^*)$ will be symmetric functions of $\Delta\rho^*$ as they are in the lattice gas. Unfortunately, there exists very little experimental information concerning the density dependence of the structure factor and the correlation length of fluids (Thomas and Schmidt 1963, 1964, Chu and Lin 1970, Lin and Schmidt 1974a), but the scant available data are at least consistent with this assumption (Hanley *et al.* 1976). We shall thus assume the symmetry properties

$$\begin{aligned}\chi(-\Delta\rho^*, \Delta T^*; k) &= \chi(\Delta\rho^*, \Delta T^*; k)\\ \xi(-\Delta\rho^*, \Delta T^*) &= \xi(\Delta\rho^*, \Delta T^*)\end{aligned} \tag{4.5.14}$$

to be valid for real fluids, but more accurate experimental information is desirable.

It is now postulated that the structure factor, for densities and temperatures sufficiently close to the critical point and for wavenumbers k small relative to the inverse molecular interaction range, is a generalized homogeneous function of its variables (Hankey and Stanley 1972)

$$\chi^*(\lambda^{b_\rho}\Delta\rho^*, \lambda^{b_T}\Delta T^*; \lambda^{b_k}k) = \lambda\chi^*(\Delta\rho^*, \Delta T^*; k). \tag{4.5.15}$$

This homogeneity property was first formulated by Kadanoff (1966) for the correlation function of the Ising model. It follows from the renormalization group theory in the same manner as shown in Section 4.2.7 for the free energy (Niemeijer and Van Leeuwen 1976). On evaluating (4.5.15) in the limit $k \to 0$, we see that the homogeneity postulate for the thermodynamic functions is recovered if the exponents b_ρ and b_T are identified with

$$b_\rho = -\frac{a_0}{1-2a_\rho} = -\frac{\beta}{\gamma} = -\frac{1}{\delta-1} \qquad b_T = -\frac{a_T}{1-2a_\rho} = -\frac{1}{\gamma}. \tag{4.5.16}$$

Substitution of (4.5.15) into (4.5.7) yields, after proper redefinition of the parameter λ,

$$\xi(\lambda^{-b_\rho/b_k}\Delta\rho^*, \lambda^{-b_T/b_k}\Delta T^*) = \lambda\xi(\Delta\rho^*, \Delta T^*). \tag{4.5.17}$$

Thus the correlation length ξ is a generalized homogeneous function of the thermodynamic variables $\Delta\rho^*$ and ΔT^* as we found earlier for the singular part of the thermodynamic properties. Along the critical isochore $\Delta\rho^* = 0$ and along the coexistence curve $\Delta\rho^*_{\text{cxc}} = \pm B\,|\,\Delta T^*\,|^{\beta}$ we find in analogy with (4.2.45) and (4.2.46)

$$\begin{aligned}\xi(0, \Delta T^*) &= \xi(0, 1)(\Delta T^*)^{-b_k/b_T} & (\Delta T^* \geqslant 0)\\ \xi(\Delta\rho^*_{\text{cxc}}, \Delta T^*) &= \xi(B, -1)\,|\,\Delta T^*\,|^{-b_k/b_T} & (\Delta T^* \leqslant 0).\end{aligned}$$

We thus recover the power laws (4.5.8) with

$$\nu = \nu' = b_k/b_T. \tag{4.5.18}$$

The homogeneity postulate (4.5.15) reads in terms of the exponents β, γ, ν adopted for the power laws

$$\chi(\lambda^{-\beta/\gamma}\Delta\rho^*, \lambda^{-1/\gamma}\Delta T^*; \lambda^{-\nu/\gamma}k) = \lambda\chi(\Delta\rho^*, \Delta T^*; k). \tag{4.5.19}$$

Taking $\chi^{\nu/\gamma} = k$, we conclude that at the critical point $\chi(0, 0; k) = k^{-\gamma/\nu}\chi(0, 0, 1)$. Comparison with (4.5.10) yields the exponent relation (Fisher 1964)

$$\gamma = \nu(2-\eta) \tag{4.5.20}$$

4.5.3 Scaling laws and hypothesis of three-scale-factor universality

A generalized homogeneous function of three variables can be scaled to become a function of only two variables. To deduce a scaling law for the structure factor we take $\lambda^{\beta/\gamma} = \Delta\rho^*$ in

(4.5.19) and obtain

$$\chi^*(\Delta\rho^*, \Delta T^*; k) = |\Delta\rho^*|^{-\gamma/\beta}\chi^*(1, x; y) \tag{4.5.21}$$

where $x = \Delta T^*/|\Delta\rho^*|^{1/\beta}$ is the thermodynamic scaling variable introduced earlier and $y = k|\Delta\rho^*|^{-\nu/\beta}$ is a new scaling variable. In the limit $k \to 0$

$$\chi^*(\Delta\rho^*, \Delta T^*; 0) = |\Delta\rho^*|^{-\gamma/\beta}D^{-1}X(x/x_0) \tag{4.5.22}$$

where $X(x/x_0) = [\delta h(x/x_0) - (x/\beta x_0)h'(x/x_0)]^{-1}$ is the scaling function for the isothermal compressibility introduced in Table 4.3.1.

The correlation length ξ satisfies a scaling law which may be written in the form

$$\xi(\Delta\rho^*, \Delta T^*) = |\Delta\rho^*|^{-\nu/\beta}\xi_{0,c}\Xi(x/x_0) \tag{4.5.23}$$

where, in accordance with (4.5.7),

$$[\xi_{0,0}\Xi(x/x_0)]^2 = -\lim_{y\to 0}\frac{\partial \ln \chi^*(1, x; y)}{\partial y^2}. \tag{4.5.24}$$

The factor $\xi_{o,c}$ is the amplitude of the power law for the correlation length along the critical isotherm

$$\xi(\Delta\rho^*, 0) = \xi_{0,c}|\Delta\rho^*|^{-\nu/\beta} \quad (\Delta T^* = 0). \tag{4.5.25}$$

It follows from (4.5.22) and (4.5.23) that the scaling law (4.5.21) for the structure factor also can be written in the form

$$\chi^*(\Delta\rho^*, \Delta T^*; k) = \chi_T^*(\Delta\rho^*, \Delta T^*)\,Y\left(\frac{x}{x_0}; k\xi\right). \tag{4.5.26}$$

This formulation has the advantage that the new scaling function $Y(u, v)$ satisfies the boundary conditions

$$\lim_{v\to 0} Y(u; v) = 1 \tag{4.5.27a}$$

$$\lim_{v\to 0}\frac{\partial Y(u; v)}{\partial v^2} = -1. \tag{4.5.27b}$$

Alternate forms for the scaling law of the structure factor are

$$\chi^*(\Delta\rho^*, \Delta T^*; k) = \xi^{2-\eta}\Phi_\chi\left(\frac{x}{x_0}; k\xi\right)$$
$$\chi^*(\Delta\rho^*, \Delta T^*; k) = k^{\eta-2}\Psi_\chi\left(\frac{x}{x_0}; k\xi\right) \tag{4.5.28}$$

where the functions $\Phi_\chi(x/x_0; k\xi)$ and $\Psi_\chi(x/x_0; k\xi)$ can be readily related to the scaling functions introduced above.

The structure factor can be determined experimentally as a function of ξ and k by measuring the intensity of scattered light as a function of temperature and scattering angle. The wavenumber k is related to the scattering angle θ by the Bragg condition $k = 2k_0\sin(\theta/2)$, where k_0 is the wavenumber of the incident light. If the observed scattered light intensity $I(k) \propto \chi(k)$, divided by the extrapolated intensity $I(0)$ at zero scattering angle, is plotted as a function of $k\xi$, it follows from the scaling law (4.5.26) that data obtained along a curve of constant x, such as

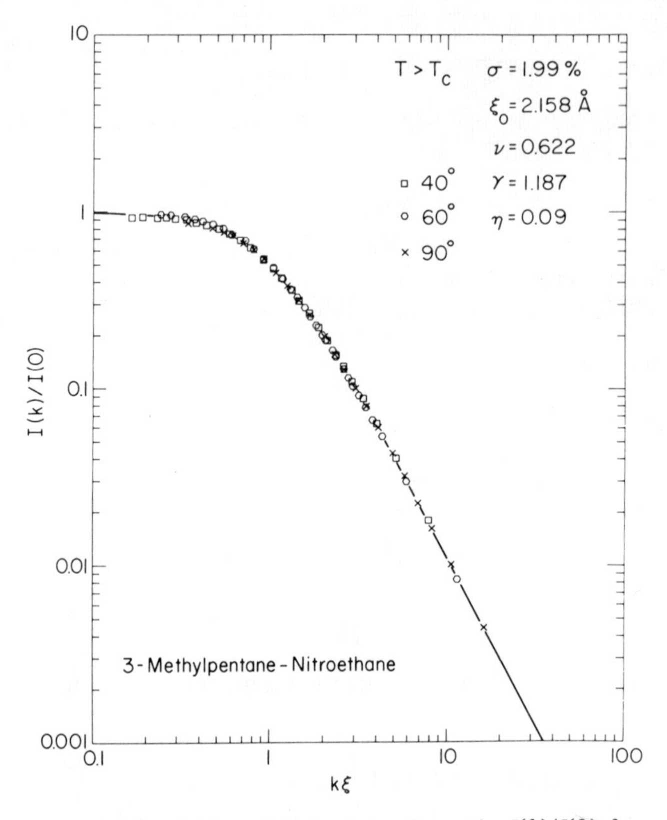

Figure 4.5.1 Scattered light intensity ratio $I(k)/I(0)$ for 3-methylpentane-nitroethane at the critical concentration as a function of $k\xi$

the critical isochore, critical isotherm or coexistence curve, should collapse onto one single curve. In Figure 4.5.1 we present an example of such a plot based on light scattering data obtained at three scattering angles (40°, 60°, 90°) for the binary liquid 3-methylpentane-nitroethane at the critical concentration as a function of temperature (Chang *et al.* 1972).

In Sections 4.2.7 and 4.3.5 we formulated the hypothesis of universality of thermodynamic behaviour near the critical point. This same principle, when extended to the correlation function, asserts that for all systems which differ only with respect to irrelevant parameters in the hamiltonians, the scaling function $\chi^*(1,x;y)$ introduced in (4.5.21) will be the same, except for three adjustable factors which define the scale of χ^*, x and y (Kadanoff 1971, Ferer and Wortis 1972). For these scale factors we may choose the power law amplitudes D, x_0 and $\xi_{0,c}$ (or ξ_0). Thus the hypothesis of three-scale-factor universality predicts that not only the function $h(x/x_0)$ introduced in (4.3.11) for the equation of state, but also the scaling functions $Y(x/x_0; k\xi)$ and $\Xi(x/x_0)$ for the structure factor and correlation length, will be the same for all systems within one universality class.

4.5.4 Correlation scaling function and correlation function exponent values

In order to reduce the correlation function exponents ν and η and the correlation length ξ from experimental data and to make a quantitative analysis of the universality hypothesis, one needs an explicit functional form for the correlation scaling function $Y(x/x_0; y)$ as a function of

$y = k\xi$. For small values of y the inverse scaling function has a Taylor series expansion in terms of y^2 (Fisher 1964)

$$Y^{-1}(x/x_0; y) = 1 + y^2 + O(y^4) \quad (y \ll 1) \tag{4.5.29}$$

where the first two terms follow from the boundary conditions (4.4.27). For very large values of the variable y the scaling function Y is expected to have the form (Fisher and Langer 1968, Stell 1968, Brézin *et al.* 1974a,b, Fisher and Aharony 1974)

$$Y\left(\frac{x}{x_0}; y\right) = \frac{C_1(x/x_0)}{y^{2-\eta}}\left(1 + \frac{C_2(x/x_0)}{y^{(1-\alpha)/\nu}} + \frac{C_3(x/x_0)}{y^{1/\nu}}\right) \quad (y \gg 1) \tag{4.5.30}$$

where C_1, C_2 and C_3 are (universal) functions of the thermodynamic scaling variable x/x_0.

The early experimental light scattering work has often been interpreted using either the Ornstein–Zernike scattering function

$$Y_{OZ} = \frac{1}{1 + y^2} \tag{4.5.31}$$

or the simple Fisher scattering function

$$Y_F = \frac{1}{(1 + y^2)^{1-\eta/2}} \tag{4.5.32}$$

for all values of y (Calmettes *et al.* 1972, Chu 1972, Lai and Chen 1972, Volochine 1972). However, the Ornstein–Zernike scattering function (4.5.31) implies that $\eta = 0$, which is known to be incorrect from the theoretical model results, while the simple Fisher scattering function (4.5.32) cannot accommodate simultaneously the correct amplitudes of the leading terms in the small and large y expansions. A proposed scattering function is not satisfactory unless it provides a method of interpolating between the asymptotic forms (4.5.29) and (4.5.30). The problem is compounded by the fact that the scaling function Y is not only a function of $k\xi$, but also of the thermodynamic scaling variable x/x_0. In practice one tries to formulate a scattering function that accommodates the known numerical results for the 2- and 3-dimension Ising model.

For the 3-dimensional Ising model Tarko and Fisher (1975) proposed

$$Y_{TF}\left(\frac{x}{x_0}; y\right) = \frac{(1 + \Phi^2 y^2)^{\eta/2}}{1 + \psi y^2}\left[1 - \Lambda + \Lambda \frac{(1 + \psi y^2/4)}{2 \ln \Omega} \ln\left(\frac{\Omega^2 + \psi y^2/4}{1 + \psi y^2/4}\right)\right] \tag{4.5.33}$$

where the parameters Φ, Λ and Ω are (universal) functions of x/x_0 and where

$$\psi\left(\frac{x}{x_0}\right) = \frac{1 + \eta\Phi^2/2}{1 - (\Lambda/4)[1 - (\Omega^2 - 1)/2\Omega^2 \ln \Omega]}\,.$$

Taking $\Lambda(\infty) = 0$, this form reduces at the critical isochore $x = \infty$ to (Fisher and Burford 1967)

$$Y_{FB}(\infty; y) = \frac{(1 + \Phi^2(\infty) y^2)^{\eta/2}}{1 + \psi(\infty) y^2} \tag{4.5.34}$$

where $\psi(\infty) = 1 + \tfrac{1}{2}\eta\Phi^2(\infty)$. For large values of y the leading term in the asymptotic expansion (4.5.30) is recovered with $C_1(\infty) = \phi^\eta(\infty)/[1 + \tfrac{1}{2}\eta\phi^2(\infty)]$. However, the Fisher–Burford and Tarko–Fisher scattering functions (4.5.34) and (4.5.33) still do not reproduce the higher order terms in the asymptotic expansion (4.5.30). A quantitative analysis of the limitations of these scattering functions by testing them on the exactly-solved 2-dimensional Ising model was made by Tracy and McCoy (1975). Lin and Schmidt (1974a) have analysed their X-ray scattering data for argon near the critical point in terms of the Fisher–Burford and Tarko–Fisher scattering functions. Swinney and Saleh (1973) evaluated the effect of the Fisher–Burford scattering function on the theoretical expression for the decay rate of the order parameter in fluids. Tartaglia and Thoen (1975) tried to discriminate between the various scattering functions from an analysis of sound absorption and dispersion data near the critical point of xenon.

Recently, Bray (1976) proposed a scattering function which in two dimensions does reproduce the known Ising model values for all $k\xi$ to within 0.03%. It was obtained by modifying an earlier scattering function of Ferrell and Scalapino (1975) so as to incorporate the Fisher–Langer form (4.5.30) for large values of $y = k\xi$. For the 3-dimensional Ising model in zero field which corresponds to the critical isochore of the lattice gas, this scattering function is defined through the equations

$$Y^{-1}(\infty; y) = 1 + y^2 f\left(\frac{y^2}{9}\right) \Big/ f\left(-\frac{1}{9}\right) \tag{4.5.35a}$$

$$f(z) = \frac{2}{\pi} \sin\left(\frac{\pi\eta}{2}\right) \int_{w_0}^{\infty} \frac{dw}{w^{\eta}} \frac{wF(w)}{w^2 + z} \tag{4.5.35b}$$

$$F(w) = \frac{P + Q\cot(\pi\eta/2)}{P^2 + Q^2} \tag{4.5.35c}$$

with

$$P(w) = 1 - \frac{C_2}{(3w)^{(1-\alpha)/\nu}} \sin\left(\frac{\pi}{2\nu}\right) + \frac{C_3}{(3w)^{1/\nu}} \cos\left(\frac{\pi}{2\nu}\right)$$

$$Q(w) = \frac{C_2}{(3w)^{(1-\alpha)/\nu}} \cos\left(\frac{\pi}{2\nu}\right) - \frac{C_3}{(3w)^{1/\nu}} \sin\left(\frac{\pi}{2\nu}\right).$$

The constants C_2 and C_3 are to be identified with the coefficients $C_2(\infty)$ and $C_3(\infty)$ in the Fisher–Langer form (4.5.30) for large values of y, and must be the same for all systems within one universality class. The integration cut-off constant w_0 in (4.5.35b) is related to the coefficient $C_1(\infty)$ by the condition

$$C_1(\infty) = 3^{-\eta} f(-1/9). \tag{4.5.35d}$$

The value of this coefficient estimated from series expansions for the 3-dimensional Ising model is 0.90 ± 0.01 (Ritchie and Fisher 1972).

Many light scattering data and X-ray scattering data are obtained under conditions for which $k\xi \ll 1$. One may then neglect the higher order terms in the expansion (4.5.29) and analyse the data in terms of

$$\chi(k) = \frac{\chi(0)}{1 + k^2\xi^2}. \tag{4.5.36}$$

Thus if the inverse scattering intensity $I^{-1}(k) \propto \chi^{-1}(k)$ is plotted as a function of k^2, at each temperature and density the data will fall on a straight line whose intercept is proportional to the inverse compressibility $\chi^{-1}(0)$ while the ratio of slope over intercept yields the correlation length ξ^2. Such a plot is usually referred to as an Ornstein–Zernike plot. According to the Ornstein–Zernike theory the data would follow a straight line for all values of $k\xi$; since the exponent of η is small the deviations from linear behaviour observed in real fluids are very small.

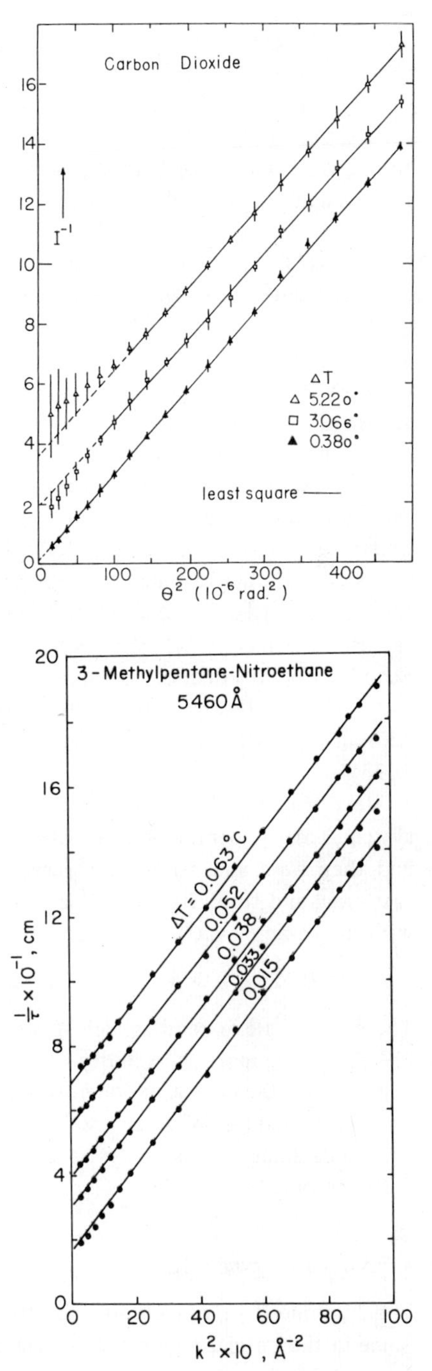

Figure 4.5.2 Reciprocal relative scattering intensity as a function of $(k/k_0)^2 \approx \theta^2$ for carbon dioxide at the critical density and at various values of $\Delta T = T - T_c$. (From Chu and Lin 1970, reproduced by permission of the American Institute of Physics)

Figure 4.5.3 Inverse scattering intensity in turbidity units as a function of k^2 for 3-methylpentane-nitroethane at the critical concentration and at various values of $\Delta T = T - T_c$ (McIntyre and Sengers 1968)

We show examples of such O–Z plots in Figure 4.5.2 for carbon dioxide near the critical point based on X-ray scattering data (Chu and Lin 1970) and in Figure 4.5.3 for the binary liquid 3-methylpentane-nitroethane near the critical mixing point based on light scattering data (Wims 1967, McIntyre and Sengers 1968).

The correlation length has been determined experimentally for such gases as ^{3}He (Ohbayashi and Ikushima 1975), ^{4}He (Tominaga 1974, Kagoshima *et al.* 1973), Ar (Lin and Schmidt

1974a), Xe (Smith *et al.* 1971), CO_2 (Chu and Lin 1970, Lunacek and Cannell 1971), SF_6 (Puglielli and Ford 1970, Cannell 1975) and for a number of binary liquids reviewed by Chu (1972), Anisimov (1975) and Scott (1977). In interpreting the experimental results one must realize that there exists a strong correlation between the exponent ν and the amplitude ξ_0 of the power law (4.5.8). Although a variety of values have been reported for the exponent ν the more reliable data are all consistent with $\nu = 0.63 \pm 0.02$ for gases as well as binary liquids.

Having determined in addition the exponent γ from the zero-angle scattering data $\lim_{k\to 0} \chi(k)$, one may then infer a value for the exponent η from the Fisher relation (4.5.20). Typical values thus obtained are $\eta = 0.07 \pm 0.04$ for CO_2 (Lunacek and Cannell 1971) and $\eta = 0.03 \pm 0.03$ for SF_6 (Cannell 1975), but they are obviously very sensitive to the values adopted for γ and ν.

In order to obtain more detailed information, one needs to measure the intensity of scattered light over a larger range of $k\xi$ values. For a given wavelength and a given range of scattering angles this can be done in principle by approaching the critical point more closely. However, here one runs into major complications due to multiple scattering and attenuation of the light beam (Chalyi 1969, White and Maccabee 1971, Oxtoby and Gelbart 1974, Bray and Chang 1975, Boots *et al.* 1976, Moldover *et al.* 1976).

As an alternative one can use X-ray scattering or neutron scattering which probe the fluctuations at larger values of k. From neutron scattering experiments Warkulwiz *et al.* 1974) reported a value $\eta = 0.11 \pm 0.03$ for neon and from X-ray scattering experiments data Lin and Schmidt (1974b) deduced a value $\eta = 0.10 \pm 0.05$ for argon. However, there exists some doubt as to whether the structure factor can still be represented by the simple scaling law (4.5.26) without correction terms at the values of k covered by these experiments (Mozer 1976, Schmidt 1976).

Very recently Chang *et al.* (1976) obtained an accurate set of light scattering data for the system 3-methylpentane-nitroethane. This system has a relatively small scattering cross-section and they were able to cover a range $0.18 < k\xi < 26$ with only minor corrections for double scattering and turbidity. The data could be well represented by the scattering function (4.5.35) with the parameters $C_2 = 1.773$, $C_3 = -2.745$, $w_0 = 1$ and the exponent values

$$\nu = 0.625 \pm 0.005, \quad \eta = 0.016 \pm 0.014. \tag{4.5.37}$$

These values are in good agreement with the exponent values (4.5.12) calculated by Baker *et al.* (1976) and support the validity of the universality hypothesis for the correlation function. It appears that the critical exponent values (4.2.78) and (4.5.12) calculated theoretically from the Landau–Ginzburg–Wilson model are in slightly better agreement with the experimental data for fluids than the estimates (4.2.64) and (4.5.11) deduced from series expansions for the 3-dimensional Ising model.

4.5.5 Hyperscaling relations

The hypothesis of universality has its origin in the long range character of the fluctuations. Since in the critical region the correlation length ξ becomes much larger than the range of the molecular interaction, it is expected that the singular nature of the cooperative phenomena becomes independent of the detailed shape of the molecules. The correlation length ξ was defined in (4.5.6). For distances much smaller than ξ the density fluctuations are highly correlated, while for distances much larger than ξ the density fluctuations become uncorrelated. Hence, the correlation length ξ may be visualized as the radius of droplet-like conglomerates referred to as clusters (Widom 1974).

The singular thermodynamic behaviour is characterized by two critical exponents among the thermodynamic exponents α, β, γ and δ. The long-range behaviour of the correlation function is characterized by the two exponents ν and η. If the singular behaviour of the system is completely determined by the long-range character of the correlation function, one would expect that the thermodynamic exponents are related to the correlation function exponents. One such a relation is the Fisher relation $\gamma = \nu(2-\eta)$. However, a second relation is needed to relate the thermodynamic exponents and the correlation function exponents uniquely. Such a relation may be obtained by an argument first proposed by Widom (1965b, 1974).

The energy associated with the spontaneous density fluctuations that extend over a distance ξ will be of order $k_B T \approx k_B T_c$. Hence the free energy density associated with these fluctuations will be of order $k_B T_c/\xi^d$. One expects this energy to be solely responsible for the singular contribution A_{sing}/V to the Helmholtz free energy density. Thus $A_{sing}/V = P_c A^*_{sing}$ (cf. 4.2.2) is assumed to be proportional to $k_B T_c/\xi^d$.

It is convenient to introduce the dimensionless quantity

$$Q^d \equiv \frac{|A_{sing}/V|\,\xi^d}{k_B T_c} = \frac{P_c\,|A^*_{sing}|\,\xi^d}{k_B T_c} \tag{4.5.38}$$

which, according to Widom's arguments, is expected to remain finite at the critical point. Since along each curve of fixed $x = \Delta T^*/|\Delta\rho^*|^{1/\beta}$, A^*_{sing} varies as $|\Delta T^*|^{2-\alpha}$ and ξ as $|\Delta T^*|^{-\nu}$, this assumption implies the exponent relation

$$\nu d = 2 - \alpha. \tag{4.5.39}$$

When this relation is combined with the other exponent relations one also obtains (Fisher 1967a, Stell 1970)

$$\frac{2-\eta}{d} = \frac{\delta - 1}{\delta + 1}. \tag{4.5.40}$$

The relations (4.5.39) and (4.5.40) are sometimes referred to as hyperscaling relations. They contain the dimensionality d and are expected to be valid in those dimensions ($d \geqslant 4$) where the correlation function becomes sufficiently long range. The hyperscaling relations are satisfied for the 3-dimensional Ising model. The validity of the hyperscaling relations in two and three dimensions is implicitly assumed in the renormalization group theory of critical phenomena. Hence they are automatically satisfied by the numerical estimates (4.2.78) and (4.5.12) for the Landau–Ginzburg–Wilson model. The problem is, however, that they are not satisfied within error by the series expansion estimates for the 3-dimensional Ising model (Baker 1976). The origin of this discrepancy has not yet been resolved. The hyperscaling relations are satisfied by the fluid critical exponents within the accuracy with which they are currently known experimentally.

4.5.6 Hypothesis of two-scale-factor universality

According to the hypothesis of universality formulated in Section 4.5.3, the scaling function for the structure factor is a universal function of the scaling variables $\Delta T^*/|\Delta\rho^*|^{1/\beta}$ and $k\,|\Delta\rho^*|^{-\nu/\beta}$ except for three scale factors D, x_0 and $\xi_{0,c}$. The number of independent scale factors can be further reduced by the hypothesis of two-scale-factor universality. It follows from the assertion, proposed by Stauffer *et al.* (1972), that the dimensionless quantity Q, defined in (4.5.38), is a universal function $Q(x/x_0)$ of the thermodynamic scaling variable x/x_0; that is, along any given curve x/x_0 = constant, the quantity Q should assume the same value for all systems within one universality class. If we now substitute into (4.5.38) the scaled expressions $A^*_{sing} = |\Delta\rho^*|^{(2-\alpha)/\beta} D a(x/x_0)$ and $\xi = |\Delta\rho^*|^{-\nu/\beta}\xi_{0,c}\Xi(x/x_0)$ in accordance with

(4.3.10) and (4.5.23), and assume the validity of the hyperscaling relations, this hypothesis of universal $Q(x/x_0)$ implies

$$\frac{P_c}{k_B T_c} D\xi_{0,c}^d = \text{universal constant} \tag{4.5.41}$$

yielding a relation between the scale factors D and $\xi_{0,c}$. The hypothesis of two-scale-factor universality does follow from the renormalization group theory, as recently shown by Hohenberg *et al.* (1976).

Since the Helmholtz free energy density is usually not measured directly, the assumption is often restated in terms of the specific heat density C_V/V. Universality of Q implies that also the combination $\xi^d \mid \Delta T^* \mid^2 C_V/V$ is universal. Empirical evidence in support of this hypothesis for fluids was presented by Bauer and Brown (1975).

In principle, the hypothesis of two-scale-factor universality could be investigated by comparing the experimental data for C_V and ξ for different fluids at the same $\Delta\rho^*$ and ΔT^* without taking recourse to a power law analysis (Stauffer *et al.* 1972, Bauer and Brown 1975). Here, however, we investigate the relations between the power law amplitudes implied by the hypothesis. In the absence of experimental data for the correlation length amplitude $\xi_{0,c}$ along the critical isotherm, we consider the value of Q along the critical isochore $x = \infty$

$$Q(\infty) = \left(\frac{P_c}{k_B T_c} \frac{A^+}{(2-\alpha)(1-\alpha)\alpha}\right)^{1/3} \xi_0 = \text{constant} \tag{4.5.42a}$$

or, in terms of the constants of the parametric equations of state in Table 4.3.2

$$Q(\infty) = \left(\frac{P_c}{k_B T_c} ak \mid f_0 \mid\right)^{1/3} \xi_0 = \text{constant}. \tag{4.5.42b}$$

The hypothesis of two-scale-factor universality presupposes that the experimental data for different fluids can be represented by a power law with universal exponents. As discussed in Section 4.3.8 the range of simple asymptotic behaviour of fluids near the gas–liquid critical point is restricted to $\mid \Delta T^* \mid < 10^{-4}$. In this range insufficient experimental correlation length data are available to investigate the validity of the hypothesis with any precision.

We have evaluated the quantity $Q(\infty)$ for a number of gases in an extended temperature range using the effective fluid exponents (4.3.29) adopted in Table 4.3.4. The results are presented in Table 4.5.1. The free energy amplitude akf_0 was calculated from the parameters presented in Table 4.3.4. Substitution of the effective thermodynamic exponents (4.3.29) into the hyperscaling relation (4.5.39) yields an effective value $\nu = 0.633$ for the correlation length exponent ν. Estimates for the corresponding correlation length amplitude ξ_0 were obtained as follows. For ^4He and Ar we took the values reported by Tominaga (1974) and by Lin and Schmidt (1974a), respectively. For Xe we took $\xi_0 = (2.0 \pm 0.2)$ Å based on the data of Smith *et al.* (1971) as reinterpreted by Swinney and Henry (1973). A re-analysis of the X-ray scattering

Table 4.5.1 Test of two-scale-factor universality for gases

	$(P_c/k_B T_c)^{1/3}$ (Å^{-1})	$(akf_0)^{1/3}$	ξ_0 (Å)	$Q(\infty)$
^4He	0.1470	1.46	2.2 ± 0.6	0.47 ± 0.13
Ar	0.1327	2.15	1.7 ± 0.2^5	0.48 ± 0.07
Xe	0.1135	2.15	2.0 ± 0.2	0.49 ± 0.05
CO_2	0.1207	2.44	1.55 ± 0.1	0.46 ± 0.03
SF_6	0.0949	2.42	1.9 ± 0.1	0.45 ± 0.03

data of Chu and Lin (1970) for CO_2 yields $\xi_0 = (1.6 \pm 0.1)$ Å, while Lunacek and Cannell (1971) report $\xi_0 = (1.50 \pm 0.09)$ Å. From the work of Cannell (1975) we estimate for SF_6 $\xi_0 = (1.9 \pm 0.1)$ Å, but there are some differences from the data of Puglielli and Ford (1970). In no case is the correlation length on an absolute basis known with an accuracy better than 10%, so that we cannot expect to check two-scale-factor universality better than to within 10%. Nevertheless, it appears from Table 4.5.1 that the quantity $Q(\infty) = 0.46 \pm 0.04$ is remarkably constant for these five gases. The hypothesis of two-scale-factor universality would thus appear to be consistent with the current experimental evidence. The corresponding value of $Q(\infty)$ for the 3-dimensional Ising model is 0.43 (Stauffer *et al.* 1972), but such a comparison is of dubious value since we have used exponent values that differ from those of the Ising model.

4.5.7 Correlation length and equation of state

If we adopt the effective universal value $Q(\infty) = 0.46$ for all gases, we can predict an effective value for the correlation length amplitude ξ_0 from the equation of state parameters presented in Table 4.3.4: the values thus estimated for ξ_0 are given in last column. In order to calculate the correlation length at densities other than the critical density we need to formulate a functional form for the scaling function $\Xi(x/x_0)$ introduced in (4.5.23). For the 3-dimensional Ising model this question has been considered by Tarko and Fisher (1975).

In analogy to the formulation used for the singular contributions to the thermodynamic properties, the required analyticity properties of ξ in the one-phase region away from the critical point can be automatically ensured by a parametric representation

$$\begin{aligned} \Delta T^* &= r(1 - b^2\theta^2) \\ \Delta\rho^* &= kr^\beta\theta(1 + c\theta^2) \\ \xi &= \xi_0 r^{-\nu} g(\theta) \end{aligned} \qquad (4.5.43)$$

where $g(\theta)$ is an analytic function of θ which must be even in θ. The case $c = 0$ corresponds again to the linear model and the case $c \neq 0$ corresponds again to the cubic model. In order to reproduce the correct power law amplitudes along the critical isochore and the coexistence curve, the function $g(\theta)$ must satisfy the boundary conditions

$$g(0) = 1 \quad g(\pm 1) = \xi_0'/\xi_0. \qquad (4.5.44)$$

For the 3-dimensional Ising model $\xi_0'/\xi_0 = 0.51$ (Tarko and Fisher 1975). For fluids the ratio ξ_0'/ξ_0 is only known with very limited accuracy. The most careful study of this ratio was made by Lin and Schmidt (1974c) with the result $\xi_0'/\xi_0 = 0.49 \pm 0.05$. For ^{4}He Tominaga (1974) reported $\xi_0'/\xi_0 = 0.45 \pm 0.1$.

A more practical approach is to relate the correlation length to the isothermal compressibility. It follows from (4.5.20) and (4.5.23) that the correlation length can be written in the form

$$\xi = \xi_0 R(\theta)(\Gamma^{-1}\chi_T^*)^{1/(2-\eta)} \qquad (4.5.45)$$

where R is a universal function of x/x_0 and, hence, a universal function of the parametric variable θ. On the critical isochore and coexistence curve it is equal to

$$R(0) = 1 \quad R(\pm 1) = \frac{\xi_0'}{\xi_0}\left[\frac{\Gamma}{\Gamma'}\right]^{1/(2-\eta)}. \qquad (4.5.46)$$

For the 3-dimensional Ising model $R(\pm 1) = 1.17$ (Tarko and Fisher 1975). The proportionality constant between ξ and $\chi_T^{*1/(2-\eta)}$ along the critical isochore and coexistence curve of argon was determined by Lin and Schmidt (1974a) with a precision of about one percent; from their work we conclude $R(\pm 1) = (4.04 \pm 0.05)/(3.45 \pm 0.03) = 1.17 \pm 0.03$, in good agreement with the theoretical value $R(\pm 1) = 1.17$ for the Ising model. The simplest polynomial accommodating this boundary condition would be

$$R(\theta) = 1 + 0.17\theta^2 . \qquad (4.5.47)$$

In practical calculations of the correlation length of fluids $R(\theta)$ is usually assumed to be independent of θ. In this approximation

$$\xi = \xi_0 (\Gamma^{-1} \chi_T^*)^{1/(2-\eta)} \qquad (4.5.48)$$

which for the restricted cubic model reduces to

$$\xi = \xi_0 r^{-\nu} . \qquad (4.5.49)$$

This assumption is not strictly correct as suggested by the Ising model and confirmed by the work of Lin and Schmidt. The approximate equation (4.5.48) reproduces correctly the behaviour of the correlation length along the critical isochore, but not along the coexistence curve. When it is used in conjunction with the equation of state parameters given in Table 4.3.4, it yields $\xi_0'/\xi_0 = (\Gamma'/\Gamma)^{1/(2-\eta)} = 0.46$ which is 10% lower than the corresponding value for the Ising model, but well within the precision with which this ratio is currently known experimentally for fluids.

Acknowledgements

The authors are indebted to Drs A. J. Bray, R. F. Chang, R. W. Gammon, M. S. Green, S. C. Greer, R. J. Hocken, P. C. Hohenberg, and M. R. Moldover for many stimulating discussions and valuable comments. The authors have benefited from a close collaboration with the late Mr T. A. Murphy. Part of the work was supported by NASA Grant NGR-21-002-344, by the U.S. Office of Standard Reference Data, and by the Center of Material Research at the University of Maryland. Some of the material was gathered while J.V.S. was supported by the Ir. Cornelis Gelderman Fund of the Technological University at Delft.

References

Aharony, A. (1976), In C. Domb and M. S. Green (Eds.), *Phase Transitions and Critical Phenomena*, Vol. 6, Academic Press, New York. Chap. 6.

Aharony, A. and P. C. Hohenberg (1976), *Phys. Rev.*, **B13**, 3081.

Alekhin, A. D., A. Z. Golik, N. P. Krupskii and A. V. Chalyi (1969a). *Ukrainian Phys. J.*, **13**, 1472.

Alekhin, A. D., A. Z. Golik, N. P. Krupskii, A. V. Chalyi and Yu. I. Shimanskii (1969b). *Ukrainian Phys. J.*, **13**, 1118.

Amirkhanov, Kh. I. and A. M. Kerimov (1963). *Teploenergetika*, **10** (8), 64; (9), 61.

Anisimov, M. A. (1975). *Sov. Phys. Uspekhi*, **17**, 722.

Anisimov, M. A., A. T. Berestov, L. S. Veksler, B. A. Koval'chuk and V. A. Smirnov (1974). *Sov Phys. JETP*, **39**, 359.

Artyukhovskaya, L. M., E. T. Shimanskaya and Yu. I. Shimanskii (1971). *Sov. Phys. JETP*, **32**, 375.

Baehr, H. D. (1963a). *Forsch. Ingenieurw.*, **29**, 143.

Baehr, H. D. (1963b). *Brennst.-Wärme-Kraft*, **15**, 514.

Baker, G. A. (1977). *Phys. Rev.*, **B15**, 1552

Baker, G. A., B. G. Nickel, M. S. Green and D. I. Meiron (1976). *Phys. Rev. Lett* , **36**, 1351.

Balzarini, D. A. (1974). *Can J. Phys.*, **52**, 499.

Balzarini, D. and K. Ohrn (1972). *Phys. Rev. Lett.* **29**, 840.
Barieau, R. E. (1966). *J. Chem. Phys*,. **45**, 3175.
Barieau, R. E. (1968). *J. Chem. Phys.*, **49**, 2279.
Barmatz, M., P. C. Hohenberg and A. Kornblit (1975). *Phys. Rev.*, **B12**, 1947. Erratum: *Phys. Rev.*, **B13**, 474.
Bauer, H. C. and G. R. Brown (1975). *Phys. Lett.*, **51A**, 68.
Behringer, R. P., T. Dorion and H. Meyer (1976). *J. Low Temp. Phys.*, **24**, 315.
Betts, D. D., A. J. Guttmann and G. S. Joyce (1971). *J. Phys. C.*, **4**, 1994.
Blagoi, Yu. P., V. I. Sokhan and L. A. Pavlichenko (1970). *Sov. Phys. JETP* Lett., **11**, 190.
Boots, H. M. J., D. Bedeaux and P. Mazur (1976). *Physica*, **84A**, 217.
Bray, A. J. (1976). *Phys. Rev. Lett.*, **36**, 285; *Phys. Rev.*, **B14**, 1248.
Bray, A. J. and R. F. Chang (1975). *Phys. Rev.*, **A12**, 2594.
Brézin, E., D. J. Amit and J. Zinn-Justin (1974a). *Phys. Rev. Lett.*, **32**, 151.
Brézin, E., J. C. le Guillou and J. Zinn-Justin (1974b). *Phys. Rev. Lett.*, **32**, 473.
Brown, G. R. and H. Meyer (1972). *Phys. Rev.*, **A6**, 364.
Calmettes, P., I. Laguës and C. Laj (1972). *Phys. Rev. Lett.*, **28**, 478.
Camp, W. J., D. M. Saul, J. P. Van Dyke and M. Wortis (1976), *Phys. Rev.*, **B14**, 3990.
Cannell, D. S. (1975). *Phys. Rev.*, **A12**, 225.
Cernuschi, F. and H. Eyring (1939). *J. Chem. Phys.*, **7**, 547.
Chalyi, A. V. (1969). *Ukrainian Phys. J.*, **13**, 828.
Chalyi, A. V. and A. D. Alekhin (1971). *Sov. Phys. JETP*, **32**, 181.
Chang, R. F., P. E. Keyes, J. V. Sengers and C. O. Alley (1972). *Ber. Bunsenges. physik. Chemie*, **76**, 260.
Chang, R. F., H. Burstyn, J. V. Sengers and A. J. Bray (1976), *Phys. Rev. Lett.*, **37**, 1481.
Chase, C. E. and G. O. Zimmerman (1973). *J. Low Temp. Phys.*, **11**, 551.
Chu, B. (1972). *Ber. Bunsenges. physik. Chemie*, **76**, 202.
Chu, B. and J. S. Lin (1970). *J. Chem. Phys.*, **53**, 4454. Erratum: *J. Chem. Phys.*, **55**, 2004.
Cornfeld, A. B. and H. Y. Carr (1972). *Phys. Rev. Lett.*, **29**, 28.
Dalton, N. W. and D. W. Wood (1969). *J. Math. Phys.*, **10**, 1271.
D'Arrigo, G., L. Mistura and P. Tartaglia, (1975). *Phys. Rev.*, **A12**, 2587.
De Boer, J., (1949). *Rep. Prog. Phys.*, **12**, 305.
De Swaan Arons, J. and G. A. M. Diepen (1966). *J. Chem. Phys.*, **44**, 2322.
Di Castro, C., G. Jona-Lasinio and L. Peliti (1974). *Annals of Physics*, **87**, 327.
Dobbs, B. C. and P. W. Schmidt (1973). *Phys. Rev.*, **A7**, 741.
Doiron, T., R. P. Behringer and H. Meyer (1976), *J. Low Temp. Phys.*, **24**, 345.
Domb, C. (1974). In C. Domb and M. S. Green (Eds.), *Phase Transitions and Critical Phenomena*, Vol. 3. Academic Press, New York. Chap. 6.
Domb, C. and D. L. Hunter (1965). *Proc. Phys. Soc.*, **86**, 1147.
Edwards, C., J. A. Lipa and M. J. Buckingham (1968). *Phys. Rev. Lett.*, **20**, 496.
Estler, W. T., R. Hocken, T. Charlton and L. R. Wilcox (1975). *Phys. Rev.*, **A12**, 2118.
Fabelinskii, I. L. (1968). *Molecular Scattering of Light*, Plenum Press, New York.
Farrell, R. A. and P. H. E. Meijer (1972). *Phys. Rev.*, **B5**, 3747.
Ferer, M. and M. Wortis (1972). *Phys. Rev.*, **B6**, 3426.
Ferrell, R. A. and D. J. Scalapino (1975). *Phys. Rev. Lett.*, **34**, 200.
Fisher, M. E. (1964). *J. Math. Phys.*, **5**, 944.
Fisher, M. E. (1966). In M. S. Green and J. V. Sengers (Eds.), *Critical Phenomena*, NBS Misc. Publ. 273, U.S. Govt. Printing Office, Washington, D.C. p. 21.
Fisher, M. E. (1967a). *Rep. Prog. Phys.*, **30** (II), 615.
Fisher, M. E. (1967b). *J. Appl. Phys.*, **38**, 981.
Fisher, M. E. (1971). In M. S. Green (Ed.), Critical Phenomena, *Proc. Intern. School of Physics, 'Enrico Fermi', Course LI*. Academic Press, New York. p. 1.
Fisher, M. E. (1974). *Rev. Mod. Phys.*, **46**, 597.
Fisher, M. E and A. Aharony(1974). *Phys. Rev.*, **B10**, 2818.
Fisher, M. E. and R. J. Burford (1967). *Phys. Rev.*, **156**, 583.
Fisher, M. E. and J. S. Langer (1968). *Phys Rev. Lett.*, **20**, 665.
Gaunt, D. S. and A. J. Guttmann (1974). In C. Domb and M. S. Green (Eds.), *Phase Transitions and Critical Phenomena*, Vol. 3. Academic Press, New York. Chap. 4.
Gaunt, D. S. and M. F. Sykes (1972). *J. Phys.* C., **5**, 1429.
Gaunt, D. S. and M. F. Sykes (1973). *J. Phys.* A., **6**, 1517.

Giglio, M. and A. Vendramini (1975). *Phys. Rev. Lett.*, **35**, 168.
Golik, A. Z., A. D. Alekhin, N. P. Krupskii, A. V. Chalyi and Yu. I. Shimanskii (1969). *Ukrainian Phys.*, **14**, 475.
Golner, G. R. and E. K. Riedel (1975). *Phys. Rev. Lett.*, **34**, 856.
Golner, G. R. and E. K. Riedel (1976), *Phys. Lett.*, **58A**, 11.
Gopal, E. S. R., R. Ramachandra and P. Chandra Sekhar (1973), *Pramāna*, **1**, 260.
Gopal, E. S. R., R. Ramachandra, P. Chandra Sekhar and S. V. Subramanyam (1974), *Phys. Rev. Lett.*, **32**, 284.
Gouy, G. (1880). *Comptes Rendus Acad. Sci. (Paris)*, **90**, 307.
Gouy, G. (1892). *Comptes Rendus Acad. Sci. (Paris)*, **115**, 720.
Green, H. S. and C. A. Hurst (1964). *Order–Disorder Phenomena*, Wiley, New York.
Greer, S. C. (1976). *Phys. Rev.*, **A14**, 1770.
Greer, S. C., Th. E. Block and C. M. Knobler (1975). *Phys. Rev. Lett.*, **34**, 250.
Griffiths, R. B. (1965a). *J. Chem. Phys.*, **43**, 1958.
Griffiths, R. B. (1965b). *Phys. Rev. Lett.*, **14**, 623.
Griffiths, R. B. (1967). *Phys. Rev.*, **158**, 176.
Griffiths, R. B. (1970). *Phys. Rev. Lett.*, **24**, 1479.
Griffiths, R. B. (1972). In C. Domb and M. S. Green (Eds.), *Phase Transitions and Critical Phenomena*, Vol. 1. Academic Press, New York. Chap. 2.
Griffiths, R. B. and J. C. Wheeler (1970). *Phys. Rev.*, **A2**, 1047.
Gulari, E. C. and C. J. Pings (1973). Private communication.
Habgood, H. W. and G. W. Schneider (1954). *Can. J. Chem.*, **32**, 98.
Hankey, A. and H. E. Stanley (1972). *Phys. Rev.*, **B6**, 3515.
Hanley, H. J. M., J. V. Sengers and J. F. Ely (1976). In T. K. Chu and P. G. Klemens (Eds.), *Proc. 14th Intern. Conference on Thermal Conductivity*. Plenum Press, New York. p. 383.
Hastings, J. R. and J. M. H. Levelt Sengers (1976). To be published.
Heller, P. (1967) *Rep. Prog. Phys.*, **30** (II), 731.
Hemmer, P. C., M. Kac and G. E. Uhlenbeck (1964). *J. Math. Phys.*, **5**, 60.
Hicks, C. P. and C. L. Young (1975). *Chem. Revs.*, **75**, 119.
Hildebrand, J. H., J. M. Prausnitz and R. L. Scott (1970). *Regular and Related Solutions*, Van Nostrand Reinhold, New York.
Hirschfelder, J. O., C. F. Curtiss and R. B. Bird (1954). *Molecular Theory of Gases and Liquids*, Wiley, New York.
Ho, J. T. and J. D. Litster (1969). *Phys. Rev. Lett.*, **22**, 603.
Ho, J. T. and J. D. Litster (1970). *Phys. Rev.*, **B2**, 4523.
Hocken, R. and M. R. Moldover (1976). *Phys. Rev. Lett.*, **37**, 29.
Hohenberg, P. C., A. Aharony, B. I. Halperin and E. D. Siggia (1976). *Phys. Rev.*, **B13**, 2986.
Hohenberg, P. C. and M. Barmatz (1972). *Phys. Rev.*, **A6**, 289.
Huang, C. C. and J. T. Ho (1973). *Phys. Rev.*, **A7**, 1304.
Hubbard, J. and P. Schofield (1972). *Phys. Lett.*, **40A**, 245.
Ising, E. (1925). *Z. Phys.*, **31**, 253.
Jasnow, D. and M. Wortis (1968). *Phys. Rev.*, **176**, 739.
Josephson, B. D. (1969). *J. Phys.* C., **2**, 1113.
Kac, M., G. E. Uhlenbeck and P. C. Hemmer (1963). *J. Math. Phys.*, **4**, 216.
Kadanoff, L. P. (1966). *Physics*, **2**, 263.
Kadanoff, L. P., (1971). In M. S. Green (Ed.), Critical Phenomena, *Proc. Intern. School of Physics, 'Enrico Fermi,' Course LI*. Academic Press, New York. p. 100.
Kadanoff, L. P. (1976). In C. Domb and M. S. Green (Eds.), *Phase Transitions and Critical Phenomena*, Vol. 5A. Academic Press, New York. Chap. 1.
Kadanoff, L. P., A. Houghton and M. C. Yalabik (1976). *J. Stat. Phys.*, **14**, 171. Erratum, **15**, 263.
Kagoshima, S., K. Ohbayashi and A. Ikushima (1973). *J. Low Temp. Phys.*, **11**, 765.
Kay, W. B. (1968). *Accounts Chem. Res.*, **1**, 344.
Kierstead, H. A. (1973). *Phys. Rev.*, **A7**, 242.
Klein, M. J. and L. Tisza (1949). *Phys. Rev.*, **76**, 1861.
Krupskii, N. P. and Yu. I. Shimanskii (1972). *Sov. Phys. JETP.*, **35**, 561.
Lai, C. C. and S. H. Chen (1972). *Phys. Lett.*, **41A**, 259.
Lebowitz, J. L. (1974). *Physica*, **73**, 48.
Lee, T. and C. N. Yang (1952). *Phys. Rev.*, **87**, 410.
Lentini, E. and M. Vicentini-Missoni (1973). *J. Chem. Phys.*, **58**, 91.

Leung, S. S. and R. B. Griffiths (1973). *Phys. Rev.*, **A8**, 2670.
Levelt Sengers, J. M. H. (1970). *Ind. Eng. Chem. Fund.*, **9**, 470.
Levelt Sengers, J. M. H. (1974). *Physica*, **73**, 73.
Levelt Sengers, J. M. H. (1975). In B. Le Neindre and B. Vodar (Eds.), *Experimental Thermodynamics*, Vol. II. Butterworths, London. Chap. 14.
Levelt Sengers, J. M. H. (1976). *Physica*, **82A**, 319.
Levelt Sengers, J. M. H., W. L. Greer and J. V. Sengers (1974). In K. D. Timmerhaus (Ed.), *Advances in Cryogenic Engineering*, **19**. Plenum Press, New York. p. 358.
Levelt Sengers, J. M. H., W. L. Greer and J. V. Sengers (1976). *J. Phys. Chem. Ref. Data*, **5**, 1.
Levelt Sengers, J. M. H. and J. V. Sengers (1975). *Phys. Rev.*, **A12**, 2622.
Levelt Sengers, J. M. H., J. Straub and M. Vicentini-Missoni (1971). *J. Chem. Phys.*, **54**, 5034.
Ley-Koo, M. and M. S. Green (1976). preprint.
Liberman, D. A. (1966). *J. Chem. Phys.*, **44**, 419.
Lin, J. S. and P. W. Schmidt (1974a), *Phys. Rev.*, **A10**, 2290.
Lin, J. S. and P. W. Schmidt (1974b). *Phys. Lett.*, **48A**, 75.
Lin, J. S. and P. W. Schmidt (1974c). *Phys. Rev. Lett.*, **33**, 1265.
Lipa, J. A., C. Edwards and M. J. Buckingham (1970). *Phys. Rev. Lett.*, **25**, 1086.
Lorentzen, H. L. (1953). *Acta Chem. Scand.*, **7**, 1335.
Lorentzen, H. L. and B. B. Hansen (1966). In M. S. Green and J. V. Sengers (Eds.), *Critical Phenomena*, NBS Misc. Publ. 273, U.S. Govt. Printing Office, Washington, D.C. p. 213.
Lunacek, J. H. and D. S. Cannell (1971). *Phys Rev. Lett.*, **27**, 841.
Ma, S. K. (1973). *Rev. Mod. Phys.*, **45**, 589.
Ma, S. K. (1976), *Modern Theory of Critical Phenomena*, W. A. Benjamin, Reading, Mass.
McIntyre, D. and J. V. Sengers (1968). In H. N. V. Temperley, J. S. Rowlinson and G. S. Rushbrooke (Eds.), *Physics of Simple Liquids*, North-Holland, Amsterdam. Chap. 11.
Maisano, G., P. Migliardo and F. Wanderlingh (1976), *Optics Comm.*, **19** 155.
Meijer, P. H. E. and R. A. Ferrell (1973), *Phys. Rev.*, **B12**, 243.
Mermin, N. D. and J. J. Rehr (1971). *Phys. Rev. Lett.*,**26**, 1155.
Michaels, S., M. S. Green and S. Y. Larsen (1970). *Equilibrium Critical Phenomena in Fluids and Fluid Mixtures: A Comprehensive Bibliography with Key-Word Descriptors*, NBS Special Publ. 327, U.S. Govt. Printing Office, Washington, D.C.
Michels, A., Blaise B. and C. Michels (1937). *Proc. Roy. Soc., London*, **A160**, 358.
Miloševic, S. and H. E. Stanley (1972). *Phys. Rev.*, **B6**, 1002.
Mistura, L. (1971). *J. Chem. Phys.*, **55**, 2375.
Moldover, M. R. (1969). *Phys. Rev.*, **182**, 342.
Moldover, M. R. and J. Gallagher (1976). to be published.
Moldover, M. R., R. J. Hocken, R. W. Gammon and J. V. Sengers (1976). *Overviews and Justifications for Low Gravity Experiments on Phase Transition and Critical Phenomena in Fluids*, NBS Technical Note 925, U.S. Gov't Printing Office, Washington, D.C.
Moore, M. A., D. Jasnow and M. Wortis (1969). *Phys. Rev. Lett.*, **22**, 940.
Morrison, G. and C. M. Knobler (1976), *J. Chem. Phys.*, **65**, 5507.
Mozer, B. (1976). Private communication.
Mulholland, G. W. (1973). *J. Chem. Phys.*, **59**, 2738.
Murphy, T. A., (1975). Private communication.
Murphy, T. A., J. V. Sengers and J. M. H. Levelt Sengers (1973). In P. E. Liley (Ed.), *Proc. 6th Symp. on Thermophysical Properties*, American Society of Mechanical Engineers, New York. p. 180.
Murphy, T. A., J. V. Sengers and J. M. H. Levelt Sengers (1975). In P. Bury, H. Perdon and B. Vodar (Eds.), *Proc 8th Intern. Conf. on the Properties of Water and Steam*, Editions Européennes Thermiques et Industries, Paris. p. 603.
Naumenko, Zh. P., E. T. Shimanskaya and Yu. I. Shimanskii (1967). *Ukrainian Phys.* **12**, 144.
Niemeijer, Th. and J. M. J. Van Leeuwen (1976). In C. Domb and M. S. Green (Eds.), *Phase Transitions and Critical Phenomena*, Vol. 6. Academic Press, New York. Chap. 7.
Ohbayashi, K. and A. Ikushima (1975). *J. Low Temp. Phys.*, **19**, 449.
Onsager, L. (1944). *Phys. Rev.*, **65**, 117.
Oxtoby, D. W. and W. M. Gelbart (1974). *Phys. Rev.*, **A10**, 738.
Palmer, H. B. (1954). *J. Chem. Phys.*, **22**, 625.
Peierls, R. (1936). *Proc. Camb. Phil. Soc.*, **32**, 477.
Plank, R. (1936). *Forsch. Geb. Ingenieurw.*, **7**, 161.

Prins, C. (Ed.) (1974). Proceedings Van der Waals Centennial Conference on Statistical Mechanics, *Physica* **73**, No. 1. North-Holland, Amsterdam.
Puglielli, V. G. and N. C. Ford (1970). *Phys. Rev. Lett.*, **25**, 143.
Rathjen, W. and J. Straub (1973). In *International Institute of Refrigeration, Meeting of Commission B7 (Zürich)*. p. 129.
Rehr, J. J. and N. D. Mermin (1973). *Phys. Rev.*, **A8**, 472.
Ritchie, D. S. and M. E. Fisher (1972). *Phys. Rev.*, **B5**, 2668.
Rivkin, S. L. and T. S. Akhundov (1962). *Teploenergetika*, **9**(1), 57.
Rivkin, S. L. and T. S. Akhundov (1963). *Teploenergetika*, **10**(9), 66.
Rivkin, S. L., T. S. Akhundov, E. K. Kremnevskaia and N. N. Assadulaieva (1966). *Teploenergetika*, **13**(4), 59.
Rivkin, S. L. and G. V. Troianovskaia (1964). *Teploenergetika*, **11**(10), 72.
Roach, P. R. (1968). *Phys. Rev.*, **170**, 213.
Rowlinson, J. S. (1969). *Liquids and Liquid Mixtures*, 2nd ed. Butterworths, London.
Rushbrooke, G. S. (1963). *J. Chem. Phys.*, **39**, 842.
Rushbrooke, G. S. (1965). *J. Chem. Phys.*, **43**, 3439.
Saam, W. F. (1970). *Phys. Rev.*, **A2**, 1461.
Saul, D. M., M. Wortis and D. Jasnow (1975). *Phys. Rev.*, **B11**, 2571.
Schmidt, E. H. W. (1966). In M. S. Green and J. V. Sengers (Eds.), *Critical Phenomena*, NBS Misc. Publ. 273, U.S. Govt Printing Office, Washington, D.C. p. 13.
Schmidt, E. H. W. and K. Traube (1962). In J. F. Masi and D. H. Tsai (Eds.), *Proc. 2nd Symposium on Thermophysical Properties*, American Society of Mechanical Engineers, New York. p. 193.
Schmidt, H. H. (1971). *J. Chem. Phys.*, **54**, 3610.
Schmidt, P. W. (1976). private communication.
Schneider, G. M. (1970). In I. Prigogine and S. A. Rice (Eds.), *Advances in Chemical Physics*, Vol. 17. Wiley, New York. p. 1.
Schneider, G. M. (1972). *Ber. Bunsenges. physik. Chemie*, **76**, 325.
Schofield, P. (1969). *Phys. Rev. Lett.*, **22** 606.
Schofield, P., J. D. Litster and J. T. Ho (1969). *Phys. Rev. Lett.*, **23**, 1098.
Schroer,B. (1974). *Revista Brasileira Física*, **4**, 323.
Scott, R. L., (1972). *Ber. Bunsenges. physik. Chemie*, **76**, 296.
Scott, R. L. (1977). In M. L. McGlashan (Ed.), *Chemical Society Specialist Reports on Thermodynamics of Mixtures*, The Chemical Society, London.
Sengers, J. V. (1971). In M. S. Green (Ed.), Critical Phenomena, *Proc. Intern. School of Physics 'Enrico Fermi,' Course LI*. Academic Press, New York. p. 445.
Sengers, J. V. (1972). *Ber. Bunsenges. physik. Chemie*, **76**, 234.
Sengers, J. V. (1973). In J. Kestin (Ed.), *Transport Phenomena – 1973*, AIP Conf. Proc. No. 11. American Institute of Physics, New York. p. 229.
Simon, M., A. A. Fannin and C. M. Knobler (1972). *Ber Bunsenges. physik. Chemie*, **76**, 321.
Smith, I. W., M. Giglio and G. B. Benedek (1971). *Phys. Rev. Lett.*, **27**, 1556.
Stanley, H. E. (1971). *Introduction to Phase Transitions and Critical Phenomena*, Oxford Univ. Press, New York and Oxford.
Stanley, H. E. (1972). In R. N. Sen and C. Weil (Eds.), *Statistical Mechanics and Field Theory*, Israel Univ. Press, Jerusalem. p. 225.
Stanley, H. E. (1973). *Cooperative Phenomena Near Phase Transitions*, MIT Press, Cambridge, Mass.
Stauffer, D., M. Ferer and M. Wortis (1972). *Phys. Rev. Lett.*, **29**, 345.
Stein, A. and G. F. Allen (1974). *J. Phys. Chem. Ref. Data*, **2**, 443.
Stell, G. (1968). *Phys. Lett.*, **A27**, 550.
Stell, G. (1970). *Phys. Rev. Lett.*, **24**, 1343.
Straub, J. (1967). *Chem.-Ing.-Techn.*, **5/6**, 291.
Streett, W. B. (1965). *J. Chem Phys.*, **42**, 500.
Streett, W. B. (1967). *J. Chem. Phys.*, **46**, 3282.
Streett, W. B. (1974). *Can. J. Chem. Eng.*, **52**, 92.
Swinney, H. L. and D. L. Henry (1973). *Phys. Rev.*, **A8**, 2586.
Swinney, H. L. and B. E. A. Saleh (1973). *Phys. Rev.*, **A7**, 747.
Tarko, H. B. and M. E. Fisher (1975). *Phys. Rev.*, **B11**, 1217.

Tartaglia, P. and J. Thoen (1975). *Phys. Rev.*, **A11**, 2061.
Teichner, G. (1904). *Annalen der Physik*, **13**, 595.
Thoen, J. and C. W. Garland (1974). *Phys. Rev.*, **A10**, 1311.
Thomas, J. E. and P. W. Schmidt (1963). *J. Chem. Phys.*, **39**, 2506.
Thomas, J. E. and P. W. Schmidt (1964). *J. Am. Chem. Soc.*, **86**, 3554.
Thompson, C. J. (1976). *J. Phys.*, **A9**, L25.
Tisza, L. (1961). *Annals of Physics*, **13** 1.
Tisza, L. and C. E. Chase (1965). *Phys. Rev. Lett.*, **15**, 4.
Tominaga, A. (1974). *J. Low Temp. Phys.*, **16**, 571.
Tracy, C. A. and B. M. McCoy (1975). *Phys. Rev.*, **B12**, 368.
Trappeniers, N. J. and J. A. Schouten (1974). *Physica*, **73**, 527, 539, 546.
Uhlenbeck, G. E., P. C. Hemmer and M. Kac (1963). *J. Math. Phys.*, **4**, 229.
Van der Waals, J. D. (1873). *Over de continuiteit van den gas – en vloeistoftoestand, Doctoral dissertation*, University of Leiden.
Van der Waals, J. D. (1893). *Verhandelingen, Kon. Akad. van Wetenschappen, Amsterdam*, **1**, No. 8.
Van der Waals, J. D. (1894). *Z. physik. Chem.*, **13**, 657.
Van der Waals, J. D. (1900). *Die Kontinuität des gasförmigen und flussigen Zustandes*, II. (Verlag J. A. Barth, Leipzig.)
Van Kampen, N. G. (1964). *Phys. Rev.*, **135**, A362.
Van Laar, J. J. (1912). *Proc. Section of Sciences, Kon. Akad. van Wetenschappen, Amsterdam*, **14** (II), 1091.
Van Leeuwen, J. M. J. (1975). In E. G. D. Cohen (Ed.), *Fundamental Problems in Statistical Mechanics*, III. North-Holland, Amsterdam. p. 81.
Verschaffelt, J. E. (1900). *Proc. Section of Sciences, Kon. Akad. van Wetenschappen, Amsterdam*, **2**, 588.
Verschaffelt, J. E. (1901), *Arch. Néerl.*, (2) **6**, 650.
Verschaffelt, J. E. (1904). *Arch. Néerl.*, (2) **9**, 125.
Verschaffelt, J. E. (1905). *Proc. Section of Sciences, Kon. Akad. van Wetenschappen, Amsterdam*, **7**, 474.
Verschaffelt, J. E. (1923). *Arch. Néerl.*, (3) **6**, 153.
Vicentini-Missoni, M., J. M. H. Levelt Sengers and M. S. Green (1969a) *J. Res. Natl. Bur. Stand. (U.S.)*, **73A**, 563.
Vicentini-Missoni, M., J. M. H. Levelt Sengers and M. S. Green (1969b). *Phys. Rev. Lett.*, **22**, 389.
Volochine, B. (1972). *Ber. Bunsenges. physik. Chemie*, **76**, 217.
Voronel, A. V. and M. Sh. Giterman (1965). *Sov. Phys. JETP*, **21**, 958.
Wallace, B. and H. Meyer (1970). *Phys. Rev.*, **A2**, 1563, 1610.
Wallace, B. and H. Meyer (1972), *Phys. Rev.*, **A5**, 953.
Wallace, D. J. (1976). In C. Domb and M. S. Green (Eds.), *Phase Transitions and Critical Phenomena*, Vol. 6. Academic Press, New York. Chap. 5.
Wallace, D. J. and R. K. P. Zia (1974). *J. Phys.* C., **7**, 3480.
Warkulwiz, V. P., B. Mozer and M. S. Green (1974). *Phys. Rev. Lett.*, **32**, 1410.
Watson, P. G. (1969). *J. Phys.* C., **2**, 1883, 2158.
Weber, L. A. (1970). *Phys. Rev.* **A2**, 2379.
Wegner, F. (1972). *Phys. Rev.*, **B5**, 4529.
Weiner, J., K. H. Langley and N. C Ford (1974). *Phys. Rev. Lett.*, **32**, 879.
White, J. A. and B. S. Maccabee (1971). *Phys. Rev. Lett.*, **26**, 1468.
White, J. A. and B. S. Maccabee (1975). *Phys. Rev.*, **A11**, 1706.
Widom, B. (1965a), *J. Chem. Phys,.* **43**, 3898.
Widom, B. (1965b). *J. Chem. Phys.*, **43**, 3892.
Widom, B. (1974). *Physica*, **73**, 107.
Widom, B. and J. S. Rowlinson (1970). *J. Chem. Phys.* **52**, 1670.
Wilcox, L. R. and D. Balzarini (1968). *J. Chem. Phys.*, **48**, 753.
Wilson, K. G. (1971). *Phys. Rev.*, **B4**, 3174, 3184.
Wilson, K. G. and J. Kogut (1974). *Physics Reports*, **12C**, 75.
Wims, A. M. (1967). *Critical Phenomena of 3-Methylpentane-Nitroethane*, Ph.D. Thesis, Dept. of Chemistry, Howard University, Washington, D.C.

Wohl, A. (1914). *Z. physik. Chemie,* **87**, 1.
Wu, S. Y. and C. J. Pings (1976). to be published.
Yang, C. N. and C. P. Yang (1964). *Phys. Rev. Lett.*, **13**, 303.
Yvon, J. (1937). *Actualités scientifiques et industrielles,* Nos. 542 and 543. (Hermann, Paris.)
Zollweg, J. A. and G. W. Mulholland (1972). *J. Chem. Phys.*, **57**, 1021.

Chapter 5

Critical Dynamics

KYOZI KAWASAKI† and JAMES D. GUNTON‡

†*Department of Physics, Faculty of Science, Kyushu University Fukuoka 812, Japan*
‡*Department of Physics, Temple University, Philadelphia, Pa. 19122, U.S.A.*

5.1. Introduction

Experimental investigations of non-equilibrium aspects of critical phenomena started shortly after the discovery by Andrews of the critical point of carbon dioxide in 1869. In 1882 Warburg and von Babo carefully investigated the behaviour of the shear viscosity near the critical point of the same substance, and in 1901 Friedlander reported on anomalous increases of the shear viscosity near the critical consolute points of isobutyric–water and phenol–water solutions. Since that early beginning there has been a steady accumulation of experimental studies of critical anomalies in various transport coefficients such as shear viscosity, thermal conductivity, mutual and self diffusion coefficients, and sound absoption.

Many of these investigations have been marred, however, by enormous experimental difficulties. For instance, small uncertainties in measuring the pressure or the temperature near a liquid–gas critical point result in large errors in the density, i.e. in the order parameter. Gravity thus has an enormous effect near the critical point. Hydrodynamic instabilities such as the convective instability further complicate the problem. It has only been quite recently that these problems have been sorted out and corrected. For further details on this aspect of the problem the reader is referred to the excellent reviews by Sengers.[1] It is also worth noting that a class of experimental investigations very much suited for critical phenomena is that of neutron and light scattering studies, in which time correlation functions of the Fourier components of certain density variables are measured directly. This has the advantage of yielding the frequency and wavenumber-dependent transport coefficients without any macroscopic disturbance of the system.

Turning to theoretical aspects of dynamic critical phenomena, it is fair to say that no successful quantitative theory existed before, say, the 1960s – and for a good reason, namely that it took a long time to formulate successfully the transport theory for dense systems. (By dense systems we mean systems which cannot be represented as an assembly of almost independent particles, quasi-particles or other elementary excitations.) This was due to the fact that the intrinsically many-body aspect of the problem had been hopelessly intermingled with the task of properly formulating the problem. These two aspects were finally sorted out in the time correlation function approach to transport theory which was completed in the 1960s.[2] Even after these developments, however, it took some years for the nature of critical dynamics to be understood. Initially many attempts were made to obtain a completely microscopic, first principle treatment, using the time correlation formulation in combination with various many-body theoretical techniques. However, such attempts were not successful for several reasons, which include the following. To begin with, any correct theory of the critical anomalies must properly incorporate the hydrodynamic modes of a system; many of the microscopic approaches did not do so. Secondly, the use of many-body techniques is often accompanied by an obscuring of the essential physics of critical dynamics, in which case the problem becomes hopelessly difficult. It is therefore understandable that these early efforts failed. This is not to say, of course, that all attempts at a microscopic understanding of critical dynamics are worthless. Rather, we are merely emphasizing the fact that, as we shall discuss below, we now have a body of theories which on a more macroscopic basis have successfully accounted for most experimental observations, and which thus should serve as a useful guide for constructing a truly successful microscopic theory of critical dynamics. It is unfortunate that this last point does not yet appear to be fully appreciated by those who attempt microscopic theories.

The first step toward a successful attack on critical dynamics was made in 1962 by Fixman,[3] who adopted the view that one should focus on those large-scale fluctuations which persist for a long time and on the interactions among them, rather than on the microscopic aspects of the problem. This view was incorporated into the calculation of time correlation functions by Kawasaki[4] and by Zwanzig and his co-workers.[5] This work was then combined with the static scaling laws[6] by Kadanoff and Swift[7] to yield correct critical exponents for transport anomalies. These approaches later developed into the so-called mode coupling theory.

Another step forward was made by the development of dynamical scaling,[8] whose origins lie in the static formulation of equilibrium critical phenomena.[6] The most important physical picture of critical dynamics to emerge from all of these developments is that, for the cases where transport coefficients diverge, there exists a class of variables including the local order parameter of the transition which constitutes an autonomous dynamic system with its own characteristic length and time scales. This in fact can lead to certain simplifications of the problem when one is near a critical point, quite contrary to earlier expectations.

More recently, the phenomenal success of the renormalization group approach to equilibrium critical phenomena[9] has given rise to a dynamic renormalization group approach to critical dynamics,[10] and has thus given new impetus to this field. Nevertheless, a complete understanding of critical dynamics on a microscopic level still appears to be a long way off.

In this chapter we will be mainly concerned with those successful theories of critical dynamics mentioned above, namely,

(i) dynamical scaling (Section 5.3)
(ii) mode coupling (or kinetic equation) theory (Sections 5.5, 5.6 and 5.8)
(iii) the dynamic renormalization group approach (Section 5.7).

The discussion of these will be preceded by a unified presentation of earlier thermodynamic treatments of critical dynamics, i.e. of the so-called conventional theory of critical slowing-down (Section 5.2). In Section 5.9 we discuss some little-explored areas such as nucleation and spinodal decomposition near a critical point. In Section 5.10 we briefly touch on some topics which are otherwise left out of this article, namely the kinetic Ising and TDGL models and the problem of the long time tail in transport theory. An exact formulation of a dynamic renormalization group is presented in the Appendix.

Finally, we should note that in this review we have given more emphasis to the developments of the main ideas of critical dynamics at the expense of thorough discussions of applications to various systems and of technical details. For these, the reader is referred to a review by one of the authors.[11]

5.2 Thermodynamic theory of critical dynamics

In thy early days before transport theories of dense systems were well-formulated, attempts were made to understand dynamic critical phenomena such as the slowing-down of spin diffusion in ferromagnets and the anomalous sound absorption in liquid helium near the λ point on a purely thermodynamic basis. As these thermodynamic theories serve as a useful qualitative introduction to critical dynamics, we present a unified treatment of them here. We are concerned with the dynamical behaviour of an appropriate order parameter M which varies in time under the influence of an effective potential energy $\Psi(M)$ which is taken to be the Landau-type thermodynamic potential

$$\Psi(M) = AM^2 + BM^4 \tag{5.2.1}$$

where

$$A = a(T - T_c) \tag{5.2.2}$$

a and B being finite constants at T_c. In order to introduce dynamics we consider a 'kinetic energy' associated with the time derivative of M of the form $\frac{1}{2}\mu\dot{M}^2$ where the 'mass' μ is assumed to be finite at T_c. We then form a Langrangian

$$\mathscr{L} = \tfrac{1}{2}\mu\dot{M}^2 - \Psi(M). \tag{5.2.3}$$

This treatment, however, is still incomplete since the dynamics of the order parameter will inevitably be influenced by its coupling with the remaining degrees of freedom, which produces a dissipative effect. Hence we also introduce the Rayleigh dissipation function $\mathscr{F}$ of the form

$$\mathscr{F} = \tfrac{1}{2}\zeta\dot{M}^2 \tag{5.2.4}$$

where the kinetic coefficient ζ is assumed to be finite at T_c. The equation of motion for M then readily follows from the quation of motion

$$\frac{\mathrm{d}}{\mathrm{d}t}\frac{\partial\mathscr{L}}{\partial\dot{M}} - \frac{\partial\mathscr{L}}{\partial M} + \frac{\partial\mathscr{F}}{\partial\dot{M}} = 0 \tag{5.2.5}$$

that is,

$$\mu\ddot{M} + \zeta\dot{M} + 2AM + 4BM^3 = 0 \tag{5.2.6}$$

In this conventional theory the critical dynamical anomaly is entirely ascribed to the vanishing of A at T_c, and thus (5.2.6) may be viewed as a natural generalization of the Landau theory of phase transitions[6] to critical dynamics.

Instead of attempting to find a general solution of (5.2.6), we will limit ourselves to considering a small deviation ΔM from the quilibrium value M_0. Hence we find

$$\mu\Delta\ddot{M} + \zeta\Delta\dot{M} = -\,[2A + 12BM_0^2]\,\Delta M.$$

If we note that in the Landau theory the coefficient $2A + 12BM_0^2$ is just the inverse susceptibility, χ^{-1}, this equation becomes

$$\mu\Delta\ddot{M} + \zeta\Delta\dot{M} = -\chi^{-1}\Delta M \tag{5.2.7}$$

where the critical anomaly enters through the divergence of χ at T_c. We find a solution of (5.2.7) of the form ΔM = const. x $\exp(i\omega t)$ with

$$\omega = \frac{1}{2}\left[i\frac{\zeta}{\mu} \pm \left(-\frac{\zeta^2}{\mu^2} + \frac{4}{\mu\chi}\right)^{1/2}\right]. \tag{5.2.8}$$

Let us now discuss certain special cases of this general result.

(i) $\zeta = 0$ (no dissipation).

Equation (5.2.8) then becomes

$$\omega = \pm\,(\mu\chi)^{-1/2}. \tag{5.2.9}$$

Here ω is real and vanishes as $\chi^{-1/2} \propto |T - T_c|^{1/2}$ at T_c. This behaviour is commonly known as a softening of the mode and is encountered in displacive ferroelectric transitions and certain structural phase transitions.[12]

(ii) $\mu = 0$ (simple relaxation)

Here we have, excluding a solution with an infinite decay rate,

$$\omega = i/\zeta\chi. \tag{5.2.10}$$

That is, the relaxation time $\zeta\chi$ blows up like χ at T_c. This behaviour is commonly known as critical slowing-down, and qualitatively explains the behaviour observed at magnetic phase transitions, order-disorder type phase transitions, etc.[13] In particular, for $T < T_c$, we have

$$\zeta\Delta\dot{M} = -\,8BM_0^2\Delta M \tag{5.2.11}$$

with the relaxation time $\zeta/8BM_0^2$, where in the Landau theory $M_0^2 \propto (T_c - T)$. This is the case first considered by Landau and Khalatnikov[14] for the problem of sound absorption in liquid Helium.

(iii) $\mu, \zeta \neq 0$

Here, if χ is allowed to vary from zero to infinity, we will first observe a damped oscillation with a constant damping rate $\zeta/2\mu$ and an angular frequency $[(\mu\chi)^{-1} - (\zeta/2\mu)^2]^{1/2}$ which decreases and finally vanishes at $\chi = 4\mu/\zeta^2$ (soft mode). After that we observe two purely relaxational behaviours, one with an increasing and another with a decreasing relaxational rate. The latter then exhibits the critical slowing-down of (ii).

The thermodynamic approach described above has the merit of simplicity, and yet successfully incorporates an important ingredient of critical dynamics, i.e. effects arising from the vanishing of the thermodynamic driving force, which is proportional to χ^{-1}. However, it is a severe limitation of this theory that nothing can be said regarding possible critical anomalies of μ and ζ, especially the latter. In fact, we now know that the kinetic coefficient ζ indeed diverges at T_c in many instances.[1,15] The fluctuation dissipation theorem,[2] in which ζ is expressed as a time integral of a time correlation function of fluctuations, then tells us that there will be a critical anomaly in the temporal behaviour of the fluctuations. Such an anomaly can be inferred, in fact, from the above thermodynamic theory, which however treats the anomaly in a rather incomplete fashion.

5.3 Dynamic scaling

One of the most successful ideas in critical phenomena is the concept of scaling[6] in which all of the critical anomaly can be attributed to the range of correlation of the order-parameter fluctuations, ξ, which increases indefinitely near a critical point. Thus the average square of a Fourier component $M_\mathbf{q}$ of the local order-parameter fluctuation is assumed to take the simple form

$$\langle |M_\mathbf{q}|^2 \rangle = q^{-2+\eta} f(q\xi). \tag{5.3.1}$$

The idea of scaling has been extended by Farrell and his co-workers and by Halperin and Hohenberg[8] to the time correlation function of $M_\mathbf{q}$ as

$$\langle M_\mathbf{q}(t) M_{-\mathbf{q}}(0) \rangle = q^{-2+\eta} F(t\Gamma(q), q\xi) \tag{5.3.2}$$

where the inverse characteristic time $\Gamma(q)$ takes the scaling form,

$$\Gamma(q) = q^z \omega(q\xi). \tag{5.3.3}$$

This hypothesis is called dynamical scaling.

Although the dynamical critical exponent z as well as the forms of the scaling functions F and ω remain unknown, some information can be gained by assuming the continuity of (5.3.2) in the space of k and ξ^{-1}, except for a singular point $k = \xi^{-1} = 0$. In particular, when the ordered phase results from a spontaneous breakdown of continuous symmetry, the exponent z is determined from the critical anomaly of the associated Goldstone mode† frequency which, in fact, is an equilibrium quantity, such as the speed of sound. The values of z thus obtained are summarized in Table 5.3.1. The predictions of dynamical scaling have been tested in a variety of systems with remarkable success.[8,11]

Nevertheless, dynamical scaling is still just an attractive hypothesis and has certain shortcomings which can be summerized as follows:

(i) The exponent z cannot be determined by the hypothesis alone, and hence in the absence of a Goldstone mode z remains undetermined. Also there is no allowance for possible different values of z above and below the transition.

(ii) There is no way of actually determining the scaling functions F and ω.

(iii) Critical dynamical behaviour is often observed for variables other than just the order

†This is the name given for low frequency modes that occur in ordered phases due to the spontaneous breakdown of continuous symmetry. Examples are the second sound in superfluids and the spin waves in magnets.

Table 5.3.1 Dynamical critical exponent z predicted by dynamical scaling

System	Isotropic ferromagnet	Isotropic antiferromagnet and planar ferro- and antiferromagnet	Superfluid helium
z	$\frac{d+2-\eta}{2}$	$\frac{d}{2}$	$\frac{d}{2}+\frac{\alpha}{2\nu}$

parameter: as for example the total magnetization of an antiferromagnet, whose order para meter is the sublattice magnetization. In some cases dynamical scaling can be extended to cove these variables. However, the theory does not tell us how far such extensions are allowed.

Under these circumstances we are compelled to look into the dynamics of critical fluctua tions on a more fundamental level, which leads us to the question of the relevant time and length scales governing the critical dynamics.

5.4 Hierarchical structure of time and length scales

The regular hydrodynamical behaviour of systems composed of a great number of molecules is a consequence of the existence of two types of motions with widely separated time and length scales, one macroscopic and another microscopic. The discussions of the foregoing sections suggest that near a critical point the motion associated with certain degrees of freedom slows down enormously, which, combined with the indefinite increase of the correlation range of critical fluctuations ξ, destroys the simple picture of widely separated scales. Here it is instructive to remind ourselves of the situation encountered in a dilute gas with mean free path L, mean free time L/v (where v is the average thermal velocity), and molecular diameter $\sigma(\ll L)$.[16] The most detailed description valid for all spatial and temporal scales is in terms of a Liouville equation for the distribution function of the entire system. However, if one is interested in the behaviour occurring on macroscopic spatial and temporal scales, $\Omega(\gg L)$ and Ω/v, respectively, the most sensible procedure is to first derive a macroscopic hydrodynamic

Table 5.4.1 Hierarchical structure. r_0 is an appropriate microscopic length, such as the lattice spacing, and τ_0 an appropriate microscopic time such as the inverse exchange frequency of a magnet. z is the dynamic critical exponent and z' is an appropriate positive exponent

Regimes	Microscopic	Kinetic	Macroscopic
Gas			
Length scale	σ	L	Ω
Time scale	σ/v	L/v	Ω/v
Governing equation	Liouville	Boltzmann	Hydrodynamic
Critical dynamics			
Length scale	r_0	ξ	Ω
Time scale	τ_0	$\tau_0(\xi/r_0)^z$	$\tau_0(\Omega/r_0)^{z'}$
Governing equation	Liouville	Kinetic	Macroscopic

equation which one then uses as a starting point. If, however, we increase L indefinitely by diluting the gas with Ω fixed, we will inevitably reach a regime (the Knudsen regime) where hydrodynamics loses its validity and yet we can nevertheless still dispense with the full details of the Liouville equation. This is where a kinetic equation (here the Boltzmann equation), intermediate between the fully microscopic and the fully macroscopic, is appropriate, where the relevant scales are L and L/v. On the other hand, if we go on to another limit of decreasing L indefinitely until it approaches σ, the kinetic equation loses its validity and the full complexity of the Liouville equation has to be contended with. This is why the transport theory of dense fluids in general is so difficult.[17] All these ideas are, of course, well-known and the reason we have discussed them at length is the fact that even in dense systems a remarkable simplicity arises if we are near a critical point because the correlation length ξ and the time scale τ_ξ associated with the critical slowing-down play a role analogous to that of L and L/v for a dilute gas. This suggests the existence of a sort of kinetic equation for critical dynamics, and this is the topic of the next section. In Table 5.4.1 we summarize the hierarchical structure.

5.5 Kinetic equation

To obtain a kinetic equation we first need to know what variables occur in it. Such variables must describe some gross features of the system appropriate to the scales concerned, yet we want them to be complete in the sense that no memory terms appear in the kinetic equation from which macroscopic hydrodynamics must follow. The first candidate that comes to our mind is the local order parameter $M(\mathbf{r})$. This alone, however, is inadequate since hydrodynamic equations often involve other variables. Thus we supplement $M(\mathbf{r})$ with other density variables which are conserved and hence are slowly varying. In this way we form a complete set of gross variables which we denote by $\{a_\alpha(\mathbf{r})\}$ or $\{a_{\alpha\mathbf{k}}\}$ or simply by $\{a_i\}$. The 'grossness' is taken into account by supposing either that $a_\alpha(\mathbf{r})$ is defined only for cells which contain a great number of molecules, yet which are macroscopically infinitesimal, or by restricting the wavenumbers k to be smaller than a cut-off Λ greater than ξ^{-1} but smaller than the inverse of a microscopic length. Examples are (i) the local spin density $S(\mathbf{r})$ of a ferromagnet and (ii) the local hydrodynamic variables of a fluid such as the local density, etc.

Once the gross variables are identified, our next task is to derive a kinetic equation which is satisfied by them. The completeness of the set of gross variables implies that the degrees of freedom of the system not included in the gross variables perform rapid irregular motion, and hence act as a random force on the slowly varying gross variables. The situation is very much analogous to the one-dimensional Brownian motion of a heavy particle of mass m located at x with momentum p moving through a viscous fluid under the influence of an external force field $\mathbf{V}$. This motion is described by the following Langevin equation

$$\dot{x} = m^{-1}p \tag{5.5.1a}$$

$$\dot{p} = -m^{-1}\gamma p + \mathbf{V}(x) + \mathbf{f} \tag{5.5.1b}$$

where γ is the friction constant and $\mathbf{f}$ is the random force acting on the Brownian particle. The fluctuation-dissipation theorem is

$$\langle \mathbf{f}(t)\mathbf{f}(t') \rangle = 2k_B T\gamma\, \delta(t-t'). \tag{5.5.2}$$

where an angular bracket denotes an equilibrium average. Let us now rewrite (5.5.1) as one equation by introducing the following notation:

$$a_1 \equiv x \qquad a_2 \equiv p \tag{5.5.3a}$$

$$v_1(\{a\}) \equiv m^{-1}p \quad v_2(\{a\}) \equiv \mathbf{V}(x) \tag{5.5.3b}$$

$$f_1(t) \equiv 0 \quad f_2(t) \equiv f(t) \tag{5.5.3c}$$

$$F_1(\{a\}) \equiv 0 \quad F_2(\{a\}) \equiv -p/mT = -\frac{\partial}{\partial a_2}\frac{a_2^2}{2mT} \tag{5.5.3d}$$

$$\zeta_{11}^0 = \zeta_{12}^0 = \zeta_{21}^0 = 0 \quad \zeta_{22}^0 \equiv T\gamma. \tag{5.5.3e}$$

Thus, (5.5.1) reduces to

$$\dot{a}_i = v_i(\{a\}) + \sum_i \zeta_{ij}^0 F_j(\{a\}) + f_i \tag{5.5.4}$$

with

$$\langle f_i(t) f_j(t') \rangle = 2k_B \zeta_{ij}^0 \delta(t - t'). \tag{5.5.5}$$

Equation (5.5.4) when extended to an arbitrary number of gross variables $\{a\}$ is more general than suggested by its derivation, and is taken to be a general form of our kinetic equation. In this case ζ_{ij}^0 is the bare Onsager kinetic coefficient, $F_j(\{a\})$ is the thermodynamic driving force for a_i, and $v_i(\{a\})$ is the instantaneous average rate of change of $\{a\}$ when all the irreversible effects arising from the random force f_i are absent. Thus $v_i(\{a\})$ is obtained by averaging (5.5.4) in an initial local equilibrium state with fixed $\{a\}$ before the irreversible effects start to build up, that is, with $\zeta_{ij}^0 = 0$. This is denoted as

$$v_i(\{a\}) = \langle \dot{a}_i; \{a\} \rangle \tag{5.5.6}$$

where

$$\langle X; \{a\} \rangle \equiv \frac{\int X(\mathbf{x})\delta(A(\mathbf{x}) - a)D_e(\mathbf{x})d\mathbf{x}}{\int \delta(A(\mathbf{x}) - a)D_e(\mathbf{x})d\mathbf{x}}. \tag{5.5.7}$$

Here $\mathbf{x}$ is a point in the phase space, $D_e(\mathbf{x})$ is the equilibrium phase space distribution function, and $\delta(A(\mathbf{x}) - a)$ is the product of delta functions of all the gross variables $\delta(A_i(\mathbf{x}) - a_i)$ where $A_i(\mathbf{x})$ is the phase space function for the a_i.

Although the kinetic equation (5.5.4) is derived here in an entirely heuristic manner, it can be obtained also by starting from an exact identity satisfied by $A_i(\mathbf{x}, t) = \exp(it\mathscr{S})A_i(\mathbf{x})$ where $\mathscr{S}$ is the Liouville operator, which takes the form of a generalized nonlinear Langevin equation with memory.[11,18] The assumption needed to obtain (5.5.4) is that the time correlation function of the random forces is equal to a delta function of the time difference with the coefficient independent of the values of $\{A_i\}$. In what follows, we also make the usual assumption that the random force obeys a Gaussian stochastic process. (For the case of a Brownian motion, it can be shown that the assumption that correlation of many random forces has a cluster property with respect to time leads to a Gaussian stochastic process for random forces.[19])

A useful expansion formula for (5.5.6) can be found by introducing a complete orthonormal set $\{\Phi_n(\{a\})\}$ which satisfies

$$\langle \Phi_m^*(\{a\})\Phi_n(\{a\}) \rangle = \delta_{mn} \tag{5.5.8}$$

$$\sum_n \Phi_n(\{a\})\Phi_n^*(\{a'\}) = g_e(\{a\})^{-1}\delta(a - a') \tag{5.5.9}$$

where $g_e(\{a\}) \equiv \int \delta(a - A(\mathbf{x}))D_e(\mathbf{x})d\mathbf{x}$ is the equilibrium probability distribution of $\{a\}$. Thus, we have

$$v_j(\{a\}) = \sum_n v_j^{(n)} \Phi_n(\{a\}) \tag{5.5.10}$$

$$v_j^{(n)} \equiv k_{\mathrm{B}}T \langle \{A, \Phi_n^*(\{A\})\}\rangle \tag{5.5.11}$$

where $\{x, y\}$ denotes the Poisson bracket. If $g_{\mathrm{e}}(\{a\})$ is Gaussian, namely,

$$g_{\mathrm{e}}(\{a\}) = \text{const.} \times \exp[-\Sigma \mid a_i \mid^2/2\chi_i] \tag{5.5.12}$$

$\Phi_n(\{a\})$ takes the form of multi-variable Hermite polymomials. The usual mode coupling theory then involves only the linear and quadratic terms of this expansion, in which case we obtain (with $\langle a_i \rangle = 0$ and $\langle a_i a_j^* \rangle = \chi_i \delta_{ij}$)

$$v_j(\{a\}) = \sum_l \mathrm{i}\omega_{jl} a_l + \frac{\mathrm{i}}{2} \sum_{lm} \mathscr{V}_{jlm}(a_l a_m - \langle a_l a_m \rangle) \tag{5.5.13}$$

where

$$\omega_{jl} = -\mathrm{i}k_{\mathrm{B}}T \langle \{A_j, A_l^*\}\rangle / \chi_l \tag{5.5.14}$$

$$\mathscr{V}_{jlm} = -\mathrm{i}k_{\mathrm{B}}T \langle \{A_j, A_l^* A_m^*\}\rangle / \chi_l \chi_m . \tag{5.5.15}$$

In this case we also have

$$F_i(\{a\}) = -\frac{k_{\mathrm{B}}}{\chi_i} a_i \tag{5.5.16}$$

and the kinetic equation (5.5.4) becomes

$$\dot{a}_j = \sum_l (\mathrm{i}\omega_{jl} - k_{\mathrm{B}}\zeta_{jl}^0/\chi_l) a_l + \frac{\mathrm{i}}{2} \sum_{lm} \mathscr{V}_{jlm}(a_l a_m - \langle a_l a_m \rangle) + f_i . \tag{5.5.17}$$

Note that from (5.5.14) and (5.5.15) we find the following relationship among the ω and $\mathscr{V}$

$$\frac{1}{\chi_j}\omega_{jl} + \frac{1}{\chi_l}\omega_{l^*j^*} = 0 \tag{5.5.18a}$$

$$\frac{1}{\chi_j}\mathscr{V}_{jlm} + \frac{1}{\chi_l}\;{}_{l^*j^*m} + \frac{1}{\chi_m}\mathscr{V}_{m^*j^*l} = 0. \tag{5.5.18b}$$

Since the set of variables $\{A_i\}$ consists of the long wavelength Fourier components of the density variables, the equal-time correlations of $\{A_i\}$ as well as the Poisson brackets among the members of $\{A\}$ can often be related to thermodynamic quantities like the magnetic susceptibility, whose critical anomalies can be studied within the framework of equilibrium statistical mechanics. On the other hand the set $\{f_j\}$ contains only those degrees of freedom that are left out of $\{A\}$. Hence if $\{A\}$ includes all the long wavelength, slowly varying variables, $\{f\}$ contains only the short wavelength, rapidly varying components. Hence the members of $\{f\}$ and their statistical properties, in particular ζ_{ij}^0, are not affected by second order phase transitions. In this manner the critical anomalies of (5.5.16) are entirely attributed to those of the static properties. This separation of critical and non-critical properties is the feature that makes the kinetic equation approach most suitable for critical dynamics.

The kinetic equation (5.5.4) also can be cast into the following alternative stochastic equation for the probability distribution function $g(\{a\}, t)$ of the gross variables, provided that the random force obeys a Gaussian process:[20]

$$\frac{\partial}{\partial t}g(\{a\},t) = -\sum_i \frac{\partial}{\partial a_i}[v_i(\{a\}) + \sum_i \zeta^0_{ij}F_j(\{a\})]g(\{a\},t) + \sum_{ij}\frac{\partial}{\partial a_i}k_B\zeta^0_{ij}\frac{\partial}{\partial a_j}g(\{a\},t). \quad (5.5.19)$$

For $v_i(\{a\})$ and $F_i(\{a\})$ given by (5.5.13) and (5.5.15), respectively, the conditions (5.5.18) are sufficient to give a Gaussian equilibrium distribution.

We will now illustrate the kinetic equation for the cases of the isotropic Heisenberg ferromagnet and for a classical liquid.

(a) Isotropic Heisenberg ferromagnet in the paramagnetic phase[11,21]

Here the gross variables are Fourier components of the spin density $S^\alpha_\mathbf{k}$, $\alpha = x,y,z$ and we have, denoting $\chi_\mathbf{k} = \langle |S^x_\mathbf{k}|^2 \rangle$ and the system volume by V,

$$\dot{\mathbf{S}}_\mathbf{q} = -q^2\frac{\zeta^0_q}{\chi_\mathbf{q}}\mathbf{S}_\mathbf{q} + \frac{k_BT}{2V^{1/2}}\sum_\mathbf{k}\left(\frac{1}{\chi_\mathbf{k}} - \frac{1}{\chi_{\mathbf{q}-\mathbf{k}}}\right)\mathbf{S}_\mathbf{k} \times \mathbf{S}_{\mathbf{q}-\mathbf{k}} + \mathbf{f}_\mathbf{q} \quad (5.5.20)$$

$$\langle f^\alpha_\mathbf{q}(t) f^\beta_\mathbf{q}(t')^* \rangle = 2q^2\zeta^0_q\delta_{\alpha\beta}\delta(t-t') \quad (5.5.21)$$

(b) Classical liquid[11,21]

Here to a good approximation we can ignore the rapidly varying sound wave modes and can thus limit the gross variables to the local order parameter, with Fourier components $c_\mathbf{k}$ and the transverse local velocity with Fourier components $u^\alpha_\mathbf{k}$. Thus, we have with $\chi_\mathbf{k} = \langle |c_\mathbf{k}|^2 \rangle$,

$$\left(\frac{\partial}{\partial t} + q^2\frac{\zeta^0_q}{\chi_\mathbf{q}}\right)c_\mathbf{q} = -\frac{\mathrm{i}}{V^{1/2}}\sum_\mathbf{k}\mathbf{k}\,.\,\mathbf{u}_{\mathbf{q}-\mathbf{k}}c_\mathbf{k} + \mathbf{f}^c_\mathbf{q} \quad (5.5.22)$$

$$\left(\frac{\partial}{\partial t} + q^2\frac{\eta^0}{\rho}\right)\mathbf{u}_\mathbf{q} = -\mathrm{i}\frac{k_BT}{2\rho V^{1/2}}\sum_\mathbf{k}\left(\frac{1}{\chi_\mathbf{k}} - \frac{1}{\chi_{\mathbf{q}-\mathbf{k}}}\right)\left(\mathbf{k} - \frac{\mathbf{q}}{q^2}(\mathbf{q}\,.\,\mathbf{k})\right)c_\mathbf{k}c_{\mathbf{q}-\mathbf{k}} + \mathbf{f}^u_\mathbf{q} \quad (5.5.23)$$

and

$$\langle f^c_\mathbf{q}(t) f^c_\mathbf{q}(t')^* \rangle = 2q^2\zeta^0_q\delta(t-t') \quad (5.5.24)$$

$$\langle f^{u\alpha}_\mathbf{q}(t) f^{u\beta}_{-\mathbf{q}}(t')^* \rangle = 2q^2\frac{\eta^0}{\rho}(\delta_{\alpha\beta} - q^{-2}q_\alpha q_\beta)\delta(t-t') \quad (5.5.25)$$

where ρ is the average mass density.

5.6 Time correlation functions

Having obtained the kinetic equation, the next task is to deduce from it the macroscopic behaviour near the critical point. Since the usual hydrodynamics is likely to become invalid in this region, the macroscopic behaviour must be expressed in a more general language. The time correlation functions seem to be most suitable for this purpose because (i) the time evolution of a time correlation function is the same as the time evolution of a small deviation of an averaged gross variable from its equilibrium value, and (ii) a time correlation function is directly accessible to experimental observation (e.g. as in neutron and light scattering experiments). Hence in the following we consider the problem of using the stochastic equation (5.5.17) to determine the behaviour of the set of time correlation functions $\{G_{jl}(t)\}$ defined by

$$G_{jl}(t) = \langle a_j(t)a_l^*(0)\rangle/\chi_l. \tag{5.6.1}$$

The problem would be very simple if the mode coupling term $\mathscr{V}_{jlm}$ could be ignored, for in that case the kinetic equation reduces to

$$\dot{a}_j^0 = \sum_l (\mathrm{i}\omega_{jl} - k_\mathrm{B}\zeta_{jl}^0/\chi_l)a_l^0 + f_j. \tag{5.6.2}$$

The stochastic process (5.6.2) is Gaussian because $\{f\}$ is Gaussian and the equilibrium distribution function (5.5.12) is of the Gaussian form. In the following we make the simplifying assumptions: $\omega_{jl} = \omega_j\delta_{jl}$ and $\zeta_{jl}^0 = \zeta_j^0\delta_{jl}$. Thus it is enough to know in this case that for $t \geqslant t' \geqslant 0$

$$G_{jl}^0(t-t') \equiv \langle a_j^0(t)a_l^{0*}(t')\rangle/\chi_l = G_j^0(t-t')\delta_{jl} \tag{5.6.3}$$

with

$$G_j^0(t) \equiv \exp[(\mathrm{i}\omega_j - \gamma_j^0)t] \tag{5.6.3a}$$

We now use the above to develop a method which incorporates the effects of mode coupling in $G_{jl}(t)$. First note that with the aid of these auxilliary variables $\{a^0\}$, (5.5.17) can be cast into the following integral form

$$a_j(t) = a_j^0(t) + \frac{\mathrm{i}}{2}\sum_{lm}\int_0^t \mathrm{d}s\,\exp[(\mathrm{i}\omega_j - \gamma_j^0)(t-s)]\mathscr{V}_{jlm}[a_l(s)a_m(s) - \langle a_la_m\rangle]. \tag{5.6.4}$$

Next we use (5.5.17) and (5.6.1) to derive an equation for $G_{jl}(t)$:

$$\frac{\mathrm{d}}{\mathrm{d}t}G_{jr}(t) = (\mathrm{i}\omega_j - \gamma_j^0)G_{jr}(t) + \frac{\mathrm{i}}{2}\sum_{lm}\mathscr{V}_{jlm}\langle a_l(t)a_m(t)a_r^*(0)\rangle. \tag{5.6.5}$$

We then substitute (5.6.4) for the $\{a(t)\}$ that occur on the right-hand side of (5.6.5) and expand the second term in powers of the $\mathscr{V}$, using (5.6.3) and noting that $\{a^0\}$ is Gaussian. The first non-vanishing contribution of this second term appears in the second order in $\mathscr{V}$, which is

$$-\frac{1}{2}\sum_{lm}\int_0^t \mathrm{d}s\,\mathscr{V}_{jlm}\mathscr{V}_{lm^*r}\chi_m\chi_r G_l^0(s)G_m^0(s)G_r^0(t-s). \tag{5.6.6}$$

In the following it is convenient to represent (5.6.6) by a diagram where a solid line and a vertex represent the free propagator, the G^0, and the mode coupling coefficient, the $\mathscr{V}$, respectively. This is in fact just the second order self-energy diagram. In order to find the second order correction to transport coefficients, we sum up all the terms of the type shown in Figure 5.6.1. The net result is to replace $G_r^0(t-s)$ in (5.6.6) by the exact propagator, and we obtain

$$\frac{\mathrm{d}}{\mathrm{d}t}G_{jr}(t) = (\mathrm{i}\omega_j - \gamma_j^0)G_{jr}(t) - \int_0^t \Xi_{jl}^{(2)}(s)G_{lr}(t-s)\mathrm{d}s \tag{5.6.7}$$

with the following memory kernel:

$$\begin{aligned}\Xi_{jl}^{(2)}(t) &\equiv \sum_{mn}\mathscr{V}_{jmn}\mathscr{V}_{mn^*l}\chi_n\chi_l G_m^0(t)G_n^0(t)\\ &= \frac{1}{2}\sum_{mn}\frac{\chi_m\chi_n}{\chi_l}\mathscr{V}_{jmn}\mathscr{V}_{lmn}^*G_m^0(t)G_n^0(t).\end{aligned} \tag{5.6.8}$$

In the second step we have used (5.5.18b) and the following relation (which follows from

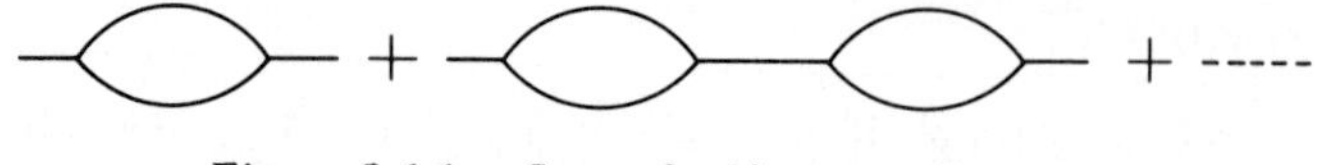

Figure 5.6.1 Sum of self-energy diagrams

(5.5.17) by taking the complex conjugate):

$$\mathscr{V}_{jlm} = -\mathscr{V}^*_{j^*l^*m^*}. \tag{5.6.9}$$

If the memory kernel $\Xi^{(2)}_{jl}(t)$ decays to zero in a time short compared with the time in which $G_{jr}(t)$ changes appreciably, we are allowed to make the 'Markoff' approximation for (5.6.7), in which the variable s in $G_{lr}(t-s)$ is dropped and the upper limit of the time integral is extended to infinity. Thus, we obtain

$$\frac{\mathrm{d}}{\mathrm{d}t} G_{jr}(t) = \mathrm{i}\omega_j G_{jr}(t) - \sum_l \frac{\zeta_{jl}}{\chi_l} G_{lr}(t) \tag{5.6.10}$$

with

$$(\zeta^0_{jl} = \gamma^0_j \delta_{jl})$$

$$\zeta_{jl} \equiv \zeta^0_{jl} + \Delta\zeta^{(2)}_{jl} \tag{5.6.11}$$

$$\Delta\zeta^{(2)}_{jl} \equiv \frac{\chi_l}{k_\mathrm{B}} \int_0^\infty \Xi^{(2)}_{jl}(t)\mathrm{d}t. \tag{5.6.12}$$

It is now evident that the nonlinear term in the kinetic equation (5.5.17) gives rise to an additional contribution to the transport coefficients that appear in the macroscopic equations of motion for the averages of the gross variables. Indeed, what are observed as transport coefficients in macroscopic measurements are the ζ_{jl} and not the ζ^0_{jl}. The effect described above is now commonly called the renormalization of transport coefficients and ζ^0_{jl} is called the bare transport coefficient (adopting the terminology of quantum field theory).

The second order calculation can be improved in various ways; the one that immediately comes to mind is to make the calculation self-consistent by replacing the free propagators in the self-energy by the exact ones. The result is to replace (5.6.7) by the following expression:

$$\frac{\mathrm{d}}{\mathrm{d}t} G_{jr}(t) = (\mathrm{i}\omega_j - \gamma^0_j) G_{jr}(t)$$

$$-\frac{1}{2} \sum_{lmn} \sum_{m'n'} \int_0^t \frac{\chi_{m'}\chi_{n'}}{\chi_l} \mathscr{V}_{jmn} \mathscr{V}^*_{lm'n'} G_{mm'}(s) G_{nn'}(s) G_{lr}(t-s)\mathrm{d}s. \tag{5.6.13}$$

If we are allowed to make the Markoffian approximation,† (5.6.13) reduces to the equation that determines the renormalized transport coefficients self-consistently. In particular, if a representation is used in which G_{jl} is diagonal,

$$G_{jl}(t) = \delta_{jl} \exp\{[\mathrm{i}\omega_j - \gamma_j]t\}, \tag{5.6.14}$$

†Except for some special circumstances, this approximation is unjustified near the critical point because there is no reason to believe that scales of the modes m,m' or n,n' are shorter than those of l,r in general. Thus, the Markoffian approximation here should be regarded more as a simplifying assumption. The order parameter dynamics in liquids provides an exception since the velocity modes in intermediate states decay much faster than the order parameter.

we obtain

$$\gamma_j = \gamma_j^0 + \frac{1}{2}\sum_{lm} \frac{\chi_l \chi_m}{\chi_j} \mid \mathscr{V}_{jlm} \mid^2 \frac{1}{-\mathrm{i}(\omega_l + \omega_m) + \gamma_l + \gamma_m}. \tag{5.6.15}$$

We illustrate the foregoing general theory for the two systems discussed previously, namely, magnets and classical liquids, restricting ourselves to the Markoffian approximation.

(a) Magnets

Here we have only one type of propagator if we restrict ourselves to the disordered phase,

$$G_q(t) = \langle S_{\mathbf{q}}^x(t) S_{-\mathbf{q}}^x(0) \rangle / \chi_{\mathbf{q}}, \tag{5.6.16}$$

If one assumes that $G_{\mathbf{q}}(t)$ decays as $\exp[-q^2 D_q t]$ where D_q is the $\mathbf{q}$-dependent diffusion coefficient, (5.6.15) reduces to

$$D_q = D_q^0 + \frac{(k_{\mathrm{B}}T)^2}{q^2} K_{d-1} \int_0^\Lambda \mathrm{d}k\, k^{d-1} \int_0^\pi \frac{\mathrm{d}\theta}{2\pi} \sin^{d-2}\theta F(k^2, q^2, \mathbf{k}.\mathbf{q}) \tag{5.6.17}$$

where

$$F(k^2, q^2, \mathbf{k}.\mathbf{q}) \equiv \frac{\chi_{\mathbf{k}}\chi_{\mathbf{q}-\mathbf{k}}}{\chi_{\mathbf{q}}} \frac{(\chi_{\mathbf{k}}^{-1} - \chi_{\mathbf{q}-\mathbf{k}}^{-1})^2}{k^2 D_k + (\mathbf{q}-\mathbf{k})^2 D_{|\mathbf{q}-\mathbf{k}|}}, \tag{5.6.18}$$

θ is the angle between $\mathbf{k}$ and $\mathbf{q}$ and K_d is the geometrical factor $K_d = 2^{-(d-1)-d/2}[\Gamma(d/2)]^{-1}$. To obtain this result the sum over $\mathbf{k}$ has been converted into an integral with the upper cut-off Λ. It is not possible to solve (5.6.17) analytically, but an equation similar to it was numerically solved by Résibois and Piette.[22] Their result was in semi-quantitative agreement with the observed linewidth of the quasi-elastic peak of magnetic neutron critical scattering for iron.[22]

(b) Classical liquid

Let us now turn to the classical liquid. Since the order parameter decays in a diffusive manner, we need to consider only the coupling of the order parameter to diffusive modes because the sound wave modes vary more rapidly and can be regarded as a part of the random forces. Furthermore, in the binary liquids we can ignore the coupling to the entropy fluctuation since the entropy shows only a weak critical anomaly. We are thus left with the order parameter and the transverse velocity as the important gross variables. Hence we consider the following two types of propagators,

$$G_{qc}(t) = \langle c_{\mathbf{q}}(t) c_{-\mathbf{q}}(0) \rangle / \chi_{\mathbf{q}} \tag{5.6.19}$$

$$\langle u_{\mathbf{q}}^{\alpha *}(t) u_{-\mathbf{q}}{}^{\beta}(0) \rangle = \frac{k_{\mathrm{B}}T}{\rho} [\delta_{\alpha\beta} - q^{-2} q_\alpha q_\beta] G_{qu}(t). \tag{5.6.20}$$

Making the Markoffian assumption we set $G_{qc}(\mathrm{t}) = \exp[-q^2 D_q \mathrm{t}]$ and $G_{qu}(t) = \exp[-q^2 \eta(q) t/\rho]$, and thus arrive at the following coupled set of equations which correspond to (5.6.15);

$$D_q = D_q^0 + \frac{k_{\mathrm{B}}T}{\rho} K_{d-1} \int_0^\Lambda \mathrm{d}k\, k^{d-1} \int_0^\pi \frac{\mathrm{d}\theta}{2\pi} \sin^d\theta \frac{\chi_{\mathbf{q}-\mathbf{k}}}{\chi_{\mathbf{q}}} \frac{1}{\rho^{-1} k^2 \eta_k + (\mathbf{q}-\mathbf{k})^2 D_{\mathbf{q}-\mathbf{k}}} \tag{5.6.21}$$

$$\eta_q = \eta^0 + \frac{k_B T}{2(d-1)\rho q^2} K_{d-1} \int_0^\Lambda dk\, k^{d+1} \int_0^\pi \frac{d\theta}{2\pi} (\sin^d \theta) \chi_{\mathbf{k}} \chi_{\mathbf{q}-\mathbf{k}} \left(\frac{1}{\chi_{\mathbf{k}}} - \frac{1}{\chi_{\mathbf{q}-\mathbf{k}}} \right)^2$$

$$\times \frac{1}{k^2 D_k + (\mathbf{q}-\mathbf{k})^2 D_{\mathbf{q}-\mathbf{k}}} \tag{5.6.22}$$

with

$$D_q^0 \equiv \zeta_q^0 / \chi_{\mathbf{q}}.$$

As in magnets, it has not been possible to find an analytic solution of these equations. Yet here the situation is much better than in the case of magnets in view of the rather weak critical anomaly of the shear viscosity and the rapid viscous relaxation process. This allows us to solve (5.6.21) and (5.6.22) by iteration by first ignoring the critical anomaly of the shear viscosity. The first approximation then yields,[11,20,21,23]

$$D_q = D_q^0 + \frac{k_B T}{6\pi\eta^0 q^2 \xi^3} K(q\xi) \tag{5.6.23}$$

$$K(x) \equiv \tfrac{3}{4}[1 + x^2 + (x^3 - x^{-1}) \tan^{-1} x] \tag{5.6.24}$$

$$\eta_{q=0} = \eta^0 + \frac{8\nu}{15\pi^2} \eta^0 \ln \tau \tag{5.6.25}$$

where the Ornstein–Zernike form $\chi_q = A(q^2 + \xi^{-2})^{-1}$ was assumed, with A being a constant. These results as well as those which come from taking into account various corrections such as the nonlocal nature of the shear viscosity, the deviations from the Ornstein–Zernike form for $\chi_{\mathbf{k}}$ and the vertex corrections have been successfully tested with the observed Rayleigh line width of light scattering[24] and the shear viscosity measurements[15] for both one-component liquids and binary liquid mixtures. The thermal conductivity for one-component fluids is given by ζ_q for $q\xi \ll 1$ and diverges as $\tau^{-(1-\eta)\nu}$, which is also in accord with the behaviour of the thermal conductivity found by macroscopic measurements.[15]

5.7 Renormalization group approach to critical dynamics

The recent introduction of the renormalization group ideas into equilibrium critical phenomena[9] virtually revolutionized the whole field and produced a concomitant impact in the area of critical dynamics.[10,25,26,27] Although this is not the place for a review of this work, we will present here a brief description of the basic ideas in a form suitable for our discussion of its extension to dynamics.

Theoretical efforts in the study of critical phenomena since Landau have been directed towards finding a suitable analytic starting point from which one could attack the problem of the highly singular behaviour at criticality. Wilson has shown that such a useful starting point is the construction of a supposedly analytic transformation from one system into another one, in the process of which some of the original degrees of freedom are eliminated. Take as an example a system whose Hamiltonian is $H_1(\{s\}^1)$ where $\{s\}^1$ is the entire set of the local order parameters $s(r)$ at every point, with corresponding Fourier components s_k, $0 < k < \Lambda$. The partition function Z is

$$Z = \int d\{s\}^1 \exp[-H_1(\{s\}^1)] \tag{5.7.1}$$

where the integration is over all the Fourier components $s_{\mathbf{k}}$, and the factor $k_B T$ is included in

H_1. Although a direct evaluation of (5.7.1) is prohibitively difficult in general, since it involves an enormous number of interacting degrees of freedom, one might reasonably hope to be able to perform an integration over a very small portion of these degrees of freedom, say over those s_k with $b^{-1}\Lambda < k \leqslant \Lambda$. One chooses b as some positive number slightly greater than unity, so that the number of degrees of freedom involved is rather small. Thus we may write (5.7.1) as

$$Z = \int \mathrm{d}\{s\}^2 \exp[-\tilde{H}_2(\{s\}^2)] \tag{5.7.2}$$

$$\exp[-\tilde{H}_2(\{s\}^2)] \equiv \int \mathrm{d}\{s\}^{1-2} \exp[-\tilde{H}_1(\{s\}^1)] \tag{5.7.3}$$

where $\{s\}^2$ and $\{s\}^{1-2}$ contain s_k with $0 \leqslant k < b^{-1}\Lambda$ and $b^{-1}\Lambda < k < \Lambda$, respectively. We now perform a suitable scale change of the variables and of the system size so that $\{s\}^2$ becomes identical to $\{s\}^1$ and then $\tilde{H}_2(\{s\}^2)$ becomes $H_2(\{s\}^1)$. The transformation $H_1 \to H_2$ is a renormalization transformation R of the Hamiltonian: $H_2 = RH_1$. After applying the transformation l times, all the degrees of freedom in the original Hamiltonian with $b^{-l}\Lambda < k < \Lambda$ are eliminated from H_1 for very large l such that $b^l\Lambda^{-1}$ exceeds ξ, H_l is essentially the free energy and hence contains all the singularities associated with a critical point. The various types of criticality are then associated with various fixed point Hamiltonians H^* of the transformation: $RH^* = H^*$. The fixed point is in general unstable in the sense that except for criticality H_l deviates from H^* with increasing l. Various critical exponents characterize the rates at which deviations occur in various directions in the space of parameters entering the Hamiltonian. Suppose for example that the initial Hamiltonian differs from the fixed point Hamiltonian by a perturbation δH. Then one assumes that for sufficiently small δH one can linearize the renormalization group equations around the fixed point, i.e. that

$$R(H^* + \delta H) = R(H^*) + L(H^*)\delta H = H^* + L(H^*)\delta H \tag{5.7.4}$$

where $L(H^*)$ is a linear operator. Then if one expands δH in terms of the eigenfunctions $\{O_i\}$ of $L(H^*)$, one finds that $H_l = R^l(H^* + \delta H)$ is given by

$$H_l = H^* + \Sigma \mu_i b^{l y_i} O_i \tag{5.7.5}$$

where the eigenvalues $\{\lambda_i\}$ of $L(H^*)$ are related to the exponents $\{y_i\}$ for the scaling fields $\{\mu_i\}$ by $\lambda_i = b^{y_i}$. Those operators O_i for which the corresponding y_i are greater (less) than zero are termed relevant (irrelevant) since an application of R^l causes an increase (decrease) in the corresponding μ_i and therefore causes H to move away from (toward) H.* The borderline case for which the operators O_i have $y_i = 0$ is more subtle and is not discussed here. Such operators are called marginal. To be at criticality requires that all the fields μ_i which are relevant must be set equal to zero, so that the limit of $R^l(H_c)$ as $l \to \infty$ equals H^*, where H_c denotes the initial Hamiltonian at criticality. An irrelevant field need not be fixed at any particular value for the system to be at criticality. In the simplest case there are two relevant fields, one corresponding to the order parameter and one to the temperature. As an example of (5.7.5) consider the situation near a magnetic critical point with all the parameters $\{\mu_i\}$ equal to zero (e.g. $T = T_c$) except for a nonzero magnetic field. Then

$$H_l = H^* - b^{l y_1} \int s(r)\mathrm{d}r \tag{5.7.6}$$

where y_1 is related to the critical exponent η by

$$y_1 = \tfrac{1}{2}(d + 2 - \eta). \tag{5.7.7}$$

Thus the problem of determining the singular behaviour of a system near its critical point is that of determining the fixed point H^* and the eigenvalues and eigenfunctions of $L(H^*)$. Of course, in general this cannot be done exactly, and so various calculational schemes have been developed recently to achieve this goal in an approximate way. For example, one such method is the so-called ϵ-expansion, in which one expands around the Gaussian fixed point, which is the correct fixed point for $d > 4$ and calculates critical exponents as an asymptotic expansion in powers of $\epsilon = 4 - d$ around the Gaussian (mean field) values. This approach is particularly useful in discussing the problems of universality and cross-over behaviour and gives reasonable values (within 5–10% accuracy) for the critical exponents for $\epsilon = 1$. As we will see, this ϵ-expansion can also be used in critical dynamics. Other methods have also been developed to handle such problems as the two- and three-dimensional Ising model.[28] A thorough exposition of the basic ideas of the renormalization group as well as an excellent reference to theoretical work up to 1974 is given in the review article by Wilson and Kogut.[9]

We now turn to non-equilibrium aspects where the problem is completely specified by a stochastic equation of the type (5.5.19) for the probability distribution function $g(\{a\},t)$ of the gross variables. We can write this equation as (the symbol $\mathscr{L}$ should not be confused with the Lagrangian of Section 5.2)

$$\frac{\partial}{\partial t} g(\{a\}, t) = \mathscr{L}(\{a\}) g(\{a\}, t). \tag{5.7.8}$$

The role of the Hamiltonian in the equilibrium case is now played by the stochastic operator $\mathscr{L}(\{a\})$, and a renormalization transformation $\mathscr{R}$ now transforms one stochastic operator $\mathscr{L}_l$ into another one $\mathscr{L}_{l+1}$:[26] $\mathscr{L}_{l+1} = \mathscr{R}\mathscr{L}_l$. In a similar manner to the equilibrium situation one assumes the existence of a fixed point stochastic operator, $\mathscr{L}^* = \mathscr{R}\mathscr{L}^*$ and also assumes that one can linearize around this fixed point such that $\mathscr{R}(\mathscr{L}^* + \delta\mathscr{L}) = \mathscr{R}(\mathscr{L}^*) + B(\mathscr{L}^*)\delta\mathscr{L}$ etc.

The elimination of a part of the degrees of freedom is conveniently accomplished with the aid of a projection operator. The details of this dynamic renormalization group are rather involved and will be deferred to the Appendix where some illustrative examples are also provided. The stochastic operator contains a set of parameters $\{\mu_i\}$ such as the coefficients of the mode coupling terms and the 'bare' transport coefficients, and the renormalization transformation can be represented as a mapping in this space of parameters $\{\mu_i\}$. Namely, we may write

$$\mathscr{L}_l(\{a\}) = \mathscr{L}(\{a\}\{\mu_i^{(l)}\}) \tag{5.7.9}$$

$$\{\mu_i^{(l+1)}\} = \mathscr{R}\{\mu_i^{(l)}\}. \tag{5.7.10}$$

The symbol $\mathscr{R}$ used in (5.7.10) to denote the dynamical renormalization transformation is not to be confused with the similar symbol used earlier to denote the equilibrium renormalization transformation.

Two examples of (5.7.10) are given in the Appendix by (A.36) for magnets and (A.52) for classical liquids. The latter gives the transformation of the mode coupling strength $\hat{\lambda}_l$ of the order parameter and the local velocity and the 'bare' shear viscosity $\hat{\eta}_l$ and the 'bare' thermal conductivity $\hat{\xi}_l$, all of which are properly scaled, and takes the following form at criticality for the dimensionality $d = 4 - \epsilon$ $(0 < \epsilon \ll 1)$:[25]

$$\hat{\lambda}_{l+1} = b^{z(d/2)-1}\hat{\lambda}_l \tag{5.7.11a}$$

$$\hat{\xi}_{l+1} = b^{z-4}\left(\hat{\xi}_l + \frac{3}{4} K \ln b \, \frac{\hat{\lambda}_l^2}{\hat{\eta}_l}\right) \tag{5.7.11b}$$

$$\hat{\eta}_{l+1} = b^{z-2}\left(\hat{\eta}_l + \frac{1}{24} K l n b \frac{\hat{\lambda}_l^2}{\hat{\zeta}_l}\right). \tag{5.7.11c}$$

In the above, b and b^z ($b > 1$) are the scale factors of space and time, respectively, at each renormalization transformation, and $K \equiv 1/8\pi^2$. z is determined below by the fixed point condition and is in fact the dynamic critical exponent for the characteristic time of the order parameter. As in the equilibrium case we now assume the existence of the fixed point stochastic operator $\mathscr{L}^*(\{a\}) = \mathscr{L}(\{a\}\{\mu_i^*\})$ such that $\mathscr{L}^* = \mathscr{R}\,\mathscr{L}^*$. In order to obtain the fixed point solution of (5.7.11), we first note that (5.7.11) reduces to the following single equation for the ratio $\kappa_l \equiv \hat{\lambda}_l^2/\hat{\zeta}_l\hat{\eta}_l$,

$$b^{\epsilon}(\kappa_l/\kappa_{l+1}) = [1 + \tfrac{3}{4} K l n b \kappa_l]\,[1 + K l n b \kappa_l/24]. \tag{5.7.12}$$

For small ϵ where $b^{\epsilon} \simeq 1 + \epsilon l n b$, (5.7.12) gives for the fixed point κ^*,

$$\kappa^* = \frac{24}{19K}\,\epsilon. \tag{5.7.13}$$

This result, together with (5.7.11b), gives for the dynamic critical exponent z,

$$z = 4 - \frac{18}{19}\,\epsilon. \tag{5.7.14}$$

Now, $\hat{\eta}_l$ and $\hat{\zeta}_l$ are related to the original sheer viscosity η_l and ζ_l by the following scale transformations

$$\hat{\zeta}_l = b^{(z-4)l}\zeta_l \tag{5.7.15a}$$

$$\hat{\eta}_l = b^{(z-2)l}\eta_l \tag{5.7.15b}$$

where η_l and ζ_l include fluctuations with wavenumbers greater than $b^{-l}\Lambda$ in the sense of mode coupling theory. If $\hat{\zeta}_l$ tends to a finite fixed point value ζ^* for large l, ζ_l behaves as $b^{(4-z)l} = b^{18\epsilon l/19}$. That is, as l increases more fluctuation contributions are included in ζ_l. This means that the wavenumber-dependent thermal conductivity $\zeta(q)$ should behave at criticality as

$$\zeta(q) \propto q^{-18\epsilon/19}. \tag{5.7.16a}$$

Now, since $\hat{\lambda}_l \propto b^{(z-d/2-1)l} = b^{(1-17\epsilon/38)l}$ by (5.7.11a), the finiteness of κ_l and ζ_l as $l \to \infty$ implies that $\eta_l \propto b^{\epsilon l/19}$. Thus the wavenumber-dependent shear viscosity $\eta(q)$ at criticality behaves as

$$\eta(q) \propto q^{-\epsilon/19}. \tag{5.7.16b}$$

Dynamical scaling then tells us that the thermal conductivity and shear viscosity at the zero wavenumber exhibit the following critical divergences:

$$\zeta \propto \tau^{-18\epsilon\nu/19} \qquad \eta \propto \tau^{-\epsilon\nu/19}. \tag{5.7.17}$$

It is also possible to formulate the problem so that both ζ_l and η_l tend to finite fixed point values by introducing two dynamic critical exponents z_{ζ} and z_{η} which correspond to the two scales that characterize the critical dynamics here, with essentially the same results. In this case there is no single finite stochastic operator, which is also true in the original formulation since η_l blows up with increasing l. Nevertheless, the recurrence relation which corresponds to (5.7.10) can give rise to finite fixed points in this alternative formulation.[54]

5.8 Mode coupling calculations of critical transport anomalies in $4-\epsilon$ and $6-\epsilon$ dimensions

So far we have discussed two alternative procedures for elucidating the nature of dynamic critical phenomena, those being the mode coupling and renormalization group approaches. A close relationship between the two is to be expected since both approaches eliminate short wavelength fluctuations to renormalize transport coefficients. The mode coupling theory accomplishes this in one step whereas the renormalization group carries this out in small steps. The important difference, however, is one of spirit. The mode coupling theory as we have presented it in this article attempts to construct the dynamical theory on the basis of the assumed knowledge of the equilibrium aspects of the problem, whereas the renormalization group approach aims at a more ambitious task of simultaneously treating both the static and the dynamical aspects of the problem. Although the interesting question of the precise relationship between these two theories has not yet been answered in a general way, in this section we show that, at least to lowest order in the parameter $\epsilon = d_c - d$ (where the 'conventional theory' is expected to be valid for the dimensionality $d > d_c$), the two approaches yield identical results for the dynamical critical exponents. We will demonstrate this for the isotropic Heisenberg ferromagnet and for a classical liquid, but the same procedure can be applied to several other models of critical dynamics.[29] The calculations explicitly show that the coupling between the modes leads to critical anomalies in the kinetic coefficients which, as mentioned in Section 5.2, is the reason that conventional theory is invalid for $d < d_c$.

It is convenient for our present purpose to use the language of propagators, $G_{jl}(t)$, introduced in Section 5.6. These propagators satisfy a set of coupled nonlinear integral equations (5.6.13) which reduce, upon making the Markoffian approximation, to (5.6.15) whose solution yields the renormalized transport coefficients. Although these coupled equations are in general quite complicated to solve, as noted earlier, we will now show that for the special case of $\epsilon = d_c - d \ll 1$ they can be reduced to ordinary differential equations whose solution can then easily be found. For convenience we will deal with the small momentum, zero frequency, critical point behaviour of the transport coefficients.

5.8.1 *Isotropic Heisenberg model*

Our starting point is (5.6.17) for the q-dependent spin diffusion constant. For $d > 6$ the mode coupling contribution to $D_{\mathbf{q}}$ can be shown to be 'negligible', so that conventional theory is valid. We now consider the situation for d slightly less than $d_c = 6$, to lowest order in the expansion parameter $\epsilon = 6 - d$, with $0 < \epsilon \ll 1$. Although this calculation is not very realistic for the physically interesting case of $d = 3$, the results are at least of qualitative validity there.

We first note that for $\epsilon \ll 1$ the function $F(k^2, q^2, \mathbf{k}\cdot\mathbf{q})$ given by (5.6.18) reduces to

$$F(k^2, q^2, \mathbf{k}\cdot\mathbf{q}) = \frac{q^2}{k^2 \cdot (\mathbf{q}-\mathbf{k})^2}\,\frac{(2\mathbf{q}\cdot\mathbf{k} - q^2)^2}{k^2 D_{\mathbf{k}} + (\mathbf{q}-\mathbf{k})^2 D_{\mathbf{q}-\mathbf{k}}} \tag{5.8.1}$$

since $\chi_k(T = T_c) \simeq k^{-2}$ to $O(\epsilon^2)$ in an appropriate dimensionless unit. Our basic approximation to (5.6.17) is then based on the observation that for small ϵ the wavenumbers much greater than q give the dominant contribution to the second term of (5.6.17). Therefore we evaluate this term by taking the $q \to 0$ limit of the integrand and introducing a lower cutoff in the integral over k which is proportional to q. We are now in a position to perform the angular integration to lowest order in ϵ (i.e. with $d = 6$), and thus obtain the following expression for the (renormalized) kinetic coefficient $\zeta(q) = D_q/q^2$,

$$\zeta(q) = \zeta^0 + g^2 \int_q^\Lambda dk/k^{1+\epsilon}\zeta(k) \tag{5.8.2}$$

where $g^2 = (k_B T)^2/(192\pi^3)$. Upon differentiating (5.8.2) with respect to $x \equiv q^{-\epsilon}$, we thus obtain the differential equation

$$\frac{\partial \zeta}{\partial x} = \frac{g^2}{\epsilon}\frac{1}{\zeta} \tag{5.8.3}$$

whose solution is

$$\zeta(x) = \left(\zeta^2(x_1) + \frac{2g^2}{\epsilon}(x - x_1)\right)^{1/2} \tag{5.8.4}$$

where x_i is an arbitrary upper limit of integration. Hence we see that the coupling between the spin modes as given in (5.6.17) leads to a diverging kinetic coefficient, $\zeta \sim q^{-\epsilon/2}$, thus invalidating a basic assumption of the conventional thermodynamic theory†. The dynamic critical exponent z which is defined by $q^2 D_q \sim q^z$ is therefore given to order ϵ by $z = 4 - \epsilon/2 = (d+2)/2$, a result which agrees with the renormalization group calculations.

We also note that the same procedure can be used to evaluate the $q = 0$, $T > T_c$ behaviour of D_q. In this case one introduces a lower cutoff which is proportional to ξ^{-1} and uses $\chi_k^{-1} \simeq \xi^{-2}$ for $k = 0$. By an analogous argument to that given above one then finds that

$$q^2 D_q \xrightarrow[q \to 0]{} q^2 \left(\frac{2g^2}{\epsilon}\right)^{1/2} \xi^{-(2-\epsilon/2)} \tag{5.8.5}$$

which agrees with the dynamical scaling and renormalization group calculations.[27]

5.8.2 *Classical liquids*

The treatment of classical liquids is completely analogous to that of the ferromagnet, but the results differ in one important aspect which we will discuss later, namely in the role played by a certain 'constant of the motion' of the differential equations. We start with the pair of coupled equations (5.6.21) and (5.6.22) for the diffusion coefficient D_q and shear viscosity η_q. Following the same procedure for evaluating the small q behaviour of these expressions as used for the ferromagnet we obtain from (5.6.21) and (5.6.22) the equations

$$\zeta(q) = \zeta^0 + \frac{3}{4}\frac{(k_B T)^2}{\rho} K \int_q^\Lambda \frac{dk}{k^{1+\epsilon}\eta(k)} \tag{5.8.6}$$

and

$$\eta(q) = \eta^0 + \frac{(k_B T)^2}{\rho}\frac{K}{24}\int_q^\Lambda \frac{dk}{k^{1+\epsilon}\zeta(k)} \tag{5.8.7}$$

where $D(q) = q^2\zeta(q)$ and $\epsilon = 4 - d$. Thus we find for the classical liquid (for $0 < \epsilon \ll 1$) the coupled pair of differential equations

$$\frac{d}{dx}\zeta = \frac{3}{4}\frac{g^2}{\epsilon\eta} \tag{5.8.8}$$

†In the case when $d > d_c (\epsilon < 0)$ we see from (5.8.4) that the kinetic coefficient ζ approaches a finite constant as $q \to 0$, so that the conventional theory is valid.

$$\frac{\mathrm{d}}{\mathrm{d}x}\eta = \frac{g^2}{24\epsilon}\,\frac{1}{\zeta} \tag{5.8.9}$$

where $x = q^{-\epsilon}$ and $g^2 = K(k_B T)^2/\rho$. Since it follows from (5.8.8) and (5.8.9) that

$$\frac{\mathrm{d}}{\mathrm{d}x}(\eta \cdot \zeta) = \frac{19}{24}\,\frac{g^2}{\epsilon} \tag{5.8.10}$$

one immediately obtains

$$\eta(x)\zeta(x) = \eta(x_1)\zeta(x_1) + \frac{19}{24}\,\frac{g^2}{\epsilon}(x - x_1). \tag{5.8.11}$$

Thus with the use of (5.8.11) one can eliminate $\zeta(x)$, say, in (5.8.9) and solve for $\eta(x)$, with the resulting solutions

$$\eta(x) = \eta(x_1)\left(1 + \frac{19}{24}\,\frac{g^2}{\epsilon\eta(x_1)\zeta(x_1)}(x - x_1)\right)^{1/19} \tag{5.8.12}$$

$$\zeta(q) = \zeta(q_1)\left(1 + \frac{19}{24}\,\frac{g^2}{\epsilon\eta(x_1)\zeta(x_1)}(x - x_1)\right)^{18/19}. \tag{5.8.13}$$

Hence in the limit as $q \to 0$ one has

$$\eta(q) \sim \left(\frac{19}{24}\,\frac{g^2}{\epsilon}\right)^{1/19} Cq^{-\epsilon/19} \tag{5.8.14}$$

$$\zeta(q) \sim \left(\frac{19}{24}\,\frac{g^2}{\epsilon}\right)^{18/19} \frac{1}{C} q^{-(18/19)\epsilon} \tag{5.8.15}$$

where $C \equiv \eta^{18/19}(q_1)/\zeta^{1/19}(q_1)$.

The dynamical critical exponents given in (5.8.14) and (5.8.15) agree with the renormalization group calculations of the preceding section. An interesting difference between these results for the liquid and those for other models of critical dynamics, such as the Heisenberg ferromagnet, is the dependence of the amplitudes of the shear viscosity, $\eta(q)$, and thermal conductivity, $\zeta(q)$, on the 'bare' coefficients $\eta(q_1)$ and $\zeta(q_1)$ through the factor C. However, this ratio of the bare coefficients, C, is in fact a kind of constant of the motion independent of q_1, as it should be. In the other models considered so far, such constants of the motion do not affect the amplitudes. Hence the binary liquid discussed here differs from the other models in that the mode coupling equation (or the renormalization group equation) itself is not sufficient to completely determine the diverging parts of the transport coefficients, in that it has to be supplemented with the initial data for C. The difference between magnets and liquid illustrates the fact that broadly speaking there are two types of critical dynamics which we have previously called the first and second classes.[11] For the first class systems, the long wavelength fluctuations of a set of gross variables including the order parameter constitute an asymptotically closed dynamical system in the sense that the bare Onsager kinetic coefficients ζ_{ij}^0 in (5.5.4) and (5.5.5) which arise from short wavelength fluctuations can be set to zero afterwards without affecting the diverging parts of the transport coefficients. Systems with Goldstone modes belong to this class, where dynamic critical exponents are obtained from dynamic scaling. For second class systems, short wavelength fluctuations cannot be entirely ignored in determining the diverging parts of transport coefficients, and dynamical scaling is not enough to find a dynamical critical exponent. In one word, first class systems are more 'mechanical' whereas second class systems are more 'thermal' in character.

[illegible] Summary of results for the small q, $T - T_c$ behaviour of the decay rate $\Gamma_j = A_j q^{z_j}$. For the Heisenberg model $\epsilon = 6 - d$; otherwise $\epsilon = 4 - d$. The notation used for the first four models is similar to that of Kawasaki[21] while that of the last model is that of Gunton and Kawasaki.[29] χ is the parallel susceptibility per unit volume. (Reproduced by permission of The Institute of Physics)

Model	Gross variables	γ_j	Dynamical critical exponent	Critical amplitude
Heisenberg	$a_1 = s^x_{\mathbf{q}}$	$\gamma_1 = q^2L^0/\chi_{1q}$	$z_1 = 4 - \frac{\epsilon}{2} = \frac{d+2}{2}$	$A_1 = \sqrt{(2g^2/\epsilon)}$ $g^2 = (k_B T v_0)^2/192\pi^3$
Planar ferromagnet	$a_2 = s^0_{\mathbf{q}}$	$\gamma_2 = q^2L^0$	$z_1 = z_2 = 2 - \frac{\epsilon}{2}$	$A_1 = \sqrt{(2g^2/\epsilon)}$
	$a_1 = s^+_{\mathbf{q}}$	$\gamma_1 = L^{\perp}/\chi_{1q}$	$= \frac{d}{2}$	$A_2 = A_1$ $g^2 = \frac{(k_B T)^2}{16\pi^2\chi_\parallel}$
Superfluid ^{4}He	$a_1 = \delta\psi^+_{\mathbf{q}}$	$\gamma_1 = L^0/\chi_{1q}$	$z_1 = z_2 = 2 - \frac{\epsilon}{2} + \frac{\alpha}{2\nu}$	$A_1 = \sqrt{\left(\frac{10g^2}{3\epsilon}\right)}$
	$a_2 = \delta s_{\mathbf{q}}$	$\gamma_2 = \frac{q^2\lambda}{\rho C_p(q)}$	$= \frac{1}{2}\left(d + \frac{\alpha}{\nu}\right)$ $(\alpha/\nu \simeq \epsilon/5)$	$A_2 = A_1/2$ $g^2 = \frac{k_B(sT)^2}{16\pi^2\rho}$
Binary liquid	$a_1 = \delta v^\alpha_{\mathbf{q}}$	$\gamma_1 = \eta^0 q^2/\rho$	$z_1 = 2 - \epsilon/19$	$A_1 = \left[\frac{19k_B T}{192\pi^2\rho\epsilon}\right]^{1/19} \frac{\eta(q_1)^{18/19}}{\lambda(q_1)^{1/19}}$
	$a_2 = \delta c_{\mathbf{q}}$	$\gamma_2 = q^2D^0$	$z_2 = 4 - \frac{18\epsilon}{19}$	$A_2 = \left[\frac{19k_B T}{192\pi^2\rho\epsilon}\right]^{18/19} \frac{\lambda(q_1)^{1/19}}{\eta(q_1)^{18/19}}$
^{3}He–^{4}He tricritical	$a_1 = \delta\psi^+_{\mathbf{q}}$ $a_2 = \delta x_{\mathbf{q}}$ $a_3 = \delta s_{\mathbf{q}}$	$\gamma_1 = L^0_\psi/\chi_{1q}$ $\gamma_2 = Dq^2$ $\gamma_3 = \frac{\kappa_s q^2}{\rho C_{px}(q)}$	$z_1 = z_2 = 2 - \frac{\epsilon}{2}$ $= \frac{d}{2}$ $z_3 = 2 - \frac{\epsilon}{2} + \frac{\alpha_t}{\nu_t} = \frac{d}{2} + \frac{\alpha_t}{\nu_t}$ $(\alpha_t/\nu_t = 1)$	$A_1 = A_2 = \sqrt{(2g^2/\epsilon)}$ $A_3 = \frac{1}{16\pi^2}\left[\frac{m_4 x k_B T}{\rho}\right]\sqrt{\left(\frac{2}{\epsilon g^2}\right)}$ $g^2 = \frac{1}{16\pi^2}\frac{k_B(m_4 s_4 T)}{\rho}$

We should note that the procedure outlined here can be applied to other dynamical models, such as for the planar ferromagnet and the ^3He–^4He mixture near its tricritical point. The results of these calculations are summarized in Table 5.8.1.

As we have suggested earlier there is an even closer relationship between the mode coupling and the renormalization group approaches than shown here for the calculation of the critical exponents. For example, one can show that the renormalization group equations of Halperin, Hohenberg and Siggia[25] for the binary liquid can be converted into exactly the same differential equations as our equations (5.8.8) and (5.8.9). In order to see this relationship, we first write (5.7.11) in a differential form by introducing $q = b^l$ and $\hat{\lambda}_l = \hat{\lambda}(q)$, etc. and choosing $b - 1$ to be infinitesimal, $b = 1 + q^{-1}\delta q$, where we have used a dimensionless unit with $\Lambda = 1$. (5.7.11) then becomes

$$\frac{d\hat{\lambda}}{dq} = \frac{z - 3 + \epsilon/2}{q}\hat{\lambda} \tag{5.8.16}$$

$$\frac{d\hat{\xi}}{dq} = \frac{z - 4}{q}\hat{\xi} + \frac{3K}{4q}\frac{\hat{\lambda}^2}{\hat{\eta}} \tag{5.8.17}$$

$$\frac{d\hat{\eta}}{dq} = \frac{z - 2}{q}\hat{\eta} + \frac{K}{24q}\frac{\hat{\lambda}^2}{\hat{\xi}}. \tag{5.8.18}$$

These equations reduce to (5.8.8) and (5.8.9) by the following transformations

$$\hat{\lambda} \equiv q^{(z-3+\epsilon/2)}\lambda \qquad \hat{\xi} = q^{z-4}\zeta \qquad \hat{\eta} = q^{z-2}\eta \tag{5.8.19}$$

$$x = q^{\epsilon}. \tag{5.8.20}$$

This means that provided the initial data are the same, $\hat{\zeta}(k)$ and $\hat{\eta}(k)$ of the renormalization group calculation are related to $\zeta(k)$ and $\eta(k)$ of the mode coupling theory via

$$\hat{\zeta}(q) = q^{z-4}\zeta(1/q) \tag{5.8.21}$$

$$\hat{\eta}(q) = q^{z-2}\eta(1/q), \tag{5.8.22}$$

The factors q^{z-4} and q^{z-2} come from the scale change in the renormalization group. If we put aside those factors, $\hat{\zeta}(q)$ and $\hat{\eta}(q)$ change in the opposite manner to $\zeta(q)$ and $\eta(q)$ as q increases. As q increases, the renormalization group operation eliminates more degrees of freedom resulting in increases of $q^{4-z}\hat{\zeta}$ and $q^{2-z}\hat{\eta}$. On the other hand, (5.8.6) and (5.8.7) tell us that as q increases fewer fluctuation degrees of freedom are incorporated into $\zeta(q)$ and $\eta(q)$ which hence decrease. The precise nature of the mode coupling and renormalization group relationship, however, needs further investigation based on exact dynamical renormalization group equations such as the one given in the Appendix.

Recently the mode coupling equations (5.6.21) and (5.6.22) for $d = 3$ at the criticality have been solved approximately by T. Ohta[55,56] who finds that $\eta(q) \propto q^{-8/15\pi^2}$.

5.9 Problems far from equilibrium

Until now we have limited ourselves to those aspects of critical dynamics in which a deviation from thermal equilibrium is small enough to be treated as a first order perturbation. On the other hand, it is widely recognized that even a small perturbation such as gravity can have drastic effects in the close vicinity of a critical point.[1,15] Hence this class of problems has to be well understood before a complete understanding of critical dynamics can be claimed. In this section we deal with some aspects of this little-explored area.

There are problems far from equilibrium such as those associated with metastable states whose understanding on a fundamental level is very incomplete. Here we wish to stress that a considerable simplification results for a system near its critical point. This case may thus serve as a useful and realistic model for investigating behaviour far from equilibrium. The reason is that in such cases the expected universality of critical dynamics allows us to ignore the complicated and irrelevant microscopic details of the dynamics, and to focus rather on processes occurring on the semimacroscopic level,[30,31] e.g. the level of the kinetic equation of Section 5.5.

We begin by noting that the order parameter is the most sensitive of the gross variables to a perturbation near a critical point and in many instances is also the most slowly varying of these variables. Thus we will formulate a theory that allows for a large deviation from equilibrium for those degrees of freedom associated with the local order parameter, where other degrees of freedom are assumed to adiabatically follow its slow variation. Such a theory can be succinctly couched in terms of a stochastic equation of the type (5.5.19) which we write as

$$\frac{\partial}{\partial t} g(\{a\}, t) = \mathscr{L}(\{a\}) g(\{a\}, t). \tag{5.9.1}$$

We now divide the set of gross variables $\{a\}$ into the local order parameter $\{c\}$ and the remainder $\{b\}$. Correspondingly, the stochastic operator $\mathscr{L}$ is written as

$$\mathscr{L}(\{a\}) = \mathscr{L}_0(\{c\}) + \mathscr{L}_b(\{b\}) + \mathscr{L}'(\{c\}\{b\}). \tag{5.9.2}$$

where $\mathscr{L}'$ is the interaction between $\{c\}$ and $\{b\}$. We are interested here in the reduced distribution function for $\{c\}$;

$$g(\{c\}, t) \equiv \int g(\{c\}\{b\}, t) \mathrm{d}\{b\}. \tag{5.9.3}$$

The formally exact closed equation for $g(\{c\},t)$ can be obtained with the aid of a projection operator technique such as that of the Appendix. We shall be content with the second order perturbation theory with respect to $\mathscr{L}'$ where we assume that the interaction between $\{c\}$ and $\{b\}$ is small. Taking into account the assumption that $\{c\}$ varies rather slowly in comparison with $\{b\}$, the approximate equation for $g(\{c\},t)$ becomes

$$\frac{\partial}{\partial t} g(\{c\}, t) = [\mathscr{L}_0(\{c\}) + \mathscr{L}_1(\{c\})] g(\{c\}, t) \tag{5.9.4}$$

$$\mathscr{L}_1(\{c\}) \equiv -\langle \mathscr{L}' \mathscr{L}b^{-1} \mathscr{L}' \rangle_b \tag{5.9.5}$$

where $\langle \ldots \rangle_b$ denotes the average over the equilibrium distribution function of $\{b\}$ alone.

As an example, consider a classical liquid where $\{b\}$ consists of the local transverse velocity field and the stochastic equation equivalent to (5.5.19) is given by (A.40)–(A.44) in the Appendix (where the superscript α should be dropped) and $\lambda = (k_B T/\rho)^{1/2}$. Eliminating the velocity field $\mathbf{j}$ we obtain the following stochastic equation for the order parameter distribution function $g(\{c\},t)$,[32]

$$\frac{\partial}{\partial t} g(\{c\}, t) = \mathscr{L}(\{c\}) g(\{c\}, t) \tag{5.9.6}$$

where $\mathscr{L} = \mathscr{L}_0 + \mathscr{L}_1$ is given by

$$\mathscr{L}(\{c\}) \equiv \iint \mathrm{d}\mathbf{r}\, \mathrm{d}\mathbf{r}' \frac{\delta}{\delta c(\mathbf{r})} \mathscr{G}(\mathbf{rr}', \{c\}) \left(\frac{\delta}{\delta c(\mathbf{r}')} + \frac{\delta \Psi(\{c\})}{\delta c(\mathbf{r}')} \right) \tag{5.9.7}$$

and k_BT has been set equal to unity. Here $\Psi(\{c\})$ is the thermodynamic potential associated with the fluctuation $\{c\}$ and

$$\mathcal{G}(\mathbf{rr}', \{c\}) = -\zeta^0 \nabla^2 \delta(\mathbf{r} - \mathbf{r}') + 2\frac{\partial c(r)}{\partial \mathbf{r}} \mathbf{T}(\mathbf{r} - \mathbf{r}') \cdot \frac{\partial c(r')}{\partial \mathbf{r}'} \tag{5.9.8}$$

with $\mathbf{T}(\mathbf{r})$ the Oseen tensor given by

$$[\mathbf{T}(\mathbf{r})]_{\alpha\beta} = \frac{1}{8\pi\eta}\left(\frac{1}{r}\delta_{\alpha\beta} + \frac{r_\alpha r_\beta}{r^3}\right). \tag{5.9.9}$$

$\mathcal{G}(\mathbf{rr}', \{c\})$ is the generalized nonlocal diffusion 'coefficient' in the function space $\{c\}$ and the nonlocality (the second term) is brought about by the effective nonlocal interaction of order parameter fluctuations mediated by the fluctuating velocity field. This interaction is an analogue of the hydrodynamic interaction among moving segments in high polymer solutions.

The stochastic equation obtained here constitutes the basic equation from which to start an investigation of order parameter dynamics in a general manner. However, it is not amenable to direct analysis in view of its highly complex character. On the other hand, we often encounter the situation where a small part $\{c^u\}$ of the degrees of freedom $\{c\}$ changes much more slowly than the remaining degrees of freedom, $\{c^s\}$. For instance, for a nucleation near a critical point, $\{c^u\}$ is the nucleation coordinate. Thus as $\{c^u\}$ changes slowly in time, $\{c^s\}$ adiabatically follows $\{c^u\}$. We may then approximate $g(\{c\},t)$ as

$$g(\{c\}, t) = P_l(\{c^s\} \mid \{c^u\}) g_u(\{c^u\}, t) \tag{5.9.10}$$

where

$$g_u(\{c^u\}, t) \equiv \int g(\{c\}, t) \mathrm{d}\{c^s\} \tag{5.9.11}$$

and P_l is the local equilibrium distribution function of $\{c^s\}$ for fixed $\{c^u\}$. We thus find for the case in which $\{c^u\}$ is a discrete set of variables c_i^u,

$$\frac{\partial}{\partial t} g_u(\{c^u\}, t) = \mathcal{L}_u(\{c^u\}) g_u(\{c^u\}, t) \tag{5.9.12}$$

where

$$\mathcal{L}_u(\{c^u\}) \equiv \sum_{ij} \frac{\partial}{\partial c_i^u} M_{ij}(\{c^u\}) \left(\frac{\partial}{\partial c_i^u} + \frac{\partial \Psi_u(\{c^u\})}{\partial c_j^u}\right) \tag{5.9.13}$$

$$M_{ij}(\{c^u\}) \equiv \iint \mathrm{d}r\, \mathrm{d}r' \int \mathrm{d}\{c^s\} \frac{\delta c_i^u}{\delta c(r)} \frac{\delta c_j^u}{\delta c(r')} (rr'; \{c\}) P_l(\{c^u\} \mid \{c^s\}) \tag{5.9.14}$$

and

$$\exp[-\Psi_u(\{c^u\})] \equiv \int \mathrm{d}\{c^s\} \exp[-\Psi(\{c^u\}\{c^s\})]. \tag{5.9.15}$$

In order to illustrate the usefulness of this general theory we discuss in some detail the nucleation near a critical point where $\{c^u\}$ is chosen to be the nucleation coordinate R, i.e. the radius of the nucleation droplet. (This problem has been considered also by Kiang *et al.*[33] and by Langer and Turski[34] from different points of view to that given here.) The stochastic equation (5.9.1) then reduces to

$$\frac{\partial}{\partial t} g(R, t) = -\frac{\partial}{\partial R} J(R, t) \tag{5.9.16}$$

where

$$J(R, t) = -M(R)\left(\frac{\partial}{\partial R} + \frac{\partial \Psi_u(R)}{\partial R}\right) g(R, t) \tag{5.9.17}$$

is the probability current of nucleation.

In the standard theory of nucleation[34] one assumes a kind of steady state in which nuclei larger than the critical nucleus are removed from and fed into the system in the form of the metastable phase. The steady state solution of (5.9.16) corresponds to the constant current $J(R) = J^*$ with[35]

$$J^* = M(R^*)(2\pi\chi_u)^{-1/2} \exp(-\Delta\Psi) \tag{5.9.18}$$

where R^* and $\Delta\Psi$ are, respectively, the radius and the activation free energy of a critical nucleus in which contributions from thermal fluctuations of $\{c^s\}$ are already included. We have also expanded $\Psi_u(R)$ near its maximum at $R = R^*$ as

$$\Psi_u(R) = \Psi_u(R^*) - 2\chi_u^{-1}(R - R^*)^2 \quad (\chi_u > 0). \tag{5.9.19}$$

$M(R^*)$ now becomes

$$M(R^*) = \iint \mathrm{d}\mathbf{r}\, \mathrm{d}\mathbf{r}' \left(\frac{\delta R}{\delta c(\mathbf{r})}\right)^* \left(\frac{\delta R}{\delta c(\mathbf{r}')}\right)^* \mathscr{G}^*(\mathbf{rr}') \tag{5.9.20}$$

with

$$\mathscr{G}^*(\mathbf{rr}') \equiv -\zeta^0 \nabla^2 \delta(\mathbf{r} - \mathbf{r}') + 2\mathrm{T}(\mathbf{r} - \mathbf{r}') : \left(\frac{\partial c^*(r)}{\partial \mathbf{r}} \frac{\partial c^*(r')}{\partial \mathbf{r}'} + \frac{\partial^2}{\partial \mathbf{r} \partial \mathbf{r}'} \chi_l(\mathbf{rr}')\right) \tag{5.9.21}$$

and

$$\chi_l(\mathbf{rr}') \equiv \langle [c(\mathbf{r}) - c^*(\mathbf{r})]\, [c(\mathbf{r}') - c^*(\mathbf{r}')] \rangle_l. \tag{5.9.22}$$

In the above an asterisk means that the expression is evaluated in a local equilibrium state with a critical nucleus centred at $r = 0$, and $\langle \ldots \rangle_l$ denotes a local equilibrium average with the value of R fixed at R^* with R expressed as a suitable function of $c(r)$. For example, this functional may be $\mathrm{R} = \{\tfrac{3}{4}\int [c(\mathbf{r}) - c_0]\, \mathrm{d}\mathbf{r}/\pi\Delta c\}^{1/3}$ where c_0 is the value of $c(\mathbf{r})$ outside the nucleus and Δc is the difference of c in the two coexising phases.

A simple expression for $M(R^*)$ can be obtained by noting that the growth rate of a critical nucleus κ is obtained as the largest eigenvalue of the following equation for $\delta c(\mathbf{r}) = \bar{c}(\mathbf{r}) - c^*(\mathbf{r})$, $\bar{c}(\mathbf{r})$ being the non-equilibrium average of $c(\mathbf{r})$:[31]

$$\kappa \delta c(r) = -\int \mathrm{d}\mathbf{r}' \mathscr{G}^*(\mathbf{rr}') \frac{\mathrm{d}\Psi_u(R)}{\mathrm{d}R} \left(\frac{\delta R}{\delta c(r')}\right)^*. \tag{5.9.23}$$

This result follows from (5.9.6) by replacing $g(\{c\},t)$ by its value in a local equilibrium state with R fixed at some value slightly different from R^* (a special case of (5.9.10)) and by finding the equation for the average $\bar{c}(\mathbf{r})$ where the time rate of change of $\delta c(\mathbf{r})$ is replaced by $\kappa \delta c(\mathbf{r})$. We have further assumed that for $c(\mathbf{r})$ near $c^*(\mathbf{r})$, the part of $\Psi(\{c\})$ that depends upon R is given by $\Psi_u(R)$. Multiplying both sides of (5.9.23) by $\delta\Psi(\{c\})/c(r) \simeq \mathrm{d}\Psi_u(R)/\mathrm{d}R \cdot (\delta R/\delta c(r))^*$ and integrating over $\mathbf{r}$, we finally obtain, noting (5.9.19) and that

$\int [\partial c(r)/\partial R]\, [\delta R/\delta c(r)]\, d\mathbf{r} = 1,$

$$M(R^*) = \kappa \chi_u. \tag{5.9.24}$$

With this result the total nucleation rate I is given by[35]

$$I = \Omega \kappa (\chi_u/2\pi)^{1/2} \exp(-\Delta\Psi) \tag{5.9.25}$$

where Ω is the volume of the function space of $c(\mathbf{r})$ due to translation of a nucleus and is given by

$$\Omega = \left[\frac{1}{3} \int (\nabla c^*)^2 \, d\mathbf{r} \right]^{3/2} V \tag{5.9.26}$$

with V the system volume. On the other hand κ has been found[31] from an equation equivalent to (5.9.23) for the case where $R^* \gg \xi$ and $\Psi(\{c\})$ is of the square-gradient type

$$\Psi(\{c\}) = \int [\tfrac{1}{2}\bar{K}(\nabla c)^2 + \phi(c)]\, dr \tag{5.9.27}$$

$\bar{K}$ being a constant related to the surface tension σ by

$$\sigma = \bar{K} \int_0^\infty \left(\frac{dc^*(r)}{dr} \right)^2 dr. \tag{5.9.28}$$

The result is

$$\kappa = 2\sigma\zeta/(R^*)^3(\Delta c)^2 \tag{5.9.29}$$

where ζ is the fully renormalized kinetic coefficient and Δc is the difference of c in the two coexisting phases.

This is not the place to discuss in detail the comparison of the present theory with experiments, but we merely note the following. A result very similar to (5.9.25) was also obtained by Langer and Turski[30] and was found to provide a somewhat better explanation of the recently observed nucleation rate in a binary critical mixture[36] than the classical Becker–Döring theory.[34] On the other hand, very recently an anomalously large supercooling (about three times the normal value) was observed for CO_2 near its critical point ($\tau \sim 10^{-3}$)[37] and no theory, including the one presented here, is able to account for this anomaly. Near the critical point, metastable states appear to be much more stable than all the current theories predict.

Let us now turn to another related topic, namely, spinodal decomposition. This is the process that takes place when the system is suddenly brought from a disordered stable phase into a state which is thermodynamically unstable against separation into coexisting phases. This process has been commonly observed in alloys and glasses[38] as well as in some polymer solutions[39] where the time scale of decomposition is slow enough for observation. In liquids, the relevant time scale is normally too short for such a process to be observed. However, in the close vicinity of a critical point the critical slowing-down can bring the time scale to the right size for observation, as was recently verified by Huang *et al.* who detected the spinodal decomposition of a binary liquid by light scattering.[40]

For alloys and glasses for which we may use (5.9.6) without the second term of (5.9.8), the initial stage of the decomposition can be described by the linear theory.[38] This gives for the average square fluctuation of $c_\mathbf{k}$ at time t,

$$\langle |c_\mathbf{k}|^2 \rangle_t = \langle |c_\mathbf{k}|^2 \rangle_0 \exp(2R_k t) \tag{5.9.30}$$

with the initial growth rate of the fluctuation ($k \leqslant [-\phi''(\bar{c}_0)/\bar{K}]^{1/2}$)

$$R_k = \zeta_0 k^2[-\phi''(\bar{c}_0) - \bar{K}k^2] \tag{5.9.31}$$

where $\bar{c}_0$ is the initial average value of c for which the second derivative ϕ'' is negative.

One of the interesting questions in the spinodal decomposition of a liquid is whether or not the renormalization of transport coefficients affects ζ_0 in the initial growth rate (5.9.31), and if it does, in what manner. The above-mentioned experimental result[40] appears to suggest that ζ_0 is fully renormalized here as well. However, from a theoretical point of view this is not so obvious. To see this, we consider the effect of the second term of (5.9.8) which contributes $\mathscr{L}_r(\{c\})$ to the stochastic operator $\mathscr{L}$ where

$$\mathscr{L}_r(\{c\}) = 2\iint \mathrm{d}\mathbf{r}\mathrm{d}\mathbf{r}' \frac{\delta}{\delta c(\mathbf{r})} \frac{\partial c(\mathbf{r})}{\partial \mathbf{r}} \cdot \mathbf{T}(\mathbf{r}'\mathbf{r}') \frac{\partial c(\mathbf{r}')}{\partial \mathbf{r}'} \left(\frac{\delta}{\delta c(\mathbf{r}')} + \frac{\delta \Psi(\{c\})}{\delta c(\mathbf{r}')} \right). \tag{5.9.32}$$

Then it is easy to see that (5.9.32) is not affected by replacing Ψ by the following expression,

$$\Psi(\{c\}) + \int \phi_1(c(\mathbf{r}))\mathrm{d}\mathbf{r} \tag{5.9.33}$$

where $\phi_1(c)$ is an arbitrary function, if one notes the fact that $(\partial/\partial\mathbf{r}') \cdot \mathbf{T}(\mathbf{r}-\mathbf{r}') = 0$. This means that in the absence of ζ_0 the stochastic equation (5.9.6) allows a class of steady state distribution functions $g(\{c\})$ of the form

$$g(\{c\}) = \text{const.} \times \exp\left(-\Psi(\{c\}) - \int \phi_1(c(\mathbf{r}))\mathrm{d}\mathbf{r} \right) \tag{5.9.34}$$

in addition to the equilibrium one. Namely, $\mathscr{L}_r$ by itself cannot bring the system to its equilibrium state $g_e(\{c\}) = \text{const.} \times \exp[-\Psi(\{c\})]$ after quenching.† The difficulty described here has subsequently been resolved by the introduction of an effective renormalized transport coefficient which increases in time in some initial time.[57]

5.10 Related developments

The present article has been mainly concerned with those theories that explicitly take into account, in one way or another, the characteristic time scale which governs the dynamics of critical fluctuations. However, there are other approaches to critical dynamics which we have not discussed and there are also developments in areas other than critical dynamics which are closely related to the context of this article. We will briefly mention some of these below.

5.10.1 Kinetic Ising models and TDGL models

Although the Ising model itself has no dynamics, it is possible to construct a kinetic Ising model in which spins can flip with appropriate probabilities in such a way as to ensure that the final equilibrium state is that of the Ising model. Since Glauber[42] introduced the model in 1963, it has been generalized in various ways and has been studied extensively.[43]

It is often convenient to replace discrete spins on a lattice by a continuous field of the local order parameter, say, $s(\mathbf{r})$. Then the model takes, for example, the simple form of the following stochastic equation for the distribution functional $g(\{s\},t)$:

$$\frac{\partial}{\partial t} g(\{s\}, t) = \zeta_0 \int \mathrm{d}r \frac{s}{\delta s(r)} \left(\frac{s}{\delta s(r)} + \frac{\delta \Psi(\{s\})}{\delta s(r)} \right) g(\{s\}, t) \tag{5.10.1}$$

†One of the authors (K.K.) is indebted to Dr Tomoji Yamada for a discussion which led him to this observation.

where ζ_0 is the 'bare' kinetic coefficient which is assumed to be finite. Ψ is the thermodynamic potential for a given $\{s\}$ which takes the following form:

$$\Psi(\{s\}) = \tfrac{1}{2}\int \mathrm{d}r[(\nabla s)^2 + r_0 s^2 + 2u_0 s^4]. \tag{5.10.2}$$

This model is often referred to as a time-dependent Ginzburg–Landau (TDGL) model.

Originally these models were considered to be microscopic models for which the conventional theory of critical slowing-down holds. However, subsequent studies based on series expansion techniques[44] and the renormalization group approach[10,45] revealed that for models such as (5.10.1) in which the order parameter is not conserved, the renormalized kinetic coefficient ζ in fact tends to zero at the critical point, where the renormalization arises from the nonlinear mode coupling due to the last term of (5.10.2). At the moment our understanding of this vanishing is less complete than of the cases discussed in the preceding sections. However, we do know that the mode coupling of a dissipative type such as the one encountered here tends to decrease a transport coefficient,[46] in contrast to the mode coupling of a streaming type with which we have been mainly concerned in this article. From this point of view the vanishing of ζ is not totally unexpected.

Indeed, the most natural way to understand this problem is in terms of a multiplicative renormalization of the life-time where the linearized macroscopic equation of motion for the average s_t of $s(\mathbf{r})$ takes the form

$$\frac{\mathrm{d}}{\mathrm{d}t}s_t = \zeta_0[h_t - \chi^{-1}s_t] - \int_0^\infty \hat{\psi}(s)s_{t-s}\mathrm{d}s \tag{5.10.3}$$

where χ is the magnetic susceptibility and $\hat{\psi}(t)$ is a sort of memory kernel. Fourier transforming (5.10.3) we find for the dynamical susceptibility $\chi(\omega)$ the form

$$\chi(\omega) = \{1 - \mathrm{i}\omega\tau_c[1 + \psi(\omega)]\}^{-1}\chi \tag{5.10.4}$$

where $\tau_c = \chi/\zeta_0$ is the lifetime predicted by the conventional theory and $\psi(\omega)$ is the one-sided Fourier transform of $\hat{\psi}(t)$. Thus, for small ω the apparent lifetime τ is given by $\tau_c[1 + \psi(0)]$. It is possible to find an exact time correlation function expression for $\psi(0)$ and to prove that $\psi(0) \geqslant 0$. Near T_c, the renormalization factor $1 + \psi(0)$ is likely to diverge at T_c for the model system considered here. Further details will be published by one of the authors (K.K.) elsewhere.

5.10.2 Long time tail problem

The fundamental assumption of the correlation function method of transport theory is that the time correlation functions which enter the expressions for the transport coefficients should vanish in a time of microscopic size which is much shorter than a time in which macroscopic variations occur.[2] This assumption is now in serious doubt. The first indication of this difficulty appeared in 1967 when Yamada and Kawaski[47] found that mode coupling contribution to the bulk viscosity remains important even away from a critical point, and in fact gives rise to a divergence in two dimensions. Similar effects were found independently in 1970 by Alder and Wainwright[48] in their molecular dynamic calculations of the self-diffusion coefficient in two and three dimensional fluids. For a d-dimensional fluid it is believed that the time correlation functions which yield the transport coefficients have long time tails which behave as $t^{-d/2}$, and which thus result in logarithmic divergences in transport coefficients in two dimensions. The physical origin of these long time tails is precisely the same as in critical dynamics: the coupling between modes produces slowly varying components in the fluxes which enter the time correlation functions of transport coefficients both near as well as away from a critical point. In three dimensions, although no divergence difficulty arises, the long time tails reveal themselves in macroscopic hydrodynamics through a singular dependence of the effective transport coefficients on frequency and wavenumber[49] as well as on the thermodynamic driving forces.[50] These long time tails have to be taken into account in evaluating transport coefficients.[51] However, precise estimates of long time tail contributions are difficult to make because the results depend sensitively on the short wavelength cut-offs. Hence to obtain reliable estimates some way must be found to get around this problem. For example, one approach might be to devise some kind of interpolation scheme between the limits of long and short times.

Finally we should also mention studies of similar long time tail problems for dense gases. Since this problem can be attacked by starting from the microscopic Liouville equation, such studies appear to be quite useful for examining the microscopic basis of the mode coupling ideas.[52]

For further details, the reader is referred to the review by Pomeau.[53]

Appendix. Exact formulation of a dynamic renormalization group

This Appendix is devoted to an exact renormalization group formulation of critical dynamics. Since critical dynamics is characterized not only by anomalies in the values of transport coefficients but also by nonlocal spatio-temporal structures in them, we start from a stochastic equation for the probability distribution function $g^1(\{a\}^1, t)$ of a set of the gross variables $\{a\}^1$ that includes such nonlocalities. The renormalization group operation eliminates certain 'short wavelength' parts $\{a\}^{1-2}$, and we are left with a stochastic equation for the probability distribution function $g^2(\{a\}^2, t)$ of the remaining gross variables $\{a\}^2$. Explicitly, these stochastic equations take the following form with $\alpha = 1,2$ and $t > 0$:

$$\frac{\partial}{\partial t} g^\alpha(\{a\}^\alpha, t) = \int_0^t ds \int d\{a'\}^\alpha \mathscr{H}^\alpha(\{a\}^\alpha, \{a'\}^\alpha; s) g^\alpha(\{a'\}^\alpha, t - s) \tag{A.1}$$

where $\mathscr{H}^\alpha$ is the stochastic operator which contains memory effects. The procedure we follow is similar to that of Kuramoto,[26] but will be carried out in a formally exact manner. (Kuramoto did not include memory effects in the starting stochastic equation.) The renormalization group transformation first eliminates a 'short wavelength' part of $\{a\}^1$ which results in a relationship between $\mathscr{H}^1$ and $\mathscr{H}^2$. Then, after an appropriate scale transformation, this relationship becomes a recurrence formula for reduced Hamiltonians in the renormalization group approach to equilibrium critical phenomena discussed in Section 5.7.

Let us first introduce projection operators $\mathscr{P}$ and $\mathscr{Q} \equiv 1 - \mathscr{P}$ which are defined for an arbitrary function $X(\{a\}^1)$ by

$$\mathscr{P} X(\{a\}^1) g_e^1(\{a\}^1) = g_e^1(\{a\}^1) \langle X(\{a\}^1) \rangle^{1-2} \tag{A.2}$$

where g_e^α denotes the equilibrium probability distribution function for $\{a\}^\alpha$, and $\langle \ldots \rangle^{1-2}$ denotes the partial average

$$\langle X(\{a\}^1) \rangle^{1-2} \equiv \frac{1}{g_e^2(\{a\}^2)} \int X(\{a\}^1) g_e^1(\{a\}^1) d\{a\}^{1-2}. \tag{A.3}$$

In particular, we have

$$\mathscr{P} g^1(\{a\}^1, t) = [g_e^1(\{a\}^1) / g_e^2(\{a\}^2)] g^2(\{a\}^2, t). \tag{A.4}$$

Physically, $\mathscr{P}$ projects onto a local equilibrium state for the variables $\{a\}^{1-2}$ with the values of $\{a\}^2$ fixed. We than derive the stochastic equation for g^2 starting from that for g^1. The procedure is quite analogous to that of Zwanzig[41] who obtained a master equation from the Liouville equation, except that here all the time evolution operators contain memory effects. First one operates with $\mathscr{Q}$ on (A.1) with $\alpha = 1$ and solves it for $\mathscr{Q}g^1$ in terms of $\mathscr{P}g^1$ with the initial condition $\mathscr{Q}g^1(t = 0)$. Next operate with $\mathscr{P}$ on (A.1) with $\alpha = 1$. The right-hand side is then expressed in terms of $\mathscr{P}g^1$ and $\mathscr{Q}g^1$. For $\mathscr{Q}g^1$ we substitute the expression obtained previously in terms of $\mathscr{P}g^1$. Since the time zero can be taken to be in the distant past, any initial deviation from local equilibrium will be washed out, and hence the term with $\mathscr{Q}g^1(t = 0)$ is dropped. In this manner we derive a closed non-Markoffian stochastic equation for g^2, that is, (A.1) with $\alpha = 2$, where the new stochastic operator $\mathscr{H}^2$ is now expressed in terms of $\mathscr{H}^1$ as follows:

$$\mathscr{H}^2(\{a\}^2, \{a'\}^2; t) \equiv \int d\{a\}^{1-2} \int d\{a'\}^{1-2} [\mathscr{H}^1(\{a\}^1, \{a'\}^1; t)$$

$$+ \int d\{a''\}^1 \int_0^t ds \int_0^{t-s} ds' \mathscr{H}^1(\{a\}^1, \{a''\}^1; s) \mathscr{Q}(\{a''\}^1)$$

$$\times U(t-s-s'; \{a''\}^1) \mathscr{Q}(\{a''\}^1) \mathscr{H}^1(\{a''\}^1, \{a'\}^1; s')] \frac{g_e^1(\{a'\}^1)}{g_e^2(\{a'\}^2)}. \qquad (A.5)$$

Here the time evolution operator U satisfies the following equation where X is an arbitrary function:

$$\frac{\partial}{\partial t} U(t, \{a\}^1) X(\{a\}^1) = \int_0^t ds \int d\{a'\}^1 \mathscr{Q}(\{a\}^1) \mathscr{H}^1(\{a\}^1, \{a'\}^1; s) U(t-s, \{a'\}^1) X(\{a'\}^1) \qquad (A.6)$$

with

$$U(0, \{a\}^1) X(\{a\}^1) = X(\{a\}^1). \qquad (A.7)$$

Now the general form of the stochastic operator can be inferred from the exact master equation obtained by Zwanzig[41] some years ago, so that, we can write, for $\alpha = 1,2$,

$$\mathscr{H}^\alpha(\{a\}^\alpha, \{a'\}^\alpha; t) = -2\sum_i^\alpha \frac{\partial}{\partial a_i} v_i^\alpha(\{a\}^\alpha) \delta^\alpha(a - a') \delta(t) \qquad (A.8)$$

$$+ \sum_i^\alpha \sum_j^\alpha \frac{\partial}{\partial a_i} K_{ij}^\alpha(t, \{a\}^\alpha, \{a'\}^\alpha) g_e^\alpha(\{a\}^\alpha) g_e^\alpha(\{a'\}^\alpha) \frac{\partial}{\partial a_j'^*} \frac{1}{g_e^\alpha(\{a'\}^\alpha)}$$

where $\delta^\alpha(a - a')$ is the product of delta functions $\delta(a_i - a_i')$ over the set $\{a\}^\alpha$ and Σ_i^α is the sum over the set $\{a\}^\alpha$. v_i^α has the physical meaning of the instantaneous rate of change of a_i and K_{ij}^α is the memory function. The relationships between v^2 and K^2, and v^1 and K^1 are obtained by substituting (A.7) into (A.4). Integrating (A.4) over an infinitesimal time interval (0,0+) we find

$$v_i^2(\{a\}^2) = \langle v_i^1(\{a\}^1) \rangle^{1-2}. \qquad (A.9)$$

This result also follows immediately from the microscopic expression for $v_i^\alpha(\{a\}^\alpha)$.

The relationship between K^2 and K^1 is more complicated, and its derivation is outlined here. We operate with (A.5) on $a_j'^* g_e^2(\{a'\}^2)$ and use (A.9) and the equilibrium condition

$$\sum_i^{\alpha} \frac{\partial}{\partial a_i} v_i^{\alpha}(\{a\}^{\alpha}) g_e^{\alpha}(\{a\}^{\alpha}) = 0 \quad (\alpha = 1, 2). \tag{A.10}$$

After some algebra we then obtain

$$\begin{aligned} K_{ij}^2(t, \{a\}^2, \{a'\}^2) &= \langle K_{ij}^1(t, \{a\}^1, \{a'\}^1) \rangle_{(a)(a')}^{1-2} \\ &+ \int d\{a''\}^1 \int_0^t ds \int_0^{t-s} ds' \; \mathscr{I}_i(s, \{a\}^2, \{a''\}^1) \mathscr{Q}(\{a''\}^1) \\ &\times U(t-s-s'; \{a''\}^1) \mathscr{Q}(\{a''\}^1) \tilde{\mathscr{I}}_j(s', \{a'\}^2, \{a''\}^1) g_e^1(\{a''\}^1) \end{aligned} \tag{A.11}$$

where

$$\begin{aligned} \mathscr{I}_i(t, \{a\}^2, \{a'\}^1) &\equiv 2v_i^1(\{a'\}^1) \frac{\delta^2(a-a')}{g_e^2(\{a\}^2)} \delta(t) \\ &+ \sum_j^1 \left(\frac{\partial}{\partial a_j'} + \frac{\partial \ln g_e^1(\{a'\}^1)}{\partial a_j'} \right) \langle K_{ij}^1(t, \{a\}^1, \{a'\}^1) \rangle_{(a)}^{1-2} \end{aligned} \tag{A.12}$$

$$\begin{aligned} \tilde{\mathscr{I}}_i(t, \{a\}^2, \{a'\}^1) &\equiv 2v_i'(\{a'\}^1)^* \frac{\delta^2(a-a')}{g_e^2(\{a\}^2)} \delta(t) \\ &- \sum_j^1 \left(\frac{\partial}{\partial a_j'} + \frac{\partial \ln g_e^1(\{a'\}^1)}{\partial a_j'} \right) \langle K_{ji}^1(t, \{a'\}^1, \{a\}^1) \rangle_{(a)}^{1-2} \end{aligned} \tag{A.13}$$

and $\langle \ldots \rangle_{(a)}^{1-2}$ here is the partial average over $\{a\}^{1-2}$, etc. In arriving at (A.11) we have made use of the fact that as far as the final stochastic equation for $g^2(\{\alpha\}^2, t)$ is concerned, $K_{ij}^2(t, \{a\}^2, \{a'\}^2)$ is arbitrary up to an additive term of the following form:

$$\left\langle \sum_n^1 v_n^1(\{a'\}^1) \frac{\partial X(\{a'\}^1, \{a\}^1)}{\partial a_n'^1} \right\rangle^{1-2}$$

Note the different signs in front of the second terms in these expressions, which are responsible for the different signs of the mode coupling contributions to transport coefficients due to streaming and dissipative nonlinearities, respectively.

From the definition of $\mathscr{P}$, (A.2), we can easily show that

$$\int d\{a\}^1 Y(\{a\}^1) \mathscr{P} X(\{a\}^1) g_e^1(\{a\}^1) = \int d\{a\}^1 X(\{a\}^1) \mathscr{P} Y(\{a\}^1) g_e^1(\{a\}^1). \tag{A.14}$$

Therefore, by (A.2) and (A.14), we see that in (A.11) $\mathscr{I}_i$ and $\tilde{\mathscr{I}}_j$ can be replaced, respectively, by

$$\mathscr{I}_i(s, \{a\}^2, \{a''\}^1) - \langle \mathscr{I}_i(s, \{a\}^2, \{a''\}^1) \rangle_{(a'')}^{1-2} \tag{A.15}$$

$$\tilde{\mathscr{I}}_j(s, \{a'\}^2, \{a''\}^1) - \langle \tilde{\mathscr{I}}_j(s, \{a'\}^2, \{a''\}^1) \rangle_{(a'')}^{1-2}. \tag{A.16}$$

The result (A.11) can thus be interpreted as follows: elimination of 'short wavelength parts' produces a new memory kernel which consists of (i) the average of the old memory kernel over the eliminated degrees of freedom and (ii) another term involving a correlation of 'random forces', (A.15) and (A.16), of the eliminated degrees of freedom.

The time evolution operator $U(t)$ for 'random forces' arises from the time-displacement stochastic operator $\mathscr{H}^1$ from which the 'long wavelength part' is removed by projection. This has the effect of removing those intermediate states in which the degrees of freedom $\{a\}^{1-2}$

are in partial equilibrium with fixed $\{a\}^2$. This can be more clearly seen by rewriting (A.6) in the following explicit form:

$$\frac{\partial}{\partial t} U(t, \{a\}^1)X(\{a\}^1) = \int_0^t \mathrm{d}s \int \mathrm{d}\{a'\}^1 \hat{\mathscr{H}}^1(\{a\}^1, \{a'\}^1; s)U(t-s, \{a'\}^1)X(\{a'\}^1) \quad (A.17)$$

with

$$\hat{\mathscr{H}}^1(\{a\}^1, \{a'\}^1; t) \equiv \hat{\mathscr{H}}^1(\{a\}^1, \{a'\}^1; t) - \frac{g_e^1(\{a\}^1)}{g_e^2(\{a\}^2)} \int \mathrm{d}\{a\}^{1-2} \mathscr{H}^1(\{a\}^1, \{a'\}^1; t). \quad (A.18)$$

The relationships (A.9) and (A.11) plus appropriate scale transformations that ensure the existence of a fixed point constitute an exact renormalization group transformation of the stochastic equation.

If we are content with a crude approximation, we can assume for K_{ij},

$$K_{ij}^{\alpha}(t, \{a\}^{\alpha}, \{a'\}^{\alpha}) = 2\zeta_{ij}^{\alpha}\delta(t)\delta^{\alpha}(a - a')/g_e^{\alpha}(\{a\}^{\alpha}) \quad (A.19)$$

where ζ_{ij}^{α} is a kinetic coefficient which is independent of $\{a\}$, and then integrate (A.11) over time and average over $g_e^2(\{a\}^2)$. Thus (A.11) reduces after some algebra to

$$\zeta_{ij}^2 = \zeta_{ij}^1 + \int_0^{\infty} \mathrm{d}t \int \mathrm{d}\{a\}^1 J_i(\{a\}^1)\mathscr{Q}(\{a\}^1)U(t; \{a\}^1)\mathscr{Q}(\{a\}^1)\tilde{J}_j(\{a\}^1)g_e^1(\{a\}^1) \quad (A.20)$$

where

$$J_i(\{a\}^1) \equiv v_i^1(\{a\}^1) - \sum_j{}^1 \zeta_{ij}^1 \frac{\partial \Phi^1(\{a\}^1)}{\partial a_j} \quad (A.21)$$

$$\tilde{J}_i(\{a\}^1) \equiv v_i^1(\{a\}^1)^* + \sum_j{}^1 \zeta_{ji}^1 \frac{\partial \Phi^1(\{a\}^1)}{\partial a_j} \quad (A.22)$$

$$\Phi^{\alpha}(\{a\}^{\alpha}) \equiv \ln g_e^{\alpha}(\{a\}^{\alpha}). \quad (A.23)$$

Again the projection operator $\mathscr{Q}(\{a\}^1)$ picks up only those parts of the 'random forces' J_i and $\tilde{J}_j$ that involve the variables $\{a\}^{1-2}$.

The remaining difficulty in evaluating the second term of (A.18) comes from the fact that $U(t)$ is governed by the projected-out part of the stochastic operator $\hat{\mathscr{H}}^1$, (A.18). At present we have not yet worked out a generally satisfactory method to handle this problem. However, if we adopt an approximation in which those modes of $\{a\}^{1-2}$ which are excited in intermediate states propagate freely (i.e. no vertex corrections for these modes), and hence intermediate states are always out of local equilibrium states with respect to $\{a\}^{1-2}$, which are compatible with fixed values of $\{a\}^2$, $\hat{\mathscr{H}}^1$ can be replaced by $\mathscr{H}^1$ in (A.17). We use this approximation below.

In the following, we apply the general theory to two simple examples. The treatments of these examples are incomplete and do not explore fully the general theory presented above. In particular the non-Gaussian nature of the probability distribution function of the order parameter and the non-local nature of transport coefficients are not properly treated. Nevertheless they serve as useful illustrations of the general formalism.

(a) Heisenberg ferromagnet

Adopting the approximation (A.19) with $\zeta_q^{\alpha} = q^2\Gamma^{\alpha}$, Γ^{α} being the Onsager kinetic coefficient associated with diffusion, the stochastic equation is

$$\frac{\partial}{\partial t}g^\alpha(\{S\}^\alpha,t) = \left(-\sum_q{}^\alpha \frac{\partial}{\partial \mathbf{S_q}}\cdot[\mathbf{v}_\mathbf{q}^\alpha(\{S\}^\alpha)+q^2\Gamma^\alpha \mathbf{F}_\mathbf{q}^\alpha(\{S\}^\alpha)] + \sum_q{}^\alpha q^2\Gamma^\alpha \frac{\partial}{\partial \mathbf{S_q}}\cdot\frac{\partial}{\partial \mathbf{S_{-q}}}\right)g^\alpha(\{S\}^\alpha,t) \quad \alpha=1,2 \tag{A.24}$$

where we put

$$\mathbf{v}_\mathbf{q}^\alpha(\{S\}^\alpha) = \frac{\lambda^\alpha}{V^{1/2}}\sum_\mathbf{k}{}^\alpha \mathbf{S_{q-k}}\times\frac{\partial\Phi^\alpha(\{S\}^\alpha)}{\partial \mathbf{S_k}} \tag{A.25}$$

$$\mathbf{F}_\mathbf{q}^\alpha(\{S\}^\alpha) = -\frac{\partial}{\partial \mathbf{S_{-q}}}\Phi^\alpha(\{S\}^\alpha) \tag{A.26}$$

and Φ^α is the thermodynamic potential for the variables $\{S\}^\alpha$ which we assume to be Gaussian (hence this treatment is incomplete for $d < 4$).

$$\Phi^\alpha = \frac{1}{2}\sum_\mathbf{q}{}^\alpha \frac{1}{\chi_\mathbf{q}^\alpha}\mathbf{S_q}\cdot\mathbf{S_{-q}}. \tag{A.27}$$

First note that the relation (A.9) requires $\lambda^2 = \lambda^1$. Next, (A.20) becomes

$$\Gamma^2 = \Gamma^1 + \frac{(\lambda^1)^2}{V}\int_0^\infty \mathrm{d}t\int \mathrm{d}\{S\}^1 q^{-2}\sum_\mathbf{k}{}^1\sum_\mathbf{k'}{}^1\left(\mathbf{S_{q-k}}\times\frac{\partial\Phi^1}{\partial\mathbf{S_k}}\right)\mathscr{Q}\,U(t)\,\mathscr{Q}\left(\mathbf{S_{-q+k'}}\times\frac{\partial\Phi^1}{\partial\mathbf{S_{-k'}}}\right)g_\mathrm{e}^1(\{S\}^1) \tag{A.28}$$

since the second terms of (A.21) and (A.22) drop out because of the presence of the operator $\mathscr{Q}$ here. Ignoring the interaction between two modes in intermediate states, (A.28) reduces for small q to

$$\begin{aligned}\Gamma^2 &= \Gamma^1 + \frac{3(\lambda^1)^2}{\Gamma^1}\int_\mathbf{k}^{(1-2)}\frac{\chi_\mathbf{k}^1\chi_\mathbf{q-k}^1}{\mathbf{k}^2/\chi_\mathbf{k}^1+(\mathbf{q}-\mathbf{k})^2/\chi_\mathbf{q-k}^1}\frac{1}{q^2}\left(\frac{1}{\chi_\mathbf{k}^1}-\frac{1}{\chi_\mathbf{q-k}^1}\right)^2\\ &\cong \Gamma^1 + \frac{3}{2d}\frac{(\lambda^1)^2}{\Gamma^1}\int_\mathbf{k}^{(1-2)}\frac{(\partial\chi_\mathbf{k}/\partial k)^2}{k^2\chi_\mathbf{k}}\end{aligned} \tag{A.29}$$

where

$$\int_\mathbf{k}^{(1-2)} = \frac{1}{(2\pi)^d}\int^{(1-2)}\mathrm{d}\mathbf{k} \tag{A.30}$$

and the superscript (1–2) indicates the range of integral $\Lambda/b < |\mathbf{k}| < \Lambda$.

We now consider the following scale transformation

$$S_\mathbf{k} \to b^{1-\eta/2}S_{b\mathbf{k}} \qquad t \to b^z t \tag{A.31}$$

z being a dynamical critical exponent. We easily find that

$$\Phi^2(\{b^{1-\eta/2}S_{b\mathbf{k}}\},\xi,V) = \Phi^1(\{S_\mathbf{k}\},\xi/b,V/b^d) \tag{A.32}$$

$$v_\mathbf{q}^2(\{b^{1-\eta/2}S_{b\mathbf{k}}\},\xi,V,\lambda^2) = b^{-d/2}v_{b\mathbf{q}}^1(\{S_\mathbf{k}\},\xi/b,V/b^d,\lambda^2). \tag{A.33}$$

The stochastic operator on the right-hand side of (A.24) which we denote as $\mathscr{L}^\alpha$ then is seen to transform as

$$\mathscr{L}^2(\{b^{1-\eta/2}S_{b\mathbf{k}}\},\xi,V,\lambda^2,\Gamma^2) = b^{-z}\mathscr{L}^1(\{S_\mathbf{k}\},\xi/b,V/b^d,b^{z-x}\lambda^2,b^{z-4+\eta}\Gamma^2) \tag{A.34}$$

where

$$x \equiv (d + 2 - \eta)/2. \tag{A.35}$$

Thus, if we require that the form of the stochastic equation remains the same after the renormalization group transformation, we obtain the following transformation properties of the parameters λ and $\Gamma(\lambda_l = \lambda^1 = \lambda^2, \lambda_{l+1} = b^{z-x}\lambda^2$ and $\Gamma_{l+1} = b^{z-4+\eta}\Gamma^2, \Gamma_l = \Gamma^1)$:

$$\lambda_{l+1} = b^{z-x}\lambda_l \tag{A.36a}$$

$$\Gamma_{l+1} = b^{z-4+\eta}\left(\Gamma_l + \frac{3}{2d}\,\frac{\lambda_l^2}{\Gamma_l}\int_{\mathbf{k}}^{(1-2)} \frac{(\partial\chi_{\mathbf{k}}/\partial k)^2}{k^2\chi_{\mathbf{k}}}\right). \tag{A.36b}$$

These equations constitute the dynamic renormalization group equations. In particular, at criticality where $\chi_{\mathbf{k}} = \chi_{\mathbf{k}}^* = k^{-2+\eta}$, (A.36) determine the values of z and $\lambda = \lambda^*$ at the fixed points, where Γ_l is set equal to unity by an appropriate choice of the unit of time. Two fixed point solutions are found.

(i) Trivial fixed point (conventional theory)

$$\lambda^* = 0 \text{ and } z = 4 - \eta \tag{A.37}$$

(ii) Non-trivial fixed point

$$z = x \tag{A.38}$$

$$\lambda^* = \pm\left((1 - b^{(d-6+\eta)/2})\Big/\frac{3}{2d}\int_{\mathbf{k}}^{(1-2)} \frac{(\partial\chi_{\mathbf{k}}/\partial k)^2}{k^2\chi_{\mathbf{k}}}\right)^{1/2}. \tag{A.39}$$

The value of λ at the nontrivial fixed point is real only for the dimensionality less than 6, suggesting its instability for $d > 6$ as was explicitly verified by Ma and Mazenko.[27] For d near 6 these results reduce to those due to Ma and Mazenko[27] quoted in Section 5.7.

(b) Classical liquids

Again using the approximation (A.19) the stochastic equation corresponding to (A.24) is, denoting $\bar{\eta}^\alpha \equiv \eta^\alpha/k_B T$ and $\mathbf{j}^\alpha \equiv (\rho/k_B T)^{1/2}\mathbf{u}^\alpha$,

$$\frac{\partial}{\partial t} g^\alpha(\{c\}^\alpha, \{\mathbf{j}\}^\alpha, t) = \mathscr{L}^\alpha(\{c\}^\alpha, \{\mathbf{j}\}^\alpha) g^\alpha(\{c\}^\alpha, \{\mathbf{j}\}^\alpha, t) \tag{A.40}$$

with

$$\mathscr{L}^\alpha(\{c\}^\alpha\{\mathbf{j}\}^\alpha) = \zeta^\alpha \sum_{\mathbf{q}}^\alpha q^2 \frac{\partial}{\partial c_{\mathbf{q}}}\left(\frac{\partial}{\partial c_{-\mathbf{q}}} + \frac{\partial\Phi^\alpha}{\partial c_{-\mathbf{q}}}\right) + \bar{\eta}^\alpha \sum_{\mathbf{q}}^\alpha \frac{\partial}{\partial \mathbf{j}_{\mathbf{q}}}\cdot(q^2\mathbf{1} - \mathbf{q}\mathbf{q})\cdot\left(\frac{\partial}{\partial \mathbf{j}_{-\mathbf{q}}} + \frac{\partial\Phi^\alpha}{\partial \mathbf{j}_{-\mathbf{q}}}\right)$$
$$- \sum_{\mathbf{q}}^\alpha \left[\frac{\partial}{\partial c_{\mathbf{q}}} v_{\mathbf{q}c}^\alpha + \frac{\partial}{\partial \mathbf{j}_q}\cdot \mathbf{v}_{\mathbf{q}j}^\alpha\right] \tag{A.41}$$

$$v_{\mathbf{q}c}^\alpha \equiv -\mathrm{i}\frac{\lambda^\alpha}{V^{1/2}}\,\mathbf{q}\cdot\frac{\partial\Phi}{\partial \mathbf{j}_{k-q}} c_k \tag{A.42}$$

$$v_{\mathbf{q}j}^{\alpha} \equiv \mathrm{i}\frac{\lambda^{\alpha}}{V^{1/2}}\left(\mathbf{k}-\frac{\mathbf{q}\cdot\mathbf{k}}{q^2}\mathbf{q}\right)c_{\mathbf{k}}\frac{\partial\Phi^{\alpha}}{\partial c_{\mathbf{k}-\mathbf{q}}} \tag{A.43}$$

$$\Phi^{\alpha}(\{c\}^{\alpha}, \{\mathbf{j}\}^{\alpha}) = \frac{1}{2}\sum_{\mathbf{k}}^{\alpha}\left(\frac{|c_{\mathbf{k}}|^2}{\chi_{\mathbf{k}}^{\alpha}} + |\mathbf{j}_{\mathbf{k}}|^2\right). \tag{A.44}$$

Here λ^{α} is the strength of the mode coupling term and its initial value is $(k_B T/\rho)^{1/2}$, as can be verified by looking at (5.5.20)–(5.5.23). From (A.9) we again have $\lambda^1 = \lambda^2$. After making the same approximations that led to (A.29), the relation (A.20) in the present case reduces to

$$\zeta^2 = \zeta^1 + \frac{(\lambda^1)^2}{\bar{\eta}^1}(1-d^{-1})\int_{\mathbf{k}}^{(1-2)}\chi_{\mathbf{k}}/k^2 \tag{A.45}$$

$$\bar{\eta}^2 = \bar{\eta}^1 + \frac{(\lambda^1)^2}{(d-1)\zeta^1}\int_{\mathbf{k}}^{(1-2)}\left[1-\left(\frac{\mathbf{q}\cdot\mathbf{k}}{qk}\right)^2\right]\left(\frac{\mathbf{q}\cdot\mathbf{k}}{q}\right)^2\chi_{\mathbf{k}}^3. \tag{A.46}$$

The scale in the present case is

$$c_{\mathbf{k}} \to bc_{b\mathbf{k}} \qquad \mathbf{j}_{\mathbf{k}} \to \mathbf{j}_{b\mathbf{k}} \qquad t \to b^z t. \tag{A.47}$$

where the critical exponent η is taken to be zero. We then find

$$\Phi^2(\{bc_{b\mathbf{k}}\}, \{\mathbf{j}_{b\mathbf{k}}\}, \xi, V) = \Phi^1(\{c_{\mathbf{k}}\}, \{\mathbf{j}_{\mathbf{k}}\}, \xi/b, V/b^d) \tag{A.48}$$

$$v_{\mathbf{q}c}^2(\{bc_{b\mathbf{k}}\}, \{\mathbf{j}_{b\mathbf{k}}\}, \xi, V, \lambda^2) = b^{-d/2}v_{b\mathbf{q},c}(\{c_{\mathbf{k}}\}, \{\mathbf{j}_{\mathbf{k}}\}, \xi/b, V/b^d, \lambda^2) \tag{A.49}$$

$$\mathbf{v}_{\mathbf{q}j}^2(\{bc_{b\mathbf{k}}\}, \{\mathbf{j}_{b\mathbf{k}}\}, \xi, V, \lambda^2) = b^{-d/2-1}\mathbf{v}_{b\mathbf{q},\mathbf{j}}(\{c_{\mathbf{k}}\}, \{\mathbf{j}_{\mathbf{k}}\}, \xi/b, V/b^d, \lambda^2). \tag{A.50}$$

Thus, the stochastic operator transforms as

$$\begin{aligned}&\mathscr{L}^2(\{bc_{b\mathbf{k}}\}, \{\mathbf{j}_{b\mathbf{k}}\}, \zeta, V, \lambda^2, \zeta^2, \bar{\eta}^2)\\ &\quad = b^{-z}\mathscr{L}^1(\{c_{\mathbf{k}}\}, \{j_{\mathbf{k}}\}, \xi/b, V/b^d, b^{z-d/2-1}\lambda^2, b^{z-4}\zeta^2, b^{z-2}\bar{\eta}^2).\end{aligned} \tag{A.51}$$

Requiring that the form of the stochastic equation does not change under the renormalization transformation, we find the following transformation rule for the parameters:

$$\lambda_{l+1} = b^{z-d/2-1}\lambda_l \tag{A.52a}$$

$$\zeta_{l+1} = b^{z-4}\left[\zeta_l + \frac{\lambda_l^2}{\bar{\eta}_l}\left(1-\frac{1}{d}\right)\int_{\mathbf{k}}^{(1-2)}\frac{\chi_{\mathbf{k}}}{k^2}\right] \tag{A.52b}$$

$$\bar{\eta}_{l+1} = b^{z-2}\left\{\bar{\eta}_l + \frac{\lambda_l^2}{(d-1)\zeta_l}\int_{\mathbf{k}}^{(1-2)}\left[1-\left(\frac{\mathbf{q}\cdot\mathbf{k}}{qk}\right)^2\right]\left(\frac{\mathbf{q}\cdot\mathbf{k}}{q}\right)^2\chi_{\mathbf{k}}^3\right\}. \tag{A.52c}$$

Finally we remark that the present formulation presupposes the existence of one time scale for critical dynamics. Otherwise, there is no single finite fixed point stochastic operator in contrast to the suggestion of Kuramoto.[26] Nevertheless, our general treatment can be extended to more general situations with multiple time scales although we do not go into details here.

References

1. J. V. Sengers (1966). In M. S. Green and J. V. Sengers (Eds), *Critical Phenomena*, NBS Miscellaneous Publication 273.
   J. V. Sengers and J. M. H. Levelt-Sengers, Chapter 4 of this book.

2. R. Zwanzig (1965). *Ann. Rev. Phys. Chem.*, **16**, 67.
R. Kubo (1966). *Rep. Prog. Phys.*, **29/1**, 255.
3. M. Fixman (1962). *J. Chem. Phys.*, **36**, 310.
4. K. Kawasaki (1966). *Phys. Rev.*, **150**, 291.
K. Kawasaki and M. Tanaka (1967). *Proc. Phys. Soc. (London)*, **90**, 791.
5. R. D. Mountain and R. W. Zwanzig (1968). *J. Chem. Phys.*, **48**, 1451.
6. H. E. Stanley (1971). *Introduction to Phase Transitions and Critical Phenomena*, Clarendon Press, Oxford.
7. L. P. Kadanoff and J. Swift (1968). *Phys. Rev.*, **166**, 89.
J. Swift (1968). *Phys. Rev.*, **173**, 257.
8. R. A. Farrell, N. Menyhárd, H. Schmidt, F. Schwabl and P. Szépfalusy (1968). *Ann. Phys. (N.Y.)*, **47**, 565.
B. I. Halperin and P. C. Hohenberg (1969). *Phys. Rev.*, **177**, 952.
9. K. G. Wilson and J. Kogut (1974). *Physics Reports*, **12C**, 75.
10. B. I. Halperin (1973), In B. Lundqvist and S. Lundqvist, *Collective Properties of Physical Systems, Nobel Symposium* 24, Academic Press, New York.
11. K. Kawasaki (1976). In C. Domb and M. S. Green (Eds), *Phase Transitions and Critical Phenomena*, Vol. 5A, Academic Press, London.
12. See, e.g., H. Thomas (1971). In E. J. Samuelsen, E. Andersen and J. Feder, *Structural Phase Transitions and Soft Modes.* Universitetsforlaget, Oslo.
13. L. van Hove (1954). *Phys. Rev.*, **95** 1374.
14. L. D. Landau and I. M. Khalatnikov (1954). *Dokl. Akad. Nauk, SSSR*, **96**, 469.
15. J. V. Sengers (1971) In M. S. Green (Ed.), *Proceedings of the International Conference of Physics Enrico Fermi Course LI*, Academic Press, New York and London; *AIP Conference Proceedings, No. 11*, American Institute of Physics, New York (1973).
16. G. E. Uhlenbeck and G. W. Ford (1963). *Lectures in Statistical Mechanics*, Providence, R.I., American Mathematical Society.
17. See, e.g., S. A. Rice and P. Gray (1965). *The Statistical Mechanics of Simple Liquids*, Wiley, New York.
18. K. Kawasaki (1973). *J. Phys.* A, **6**, 1289.
19. G. E. Uhlenbeck and L. S. Ornstein (1930). *Phys. Rev.*, **36**, 823.
20. K. Kawasaki (1971). In M. S. Green (Ed.). *Proceedings of the International Conference of Physics Enrico Fermi Course LI.* Academic Press, New York and London.
21. K. Kawasaki (1969). *Phys. Lett.*, **30A**, 325; *Ann. Phys. (N.Y.)*, **61**, 1 (1970).
22. P. Résibois and C. Piette (1970). *Phys. Rev. Lett.*, **24**, 514.
For an experimental test, see G. Parette and R. Kahn (1971). *J. de Phys.* (*Paris*), **32**, 447.
23. R. A. Farrell (1970). *Phys. Rev. Lett.*, **24**, 1169.
24. There is an extensive literature on this subject. For a review, see e.g. H. L. Swinney and D. L. Henry (1973). *Phys. Rev.* A, **8**, 2586.
25. B. Halperin, P. C. Hohenberg and E. Siggia (1974). *Phys. Rev. Lett.*, **32**, 1289.
26. Y. Kuramoto (1974). *Prog. Theor. Phys. (Kyoto)*, **51**, 1712.
M. Suzuki (1973). *Prog. Theor. Phys. (Kyoto)*, **50**, 393, 1767.
27. S. Ma and G. F. Mazenko (1974). *Phys. Rev. Lett.*, **33**, 1384.
28. Th. Niemeijer and J. M. J. van Leeuwen (1974). *Physica*, **71**, 17.
Th. Niemeijer and J. M. J. van Leeuwen (1973). *Phys. Rev. Lett.*, **31**, 1411.
S. Hsu, Th. Niemeijer and J. D. Gunton (1975). *Phys. Rev.*, **B11**, 2699.
L. P. Kadanoff and A. Houghton (1975). *Phys. Rev.*, **B11**, 377.
G. R. Golner and E. K. Riedel (1975). *Phys. Rev. Lett.*, **34**, 856.
J-L. Colot, J. A. C. Loodts and R. Brout (1975). *J. Phys.* A, **8**, 594.
29. J. D. Gunton and K. Kawasaki (1975). *J. Phys.* A, **8**, L9.
30. J. S. Langer and L. A. Turski (1973). *Phys. Rev.*, A, **8**, 3230.
31. K. Kawasaki (1975). *J. Stat. Phys.*, **12**, 365.
32. K. Kawasaki (1973). In H. Haken (Ed.), *Synergetics*, B. G. Teubner, Stuttgart.
33. D. Stauffer and C. S. Kiang, to be published in A. C. Zettlemoyer (Ed.), *Nucleation II.*
34. See e.g. W. J. Dunning (1969). In A. C. Zettlemoyer (Ed.), *Nucleation*, M. Dekker, New York.
35. J. S. Langer (1969). *Ann. Phys. (N.Y.)*, **54**, 258.
36. J. S. Huang, S. Vernon and N. C. Wong (1974). *Phys. Rev. Lett.*, **33**, 140.
37. J. S. Huang, W. I. Goldburg and M. R. Moldover (1975). *Phys. Rev. Lett.*, **11**, 639.

38. J. W. Cahn (1971). In R. E. Mills, E. Ascher and R. I. Jaffee (Eds), *Critical Phenomena in Alloys, Magnets, and Superconductors,* McGraw-Hill, New York.
39. J. J. van Aartsen and C. A. Smolders (1970). *Eur. Polym. J.*, **6**, 1105.
C. A. Smolders, J. J. van Aartsen and A. Steenberger (1971). *Kolloid-Z, Z. Polym.*, **243**, 14.
P. T. Van Emmeric and C. A. Smolders (1973). *Eur. Polym. J.*, **9**, 309.
40. J. S. Huang, W. I. Goldburg, and A. W. Bjerkaas (1974). *Phys. Rev. Lett.*, **32**, 921.
A. J. Schwartz, J. S. Huang and W. I. Goldburg, to be published.
41. R. W. Zwanzig (1961). *Phys. Rev.*, **124**, 983.
42. R. J. Glauber (1963). *J. Math. Phys.*, **4**, 294.
43. K. Kawasaki (1972). In C. Domb and M. S. Green (Eds), *Phase Transitions and Critical Phenomena,* Vol. II. Academic Press, London.
44. H. Yahata and M. Suzuki (1969). *J. Phys. Soc. (Japan)*, **27**, 1421.
H. Yahata (1971). *J. Phys. Soc. (Japan)*, **30**, 657.
45. B. I. Halperin, P. C. Hohenberg and S. Ma (1972). *Phys. Rev. Lett.*, **29**, 1548.
B. I. Halperin, P. C. Hohenberg and S. Ma (1974). *Phys. Rev.*, **B10**, 139.
M. Suzuki and G. Igarashi (1973). *Prog. Theor. Phys.*, **49**, 1070.
M. Suzuki (1973). *Phys. Lett.*, **A43**, 245.
M. Suzuki (1973). *Prog. Theor. Phys.*, **50**, 1769.
M. Suzuki and F. Tanaka (1974). *Prog. Theor. Phys.*, **52**, 722.
H. Yahata (1974). *Prog. Theor. Phys.*, **51**, 2003.
H. Yahata (1974). *Prog. Theor. Phys.*, **52**, 871.
C. De Dominicis, E. Brezin and J. Zinn-Justin (1975). *Saclay Preprint DPH-T/75/17.*
K. Kawasaki (1974). *Prog. Theor. Phys.*, **52**, 359.
46. K. Kawasaki (1973). *J. Phys.* A., **6**, L1.
47. T. Yamada and K. Kawasaki (1967). *Prog. Theor. Phys. (Kyoto)*, **38**, 1031.
48. B. J. Alder and T. E. Wainwright (1970). *Phys. Rev.*, **A1**, 18.
49. M. H. Ernst and J. R. Dorfman (1972). *Physica,* **61**, 157.
50. K. Kawasaki and J. D. Gunton (1973). *Phys. Rev.*, **A8**, 2048.
51. D. Levesque, L. Verlet and J. Kurkijarvi (1973). *Phys. Rev.*, **A7**, 1690.
52. J. R. Dorfman and E. G. D. Cohen (1972). *Phys. Rev.*, **A6**, 776.
G. F. Mazenko (1974). *Phys. Rev.*, **A9**, 360.
53. Y. Pomeau and P. Résibois (1975). *Phys. Rept.*, **19C**, 64.
54. K. Kawasaki and J. D. Gunton (1976). *Phys. Rev.*, **B13**, 4658.
55. T. Ohta (1975). *Prog. Theor. Phys. (Kyoto)*, **54**, (1566).
56. T. Ohta and K. Kawasaki (1976). *Prog. Theor. Phys. (Kyoto)*, **55**, 1384.
57. K. Kawasaki, *Prog. Theor. Phys.*, (*Kyoto*), (to be published).

Chapter 6

The Structure of Quantum Fluids

C. E. CAMPBELL

School of Physics and Astronomy
University of Minnesota, Minneapolis, Minnesota 55455

6.1 Introduction

Historical background

The element helium was discovered first in 1869 by spectral analysis of the solar corona; terrestrial helium was not discovered until 1895. At that time a substantial effort was being made to liquefy hydrogen, most notably in the laboratories of Dewar in London, Olszewski in Cracow, and Kamerlingh Onnes in Leiden. Olszewski attempted to liquefy helium in 1896, even before Dewar has succeeded in liquefying hydrogen (May 1898). He reached a temperature of 9 °K and saw no sign of liquefaction: it was only through the monumental efforts of Onnes that helium liquefaction was finally achieved, in July 1908.[1] An interesting account of these efforts is given in Mendelssohn's *The Quest for Absolute Zero.*[2]

In the same experiment in which helium was first liquefied, Onnes allowed the liquid to boil under reduced pressure in order to see the solid form. Although he obtained a temperature of about 1.7 °K and a pressure of one-tenth atmosphere, he saw no sign of the solid. The failure of repeated attempts to solidify helium under its own vapour pressure by lowering the temperature led Onnes to speculate that helium would remain liquid when cooled to absolute zero. Keesom provided the first step in the verification of this speculation in 1926 by solidifying helium under pressure.[3] Ensuing measurements showed that the liquid–solid equilibrium curve extrapolates to finite pressure (25 atm) at absolute zero, verifying Onnes' speculation: helium can be solidified only by applying finite pressure, even at $T = 0$.

Quantum theory of corresponding states

Schematic representations of the phase diagram of ^{4}He and ^{3}He are shown in Figure 6.1.1 along with a comparison to the prediction of the classical theory of corresponding states. The failure of the classical theory is evidence for the importance of quantum mechanical effects in helium. In short, the fluidity of helium at low temperatures is due to the large zero point motion. The light mass and the small interaction strength means that the zero point energy is comparable to the potential energy of the system, in contrast with the other noble gases.

The effect of quantum mechanics was introduced into the theory of corresponding states by de Boer.[4] He defined the quantum parameter Λ^*:

$$\Lambda^* = (h^2/m\epsilon\sigma^2)^{1/2} \tag{6.1.1}$$

where h is Planck's constant, m is the mass of the atom, and ϵ and σ are respectively measures of the strength and range of the interaction potential. Λ^* is a measure of the ratio of the de Broglie wavelength to the diameter of the atom, and is around 3 for the helium liquids,

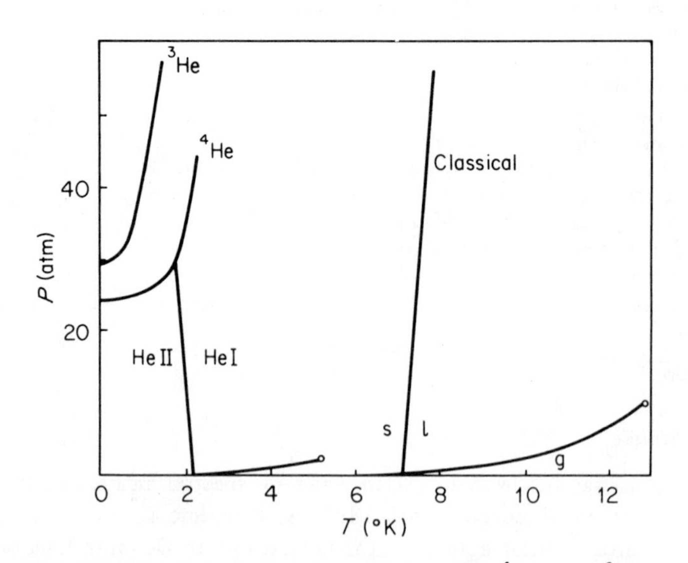

Figure 6.1.1 Schematic phase diagram of ^{4}He and ^{3}He compared with the prediction of the classical theory of corresponding states. (From de Boer, reproduced by permission of Academic Press Inc.)

approximately 30 times its value for other noble gases. By including H_2, HD, and D_2 (with Λ^* between 1 and 2) in the quantum theory of corresponding states, de Boer correctly predicted the vapour pressure curve for the much rarer isotope ^{3}He before a sufficient quantity of it became available to liquefy in 1949.

Actually, the difference in mass between ^{4}He and ^{3}He is not sufficient to explain completely the differences in the phase diagrams of these two isotopes. It is also necessary to incorporate the statistics: ^{4}He atoms are bosons, while ^{3}He atoms are fermions. The most spectacular manifestation of these differences is in the appearance of a fourth state of matter, known as the superfluid state, which we discuss further below. The location of the other three phases is also influenced very much by the statistics. The reason is simple to see: ^{3}He must satisfy the Pauli exclusion principle, which raises its energy relative to a hypothetical boson ^{3}He system. The change is substantial in the liquid and vapour phase, being comparable to the energy of a non-interacting Fermi gas of the same mass and density. In the solid phase however, the particles are localized at separate sites with only a weak overlap, so that the energy of the solid is nearly independent of statistics. Hence the difference between the energy of the liquid and the solid phase at a given density should be less for a system satisfying Fermi statistics than the same one satisfying Bose statistics (assuming the liquid has lower energy). This point is well illustrated by the recent calculation of Nosanow, Parish and Pinski.[6] They approached the quantum theory of corresponding states at $T = 0$ by calculating the energy of the liquid and solid for a system of N particles interacting via the Lennard–Jones potential

$$v(r) = \epsilon v^*(r/\sigma) \tag{6.1.2}$$

where

$$v^*(x) = 4(x^{-12} - x^{-6}). \tag{6.1.3}$$

The reduced variables of the system are the reduced volume

$$V^* = V/N\sigma^3 \tag{6.1.4}$$

the reduced pressure

$$P^* = P/\sigma^3\epsilon \tag{6.1.5}$$

and the reduced energy

$$E^* = E/N\epsilon. \tag{6.1.6}$$

The energy is calculated as a function of V^* and Λ^* by approximating the state of the system by a variationally determined Jastrow function with symmetry and density profile appropriate for the state of interest; the details of this type of calculation are discussed in Section 6.2. The results of these calculations are shown in Figure 6.1.2 for several values of the quantum parameter η defined by

$$\eta = (\Lambda^*/2\pi)^2 \tag{6.1.7}$$

The values of η for ^{3}He and ^{4}He are

$$\eta = 0.2409,\ ^3\text{He}$$

$$\eta = 0.1815,\ ^4\text{He}$$

which are obtained by using the de Boer–Michels parameters

$$\epsilon = 10.22\ ^\circ\text{K}$$

$$\sigma = 2.556\ \text{Å} \tag{6.1.8}$$

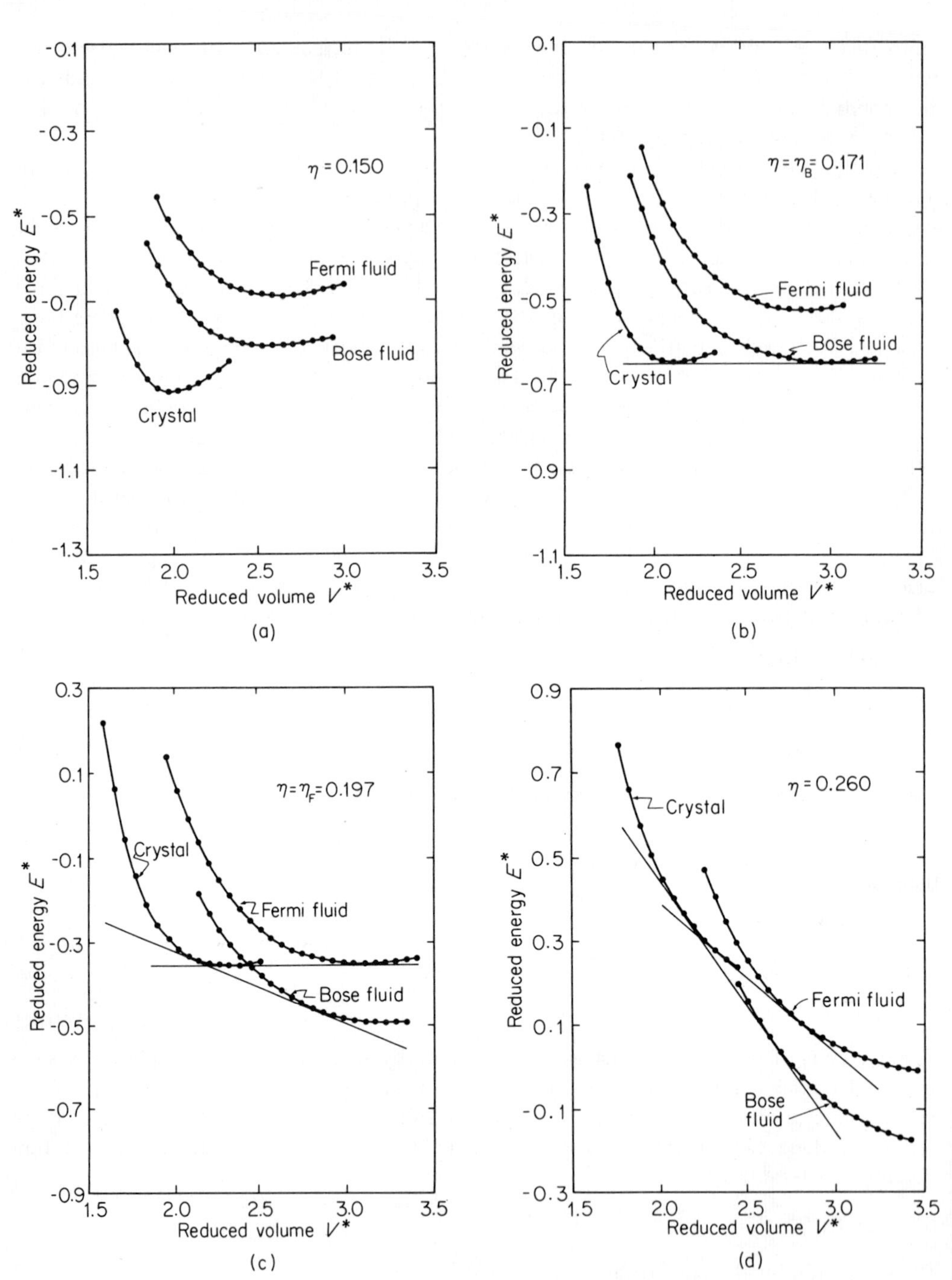

Figure 6.1.2 Reduced ground-state energy E^* as a function of the reduced volume V^* for the liquid and crystalline phases of a Lennard–Jones system with several values of the quantum parameter η: (a) $\eta = 0.150$; (b) $\eta = 0.171$, the critical value for Bose statistics, where the liquid and crystalline phases of the boson system coexist; (c) $\eta = 0.197$, the critical value for Fermi statistics; and (d) $\eta = 0.260$, above the critical value for both statistics, requiring a finite pressure to solidify in each case. (From Nosanow, Parish and Pinski,[6] reproduced by permission of the American Institute of Physics)

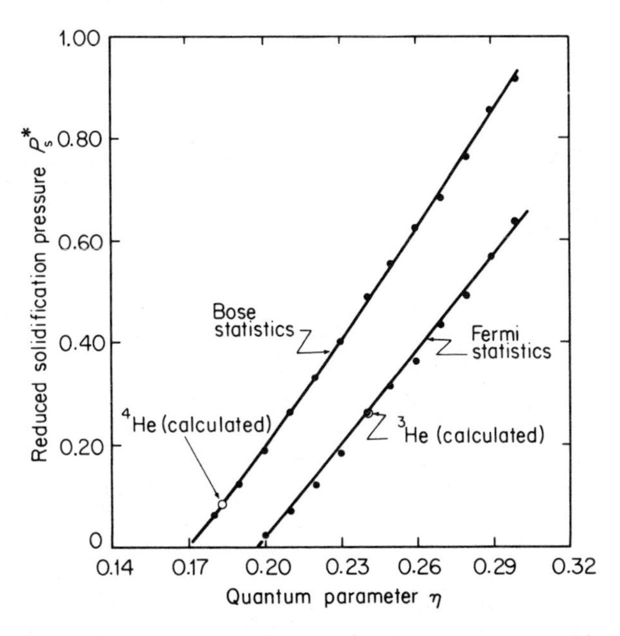

Figure 6.1.3 The reduced solidification pressure P_s^* as a function of the quantum parameter η for Bose and Fermi Lennard–Jones systems. (From Nosanow, Parish and Pinski,[6] reproduced by permission of the American Institute of Physics)

Figure 6.1.2a is at a sufficiently small value of η that the stable phase for both statistics is the solid. Increasing the value of η to $\eta_B = 0.171$ produces a system for which the Bose fluid can coexist in equilibrium with the solid (Figure 6.1.2b). For larger values of η (e.g. Figure 6.1.2c and 6.1.2d) the liquid is in the stable phase at zero pressure and the solid occurs at a finite pressure. For the Fermi system, however, the critical value of η is at the larger value $\eta_F = 0.197$, shown in Figure 6.1.2c. The solidification pressure as a function of η for Bose statistics and Fermi statistics is shown in Figure 6.1.3. Note that these calculations indicate that ^{4}He would be a solid if it were a fermion system. On the other hand, ^{3}H would require twice the pressure to solidify if it were a boson system.

Increasing Λ^* (i.e. η) results in a larger zero-point energy and should eventually make the vapour phase the lowest energy phase even at $T = 0$. This was demonstrated by Miller, Nosanow and Parish in an extension of the quantum corresponding states calculation.[7] Their results are summarized in Figure 6.1.4, which shows E^* as a function of $\rho^* = 1/V^*$. Note that for the boson system there is a critical value $\eta = \eta_{CB}$ above which the system is under finite pressure at all densities investigated and is, therefore, in a vapour phase, while for lower values of η there is a self-bound liquid at finite density under zero pressure. The fermion system has a somewhat more complicated behaviour, having two critical values of η. Below the first critical value ($\eta = 0.269$ in this calculation) there is a self-bound liquid. Above this value of η the low density phase is a vapour, but the liquid can be produced by increasing the pressure. Thus there is a critical pressure at which the liquid and vapour can coexist. For values of η larger than η_{CF}, the system is always in the vapour phase.

The difference between the fermion system, where the liquid and vapour can coexist for a range of the parameters, and the boson system, for which they cannot, can be explained by the density dependence of the Fermi energy.[7] While the energy of the boson system can be

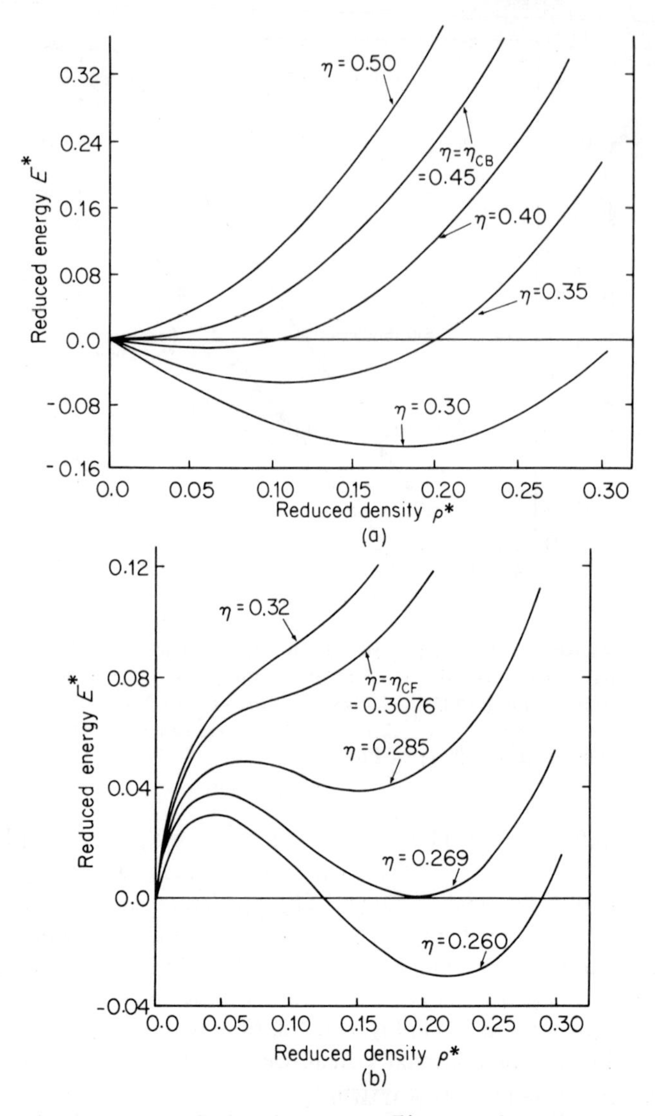

Figure 6.1.4 Reduced energy, E^*, as a function of the reduced density, ρ^*, for a low-density Lennard–Jones system at several values of the quantum parameter η: (a) Bose statistics, and (b) Fermi statistics. (From Miller, Nosanow and Parish,[7] reproduced by permission of the American Institute of Physics)

expanded in integral powers of the density, the fermion system has the additional Fermi energy which is proportional to $\rho^{*2/3}$. This produces a double tangent construction in the energy and consequently the two phase region.

While the quantum parameter Λ^* for the helium systems is not large enough to see this quantum vapour effect, it evidently is large enough in the experimentally accessible system of helium monolayers adsorbed on a smooth substrate. Our calculations show that, at $T = 0$, ^{4}He forms a weakly bound liquid in this system,[8,9,10] while ^{3}He is a vapour at all densities below solidification.[10] Note that since the Fermi energy is proportional to the density in two dimensions, one would not expect to see a two-phase liquid–vapour region even for the fermion system.

Experiments which have been performed on helium monolayers adsorbed on basal-plane oriented graphite (a very smooth substrate) show no evidence for the existence of the liquid for ^{3}He but do show an anomaly interpreted as an indication of liquefaction of ^{4}He.[11]

The superfluid states

While many experiments were performed on liquid ^{4}He at and below 2 °K in the quarter century after it was first liquefied, no observation was made of the fact that it existed in a fourth state of matter (other than gas, liquid and solid) at these temperatures until Wolfke and Keesom in 1928 observed an anomaly in the dielectric constant just below 2.2 °K under the vapour pressure. They suggested that this anomaly signalled a phase transition[12] and referred to the high temperature phase as He I and the low temperature phase as He II.

Heat capacity experiments showed an anomaly at the same temperature under vapour pressure.[13] The shape of the anomaly resembles the Greek letter λ and has resulted in the transition being called the λ-transition, and the temperature of the transition being referred to as T_λ. Measurements of the heat capacity under pressure revealed a line of λ-transitions separating He I from He II, beginning on the vapour pressure curve and ending on the liquid–solid curve (Figure 6.1.1)[13] Consequently, instead of having no triple point, liquid ^{4}He has two triple points.

It was several years before the peculiar properties of He II which gave it the name superfluid were observed. (Indeed, one of the earliest conjectures was that He II is a crystalline phase.)[2] At about the same time a phenomenological model known as the two-fluid model was being developed which codified the existing experimental data below the λ point and predicted some other results soon to be verified (e.g. the existence of second sound). Tisza's formulation of the two-fluid model[14] was motivated by London's suggestion that the existence of He II is closely related to the Bose–Einstein condensation of an ideal Bose gas.[15,16] Above the Bose–Einstein condensation temperature , T_{BE}, the single-particle states all have occupation number of order unity:

$$N_{\mathbf{k}} = \langle a_{\mathbf{k}}^\dagger a_{\mathbf{k}} \rangle = O(N^0) \tag{6.1.9}$$

where N is the number of particles in the system. Below T_{BE}, the number of particles in the zero-momentum single-particle state takes on a macroscopic value:

$$N_0(T) = \langle a_0^\dagger a_0 \rangle = N\left[1 - \left(\frac{T}{T_{BE}}\right)^{3/2}\right] \qquad T < T_{BE} \tag{6.1.10}$$

where

$$T_{BE} = \left(\frac{2\pi\hbar^2}{mk_B}\right)\left(\frac{N}{2.612\,\Omega}\right)^{3/2} \tag{6.1.11}$$

where Ω is the volume of the system and k_B is Boltzmann's constant. Thus $N_0(T)$ increases to N as T decreases to 0. London suggested that, since liquid ^{4}He is a low density collection of bosons, although interacting, some of the features of this condensation are retained. The two-fluid model postulated the existence of two interpenetrating fluids, the superfluid and the normal fluid. Their mass densities, ρ_s and ρ_n respectively, total to the mass density of the liquid

$$\rho_s + \rho_n = \rho. \tag{6.1.12}$$

The superfluid density increases from its value of zero at and above T_λ to ρ at $T = 0$, similar to the condensate fraction $n_0(T) = N_0(T)/N$. It was also postulated that each 'fluid' has a velocity field, $\mathbf{v}_s$ and $\mathbf{v}_n$ respectively so the total mass current density $\mathbf{J}$ is given by

$$\mathbf{J} = \rho_s \mathbf{v}_s + \rho_n \mathbf{v}_n. \tag{6.1.13}$$

Finally it was postulated that the superfluid carries no entropy, which is to suggest that it represents a single state. Clearly the $T \to 0$ limit implies that this single state is the ground state of the system.

The relationship between the macroscopic condensate fraction $n_0(T)$ and the superfluid density $\rho_s(T)$ remains a subject of continued interest. On the one hand it seems clear that $\rho_s(T)$ tends to $\rho n_0(T)$ as the interaction is turned off. On the other hand it is reasonable to expect the condensate to remain macroscopic as a weak interaction is turned on. Indeed, that is the basis of Bogoliubov's pioneering microscopic theory of the weakly-interacting Bose gas.[17] It should be noted, however, that the superfluid density is not given by the condensate fraction in an interacting system. That is clear from the fact that the ground state of the interacting system is different from the ideal Bose gas so that $N_\mathbf{k} \neq 0$ and $n_0 < 1$ at $T = 0$. In contrast, $\rho_s = \rho$ at $T = 0$. Indeed, numerous calculations of the condensate fraction in the ground state of liquid ^{4}He indicate that $n_0 \simeq 0.1$,[18,19,20] and recent inelastic thermal neutron scattering indicates that $n_0 = 0.02 \pm 0.02$[21] or smaller.[22] This suggests that the relationship between n_0 and ρ_s is, at the most, qualitative.

Before continuing with the introduction, we elaborate here on the interesting question of the existence of the condensate in the interacting boson system. Toward that end we define the reduced density matrices for the system by

$$\Gamma_n(\mathbf{r}_1 \ldots \mathbf{r}_n; \mathbf{r}'_1 \ldots \mathbf{r}'_n) = \frac{N!}{(N-n)!n!}$$

$$\times \int d\mathbf{x}_{n+1} \ldots d\mathbf{x}_N \Gamma_N(\mathbf{r}_1 \ldots \mathbf{r}_n, \mathbf{x}_{n+1} \ldots \mathbf{x}_N; \mathbf{r}'_1 \ldots \mathbf{r}'_n, \mathbf{x}_{n+1} \ldots \mathbf{x}_N) \tag{6.1.14}$$

where the Γ_N is given by

$$\Gamma_N = \frac{1}{Z} \sum_\alpha \exp(-\beta E_\alpha) \Psi^*_\alpha(\mathbf{r}_1 \ldots \mathbf{r}_N) \Psi_\alpha(\mathbf{r}'_1 \ldots \mathbf{r}'_N) \tag{6.1.15}$$

at temperature

$$T = 1/k_B\beta \tag{6.1.16}$$

and Ψ_α are the eigenstates and E_α the corresponding eigenvalues of the system. Z is the partition function. At $T = 0$ only the ground state is involved:

$$\Gamma_N = \Psi^*_0(\mathbf{r}_1 \ldots \mathbf{r}_N) \Psi_0(\mathbf{r}'_1 \ldots \mathbf{r}'_N) \tag{6.1.17}$$

For boson systems the presence of a condensate is revealed by considering the single particle density matrix $\Gamma_1(\mathbf{r},\mathbf{r}')$ which satisfies[23]

$$\int \Gamma_1(\mathbf{x}, \mathbf{x}') \psi_\lambda(\mathbf{x}) \, d\mathbf{x} = N_\lambda \psi_\lambda(\mathbf{x}') \tag{6.1.18}$$

For a translationally invariant system, the ψ_λ are single-particle plane wave states

$$\psi_\lambda(\mathbf{x}) = \exp(i\mathbf{k}_\lambda \cdot \mathbf{x})/\sqrt{\Omega} \tag{6.1.19}$$

and N_λ is the occupation number of the single-particle state $\mathbf{k}_\lambda$:

$$N_\lambda = \langle a^\dagger_{\mathbf{k}\lambda} a_{\mathbf{k}\lambda} \rangle \tag{6.1.20}$$

Γ_1 may be written in terms of its eigenfunctions

$$\Gamma_1(\mathbf{x}, \mathbf{x}') = \sum_\lambda N_\lambda \psi^*_\lambda(\mathbf{x})\psi_\lambda(\mathbf{x}'). \tag{6.1.21}$$

Bose–Einstein condensation occurs when one of the eigenvalues, say N_{λ_0}, becomes macroscopic:

$$N_{\lambda_0} = O(N). \tag{6.1.22}$$

Then

$$n_0(T) = \lim_{N\to\infty} N_{\lambda_0}(T)/N. \tag{6.1.23}$$

A common statement of this property is that the system has off-diagonal long range order (ODLRO)[24], since the off-diagonal components of the density matrix remain finite under infinite separation:

$$\Gamma_1(\mathbf{x}, \mathbf{x}') \xrightarrow[|\mathbf{x}-\mathbf{x}'|\to\infty]{} N_{\lambda_0} \psi^*_{\lambda_0}(\mathbf{x})\, \psi_{\lambda_0}(\mathbf{x}'). \tag{6.1.24}$$

It is this long range coherence of the system which may account for the macroscopic quantum properties associated with superfluidity. This connection, however, has not been rigorously established. Indeed, it has been argued that a finite condensate fraction has not been measured in ^{4}H.[22] Furthermore, the theoretical calculations of a finite condensate in the ground state of liquid ^{4}He have been based upon Jastrow approximations[18,19,20] to the ground state wave function

$$\Psi_J(\mathbf{r}_1 \ldots \mathbf{r}_N) = \prod_{i<j}^{N} \exp[\tfrac{1}{2}u(r_{ij})] \tag{6.1.25}$$

It has been shown by Reatto, however, that a Jastrow function always has a condensate.[25] Thus the trial functions used to estimate the condensate do not include the possibility of the absence of a condensate, and cannot shed much light on the question of its existence.

An associated problem is whether a boson system which is a fluid in its ground state must have a finite condensate. An immediate counter-example exists – the interacting Bose gas in one dimension has no condensate[26] – but three dimensions differs sufficiently from one dimension for the question still to remain. There is a two-dimensional counter-example – the two-dimensional Coulomb gas, with two-particle potential whose Fourier transform is proportional to k^{-2} (note that this is not a $1/r$ potential), has no condensate in its ground state. Again this is an extremely pathological example, involving in this case infinite-range forces. The pathology of the exceptions suggests that, under rather weak restrictions, a three-dimensional fluid ground state must contain a finite condensate. Furthermore, Chester has observed that even a quantum solid may have a condensate.[27]

To add to this confusion about the importance of the condensate, it should be noted that Landau arrived at the two-fluid model independently and without reference to the concept of Bose–Einstein condensation. Instead, he postulated the existence of an excitation spectrum $\epsilon(\mathbf{k})$ in terms of which the excited state energies have the form

$$E(\{n_\mathbf{k}\}) = \sum_\mathbf{k} n_\mathbf{k}\epsilon(\mathbf{k}) \tag{6.1.26}$$

where $n_\mathbf{k}$ can be any non-negative integer. The normal fluid is then just the gas of excitations present by virtue of the finite temperature, and the superfluid is, in a crude sense, what remains of the ground state.[28] Landau made the additional assumption that the λ point corresponds to the point at which the density of these excitations is *equal to* the particle density. This is a weak point in Landau's model since the concept of non-interacting elementary excitations is an approximation which breaks down badly when the number of excitations becomes comparable to the number of particles in the system. London considered this an argument in favour of Tisza's conception of the two-fluid model, since the latter has a well-defined criterion for the λ transition: the disappearance of the condensate.[16]

Landau's model was phenomenological – he had to deduce $\epsilon(\mathbf{k})$ in order to reproduce available experimental results. The microscopic theory of excited states is in a similarly ambivalent situation with regard to the role of the condensate. On the one hand there is Feynman's proof that Landau's elementary excitation spectrum is the lowest lying state, which is important both from the point of view of thermodynamics (a lower energy state would dominate the low temperature properties) and for the existence of superfluidity (a lower lying state with a quadratic momentum dependence would destroy superfluidity).[29] Feynman's proof was based only on the fluidity of the ground state and the Bose statistics, with no necessity for involving the condensate. Evidently it would work equally well for a fluid ground state without a condensate, if it exists. On the other hand, other microscopic theories based upon the perturbation theory of the weakly interacting Bose gas rely on the condensate to suppress the single-particle-like modes which would destroy superfluidity.[17,30,31]

We do not see an answer to the question of the role of the condensate in superfluid ^{4}He, but highlight it here as one of the more intriguing fundamental questions still with us in this subject.

The absence of superfluidity in liquid ^{3}He was taken by London as further evidence that its presence in ^{4}He is related to Bose–Einstein condensation.[16] ^{3}He, being composed of fermions, cannot have more than one particle occupying a single-particle state, and thus cannot have a macroscopic single particle condensate. On the other hand, another example of superfluidity is the superconductivity of electrons in a metal, which is a collection of fermions. The explanation for superfluidity in macroscopic systems of fermions was given by Bardeen, Cooper and Schrieffer (BSC) in 1957.[32] The central idea is that a macroscopic number of electrons associate themselves into the same pair state, which is then a two-body condensate. A manifestation of this pair condensation is the occurrence of ODLRO in the two-body density matrix:[24]

$$\Gamma_2(\mathbf{r}_1, \mathbf{r}_2; \mathbf{r}_1', \mathbf{r}_2') \xrightarrow[|(\mathbf{r}_1, \mathbf{r}_2)-(\mathbf{r}_1', \mathbf{r}_2')|\mapsto\infty]{} F^*(\mathbf{r}_1, \mathbf{r}_2)F^*(\mathbf{r}_1', \mathbf{r}_2'). \tag{6.1.27}$$

The condition for pair-condensation to occur is that the fermions have an attractive interaction between single-particle states near the Fermi surface. Attempts were made around 1960 to extend the BCS model to ascertain the possibility of pair-condensed states in liquid ^{3}He.[33,34,35] While pairing in superconductors occurs in relative s-wave angular momentum states, it was pointed out that the strong, short-range repulsion between helium atoms makes pairing more likely in higher relative angular momentum states; the early conjecture was pairing in relative d-wave states.[34]

The predicted maximum temperature at which this new phase should be observed ranged from millidegrees down to microdegrees above absolute zero. Experiments, however, are currently feasible only as low as about one millidegree and until recently showed no property which might be associated with pair-condensation or superfluidity. The situation changed radically several years ago when NMR experiments on liquid ^{3}He manifested anomalous behaviour at about .002 °K along the melting curve.[36] Rapidly mounting evidence makes it

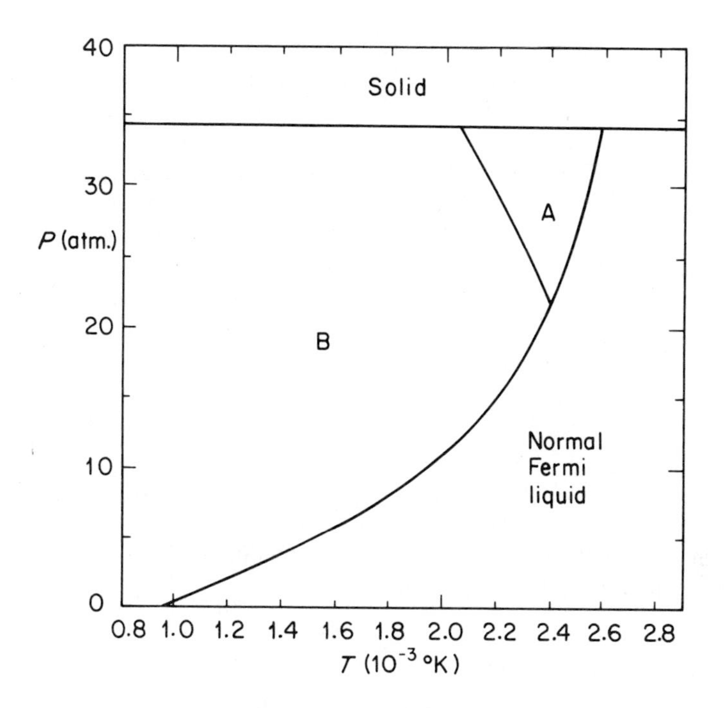

Figure 6.1.5 The phase diagram of ^{3}He at very low temperature, in zero magnetic field. (From Wheatley,[37] reproduced by permission of the American Institute of Physics)

clear that superfluid liquid ^{3}He has at long last been found. For more details, see the excellent recent review by Wheatley.[37]

The phase diagram as it is presently understood is shown in Figure 6.1.5. Note that there are two superfluid phases, referred to as ^{3}He A and ^{3}He B. While the transition from normal fluid to superfluid appears to be second order, the transition between the A and B phase is first order. In light of London's remarks concerning the importance of the absence of superfluidity in ^{3}He as evidence for the theoretical interpretation of superfluidity in ^{4}He, it is amusing to consider the possible theoretical course the subject might have taken had these experiments on ^{3}He been performed in, say, 1940.

The superfluidity in ^{3}He is clearly much more complicated than in ^{4}He or superconductors. While the early theoretical work had focused on singlet pairing (i.e. even angular momentum states), the interpretation of the experimental results, particularly the magnetic properties, requires triplet pairing (i.e. odd angular momentum states).[38] In retrospect, it seems as if this might have been expected. In the years following the early theoretical speculation, it became clear than normal liquid ^{3}He is nearly ferromagnetic, having a static susceptibility ten times that of the Pauli susceptibility.[39] This indicates a preference for spin triplet correlations, i.e. for an enhanced effective interaction between ^{3}He atoms with parallel spins.

Earlier work on superconductors by Anderson and Morel (AM) had generalized the BCS theory to include the possibility of triplet pairing.[40] The solutions of their generalized BCS model are anisotropic pair condensed states. The anisotropy is evident by the existence of two directions in space: the direction of the spin quantization axis, and the direction of the finite relative angular momentum. Balian and Werthamer (BW) followed several years later with a slightly more general theory which included the AM anisotropic solutions as well as an isotropic solution for the triplet pairing p-wave states.[41] Furthermore, they showed that the isotropic solution is energetically favoured over the AM anisotropic solution for weak-coupling systems.

Shortly after the NMR experiments,[36] Leggett pointed out that the A phase can be understood as the anisotropic triplet pairing state of Anderson and Morel.[40] This apparent contradiction of the conclusions of Balian and Werthamer was resolved by Anderson and Brinkman[42] and by Mermin and Stare,[43] who pointed out that the effective interaction between ^{3}He quasi-particles has a more complicated spin and angular momentum dependence than that assumed by BW and AM, in which case it is possible for the AM state to be stabilized relative to the isotropic BW state. Anderson and Brinkman showed that this stabilization may be accomplished by the exchange of paramagnons, which enhances the coupling between parallel spins.[42] Thus the A phase is now generally accepted to be in the ABM state (for Anderson, Brinkman and Morel), while the B phase is understood to be isotropic triplet pairing of the BW variety. The present theoretical situation is given considerable elaboration in the excellent review by Leggett.[44] The general conclusion we would like to emphasize here is that the determination of the nature of the pair-condensed state in liquid ^{3}He requires a much more detailed knowledge of the effective interactions between the quasi-particles than is required for the single-parameter weak coupling theory of superconductors.[32] We will discuss a possible procedure for accomplishing that goal in Section 6.4.

A microscopic theory of the helium liquids

The main subject of the remainder of this chapter is a description of the features of a microscopic theory of liquid ^{4}He and liquid ^{3}He. The goal of the theoretical procedures we describe is always *quantitative* predictions about these systems and/or *quantitative* agreement with experiment. Toward that goal we are interested in solving the Schrödinger equation

$$H\psi_n = E_n\psi_n \tag{6.1.28}$$

where the Hamiltonian H contains as realistic a model as available for the interaction potential between the helium atoms. Therein lies the chief difficulty: the interaction between two helium atoms contains a strong, short-range repulsion. A frequently used model is the Lennard–Jones 6–12 potential, (equation 6.1.3), although a number of other candidates exist.[45] A wave function which represents the ground state or a low excited state of a system of helium atoms must be negligible for inter-particle distances within the highly repulsive range. This precludes the direct application of the low-density weak-interaction models usually treated in many-body theory. These models do, however, provide guides to the important effects in helium. In the procedures described in the following sections, this model information is combined with a theory of the short-range correlations so that a realistic model of the states of the many-body systems of helium atoms may be studied.

Sophisticated many-body procedures have been developed to treat the short-range correlations in nuclear matter and have been generalized to the helium problem.[46] These theories have been less successful in the latter case than in the former because the densities of interest in helium are effectively higher than those of nuclear matter. There is, however, a much simpler approach to the theory of helium which has been developed in large part by Feenberg and his students since the early 1960s.[47,48] This procedure is called the *method of correlated wavefunctions,* and is the basic theoretical method discussed below. The basic idea is the same as that embodied in Feynman's elegant discussion of the low excited states of ^{4}He,[29] which is that the short-range correlations in the low-momentum eigenstates differ very little from the short-range correlations in the ground state. Feenberg took this further with two observations: first,[49] the short-range correlations are also nearly independent of the statistics; and second,[50] a mathematically complete basis can be constructed which contains these correlations at the outset.

One consequence of the first assertion is that a great deal of information about the Fermi system, liquid ^{3}He, is contained in the boson ground state of the ^{3}He Hamiltonian. To describe the Fermi system, this boson ground state is taken as a correlation operator which is applied to a complete set of model states of appropriate statistics. The choices of model states for the ^{3}He normal-liquid are the plane-wave Slater determinants, which are the eigenstates of the non-interacting Fermi system.[49] Recently, Clark and Yang[51] used the BCS[32] ground-state as a model function for the pair-condensed phase of nuclear matter (their correlation function, however, is a Jastrow function). For excited states of the boson system, liquid ^{4}He, the choice of model functions was a set of polynomials in the density fluctuation operator $\rho_{\mathbf{k}}$,[52] which is a symmetrized plane wave:

$$\rho_{\mathbf{k}} = \sum_{i=1}^{N} \exp(i\mathbf{k} \cdot \mathbf{r}_i) \tag{6.1.29}$$

for an N particle system.

The elements of the basis formed by the product of the symmetric ground-state wave-function by the model states are called the correlated wave functions. The procedure is to diagonalize simultaneously the identity and the Hamiltonian operators.[50] It is at this point that approximation schemes must be introduced. For the boson system, the density fluctuation operators $\rho_{\mathbf{k}}$ behave much as random variables. The overall approximation scheme which has been applied so far is to consider correlations between increasingly larger groups of $\rho_{\mathbf{k}}$. Since $\rho_{\mathbf{k}}$ is closely related to the creation operator for a 'Feynman phonon', this amounts to taking into account multi-phonon interactions between increasingly larger numbers of phonons. The simplest correlations are between pairs of phonons of equal and opposite momentum. All correlations of this type were treated in a theory called the *paired phonon analysis,*[52] the result of which is an excitation spectrum of non-interacting quasi-particles with energy given by the Bijl–Feynman form

$$\epsilon(\mathbf{k}) = \frac{\hbar^2 k^2}{2mS(k)} \tag{6.1.30}$$

where $S(k)$ is the liquid structure factor for the ground-state. This supplies some microscopic support for the Landau model. The next level of approximation is the inclusion of three-phonon processes, which is to say the inclusion of one phonon decaying to two or the inverse process. These processes are important in the roton region of the excitation spectrum[53] and are the same type of terms as those included in Feynam–Cohen backflow.[57] The general mathematical structure of multi-phonon processes has been given recently by Wu.[55,56]

The anti-symmetry of the fermion eigenstates makes the ^{3}He problem more difficult than ^{4}He. Matrix elements between states composed of a plane-wave Slater determinant correlated by a symmetric Bose function are difficult to calculate. A great deal of progress was made by using cluster expansion for the Slater determinants.[57] In this connection, attention was restricted to a subspace of low-energy particle-hole states, where the correlated basis was orthogonalized and then approximately diagonalized using second-order perturbation theories.[49,57] The results of this procedure applied to the ground-state are in reasonably good agreement with experiment for such quantities as energy versus density, liquid structure function, and paramagnetic susceptibility.[49,57] This procedure was also applied to calculate the interaction between two quasi-particles which can be used in turn to calculate the Landau–Fermi liquid parameters.[58] The agreement with experiment is remarkable considering approximations necessary in the calculation. Tan and Woo[59] have shown how to obtain a second-quantized Hamiltonian in the quasi-particle creation and annihilation operators which contains the multi-quasi-particle interactions.

In Section 6.4 we describe a method for treating the Fermi statistics of the ^{3}He problem without using cluster expansion.[60] Instead, we observe that expectation values between correlated model Fermi functions can be rewritten as expectation values of symmetric functions in the Bose ground-state. The reason is that the fermion expectation values always contain two anti-symmetric model functions whose combination is symmetric. We have much more confidence in procedures for calculating expectation values in Bose ground states and can appeal to techniques developed in the theory of liquid ^{4}He. Therefore we expand the symmetric operators in a complete set of symmetrized plane waves, each of which is a simple polynomial in the functions. Bose expectation values of these symmetrized plane waves have a general mathematical structure similar to that obtained by Wu[55] for multi-phonon matrix elements. Therefore we may systematically include all paired-phonon processes, then all three phonon processes, etc. in an approximation scheme similar to that applied to the boson system. We call this the *effective Bose expansion.* Preliminary application of this procedure at the paired-phonon level to the normal liquid ^{3}He system improves agreement with experiment for the ground-state energy and liquid structure factor.[60] We are, therefore, encouraged to apply this procedure to the more difficult problem of the pair-condensation in liquid ^{3}He.

As described above, the method of correlated wave functions makes extensive use of the symmetric ground-state wave function of the Hamiltonian for the system of interest. The results of the calculations using the correlated wave functions are expressed in terms of the n-body distribution functions of this Bose ground-state;[49] the value of the index n depends upon the complexity of the processes included in the approximation. The only available experimental information about these distribution functions is the liquid structure factor for liquid ^{4}He;[61,62] it is essentially the Fourier transform of the two-body distribution function (but must be corrected for the effects of finite temperature). Higher distribution functions must be calculated theoretically. Even the two-body distribution function for the *Bose* ground state of the ^{3}He Hamiltonian must be obtained theoretically. Most calculations of these distribution functions use the Jastrow approximation for the ground-state wave functions, Equation 6.1.25.[19,63–69] This is the simplest choice of trial function which can be chosen to be negligible for inter-particle distances in the high repulsive region of the two-particle potential. This is accomplished by letting $u(r)$ become large and negative for configurations where the potential is large and positive. The function $u(r)$ is chosen to minimize the expectation value of the Hamiltonian in state ψ_J. Although numerous variational calculations have been done for choices of $u(r)$ depending on a small number of variational parameters,[19,64–68] several years ago we showed how to obtain the optimum Jastrow wave function from the class of all functions of the form of ψ_J.[70] Limited computing capability forced us to introduce further approximations (discussed in Section 6.2) to carry out these unconstrained variational calculations. Nevertheless, our results for the liquid structure function using the approximations are in good agreement with experiment.[70,62]

While Jastrow functions give reasonably good account of some of the properties of liquid ^{4}He, recent interest has been focused upon enlarging the class of correlation functions to obtain improved agreement with experiment.[71–73] This is accomplished by defining an extended Jastrow function as

$$\psi(\mathbf{r}_1 \ldots \mathbf{r}_N) = \exp[\tfrac{1}{2}\chi(\mathbf{r}_1 \ldots \mathbf{r}_N)] \tag{6.1.31}$$

where χ is a real function given by

$$\chi(\mathbf{r}_1 \ldots \mathbf{r}_N) = \sum_{n=1}^{N} \frac{1}{n!} \sum_{\substack{i_1 \ldots i_n \\ i_s \neq i_t}} u_n(\mathbf{r}_{i_1} \ldots \mathbf{r}_{i_n}) \tag{6.1.32}$$

where u_n is symmetric under interchange of its arguments. We show in Section 6.2 that the exact ground state wave function of a system of bosons can be chosen in this form.

A systematic approximation scheme is now evident. The Hartree approximation for a boson system is obtained by retaining only the $u_1(\mathbf{r})$ term in χ and determining it variationally. The Jastrow approximation is obtained by retaining $u_1(\mathbf{r})$ and $u_2(\mathbf{r},\mathbf{r}')$. We have carried out calculations including $u_3(\mathbf{r}_1,\mathbf{r}_2,\mathbf{r}_3)$ and find improved agreement with the ^{4}He ground state energy and liquid structure factor.[72] Furthermore, the changes in the calculated values from the pure Jastrow function are 10% or less, indicating that convergence is rapid with increasing n.

In Section 6.2 we discuss the ground state of liquid ^{4}He based upon the extended Jastrow framework just introduced. An extensive discussion of the results of the pure Jastrow approximation is included. In Section 6.3 we discuss the low excited states of liquid ^{4}He and its response to external probes. In this section we assume that we have precise knowledge of the ^{4}He ground state wave function, and use it as a correlation factor in a set of correlated wave functions. While Sections 6.2 and 6.3 are formally independent of one another, we must rely on the procedures of Section 6.2 to obtain the information about the ground state wave function necessary to carry out the calculations described in Section 6.3.

Finally in Section,6.4 we discuss the theory of liquid ^{3}He, using the method of correlated wave functions to obtain information about the ground state (both normal and pair-condensed) and the low excited states. The correlation function in this case is the fictitious boson liquid ^{3}He ground state wave function, so we must again rely on the methods of Section 6.2 to provide the necessary information about the correlation function.

Analyses such as we have described here have also been applied to the following problems: ^{3}He–^{4}He mixtures,[74,75] adsorbed quantum fluids,[8–10,76] the structure of the free surface of liquid ^{4}He,[77] the states of ^{3}He atoms and of electrons on the free surface of liquid ^{4}He,[77,78] and the structure of a quantized vortex in liquid ^{4}He.[79] All of these problems but the first have the additional complication of non-uniform density. The scope of this chapter necessarily precludes the treatment of these topics.

There are numerous recent reviews which in sum cover most of the material necessarily omitted and some of the material included. In particular we mention the review of the structure and excitations of liquid ^{4}He by Woods and Cowley,[80] the reviews by Wheatley[37] and by Leggett[44] on superfluid ^{3}He (already mentioned above) and the forthcoming volume of reviews on liquid and solid helium edited by Bennemann and Ketterson.[81]

6.2 Static structure of liquid ^{4}He: The ground state

We begin this section with a general discussion of the ground state of a many-body boson system with a view toward calculating the properties of liquid ^{4}He. Below, we discuss the Jastrow approximation to the ground state, first describing the parametrized variational calculation and then the Euler–Lagrange approach which leads to the optimum Jastrow function in the class of all Jastrow functions. The Euler–Lagrange approach is subsequently extended to include non-Jastrow functions.

General properties of the many-boson ground state

The problem we wish to consider is the nature of the boson ground state of the Hamiltonian

$$H = T + V \tag{6.2.1}$$

where the kinetic energy operator

$$T = \sum_{i=1}^{N} \frac{-\hbar^2}{2m} \nabla_i^2 \tag{6.2.2}$$

and the potential energy is local. We will focus on a two-body potential energy of the form

$$V = \sum_{i<j}^{N} V(r_{ij}) \tag{6.2.3}$$

although other terms such as an external one-body potential or a three-body potential would not change the results of this section. The boson ground state wave function for N particles is symmetric under interchange of particle coordinates and satisfies the Schrödinger equation

$$H\Psi_0(\mathbf{r}_1 \ldots \mathbf{r}_N) = E_0\Psi_0(\mathbf{r}_1 \ldots \mathbf{r}_N) \tag{6.2.4}$$

where E_0 is the lowest eigenvalue of the boson problem. Thus

$$E_0 \leqslant \langle \Psi | H | \Psi \rangle / \langle \Psi | \Psi \rangle \tag{6.2.5}$$

where Ψ is any N-body function symmetric under interchange of particle coordinates.

It was shown by Penrose and Onsager[18] and others that the ground state wave function of a many-body boson system has two important features:

(i) It has no nodes, and consequently can be chosen to be real positive; and
(ii) It is non-degenerate.

The proof of these properties is simple and also useful for the ensuing analysis, so we include it here.

To demonstrate (i) write Ψ_0 in the form

$$\Psi_0(\mathbf{r}_1 \ldots \mathbf{r}_N) = \exp[\tfrac{1}{2}\chi(\mathbf{r}_1 \ldots \mathbf{r}_N)] \tag{6.2.6}$$

and suppose that χ may be complex:

$$\chi(\mathbf{r}_1 \ldots \mathbf{r}_N) = \chi_{\mathrm{R}}(\mathbf{r}_1 \ldots \mathbf{r}_N) + \mathrm{i}\chi_{\mathrm{I}}(\mathbf{r}_1 \ldots \mathbf{r}_N) \tag{6.2.7}$$

where χ_{R} and χ_{I} are real and symmetric. Then

$$\Psi_0^* \Psi_0 = \exp(\chi_{\mathrm{R}}) \tag{6.2.8}$$

and the total potential energy is independent of χ_{I}:

$$\frac{\langle \Psi_0 | V | \Psi_0 \rangle}{\langle \Psi_0 | \Psi_0 \rangle} = \frac{\int \mathrm{d}\mathbf{r}_1 \ldots \mathrm{d}\mathbf{r}_N \exp(\chi_{\mathrm{R}}) \sum_{i<j}^{N} V(r_{ij})}{\int \mathrm{d}\mathbf{r}_1 \ldots \mathrm{d}\mathbf{r}_N \exp(\chi_{\mathrm{R}})} \tag{6.2.9}$$

To calculate the kinetic energy, we make use of the identity

$$\begin{aligned} -\frac{\hbar^2}{2m} \int \psi^* \nabla_i^2 \psi \, \mathrm{d}\mathbf{r}_1 \ldots \mathrm{d}\mathbf{r}_N &= -\frac{\hbar^2}{2m} \int \psi \nabla_i^2 \psi^* \mathrm{d}\mathbf{r}_1 \ldots \mathrm{d}\mathbf{r}_N \\ &= \frac{\hbar^2}{2m} \int \nabla_i \psi^* \cdot \nabla_i \psi \, \mathrm{d}\mathbf{r}_1 \ldots \mathrm{d}\mathbf{r}_N \end{aligned} \tag{6.2.10}$$

so that the total kinetic energy is given by

$$\frac{\langle \psi | T | \psi \rangle}{\langle \psi | \psi \rangle} = -\frac{\hbar^2}{8m} \sum_{i=1}^{N} \frac{\int [\psi^* \nabla_i^2 \psi + \psi \nabla_i^2 \psi^* - 2\nabla_i \psi^* \cdot \nabla_i \psi] \, d\mathbf{r}_1 \ldots d\mathbf{r}_N}{\langle \psi | \psi \rangle}$$

$$= \frac{\int d\mathbf{r}_1 \ldots d\mathbf{r}_N \exp(\chi_R) \, [(-\hbar^2/8m) \sum_{i=1}^{N} \nabla_i^2 \chi_R + (\hbar^2/8m) \sum_{i=1}^{N} (\nabla_i \chi_I)^2]}{\int d\mathbf{r}_1 \ldots d\mathbf{r}_N \exp(\chi_R)}$$

(6.2.11)

Only the second term in the numerator of this expression depends on χ_I, and it is positive. Consequently, the energy is minimized by setting

$$\chi_I = 0 \tag{6.2.12}$$

which verifies (i) above.

The fact that Ψ_0 is non-degenerate follows immediately from the observation that there are no nodes. If there were another eigenstate with energy E_0, then it could be chosen orthogonal to Ψ_0. Consequently it would have a node (since Ψ_0 is positive), which means that it has a part $\chi_I \neq 0$. It follows that its energy would be lowered by setting $\chi_I = 0$, which violates the assumption that E_0 is the lowest eigenvalue.

The function $\chi = \chi_R$ can now be decomposed into fundamental n-body functions u_n, $1 \leqslant n \leqslant N$:[72]

$$\chi(\mathbf{r}_1 \ldots \mathbf{r}_N) = \sum_{n=1}^{N} \sum_{\langle i_1 \ldots i_n \rangle} u_n(\mathbf{r}_{i_1} \ldots \mathbf{r}_{i_n}) \tag{6.2.13}$$

where the notation $\langle i_1 \ldots i_n \rangle$ refers to the summation over all distinct choices of the n coordinates from the N coordinates. For simplicity we impose periodic boundary conditions. Then the liquid ground state Ψ_0 will be translationally invariant. That fact along with the reality of Ψ_0 and the condition that u_n be uniquely defined requires that u_n be:[72]

(1) real;
(2) symmetric under interchange of the particle coordinates $\mathbf{r}_i$;
(3) invariant under translation of the centre of mass variable

$$\mathbf{R} = \frac{1}{n} \sum_{i=1}^{n} \mathbf{r}_i,$$

(4) invariant under rigid rotation about the centre of mass $\mathbf{R}$;
(5) invariant under simultaneous inversion of all the coordinates, $\mathbf{r}_i \to -\mathbf{r}_i$; and
(6) short-ranged for separation of the subsets of n coordinates.

Condition (6) makes the decomposition in equation (6.2.13) unique. It excludes the possibility of symmetrized u_p type terms in u_n, $n > p$. For example, $u_3(\mathbf{r}_1, \mathbf{r}_2, \mathbf{r}_3)$ may not contain a term of the form $f(\mathbf{r}_1, \mathbf{r}_2) + f(\mathbf{r}_2, \mathbf{r}_3) + f(\mathbf{r}_3, \mathbf{r}_1)$; it belongs instead in u_2.

The variational principle, equation (6.2.9) can be used to replace the Schrödinger equation (6.2.4) by a set of simultaneous integral-differential equations for u_n:

$$\frac{\delta}{\delta u_n}(\mathbf{r}_1 \ldots \mathbf{r}_n) \left[\frac{\langle \Psi_0 | H | \Psi_0 \rangle}{\langle \Psi_0 | \Psi_0 \rangle} \right] = 0. \tag{6.2.14}$$

Of course there is no hope of solving this complete set of equations. This formulation of the problem, however, suggests a systematic set of approximations which should produce a close approach to the ground state energy while retaining a variational interpretation of the computed energy. The obvious procedure is to truncate (6.2.13) at some small value of $n = M < N$, i.e. set

$$u_n = 0 \quad n > M. \tag{6.2.15}$$

The Jastrow function (6.1.25) is obtained by choosing $M = 2$. The variational equation obtained from (6.2.14) with $n = 2$[82] is shown below in equation (6.2.52). The procedure for solving this equation, known as the paired-phonon analysis,[70] is the subject of the next section. Procedures for solving this set of equations for $M > 2$ are discussed in a subsequent section.

There are several points to be made about approximations of this sort. First, the conditions (1–6) should be retained in the truncated trial function. In particular, the proof that u_n should be real in order to minimize the energy is not restricted to the exact ground state; it applies equally well for these trial functions.

The second point to be made is that the truncation at finite n reduces the amount of information needed about the trial function in order to calculate E_0. This information appears in the form of the n-body distribution functions, g_n, defined by

$$\rho^n g_n(\mathbf{r}_1 \ldots \mathbf{r}_n) = \frac{N!}{(N-n)!} \frac{\int d\mathbf{r}_{n+1} \ldots d\mathbf{r}_N \,|\Psi_0(\mathbf{r}_1 \ldots \mathbf{r}_N)|^2}{\langle \Psi_0 | \Psi_0 \rangle} \tag{6.2.16}$$

where ρ now means the number density

$$\rho = N/\Omega \tag{6.2.17}$$

where Ω is the volume. (In Section 6.1, ρ was the mass density). g_n is proportional to the diagonal part of Γ_n, the n-body density matrix (equation 6.1.14). The function g_2 is the well-known radial distribution function g:

$$g_2(\mathbf{r}_1, \mathbf{r}_2) = g(r_{12}). \tag{6.2.18}$$

The potential energy requires only g_2 if it is a two-body potential:

$$\frac{\langle \Psi_0 | V | \Psi_0 \rangle}{\langle \Psi_0 | \Psi_0 \rangle} = \frac{\rho^2}{2} \int d\mathbf{r}_1 d\mathbf{r}_2 g_2(\mathbf{r}_1, \mathbf{r}_2)\, V(r_{12}) = \frac{N\rho}{2} \int d\mathbf{r} g(\mathbf{r}) V(\mathbf{r}). \tag{6.2.19}$$

It can be seen from equation (6.2.11) with $\chi_I = 0$ that the contribution of u_n to the kinetic energy requires only g_n:

$$\frac{\langle \Psi_0 | T | \Psi_0 \rangle}{\langle \Psi_0 | \Psi_0 \rangle} = -\frac{\hbar^2}{8m} \sum_n \frac{\rho^n}{(n-1)!} \int d\mathbf{r}_1 \ldots d\mathbf{r}_n g_n(\mathbf{r}_1 \ldots \mathbf{r}_n) \nabla_1^2 u_n(\mathbf{r}_1 \ldots \mathbf{r}_n) \tag{6.2.20}$$

The simplicity of this result is a bit misleading because of the difficulty of obtaining g_n. Some guidance in that regard may be obtained from the theory of classical fluids: in particular, the problem of obtaining the n-particle distribution g_n from the N-particle probability density $|\Psi_0|^2$ is completely equivalent to the problem of calculating the classical n-particle distribution g_n^{Cl} at some temperature T for a fictitious classical system of particles in the canonical ensemble with many-body interactions $\Phi_p(\mathbf{r}_1 \ldots \mathbf{r}_p)$ defined by

$$\Phi_p(\mathbf{r}_1 \ldots \mathbf{r}_p) = -k_B T u_p(\mathbf{r}_1 \ldots \mathbf{r}_p). \tag{6.2.21}$$

Indeed, with this equality,

$$g_n^{\mathrm{Cl}}(\mathbf{r}_1 \ldots \mathbf{r}_n | T) = g_n(\mathbf{r}_1 \ldots \mathbf{r}_n) \tag{6.2.22}$$

as can be seen immediately by noting that the Boltzmann factor for the fictitious classical system is equal to $|\Psi_0|^2$. This property is particularly useful for the Jastrow function (as is discussed in the next section) since by far the largest amount of work in the theory of classical fluids has been restricted to two-body potentials.

Finally we comment on the calculations of quantities other than the energy. It is well-known that variationally determined wave functions may produce poor results for quantities other than the variational functional (in this case the energy). The magnitude of the error is closely related to the size of the function space sampled in the variation. In this connection Euler–Lagrange procedures are far superior to limited parametrizations in determining quantities other than the energy. For example, the expression for the potential energy in equation (6.2.19) depends only on $g(r)$ over the range of the potential $V(r)$, which is generally quite small. $g(r)$, however, achieves its asymptotic value much more slowly than the potential energy, and in fact

$$g(r) \xrightarrow[r \to \infty]{} 1. \tag{6.2.23}$$

Consequently the kinetic energy (equation 6.2.20) may be reduced by expanding the range over which u_n is non-negligible. For example, it has been shown that, for a short-range two-body potential $V(r)$, the function $u_2(\mathbf{r}_1, \mathbf{r}_2)$ is long-ranged in order to properly include the zero-point motion of the phonon field.[83] While the short-ranged parametrized trial functions $u_2(\mathbf{r}_1, \mathbf{r}_2)$ fail to take this into account, the Euler–Lagrange procedure applied to an unlimited variation in u_2 produces the correct long-wavelength behaviour. This point is elaborated on below.

Our feeling is that an unlimited variation on $u_2(\mathbf{r}_1, \mathbf{r}_2)$ should produce a good approximation for any two-body property, i.e. for any quantity F which can be expressed as

$$\langle F \rangle = \int d\mathbf{r}_1 \, d\mathbf{r}_2 F(\mathbf{r}_1 \mathbf{r}_2) g_2(\mathbf{r}_1, \mathbf{r}_2).$$

This is so because of the sensitive dependence of the kinetic energy on g_2 (equation 6.2.20). One such quantity is the liquid structure function $S(k)$, which is measured by X-ray scattering[61,62] and integrated neutron scattering.[80]

$$S(k) = \frac{\langle \Psi_0 | \rho_{\mathbf{k}} \rho_{-\mathbf{k}} | \Psi_0 \rangle}{N \langle \Psi_0 | \Psi_0 \rangle} \tag{6.2.24}$$

where $\rho_{\mathbf{k}}$ is the density fluctuation operator given by equation (6.1.29). $S(k)$ may be expressed in terms of the two-particle distribution function as

$$S(k) = 1 + \frac{\rho^2}{N} \int d\mathbf{r}_1 \, d\mathbf{r}_2 \exp(i\mathbf{k} \cdot \mathbf{r}_{12}) g_2(\mathbf{r}_1, \mathbf{r}_2). \tag{6.2.25}$$

Similarly, a good approximation for the ground-state expectation value of a three-body function may require a trial function containing a carefully determined u_3. In that connection we call attention to the recent work of Berdahl on the long wavelength behaviour of g_3 in the exact ground-state, as opposed to its behaviour in a simple Jastrow approximation.[84]

The Jastrow approximation

The original motivation for the use of the Jastrow function as a trial ground state wave function for liquid ^{4}He was that it is the simplest function which produces finite matrix elements for a potential energy which has short range divergences as in helium.[85] Recall that the Jastrow function is defined by equation (6.1.25), which we repeat here

$$\psi_J(\mathbf{r}_1 \ldots \mathbf{r}_N) = \exp[\tfrac{1}{2} \sum_{i<j} u(r_{ij})]$$

where we make contact with equation (6.2.13) by identifying

$$u(r_{ij}) = u_2(\mathbf{r}_i, \mathbf{r}_j) \tag{6.2.26}$$

The function $u(r)$ is required to become increasingly negative at small r in order that the potential energy expectation value be finite.

The estimate of the ground state energy is

$$E_J = \langle \psi_J | H | \psi_J \rangle / \langle \psi_J | \psi_J \rangle \tag{6.2.27}$$

and is an upper bound on the exact ground state energy:

$$E_J \geqslant E_0. \tag{6.2.28}$$

Equations (6.2.19) and (6.2.20) can be used to express the energy in terms of the radial distribution function $g_J(r)$:

$$\frac{E_J}{N} = \frac{\rho}{2} \int \mathrm{d}\mathbf{r} g_J(r) V^*(r) \tag{6.2.29}$$

where

$$V^*(r) = V(r) - \frac{\hbar^2}{4m} \nabla^2 u(r). \tag{6.2.30}$$

The second term is from the kinetic energy and therefore is a manifestation of zero-point motion. Since $u(r)$ must be increasingly large and negative as r decreases, $V^*(r)$ has a larger core than $V(r)$. Therefore $g_J(r)$ peaks at a larger distance that the minimum in $V(r)$ (see Figure 6.2.1 for a typical g_J, V and V^*).

The liquid structure function defined in equations (6.2.24) and (6.2.25) may also be written entirely in terms of g_J:

$$S(k) - 1 = \rho \int \mathrm{d}\mathbf{r} \exp(\mathrm{i}\mathbf{k} \cdot \mathbf{r})(g_J(r) - 1) \quad k \neq 0 \tag{6.2.31}$$

where the asymptotic value subtracted from g_J in the integrand contributes nothing at $k \neq 0$. Thus the problem of calculating with a Jastrow function reduces to the determination of the corresponding radial distribution function. It is at this point that the similarity between the square of the Jastrow function and the Boltzmann factor of a classical system opens the gates to an enormous amount of previous work on classical fluids. Specifically, the determination of $g_J(r)$ from $u(r)$ is mathematically equivalent to the determination of the radial distribution function $g_{\mathrm{Cl}}(r)$ for the classical fluid interacting via the fictitious potential $\phi_{\mathrm{Cl}}(r)$ at some arbitrary temperature T such that

$$u(r) = -\phi_{\mathrm{Cl}}(r)/k_B T \tag{6.2.32}$$

Then

$$g_J(r) = g_{\mathrm{Cl}}(r) \tag{6.2.33}$$

Approximations developed in the theory of classical fluids such as BBGKY–KSA, PY, and HNC generally have a closed functional form[86]

$$g_{\mathrm{Cl}}(r) = F[-\phi/k_B T, g_{\mathrm{Cl}}; r] \tag{6.2.34}$$

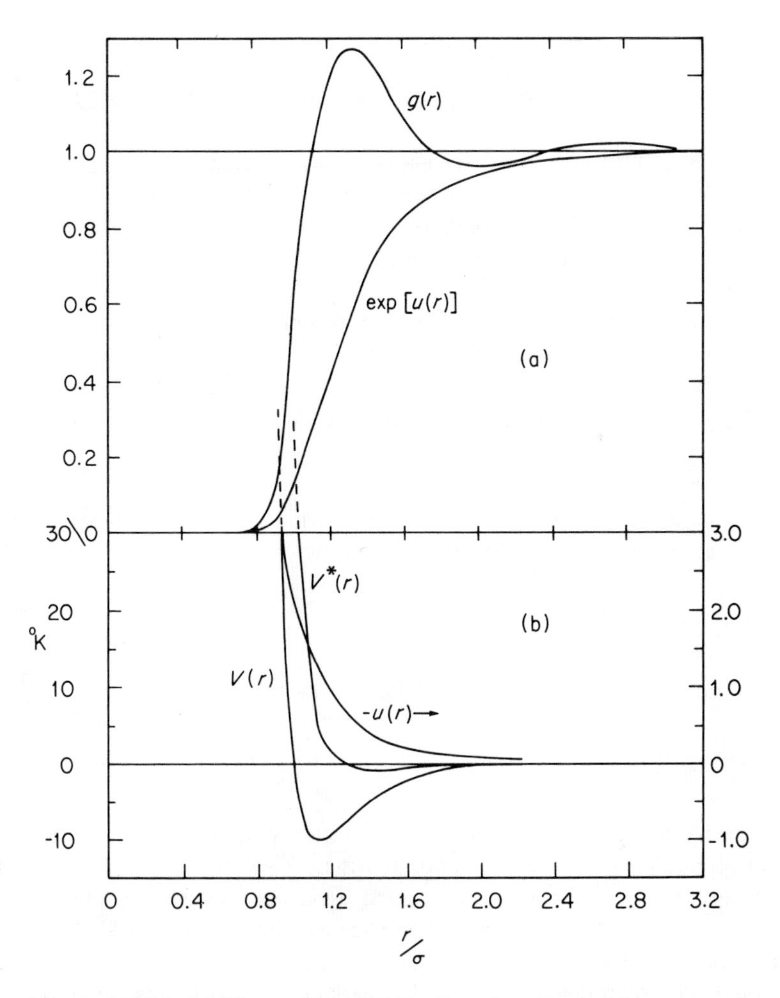

Figure 6.2.1 (a) The radial distribution function $g(r)$ compared with the Jastrow factor $\exp[u(r)]$ for the $u(r)$ of the form of equation (6.2.41) with parameters given in equation (6.2.42) obtained in a variational calculation for liquid ^{4}He at density $\rho = 0.0218$ Å^{-3}; (b) The Lennard–Jones potential $V(r)$ used in the variational calculation (equations 6.1.2 and 6.1.8) compared with $-u(r)$ and $V^*(r)$, equation (6.2.30)

which is then solved for g_{Cl}. Then the analogous approximation for the Jastrow functions is

$$g_{\mathrm{J}}(r) = F[u, g_{\mathrm{J}}; r]. \tag{6.2.35}$$

For example, the three most common approximations are[86]

$$g_{\mathrm{J}}(r) = \exp[u(r) + P(r)] \qquad \text{(HNC)} \tag{6.2.36}$$

$$g_{\mathrm{J}}(r) = [1 + P(r)]\exp[u(r)] \qquad \text{(PY)} \tag{6.2.37}$$

where

$$P(r) = \frac{1}{(2\pi)^3 \rho} \int \exp(\mathrm{i}\mathbf{k}\cdot\mathbf{r}) \frac{(S(k)-1)^2}{S(k)} \tag{6.2.38}$$

and

$$\nabla \ln g_J(r_{12}) = \nabla u(\mathbf{r}_{12}) + \rho \int d\mathbf{r}_3 g_J(\mathbf{r}_{13}) g_J(\mathbf{r}_{23}) \nabla u(\mathbf{r}_{13}) \quad \text{(BBGKY–KSA)} \tag{6.2.39}$$

Nothing has been said here yet about the choice of $u(r)$, except regarding the character of its short range behaviour. It is not possible, however, that the Jastrow function is an eigenstate of the Hamiltonian, at least for an ordinary two-body potential. The closest approach to the energy eigenvalue E_0 is obtained by minimizing E_J with respect to $u(r)$.

Parametrized Jastrow functions. We shall discuss a procedure for sampling the entire class of all real functions $u(r)$ in the variation, but first we discuss the large amount of work based upon simple parametrizations of $u(r)$.

Suppose we have a set of variational parameters $\{\xi_i\}$, such that

$$u(r) = u(r \mid \{\xi_i\}).$$

Then E_J is a function of $\{\xi_i\}$ through u. The variational procedure is to minimize E_J with respect to these parameters by solving the set of equations

$$\frac{\partial}{\partial \xi_i} E_J(\{\xi_i\}) = 0. \tag{6.2.40}$$

The earliest and most widely used form of $u(r)$ depends on two parameters, p and a:[19,66–68]

$$u(r \mid p, a) = -\left(\frac{a}{r}\right)^p. \tag{6.2.41}$$

The fact that this is a homogeneous function of r permits the use of a scaling procedure to carry out calculations simultaneously at a number of densities.[19] Except for the long range structure, this simple form has produced reasonably good agreement with experiment. It has been difficult to improve upon it much by more complicated parametrizations.[66,87] One surprising feature is that this $-u(r)$ is uniformly repulsive. The solution from the class of all Jastrow functions at the equilibrium density of ^4He is remarkably similar to this $u(r)$; in particular it is also monotonic. It might be expected *a priori* that $-u$ would have more structure, perhaps bearing some resemblance to classical fluid potentials, with a negative minimum near the nearest-neighbour distance. If not at equilibrium density, then perhaps such a minimum should develop at higher densities, particularly at densities approaching the liquid–solid phase transition; such a minimum would certainly aid localization of the particles and strengthen the maximum in $g(r)$. Arguing against the appearance of such a minimum, however, is the fact that it is the Laplacian acting on $u(r)$ which determines the kinetic energy, and increasing the structure in $u(r)$ will make the kinetic energy more positive. So it is difficult in advance to predict the behaviour of $u(r)$. The fact is that all calculations performed thus far with realistic potentials have shown a monotonic $u(r)$ at all densities including the lowest density of the solid. The only example of which we are aware giving a minimum in $-u(r)$ was Massey's BBGKY–KSA calculation based upon a 6–12 potential chosen to reproduce the binding energy and equilibrium density of ^4He (see discussion below).[64]

The calculation of E_J as a function of the variational parameters requires the determination of $g(r)$. The several methods for doing this fall into two categories: essentially exact calculations for 10^2–10^3 particles;[19,66,67] and approximate integral equations for infinite systems.[64–66,68,69]

The first of these procedures includes the use of Monte Carlo[19,66] and molecular dynamics[67] integration procedures to obtain $g(r)$ from $u(r)$ for a finite number of particles. The earliest of these calculations, done by McMillan for 32 particles using the simple trial form of equation (6.2.41),[19] has been shown to be surprisingly accurate by more extensive calculations with larger numbers of particles and more complicated trial functions u.[66,67] The alternative molecular dynamics calculation makes use of the aforementioned analogy of $u(r)$ to a classical potential $-\beta\phi_{Cl}(r)$.[67] Then $g(r)$ is calculated by following a point in phase space as a function of time whose motion is that of a classical particle at temperature $T = (k_B\beta)^{-1}$ and interaction $\phi_{Cl}(r)$. Use is made of the ergodic theorem to replace configuration averages with time averages, and the equivalence of the microcanonical ensemble to the canonical ensemble to show that the $g_{Cl}(r)$ thereby determined is equivalent to the $g(r)$ being calculated.

Comparisons between Monte Carlo and molecular dynamics calculations for the same parameters have shown that each produces the same information to the same accuracy for the same computing time. The errors in these calculations are (i) statistical, due to the finite number of configurations sampled, and (ii) surface effects due to the finite number of particles. The latter are reduced somewhat by using periodic boundary conditions to extend the finite system to an infinite volume. Thus the calculated density is uniform, whereas it would go to zero if hard wall boundary conditions were used. The uniform density is a feature of the infinite system and therefore a desired feature.

The molecular dynamics procedure has an inherent advantage for classical systems because of the ability to deduce time correlations from it. That advantage is lost on the quantum system because the propagation in time of the fictitious classical system has nothing to do with the quantum system. Furthermore, there is a practical disadvantage to the molecular dynamics system when compared to Monte Carlo, which arises from the fact that $u(r)$ is obtained by defining $\phi_{Cl}(r)$ and then establishing a velocity distribution which gives a temperature β. The temperature, however, is subject to fluctuations in the microcanonical ensemble. Consequently there is an uncertainty in what $u(r)$ actually is. There is no comparable uncertainty in the Monte Carlo calculation.

The energy as a function of density is shown in Figure 6.2.2. Figure 6.2.1(b) has a comparison between $V(r)$, $V^*(r)$ and $-u(r)$ for the optimum parameters at equilibrium density. The potential used in these calculations is the 6–12 potential, equation (6.1.3), with the de Boer–Michels parameters, equation (6.1.8). The parameters in the $u(r)$ of equation (6.2.41) which minimize the energy are

$$p = 5 \quad a = 2.965\ \text{Å}^{-1}. \tag{6.2.42}$$

The value of p is not surprising in light of the use of a 6–12 potential. The very short range behaviour of $\exp[\tfrac{1}{2}u(r)]$ should be the same as an eigenfunction obtained from the two-body Schrödinger equation with the same potential. Thus the r^{-12} behaviour forces the r^{-5} behaviour of $u(r)$.

The functions $\exp[-u(r)]$ and $g(r)$ are compared in Figure 6.2.1(a). The liquid structure function $S(k)$ obtained from $g(r)$ is compared with the low temperature experimental results of Hallock[62] and of Achter and Meyer[61] in Figure 6.2.3.

There are other two-body potentials than the Lennard–Jones potential used here which have been proposed for helium.[45] Murphy has investigated a number of these potentials by the Monte Carlo method.[88] By using the variational principle, equation (6.2.5), he was able to set limits on the parameters in some model potentials so that E_J would not be below the experimental energy at any measured density.

The uncertainties in the above results due to statistics and finite size effects can be avoided by going to approximate relations between $u(r)$ and $g(r)$, such as BBGKY–KSA, HNC and PY (equations 6.2.35–38). Introduction of these approximations, however, invalidates the

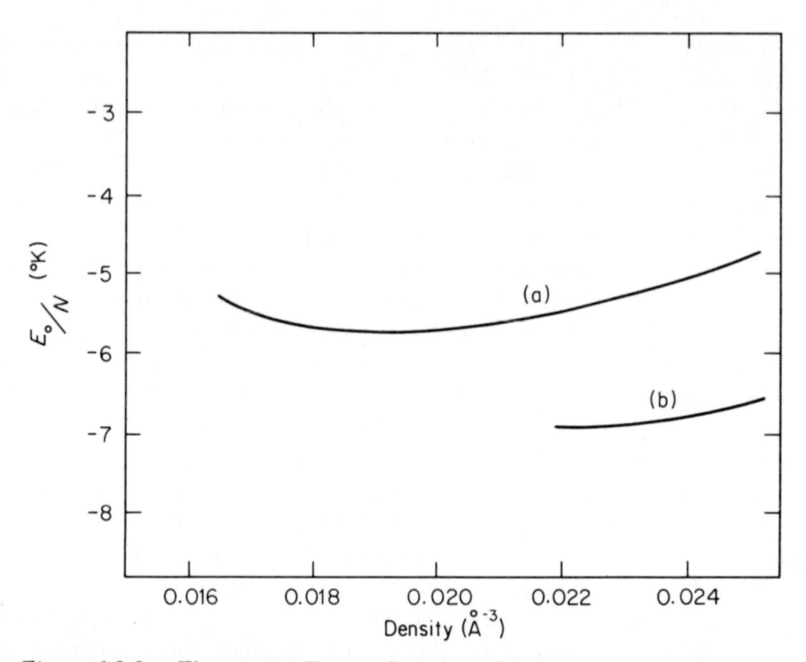

Figure 6.2.2 The energy E_0 as a function of density ρ: (a) obtained by minimizing the expectation value of the ^{4}He Hamiltonian H with the Lennard–Jones potential (equations 6.1.2 and 6.1.8) in the parametrized Jastrow function (equations 6.2.41 and 6.2.42), and (b) obtained from experiment. (From Murphy and Watts,[68] reproduced by permission of Plenum Publishing Corporation)

variational principle. It is necessary, nevertheless, to have some such principle to replace the Schrödinger equation in order that the best $u(r)$ may be chosen, and the variational principle is the only candidate. Thus the energy in the form of equation (6.2.29) is minimized as a function of $u(r)$, where $g(r)$ is given in terms of $u(r)$ by solving one of the equations (6.2.35–38) depending on the approximation chosen. This will be a useful procedure as long as the optimum choice of $u(r)$ is one for which the approximation for $g(r)$ is reasonable.

All three of these approximations are highly nonlinear integral equations for $g(r)$ as a function of $u(r)$. It is frequently more convenient to use these same equations as expressions for $u(r)$ as a function of $g(r)$. Then HNC and PY are simple integral expressions for $u(r)$, and BBGKY–KSA becomes a linear integral equation for $\nabla u(r)$. Furthermore, the variation of $u(r)$ may be replaced equivalently by the variation of $g(r)$.

The first variational procedure of this sort was done by Massey[64,47] at about the same time as McMillan's Monte Carlo calculation,[19] and the results are quite similar. Massey developed a procedure for solving the BBGKY–KSA equation for $u(r)$ from a parametrized $g(r)$, after which the energy is minimized as a function of the parameters in $g(r)$. His procedure is an iterative method taking advantage of the Feenberg–Wu extremum principle.[89,47] Since this extremum principle is of wider application than the present context, it is worthwhile deriving it here.

The Feenberg–Wu functional is defined by[89,47]

$$J[g, p^{(3)} \mid h] = -\int g(r_{12})h(r_{12})^2 \,\mathrm{d}\mathbf{r}_2 - 2\int g'(r_{12})h(r_{12})\,\mathrm{d}\mathbf{r}_2 - \frac{1}{\rho^2}\int p^{(3)}(\mathbf{r}_1, \mathbf{r}_2, \mathbf{r}_3)h(r_{12})h(r_{13})\frac{\mathbf{r}_{12}\cdot\mathbf{r}_{13}}{r_{12}r_{13}}\,\mathrm{d}\mathbf{r}_2\,\mathrm{d}\mathbf{r}_3 \tag{6.2.43}$$

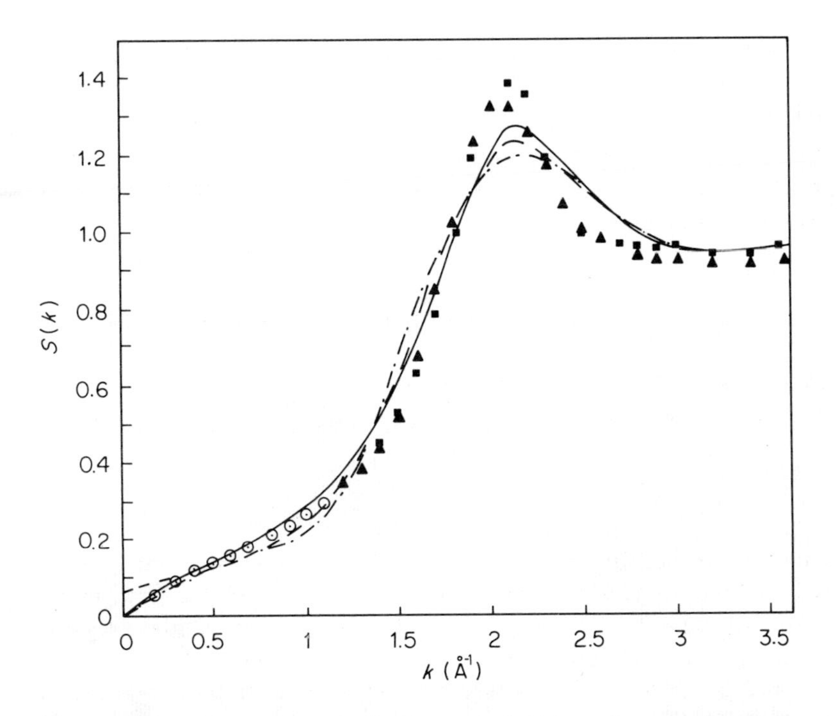

Figure 6.2.3 Comparison of the calculated and experimentally determined liquid structure, $S(k)$, at $\rho = 0.0218$ Å^{-3}: ▲ X-ray scattering at $T = 0.79$ °K;[61]. X-ray scattering at $T = 0.38$ °K;[62] ■ Neutron scattering at $T = 1.1$ °K;[115] ––– Parametrized Jastrow calculation for $g(r)$ of Figure 6.2.1; – · – · – Parametrized Jastrow calculation of Massey and Woo;[65] —— Results of the paired-phonon analysis

where $p^{(3)}$ is a three-body function, symmetric under interchange of its arguments (eventually to be identified as the three-particle distribution function or some approximation thereto). For a given g and $p^{(3)}$, the function h which extremizes this form (which is a quadratic functional of h) is the solution of an integral equation obtained by setting $\delta J = 0$, where

$$\begin{aligned}\delta J &= J[g, p^{(3)} \mid h + \delta h] - J[g, p^{(3)} \mid h]\\ &= -2 \int [g(r_{12})h(r_{12}) + g'(r_{12})]\, \delta h(r_{12}) d\mathbf{r}_2\\ &\quad - \frac{2}{\rho^2} \int p^{(3)}(\mathbf{r}_1, \mathbf{r}_2, \mathbf{r}_3) h(r_{13}) \frac{\mathbf{r}_{12} \cdot \mathbf{r}_{13}}{r_{12} r_{13}} d\mathbf{r}_3 \delta h(r_{12}) d\mathbf{r}_2 .\end{aligned} \tag{6.2.44}$$

Assuming further that $p^{(3)}$ depends only on the coordinate differences, the extremum condition produces the integral equation for h:

$$\begin{aligned}0 &= g(r_{12})h(r_{12}) + g'(r_{12})\\ &\quad + \frac{1}{\rho^2} \int p^{(3)}(\mathbf{r}_1, \mathbf{r}_2, \mathbf{r}_3) h(r_{12}) \frac{\mathbf{r}_{12} \cdot \mathbf{r}_{13}}{r_{12} r_{13}} d\mathbf{r}_3 .\end{aligned} \tag{6.2.45}$$

This is the BBGKY equation if $p^{(3)}$ and g are identified as the three-particle distribution function $\rho^3 g_3$ and the radial distribution, respectively, and h is defined by

$$h(r) = u'(r). \tag{6.2.46}$$

The extremum principle is valid independent of the choice of $p^{(3)}$. In particular, it applies to the BBGKY–KSA approximation, equation (6.2.39) where the Kirkwood superposition approximation is made for g_3:

$$g_3(\mathbf{r}_1, \mathbf{r}_2, \mathbf{r}_3) = g(r_{12})g(r_{23})g(r_{31}). \tag{6.2.47}$$

The extremum value of J is obtained by substituting the solution of (6.2.45) into (6.2.43):

$$J_{\text{ext}} = J[g, p^{(3)} \mid h^{\text{ext}}] = -\int g'(r)h^{\text{ext}}(r)\,\mathrm{d}\mathbf{r}. \tag{6.2.48}$$

Using u' in place of h^{ext}, this is just proportional to the kinetic energy per particle:

$$\frac{\langle \text{K.E.} \rangle}{\text{N}} = \frac{\hbar^2 \rho}{8m} J_{\text{ext}}. \tag{6.2.49}$$

Although the second functional derivative of J with respect to h can be either positive or negative due to the cosine which appears in the last term, experience shows that the extremum is always a maximum. This is presumably because the first term is always negative and tips the balance to a negative second derivative.

Massey used the Feenberg–Wu extremum principle to hasten the convergence of an iterative solution of the BBGKY–KSA approximation.[64,47] The variational parameters are included in $g(r)$. He found it necessary to have seven free parameters in $g(r)$ so that it has a nearest neighbour maximum followed by a minimum. That amount of structure in $g(r)$ is necessary to produce a reasonable behaviour in $S(k)$.

Massey carried out this calculation for a Lennard–Jones 6–12 potential of the form (6.1.3) but with parameters ϵ and σ chosen to produce the experimental binding energy at the experimental equilibrium density of liquid ^{4}He.[64] The calculation was redone by Massey and Woo using the De Boer–Michels parameters, equation (6.1.8).[65] Their energy–vs. density curve is quite close to the Monte Carlo and molecular dynamics results, with a minimum energy per particle of –6.06 °K at the calculated equilibrium density 0.0229 Å^{-3}.

It should be noted, however, that in spite of the complicated parametrization of $g(r)$, their $u(r)$ is always found to be monotonic at the parameter values which minimize the energy.[65] This partly overcomes the advantages of solving the BBGKY–KSA for $u(r)$ from the given $g(r)$ in comparison to solving it for $g(r)$ given a simply parametrized $u(r)$ such as equation (6.2.41). While the former procedure is much faster for finding the (u,g) pair the parameter space is so much larger that the overall effort is comparable to the Monte Carlo or molecular dynamics calculation.

Other approximations such as HNC and PY may be used in a similar fashion.[66,68] The usual procedure is to use the simple parametrization of $u(r)$ equation (6.2.41). The results seem to be characterized by an HNC energy which is above the Monte Carlo energy, a PY energy which is below the Monte Carlo energy, and a BBGKY energy which is very comparable:

$$E_{\text{HNC}} > E_{\text{MC}} \approx E_{\text{BBGKY}} > E_{\text{PY}}$$

Table 6.2.1 compares the energies obtained for the various approximation to $u(r)$ for Massey and Woo's $g(r)$ at $\rho = 0.0218$ Å^{-3}. The different u are shown in Figure 6.2.4. For further comparisons see References 90, 68, and 9. The discrepancy between the calculated energies increases as the density increases. Since the energy is composed of two terms which largely cancel one another, it is a more sensitive function of the approximation than the structure.

Table 6.2.1 Kinetic energy per particle T/N for a fixed $g(r)$ and several approximations for $u(r)$ (see Figure 6.2.4 for $u(r)$)

Approximation	T/N (°K)
BBGKY–KSA	14.39
HNC	16.87
PY	14.00
Molecular dynamics	14.0 ± 0.1

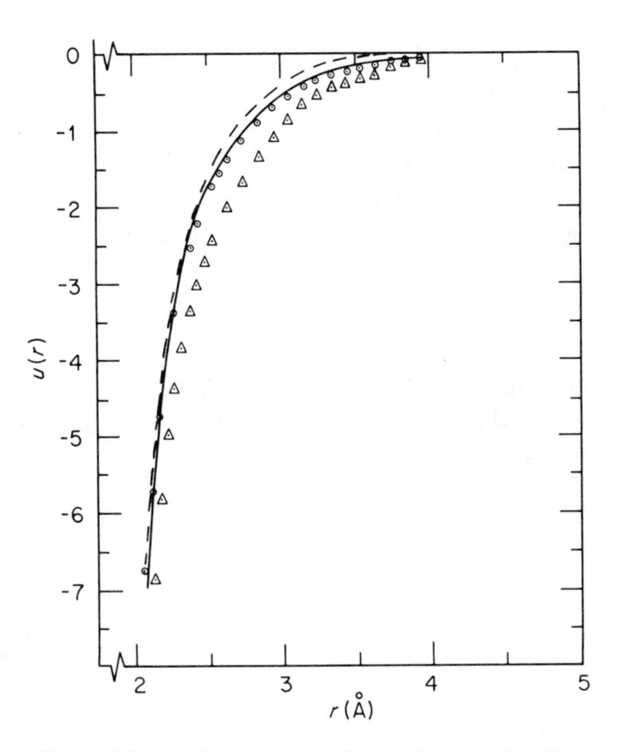

Figure 6.2.4 Comparison of several approximations for the Jastrow function $u(r)$ from the same radial distribution at $\rho = 0.0218$ Å^{-3}: —— exact (by molecular dynamics); ––– BBGKY with the Kirkwood superposition approximation; ● Percus–Yevick approximation; △ hypernetted chain approximation

Advantage has been taken of the similarity between HNC and PY calculations to hybridize the two into approximations in better agreement with the Monte Carlo calculations.[90,91]

Improvements upon HNC and PY are available. In particular, these approximations can be obtained as the first term of functional expansions, and perhaps improved by keeping higher terms.[86] This leads to the PY II and HNC II approximation. The results of calculations with the PY II approximation are very similar to the Monte Carlo results, but are expensive to do because of the complexity of the equation.[66]

Functional variations. Although the parameter variations described above have led to a reasonably good representation of the experimental energy and structure function, improvements may be expected if a wider space of variational functions can be tested. This can be accomplished in two steps: find the best Jastrow function from amongst all possible Jastrow functions (not just those limited by a specific parametrization); and expand the class of trial functions to include non-Jastrow functions.

The variational problem within the Jastrow function space is to solve the equation

$$\frac{\delta}{\delta u(r)} \frac{\langle \psi_J | H | \psi_J \rangle}{\langle \psi_J | \psi_J \rangle} = 0 \tag{6.2.50}$$

where ψ_J is the Jastrow function defined in equation (6.1.25). Using (6.2.29) and (6.2.30) this gives the equation

$$\frac{\hbar^2}{4m} \nabla^2 g(r) = \int V^*(\mathbf{r}') \frac{\delta g(\mathbf{r}')}{\delta u(\mathbf{r})} \, d\mathbf{r}'. \tag{6.2.51}$$

Using the definition of $g(r)$ in terms of $u(r)$ given by (6.2.16) and (6.2.18) with ψ_J replacing Ψ_0 to calculate the functional derivative gives[82]

$$\frac{\hbar^2}{4m} \nabla^2 g(r_{12}) = g(r_{12}) V^*(r_{12}) + 2\rho \int V^*(r_{31}) g_3(\mathbf{r}_1, \mathbf{r}_2, \mathbf{r}_3) d\mathbf{r}_3$$
$$+ \frac{\rho^2}{2} \int V^*(r_{34}) [g_4(\mathbf{r}_1 \mathbf{r}_2 \mathbf{r}_3 \mathbf{r}_4) - g(\mathbf{r}_{12}) g(\mathbf{r}_{34})] \, d\mathbf{r}_3 d\mathbf{r}_4 . \tag{6.2.52}$$

A convenient reformulation of this equation is obtained by defining the generalized normalization integral $I(\alpha)$ which may be used to generate the energy expectation value:[70]

$$I(\alpha) = \langle \psi(\alpha) | \psi(\alpha) \rangle \tag{6.2.53}$$

where

$$\psi(\alpha) = \psi \prod_{i<j}^{N} \exp[\alpha V^*(r_{ij})/2] \tag{6.2.54}$$

Then

$$E = \frac{d}{d\alpha} \ln I(\alpha) \bigg|_{\alpha=0} \tag{6.2.55}$$

Making use of these definitions the Fourier transform of equation (6.2.52) brings the variational equation to the form[52]

$$\frac{\hbar^2 k^2}{4m} \left[1 - S(k) \right] = \frac{d}{d\alpha} S(k | \alpha) \bigg|_{\alpha=0} \equiv \mathcal{H}(k) \tag{6.2.56}$$

where $S(k | \alpha)$ is obtained from equation (6.2.24) with $\psi(\alpha)$ in place of Ψ_0. Consequently,

$$\mathcal{H}(k) = \frac{1}{N} \frac{\left\langle \psi \left| \rho_{\mathbf{k}} \rho_{-\mathbf{k}} \left[\sum_{i<j}^{N} V^*(r_{ij}) - E \right] \right| \psi \right\rangle}{\langle \psi | \psi \rangle} \tag{6.2.57}$$

To solve these equations, we begin with some Jastrow function ψ_0 which does not satisfy

(6.2.56) and then we try to improve ψ_0 by an iterative procedure until this variational extremum condition is satisfied.[70] Then

$$\psi_0 = \prod_{i<j}^{N} \exp[u_0(r_{ij})/2] \tag{6.2.58}$$

and we are looking for

$$\psi = \prod_{i<j}^{N} \exp[u(r_{ij})/2] \tag{6.2.59}$$

which satisfies (6.2.56), where

$$u(r) = u_0(r) + \Delta u(r). \tag{6.2.60}$$

Thus

$$\psi = \psi_0 \exp\left[\frac{1}{2} \sum_{i<j}^{N} \Delta u(r_{ij})\right]. \tag{6.2.61}$$

The beginning function $u_0(r)$ should be the best one available from previous variational calculations, such as those discussed in the last section.

Defining the Fourier transform of $\Delta u(r)$ by

$$\Delta u(r) = \frac{1}{N} \sum_{\mathbf{k}} c_{\mathbf{k}} \exp(i\mathbf{k} \cdot \mathbf{r}) \tag{6.2.62}$$

the function in the exponential in (6.2.61) is

$$\begin{aligned} \sum_{i<j}^{N} \Delta u(r_{ij}) &= \frac{1}{2} \sum_{\mathbf{k}} c_{\mathbf{k}} \left(\frac{\rho_{\mathbf{k}}\rho_{-\mathbf{k}}}{N} - 1\right) \\ &= \sum_{\mathbf{k},\mathbf{k}_x>0} c_{\mathbf{k}} \left(\frac{\rho_{\mathbf{k}}\rho_{-\mathbf{k}}}{N} - 1\right) \end{aligned} \tag{6.2.63}$$

where $\rho_{\mathbf{k}}$ is the density fluctuation operator defined in equation (6.1.29). The second line is obtained by using the fact that Δu is real, and $\rho_{\mathbf{k}}^* = \rho_{-\mathbf{k}}$; the restriction $\mathbf{k}_x > 0$ means that when $\mathbf{k}$ is included in the sum, $-\mathbf{k}$ is not. Then our trial function ψ takes the form

$$\psi = \psi_0 \prod_{\mathbf{k},\mathbf{k}_x>0} \exp\left[\frac{1}{2} c_{\mathbf{k}} \left(\frac{\rho_{\mathbf{k}}\rho_{-\mathbf{k}}}{N} - 1\right)\right] \tag{6.2.64}$$

and our objective is now to minimize the energy with respect to the function $c_{\mathbf{k}}$. Before solving that problem, note that the present formulation permits the simplest expression of the variational conditions (6.2.56), which must be equivalent to stability with respect to an infinitesimal variation in $c_{\mathbf{k}}$ That is, (6.2.56) is equivalent to

$$\begin{aligned} 0 &= \frac{\partial}{\partial c_{\mathbf{k}}} \left[\frac{\langle \psi | H | \psi \rangle}{\langle \psi | \psi \rangle}\right] \\ &= \frac{\langle \psi | \rho_{\mathbf{k}}\rho_{-\mathbf{k}}(H - E) | \psi \rangle}{\langle \psi | \psi \rangle} + \frac{\langle \psi | (H - E)\rho_{\mathbf{k}}\rho_{-\mathbf{k}} | \psi \rangle}{\langle \psi | \psi \rangle} \end{aligned} \tag{6.2.65}$$

Reality permits this to be rewritten as

$$\frac{\langle \psi \mid (H-E)\rho_{\mathbf{k}}\rho_{-\mathbf{k}} \mid \psi \rangle}{\langle \psi \mid \psi \rangle} = 0. \tag{6.2.66}$$

That this is equivalent to (6.2.56) is established by calculating the commutator of the Hamiltonian with $\rho_{\mathbf{k}}\rho_{-\mathbf{k}}$.

To calculate the energy as a function of $c_{\mathbf{k}}$, note that V^* may be written as

$$V^*(r) = V_0^*(r) - \frac{\hbar^2}{4m} \nabla^2 \Delta u(r) \tag{6.2.67}$$

where

$$V_0^*(r) = V(r) - \frac{\hbar^2}{4m} \nabla^2 u_0(r). \tag{6.2.68}$$

Then the generalized normalization integral becomes

$$I(\alpha) = I_0(\alpha)\left(\prod_{\mathbf{k},k_x>0} \exp[-c_{\mathbf{k}}(\alpha)]\right) \frac{\left\langle \psi_0(\alpha) \middle| \prod_{\mathbf{k},k_x>0} \exp[c_{\mathbf{k}}(\alpha)\rho_{\mathbf{k}}\rho_{-\mathbf{k}}/N] \middle| \psi_0(\alpha) \right\rangle}{\langle \psi_0(\alpha) \mid \psi_0(\alpha) \rangle} \tag{6.2.69}$$

where $I_0(\alpha)$ and $\psi_0(\alpha)$ are given by equations (6.2.53) and (6.2.54), respectively, with ψ_0 replacing ψ and

$$c_{\mathbf{k}}(\alpha) = \left(1 + \alpha \frac{\hbar^2 k^2}{4m}\right) c_{\mathbf{k}}. \tag{6.2.70}$$

Density fluctuation cumulant analysis and a linked cluster expansion[92]. To evaluate $I(\alpha)$ it is necessary to obtain the last factor on the right-hand side of equation (6.2.69), which has the form

$$J(y) = \left\langle \prod_{\mathbf{k},k_x>0} \exp(y_{\mathbf{k}}\rho_{\mathbf{k}}\rho_{-\mathbf{k}}) \right\rangle_0 \tag{6.2.71}$$

where $\langle \ldots \rangle_0$ refers to a normalized expectation value (in the present case with respect to ψ_0 or $\psi_0(\alpha)$). $J(y)$ may be evaluated as a linked cluster expansion in terms of $y_{\mathbf{k}}$ and the cumulants of the density fluctuation operators. To define these cumulants, first define the average value of a product of $\rho_{\mathbf{k}}$ by

$$I_n(\mathbf{k}_1 \cdots \mathbf{k}_n) = \left\langle \prod_{i=1}^{n} \rho_{\mathbf{k}_i} \right\rangle_0 \tag{6.2.72}$$

Then the n-th-cumulant, $U_n^0(\mathbf{k}_1 \ldots \mathbf{k}_n)$ is defined by the first n equations of the following sequence:[55]

$$\begin{aligned} I_1(\mathbf{k}_1) &= U_1^0(\mathbf{k}_1) \\ I_2(\mathbf{k}_1\,\mathbf{k}_2) &= U_1^0(\mathbf{k}_1)\, U_1^0(\mathbf{k}_2) + U_2^0(\mathbf{k}_1, \mathbf{k}_2) \\ I_3(\mathbf{k}_1\,\mathbf{k}_2\,\mathbf{k}_3) &= U_1^0(\mathbf{k}_1)U_1^0(\mathbf{k}_2)U_1^0(\mathbf{k}_3) + U_1^0(\mathbf{k}_1)U_2^0(\mathbf{k}_2\,\mathbf{k}_3) + U_1^0(\mathbf{k}_2)U_2^0(\mathbf{k}_1\,\mathbf{k}_3) \\ &\quad + U_1^0(\mathbf{k}_3)U_2^0(\mathbf{k}_1\,\mathbf{k}_2) + U_3^0(\mathbf{k}_1\mathbf{k}_2\mathbf{k}_3) \end{aligned} \tag{6.2.73}$$

For a uniform system, I_n and U_n^0 are zero unless the sum of their momenta is zero. Then, for example,

$$U_1^{(0)}(\mathbf{k}) = N\delta_{\mathbf{k},0} \tag{6.2.74}$$

and

$$U_2^0(\mathbf{k}_1, \mathbf{k}_2) = N\delta_{\mathbf{k}_1, -\mathbf{k}_2} S_0(\mathbf{k}_1) \tag{6.2.75}$$

where $S_0(k)$ is the liquid structure function for the distribution over which the average is calculated:

$$S_0(\mathbf{k}) = \langle \rho_{\mathbf{k}} \rho_{-\mathbf{k}} \rangle_0 / N \tag{6.2.76}$$

One of the main advantages of a cumulant decomposition is that it permits the explicit identification of the order of magnitude in the number of particles, N. Wu showed that the cumulants U_n^0 are of order N; thus the leading term in powers of N for I_n will be the term with the largest number of factors U_i^0.[55] Wu's demonstration that the U's are extensive is based upon an analysis of the n-particle distribution function, defined as in equation (6.2.16):

$$g_n(\mathbf{r}_1 \ldots \mathbf{r}_n) = \frac{N!}{(N-n)!} \rho^{-n} \int W_N(\mathbf{r}_1 \ldots \mathbf{r}_N)\, \mathrm{d}\mathbf{r}_{n+1} \ldots \mathrm{d}\mathbf{r}_N \tag{6.2.77}$$

where W_N is the N-body probability density over which the averages are calculated:

$$\langle A \rangle_0 = \int A W_N(\mathbf{r}_1 \ldots \mathbf{r}_N)\, \mathrm{d}\mathbf{r}_1 \ldots \mathrm{d}\mathbf{r}_N \tag{6.2.78}$$

In the present case W_N is just the normalized square of a wave function. An infinite homogeneous and isotropic system such as a liquid is expected to have correlations which are short-ranged for separation of subsets of the n coordinates. Thus we assume the limiting behaviour

$$g_n(\mathbf{r}_1 \ldots, \mathbf{r}_p, \mathbf{r}_{p+1}, \ldots, \mathbf{r}_n) \longrightarrow g_p(\mathbf{r}_1 \ldots \mathbf{r}_p) g_n(\mathbf{r}_{p+1} \ldots \mathbf{r}_n) \tag{6.2.79}$$

as $r_{ij} \to \infty$ for every i and j satisfying

$$1 \leqslant i \leqslant p < j \leqslant n.$$

This suggests an additive cluster decomposition whose formal structure is that of the Ursell–Mayer cluster expansion in classical statistical mechanics. The cluster functions f_n are defined as follows:

$$\begin{aligned}
g_1(\mathbf{r}_1) &= f_1(\mathbf{r}_1) \\
g_2(\mathbf{r}_1\, \mathbf{r}_2) &= f_1(\mathbf{r}_1) f_1(\mathbf{r}_2) + f_2(\mathbf{r}_1\, \mathbf{r}_2) \\
g_3(\mathbf{r}_1\, \mathbf{r}_2\, \mathbf{r}_3) &= f_1(\mathbf{r}_1) f_1(\mathbf{r}_2) f_1(\mathbf{r}_3) + f_1(\mathbf{r}_1) f_2(\mathbf{r}_2, \mathbf{r}_3) + f_1(\mathbf{r}_2) f_2(\mathbf{r}_1\, \mathbf{r}_3) \\
&\quad + f_1(\mathbf{r}_3) f_2(\mathbf{r}_1\, \mathbf{r}_2) + f_3(\mathbf{r}_1\, \mathbf{r}_2\, \mathbf{r}_3) \\
&\quad \vdots
\end{aligned} \tag{6.2.80}$$

The cluster condition (6.2.79) implies that the f functions are short range functions under separation of subsets of the arguments. In particular,

$$\lim_{r_{ij} \to \infty} f_n(\mathbf{r}_1 \ldots \mathbf{r}_n) = 0 \quad 1 \leqslant i < j \leqslant n. \tag{6.2.81}$$

It is this short-ranged property of f_n which leads to the conclusion that $U_p^0 = O(N)$.[55] To show this, it is necessary to express the U's in terms of the f's. Toward that end, we need the Fourier transforms of f and g:

$$F_n(\mathbf{k}_1 \ldots \mathbf{k}_n) = \rho^n \int f_n(\mathbf{r}_1 \ldots \mathbf{r}_n) \prod_{i=1}^{n} \exp(i\mathbf{k}_i \cdot \mathbf{r}_i)\, d\mathbf{r}_1 \ldots d\mathbf{r}_n \tag{6.2.82}$$

$$G_n(\mathbf{k}_1 \ldots \mathbf{k}_n) = \rho^n \int g_n(\mathbf{r}_1 \ldots \mathbf{r}_n) \prod_{i=1}^{n} \exp(i\mathbf{k}_i \cdot \mathbf{r}_i)\, d\mathbf{r}_1 \ldots d\mathbf{r}_n. \tag{6.2.83}$$

Note that the short-ranged nature of f_n implies that

$$F_n = O(N) \tag{6.2.84}$$

since the only integral in the integration in equation (6.2.82) which is free to range over the entire volume is the 'centre of mass' integration, or, equivalently, the last integration done.

The functions I_n can be expressed in terms of the G by using the definition (6.1.29) of $\rho_\mathbf{k}$ in equation (6.2.72) for I_n. The result is[55]

$$\begin{aligned} I_1(\mathbf{k}_1) &= G_1(\mathbf{k}_1) \\ I_2(\mathbf{k}_1\, \mathbf{k}_2) &= G_1(\mathbf{k}_1 + \mathbf{k}_2) + G_2(\mathbf{k}_1, \mathbf{k}_2) \\ I_3(\mathbf{k}_1\, \mathbf{k}_2\, \mathbf{k}_3) &= G_1(\mathbf{k}_1 + \mathbf{k}_2 + \mathbf{k}_3) + G_2(\mathbf{k}_1, \mathbf{k}_2 + \mathbf{k}_3) + G_2(\mathbf{k}_2, \mathbf{k}_1 + \mathbf{k}_3) \\ &\quad + G_2(\mathbf{k}_3, \mathbf{k}_1 + \mathbf{k}_2) + G_3(\mathbf{k}_1, \mathbf{k}_2, \mathbf{k}_3) \\ &\vdots \\ I_n(\mathbf{k}_1 \ldots \mathbf{k}_n) &= \Sigma\, G_l\,(\text{all distinct partitions of the } \mathbf{k}). \end{aligned} \tag{6.2.85}$$

To complete the proof, a short-hand notation is introduced for the two different decompositions of I_n. Equation (6.2.73) becomes

$$I = \Lambda U^0 \tag{6.2.86}$$

and equation (6.2.85) becomes

$$I = \Gamma G. \tag{6.2.87}$$

Note that the formal relationship between I and U^0 is the same as that between g and f:

$$g = \Lambda f \tag{6.2.88}$$

and, consequently,

$$G = \Lambda F. \tag{6.2.89}$$

Then equation (6.2.87) becomes

$$I = \Gamma \Lambda F. \tag{6.2.90}$$

We now use the identity

$$\Gamma \Lambda = \Lambda \Gamma \tag{6.2.91}$$

which is readily verified by noting that every term in $\Gamma\Lambda F$ is in $\Lambda\Gamma F$ and *vice versa.* Consequently, the two expressions for I can be equated to give

$$\Lambda U^0 = \Lambda \Gamma F. \tag{6.2.92}$$

Then

$$U^0 = \Gamma F \tag{6.2.93}$$

that is,

$$
\begin{aligned}
U_1^0(\mathbf{k}_1) &= F_1(\mathbf{k}_1)\\
U_2^0(\mathbf{k}_1\,\mathbf{k}_2) &= F_1(\mathbf{k}_1+\mathbf{k}_2)+F_2(\mathbf{k}_1,\mathbf{k}_2)\\
U_3^0(\mathbf{k}_1\,\mathbf{k}_2\,\mathbf{k}_3) &= F_1(\mathbf{k}_1+\mathbf{k}_2+\mathbf{k}_3)+F_2(\mathbf{k}_1,\mathbf{k}_2+\mathbf{k}_3)+F_2(\mathbf{k}_2,\mathbf{k}_3+\mathbf{k}_1)\\
&\quad +F_2(\mathbf{k}_3,\mathbf{k}_1+\mathbf{k}_2)+F_3(\mathbf{k}_1,\mathbf{k}_2,\mathbf{k}_3)\\
&\ \vdots
\end{aligned}
\tag{6.2.94}
$$

Thus U_n^0 is a linear combination of F_i, $1 \leqslant i \leqslant n$, and is therefore $O(N)$.

Returning now to the original task, which is the evaluation of $J(y)$ (equation 6.2.71), we have shown elsewhere that $\ln J(y)$ can be written as a linked cluster series in the cumulant functions U_n^0:[92]

$$
\ln J(y) = \sum_p \frac{1}{p!} \sum_{j=1}^{p} \left(\sum_{\mathbf{k}_j, \mathbf{k}_{jx} > 0} y_{\mathbf{k}_j} \right) \sum_{\substack{n_1 \leqslant n_2 \ldots \leqslant n_l \\ \Sigma n_\alpha = 2p}} [\prod_\alpha U_{n_\alpha}^0]_c\{(\mathbf{k}_1, -\mathbf{k}_1), \ldots, (\mathbf{k}_p, -\mathbf{k}_p)\}
\tag{6.2.95}
$$

where we have introduced a notation to represent a connected product of U:

$$
[\prod_j U_{n_j}^0]_c\{(\mathbf{h}_1 \ldots \mathbf{h}_{p_1})(\mathbf{k}_1 \ldots \mathbf{k}_{p_2}) \ldots (\mathbf{l}_1 \ldots \mathbf{l}_{p_n})\}
\tag{6.2.96}
$$

represents the sum over all distinct partitions of the momenta in braces into arguments of the $U_{n_j}^0$ factors in such a way that the product over j is connected by subsets of momenta in parentheses. To illustrate,

$$
\begin{aligned}
&[u_2^0 U_4^0]_c\{(\mathbf{h}_1^{(1)}\mathbf{h}_2^{(1)}), (\mathbf{h}_1^{(2)}\mathbf{h}_2^{(2)}), (\mathbf{h}_1^{(3)}\mathbf{h}_2^{(3)})\}\\
&= \sum_{1 \leqslant i < j \leqslant 3} \sum_{\alpha_i \neq \beta_i} \sum_{\alpha_j \neq \beta_j} U_2^0(\mathbf{h}_{\alpha_i}^{(i)} \mathbf{h}_{\alpha_j}^{(j)}) U_4^0(\mathbf{h}_{\beta_j}^{(j)} \mathbf{h}_{\beta_i}^{(i)} \mathbf{h}_1^{(k)} \mathbf{h}_2^{(k)})
\end{aligned}
\tag{6.2.97}
$$

where the horizontal brackets aid the eye in observing that all terms are connected. Note that a disconnected term which is absent from this expression has the form

$$
U_2^0(\mathbf{h}_1^{(i)} \mathbf{h}_2^{(i)}) U_4^0(\mathbf{h}_1^{(j)} \mathbf{h}_2^{(j)} \mathbf{h}_1^{(k)} \mathbf{h}_2^{(k)})
\tag{6.2.98}
$$

To simplify the display of this expression, a diagrammatic representation is defined in Figure 6.2.5; a dashed line represents a momentum, $y_\mathbf{k}$ is represented by a dot connected to

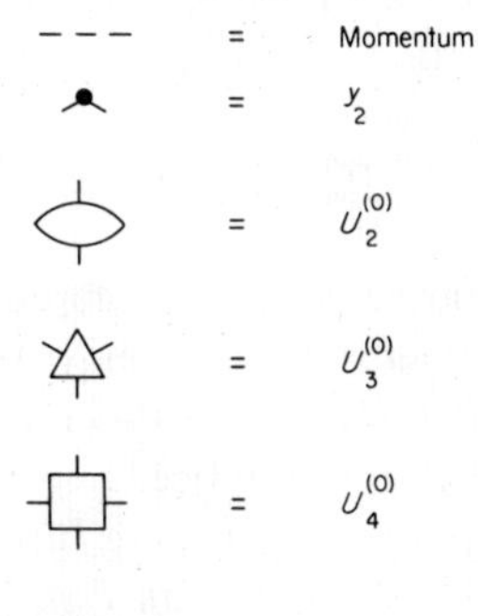

Figure 6.2.5 The elements of a cumulant expansion. (From Campbell,[92] reproduced by permission of the American Institute of Physics)

Momentum
y_2
$U_2^{(0)}$ $U_3^{(0)}$ $U_4^{(0)}$

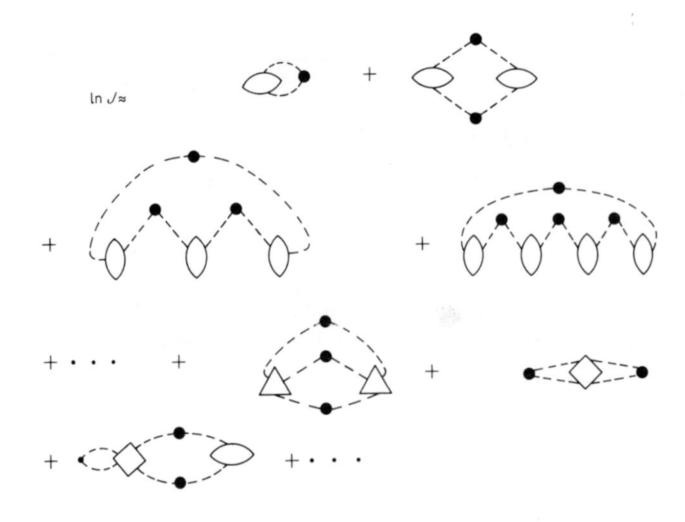

Figure 6.2.6 A schematic expansion of the logarithm of the normalization integral. Momentum must be assigned to the dashed lines and summed over. Appropriate combinational factors must be assigned to each term. (See discussion in text.) (From Campbell,[92] reproduced by permission of the American Institution of Physics)

two momenta $(\mathbf{k}, -\mathbf{k})$, and an m-sided polygon represents U_m^0. All momentum lines run from a dot to a polygon side. Then a connected diagram is a diagram which is topologically connected. The diagrammatic expression for $\ln J$ is a sum over all connected diagrams with no free momentum lines; the first few terms are shown in Figure 6.2.6. The counting factor for each diagram is chosen so that any distribution of the momentum variables which is not equivalent to the first by a permutation of the U_m^0 or the arguments of a U_m^0 or a combination of both must be considered a new term. The factor $1/p!$ must also be retained.

These expansions lend themselves to systematic approximation because each topologically distinct diagram is of the same order in N when the sum over momenta is included. Thus any partial summation has the schematic result

$$\langle \exp[O(N)] \rangle = \exp[O(N)] \tag{6.2.99}$$

Paired phonon analysis[52,70] We now approximate $\ln J(y)$ by retaining only the contributions of U_2^0; i.e. we set

$$U_n^0 = 0, \quad n \geqslant 3. \tag{6.2.100}$$

Having made this approximation, we can obtain algebraically the expression for $c_{\mathbf{k}}$ which minimizes the energy (i.e. the solution of equation (6.2.66) or, equivalently, (6.2.56) or (6.2.52)). The errors introduced by the approximation (6.2.100) can be corrected by iterating the entire procedure, with u_0 replaced by $u_0 + \Delta u$. The equations obtained from this approximation were first obtained in a somewhat different fashion in a procedure known as the *paired phonon analysis.*[52,70] The name is derived from the fact that, as we shall see in Section 6.3, the $\rho_{\mathbf{k}}$ operator is the phonon creation operator in the long wavelength limit. Retaining only U_2^0 means that these operators always must be paired to give a contribution.

With the approximation (6.2.100) the diagrammatic expansion for ln$J$ contains only the cyclic diagrams, the first few of which are the first four terms in Figure 6.2.6. In that case, the diagram rules simplify considerably. The expression for ln$J$ is

$$\ln J(y) = \sum_{p=1}^{\infty} \frac{1}{p!} \sum_{i=1}^{p} \Big(\sum_{\mathbf{k}_i, \mathbf{k}_{ix} > 0} y_{\mathbf{k}_i} \Big) [(U_2^0)^p]_c \{(\mathbf{k}_1, -\mathbf{k}_1) \ldots (\mathbf{k}_p, -\mathbf{k}_p)\}. \tag{6.2.101}$$

To simplify this expression, assign momenta to the specific dashed lines in the figure. The restriction to half of k-space denoted by $\mathbf{k}_{ix} > 0$ can be removed by noting that each pair $(\mathbf{k}_i, -\mathbf{k}_i)$ gives rise to two equivalent terms, except the first which must have a factor ½ to remove the restriction. The factor $1/p!$ becomes $1/p$ because of the $(p-1)!$ non-cyclic permutations which reproduce the term after a change of variables.

Then

$$\ln J(y) = \frac{1}{2} \sum_{p=1}^{\infty} \sum_{\mathbf{k}_1 \ldots \mathbf{k}_p} y_{\mathbf{k}_1} U_2^0(\mathbf{k}_1, -\mathbf{k}_1) y_{\mathbf{k}_2} U_2^0(\mathbf{k}_2, -\mathbf{k}_2) \ldots y_{\mathbf{k}_p} U_2^0(\mathbf{k}_p, -\mathbf{k}_p). \tag{6.2.102}$$

The new liquid structure function is defined by

$$S(k) = U_2(\mathbf{k}, -\mathbf{k})/N \tag{6.2.103}$$

for a translationally invariant system, where U_2 may be obtained in general by

$$\frac{\partial}{\partial y_{\mathbf{k}}} \ln J(y) = U_2(\mathbf{k}, -\mathbf{k}). \tag{6.2.104}$$

Using the approximate result (6.2.102), this becomes a Dyson-like equation

$$U_2(\mathbf{k}, -\mathbf{k}) = U_2^0(\mathbf{k}, -\mathbf{k}) + \sum_{\mathbf{q}} U_2^0(\mathbf{k}, \mathbf{q}) y_{\mathbf{q}} U_2(\mathbf{q}, -\mathbf{k}). \tag{6.2.105}$$

Equations (6.2.102) and (6.2.105) simplify considerably when ψ_0 is translationally invariant so that equation (6.2.75) holds, eliminating intermediate momentum sums. Then equation (6.2.102) becomes

$$\begin{aligned} \ln J(y) &= \sum_{p=1}^{\infty} \frac{1}{p} \sum_{\mathbf{k}, \mathbf{k}_x > 0} y_{\mathbf{k}}^p U_2^0(\mathbf{k}, -\mathbf{k})^p \\ &= - \sum_{\mathbf{k}, \mathbf{k}_x > 0} \ln [1 - y_{\mathbf{k}} U_2^0(\mathbf{k}, -\mathbf{k})] \end{aligned} \tag{6.2.106}$$

so that

$$J(y) = \prod_{\mathbf{k}, \mathbf{k}_x > 0} \frac{1}{1 - y_{\mathbf{k}} U_2^0(\mathbf{k}, -\mathbf{k})} = \prod_{\mathbf{k}, \mathbf{k}_x > 0} \frac{1}{1 - c_{\mathbf{k}} S_0(k)} \tag{6.2.107}$$

and equation (6.2.105) becomes

$$U_2(\mathbf{k}, -\mathbf{k}) = \frac{U_2^0(\mathbf{k}, -\mathbf{k})}{1 - y_{\mathbf{k}} U_2^0(\mathbf{k}, -\mathbf{k})} \tag{6.2.108}$$

or, equivalently,

$$S(k) = \frac{S_0(k)}{1 - c_{\mathbf{k}} S_0(k)}. \tag{6.2.109}$$

With these expressions in hand, the energy may be calculated as a function of $c_{\mathbf{k}}$ by making use of this result in the expression (6.2.69) for $I(\alpha)$; this in turn gives the energy by equation (6.2.55). Then we must make the identification

$$y_{\mathbf{k}} = c_{\mathbf{k}}(\alpha)/N \tag{6.2.110}$$

where $c_{\mathbf{k}}(\alpha)$ is defined in equation (6.2.70). Then

$$I(\alpha) = I_0(\alpha) \prod_{\mathbf{k}, k_x > 0} \frac{\exp[-c_{\mathbf{k}}(\alpha)]}{1 - c_{\mathbf{k}}(\alpha) S_0(k, \alpha)} \tag{6.2.111}$$

and, from equation (6.2.55),

$$E = E_0 - \sum_{\mathbf{k}, k_x > 0} c_{\mathbf{k}} \left(\frac{\hbar^2 k^2}{4m} - \frac{[(\hbar^2 k^2/4m) S_0(k) + \mathscr{H}(k)]}{1 - c_k S_0(k)} \right)$$

$$= E_0 + \sum_{\mathbf{k}, k_x > 0} \delta E_{\mathbf{k}} \tag{6.2.112}$$

where $\mathscr{H}(k)$ is the derivative of $S_0(k,\alpha)$ with respect to α, defined in equation (6.2.57) with ψ replaced by ψ_0, and E_0 is the energy expectation value in state ψ_0:

$$E_0 = \frac{\langle \psi_0 | H | \psi_0 \rangle}{\langle \psi_0 | \psi_0 \rangle} = \frac{\mathrm{d}}{\mathrm{d}\alpha} \ln I_0(\alpha) \bigg|_{\alpha=0} \tag{6.2.113}$$

For a particular $\mathbf{k}$ the only term containing $c_{\mathbf{k}}$ is $\delta E_{\mathbf{k}}$. Minimizing $\delta E_{\mathbf{k}}$ with respect to $c_{\mathbf{k}}$ gives

$$c_{\mathbf{k}} = \frac{1 - \sqrt{[S_0(k) + (4m/\hbar^2 k^2)\mathscr{H}(k)]}}{S_0(k)} \tag{6.2.114}$$

which gives for the energy shift

$$\delta E_k = -\frac{\hbar^2 k^2}{4m S_0(k)} \{1 - \sqrt{[S_0(k) + (4m/\hbar^2 k^2)\mathscr{H}(k)]}\}^2 . \tag{6.2.115}$$

Note that this is negative, so that the extremum is indeed a minimum. The liquid structure function corresponding to the improved wave function,

$$S(k) = \frac{1}{N} \frac{\langle \psi | \rho_{\mathbf{k}} \rho_{-\mathbf{k}} | \psi \rangle}{\langle \psi | \psi \rangle} \tag{6.2.116}$$

may be evaluated by using equation (6.2.114) in equation (6.2.109) to obtain

$$S(k) = \frac{S_0(k)}{\sqrt{[S_0(k) + (4m/\hbar^2 k^2)\mathscr{H}(k)]}} \tag{6.2.117}$$

Note that by using S to eliminate the square root in equation (6.2.115), δE_k takes on the simpler form

$$\delta E_k = -\frac{\hbar^2 k^2}{4m S_0(k)} \left(1 - \frac{S_0(k)}{S(k)}\right)^2 . \tag{6.2.118}$$

To complete this stage of the analysis, it is necessary to calculate $\mathscr{H}(k)$. Assuming for the moment that $\mathscr{H}(k)$ has been evaluated for a particular ψ_0, equation (6.2.114) for $c_{\mathbf{k}}$ can be Fourier transformed to give equation (6.2.60). Then, using $u = u_0 + \Delta u$, the new energy and liquid structure function can be calculated directly from the new wave function and compared to $S(k)$ and

$$E = E_0 + \sum_{\mathbf{k}, k_x > 0} \delta E_k \tag{6.2.119}$$

$$= E_0 + \delta E_{\text{PPA}}$$

obtained in the above analysis. Some discrepancy in these two evaluations may exist due to the approximation (6.2.100) in the determination of $c_{\mathbf{k}}$. Nevertheless, if the initial wave function ψ_0 is chosen carefully, the directly calculated energy should be lower than E_0 and not far from the predicted E. The new wave function can be subjected to a repeat of the above analysis to further improve the energy.

The most direct and accurate procedure for calculating $\mathscr{H}(k)$ is to evaluate equation (6.2.57) by using Monte Carlo or molecular dynamics integration routines for a finite system of particles.[70] This method is complicated, however, by the fact that $\mathscr{H}(k)$ is a fluctuation, being the difference of two terms of order N to give a term of order unity. This magnifies the statistical errors and requires a more careful (and expensive) calculation than the determination of E_0 and $S_0(k)$. Nevertheless, we have done such a calculation for a system of 49 helium atoms in a two-dimensional box with encouraging results, and plan to carry the calculation out in three dimensions in the near future.[93]

Approximations for $\mathscr{H}(k)$ in an infinite system may be obtained by using any of the approximate relations between $u(r)$ and $g(r)$ of the form of equation (6.2.35). In particular, $\mathscr{H}(k)$ is the Fourier transform of $\mathrm{d}g_0/\mathrm{d}\alpha\,|_{\alpha=0}$:

$$\mathscr{H}(k) = \rho \int \exp(\mathrm{i}\mathbf{k}\cdot\mathbf{r}) \left.\frac{\mathrm{d}g_0(r|\alpha)}{\mathrm{d}\alpha}\right|_{\alpha=0} \mathrm{d}\mathbf{r} \tag{6.2.120}$$

where $g_0(r\,|\,\alpha)$ is the radial distribution function for $\psi_0(\alpha)$. Then using the α-dependent u_0 in equation (6.2.35) and differentiating with respect to α gives an equation for $\mathrm{d}g_0/\mathrm{d}\alpha$:

$$\left.\frac{\mathrm{d}}{\mathrm{d}\alpha} g_0(r\,|\,\alpha)\right|_{\alpha=0} = \left.\frac{\partial}{\partial\alpha} F(u_0 + \alpha V^*, g_0(\alpha); r)\right|_{\alpha=0} \tag{6.2.121}$$

The HNC and PY approximations give particularly simple integral equations for $\mathscr{H}(k)$ by this procedure.[70] Fourier transforming equation (6.2.121), the HNC and PY approximations give a linear integral equation

$$L(k) = V_k^* + \frac{1}{(2\pi)^3\rho} \int (1 - M(\mathbf{k}+\mathbf{h}))(1 - S_0(h)^2) L(h)\,\mathrm{d}\mathbf{h} \tag{6.2.122}$$

where

$$L(k) = \frac{\mathscr{H}(k)}{S_0(k)^2} \tag{6.2.123}$$

$$V_k^* = \rho \int g_0(r) V^*(r) \exp(\mathrm{i}\mathbf{k}\cdot\mathbf{r})\,\mathrm{d}\mathbf{r} \tag{6.2.124}$$

and

$$M(k) = \begin{cases} S_0(k), & \text{HNC} \\ 1 + \rho \int (\exp[u_0(r)] - 1) \exp(\mathrm{i}\mathbf{k}\cdot\mathbf{r})\,\mathrm{d}\mathbf{r} & \text{PY.} \end{cases} \tag{6.2.125}$$

It should not be surprising that these two approximations produce very similar results. Both integral equations were solved by Campbell and Feenberg,[70] using as input the Massey and Woo

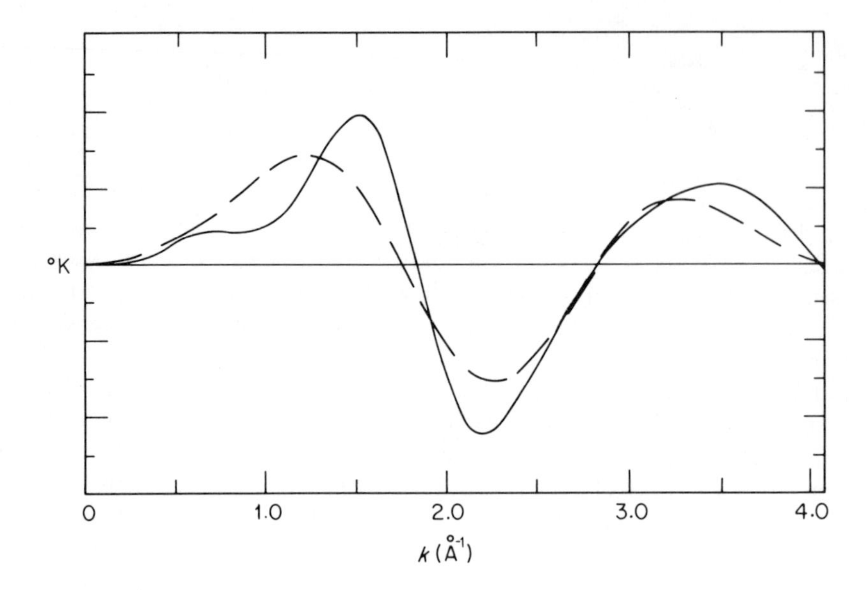

Figure 6.2.7 The hypernetted chain approximation for $\mathscr{H}(k)$ (solid line) and $(\hbar^2 k^2/4m)[1 - S(k)]$ (dashed line). These two curves would coincide if the energy expectation value were an extremum

$g_0(r)$.[65] The results at experimental equilibrium density, $\rho_0 = 0.0218$ Å^{-3}, are

$$\begin{aligned} \delta E_0/N &= -0.69\ ^\circ\text{K} \quad \text{HNC} \\ \delta E_0/N &= -0.68\ ^\circ\text{K} \quad \text{PY} \end{aligned} \tag{6.2.126}$$

which give the ground state energy per particle of –6.66 °K and –6.65 °K, respectively, when added to Massey and Woo's energy. This brings the calculated binding to within 0.5 °K of the experimental value. $\mathscr{H}(k)$ and $(\hbar^2 k^2/4m)(1 - S_0(k))$ are shown in Figure 6.2.7 for the HNC approximation. If the variational condition equation (6.2.56) were satisfied, these two curves would coincide. The liquid structure function $S(k)$ is compared to Massey and Woo's $S(k)$ and the results of X-ray and neutron scattering in Figure 6.2.3. While the improvement is significant, particulary around 1 Å^{-1} the maximum in $S(k)$ stills falls somewhat short. Of course, the energy is still somewhat larger than experiment. Both of these shortcomings should be improved by making non-Jastrow corrections to the wave function. That is indeed what happens, as will be discussed below.

Self-consistent approximations It should be pointed out that the calculation just described using HNC or PY approximations for $\mathscr{H}(k)$ is not self-consistent and cannot be iterated. The inconsistency is illustrated by the fact that, had we begun with the optimum $u(r)$ and had access to the corresponding $g(r)$ without approximations, the variational condition (6.2.56) would not be satisfied if to calculate $\mathscr{H}(k)$ we had to use some approximation (such as HNC or PY, as above). Thus a negative energy shift would have been predicted, when, by virtue of the fact that we began with the variationally best $u(r)$, the actual energy change would need to be positive. Thus, until $\mathscr{H}(k)$ is calculated without resorting to these approximations, the energy shift and change in $S(k)$ must be viewed as estimates only.

There is an alternative course, which is to minimize the energy while staying entirely within one of the approximations. That is, the energy is defined by equation (6.2.29), which we repeat here.[94]

$$\frac{E}{N}=\frac{\rho}{2}\int \mathrm{d}\mathbf{r}g(r)V^*(r) \tag{6.2.127}$$

where

$$V^*(r)=V(r)-\frac{\hbar^2}{4m}\nabla^2 u(r). \tag{6.2.128}$$

Now, however, we require $u(r)$ to be related to $g(r)$ by the approximate equation (e.g. HNC). The Euler–Lagrange equation is still given by equation (6.2.51), but now the functional derivative is calculated consistently within the approximation, namely,

$$\frac{\delta g(r)}{\delta u(r')}=\frac{\delta F(u,g;r)}{\delta u(r')} \tag{6.2.129}$$

when the approximation of interest is given by equation (6.2.39).

We first consider the HNC equation. Because of the form of the equation, it is easier (but equivalent) to minimize the energy with respect to $g(r)$:

$$\frac{\delta E}{\delta g(r)}=0 \tag{6.2.130}$$

which gives

$$V^*(r)-\frac{\hbar^2}{4m}\int \frac{\delta u(r')}{\delta g(r)}\nabla^2 g(r')\,\mathrm{d}\mathbf{r}'=0. \tag{6.2.131}$$

The HNC relation (6.2.36) gives

$$\frac{\delta u(r)}{\delta g(r')}=\delta(\mathbf{r}-\mathbf{r}')\left(\frac{1}{g(r)}-1\right)+\frac{1}{(2\pi)^3}\int \frac{\exp[\mathrm{i}\mathbf{k}\cdot(\mathbf{r}-\mathbf{r}')]}{S(k)^2}\,\mathrm{d}\mathbf{k}. \tag{6.2.132}$$

Substituting this into (6.2.131) gives the HNC variational extremum condition

$$0=V^*(r)-\frac{\hbar^2}{4m}\left(\frac{1}{g(r)}-1\right)\nabla^2 g(r)-\frac{1}{(2\pi)^3\rho}\int \frac{\hbar^2k^2}{4m}\,\frac{1-S(k)}{S(k)^2}\exp(\mathrm{i}\mathbf{k}\cdot\mathbf{r})\,\mathrm{d}\mathbf{k}. \tag{6.2.133}$$

By Fourier transforming this equation and comparing to the HNC expression for $\mathscr{H}(k)$ (equations 6.2.122–125), we obtain the result that the HNC optimization condition (6.2.133) is equivalent to

$$\mathscr{H}(k)=\frac{\hbar^2k^2}{4m}[1-S(k)] \tag{6.2.134}$$

i.e. it is formally identical to the general optimization condition, equation (6.2.56). Thus the paired-phonon analysis may be carried out self-consistently within the HNC approximation to obtain the optimum HNC $u(r)$. In contrast, a similar analysis shows that the other two approximations we have discussed, PY and BBGKY–KSA, do not possess this self-consistency. We comment further on this below.

The procedure for using the paired-phonon analysis to find the variationally optimum HNC $u(r)$ is the same as discussed in the last section for the more general case. That is, the new $u(r)$ obtained from the paired-phonon analysis by using equation (6.2.114) in equations (6.2.62) and (6.2.60) will have a lower HNC energy, and can be used as a new starting point for the paired-phonon analysis; this procedure is repeated until the optimization condition (6.2.134) is satisfied.

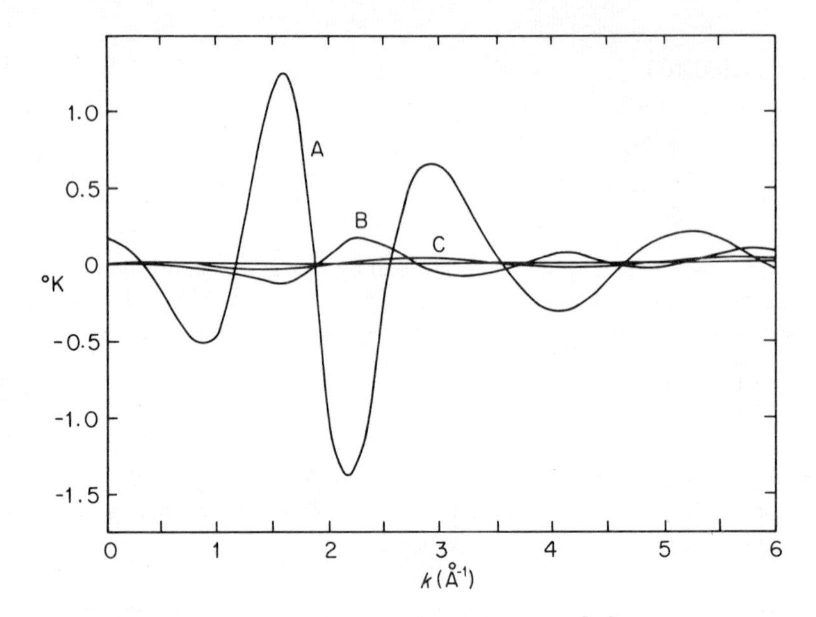

Figure 6.2.8 The difference between $\mathscr{H}(k)$ and $(\hbar^2 k^2/4m)[1 - S(k)]$ for several iterations of the self-consistent paired-phonon analysis within the hypernetted chain approximation: A, first iteration; B, second iteration; C, fourth iteration

This HNC optimization has been carried out at the equilibrium density of ^{4}He.[95] We begin with the optimum $u_0(r)$ of the form (equation 6.2.41) obtained by Murphy and Watts[68] within the HNC approximation. Those optimum parameters are

$$a = 3.0 \text{ Å}, \quad p = 5. \tag{6.2.135}$$

Figure 6.2.8 shows the results of the four iterations of the paired-phonon analysis for $\mathscr{H}(k) - (\hbar^2 k^2/4m)[1 - S(k)]$. The initial liquid structure function and the result after each iteration are given in Table 6.2.2. The convergence is rapid. The improved structure functions are nearly identical to S_{PPA} of Figure 6.2.3 and bear the same resemblance to experiment. Table 6.2.3 compares the predicted energy with the directly calculated energy.

Although it should be clear from Figure 6.2.8 that the optimum HNC $u(r)$ has been found, it is nevertheless useful to demonstrate that the final result is independent of the choice of the initial $u_0(r)$. Toward that end, Chang has chosen a different $u_0(r)$, with parameters

$$a = 2.8 \text{ Å} \quad p = 6 \tag{6.2.136}$$

and repeated the above analysis. The two different initial u_0 and the final u are compared in Figure 6.2.9. The energies of each iteration are shown in Table 6.2.3. Note that the initial energy lies above the initial energy obtained from the parameters (6.2.135). The final energies are nearly the same, however and would be equal in a more accurate numerical calculation after more iterations. The final $u(r)$ and $S(k)$ are indistinguishable on the scale of the figures.

Clearly, we may conclude that the final solution is independent of the choice of $u_0(r)$. It should be pointed out, however, that an extremely bad choice of $u_0(r)$, requiring a large shift in energies, might not converge via this procedure.

The long range structure in the improved $u(r)$ as indicated by the small k behaviour of $S(k)$ is particularly noteworthy. It is well known from the theory of classical fluids that a

Table 6.2.2 The liquid structure function $S(k)$ at $\rho = 0.02185$ Å^{-3} for several iterations of the self-consistent paired-phonon analysis within the hypernetted chain approximation. $n = 0$ is the initial structure factor from $u(r) = -(a/r)^5$ at the optimum value of a

k(Å^{-1})	$n = 0$	1	2	3
0.2	0.096	0.071	0.073	0.072
0.4	0.115	0.125	0.126	0.126
0.6	0.146	0.173	0.174	0.174
0.8	0.194	0.227	0.227	0.227
1.0	0.269	0.294	0.295	0.295
1.2	0.383	0.387	0.388	0.388
1.4	0.551	0.521	0.523	0.523
1.6	0.774	0.716	0.719	0.718
1.8	1.008	0.974	0.975	0.974
2.0	1.170	1.207	1.203	1.204
2.2	1.209	1.262	1.257	1.258
2.4	1.160	1.178	1.176	1.177
2.8	1.019	1.003	1.005	1.005
3.2	0.953	0.945	0.947	0.946
3.6	0.959	0.960	0.960	0.960
4.0	0.992	0.997	0.996	0.996

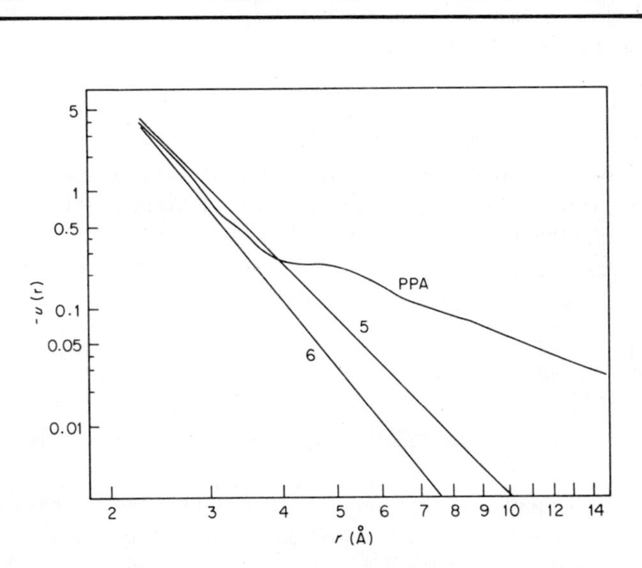

Figure 6.2.9 The final $u(r)$ obtained from the self-consistent paired-phonon analysis (ppA) compared with two different choices of initial $u(r)$ ($n = 5$ and $n = 6$). The two initial choices produce the same final result

short-ranged interaction potential produces a finite value of $S(k)$ in the limit $k \to 0$.[96] Specifically,

$$S_{\text{Cl}}(k) \xrightarrow[k \to 0]{} k_\beta T \rho \kappa \tag{6.2.137}$$

where κ is the compressibility of the classical fluid. Mathematically, this implies that the initial

Table 6.2.3 The energy E (°K per particle) at the end of each iteration of the paired-phonon analysis within the HNC approximation with two different initial functions $u(r) = -(a/r)^p$. The predicted energy shift δE_{PPA} (equation 6.2.119) is also shown

Iteration		0	1	2	3	4	5	6	7
$p = 5$	E	−4.414	−4.609	−4.615	−4.617	−4.6165			
	δE_{PPA}	—	−0.214	−0.004	−0.001	−0.0006			
$p = 6$	E	−4.205	−4.522	−4.527	−4.540	−4.542	−4.544	−4.549	−4.552
	δE_{PPA}		−0.351	−0.015	−0.0040	−0.0027	−0.0022	−0.0019	−0.001

$S_0(k)$, obtained from the short ranged $u_0(r)$ with parameters (6.2.135) will also have a finite value as k goes to zero. That behaviour is verified in Table 6.2.2 and Figure 6.2.3. On the other hand, sum rule arguments may be used to show that, at $T = 0$,

$$\lim_{k\to 0} S(k) = \frac{\hbar k}{2mc} \tag{6.2.138}$$

where c is the velocity of sound.[47] Note that the paired-phonon analysis puts this behaviour in automatically, as evidenced in Table 6.2.2. The relationship of the slope to the velocity of sound within the class of Jastrow functions has not been proven rigorously. The numerical value obtained from the last $S(k)$ is

$$c = 206 \text{ m/sec} \tag{6.2.139}$$

which is reasonably close to the experimental value of 238 m/sec. The linear behaviour is obtained immediately in the first iteration because a short-ranged $u_0(r)$ gives an $\mathscr{H}(k)$ which is constant as k tends to zero. Then the denominator on the right-hand side of equation (6.2.117) diverges as k^{-1}, putting the linear behaviour into $S(k)$.[97]

It must also be concluded that the new $u(r)$ giving a linear $S(k)$ is not short ranged. Indeed, equation (6.2.114) shows that $c_k \sim k^{-1}$ at small k; consequently

$$\Delta u(r) \underset{r\to\infty}{\propto} \frac{1}{r^2}. \tag{6.2.140}$$

The log–log plot of $u(r)$ in Figure 6.2.9 (and for comparison $u_0(r)$) shows this long-range behaviour.

The results of this calculation at densities other than equilibrium will be discussed below.

We observed above that the HNC is a self-consistent approximation within the paired-phonon analysis, while the PY and BBGKY–KSA approximations are not. The condition that an approximation be self-consistent is that the functional derivative of $g(r)$ with respect to $u(r')$ be symmetric:[91]

$$\frac{\delta g(r)}{\delta u(r')} = \frac{\delta g(r')}{\delta u(r)}. \tag{6.2.141}$$

This symmetry is obtained from any approximation which can be expressed as an approximation for the normalization integral (or in classical systems, the partition function):[98]

$$I = \int \prod_{i<j} \exp[u(r_{ij})]\, d\mathbf{r}_1 \ldots d\mathbf{r}_N. \tag{6.2.142}$$

Then

$$\rho^2 g(r) = \frac{\delta}{\delta u(r)} \ln I \tag{6.2.143}$$

and

$$\frac{\delta g(r)}{\delta u(r')} = \frac{1}{\rho^2} \frac{\delta^2}{\delta u(r')\delta u(r)} \ln I = \frac{\delta g(r')}{\delta u(r)}. \tag{6.2.144}$$

This provides a restriction which should be imposed on any approximation applied to the paired-phonon analysis. We have not investigated the more complicated approximations such as HNC II and PY II to see if they satisfy the self-consistency.

Beyond Jastrow functions

Improvements upon the results of variational calculations within the space of Jastrow functions require an expanded space of trial functions. This has led to the suggestion that three-body factors be included in the trial function:[71–73]

$$\psi_0 = \prod_{i<j} \exp[\tfrac{1}{2}u(r_{ij})] \prod_{i<j<k} \exp[\tfrac{1}{2}u_3(\mathbf{r}_i, \mathbf{r}_j, \mathbf{r}_k)]. \tag{6.2.145}$$

The logarithm of this function has the extended Jastrow form of equation (6.2.13) with

$$u_n = 0 \quad n \geqslant 4. \tag{6.2.146}$$

Therefore the restrictions imposed on u_n in Section 2.1 also apply to u and u_3 in this trial function.

The only treatment of u_3 in an accurate calculation presently available is the parametrized variational calculation of Woo and Coldwell. They chose u_3 to have the form[99]

$$u_3(\mathbf{r}_1, \mathbf{r}_2, \mathbf{r}_3) = \left(\frac{b}{R}\right)^n \tag{6.2.147}$$

where

$$R^2 = r_{12}^2 + r_{23}^2 + r_{31}^2. \tag{6.2.148}$$

This choice was made because it favours short-range hexagonal coordination. They carried the calculation out by Monte Carlo integration for two-dimensional liquid ^{4}He and found that the parameters which minimize the energy sharpen the structure as measured by the liquid structure function.

Clearly, the possibilities for parametrized trial functions of the extended Jastrow form (equation 6.2.13) are enormous, and we have little intuition to guide us. More importantly, even for a simple parametrization such as (6.2.147) it is necessary to obtain the three-body distribution function in order to calculate the kinetic energy (equation 6.2.20). That is an expensive proposition for a single choice of u_3. The necessity of carrying the calculation out for a range of parameters makes the expense prohibitive. Clearly, then, a procedure such as the paired-phonon analysis is needed to make such a calculation feasible. With some luck, one can begin with the optimized Jastrow function u_2 and predict a u_3 in a manner similar to the prediction of Δu_2 above.

We have made the first tentative steps in that direction and have obtained encouraging results for u_3.[72] The procedure we use parallels the paired-phonon analysis. There are some differences between $n = 2$ and $n > 2$ which require a separate treatment of the latter.

As above, we suppose that we have a wave function ψ_0 which does not satisfy the variational condition (6.2.14) for a particular $n > 2$. Then we try to find a function $\Delta u_n(\mathbf{r}_1 \ldots \mathbf{r}_n)$ for which

$$\psi = \psi_0 \exp[\tfrac{1}{2} \sum_{\langle i_1 \ldots i_n \rangle} \Delta u_n(\mathbf{r}_1 \ldots \mathbf{r}_n)] \tag{6.2.149}$$

satisfies (6.2.14) for n.

The variational equations (6.2.14) are complicated to deal with in configuration space but take on a much simpler form in momentum space. To reformulate them in momentum space we define the Fourier transform of Δu_n by

$$\Delta u_n(\mathbf{r}_1 \ldots \mathbf{r}_n) = N^{1-n} \sum_{\mathbf{k}_1 \ldots \mathbf{k}_n} c_n(\mathbf{k}_1 \ldots \mathbf{k}_n) \prod_{i=1}^{n} \exp(\mathrm{i}\mathbf{k}_i \cdot \mathbf{r}_i). \tag{6.2.150}$$

Then the argument of the exponential factor in (6.2.149) becomes

$$\frac{1}{2N^{n-1}} \sum_{\langle i_1 \ldots i_n \rangle} \Delta u_n(\mathbf{r}_{i_1} \ldots \mathbf{r}_{i_n}) = \frac{1}{2n!N^{n-1}} \sum_{\mathbf{k}_1 \ldots \mathbf{k}_n} \rho^{(n)}(\mathbf{k}_1 \ldots \mathbf{k}_n) c_n(\mathbf{k}_1 \ldots \mathbf{k}_n) \tag{6.2.151}$$

where $\rho^{(n)}(\mathbf{k}_1 \ldots \mathbf{k}_n)$ is a symmetrized N-body plane wave:

$$\rho^{(n)}(\mathbf{k}_1 \ldots \mathbf{k}_n) = \sum_{\substack{i_1 \ldots i_n \\ i_s \neq i_t \,(\text{all } st)}}^{N} \prod_{i=1}^{n} \exp(\mathrm{i}\mathbf{k}_i \cdot \mathbf{r}_{i_p}) \tag{6.2.152}$$

The functional derivatives in the variational equations (6.2.14) can now be replaced by derivatives with respect to c_n, in which case the variational equations become

$$\langle \psi | (H - E)\rho^{(n)}(\mathbf{k}_1 \ldots \mathbf{k}_n) | \psi \rangle = 0 \tag{6.2.153}$$

where

$$E = \frac{\langle \psi | H | \psi \rangle}{\langle \psi | \psi \rangle}. \tag{6.2.154}$$

Since the exact ground state wave function satisfies the Schrödinger equation, it also must satisfy (6.2.153) because $H - E$ operating to the left will give zero. Conversely, a state which satisfies (6.2.153) for all n and all sets $\{\mathbf{k}_1 \ldots \mathbf{k}_n\}$ must be an eigenstate of H with energy eigenvalue E since the $\rho^{(n)}(\mathbf{k}_1 \ldots \mathbf{k}_n)$ are a complete set of plane waves. The ground state corresponds to the lowest value of E; equivalently, the ground-state wave function is the real, positive solution of the complete set of equations, (6.2.153). As above, it is more convenient for the determination of Δu_n which solves (6.2.152) to calculate the energy from the generalized normalization integral:

$$E = \frac{\mathrm{d}}{\mathrm{d}\alpha} \ln I(\alpha) \bigg|_{\alpha=0} = \frac{\langle \psi | V^* | \psi \rangle}{\langle \psi | \psi \rangle} \tag{6.2.155}$$

where, as before

$$I(\alpha) = \langle \psi(\alpha) | \psi(\alpha) \rangle \tag{6.2.156}$$

but now

$$\psi(\alpha) = \psi \exp(\tfrac{1}{2}\alpha V^*) \tag{6.2.157}$$

where

$$V^*(\mathbf{r}_1 \ldots \mathbf{r}_N) = V(\mathbf{r}_1 \ldots \mathbf{r}_N) - \frac{\hbar^2}{8m} \sum_{i=1}^{N} \nabla_i^2 \ln \psi^2$$

$$= \sum_{i<j}^{N} V(r_{ij}) - \frac{\hbar^2}{8m} \sum_{n} \sum_{\langle i_1 \ldots i_n \rangle} \left[\sum_{j=1}^{n} \nabla_{i_j}^2 u_n(\mathbf{r}_{i_1} \ldots \mathbf{r}_{i_n}) \right] \tag{6.2.158}$$

where

$$u_n = u_n^0 + \Delta u_n. \tag{6.2.159}$$

Note then, that the variational equation may be rewritten in a third form:

$$0 = \frac{\partial}{\partial c_n(\mathbf{k}_1 \ldots \mathbf{k}_n)} \frac{\langle \psi \mid V^* \mid \psi \rangle}{\langle \psi \mid \psi \rangle}$$

$$= \frac{\langle \psi \mid (V^* - E)\rho^{(n)}(\mathbf{k}_1 \ldots \mathbf{k}_n) \mid \psi \rangle}{\langle \psi \mid \psi \rangle} + \sum_{i=1}^{n} \frac{\hbar^2}{8m} k_i^2 \frac{\langle \psi \mid \rho^{(n)}(\mathbf{k}_1 \ldots \mathbf{k}_n) \mid \psi \rangle}{\langle \psi \mid \psi \rangle} \tag{6.2.160}$$

where the first term comes from $\partial \psi / \partial c_n$ and the second term comes from $\partial V^* / \partial c_n$. In a notation similar to that of equation (6.2.56) and (6.2.57) we define

$$\mathcal{H}_n(\mathbf{k}_1 \ldots \mathbf{k}_n) = \frac{1}{N} \frac{\mathrm{d}}{\mathrm{d}\alpha} G_n(\mathbf{k}_1 \ldots \mathbf{k}_n \mid \alpha) \bigg|_{\alpha=0}$$

$$= \frac{\langle \psi \mid (V^* - E)\rho^{(n)}(\mathbf{k}_1 \ldots \mathbf{k}_n) \mid \psi \rangle}{N \langle \psi \mid \psi \rangle} \tag{6.2.161}$$

where G_n is defined in equation (6.2.85), and by using (6.2.152),

$$G_n(\mathbf{k}_1 \ldots \mathbf{k}_n) = \frac{\langle \psi \mid \rho^{(n)}(\mathbf{k}_1 \ldots \mathbf{k}_n) \mid \psi \rangle}{\langle \psi \mid \psi \rangle}. \tag{6.2.162}$$

Then the third form of the variational equations (6.2.160) is written as

$$\mathcal{H}_n(\mathbf{k}_1 \ldots \mathbf{k}_n) + \frac{\hbar^2}{8mN} \sum_{i=1}^{N} k_i^2 \, G_n(\mathbf{k}_1 \ldots \mathbf{k}_n) = 0. \tag{6.2.163}$$

To solve this equation for c_n, we write $I(\alpha)$ as a function of c_n:

$$I(\alpha) = I_0(\alpha) \frac{\langle \psi_0(\alpha) \mid \exp\{(N^{1-n}/n!) \sum_{\mathbf{k}_1 \ldots \mathbf{k}_n} c_n(\mathbf{k}_1 \ldots \mathbf{k}_n \mid \alpha)\rho^{(n)}(\mathbf{k}_1 \ldots \mathbf{k}_n)\} \mid \psi_0(\alpha) \rangle}{\langle \psi_0(\alpha) \mid \psi_0(\alpha) \rangle} \tag{6.2.164}$$

where

$$c_n(\mathbf{k}_1 \ldots \mathbf{k}_n \mid \alpha) = c_n(\mathbf{k}_1 \ldots \mathbf{k}_n) \left(1 + \alpha \frac{\hbar^2}{8m} \sum_{i=1}^{n} k_i^2 \right). \tag{6.2.165}$$

Then the calculation of the energy is reduced to a calculation of an expression of the form

$$J(y) = \langle \exp \{ \sum_{\mathbf{k}_1 \ldots \mathbf{k}_n} y_n(\mathbf{k}_1 \ldots \mathbf{k}_n)\rho^{(n)}(\mathbf{k}_1 \ldots \mathbf{k}_n)\} \rangle. \tag{6.2.166}$$

where $\langle \ldots \rangle$ is a normalized expectation value. To evaluate $J(y)$, we could develop a linked-cluster cumulant expansion similar to that of described above. The correct procedure is to express $\rho^{(n)}(\mathbf{k}_1 \ldots \mathbf{k}_n)$ as an n-th-order polynomial in the density fluctuation functions $\rho_{\mathbf{k}}$.

The leading term in such an expression is always $\Pi_{i=1}^{n} \rho_{\mathbf{k}_i}$. The easier expression to write down is the one for a product of some $\rho_{\mathbf{k}}$ as a linear combination of $\rho^{(n)}$:

$$
\begin{aligned}
\rho_{\mathbf{k}_1} &= \rho^{(1)}(\mathbf{k}_1) \\
\rho_{\mathbf{k}_1}\rho_{\mathbf{k}_2} &= \rho^{(1)}(\mathbf{k}_1 + \mathbf{k}_2) + \rho^{(2)}(\mathbf{k}_1, \mathbf{k}_2) \\
\rho_{\mathbf{k}_1}\rho_{\mathbf{k}_2}\rho_{\mathbf{k}_3} &= \rho^{(1)}(\mathbf{k}_1 + \mathbf{k}_2 + \mathbf{k}_3) + \rho^{(2)}(\mathbf{k}_1, \mathbf{k}_2 + \mathbf{k}_3) + \rho^{(2)}(\mathbf{k}_2, \mathbf{k}_1 + \mathbf{k}_3) + \rho^{(2)}(\mathbf{k}_3, \mathbf{k}_1 + \mathbf{k}_2) \\
&\quad + \rho^{(3)}(\mathbf{k}_1, \mathbf{k}_2, \mathbf{k}_3) \\
&\vdots
\end{aligned}
\tag{6.2.167}
$$

Then the product of density fluctuation operators is related to the symmetrized plane waves $\rho^{(n)}$ by the operation Γ defined in equations (6.2.86) and (6.2.87). Indeed, equation (6.2.87) is obtained from (6.2.167) by taking expectation values of both sides of equation (6.2.167). The general expression for $\rho^{(n)}$ in terms of the $\rho_{\mathbf{k}}$ is given elsewhere.[92] Here we note the first three entries, which is all we will have practical application for now:

$$
\begin{aligned}
\rho^{(1)}(\mathbf{k}_1) &= \rho_{\mathbf{k}_1} \\
\rho^{(2)}(\mathbf{k}_1, \mathbf{k}_2) &= \rho_{\mathbf{k}_1}\rho_{\mathbf{k}_2} - \rho_{\mathbf{k}_1 + \mathbf{k}_2} \\
\rho^{(3)}(\mathbf{k}_1, \mathbf{k}_2, \mathbf{k}_3) &= \rho_{\mathbf{k}_1}\rho_{\mathbf{k}_2}\rho_{\mathbf{k}_3} - \rho_{\mathbf{k}_1}\rho_{\mathbf{k}_2 + \mathbf{k}_3} - \rho_{\mathbf{k}_2}\rho_{\mathbf{k}_1 + \mathbf{k}_3} - \rho_{\mathbf{k}_3}\rho_{\mathbf{k}_1 + \mathbf{k}_2} + 2\rho_{\mathbf{k}_1 + \mathbf{k}_2 + \mathbf{k}_3}.
\end{aligned}
\tag{6.2.168}
$$

The general expression has alternating sign according to the order of the term in $\rho_{\mathbf{k}}$ and the coefficient is generally different from 1. Our generalization of the linked-cluster cumulant expansion can handle exponentiated polynomials of $\rho_{\mathbf{k}}$ of arbitrary order, but the results are complicated and very tedious to present.[92] We shall instead present a simpler version[72] by focusing on one set of wave vectors $(\mathbf{k}_1 \ldots \mathbf{k}_n)$ and require this set to have no subset which sums to zero:

$$
\sum_{j=1}^{p} \mathbf{k}_{i_j} \neq 0 \quad p < n \tag{6.2.169}
$$

for all subsets. For such a set of $\mathbf{k}$ vectors with no subset momentum conservation, equation (6.2.89) shows that

$$
G_n(\mathbf{k}_1 \ldots \mathbf{k}_n) = F_n(\mathbf{k}_1 \ldots \mathbf{k}_n) = O(N). \tag{6.2.170}
$$

We can remove restriction (6.2.169) in the end since the omitted terms contribute lower orders in N.

Then we define a generating function

$$
J_n(\mathbf{k}_1 \ldots \mathbf{k}_n | y) = \left\langle \exp\left(\frac{y}{2N^{n-1}} [\rho^{(n)}(\mathbf{k}_1 \ldots \mathbf{k}_n) + \rho^{(n)}(-\mathbf{k}_1 \ldots -\mathbf{k}_n)] \right) \right\rangle \tag{6.2.171}
$$

and a y-dependent expectation value

$$
\langle A \rangle_y = \frac{\langle A \exp\{(y/2N^{n-1})[\rho^{(n)}(\mathbf{k}_1 \ldots \mathbf{k}_n) + \rho^{(n)}(-\mathbf{k}_1 \ldots -\mathbf{k}_n)]\} \rangle}{J_n(\mathbf{k}_1 \ldots \mathbf{k}_n | y)} \tag{6.2.172}
$$

We wish to obtain an alternative, tractable expression for J_n. We first note that $\ln J_n$ is a generating function for the real part of F_n:

$$\frac{d}{dy}\ln J_n(\mathbf{k}_1 \ldots \mathbf{k}_n \mid y) = \frac{\mathrm{Re}\, F_n(\mathbf{k}_1 \ldots \mathbf{k}_n \mid y)}{N^{n-1}}. \tag{6.2.173}$$

Repeated differentiation with respect to y will eventually lead to a closed set of equations due to the clustering properties of the products. To see this, the next derivative gives

$$N^{n-1}\frac{d^2}{dy^2}\ln J_n = \frac{d}{dy}\mathrm{Re}\, F_n$$
$$= \frac{1}{4N^{n-2}}\langle[\rho^{(n)}(\mathbf{k}_1 \ldots \mathbf{k}_n) + \rho^{(n)}(-\mathbf{k}_1 \ldots -\mathbf{k}_n)]^2\rangle_y$$
$$- \frac{1}{4N^{n-2}}\langle[\rho^{(n)}(\mathbf{k}_1 \ldots \mathbf{k}_n) + \rho^{(n)}(-\mathbf{k}_1 \ldots -\mathbf{k}_n)]\rangle_y^2. \tag{6.2.174}$$

The first term on the right-hand side of this equation is a polynomial of degree $2n$ in $\rho_{\mathbf{k}}$. The evaluation of the expectation value of polynomials of this sort is simplified by making use of the properties of the $\rho_{\mathbf{k}}$ cumulants discussed above. In particular, the term of leading order in N will be the one for which the $\rho_{\mathbf{k}}$ can be grouped in the maximum number of momentum conserving factors. In this case, the result is straightforward since the leading term will be the cross term between the terms $\Pi_{i=1}^{n}\rho_{\mathbf{k}_i}$ in $\rho^{(n)}(\mathbf{k}_1 \ldots \mathbf{k}_n)$ and $\Pi_{i=1}^{n}\rho_{-\mathbf{k}_i}$ in $\rho^{(n)}(-\mathbf{k}_1 \ldots -\mathbf{k}_n)$. Note that had we not included both the $\rho^{(n)}$ in the exponential in equation (6.2.171) we would have missed this important cross-correlation.

Then equation (6.2.174) becomes

$$\frac{d}{dy}\mathrm{Re}\, F_n(\mathbf{k}_1 \ldots \mathbf{k}_n \mid y) = \frac{N}{2}\prod_i n_i!\, S(k_i \mid y)^{n_i} + O(1) \tag{6.2.175}$$

where $\Sigma_i n_i = n$ and n_i are the multiplicities of the vectors $\mathbf{k}_i$. Thus we conclude that both F_n and its derivative are $O(N)$.

To close the set of equations, we note that a similar analysis shows that

$$\frac{d}{dy}\ln S(k_i \mid y) = \frac{1}{2N^{n-1}}\Big\langle \rho_{\mathbf{k}_i}\frac{\partial}{\partial\rho_{\mathbf{k}_i}}\rho^{(n)}(\mathbf{k}_1 \ldots \mathbf{k}_n)$$
$$+ \rho_{-\mathbf{k}_i}\frac{\partial}{\partial\rho_{-\mathbf{k}_i}}\rho^{(n)}(-\mathbf{k}_1 \ldots -\mathbf{k}_n)\Big\rangle_y\left[1 + O\left(\frac{1}{N}\right)\right]. \tag{6.2.176}$$

The leading term in orders of N is contributed by a term $\rho_{\mathbf{k}_i}\rho^{(n-1)}(\mathbf{k}_1 \ldots \mathbf{k}_{i-1}\mathbf{k}_{i+1} \ldots \mathbf{k}_n)$ from $\rho^{(n)}(\mathbf{k}_1 \ldots \mathbf{k}_n)$ and its complex conjugate from $\rho^{(n)}(-\mathbf{k}_1 \ldots -\mathbf{k}_n)$. Then

$$\frac{d}{dy}\ln S(k_i \mid y) = \frac{n_i}{2N^{n-1}}\langle \rho_{\mathbf{k}_i}\rho^{(n-1)}(\mathbf{k}_1 \ldots \mathbf{k}_{i-1}\mathbf{k}_{i+1} \ldots \mathbf{k}_n)$$
$$+ \rho_{-\mathbf{k}_i}\rho^{(n-1)}(-\mathbf{k}_1 \ldots -\mathbf{k}_{i-1} - \mathbf{k}_{i+1} \ldots -\mathbf{k}_n)\rangle_y\left[1 + O\left(\frac{1}{N}\right)\right]$$
$$= O\left(\frac{1}{N^{n-2}}\right) \tag{6.2.177}$$

Then, for $n > 2$, the y may be set to zero on the right-hand side of equation (6.2.175) to leading order in N and differential equations may be integrated, giving

$$\mathrm{Re}\, F_n(\mathbf{k}_1 \ldots \mathbf{k}_n \mid y) = \mathrm{Re}\, F_n(\mathbf{k}_1 \ldots \mathbf{k}_n \mid 0) + \frac{Ny}{2}\prod_i n_i!\, S(\mathbf{k}_i \mid 0)^{n_i} \tag{6.2.178}$$

and

$$\ln J_n(\mathbf{k}_1 \ldots \mathbf{k}_n \mid y) = \ln J_n(\mathbf{k}_1 \ldots \mathbf{k}_n \mid 0) + \frac{y \operatorname{Re} F_n(\mathbf{k}_1 \ldots \mathbf{k}_n \mid 0)}{N^{n-1}}$$

$$+ \frac{y^2}{4N^{n-2}} \prod_i n_i! S(k_i \mid 0)^{n_i}. \tag{6.2.179}$$

The connection with the variational wave function is obtained by choosing

$$y = 2c_n(\mathbf{k}_1 \ldots \mathbf{k}_n \mid \alpha) \tag{6.2.180}$$

where $c_n(\alpha)$ is defined in equation (6.2.165). Then the shift in energy due to the set of vectors $(\mathbf{k}_1 \ldots \mathbf{k}_n)$ and $(-\mathbf{k}_1 \ldots -\mathbf{k}_n)$ is given by

$$\delta E_n(\mathbf{k}_1 \ldots \mathbf{k}_n) = \frac{\mathrm{d}}{\mathrm{d}\alpha} \ln J_n(\mathbf{k}_1 \ldots \mathbf{k}_n \mid 2c_n(\mathbf{k}_1 \ldots \mathbf{k}_n \mid \alpha)) \bigg|_{\alpha=0}. \tag{6.2.181}$$

Extremizing this as a function of $c_n(\mathbf{k}_1 \ldots \mathbf{k}_n)$gives

$$\delta E_n(\mathbf{k}_1 \ldots \mathbf{k}_n) = \frac{-N^{2-n}\left[\mathscr{H}_n(\mathbf{k}_1 \ldots \mathbf{k}_n) + (\hbar^2/8mN) \sum_{i=1}^{n} k_i^2 \operatorname{Re} F_n(\mathbf{k}_1 \ldots \mathbf{k}_n)\right]^2}{\left(\sum_i n_i \{[\mathscr{H}(k_i)/S(k_i)] + (\hbar^2/4m)k_i^2\}\right) \prod_i n_i! \, S(k_i)^{n_i}} \tag{6.2.182}$$

and

$$c_n(\mathbf{k}_1 \ldots \mathbf{k}_n) = \frac{-\left[\mathscr{H}_n(\mathbf{k}_1 \ldots \mathbf{k}_n) + (\hbar^2/8mN) \sum_{i=1}^{n} k_i^2 \operatorname{Re} F_n(\mathbf{k}_1 \ldots \mathbf{k}_n)\right]}{\left(\sum_i n_i \{[\mathscr{H}(k_i)/S(k_i)] + (\hbar^2/4m)k_i^2\}\right) \prod_i n_i! \, S(k_i)^{n_i}} \tag{6.2.183}$$

where $\mathscr{H}_n$ is defined in equation (6.2.161), $\mathscr{H}(\mathbf{k})$ is defined in equation (6.2.57).

The extremum is a minimum if the denominator of δE_n is positive.
Although the general case is not certain, the sign of this term is easily checked in the case that equation (6.2.56) is satisfied, i.e. when the Jastrow part of the initial function is already optimized. Then the sum in brackets in the denominator becomes

$$\sum_i n_i \frac{\hbar^2 k_i^2}{4mS(k_i)}$$

and the denominator is positive definite. Also in this case the expression for the shift in energy takes the form of second order perturbation theory

$$\delta E_n(\mathbf{k}_1 \ldots \mathbf{k}_n) = -\frac{|\langle \psi_0 | (H - E_0)\rho^{(n)}(\mathbf{k}_1 \ldots \mathbf{k}_n) | \psi_0 \rangle|^2}{\langle \psi_0 | \rho^{(n)}(\mathbf{k}_1 \ldots \mathbf{k}_n)^* (H - E_0)\rho^{(n)}(\mathbf{k}_1 \ldots \mathbf{k}_n) | \psi_0 \rangle} \tag{6.2.184}$$

for the particular choice of basis $\rho^{(n)}\psi_0$.

The results obtained thus far are exact to the leading order in N, and δE_n and c_n vanish if the optimization condition (6.2.163) is satisfied for n. The improved wave function, however, has only two terms in the Fourier transform of Δu_n:

$$\psi = \psi_0 \exp\{(2N^{n-1})^{-1} c_n(\mathbf{k}_1 \ldots \mathbf{k}_n)[\rho^{(n)}(\mathbf{k}_1 \ldots \mathbf{k}_n) + \rho^{(n)}(-\mathbf{k}_1 \ldots -\mathbf{k}_n)]\}. \tag{6.2.185}$$

We now approximate the result for the wave function ψ involving the full Fourier transform of Δu_n as defined in equations (6.2.149) and (6.2.150) by summing over the results of the single c_n:

$$\delta E = \frac{1}{2n!} \sum_{\mathbf{k}_1 \ldots \mathbf{k}_n} \delta E_n(\mathbf{k}_1 \ldots \mathbf{k}_n). \tag{6.2.186}$$

This approximation is similar to the approximation in the paired-phonon analysis obtained by setting $U_3^{\mathrm{o}}(\mathbf{k}_1\mathbf{k}_2\mathbf{k}_3) = 0$. It cannot, however, be expressed as simply in this case. For example, it is not true that our $n = 3$ result is obtained by setting $U_4^{\mathrm{o}} = 0$. Nevertheless, this does share the important feature of the paired-phonon analysis that it is a predictor–corrector procedure whose predicted shifts vanish when the exact optimization condition is satisfied. The errors in a nonzero shift due to the present approximation can be corrected by using the new wave function ψ as a starting point for a reiteration of the above procedure, where ψ is given by equation (6.2.149) using equation (6.2.183) for $c_n(\mathbf{k}_1 \ldots \mathbf{k}_n)$. In a similar approximation, the new liquid structure function $S(k)$ corresponding to ψ is obtained by integrating (6.2.177) and then summing over all wave vectors except $\mathbf{k}_i$. The result is

$$S(k) = S_0(k) \exp\left\{\frac{1}{2N^n(n-1)!} \sum_{\mathbf{l}_1 \ldots \mathbf{l}_{n-1}} c_n(\mathbf{l}_1 \ldots \mathbf{l}_{n-1}) \times \langle \psi_0 | \rho_{\mathbf{k}} \rho^{(n-1)}(\mathbf{l}_1 \ldots \mathbf{l}_{n-1}) + \rho_{-\mathbf{k}} \rho^{(n-1)}(-\mathbf{l}_1 \ldots -\mathbf{l}_{n-1}) | \psi_0 \rangle\right\}. \tag{6.2.187}$$

Even with the approximations already included in these equations, there is a major difficulty in that it is still necessary to calculate the functions F_n and $\mathscr{H}_n$. The proper way to approach this problem is by using the Monte Carlo integration techniques to calculate these functions. Short of that, some approximation which is self-consistent in the same sense that HNC was for $n = 2$, i.e. one for which the minimum energy within the approximation satisfies an optimization condition equivalent to (6.2.163), would provide useful information concerning the importance of terms u_n in the wave function. Neither of these approaches has been carried out.

An estimate of δE and the shift in $S(k)$ may be obtained by using the Feenberg convolution approximation[100] to produce manageable expressions for $\mathscr{H}_n$ and F_n. This approximation may be characterized as the simplest approximation for which the sequential relation required by the definition of the n-body distribution function, equation (6.2.16), is satisfied:

$$\rho \int g_{n+1}(\mathbf{r}_1 \ldots \mathbf{r}_{n+1}) d\mathbf{r}_{n+1} = (N-n) g_n(\mathbf{r}_1 \ldots \mathbf{r}_n). \tag{6.2.188}$$

This approximation is stated most simply in terms of the cumulants U_n. Wu has obtained the general expression for U_n in the form[55]

$$U_n^{\mathrm{C}}(\mathbf{k}_1 \ldots \mathbf{k}_n) = N\delta_{\mathbf{k}_1 + \ldots + \mathbf{k}_n, 0} \left[\prod_{i=1}^{n} S(k_i)\right](1 + A_n) \tag{6.2.189}$$

where, for example

$$A_3 = 1$$

$$A_4 = 1 + \tfrac{1}{2} \sum_{1 \leqslant i < j \leqslant 4} [S(\mathbf{k}_i + \mathbf{k}_j) - 1]$$

$$A_5 = 1 + \sum_{1 \leqslant i < j \leqslant 5} [S(\mathbf{k}_i + \mathbf{k}_j) - 1] + \sum_{1 \leqslant i < j < k \leqslant 5} [S(\mathbf{k}_i + \mathbf{k}_j) - 1][S(\mathbf{k}_i + \mathbf{k}_j + \mathbf{k}_k) - 1]. \tag{6.2.190}$$

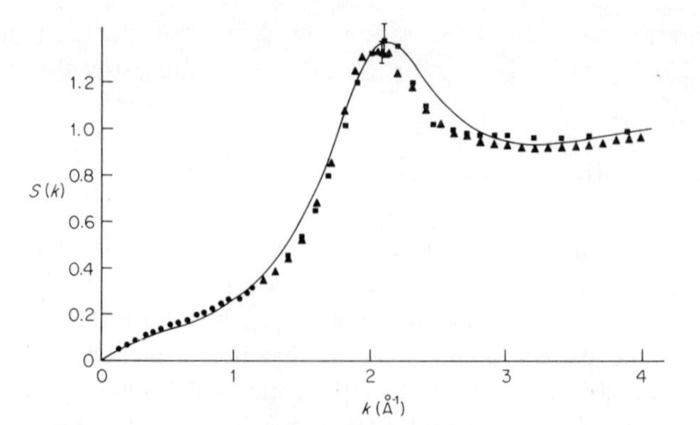

Figure 6.2.10 The liquid structure factor of liquid ^{4}He at equilibrium density calculated from an optimized trial wave function including three-body extended Jastrow factors (solid line) compared with experiment: ▲ X-ray scattering at $T = 0.79\ ^\circ$K;[61] ● X-ray scattering at $T = 0.38\ ^\circ$K;[62] ■ neutron scattering at $T = 1.1\ ^\circ$K.[115] (From Campbell,[72] reproduced by permission of North-Holland Publishing Company)

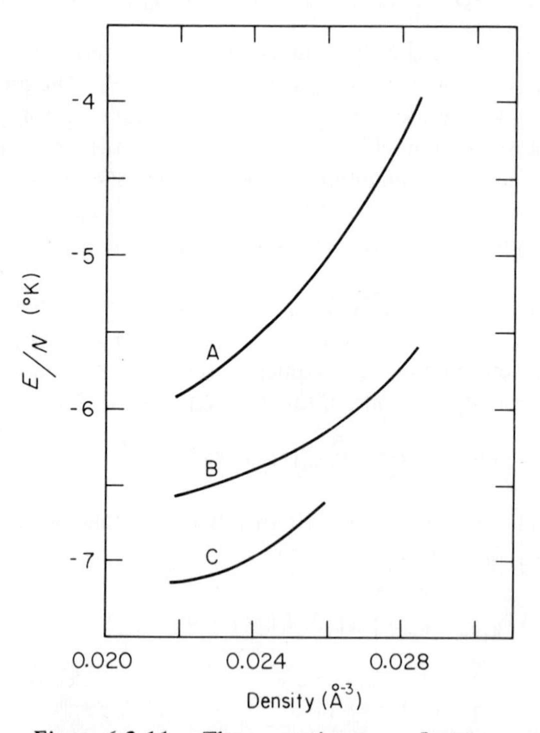

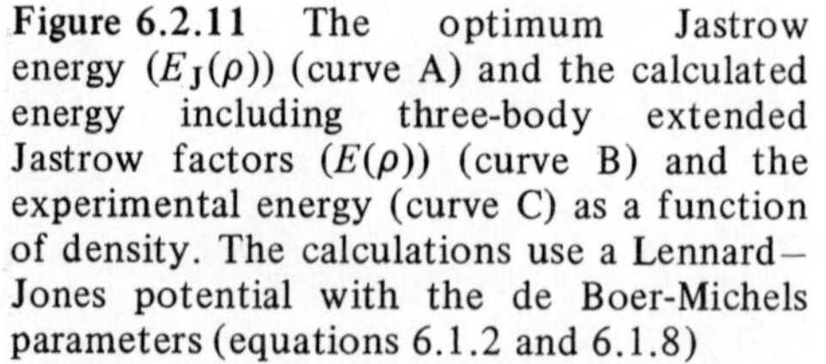

Figure 6.2.11 The optimum Jastrow energy ($E_J(\rho)$) (curve A) and the calculated energy including three-body extended Jastrow factors ($E(\rho)$) (curve B) and the experimental energy (curve C) as a function of density. The calculations use a Lennard–Jones potential with the de Boer-Michels parameters (equations 6.1.2 and 6.1.8)

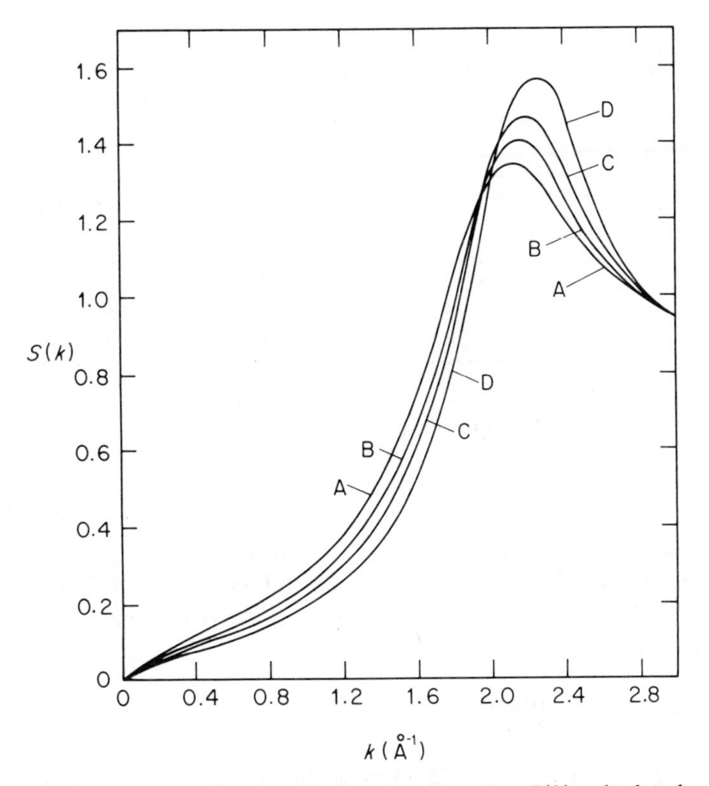

Figure 6.2.12 The liquid structure function $S(k)$ calculated from an optimized trial wavefunction including three-body extended Jastrow factors, at four densities: A, $\rho = 0.02185$ Å^{-3}; B, $\rho = 0.02388$ Å^{-3}; C, $\rho = 0.02571$ Å^{-3}; D, $\rho = 0.02844$ Å^{-3}

With this approximation, the inverse of equation (6.2.93) can be used to obtain F_n entirely in terms of $S(k)$, and $\mathscr{H}_n$ entirely in terms of $\mathscr{H}(k)$ and $S(k)$.

We have used this procedure to estimate δE_3 for liquid ^{4}He at equilibrium density, obtaining[72]

$$\delta E_3/N = -0.68\ {}^\circ\mathrm{K} \tag{6.2.191}$$

which is about 10% of the total energy. Although adding this to the best Jastrow energy exceeds the experimental binding energy by about 0.2 °K, the important point is that the order of magnitude is sufficient to account for the discrepancy between the best Jastrow energy and the experimental results.

Using this procedure to calculate $S(k)$ also provides improved agreement with experiment, as can be seen in Figure 6.2.10. Note, however, that there is little shift in the regions 1.2–1.6 Å^{-1} and 2.4–2.8 Å^{-1}, where the theory is not in such good agreement with experiment. The explanation may lie in an inadequate two-body potential or the absence of a three-body potential, or it could be that it is necessary to have u_n for $n > 3$. More work needs to be done on this problem.

The results of the paired-phonon analysis and the u_3 contribution as a function of density are shown in Figures 6.2.11 and 6.2.12. To obtain the density dependence we use the

Table 6.2.4 The estimated optimum Jastrow energy E_J and the energy including three-body extended Jastrow factors E at several densities ρ. The first column is the optimum Jastrow energy for the simply parametrized $u(r) = -(a/r)^5$ from the molecular dynamics calculation[67]

ρ(Å^{-3})	E_J^{MD}/N (°K)	E_J/N (°K)	E/N (°K)
0.02185	−5.73	−5.93	−6.59
0.02388	−5.32	−5.60	−6.48
0.02571	−4.74	−5.08	−6.21
0.02884	−3.55	−3.99	−5.62

optimized Jastrow energy E_0^{HNC} from the simple parametrized form

$$u_0(r) = -\left(\frac{a}{r}\right)^5$$

from the HNC approximation, and then apply the paired phonon analysis to this result to obtain the optimum HNC energy E_{PPA}^{HNC}. The energy shift is then added to the optimized Monte Carlo energy E_0^{MC} based upon the same simple parametrization, giving our estimate of the optimum Jastrow energy as

$$E_J = E_0^{MC} + \left(E_{PPA}^{HNC} - E_0^{HNC}\right). \tag{6.2.192}$$

Finally, the three-body energy is calculated using the expressions derived in this section, giving our estimate of the energy as

$$E(\rho) = E_J(\rho) + \delta E_3(\rho). \tag{6.2.193}$$

These results are summarized in Table 6.2.4 and Figure 6.2.11.[101] The liquid structure function including the three-body contributions using equation (6.2.187) is shown at several densities in Figure 6.2.12.

In concluding this section, it is important to make an observation about the general approach presented herein. The extended Jastrow formulation of the ground-state problem is quite general. The approximation scheme that it suggested, however, is certainly not the only possible way to proceed. As frequently happens in a series approach to a problem (in this case $\chi = \Sigma u_n$), it may be necessary at some point to introduce a resummation of the series or a partial summation including a part of every u_n. An example of an approximate ground state wave function which cannot be written as the sum of a few u_n is the pairing ground state which received attention some years ago.[102] While the pairing model cannot account for the short-range correlation induced by the short-range repulsion in helium, Hastings and Halley have suggested that it may be necessary to include pairing correlations to understand some features of liquid ^{4}He.[103] Hastings and Johnson are presently investigating a correlated pairing model for liquid ^{4}He[104] motivated by the same model applied to superfluid ^{3}He which we discuss below in Section 4. This model contains both pairing correlations and adequate short-range correlations to care for the short range repulsion.

6.3 Dynamic structure of liquid ^{4}He: The excited states

To understand the thermodynamic properties of the liquid and its response to scattering probes, it is necessary to have information about excited states as well as the ground state. In particular, the thermodynamics of the system at temperatures well below T_λ (2.17 °K) can be understood in terms of an elementary excitation model, where the energy of the excited states is labelled by set of quantum numbers $\{n_\mathbf{k}\}$:

$$E(\{n_\mathbf{k}\}) = E_0 + \sum_\mathbf{k} n_\mathbf{k}\epsilon(\mathbf{k}). \tag{6.3.1}$$

This model was proposed by Landau in 1941, who later showed that $\epsilon(\mathbf{k})$ must have a phonon region[28]

$$\epsilon_{\mathrm{Ph}}(k) = \hbar kc \tag{6.3.2}$$

and a region characterized by a quadratic minimum at a finite momentum k_0:

$$\epsilon_{\mathrm{rot}}(k) = \Delta + \frac{\hbar^2(k-k_0)^2}{2\mu}. \tag{6.3.3}$$

(This is referred to as a roton branch for largely historical reasons.) The quantum numbers $n_\mathbf{k}$ take on any non-negative integer value:

$$n_\mathbf{k} = 0,1,2,3,\ldots . \tag{6.3.4}$$

The simple form of equation (6.3.1) with (6.3.4) gives an occupation number at finite temperature T of the Bose–Einstein form

$$n_\mathbf{k}(T) = \frac{1}{\exp[\beta\epsilon(k)] - 1} \qquad \beta = 1/k_\beta T. \tag{6.3.5}$$

This model gives a low temperature heat capacity composed of two terms: a T^3 term originating from the phonon region (equation 6.3.2) whose coefficient is determined by the velocity of sound; and a $T^{3/2}\exp(-\beta\Delta)$ term arising from the roton region with a coefficient depending on the curvature of $\epsilon_{\mathrm{rot}}(k)$ as given by μ. The parameters necessary to fit the measured low temperature heat capacity are

$$\begin{aligned} \Delta/k_\mathrm{B} &= 8.6\ ^\circ\mathrm{K} \\ k_0 &= 1.9\ \text{Å}^{-1} \\ \mu &= 0.16\, m_{^4\mathrm{He}} \end{aligned} \tag{6.3.6}$$

The surprising thing about this elementary excitation spectrum is that it is not necessary to introduce any single-particle mode to account for the thermodynamic properties. That this is intrinsically connected to the Bose statistics was first demonstrated in an elegant argument by Feynman.[29] He showed that the Bose statistics implied that the lowest lying long wavelength excitation is a phonon with dispersion relation given by equation (6.3.2). It is not surprising that a phonon should be an elementary excitation of the system: this is concluded from the theory of the continuous liquid.[47] Translational symmetry also requires the existence of excited state with vanishing energy in the long wavelength limit (the Goldstone mode).[105] This mode is the phonon. The predominance of this mode is surprising however, as is the fact that the phonon dispersion relation is valid well into non-hydrodynamic wavelengths ($k > 0.5$ Å^{-1}). Numerous microscopic calculations for weakly interacting or low-density many-body boson systems have verified these results.[17,30,31,46] For example, the seminal work of Bogoliubov the weakly interacting Bose gas gave an excitation energy of the form[17]

$$\epsilon(k) = \left[\left(\frac{\hbar^2k^2}{2m}\right)^2 + 2\frac{\hbar^2k^2}{2m}\frac{N_0}{\Omega}v(k)\right]^{1/2} \tag{6.3.7}$$

where $v(k)$ is the Fourier transform of the two-body potential $V(r)$, and N_0 is the Bose–Einstein condensate, i.e. the number of particles in the single-particle $\mathbf{k} = 0$ state when the ground-state wave function is expanded in plane waves. The long wavelength behaviour of $\epsilon(k)$ is

$$\epsilon(k) \xrightarrow[k\to 0]{} \hbar ck \tag{6.3.8}$$

where

$$mc^2 = \frac{N_0 v(0)}{\Omega}. \tag{6.3.9}$$

This illustrates the point, discussed in Section 6.1, that microscopic theories such as Bogoliubov's[17] and its more complicated descendants[30,31] require the macroscopic condensate ($N_0 = O(N)$) in order to obtain the phonon. The calculated spectrum would be single-particle-like in the absence of a condensate, in contradiction to Feynman's proof[29] and the more general arguments[105] demonstrating the existence of the phonon.

Of course the Bogoliubov theory and the others are really pertubation theories, requiring the ground-state of the interacting system to be near the non-interacting ground state. Thus the condensate fraction must be near unity for the theories to be valid.

Shortly after the BCS pairing theory for superconductors was proposed[32] a similar pairing theory for the many-body boson system was suggested.[102] In principle such a model is not restricted to weakly interacting systems: the depletion of the condensate can be large, even total. As in the BCS model, the long wavelength excitation has a gap:

$$\epsilon(k) \xrightarrow[k\to 0]{} [\delta^2 + (\hbar ck)^2]^{1/2} \tag{6.3.10}$$

where δ is proportional to $(N_0/N)^{1/2}$ Because of the absence of the phonon from this theory, it has generally been felt that it is not a useful model for helium.

It is amusing to note that the same symmetry considerations which apply to helium also demonstrate the existence of a long-wavelength collective mode in chargeless fermion systems.[105] This did not, however, result in the BCS model being discarded. Instead, Anderson showed that the collective mode exists simultaneously with the gap modes.[106] Hastings and Halley have recently suggested that a similar situation may occur in the boson system, namely, the coexistence of the long-wavelength phonon and the elementary excitations (complete with gap) deduced from the pairing theory.[107]

Before discussing the consequences of their suggestion, we make a few more remarks about Feynman's long wavelength phonon.

Feynman showed that the wave function of the lowest lying excited state in the long wavelength limit has the form

$$\psi_{\mathbf{k}} = \rho_{\mathbf{k}} \Psi_0 \tag{6.3.11}$$

where Ψ_0 is the exact ground state wave function.[29] This same conclusion is reached in the theory of quantized density fluctuations in the continuum limit, where the collective coordinate is $\rho_{\mathbf{k}}$ and the excited state wave functions are hermite polynomials in $\rho_{\mathbf{k}}$ acting on the ground state wave function; the lowest excited state has the form of $\psi_{\mathbf{k}}$. The commutator

$$[[\rho_{\mathbf{k}}, H], \rho_{-\mathbf{k}}] = \frac{Nk^2}{m} \tag{6.3.12}$$

is used to obtain an evaluation of the excitation energy based upon Feynman's trial function:

$$\frac{\langle \psi_{\mathbf{k}} | H | \psi_{\mathbf{k}} \rangle}{\langle \psi_{\mathbf{k}} | \psi_{\mathbf{k}} \rangle} = E_0 + \epsilon_{\mathrm{BF}}(k) \tag{6.3.13}$$

where

$$\epsilon_{\mathrm{BF}}(k) = \frac{\hbar^2 k^2}{2mS(k)} . \tag{6.3.14}$$

The denominator contains the liquid structure factor $S(k)$ because the normalization of the state $\psi_{\mathbf{k}}$ is just $NS(k)$, (equation 6.2.24). Since this is the liquid structure factor at $T = 0$, it has the limit[47]

$$S(k) \xrightarrow[k \to 0]{} \frac{\hbar k}{2mc} . \tag{6.3.15}$$

Thus

$$\epsilon_{\mathrm{BF}}(k) \xrightarrow[k \to 0]{} \hbar kc. \tag{6.3.16}$$

Bijl has obtained the result (6.3.14) previously, although he failed to identify the denominator with the liquid structure factor.[108] The Bijl–Feynman spectrum is a variational upper-bound on the lowest excitation energy at each momentum $\hbar\mathbf{k}$:

$$\epsilon_{\mathrm{BF}}(k) \geqslant \epsilon(k). \tag{6.3.17}$$

We have just noted that equality occurs in the long wavelength limit. Much recent discussion has centred on dispersion of the phonons. This has been motivated by careful heat capacity measurements by Phillips *et al.* below 0.6 °K[109] and by measurements of ultransonic attenuation.[110] The specific heat measurements show a deviation from the T^3 behaviour, falling above the T^3 dependence at low pressure and below the T^3 dependence at elevated pressure. One suggested explanation of this is to assume the spectrum has dispersion of the form[109,111]

$$\epsilon(k) = \hbar ck(1 - \gamma\hbar^2 k^2 + \ldots). \tag{6.3.18}$$

The low temperature heat capacity anomaly would then require a negative γ at low pressures and a positive γ at higher pressures.

This form of the spectrum was originally proposed by Landau and Khalatnikov, with $\gamma > 0$.[112] In that case the spectrum curves down, and energy and momentum conservation forbid the decay of a single phonon into two phonons. However, with anomalous dispersion ($\gamma < 0$) this phonon decay does occur and will provide an anomalous ultrasonic attenuation, similar to the observed attenuation.[112] Though these experiments would seem to be good evidence for the sign and magnitude of γ,[113] there is more direct evidence available through the direct measurement of the excitation spectrum at long wavelengths which contradict these conclusions. In particular, the difference in time of flight of 30 and 90 MHz ultrasound as a function of the path length travelled gives no evidence of anomalous dispersion.[114] Furthermore, neutron scattering measurements indicate that the excitation spectrum has normal dispersion above 10^4 MHz.[115] Unfortunately, no experiments have been done which directly measure the excitation spectrum in the region 10^2–10^3 MHz, which is the region of most importance to the heat capacity at the temperatures where the anomaly is detected. The absence of anomalous dispersion in the spectrum immediately above and immediately below this experimental gap does not rule out anomalous dispersion in the gap. It is unlikely, however, that the analytic form of equation (6.3.18) can account for the dispersion in all three regions. It may be that an analytic expression is not appropriate at all.

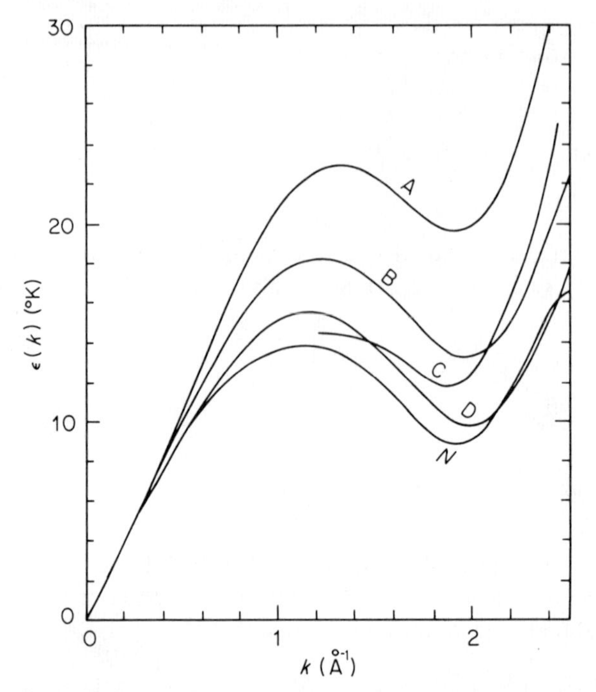

Figure 6.3.1 The elementary excitation spectrum for liquid ^{4}He at $\rho = 0.0218$ Å^{-3}: *N*, neutron scattering; A, the Bijl–Feynman approximation; B, the Jackson–Feenberg approximation; C, the Feynman–Cohen approximation[121] D, the Lee–Lee approximation.[124] (From Lee and Lee,[124] reproduced by permission of the American Institute of Physics)

An intriguing alternative explanation for these anomalous results has been suggested by Hastings and Halley based upon their phonon-augmented pairing theory discussed above.[107] They note that the estimates of the condensate fraction in helium are quite small, of the order of 2% or less.[21,22] Consequently the zero momentum gap in the excitation spectrum predicted by a pairing theory (δ of equation 6.3.10) may be quite small. They suggest that this second branch exists above the phonon branch at small momentum and merges with it at finite k. The magnitude of the gap needed to account for the low T heat capacity anomalies is in the range 0.01–0.05 °K, increasing as a function of density. They do not have a model for the coupling of these excitations to phonons and thus do not comment on the implications of this second branch for ultrasonic attenuation. Unfortunately the energies of the second branch are in the range of energies which have not been directly measured, and it is not obvious what external probe (besides temperature) would couple to them. Hastings and Halley suggests that light scattering and ultrasonic experiments may detect the existence of this branch.[107] Some resolution to the problem of the anomalies in the heat capacity and ultrasonic attenuation will occur when the experimental gap in direct measurement of the energy spectrum is filled in.

The elementary excitation spectrum at intermediate and short wavelengths is bounded by the Bijl–Feynman spectrum. Since $S(k)$ is directly measured in X-ray[61,62] and neutron scattering,[115] $\epsilon_{\rm BF}(k)$ can be compared directly to the Landau excitation spectrum. As can be seen in Figure 6.3.1, $\epsilon_{\rm BF}(k)$ does possess a phonon–roton structure, with a roton minumum near k_0, but with a gap value Δ too large by a factor of two. Note that the inequality (6.3.17) is satisfied.

Neutron scattering

As was mentioned above, direct information about the excited states of the liquid is obtained by scattering various probes from the liquid. These probes leave behind an amount of energy and momentum which is distributed in the liquid according to the coupling of the probe to the states. Cohen and Feynman pointed out that inelastic neutron scattering would provide a direct measurement of the excitation spectrum.[116] The results of the neutron scattering experiment are in very good agreement with the Landau spectrum, as can be seen in Figure 6.3.1.

We give a brief sketch here of the theory of neutron scattering.[117] The interaction between the neutron and the helium atom is extremely short-ranged and can be well-approximated by the Fermi pseudo-potential

$$V_{\text{He}-\text{n}}(\mathbf{r}) = \nu\delta(\mathbf{r}) \tag{6.3.19}$$

where $\delta(\mathbf{r})$ is the Dirac delta function. Then interaction of the neutron with the liquid is

$$H_{\text{He}-\text{n}} = \nu \sum_{i=1}^{N} \delta(\mathbf{r} - \mathbf{r}_i) = \nu\rho(\mathbf{r}) \tag{6.3.20}$$

where $\rho(\mathbf{r})$ is the density operator for the liquid. (Note that the $\mathbf{k}$ Fourier transform of $\rho(\mathbf{r})$ is the density fluctuation operator $\rho_{\mathbf{k}}$ defined in equation (6.1.29)). Using Fermi's Golden Rule to calculate transition rates in a scattering event, the double differential cross-section separates into two factors:

$$\frac{\mathrm{d}^2\sigma}{\mathrm{d}\Omega\mathrm{d}\omega} = A\,S(q,\omega) \tag{6.3.21}$$

where the first factor A is independent of the target and $S(q,\omega)$ is a response function – the dynamic structure function – characterizing the liquid only. $\mathbf{\Omega}$ is the solid angle. At $T = 0$,

$$\begin{aligned} S(q,\omega) &= \frac{\hbar}{N} \sum_n |\langle n | \rho_{\mathbf{q}} | 0 \rangle|^2 \delta(E_n - E_0 - \hbar\omega) \\ &= \frac{1}{2\pi N} \int \exp(-\mathrm{i}\omega t) \langle 0 | \rho_{-\mathbf{q}}(0)\, \rho_{\mathbf{q}}(t) | 0 \rangle\, \mathrm{d}t. \end{aligned} \tag{6.3.22}$$

The momentum change of the neutron upon scattering is $\hbar\mathbf{q}$, and its energy loss is $\hbar\omega$. The state $|n\rangle$ is an exact excited state of energy $E_0 + \hbar\omega$ and momentum $\hbar\mathbf{q}$. Thus $S(q,\omega)$ contains information about excited states which are coupled to the ground state by the density fluctuation operator $\rho_{\mathbf{q}}$. An alternative way of expressing this observation is that $S(q,\omega)$ is a measure of the overlap of an excited state of energy $E_0 + \hbar\omega$ and momentum $\hbar\mathbf{q}$ with the Feynman approximation to the low excited state wave function, $\psi_{\mathbf{q}}$ (equation,6.3.11):

$$S(q,\omega) = \hbar S(q) \sum_n |\langle n | 1 | \Psi_{\mathbf{q}} \rangle|^2\, \delta(E_n - E_0 - \hbar\omega) \tag{6.3.23}$$

where the normalized Feynman state is used here:

$$\Psi_{\mathbf{q}} = \frac{\rho_{\mathbf{q}}\Psi_0}{\sqrt{[NS(q)]}} \tag{6.3.24}$$

and Ψ_0 is the normalized ground-state wave function. In the limit that $\Psi_{\mathbf{q}}$ is an eigenstate of the Hamiltonian, $S(q,\omega)$ will be a δ function at energy $\epsilon_{\text{BF}}(q)$:

$$S(q,\omega) = \hbar S(q)\delta(\epsilon_{\text{BF}}(q) - \hbar\omega) \tag{6.3.25}$$

if

$$H\psi_{\mathbf{q}} = [E_0 + \epsilon_{BF}(q)]\ \psi_{\mathbf{q}}. \tag{6.3.26}$$

Two sum rules satisfied exactly by $S(q,\omega)$[47] are also satisfied by this approximation. The first is obtained by integrating $S(q,\omega)$ over all ω. This gives

$$\begin{aligned}\int_0^\infty S(\mathbf{q},\omega)\,\mathrm{d}\omega &= \sum_n |\langle n\,|\rho_{\mathbf{q}}|\,0\rangle|^2 \\ &= \frac{1}{N}\langle 0\,|\rho_{-\mathbf{q}}\rho_{\mathbf{q}}|\,0\rangle \\ &= S(q).\end{aligned} \tag{6.3.27}$$

Thus integrated neutron scattering gives the liquid structure function. Note that the single resonance ansatz, equation (6.3.25), is in agreement with this sum rule. The second sum rule is obtained by taking the first moment of ω with $S(q,\omega)$:

$$\begin{aligned}\int_0^\infty \hbar\omega S(\mathbf{q},\omega)\,\mathrm{d}\omega &= \frac{1}{N}\sum_n (E_n - E_0)\,|\langle n\,|\rho_{\mathbf{q}}|\,0\rangle|^2 \\ &= \frac{1}{N}\langle 0\,|\rho_{-\mathbf{q}}(H - E_0)\rho_{\mathbf{q}}|\,0\rangle \\ &= \frac{\hbar^2 q^2}{2m}\end{aligned} \tag{6.3.28}$$

where the last line is obtained by using the double commutator in equation (6.3.12). This same result is obtained from the single resonance ansatz.

Extensive neutron scattering has provided a large amount of information about the excited states of helium.[80,115] The results cannot be represented by the single resonance ansatz, but are not badly represented by a single resonance and a background:

$$S(q,\omega) = \hbar Z(q)\delta(\hbar\omega - \epsilon(q)) + S^{\mathrm{II}}(q,\omega).$$

A schematic representation for $S(q,\omega)$ is given in Figure 6.3.2. The energy $\epsilon(q)$ at the resonance is in very good agreement with the Landau excitation curve and falls below $\epsilon_{BF}(q)$ at intermediate and short wavelengths (Figure 6.3.3). The quantity $Z(q)$ is a measure of the overlap between the exact single excited state with energy $\epsilon(q)$ and Feynman's wave function:

$$\frac{Z(q)}{S(q)} = |\langle \mathbf{q}\,|\Psi_{\mathbf{q}}\rangle|^2 \tag{6.3.29}$$

where

$$H\,|\mathbf{q}\rangle = (E_0 + \epsilon(q))\,|q\rangle. \tag{6.3.30}$$

$Z(q)$ and $S(q)$ are compared in Figure 6.3.3, where it is seen that the strength of the Feynman state in $|\mathbf{q}\rangle$ is one for long wavelength and falls with increasing q. Since $S(q,\omega)$ must satisfy an overall sum rule with strength $S(q)$, the integrated background $S^{\mathrm{II}}(q)$ increases to approximately 1 above 3 Å^{-1} (Figure 6.3.3). Thus as $|\Psi_{\mathbf{q}}\rangle$ becomes an increasingly poor representation of $|\mathbf{q}\rangle$ with increasing q, other eigenstates with momentum $\hbar\mathbf{q}$ have an increasingly large overlap with $|\Psi_{\mathbf{q}}\rangle$:

$$|\Psi_{\mathbf{q}}\rangle = \sqrt{\left(\frac{Z(q)}{S(q)}\right)}\,|\mathbf{q}\rangle + \sum_n A_n(\mathbf{q})\,|\mathbf{q},n\rangle. \tag{6.3.31}$$

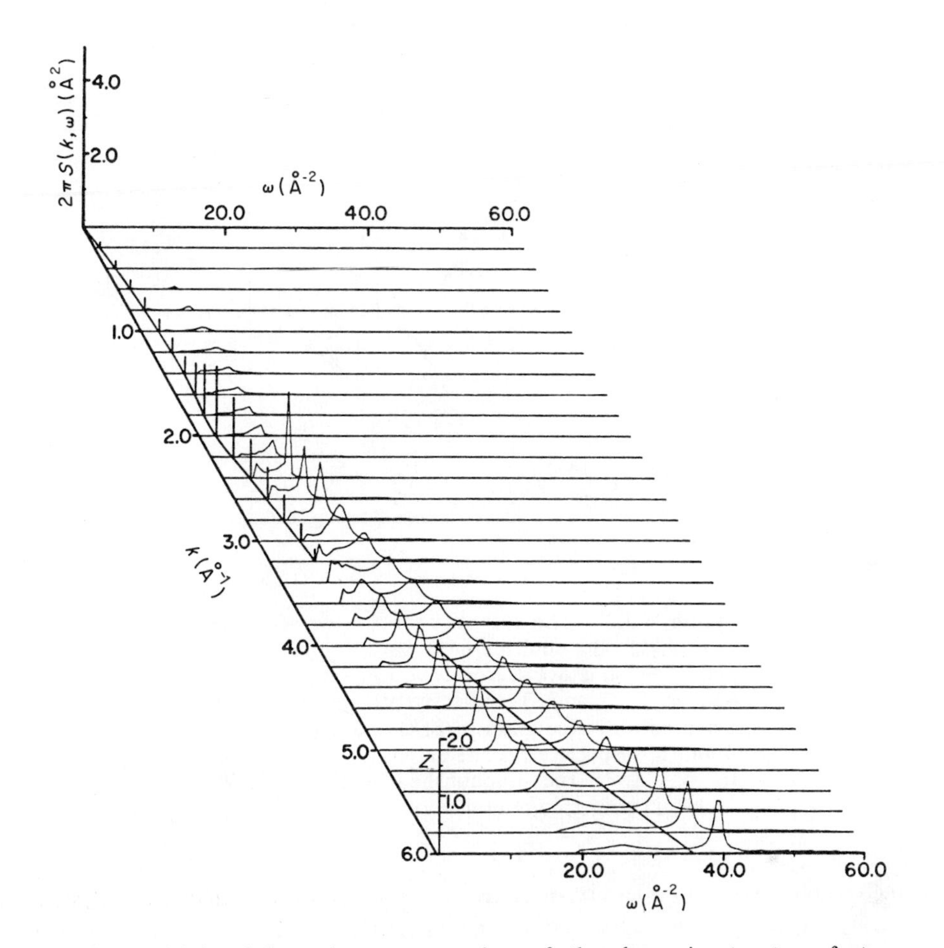

Figure 6.3.2 Schematic representation of the dynamic structure factor $S(k, \omega)$ taken from Jackson's perturbation calculation.[128] The line in the ω–k plane traces out the calculated dispersion curve, and is virtually identical to the B curve in Figure 6.3.1. The height of the peaks is to be read on the scale labelled Z. Multiply by 6.061 to convert Å^{-2} to °K. (From Jackson,[128] reproduced by permission of the American Institute of Physics)

en

$$S^{\mathrm{II}}(q, \omega) = \hbar S(q) \sum_n |A_n(q)|^2 \delta(\epsilon_n(q) - \hbar\omega) \tag{6.3.32}$$

ere

$$H|\mathbf{q}, n\rangle = (E_0 + \epsilon_n(q))|\mathbf{q}, n\rangle. \tag{6.3.33}$$

night be expected that the excited states $|\mathbf{q},n\rangle$ with the largest contribution to $S^{\mathrm{II}}(q,\omega)$ uld be the two excitation states, for which

$$\epsilon_n(\mathbf{q}) = \epsilon(\mathbf{k}) + \epsilon(\mathbf{q} - \mathbf{k}). \tag{6.3.34}$$

e that this would provide a finite strength to $S^{\mathrm{II}}(q,\omega)$ in a wide range of frequencies given all possible values of $\mathbf{k}$ in (6.3.34). It does turn out in fact that much of the structure of ω) can be understood in terms of the elementary excitation spectrum of Landau and the

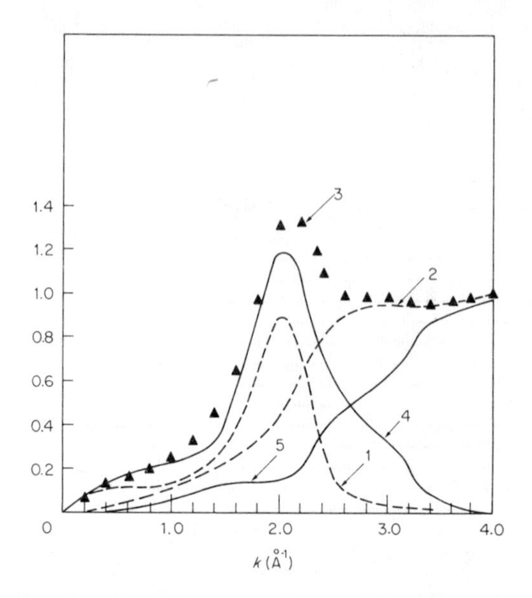

Figure 6.3.3 The intensity of the one-phonon peak $Z(k)$, the multiphonon background $S_{II}(k)$, and their sum $S(k)$, from neutron scattering[115] (curves 1, 2 and 3 respectively) and Jackson's perturbation calculation of $Z(k)$ and $S_{II}(k)$ curves 4 and 5, respectively). The extension of the curves to 4 $Å^{-1}$ is the unpublished work of H. W. Jackson (private communication)

simultaneous multi-excitation density of states. Jackson has made a detailed analysis of the joint density of states based upon the Landau spectrum and has shown that most of the features of $S^{II}(q,\omega)$ can be associated with similar features in the joint density of states.[118]

It is now known that some care must be exercised here because elementary excitations do interact, and consequently some of the states may differ radically from the multi-excitation picture.[119] It has been strongly suggested that some of the excited states correspond to bound roton pairs.[120] Some features in the neutron scattering and the Raman scattering (which we discuss below) have been attributed to these bound states.[119]

While the joint density of states gives many of the features of the neutron scattering, it does not give any of the information necessary to calculate $Z(q)$ or $A_n(q)^2$. To obtain information about these quantities, a better approximation to the single excitation wave function than $\Psi_\mathbf{q}$ must be found, and some approximation to the other excited states must be found. Feynman pointed out that a natural trial function for two excitations

$$\psi_{\mathbf{q}-\mathbf{k},\,\mathbf{k}} = \rho_{\mathbf{q}-\mathbf{k}}\rho_\mathbf{k}\Psi_0/\langle 0\,|\rho_\mathbf{k}^*\rho_{\mathbf{q}-\mathbf{k}}^*\rho_{\mathbf{q}-\mathbf{k}}\rho_\mathbf{q}\,|0\rangle \tag{6.3.35}$$

has energy expectation value

$$\langle \psi_{\mathbf{q}-\mathbf{k},\mathbf{k}}\,|H\,|\,\psi_{\mathbf{q}-\mathbf{k},\mathbf{k}}\rangle = E_0 + \epsilon_{\mathrm{BF}}(\mathbf{q}-\mathbf{k}) + \epsilon_{\mathrm{BF}}(\mathbf{k}) + O(1/\Omega). \tag{6.3.36}$$

In the long wavelength limit, this function must be orthogonalized to $\Psi_\mathbf{q}$ and therefore the strength of $\Psi_\mathbf{q}$ in the two-excitation function is zero.

To improve this model, Feynman and Cohen made use of the observation that a wave-packet built out of Feynman states (6.3.24) does not satisfy conservation of probability density as a stationary eigenstate must.[54] They showed that a dipolar backflow in the surrounding liquid is missing. To correct this they introduced a trial function with dipolar back-flow centred on each particle

$$\psi_{\mathbf{q}}^{B} = \sum_{i=1}^{N} \{\exp(i\mathbf{q}\cdot\mathbf{r}_i)\exp[i\sum_{j\neq i}\phi(\mathbf{r}_{ij})]\}\Psi_0 \tag{6.3.37}$$

where $\phi(\mathbf{r})$ is essentially a velocity potential about $\mathbf{r} = 0$. Ψ_0 is the exact ground state again. They used a dipolar form for ϕ with an adjustable strength

$$\phi(\mathbf{r}) = \frac{-A}{4\pi\rho}\frac{\mathbf{k}\cdot\mathbf{r}}{r^3} \tag{6.3.38}$$

and further simplified the calculation by expanding the exponential containing ϕ:

$$\psi_{FC,\mathbf{q}} = \sum_{i=1}^{N} \exp(i\mathbf{q}\cdot\mathbf{r}_i)\left[1 + i\sum_{j\neq i}^{N}\phi(\mathbf{r}_{ij})\right]\Psi_0. \tag{6.3.39}$$

The classical value of $A = 1$ compares favourably with their variationally best $A = 0.93$.[54]

Recently, Chester and Padmore repeated this calculation, using Monte Carlo integration to avoid many approximations in Feynman and Cohen's work.[121] Their results are compared with experiment in Figure 6.3.1. The energy of the roton minimum of 11.2 °K is an improvement over the Bijl–Feynman energy.

An alternative to the back-flow picture can be motivated by expressing the backflow function in equation (6.3.39) in terms of its Fourier transform. Then

$$\psi_{FC,\mathbf{q}} = \left\{\rho_{\mathbf{q}} - \frac{A}{N}\sum_{\mathbf{k}}\frac{\hbar^2}{2m}\frac{\mathbf{q}\cdot\mathbf{k}}{k^2}[\rho_{\mathbf{q}-\mathbf{k}}\rho_{\mathbf{k}} - \rho_{\mathbf{q}}]\right\}\Psi_0 \tag{6.3.40}$$

so that it is a mixture between the single-Feynman-phonon state $\Psi_{\mathbf{q}}$ and the two-Feynman-phonon state $\psi_{\mathbf{q}-\mathbf{k},\mathbf{k}}$ (equation 6.3.35). Such a hybridization of two ρ states with the single ρ state is to be expected by noting that the Hamiltonian connects these states:

$$\langle\Psi_{\mathbf{q}}|(H - E_0)|\psi_{\mathbf{k}-\mathbf{q},\mathbf{k}}\rangle = \frac{(\hbar^2/2m)\mathbf{q}\cdot[(\mathbf{k}-\mathbf{q})S(\mathbf{k}-\mathbf{q}) + \mathbf{k}S(\mathbf{k})]}{[NS(k)S(q)S(\mathbf{k}-\mathbf{q})]^{1/2}}. \tag{6.3.41}$$

Jackson and Feenberg[53] and Kuper[122] suggested that these two-Feynman-phonon processes might make an important contribution to the excitation spectrum. Jackson and Feenberg used Brillouin–Wigner perturbation theory to include these processes, solving the transcendental equation[53]

$$\epsilon(q) = \epsilon_{BF}(q) + \frac{1}{2}\sum_{\mathbf{k}}{}'\frac{|\langle\Psi_{\mathbf{q}}|H - E_0 - \epsilon_{BF}(q)|\psi_{\mathbf{q}-\mathbf{k},\mathbf{k}}\rangle|^2}{\epsilon(q) - \epsilon_{BF}(k) - \epsilon_{BF}(\mathbf{q}-\mathbf{k})}. \tag{6.3.42}$$

Their results, shown in Figure 6.3.1, are very similar to those of Feynman and Cohen's calculation. Recently, Lin-Liu and Woo provided a variational formulation of this problem which avoids the difficulties of a non-orthogonalized basis.[123] They propose a trial function

$$\psi = \zeta_{\mathbf{q}}|\Psi_{\mathbf{q}}\rangle + \tfrac{1}{2}\sum_{\mathbf{k}}{}'\xi_{\mathbf{q}}(\mathbf{k})|\psi_{\mathbf{q}-\mathbf{k},\mathbf{k}}\rangle \tag{6.3.43}$$

and minimize $\langle\psi|H|\psi\rangle/\langle\psi|\psi\rangle$ with respect to $\zeta_{\mathbf{q}}$ and $\xi_{\mathbf{q}}(\mathbf{k})$. The equations they obtain are

$$[\epsilon(\mathbf{q}) - \epsilon_{BF}(\mathbf{q})]\zeta_{\mathbf{q}} = \frac{1}{2}\sum_{\mathbf{k}}{}'\xi_{\mathbf{q}}(\mathbf{k})\langle\Psi_{\mathbf{q}}|H - E_0 - \epsilon(\mathbf{q})]|\psi_{\mathbf{q}-\mathbf{k},\mathbf{k}}\rangle \tag{6.3.44}$$

and

$$[\epsilon(\mathbf{q}) - \epsilon_{BF}(\mathbf{k}) - \epsilon_{BF}(\mathbf{q}-\mathbf{k})]\,\xi_{\mathbf{q}}(\mathbf{k}) = \zeta_{\mathbf{q}} \langle \Psi_{\mathbf{q}} | H - E_0 - \epsilon_{BF}(\mathbf{q}) | \psi_{\mathbf{q}-\mathbf{k},\mathbf{k}} \rangle + \frac{1}{2} \sum_{\mathbf{l}(\neq \mathbf{k},\mathbf{q}-\mathbf{k})} \xi_{\mathbf{q}}(\mathbf{l}) \langle \Psi_{\mathbf{q}-\mathbf{k},\mathbf{k}} | H - E_0 - \epsilon(\mathbf{q}) | \Psi_{\mathbf{q}-\mathbf{l},\mathbf{l}} \rangle. \tag{6.3.45}$$

The Brillouin–Wigner expression of Jackson and Feenberg is obtained from these equations by ignoring the last term in equation (6.3.45). Comparing the form of the trial wave function (6.3.43) with the Feynman–Cohen trial form equation (6.3.40), it is not at all surprising that Jackson and Feenberg obtain results very similar to Feynman and Cohen's results.

In a recent calculation, Lee and Lee have obtained further improvement with the experimental $\epsilon(q)$ by including third and fourth order terms in the Brillouin–Wigner perturbation scheme.[124] The details of this tedious calculation are not included here. We simply note that the results suggest a fairly rapid convergence of the perturbation theory approach as can be seen in Figure 6.3.1.

Experimental results for the density dependence of the roton gap parameter Δ show a marked decrease in Δ with increasing density.[125] While the calculation of Lee and Lee was only applied at equilibrium density, Bartley *et al.* have carried out the Jackson–Feenberg calculation as a function of density.[126] They used the liquid structure function calculated for the parametrized Jastrow function (equation 6.2.41) discussed in Section 6.2. Their results are compared with experimental determinations of the roton gap Δ, the roton momentum k_0, and the curvature parameter μ in Figure 6.3.4. Also shown are points at these densities obtained in the Monte Carlo calculation of the Feynman–Cohen spectrum done by Padmore and Chester.[121] This latter calculation used the same form of the Jastrow function.

Improvements in these theoretical calculations can be expected by including the three-body extended Jastrow terms (discussed in Section 6.2) in the ground-state. For example, it can be seen in Figures 6.2.10 and 6.2.12 that the three-body terms make significant contributions at the maximum in $S(k)$, which is located very near k_0. These contributions become more important with increasing density (Figure 6.2.12). The height of the first maximum in $S(k)$ plays an important role in the determination of Δ, as can be seen in the Bijl–Feynman approximation for the excitation spectrum (equation 6.3.14).

We would like to make some comments on the variational formulation of the theory of the phonon–roton spectrum proposed by Lin-Liu and Woo (equations 6.3.43–6.3.45).[123] This set of equations has several different possible types of solution. The first is an improvement on the Bijl–Feynman spectrum, which should be essentially the result of Lee and Lee[124] if the last term in (6.3.45) is retained, and will be the Jackson–Feenberg result if this term is ignored. Most of the strength of ψ will be in $\zeta_{\mathbf{q}}$ in this solution. The second set of solutions should be the scattering solutions for two-Feynman-phonons with total momentum $\hbar\mathbf{q}$. The resultant eigenvalue is the sum of the Bijl–Feynman energies:

$$\epsilon(\mathbf{q}-\mathbf{k},\mathbf{k}) = \epsilon_{BF}(\mathbf{q}-\mathbf{k}) + \epsilon_{BF}(\mathbf{k}). \tag{6.3.46}$$

The corresponding wave functions ψ, however, will contain phase shifts which affect the value of $|A_n(q)|^2$ which contribute to $S^{II}(q,\omega)$ in the neutron scattering (see equation 6.3.32) (n refers to the different choices of $\mathbf{k}$ in equation (6.3.46)). The Green function calculations of $S(q,\omega)$ based upon second-quantized representations of the phonon interactions (see below) implicitly account for these phase shifts.[127–130,119]

Finally, a possible solution of these equations is a bound state of *two-Feynman-phonons.*

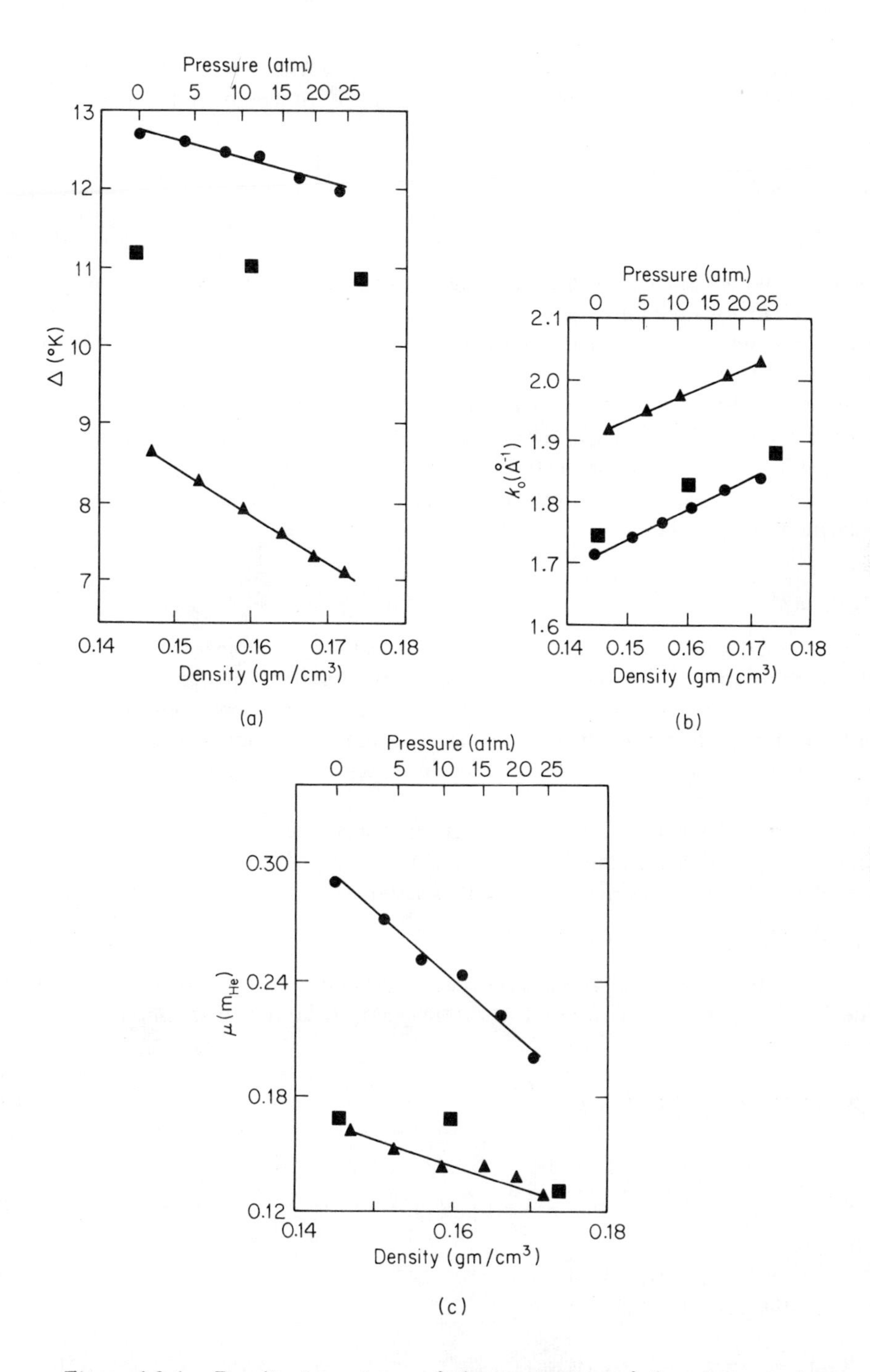

Figure 6.3.4 Density dependence of the parameters of the roton spectrum: (a) the roton gap Δ, (b) the roton momentum k_0; (c) the roton curvature parameter μ; (▲) experimental[125]; (●) Brillouin–Wigner perturbation theory using the single parameter Jastrow function (equation 6.2.41)[126]; (■) Feynman–Cohen calculation using the same Jastrow wave function.[121] (From Bartley, Robinson and Woo,[126] reproduced by permission of the Plenum Publishing Corporation)

These states would be characterized by $\xi_{\mathbf{q}}(\mathbf{k})$ nearly independent of $\mathbf{k}$ so that the phonons are localized close to one another. A bound state near the Bijl–Feynman roton minimum

$$\Delta_{BF} = \epsilon_{BF}(\mathbf{k}_0) \tag{6.3.47}$$

would have energy

$$\epsilon = 2\Delta_{BF} + \delta\epsilon. \tag{6.3.48}$$

In the next section we discuss evidence from light scattering for the existence of bound roton pairs (with a negative energy shift $\delta\epsilon$).

A calculation similar to this has been carried out by Kebukawa *et al.*[130] The matrix elements are calculated within a density fluctuation theory,[131] which is a semi-phenomenological theory since they chose the two-body potential energy to give agreement with experimental results for $\epsilon(q)$. Having made this choice, they obtain semi-quantitative agreement with the multi-phonon features in the neutron scattering $S(q,\omega)$. Of particular interest is their identification of the plateau in $\epsilon(q)$ around 3 Å^{-1} as a two-roton branch with finite total momentum.

Light scattering

Helium is a polarizable medium and consequently scatters electromagnetic radiation. The resemblance to neutron scattering is emphasized if we imagine a photon which undergoes inelastic scattering from the liquid. The energy as a function of momentum for a photon is quite different from a thermal neutron, so the excited states of the liquid which are left behind by a scattered photon will lie in an energy range different from those resulting from neutron scattering.

The response of the liquid to light is not as familiar as the response to a neutron probe (discussed above) so we shall go into it in some detail here.[132,133]

The incident light in the absence of the liquid is described by the electromagnetic field

$$\boldsymbol{\varepsilon}_0(\mathbf{r}, t) = \boldsymbol{\varepsilon}_0 \exp[\mathrm{i}(\mathbf{k}_0 \cdot \mathbf{r} - \omega_0 t)]. \tag{6.3.49}$$

The intensity of the scattered light is measured at some point $\mathbf{R}$ far from the scattering sample. The time Fourier transform of the total electromagnetic field at that distant point is given by[134]

$$\boldsymbol{\varepsilon}(\mathbf{R}, \omega) = \boldsymbol{\varepsilon}_0(\mathbf{R}, \omega) + \boldsymbol{\varepsilon}'(\mathbf{R}, \omega) \tag{6.3.50}$$

where

$$\boldsymbol{\varepsilon}'(\mathbf{R}, \omega) = \frac{\omega^2}{c^2 R^2} \exp\left(\frac{\mathrm{i}\omega R}{c}\right)\left(\overleftrightarrow{1} - \frac{\mathbf{RR}}{R^2}\right) \int \mathrm{d}t'\mathrm{d}\mathbf{r}' \exp[-\mathrm{i}(\mathbf{k} \cdot \mathbf{r}' - \omega t)]\mathbf{P}(\mathbf{r}', t') \tag{6.3.51}$$

$\mathbf{P}$ is the polarization operator and $\mathbf{k}$ is the wave vector of the scattered light. The measured intensity of scattered light of polarization $\hat{\epsilon}$ is[135]

$$I_{\hat{\epsilon}}(\mathbf{R}, \omega) = \frac{R^2 c}{\Omega(4\pi)^2 2\tau} \langle\, |\hat{\epsilon} \cdot \boldsymbol{\varepsilon}'(\mathbf{R}, \omega)|^2 \rangle \tag{6.3.52}$$

where Ω is the volume of the sample scattering light and 2τ is the duration in time of the incoming pulse of light. Then using the fact that

$$\hat{\epsilon} \cdot \left(\overleftrightarrow{1} - \frac{\mathbf{RR}}{R^2}\right) = \hat{\epsilon} \tag{6.3.53}$$

equation (6.3.52) becomes

$$I_{\hat{\epsilon}}(\mathbf{k}, \omega) = \frac{\omega^4}{(4\pi)^2 c^3} \int_{-\infty}^{\infty} dt \exp(i\omega t) \int d\mathbf{r} \int d\mathbf{r}' \exp[-i\mathbf{k} \cdot (\mathbf{r} - \mathbf{r}')] \langle \hat{\epsilon} \cdot \mathbf{P}(\mathbf{r}t)\hat{\epsilon} \cdot \mathbf{P}^{\dagger}(\mathbf{r}', 0) \rangle. \tag{6.3.54}$$

One can show that the polarization operator is given by a series[132]

$$\mathbf{P}(\mathbf{r}, t) = \mathbf{P}^{(0)}(\mathbf{r}, t) + \mathbf{P}^{(1)}(\mathbf{r}, t) + \ldots \tag{6.3.55}$$

where

$$\mathbf{P}^{(0)}(\mathbf{r}, t) = \alpha_0 \boldsymbol{\varepsilon}_0(\mathbf{r}\, t)\rho(\mathbf{r}\, t) \tag{6.3.56}$$

$$\mathbf{P}^{(1)}(\mathbf{r}, t) = \sum_{\hat{\mathrm{p}},\mathrm{p}'} \rho(\mathbf{r}\, t)\, \alpha_{\mathrm{p}}\alpha_{\mathrm{p}'} \int \overleftrightarrow{T}_{\mathrm{pp}'}(\mathbf{r} - \mathbf{r}') \cdot \boldsymbol{\varepsilon}_0(\mathbf{r}'\, t)\rho(\mathbf{r}'\, t)\, d\mathbf{r}' \tag{6.3.57}$$

where $\overleftrightarrow{T}_{\mathrm{pp}'}(\mathbf{r})$ describes the hopping rate of the p atomic excited state on a helium atom at the origin to the p′ atomic excited state on another helium atom at **r**. The atomic polarizability α_0 is given by

$$\alpha_0 = \sum_{\mathrm{p}} \alpha_{\mathrm{p}} \tag{6.3.58}$$

where

$$\alpha_{\mathrm{p}} = \frac{2\,|\,X_{\mathrm{p}}\,|^2 \epsilon_{\mathrm{p}}}{\epsilon_{\mathrm{p}}^2 - \hbar^2\omega_0^2}$$

where ϵ_{p} is the p atomic state energy and X_{p} is the matrix element of the dipole operator between the ground state and the p excited state. $\overleftrightarrow{T}_{\mathrm{pp}'}(\mathbf{r})$ has the general form

$$\overleftrightarrow{T}_{\mathrm{pp}'}(\mathbf{r}) = t_{\mathrm{d}}^{\mathrm{pp}'}(r)\left(\frac{3\mathbf{rr}}{r^2} - \overleftrightarrow{1}\right) + t_{\mathrm{s}}^{\mathrm{pp}'}(r)\,\overleftrightarrow{1} \tag{6.3.59}$$

At large r, $t_{\mathrm{d}}^{\mathrm{pp}'}(r) \to 1/r^3$ and $t_{\mathrm{s}}(r) \to 0$. There is, however, a great deal of uncertainty about these functions for small r.[132]

It is convenient to define the function

$$F_{\hat{\epsilon},\hat{\epsilon}_0}(\mathbf{q}) = \sum_{\mathrm{pp}'} \alpha_{\mathrm{p}}\alpha_{\mathrm{p}'}\hat{\epsilon} \cdot \overleftrightarrow{T}_{\mathrm{pp}'}(\mathbf{q}) \cdot \hat{\epsilon}_0 \tag{6.3.60}$$

where $\hat{\epsilon}_0$ is the polarization of the incident light and

$$\overleftrightarrow{T}_{\mathrm{pp}'}(\mathbf{q}) = \int d\mathbf{r} \exp(-i\mathbf{q} \cdot \mathbf{r})\overleftrightarrow{T}_{\mathrm{pp}'}(r). \tag{6.3.61}$$

Then the important part of $\mathbf{P}^{(1)}(\mathbf{r}, t)$ is

$$\hat{\epsilon} \cdot \mathbf{P}^{(1)}(\mathbf{r}, t) = \sum_{\mathbf{q}} \rho(\mathbf{r}\, t)\rho_{\mathbf{k}_0-\mathbf{q}}(t)F_{\hat{\epsilon}\hat{\epsilon}_0}(\mathbf{q}) \exp[i(\mathbf{q} \cdot \mathbf{r} - \omega_0 t)]\,. \tag{6.3.62}$$

Equations (6.3.62) and (6.3.56) may now be substituted into (6.3.54) to obtain the relative intensity $h_{\hat{\epsilon}}$ of light scattered through a momentum change of $\hbar\Delta\mathbf{k}$ and energy change $\hbar\Delta\omega$

given by

$$\Delta\mathbf{k} = \mathbf{k} - \mathbf{k}_0 \qquad \Delta\omega = \omega - \omega_0$$

$$h_{\hat{\epsilon}}(\Delta\mathbf{k}, \Delta\omega) = I_{\hat{\epsilon}}(\mathbf{k}, \omega)/I_0 = \frac{\omega^4 N}{(4\pi)^2 c^3}\Bigg\{\alpha_0^2(\hat{\epsilon}\cdot\hat{\epsilon}_0)^2 S(\Delta\mathbf{k}, \Delta\omega)$$

$$+ \alpha_0\hat{\epsilon}\cdot\hat{\epsilon}_0 2\mathrm{Re}\frac{1}{\Omega}\sum_{\mathbf{q}} F_{\hat{\epsilon},\hat{\epsilon}_0}(\mathbf{q}+\mathbf{k}) S_3(\mathbf{q}, -\mathbf{q}-\Delta\mathbf{k}; \Delta\mathbf{k}; \Delta\omega)$$

$$+ \frac{1}{\Omega^2}\sum_{\mathbf{q}}\sum_{\mathbf{q}'} F_{\hat{\epsilon}\hat{\epsilon}_0}(\mathbf{q}+\mathbf{k}) F_{\hat{\epsilon}\hat{\epsilon}_0}(\mathbf{q}'+\mathbf{k}) S_4(\mathbf{q}, -\mathbf{q}-\Delta\mathbf{k}; -\mathbf{q}', \mathbf{q}'+\Delta\mathbf{k}; \Delta\omega)\Bigg\} \qquad (6.3.63)$$

where $S(k, \omega)$ is the dynamic structure function measured in neutron scattering (equation 6.3.22).

$$S_3(\mathbf{k}, \mathbf{k}'; \mathbf{q}; \omega) = \frac{1}{N}\int \langle 0 \,|\, \rho_{\mathbf{k}}(t)\rho_{\mathbf{k}'}(t)\rho_{\mathbf{q}}(0) \,|\, 0\rangle \exp(-i\omega t)\, dt$$

$$= \frac{\hbar}{N}\sum_n \langle 0 \,|\, \rho_{\mathbf{k}}\rho_{\mathbf{k}'} \,|\, n\rangle\langle n \,|\, \rho_{\mathbf{q}} \,|\, 0\rangle\, \delta(E_n - E_0 - \hbar\omega)\delta_{\mathbf{k}+\mathbf{k}'+\mathbf{q},0} \qquad (6.3.64)$$

and

$$S_4(\mathbf{k}, \mathbf{k}'; \mathbf{q}, \mathbf{q}'; \omega) = \frac{1}{N}\int \langle 0 \,|\, \rho_{\mathbf{k}}(t)\rho_{\mathbf{k}'}\rho_{\mathbf{q}}(0)\rho_{\mathbf{q}'}(0) \,|\, 0\rangle \exp(-i\omega t)\, dt$$

$$= \frac{\hbar}{N}\sum_n \langle 0 \,|\, \rho_{\mathbf{k}}\rho_{\mathbf{k}'} \,|\, n\rangle\langle n \,|\, \rho_{\mathbf{q}}\rho_{\mathbf{q}'} \,|\, 0\rangle\, \delta(E_n - E_0 - \hbar\omega)\delta_{\mathbf{k}+\mathbf{k}'+\mathbf{q}+\mathbf{q}',0} \qquad (6.3.65)$$

The last term of equation (6.3.63) reduces to the results of other analyses[80] if $\overleftrightarrow{T}_{\mathrm{pp}'}(\mathbf{r})$ is taken as independent of p and p$'$, and the s wave portion is omitted. See Reference 133 for a more general analysis.

The inelastic scattering of photons from the liquid must satisfy conservation of total energy and total momentum. Consequently $\hbar\Delta\omega$ and $\hbar\Delta\mathbf{k}$ may be interpreted as the energy and momentum deposited in the liquid by the photon in the form of an excited state of the liquid. That is the source of the delta functions in equations (6.3.22), (6.3.64) and (6.3.65). The extreme difference between the dispersion relation of a photon and the excitation curve in the liquid limits the inelastic photon scattering to very small values of $\Delta k (< 10^{-3}$ Å$^{-1})$. That feature shows up in a calculation of the density of states.[136] The magnitude of $h_{\hat{\epsilon}}$ is also determined by the size of the matrix elements of the density fluctuation operators in each term. The fact that the density of states restricts $\Delta\mathbf{k}$ to long wavelengths permits us to use the analysis of the last section to assess the importance of each of the three terms in $h_{\hat{\epsilon}}$. The most important fact is that at these long wavelengths, the lowest lying excited state is given nearly exactly by the Feynman state, $\Psi_{\Delta\mathbf{k}}$ (equation 6.3.24). Then the one factor in the first and second terms of $h_{\hat{\epsilon}}$,

$$\langle 0 \,|\, \rho_{\Delta\mathbf{k}} \,|\, n\rangle \xrightarrow[\Delta\mathbf{k}\to 0]{} 0 \qquad (6.3.66)$$

for $|\, n\rangle \neq |\, \Psi_{\Delta\mathbf{k}}\rangle$.

Similarly,

$$\langle 0 \,|\, \rho_{\mathbf{q}}\rho_{-\mathbf{q}-\Delta\mathbf{k}} \,|\, n\rangle \xrightarrow[\Delta\mathbf{k}]{} 0 \qquad (6.3.67)$$

if

$$|n\rangle = |\Psi_{\Delta \mathbf{k}}\rangle.$$

Then the middle term in $h_{\hat{\epsilon}}$, given by

$$h_{\hat{\epsilon}}^{(2)}(\Delta \mathbf{k}, \Delta\omega) \propto \frac{N}{\Omega} \sum_{\mathbf{q}} F(\mathbf{q}) S_3(\mathbf{q}, -\mathbf{q} - \Delta\mathbf{k}; \Delta\mathbf{k}; \Delta\omega) \xrightarrow[\Delta\mathbf{k}\to 0]{} 0. \tag{6.3.68}$$

By the same token, the sum over intermediate states in $S(\Delta\mathbf{k},\Delta\omega)$ will include only $|\Psi_{\Delta\mathbf{k}}\rangle$ and the sum over intermediate states in S_4 will not contain $|\Psi_{\Delta\mathbf{k}}\rangle$ in the long wavelength limit.

The measurement of $S(\Delta\mathbf{k},\Delta\omega)$ by inelastic light scattering is known as Brillouin scattering. By the above arguments we see that, at these long wavelengths, $S^{\mathrm{II}}(\Delta\mathbf{k},\Delta\omega)$ is nearly completely suppressed and $S(\Delta k,\Delta\omega)$ is dominated by a single peak at

$$\Delta\omega = \Delta k c_s \tag{6.3.69}$$

where c_s is the velocity of sound in the liquid. The measurements of the Brillouin peak in liquid ^4He[137] are in good agreement with the ultrasonic measurements of the phonon dispersion curve at the same frequencies.

From the sum rule, equation (6.3.27), we see that the total intensity of the Brillouin scattering gives the liquid structure factor $S(\Delta\mathbf{k})$. In principle, our discussion thus far applies also to X-ray scattering. Since the X-ray scattering is at much shorter wavelengths (comparable to neutron scattering wavelengths) limitations on energy resolution make it only practical to measure the total intensity $S(\Delta k)$. These are the most reliable determinations of the liquid structure at short and intermediate wavelengths.[62,63]

The measurement of the third term in $h_{\hat{\epsilon}}$,

$$h_{\hat{\epsilon}}^{(3)}(\Delta\mathbf{k}, \Delta\omega) = \frac{\omega^4 N}{(4\pi)^2 c^3} \frac{1}{\Omega^2} \sum_{\mathbf{q}\mathbf{q}'} F_{\hat{\epsilon}\hat{\epsilon}_0}(\mathbf{q}+\mathbf{k}) F_{\hat{\epsilon}\hat{\epsilon}_0}(\mathbf{q}'+\mathbf{k}) S_4(\mathbf{q}, -\mathbf{q}-\Delta\mathbf{k}; -\mathbf{q}', \mathbf{q}'+\Delta\mathbf{k}; \Delta\omega) \tag{6.3.70}$$

has been the subject of considerable effort recently.[138,139,120] Halley pointed out in 1969 that Raman scattering would be a particularly useful technique for studying higher frequency excitations in helium.[130] This second order scattering may create two excitations of nearly equal but opposite momentum, so that the total energy $\hbar\Delta\omega$ may be of the order of 10 °K. Furthermore, Halley pointed out that the joint density of states diverges at energies equal to twice the roton energy and twice the maximum energy of the excitation spectrum, and that these features should be evident in the structure of $h^{(3)}$.[136] The joint density of states is shown in Figure 6.3.5. Note that there is also a singularity above 3 Å^{-1} due to the plateau in the elementary excitation curve as measured by neutron scattering.[140]

The lower resolution Raman scattering results are shown in Figure 6.3.6. There is a single peak near twice the roton minimum but the remaining structure from the density of states is absent. The difference between this and Figure 6.3.5 may be attributed to the strength of the coupling of the ground state to the excited states through two density fluctuation operators (equation 6.3.6) and the coupling of light to the liquid as expressed by the functions $F(q)$ in equation (6.3.70).

In some models it has been assumed that the overall structure can be understood in terms of non-interacting elementary excitations.[134,141] In such models the time Fourier transform of

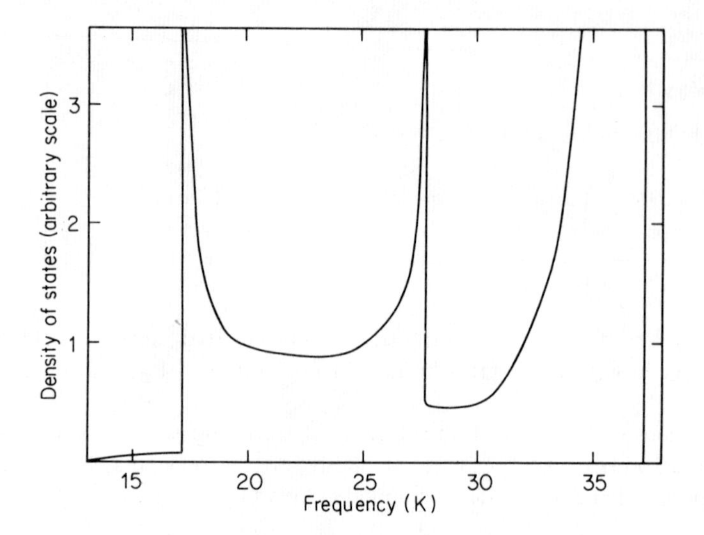

Figure 6.3.5 The calculated two-excitation density of states using the observed dispersion relation $\epsilon(q)$ for $0 < q < 3.6$ Å^{-1}. The singularities occur at the energies of the roton minimum, the maximum for $q \simeq 1$ Å^{-1} and the plateau at $q \gtrsim 3$ Å^{-1}. (From Woods and Cowley,[80] reproduced by permission of The Institute of Physics)

S_4 decouples in a simple manner

$$S_4(\mathbf{q}, -\mathbf{q} - \Delta\mathbf{k}; -\mathbf{q}', \mathbf{q}' + \Delta\mathbf{k}; t) = \frac{1}{N}\langle \rho_{\mathbf{q}}(t)\rho_{-\mathbf{q}-\Delta\mathbf{k}}(t)\rho_{-\mathbf{q}'}(0)\rho_{\mathbf{q}'+\Delta\mathbf{k}}(0)\rangle$$

$$\Rightarrow \frac{1}{N}\langle \rho_{\mathbf{q}}(t)\rho_{-\mathbf{q}}(0)\rangle\langle \rho_{-\mathbf{q}-\Delta\mathbf{k}}(t)\rho_{\mathbf{q}+\Delta\mathbf{k}}(0)\rangle\,[\delta_{\mathbf{q},\mathbf{q}'} + \delta_{\mathbf{q},-\mathbf{q}'-\Delta\mathbf{k}}]$$

$$= NS(\mathbf{q}, t)S(-\mathbf{q} - \Delta\mathbf{k}, t)[\delta_{\mathbf{q},\mathbf{q}'} + \delta_{\mathbf{q},-\mathbf{q}'-\Delta\mathbf{k}}] \qquad (6.3.71)$$

where $S(\mathbf{q},t)$ is the time Fourier transform of $S(\mathbf{q},\omega)$. Thus the neutron scattering measurement of $S(\mathbf{q},\omega)$ can be used to predict the Raman scattering results. Cowley's calculation of the

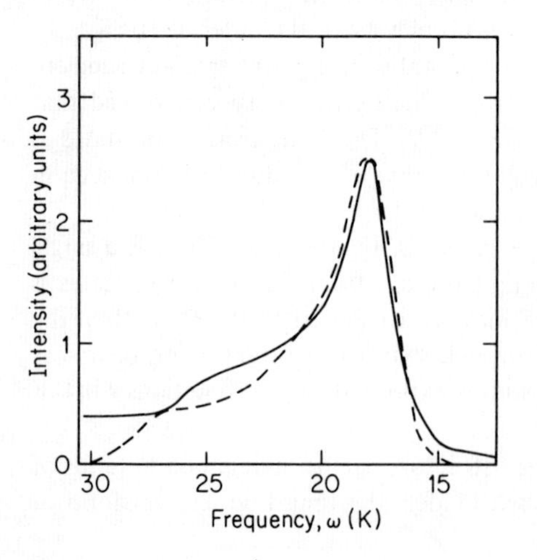

Figure 6.3.6 The Raman scattering intensity from liquid ^{4}He at low resolution.[138] The dashed line is the interacting approximation of equation (6.3.71). (From Woods and Cowley,[80] reproduced by permission of the Institute of Physics)

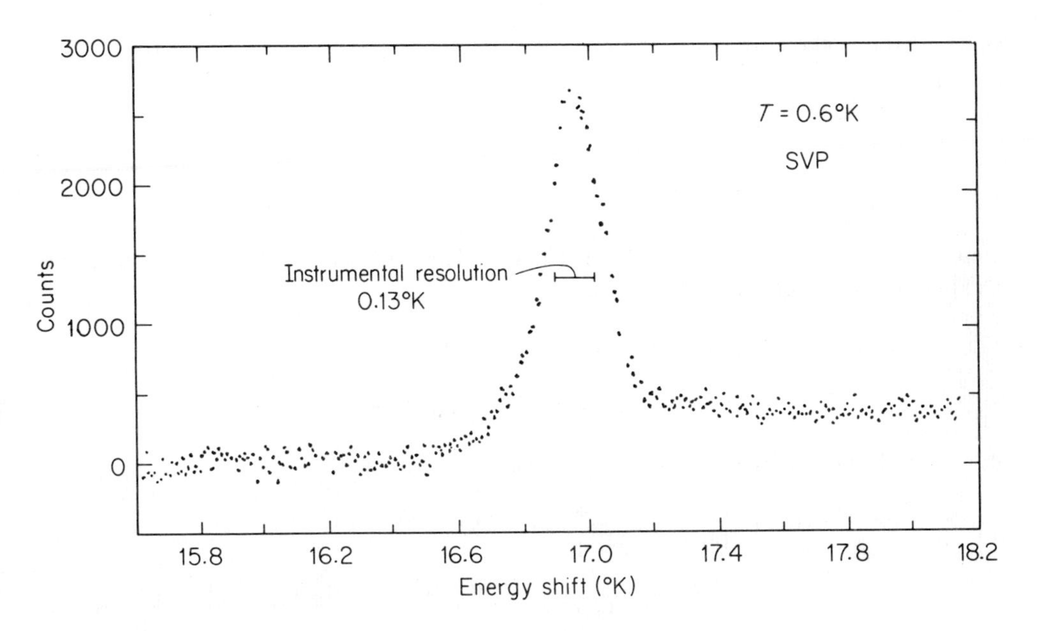

Figure 6.3.7 High resolution Raman scattering intensity. (From Murray, Woerner and Greytak,[120] reproduced by permission of The Institute of Physics)

Raman scattering using this approximation and a model for the function $F(q)$ is shown in Figure 6.3.6.[141] The high energy peaks are suppressed in agreement with experiment.

Ruvalds and Zawadowski suggested earlier that the structure of the Raman scattering could be understood in terms of the interactions of the elementary excitations.[142] While their model does not lead to any better agreement with the low resolution experiments than the non-interacting model, it was responsible for a very important prediction: an attractive interaction between the rotons will produce a two-roton bound state with energy below twice the roton minimum. The strongest evidence for existence of the bound roton pair is to be found in the structure in the high resolution Raman scattering at and below 2Δ[120,139] (see Figure 6.3.7. The peak in the highest resolution experiment occurs at 16.97 ± 0.03 °K which is significantly lower than the value of 2Δ determined by neutron scattering, 17.34 ± 0.08 °K.[120] Murray *et. al.* interpret the line shape as indication of a two-roton binding energy of 0.22 ± 0.07 °K, and the shift in the peak as a binding energy of 0.37 ± 0.1 °K.[120] Note however that there is no sign of a two-peak structure (one peak at 2Δ and the other at the bound roton energy), which can be explained by the resolution of the spectrometer.[120]

It remains possible, however, that more detailed models than those presently available may explain the peak shift and shape without the existence of bound roton pairs for reasons we discuss next.[143]

In the next section we discuss the theory of the interactions between elementary excitations. Before doing so, however, we can appeal to a previous analysis to emphasize the importance of the interactions. Toward that end we consider a sum rule on $h^{(3)}$:

$$\frac{1}{\omega^4}\int d\nu\, h^{(3)}_{\hat{\epsilon}}(\Delta\mathbf{k};\nu) = \frac{N}{(4\pi)^2 c^3}\,\frac{1}{\Omega^2}\sum_{\mathbf{q}\mathbf{q}'} F(\mathbf{q})F(\mathbf{q}')S_4(\mathbf{q}, -\mathbf{q}-\Delta\mathbf{k}, -\mathbf{q}', \mathbf{q}'+\Delta\mathbf{k}) \tag{6.3.72}$$

where the four ρ static structure factor is

$$S_4(\mathbf{k}_1\,\mathbf{k}_2\,\mathbf{k}_3\,\mathbf{k}_4) = \frac{1}{N}\langle \rho_{\mathbf{k}_1}\rho_{\mathbf{k}_2}\rho_{\mathbf{k}_3}\rho_{\mathbf{k}_4}\rangle$$

$$= \frac{1}{N} I_4(\mathbf{k}_1\,\mathbf{k}_2\,\mathbf{k}_3\,\mathbf{k}_4) \tag{6.3.73}$$

where I_n is defined in equation (6.2.72). The cumulant analysis described in Section 6.2 gives an expression for I_4 of the form

$$I_4(\mathbf{k}_1\,\mathbf{k}_2\,\mathbf{k}_3\,\mathbf{k}_4) = N^2\,[S(\mathbf{k}_1)S(\mathbf{k}_2)(\delta_{\mathbf{k}_1,-\mathbf{k}_3}\delta_{\mathbf{k}_2,-\mathbf{k}_4} + \delta_{\mathbf{k}_1,-\mathbf{k}_4}\delta_{\mathbf{k}_2,-\mathbf{k}_3})$$

$$+ S(k_1)S(k_3)\delta_{\mathbf{k}_1-\mathbf{k}_2}\delta_{\mathbf{k}_3-\mathbf{k}_4}] + U_4(\mathbf{k}_1, \mathbf{k}_2, \mathbf{k}_3, \mathbf{k}_4) \tag{6.3.74}$$

for a uniform system. Then equation (6.3.72) can be rewritten as[144]

$$\frac{1}{\omega^4}\int \mathrm{d}\nu\, h_{\hat{\epsilon}}^{(3)}(\Delta\mathbf{k}, \nu) = \frac{N}{(4\pi)^2 c^3}\left(\frac{1}{\Omega}\sum_{\mathbf{q}} F(q)^2 S(q) S(|\Delta\mathbf{k} + \mathbf{q}|)\right.$$

$$\left. + \frac{1}{\Omega^2}\sum_{\mathbf{q}\mathbf{q}'} F(q)F(q')\frac{U_4(\mathbf{q}, -\mathbf{q} - \Delta\mathbf{k}, -\mathbf{q}', \mathbf{q}' + \Delta\mathbf{k})}{N}\right) \tag{6.3.75}$$

The first term in brackets comes from the non-interacting model, equation (6.3.71). The second term represents the effects of interactions and, since U_4 is of order N equation (6.2.94), may be comparable to the first term.

Kleban and Hastings have shown that the strong short-range correlations in liquid helium require U_4 to be of significant size, and that a similar conclusion must apply to that part of U_4 which produces the second term in (3.75).[145,143] These fundamental short-range correlations should be particularly important for $\mathbf{q}$ and $\mathbf{q}'$ corresponding to wavelengths comparable to the interparticle spacing in liquid helium, which happens to be in the vicinity of the roton minimum. Furthermore, there must be important contributions for $\mathbf{q} \neq \mathbf{q}'$. These facts, together with the fact that $F(q)$ may be varying rapidly in the same wavelength region, means that the uncertainties in $F(q)$ cannot be rendered insignificant by looking at a small energy interval as argued in Reference 120. This casts some doubt on whether the high resolution Raman experiments can be regarded as providing conclusive evidence for the existence of a two-roton bound state. Kleban and Hastings suggest that the impressive agreement of the non-interacting model with the shape of the low-resolution experiment may be due to a fortuitous cancellation of the unknown short-range structure in U_4 and $F(q)$. A calculation of U_4 would provide some insight into the size of these short range effects.

The correlated basis

In this section we describe a theory of the excited states of ^{4}He with a view toward deriving some information about the interactions between final states in neutron and Raman scattering. The ultimate goal should be to have a unified theory capable of describing these two experiments.

The many-body theory we use is the method of correlated wave functions developed by Feenberg and his collaborators to describe the low excited states of the helium liquids.[47] We will express the results of the theory as a quasi-particle Hamiltonian having a form similar to several other theories based upon the weakly interacting or low density Bose gas.

The motivation for the correlated basis lies in an observation already noted, made originally by Feynman: the excitation energy of the state $\rho_{\mathbf{k}}\rho_{\mathbf{l}}\Psi_0$ is $\epsilon_{BF}(\mathbf{k}) + \epsilon_{BF}(\mathbf{l}) + O(1/N)$ (equation 6.3.36), where Ψ_0 is the exact ground state wave function and ϵ_{BF} is the Bijl–Feynman energy (equation 6.3.14).[29] This suggests that a multi-Feynman-phonon wave function is of the form

$$|n_{\mathbf{k}_1} n_{\mathbf{k}_2} \ldots n_{\mathbf{k}_p}) = \left[\prod_{i=1}^{p} \rho_{\mathbf{k}_i}^{n_{\mathbf{k}_i}}\right] \Psi_0. \tag{6.3.76}$$

Indeed, we shall see shortly that this state has energy expectation value

$$\frac{(n_{\mathbf{k}_1} \ldots n_{\mathbf{k}_p} | (H - E_0) | n_{\mathbf{k}_1} \ldots n_{\mathbf{k}_p})}{(n_{\mathbf{k}_1} \ldots n_{\mathbf{k}_p} | n_{\mathbf{k}_1} \ldots n_{\mathbf{k}_p})} = \sum_{i=1}^{p} n_{\mathbf{k}_i} \epsilon_{BF}(\mathbf{k}_i) + O\left(\frac{\Sigma n}{N}\right) \tag{6.3.77}$$

providing that no subset of the momenta in equation (6.3.76) add to zero. This condition is actually an annoying complication due entirely to the fact that the states (6.3.76) do not form an orthogonal set. The simplest example is the following: the state $\rho_{\mathbf{k}}\rho_{-\mathbf{k}}\Psi_0$ is a state of total momentum zero which we would like to associate with excitation energy $\epsilon_{BF}(\mathbf{k}) + \epsilon_{BF}(-\mathbf{k})$. At the same time, to be a sensible approximation to an excited state, it should be orthogonal to the ground state, which it is not since

$$(\Psi_0 | \rho_{\mathbf{k}}\rho_{-\mathbf{k}} | \Psi_0) = NS(k). \tag{6.3.78}$$

This is taken care of by defining the orthogonal state

$$| n_{\mathbf{k}} = 1, n_{-\mathbf{k}} = 1 \rangle = \frac{[\rho_{\mathbf{k}}\rho_{-\mathbf{k}} - NS(k)]\ \Psi_0}{[\langle \Psi_0 | (\rho_{\mathbf{k}}\rho_{-\mathbf{k}} - NS(k))^2 | \Psi_0 \rangle]^{1/2}}. \tag{6.3.79}$$

With this definition, we obtain

$$\langle n_{\mathbf{k}} = 1, n_{-\mathbf{k}} = 1 | (H - E_0) | n_{\mathbf{k}} = 1, n_{-\mathbf{k}} = 1 \rangle = \epsilon_{BF}(\mathbf{k}) + \epsilon_{BF}(-\mathbf{k}) + O\left(\frac{1}{N}\right). \tag{6.3.80}$$

Before proceeding to more involved calculations, one more point needs to be made about the set of states (6.3.76): it is a complete basis. This is shown most simply by referring back to Section 6.2 where we introduced the symmetrized plane waves $\rho^{(n)}(\mathbf{k}_1 \ldots \mathbf{k}_n)$ (equation 6.2.152). These symmetrized plane waves form a complete set of functions for N bosons, since they are composed of all the eigenstates of N non-interacting bosons. We pointed out that the $\rho^{(n)}$ can be expressed as an n-th order polynomial in $\rho_{\mathbf{k}}$ (equation 6.2.168), so that the set of functions $\Pi_{i=1}^{n}\rho_{\mathbf{k}_i}$ is also complete. In fact, it is over-complete, since we can allow $n > N$ in this product, but not in $\rho^{(n)}$; this means that, strictly speaking, we should be using the functions $\rho^{(n)}$, but the final results will be no different. The correlated basis is composed of products $\Pi_{i=1}^{n}\rho_{\mathbf{k}_i}$ multiplied on the ground state wave function. Since Ψ_0 can be chosen to be positive-definite, its inclusion as a factor in the basis does not alter their completeness property. Henceforth, Ψ_0 will be viewed as a renormalized vacuum for our excited states and will be labelled by the ket $|0\rangle$:

$$\Psi_0 = |0\rangle \tag{6.3.81}$$

The procedure now will be:

(1) orthonormalize the basis;
(2) calculate Hamiltonian matrix elements; and
(3) express the results as a second quantized quasi-particle Hamiltonian.

Before proceeding with steps (1) and (2), we can formally introduce the second quantized notation. Step (1) will produce an orthonormal basis from the set (6.3.76) which we label

$$|n_{\mathbf{k}_1} n_{\mathbf{k}_2} \ldots n_{\mathbf{k}p}\rangle \tag{6.3.82}$$

where $n_{\mathbf{k}}$ are positive integers or zero. Then we define boson creation and annihilation operators by

$$\begin{aligned} a_{\mathbf{k}}^{\dagger} \,|\, n_{\mathbf{k}}, \{n_{\mathbf{l}}\}\rangle &= \sqrt{(n_{\mathbf{k}}+1)}\,|\, n_{\mathbf{k}}+1, \{n_{\mathbf{l}}\}\rangle \\ a_{\mathbf{k}} \,|\, n_{\mathbf{k}}, \{n_{\mathbf{l}}\}\rangle &= \sqrt{n_{\mathbf{k}}}\,|\, n_{\mathbf{k}}-1, \{n_{\mathbf{l}}\}\rangle \\ a_{\mathbf{k}}\,|\,0\rangle &= 0. \end{aligned} \tag{6.3.83}$$

With these definitions, these operators satisfy the usual Bose commutation relations

$$\begin{aligned} [a_{\mathbf{k}}^{\dagger}, a_{\mathbf{q}}] &= \delta_{\mathbf{k},\mathbf{q}} \\ [a_{\mathbf{k}}, a_{\mathbf{q}}] &= 0 = [a_{\mathbf{q}}^{\dagger}, a_{\mathbf{k}}^{\dagger}]. \end{aligned} \tag{6.3.84}$$

The over-completeness of the basis is ignored by these definitions. One way to avoid the over-completeness is to supplement (6.3.83) by the additional condition

$$a_{\mathbf{k}}^{\dagger}\,|\,n_{\mathbf{q}_1} \ldots n_{\mathbf{q}_p}\rangle = 0 \tag{6.3.85}$$

if

$$\sum_{i=1}^{p} n_{\mathbf{q}_i} = N.$$

In that case the commutation relation (6.3.84) fails when acting on a state with N Feynman excitations. Those states will not enter our analysis with any significant weight, so we will ignore condition (6.3.85).

It should be emphasized that these creation and annihilation operators are not bare particle operators. We will see that they are operators which create and destroy excitations which are non-interacting in the lowest order of approximation.

There are numerous possible orthogonalization schemes which map the non-orthogonal set $|\{n_{\mathbf{k}}\})$ onto the orthonormal set $|\{n_{\mathbf{k}}\}\rangle$. The most obvious choice, however, is to take advantage of the properties of the operators shown in equations (6.3.77), (6.3.79) and (6.3.80), namely that they are closely related to creation operators for Feynman phonons. Then the procedure is to define the one-phonon state as

$$|\,n_{\mathbf{k}} = 1\,\rangle = [\rho_{\mathbf{k}}/\sqrt{[NS(k)]}]\;|\,0\,\rangle. \tag{6.3.86}$$

The space of two-phonon states is then spanned by the set of states

$$|\,\mathbf{k}, \mathbf{l}] = \left\{\rho_{\mathbf{k}}\,\rho_{\mathbf{l}} - \frac{I_3(\mathbf{k}, \mathbf{l}, -\mathbf{k}-\mathbf{l})}{NS(\mathbf{k}+\mathbf{l})}\,\rho_{\mathbf{k}+\mathbf{l}}\right\}\,|\,0\,\rangle \tag{6.3.87}$$

where the second term in braces assures that this state is orthogonal to $|\,n_{\mathbf{k}+\mathbf{l}} = 1\,\rangle$. I_3 is defined in equation (6.2.72) as

$$\langle 0\,|\,\rho_{-\mathbf{k}-\mathbf{l}}\,\rho_{\mathbf{k}}\,\rho_{\mathbf{l}}\,|\,0\rangle = I_3(\mathbf{k}, \mathbf{l}, -\mathbf{k}-\mathbf{l}). \tag{6.3.88}$$

Note, however, that these two phonon states of the same total momentum are not mutually orthogonal:

$$[\mathbf{k}-\mathbf{l},\mathbf{l}\,|\,\mathbf{k}-\mathbf{l}',\mathbf{l}'] = [I_4(-\mathbf{k},-\mathbf{l},\mathbf{k}',\mathbf{l}') - I_3^*(\mathbf{k},\mathbf{l},-\mathbf{k}-\mathbf{l})I_3(\mathbf{k},\mathbf{l}',\mathbf{k}-\mathbf{l}')]$$
$$\equiv \mathscr{M}_{\mathbf{l},\mathbf{l}'}. \tag{6.3.89}$$

This set of two-phonon states can be orthonormalized by performing a Löwdin transformation, which is defined by[146]

$$|\,n_{\mathbf{k}-\mathbf{l}} = 1, n_{\mathbf{l}} = 1\,\rangle = \sum_{\mathbf{q}} |\,\mathbf{k}-\mathbf{q},\mathbf{q}]\,(\mathscr{M}^{-1/2})_{\mathbf{q},\mathbf{l}}. \tag{6.3.90}$$

This procedure is repeated for the three-phonon space by first orthogonalizing the states $\rho_{\mathbf{k}_1}\rho_{\mathbf{k}_2}\rho_{\mathbf{k}_3}\Psi_0$ to the two-phonon and one-phonon states with the same total momentum, and then orthonormalizing the resulting three phonon states by a Löwdin procedure. By repeating this procedure, n-phonon space is separated from the n'-phonon space by a Gram–Schmidt orthogonalization procedure beginning with the Feynman wave function for the one phonon states, and then orthonormalizing the n-phonon states to one another by a Löwdin transformation.

The remaining task is the calculation of matrix elements of the Hamiltonian. This is most simply done by evaluating the Hamiltonian in the non-orthogonal basis and then transforming to the orthogonal basis.

We consider two states,

$$\begin{aligned}\psi_a &= \varphi_a\Psi_0\\ \psi_b &= \varphi_b\Psi_0\end{aligned} \tag{6.3.91}$$

and calculate the matrix element of the kinetic energy by making use of a transformation of the form of the first equality in equation (6.2.11):

$$(\psi_a\,|\,T\,|\,\psi_b) = \frac{\hbar^2}{2m}\sum_{i=1}^{N}(\Psi_0\,|\,[\nabla_i\varphi_a^*\cdot\nabla_i\varphi_b - \varphi_a^*\varphi_b\nabla_i^2]\,|\,\Psi_0). \tag{6.3.92}$$

The second term in brackets combines with the potential energy to give E_0 when Ψ_0 is the exact ground state. Then

$$(\psi_a\,|\,(H-E_0)\,|\,\psi_b) = \frac{\hbar^2}{2m}\sum_{i=1}^{N}(\Psi_0\,|\nabla_i\varphi_a^*\cdot\nabla_i\varphi_b\,|\Psi_0). \tag{6.3.93}$$

To evaluate the matrix elements in the basis (6.3.76), choose

$$\begin{aligned}\varphi_a &= \prod_{\alpha=1}^{p}\rho_{\mathbf{k}_\alpha}\\ \varphi_b &= \prod_{\beta=1}^{q}\rho_{\mathbf{l}_\beta}.\end{aligned} \tag{6.3.94}$$

Then[56]

$$\begin{aligned}(\Psi_0\,|\,\varphi_a^*(H-E_0)\varphi_b\,|\,\Psi_0) &= \sum_{\alpha\beta}\frac{\hbar^2}{2m}\mathbf{k}_\alpha\cdot\mathbf{l}_\beta\left(\Psi_0\left|\,\varphi_a^*\,\frac{\rho_{\mathbf{l}_\beta-\mathbf{k}_\alpha}}{\rho_{-\mathbf{k}_\alpha}\rho_{\mathbf{l}_\beta}}\,\varphi_b\,\right|\Psi_0\right)\\ &= \sum_{\alpha\beta}\frac{\hbar^2}{2m}\mathbf{k}_\alpha\cdot\mathbf{l}_\beta\left(\Psi_0\left|\left[\prod_{\alpha'\neq\alpha}\rho_{-\mathbf{k}'_\alpha}\right]\rho_{\mathbf{l}_\beta-\mathbf{k}_\alpha}\left[\prod_{\beta\neq\beta'}\rho_{\mathbf{l}_{\beta'}}\right]\right|\Psi_0\right).\end{aligned} \tag{6.3.95}$$

Consequently, matrix elements of the identity and Hamiltonian require only a knowledge of the functions $I_n(\mathbf{k}_1\ldots\mathbf{k}_n)$. All of the information about the potential energy has reappeared in terms of correlation functions of the $\rho_{\mathbf{k}}$ operators in the exact ground state. This is particularly convenient because the correlation functions are much better behaved than the bare potential energy. Furthermore, some of them are available from direct measurement (e.g. $S(k) = I_2(\mathbf{k}, -\mathbf{k})/N$ and a moment of I_4 is available by integrating $h_\epsilon^{(3)}(\mathbf{k},\omega)$ over ω equation (6.3.72)).

Some systematic truncation of these correlations is necessary to reduce this formalism to a practical calculation. One scheme is to refer back to the cumulants of the $\rho_\mathbf{k}$ operators, $U_n(\mathbf{k}_1 \ldots \mathbf{k}_n)$ (equation 6.2.73), and set them to zero beyond some value of n. The procedure we prefer is closely related to this truncation, but somewhat more consistent for the evaluation of the Hamiltonian. We begin by defining an operator which generates the matrix elements of the Hamiltonian when operating on the functions I_n.

First the function $I_{p,q}$ is defined as the matrix element of the identity between ψ_a and ψ_b:

$$I_{p,q}(\mathbf{k}_1 \ldots \mathbf{k}_p;\mathbf{l}_1 \ldots \mathbf{l}_q) = (\Psi_0 \mid \left(\prod_{\alpha=1}^{p} \rho_{\mathbf{k}_\alpha}\right)^* \prod_{\beta=1}^{q} \rho_{\mathbf{l}_\beta} \mid \Psi_0)$$

$$= I_{p+q}(-\mathbf{k}_1, \ldots -\mathbf{k}_p, \mathbf{l}_1 \ldots \mathbf{l}_q) \tag{6.3.96}$$

and the related cumulant function is defined as

$$U_{p_i q_j}(\mathbf{k}_{\alpha_1} \ldots \mathbf{k}_{\alpha_{p_i}}; \mathbf{l}_{\beta_1} \ldots \mathbf{l}_{\beta_{p_j}}) = U_{p_i+q_j}(-\mathbf{k}_{\alpha_1} \ldots \mathbf{k}_{\alpha_{p_i}} \mathbf{l}_{\beta_1} \ldots \mathbf{l}_{\beta_{qj}}). \tag{6.3.97}$$

Then $\hat{O}$ is an operator defined by

$$\hat{O} f_1 f_2 \ldots f_m = \sum_{i=1}^{M} f_1 f_2 \ldots f_{i-1}(\hat{O} f_i) f_{i+1} \ldots f_M \tag{6.3.98}$$

where

$$\hat{O} U_{p,q}(\mathbf{k}_1 \ldots \mathbf{k}_p;\mathbf{l}_1 \ldots \mathbf{l}_q) = \sum_{\alpha=1}^{p} \sum_{\beta=1}^{q} \frac{\hbar^2}{2m} \mathbf{k}_\alpha \cdot \mathbf{l}_\beta$$

$$\times U_{p+q-1}(-\mathbf{k}_1 \ldots -\mathbf{k}_{\alpha-1} -\mathbf{k}_{\alpha+1} \ldots -\mathbf{k}_p, \mathbf{l}_\beta - \mathbf{k}_\alpha, \mathbf{l}_1 \ldots \mathbf{l}_{\beta-1} \mathbf{l}_{\beta+1} \ldots \mathbf{l}_q) \tag{6.3.99}$$

Then the matrix element of $H - E_0$ as given in equation (6.3.95) can be written

$$(\Psi_0 \mid \varphi_a^*(H - E_0) \varphi_b \mid \Psi_0) = \hat{O} I_{p,q}(\mathbf{k}_1 \ldots \mathbf{k}_p;\mathbf{l}_1 \ldots \mathbf{l}_q) \tag{6.3.100}$$

which is then evaluated in terms of the cumulants by using the cumulant expansion for I_{p+q}, (equation 6.2.73). With these definitions, the approximation we take is to set $U_n = 0$ beyond some value of n in the cumulant expansion of I_{p+g}, and then calculate the matrix element of the Hamiltonian by operating on the truncated I_{p+q} with $\hat{O}$.

The simplest truncation is to set

$$U_n = 0 \quad n \geqslant 3. \tag{6.3.101}$$

Then the orthogonalization can be done exactly,[52] primarily because it is not necessary to use the Löwdin procedure; spaces containing different numbers of phonons are not connected by U_2. Rather than going through the analysis here, we simply state the results in the second-quantized picture developed above. The orthogonalization is equivalent to the second quantized representation of the density fluctuation operator[70]

$$\rho_\mathbf{k} = \sqrt{[(NS(k)]}\,(a_\mathbf{k}^\dagger + a_{-\mathbf{k}}). \tag{6.3.102}$$

Said another way, this equality used in equation (6.3.76) with Ψ_0 replaced by $\mid 0 \rangle$ on the right-hand side defines the transformation between the non-orthogonal set $\mid n_{\mathbf{k}_1} \ldots n_{\mathbf{k}_p})$ and the orthonormal set satisfying equation (6.3.83). The second quantized Hamiltonian in this approximation is[147]

$$H = E_0 + \sum_{\mathbf{k}} a_{\mathbf{k}}^{\dagger} a_{\mathbf{k}} \epsilon_{\mathrm{BF}}(k) \tag{6.3.103}$$

Retaining $U_3 \neq 0$ introduces terms into the Hamiltonian which are not diagonal in the second quantized representation. Equation (6.3.89) shows that the two-phonon states are no longer orthogonal, and the Löwdin transformation becomes necessary. Even without this transformation, a nonzero U_3 implies the presence of bilinear terms in the second quantized representation of $\rho_{\mathbf{k}}$:

$$\langle n_{\mathbf{k}+\mathbf{q}} = 1 \mid \rho_{\mathbf{q}} \mid n_{\mathbf{k}} = 1 \rangle = \frac{U_3(\mathbf{k}, \mathbf{q}, -\mathbf{k} - \mathbf{q})}{\sqrt{[N^2 S(k) S(\mathbf{k} + \mathbf{q})]}} \tag{6.3.104}$$

so that

$$\rho_{\mathbf{q}} = \sqrt{[NS(q)]}\, [a_{\mathbf{q}}^{\dagger} + a_{-\mathbf{q}} + \sum_{\mathbf{k}} R_{\mathbf{k},\mathbf{q}}^{(2)} a_{\mathbf{k}+\mathbf{q}}^{\dagger} a_{\mathbf{k}} + \ldots] \tag{6.3.105}$$

where

$$R_{\mathbf{k},\mathbf{q}}^{(2)} = \frac{U_3(\mathbf{k}, \mathbf{q}, -\mathbf{k} - \mathbf{q})}{\sqrt{[N^3 S(\mathbf{k} + \mathbf{q}) S(\mathbf{k}) S(\mathbf{q})]}} \,. \tag{6.3.106}$$

Moreover, the presence in $|\mathbf{k},\mathbf{l}]$ of the term $\rho_{\mathbf{k}+\mathbf{l}}$ (equation 6.3.87) implies that $\rho_{\mathbf{q}}$ must also have trilinear terms of the form $a^{\dagger}a^{\dagger}a$ and $a^{\dagger}aa$. In fact, n-phonon terms must contain a term linear in $\rho_{\mathbf{k}}$ to be orthogonal to one-phonon terms, so that the expression for $\rho_{\mathbf{q}}$ must contain terms of the form $a^{\dagger}[(a^{\dagger})^p (a)^q]a$, even when only U_2 and U_3 are non-zero. (We are spending so much effort on $\rho_{\mathbf{q}}$ because it is the operator by which the external probes of interest (neutrons and photons) couple to the system.)

The same analysis shows that the Hamiltonian must also be an infinite series in terms of the form $a^{\dagger}[(a^{\dagger})^p (a)^q]a$:

$$H = E_0 + \sum_{\mathbf{k}} \epsilon_{\mathrm{BF}}(k) a_{\mathbf{k}}^{\dagger} a_{\mathbf{k}} + \frac{1}{2} \sum_{\mathbf{k}\mathbf{l}} [a_{\mathbf{k}+\mathbf{l}}^{\dagger} a_{\mathbf{k}} a_{\mathbf{l}} g_{\mathbf{k}\mathbf{l}}^{(3)} + \text{h.c.}]$$
$$+ \frac{1}{4} \sum_{\mathbf{l}\mathbf{l}'\mathbf{k}} a_{\mathbf{l}'}^{\dagger} a_{\mathbf{k}-\mathbf{l}'}^{\dagger} a_{\mathbf{l}} a_{\mathbf{k}-\mathbf{l}}\, g_{\mathbf{k}\mathbf{l}\mathbf{l}'}^{(4)} + \ldots \tag{6.3.107}$$

where

$$g_{\mathbf{k},\mathbf{l}}^{(3)} = (\hbar^2/2m) \frac{[\mathbf{k} \cdot (\mathbf{k} + \mathbf{l}) S(\mathbf{l}) + \mathbf{l} \cdot (\mathbf{k} + \mathbf{l}) S(\mathbf{k}) - (\mathbf{k} + \mathbf{l})^2 U_3(\mathbf{k}, -\mathbf{k} - \mathbf{l}, \mathbf{l})/NS(\mathbf{k} + \mathbf{l})]}{\sqrt{[NS(k) S(l) S(\mathbf{k} + \mathbf{l})]}} \tag{6.3.108}$$

and $g^{(4)}$ is proportional to U_3/N^2. If U_4 is retained, however, it will contribute an important part to $g^{(4)}$, as can be seen from the analysis in the last section. One very convenient property of this Hamiltonian is that the **k** values for creation and annihilation operators never vanish, and all terms are normal ordered. Consequently

$$H \mid 0 \rangle = E_0 \mid 0 \rangle \tag{6.3.109}$$

i.e. the ground state is an eigenstate of the Hamiltonian.

With specific choices of the vertex strengths $g^{(3)}$ and $g^{(4)}$ this model is formally the same as that studied by several groups to explain the results of neutron and Raman scattering.[127,128,119,129] The first such calculation was done by Jackson under the assumption that $g^{(4)} = 0$, $R^{(2)} = 0$, and $g^{(3)}$ is given by equation (6.3.108).[127,128] The assumption that $R^{(2)} = 0$ permits a particularly simple Green function analysis. Jackson uses the convolution approximation for U_3 (equation 6.2.189) and calculates $S(k,\omega)$ by using the Born approximation in the Dyson equation for the Feynman phonon propagator. The results of this calculation are shown in Figure 6.3.2. The excitation spectrum he obtains is virtually identical to the Brillouin–Wigner result of Jackson and Feenberg, Figure 6.3.1. The strength of the quasiparticle peak, $Z(k)$ and the multi-excitation background $S_{II}(k)$ are compared in Figure 6.3.3. The agreement is qualitative at best.

An improved agreement with experiment might be expected by including $R^{(2)}$ and quasi-particle interactions, $g^{(4)}$. Moreover, a calculation of the Raman scattering without including $g^{(4)}$ would be unlikely to produce the shift in the peak intensity to frequencies below 2Δ, the effect which is generally attributed to the existence of two-roton bound states.

Zawadowski, Ruvalds and Solana considered a model Hamiltonian of the form (6.3.107) with $g^{(3)}$ and $g^{(4)}$ taken as adjustable constants, in an attempt to explain the structure of $S_{II}(k,\omega)$.[119] Hastings and Halley reconsidered the same model but attempted to choose the parameters $g^{(3)}$ and $g^{(4)}$ to fit both the neutron and Raman scattering data.[129] They also choose $R^{(2)} = 0$. Their conclusions are that no model Hamiltonian of the form of equation (6.3.107) with constant $g^{(3)}$ and $g^{(4)}$ can account quantitatively for either the neutron or the Raman scattering data.[129] Furthermore, momentum dependence of $g^{(3)}$ does not alter this result. They also considered a $g^{(4)}$ which restricted the interactions to the rotons. While it was possible to choose this $g^{(4)}$ to give reasonably good agreement with the light scattering, it cannot account for the strength of the multi-phonon part of the neutron scattering, $S_{II}(k,\omega)$.[129]

These calculations of Jackson,[127,128] Zawadowski, Ruvalds and Solana,[119] and Hastings and Halley[129] make it clear that a careful calculation of $g^{(3)}$ and $g^{(4)}$ from first principles is called for; moreover, it is probably necessary to include nonlinear terms such as $R^{(2)}$ in the second-quantized expression for $\rho_{\mathbf{q}}$. This latter point is somewhat disheartening, for the linear approximation (6.3.102) was largely responsible for the tractability of this problem.

6.4 Liquid ^{3}He

The Hamiltonian of N ^{3}He atoms is identical to the Hamiltonian for N ^{4}He atoms except that the mass of the ^{3}He atom is 3/4 that of the ^{4}He atom. The two-body potential is identical. Thus, if ^{3}He atoms were bosons, the qualitative results of our previous discussion would be unchanged. The lighter mass of the ^{3}He atom would increase the kinetic energy so that the binding energy of the atom would be less and the equilibrium density lower in boson liquid ^{3}He than in liquid ^{4}He. In fact, the Jastrow calculations have been done for boson liquid ^{3}He. The variational calculation of Massey and Woo (described in Section 6.2 for ^{4}He) gave an equilibrium density of 0.0157 Å^{-3} (to be compared with 0.0164 Å^{-3} as the experimental result for Fermion liquid ^{3}He, and 0.0218 Å^{-3} for liquid ^{4}He); they also obtained a ground state energy per particle of –2.09 °K for Boson liquid ^{3}He, to be compared with the experimental result of –2.52 °K for Fermion liquid ^{3}He and –7.14 °K for liquid ^{4}He.[65] The paired-phonon analysis (Section 6.2) applied to this problem shifted the energy to –2.64 °K, falling below the fermion experimental value.[70] The liquid structure function for boson liquid ^{3}He obtained in the paired-phonon analysis is compared with the experimental results for fermion liquid ^{3}He in Figure 6.4.1. They are remarkably close, and would seem to indicate that the statistics are unimportant.

However, we know of course that the statistics are very important. Normal liquid ^{3}He is a

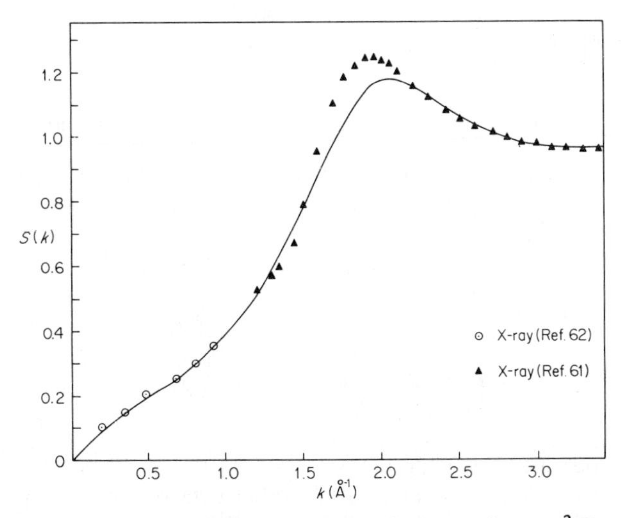

Figure 6.4.1 Comparison of the fictitious boson ^{3}He liquid structure function with the (fermion) X-ray scattering[61]

very good example of a Landau–Fermi liquid, whose thermal properties differ enormously from liquid ^{4}He. Furthermore, as we discussed in Section 6.1, it is now clear that the low temperature phase of liquid ^{3}He is a superfluid very similar to the superconducting state of electrons in metals. The transition to this phase occurs at about 0.002 °K.

Landau–Fermi liquid theory

All of the sophisticated many-body theoretical techniques[148] have been applied at one time or another to the calculation of the properties of liquid ^{3}He. Until the recent interest in the superfluid phase, the objective of these calculations has usually been to give a microscopic formulation of the Landau–Fermi liquid theory, and to account for the binding energy and structure of the liquid.

There are many accounts of Landau–Fermi liquid theory in the literature.[149,150] Here we will summarize only enough of that theory to enable us to discuss the microscopic theories. Our summary is based on the presentation of Pines and Nozières,[149] and we use their notation.

The non-interacting Fermi gas of N spin 1/2 particles has eigenstates which are the anti-symmetrized product of N eigenfunctions of the single-body Hamiltonian. For uniform systems the full Hamiltonian is

$$H^{(0)} = T = \sum_{i=1}^{N} -\frac{\hbar^2}{2m} \nabla_i^2 \tag{6.4.1}$$

with eigenstate

$$\Psi_{\mathbf{k}_1\alpha_1 \cdots \mathbf{k}_N\alpha_N} = \frac{1}{\sqrt{N!}} \sum_{\nu} (-1)^{\nu} P_{\nu} \left[\prod_{i=1}^{N} \psi_{\mathbf{k}_i\alpha_i}(x_i) \right] \tag{6.4.2}$$

where $x_i = (\mathbf{r}_i, \sigma_i)$ and $\psi_{\mathbf{k}_i\alpha_i}$ are the (plane wave) eigenstates of the single-body operator $-(\hbar^2/2m)\nabla^2$:

$$-\frac{\hbar^2}{2m} \nabla^2 \psi_{\mathbf{k}_i\alpha_i} = \frac{\hbar^2 k_i^2}{2m} \psi_{\mathbf{k}_i\alpha_i} = \epsilon^0_{\mathbf{k}_i} \psi_{\mathbf{k}_i\alpha_i} \tag{6.4.3}$$

$$\psi_{\mathbf{k}_i\alpha_i} = \frac{\exp(\mathrm{i}\mathbf{k}_i \cdot \mathbf{r})}{\Omega^{1/2}}\,\delta_{\sigma,\alpha_i} \tag{6.4.4}$$

These N-body states are most conveniently described as Slater determinants

$$\Psi_{\mathbf{k}_1\alpha_1\cdots\mathbf{k}_N\alpha_N} = \frac{1}{\sqrt{N!}}\,\mathrm{Det}\begin{vmatrix} \psi_{\mathbf{k}_1\alpha_1}(x_1) & \cdots & \psi_{\mathbf{k}_1\alpha_1}(x_N) \\ \vdots & & \vdots \\ \psi_{\mathbf{k}_N\alpha_N}(x_1) & \cdots & \psi_{\mathbf{k}_N\alpha_N}(x_N) \end{vmatrix} \tag{6.4.5}$$

and the eigenvalues of $H^{(0)}$ are

$$H^{(0)}\,\Psi_{\mathbf{k}_1\alpha_1\cdots\mathbf{k}_N\alpha_N} = \left(\sum_{i=1}^{N}\frac{\hbar^2k_i^2}{2m}\right)\Psi_{\mathbf{k}_1\alpha_1\cdots\mathbf{k}_N\alpha_N} \tag{6.4.6}$$

The ground state is obtained by filling the N lowest eigenstates. In $\mathbf{k}$ space this is described as filling two Fermi spheres (one with spin up, the other with spin down) up to the Fermi level k_F, where

$$\sum_{\substack{\mathbf{k}\\ k<k_\mathrm{F}}} 1 = \frac{N}{2} \tag{6.4.7}$$

(the 1/2 is due to spin).

It is convenient for what follows to introduce an occupation number notation, where $n_{\mathbf{k}\alpha}$ is one if the state $\psi_{\mathbf{k}\alpha}$ is occupied and zero if it is not. Then the eigenvalues of the non-interacting system are of the form

$$E(\{n_{\mathbf{k}\alpha}\}) = \sum_{\mathbf{k}\alpha} n_{\mathbf{k}\alpha}\,\epsilon_{\mathbf{k}}^0 \tag{6.4.8}$$

subject to the restrictions

$$n_{\mathbf{k}\alpha} = 0,\ 1 \tag{6.4.9}$$

$$\sum_{\mathbf{k}\alpha} n_{\mathbf{k}\alpha} = N. \tag{6.4.10}$$

The ground state is given by the distribution $n_{\mathbf{k}\alpha}^0$:

$$\begin{aligned} n_{\mathbf{k}\alpha}^0 &= 1 \quad k \leqslant k_\mathrm{F} \\ &= 0 \quad k > k_\mathrm{F} \end{aligned} \tag{6.4.11}$$

and the corresponding ground state energy is

$$E_0 = \sum_{\mathbf{k}\alpha} n_{\mathbf{k}\alpha}^0\epsilon_{\mathbf{k}}^0 = 2\sum_{k\leqslant k_\mathrm{F}}\epsilon_{\mathbf{k}}^0 = N\frac{3}{5}\left(\frac{\hbar^2k_\mathrm{F}^2}{2m}\right). \tag{6.4.12}$$

Then the excited state energies may be written as

$$E - E_0 = \sum_{\mathbf{k}\alpha}\delta n_{\mathbf{k}\alpha}\,\epsilon_{\mathbf{k}}^0 \tag{6.4.13}$$

where

$$\delta n_{\mathbf{k}\alpha} = n_{\mathbf{k}\alpha} - n_{\mathbf{k}\alpha}^{(0)} \tag{6.4.14}$$

has values $-1, 0, 1$, and must satisfy

$$\sum_{\mathbf{k}\alpha} \delta n_{\mathbf{k}\alpha} = 0 \tag{6.4.15}$$

The energies $\epsilon^0_{\mathbf{k}}$ of interest at low temperatures are located near the Fermi surface. The energy of a state on the Fermi surface, $\epsilon^0_{k_F} = \hbar^2 k_F^2/2m$, is also the energy of the last particle added to the system, and is therefore the chemical potential at $T = 0$:

$$\mu = \epsilon^0_{k_F}. \tag{6.4.16}$$

Because of the restriction (6.4.15) we may subtract μ from $\epsilon^0_{\mathbf{k}}$ in equation (6.4.13). Finally, at finite temperature we will be interested in the difference of the free energy from its ground state value $E_0 - \mu N$, so we define an excitation free energy F by

$$F - F_0 = E - E_0 = \sum_{\mathbf{k}\alpha} \delta n_{\mathbf{k}\alpha}(\epsilon^0_{\mathbf{k}} - \mu). \tag{6.4.17}$$

At finite temperature, μ is determined so that (6.4.15) is satisfied.

The finite temperature occupation of states is given by the Fermi–Dirac distribution function

$$n_{\mathbf{k}\alpha}(T) = \frac{1}{\exp[(\epsilon^0_{\mathbf{k}\alpha} - \mu)/k_B T] + 1}. \tag{6.4.18}$$

We now consider the effects of interactions. It is well known that the properties of electrons in metals (a Fermi liquid, albeit non-uniform) can be explained for the most part by a theory of non-interacting electrons, ignoring the Coulomb repulsion between the electrons. Thus the above analysis may be used to calculate the properties of that system, with an appropriate choice of excitation spectrum $\epsilon^0_{\mathbf{k}}$. The free energy F would be given by (6.4.17) with $\delta n_{\mathbf{k}\alpha}$ being defined at finite temperature through equation (6.4.18). Of course, it is well known that excitations interact with one another, in which case equation (6.4.17) may be viewed as the first term in a series expansion in the occupation numbers about their ground state value:

$$F - F_0 = \sum_{\mathbf{k}\alpha} \delta n_{\mathbf{k}\alpha}(\epsilon^0_{\mathbf{k}} - \mu) + \frac{1}{2} \sum_{\substack{\mathbf{k}\alpha \\ \mathbf{k}'\alpha'}} f_{\mathbf{k}\alpha\mathbf{k}'\alpha'} \delta n_{\mathbf{k}\alpha} \delta n_{\mathbf{k}'\alpha'} + \dots. \tag{6.4.19}$$

Landau pointed out an inconsistency in the point of view that this is an expansion by noting that, at low temperature, $\delta n_{\mathbf{k}\alpha}$ is of order $(\epsilon^0_{\mathbf{k}} - \mu)/\mu$ and therefore the first term in the expansion is actually of second order in δn. Thus a consistent treatment requires both the first and second term to be retained in leading order.

To obtain the thermodynamic properties, we can use equation (6.4.19) as a description of the eigenvalues, with $\delta n_{\mathbf{k}\alpha} = -1, 0$ or 1. These eigenvalues are then used to calculate the partition function. The results are most easily expressed in terms of the quasiparticle energy $\hat{\epsilon}_{\mathbf{k}}$ given by

$$\hat{\epsilon}_{\mathbf{k}} - \mu = (\epsilon^0_{\mathbf{k}} - \mu) + \sum_{\mathbf{k}'\alpha'} f_{\mathbf{k}\alpha\mathbf{k}'\alpha'} \delta n_{\mathbf{k}'\alpha'} \tag{6.4.20}$$

where

$$\delta n_{\mathbf{k}\alpha} = n_{\mathbf{k}\alpha}(T, \mu) - n_{\mathbf{k}\alpha}(0, \mu) \tag{6.4.21}$$

and

$$n_{\mathbf{k}\alpha}(T, \mu) = \frac{1}{\exp[(\hat{\epsilon}_{\mathbf{k}} - \mu)/k_B T] + 1}. \tag{6.4.22}$$

Note that $\hat{\epsilon}_{\mathbf{k}}$ depends on the temperature through $\delta n_{\mathbf{k}\alpha}(T)$, and thus this last equation is a set of nonlinear equations for $\delta n_{\mathbf{k}\alpha}(T)$.

At low temperatures where this theory is valid, the quasiparticle occupation $\delta n_{\mathbf{k}\alpha}$ will be nonzero only near the Fermi surface, and $f_{\mathbf{k}\alpha\mathbf{k}'\alpha'}$ need be evaluated only on the Fermi surface ($\epsilon_k^0 = \epsilon_{k'}^0 = \mu$). Consequently the properties of the system depend only on the expansion coefficients of $f_{\mathbf{k}\alpha\mathbf{k}'\alpha'}$ in Legendre polynomials on the Fermi surface:

$$f_{\mathbf{k}\mathbf{k}'} = \sum_{l=0}^{\infty} f_l P_l(\hat{\mathbf{k}} \cdot \hat{\mathbf{k}}'). \tag{6.4.23}$$

There are two independent spin components of f, which are written in the form

$$\begin{aligned} f_l^{\uparrow\uparrow} &= f_l^{\mathrm{s}} + f_l^{\mathrm{a}} \\ f_l^{\uparrow\downarrow} &= f_l^{\mathrm{s}} - f_l^{\mathrm{a}}. \end{aligned} \tag{6.4.24}$$

Finally, the effect of the interactions is also controlled by the density of states at the Fermi surface,

$$\nu(\epsilon_{\mathrm{F}}) = \Omega \frac{m^* k_{\mathrm{F}}}{\pi^2 \hbar^3} \tag{6.4.25}$$

where the effective mass m^* is given by

$$\frac{\hbar^2 k_{\mathrm{F}}}{m^*} = \nabla_{\mathbf{k}} \epsilon_{\mathbf{k}}.$$

This is measured most directly by the low temperature specific heat at constant volume

$$C_V = \frac{m^* k_{\mathrm{F}} k_{\mathrm{B}}^2 T}{3\hbar^2}. \tag{6.4.26}$$

The parameters which enter the remaining experimental quantities are best expressed in terms of the reduced interaction strengths

$$F_l = \nu(\epsilon_{\mathrm{F}}) f_l \tag{6.4.27}$$

Then the thermodynamic properties are

$$\begin{aligned} m^* &= m(1 + \tfrac{1}{3}F_1^{\mathrm{s}}) \\ C_1 &= \frac{\hbar k_{\mathrm{F}}}{m}\left(\frac{1 + F_0^{\mathrm{s}}}{3 + F_1^{\mathrm{s}}}\right)^{1/2} \\ K_T &= m\rho C_1^2 \\ \chi &= \chi_0 \frac{(1 + \tfrac{1}{3}F_1^{\mathrm{s}})}{1 + F_0^{\mathrm{a}}} \end{aligned} \tag{6.4.28}$$

where C_1, K_T, χ and χ_0 are, respectively, the velocity of first sound, the isothermal compressibility, the magnetic susceptibility, and the magnetic susceptibility of the free Fermi gas of the same mass m. The experimental values are shown in Table 6.4.1.

Table 6.4.1 Landau–Fermi liquid parameters for liquid ^{3}He from the t-matrix calculation of Østgaard,[156] the correlated basis function (CBF) calculation of Tan and Feenberg,[163] and experiment

	t-matrix	CBF	Experiment
Equilibrium density			
ρ_0 (Å^{-3})	0.0151	0.0156	0.0164
C_1 (m/sec)	175	178	180
K_T (%/atm)	4.3	3.9	3.8
χ/χ_0	10	9.2	9.2
m^*/m	2–3	2.28	3.08

Microscopic theories of normal and superfluid liquid ^{3}He

As noted above, one objective of a microscopic theory of liquid ^{3}He is a direct calculation of the quasi-particle interaction function $f_{\mathbf{k}\alpha\mathbf{k}'\alpha'}$ from a model of N ^{3}He atoms in a volume Ω interacting via a realistic two-body potential (e.g. the Lennard–Jones potential, equations 6.1.3 and 6.1.8). In the Landau–Fermi liquid theory, the Hamiltonian is replaced by a system of Fermi quasi-particles and an interaction between the quasi-particles. The interaction is then treated in mean-field, ignoring the scattering of quasi-particles from one another into new states. In principle, however, a microscopic calculation contains information about these off-diagonal scattering processes.

In the spirit of our discussion of the excitation spectrum of ^{4}He, we define a number representation for quasi-particles by introducing quasi-particle creation and annihilation operators $a^\dagger_{\mathbf{k}\sigma}$ and $a_{\mathbf{k}\sigma}$ and the corresponding Hamiltonian

$$H = E_0 + \sum_{\mathbf{k}\sigma} |\epsilon_{\mathbf{k}\sigma} - \mu| a^\dagger_{\mathbf{k}\sigma} a_{\mathbf{k}\sigma}$$

$$+ \tfrac{1}{2} \sum_{\substack{\mathbf{k}_1\sigma_1\,\mathbf{k}_2\sigma_2 \\ \mathbf{k}_3\sigma_3\,\mathbf{k}_4\sigma_4}} g_4(\mathbf{k}_1\sigma_1, \mathbf{k}_2\sigma_2, \mathbf{k}_3\sigma_3, \mathbf{k}_4\sigma_4) a^\dagger_{\mathbf{k}_1\sigma_1} a^\dagger_{\mathbf{k}_2\sigma_2} a_{\mathbf{k}_4\sigma_4} a_{\mathbf{k}_3\sigma_3} + \ldots \tag{6.4.29}$$

Here we have used the idea of a hole for $\epsilon_{\mathbf{k}\sigma} < \mu$, with energy $-(\epsilon_{\mathbf{k}\sigma} - \mu) = |\epsilon_{\mathbf{k}\sigma} - \mu|$. These creation and annihilation operators satisfy Fermi commutation relations. If we treat E_0 as the ground state energy and consider the ground state to be the vacuum of the annihilation operators

$$a_{\mathbf{k}\sigma} |\Psi_0\rangle = 0 \tag{6.4.30}$$

then we should also include the existence of collective modes by defining a set of boson creation and annihilation operators $b^\dagger_{\mathbf{k}}$ and $b_{\mathbf{k}}$ and including terms such as $\Sigma \hbar\omega_{\mathbf{k}} b^\dagger_{\mathbf{k}} b_{\mathbf{k}}$, $\Sigma g_{\mathbf{hqlk}} b^\dagger_{\mathbf{h}} a^\dagger_{\mathbf{q}\sigma} a_{\mathbf{l}\sigma'} b_{\mathbf{k}}$, etc. We will ignore these collective modes for the purposes of our discussion here.

By choosing the ground state as the vacuum, an enormous amount of the complication is hidden in the function $g_4(\mathbf{k}_1\sigma_1\mathbf{k}_2\sigma_2\mathbf{k}_3\sigma_3\mathbf{k}_4\sigma_4)$ and in the higher terms not written in equation

(6.4.29). With this form of the Hamiltonian, we have

$$H|\psi_0\rangle = E_0|\psi_0\rangle$$

$$Ha^\dagger_{\mathbf{k}\sigma}|\psi_0\rangle = [E_0 + |\epsilon_{\mathbf{k}\sigma} - \mu|]a^\dagger_{\mathbf{k}\sigma}|\psi_0\rangle$$

$$Ha^\dagger_{\mathbf{k}\sigma}a^\dagger_{\mathbf{l}\sigma'}|\psi_0\rangle = [E_0 + |\epsilon_{\mathbf{k}\sigma} - \mu| + |\epsilon_{\mathbf{l}\sigma'} - \mu|]a^\dagger_{\mathbf{k}\sigma}a^\dagger_{\mathbf{l}\sigma'}|\psi_0\rangle$$
$$+\frac{1}{2}\sum_{\mathbf{k}_1\sigma_1\mathbf{k}_2\sigma_2}[g_4(\mathbf{k}_1\sigma_1\mathbf{k}_2\sigma_2\mathbf{k}\sigma\mathbf{l}\sigma') - g_4(\mathbf{k}_1\sigma_1\mathbf{k}_2\sigma_2\mathbf{l}\sigma'\mathbf{k}\sigma)]a^\dagger_{\mathbf{k}_1\sigma_1}a^\dagger_{\mathbf{k}_2\sigma_2}|\psi_0\rangle \tag{6.4.31}$$

$|\psi_0\rangle$ and $a^\dagger_{\mathbf{k}\sigma}|\psi_0\rangle$ are eigenkets, but $a^\dagger_{\mathbf{k}\sigma}a^\dagger_{\mathbf{l}\sigma'}|\psi_0\rangle$ is not. The energy of that state is

$$\langle\psi_0|a_{\mathbf{l}\sigma'}a_{\mathbf{k}\sigma}Ha^\dagger_{\mathbf{k}\sigma}a^\dagger_{\mathbf{l}\sigma'}|\psi_0\rangle = E_0 + |\epsilon_{\mathbf{k}\sigma} - \mu| + |\epsilon_{\mathbf{l}\sigma'} - \mu|$$
$$+ g_4(\mathbf{k}\sigma\mathbf{l}\sigma'\mathbf{l}\sigma'\mathbf{k}\sigma) - g_4(\mathbf{k}\sigma\mathbf{l}\sigma'\mathbf{k}\sigma\mathbf{l}\sigma') \tag{6.4.32}$$

A microscopic evaluation of the Landau–Fermi liquid parameters is obtained by identifying

$$f_{\mathbf{k}\alpha\mathbf{k}'\alpha'} = g_4(\mathbf{k}\alpha\mathbf{k}'\alpha'\mathbf{k}\alpha\mathbf{k}'\alpha') - g_4(\mathbf{k}\alpha\mathbf{k}'\alpha'\mathbf{k}'\alpha'\mathbf{k}\alpha). \tag{6.4.33}$$

Equivalently, the Landau energy eigenvalues (6.4.19) are obtained from a reduced Hamiltonian H^{LF} which only has the diagonal terms of the interaction:

$$H^{\mathrm{LF}} = E_0 + \sum_{\mathbf{k}\sigma}|\epsilon_{\mathbf{k}\sigma} - \mu|a^\dagger_{\mathbf{k}\sigma}a_{\mathbf{k}\sigma}$$
$$+\frac{1}{2}\sum_{\mathbf{k}\sigma\mathbf{k}'\sigma'}[g_4(\mathbf{k}\sigma\mathbf{k}'\sigma'\mathbf{k}'\sigma'\mathbf{k}\sigma) - g_4(\mathbf{k}\sigma\mathbf{k}'\sigma'\mathbf{k}\sigma\mathbf{k}'\sigma')]a^\dagger_{\mathbf{k}\sigma}a_{\mathbf{k}\sigma}a^\dagger_{\mathbf{k}'\sigma'}a_{\mathbf{k}'\sigma'}. \tag{6.4.34}$$

Before discussing the microscopic calculation of g_4, it is convenient to discuss the possibility of pair-condensed states in liquid ^{3}He from the point of view of the quasiparticle Hamiltonian, equation (6.4.29). This form is postulated on the existence of a normal ground state, which is obtained from the non-interacting ground state by adiabatically switching on the interactions. The excitation spectrum is assumed continuous across the Fermi surface, as is the interaction g_4. It may be, however, that this adiabatic switching does not produce the ground state. The best known example of an interacting Fermi system whose ground state is not normal is the superconducting ground state of electrons in metals.

The pair condensation arises from an attractive interaction between normal electrons (or holes) near the Fermi surface. Cooper showed that such an attractive interaction leads to the formation of bound electron pairs (Cooper pairs) of opposite spin and momentum.[151] Bardeen, Cooper, and Schrieffer (BCS) pointed out that this signals an instability in the normal ground state in favour of Cooper pair consensation near the Fermi surface. They introduced a trial ground state of the form[32]

$$|\psi_{\mathrm{BCS}}\rangle = \prod_{\substack{\mathbf{k},k_x>0\\ \sigma\sigma'}}[1 + h_{\mathbf{k}\sigma\sigma'}a^\dagger_{\mathbf{k}\sigma}a^\dagger_{-\mathbf{k}\sigma'}]|\psi_0\rangle. \tag{6.4.35}$$

This is a slight generalization on the BCS trial function, having other than singlet pairing.[40,41] If g is attractive near the Fermi surface for pairs of opposite momentum, then a nonzero function $h_{\mathbf{k}\sigma\sigma'}$ minimizes the energy and consequently $|\psi_0\rangle$ is not the ground state. The fact that we have formulated the problem in terms of quasi-particle and quasi-hole operators does not change the BCS argument.[152] The one technical difficulty which remains is to show that $|\psi_{\mathrm{BCS}}\rangle$ has off-diagonal long-range order in the two-particle density matrix. But that follows immediately from the fact that

$$\langle \psi_{\text{BCS}} | a^\dagger_{\mathbf{k}\sigma} a^\dagger_{-\mathbf{k}\sigma'} | \psi_{\text{BCS}} \rangle \neq 0 \tag{6.4.36}$$

and the fact that these creation operators, while not bare-particle (or bare-hole) creation operators $c^\dagger_{\mathbf{k}\sigma}(c_{\mathbf{k}\sigma})$, are obtained from them adiabatically in the same sense that $|\psi_0\rangle$ is obtained adiabatically from the non-interacting ground state:

$$\langle \psi_{\text{BCS}} | a^\dagger_{\mathbf{k}\sigma} a^\dagger_{-\mathbf{k}\sigma'} | \psi_{\text{BCS}} \rangle \neq 0 \longrightarrow \langle \psi_{\text{BCS}} | c^\dagger_{\mathbf{k}\sigma} c^\dagger_{-\mathbf{k}\sigma'} | \psi_{\text{BCS}} \rangle \neq 0. \tag{6.4.37}$$

The fact that $|\psi_{\text{BCS}}\rangle$ is no longer an eigenfunction of the total number of particles means that condition (6.4.15) can only be satisfied in the expectation value by the choice of the chemical potential.

At the time of writing, the author knows of no microscopic calculation of g_4 which shows an attractive interaction between quasiparticles near the Fermi surface, although efforts in that direction are under way. Nevertheless, as we discussed in Section 6.1, the anomalous properties reported in the last several years for ^{3}He below about 2°mK indicate that the low temperature phase of ^{3}He is pair condensed.[37,44] The magnetic measurements on the system indicate that it is paired in spin triplet states, with the low temperature phase being an isotropic spin triplet and the high temperature phase being an anisotropic spin triplet. The phase diagram is shown in Figure 6.1.5.

We will not detail the experimental measurements here. Instead, we will discuss possible ways of calculating the quasiparticle interaction function g, in order to understand both the Fermi liquid properties and the superfluid properties from a microscopic point of view.

t-matrix calculations The method of Green functions has been applied to the calculation of the Landau quasi-particle interaction function f. A two-quasi-particle Green function can be defined in the form

$$\tilde{G}_{\mathbf{k}\sigma\mathbf{l}\sigma'}(t) = \langle \psi_0 | a_{\mathbf{l}\sigma'}(0) a_{\mathbf{k}\sigma}(0) a^\dagger_{\mathbf{k}\sigma}(t) a^\dagger_{\mathbf{l}\sigma'}(t) | \psi_0 \rangle \quad t<0 \tag{6.4.38}$$

where the time dependence is in the Heisenberg representation

$$a^\dagger_{\mathbf{k}\sigma}(t) = \exp(\mathrm{i}Ht/\hbar) a_{\mathbf{k}\sigma} \exp(-\mathrm{i}Ht/\hbar). \tag{6.4.39}$$

Then, using equations (6.4.32) and (6.4.33), we see that

$$f_{\mathbf{k}\sigma\mathbf{l}\sigma'} = \left[-\mathrm{i}\hbar \frac{\mathrm{d}}{\mathrm{d}t} \tilde{G}_{\mathbf{k}\sigma\mathbf{l}\sigma'}(t) \right]_{t=0^-} - |\epsilon_{\mathbf{k}\sigma} - \mu| - |\epsilon_{\mathbf{l}\sigma'} - \mu|. \tag{6.4.40}$$

Equivalently, it can be said that the time dependence of G is given primarily by

$$\tilde{G}_{\mathbf{k}\sigma\mathbf{l}\sigma'}(t) \propto \exp[\mathrm{i}(|\epsilon_{\mathbf{k}\sigma} - \mu| + |\epsilon_{\mathbf{l}\sigma'} - \mu| + f_{\mathbf{k}\sigma\mathbf{l}\sigma'})t/\hbar] \quad t<0. \tag{6.4.41}$$

Now, we can use the fact that the $a^\dagger_{\mathbf{k}\sigma}$ are quasi-particle operators for normal Fermi liquids and consequently may be obtained adiabatically from the bare particle operators $c^\dagger_{\mathbf{k}\sigma}$

$$a^\dagger_{\mathbf{k}\sigma} = A_{\mathbf{k}\sigma} c^\dagger_{\mathbf{k}\sigma} + \ldots. \tag{6.4.42}$$

Consequently, the bare two-particle Green function

$$G_{\mathbf{k}\sigma\mathbf{l}\sigma'}(t) = \langle \psi_0 | c_{\mathbf{l}\sigma'}(0) c_{\mathbf{k}\sigma}(0) c^\dagger_{\mathbf{k}\sigma}(t) c^\dagger_{\mathbf{l}\sigma'}(t) | \psi_0 \rangle \quad t<0 \tag{6.4.43}$$

should have one Fourier component located at the same frequency as $\tilde{G}$:

$$G_{\mathbf{k}\sigma\mathbf{l}\sigma'}(t) = B \exp[\mathrm{i}(|\epsilon_{\mathbf{k}\sigma} - \mu| + |\epsilon_{\mathbf{l}\sigma'} - \mu| + f_{\mathbf{k}\sigma\mathbf{l}\sigma'})t/\hbar] \qquad t<0 \tag{6.4.44}$$

where B has most of the weight in the Fourier series.

The methods of quantum field theory may be applied to the calculation of G.[148] If the bare two-body interaction $V(r)$ has a Fourier transform, then the full Hamiltonian may be written as

$$H - \mu N = \sum_{\mathbf{k}\sigma} c^{\dagger}_{\mathbf{k}\sigma} c_{\mathbf{k}\sigma}\left(\frac{\hbar^2 k^2}{2m} - \mu\right) + \frac{1}{2}\sum_{\substack{\mathbf{k}_1\sigma_1\,\mathbf{k}_2\sigma_2\\ \mathbf{k}_3\sigma_3\,\mathbf{k}_4\sigma_4}} c^{\dagger}_{\mathbf{k}_1\sigma_1} c^{\dagger}_{\mathbf{k}_2\sigma_2} c_{\mathbf{k}_4\sigma_4} c_{\mathbf{k}_3\sigma_3} V^{\mathbf{k}_1\sigma_1\,\mathbf{k}_2\sigma_2}_{\mathbf{k}_3\sigma_3\,\mathbf{k}_4\sigma_4} \tag{6.4.45}$$

where

$$V^{\mathbf{k}_1\sigma_1\,\mathbf{k}_2\sigma_2}_{\mathbf{k}_3\sigma_3\,\mathbf{k}_4\sigma_4} = \delta_{\mathbf{k}_1+\mathbf{k}_2,\,\mathbf{k}_3+\mathbf{k}_4}\,\delta_{\sigma_1,\sigma_4}\,\delta_{\sigma_2,\sigma_3}\,V(\mathbf{k}_1 - \mathbf{k}_4) = \langle \mathbf{k}_1\sigma_1\mathbf{k}_2\sigma_2 | V | \mathbf{k}_4\sigma_4\mathbf{k}_3\sigma_3 \rangle. \tag{6.4.46}$$

If G is calculated in a Hartree–Fock approximation, the time dependence is given by

$$\hbar\omega_{\mathbf{kl}} = \frac{\hbar^2 k^2}{2m} - \mu + \frac{\hbar^2 l^2}{2m} - \mu + \langle \mathbf{k}\sigma\mathbf{l}\sigma' | V | \mathbf{k}\sigma\mathbf{l}\sigma' \rangle - \langle \mathbf{k}\sigma\mathbf{l}\sigma | V | \mathbf{l}\sigma'\mathbf{k}\sigma \rangle \delta_{\sigma\sigma'} \tag{6.4.47}$$

and the last two terms are the Hartree–Fock approximation for $f_{\mathbf{k}\sigma\mathbf{l}\sigma'}$.

The interaction between ^{3}He atoms cannot be treated in this manner because of the strong, short-range repulsion. These matrix elements of V either do not exist (e.g. the 6–12 potential possesses no Fourier transform) or are so large that this mean field approach is not valid. Brueckner developed the t-matrix method to treat similar difficulties in the theory of nuclear matter.[153] In his procedure, the potential V is replaced by a complicated many-body operator t, whose effect is to sum the repeated virtual scattering of a pair of particles in states **k** and **l**, thereby including an infinite subset of the terms from perturbation theory. Then

$$\hbar\omega_{\mathbf{kl}} = \frac{\hbar^2 k^2}{2m} - \mu + \frac{\hbar^2 l^2}{2m} - \mu + \langle \mathbf{k}\sigma\mathbf{l}\sigma' | \hat{t} | \mathbf{k}\sigma\mathbf{l}\sigma' \rangle - \langle \mathbf{k}\sigma\mathbf{l}\sigma' | \hat{t} | \mathbf{l}\sigma'\mathbf{k}\sigma \rangle. \tag{6.4.48}$$

The full definition of the t-matrix may be found in References (148) and (153). The quasi-particle interaction function is approximately given by the last two terms. The t-matrix calculation was applied to ^{3}He by Brueckner and Gammel,[154] and in further refinements by L. Campbell and Brueckner,[155] and most recently by Østgaard.[156] They found it necessary to use a different potential (the Yntema–Schneider potential) to obtain a negative energy for the ground state energy. Østgaard's calculated Landau–Fermi liquid properties may be compared with the experimental results in Table 6.4.1.

The correlated theory of the normal liquid Our discussion above implies a treatment of ^{3}He similar to the treatment of ^{4}He, where the problem divides neatly into two tasks: (i) calculate the ground state wave function and the corresponding distribution functions; and (ii) treat the ground state as the N-particle vacuum and determine operators which create excited states out of this vacuum. For fermions, step (i) is complicated enormously by the anti-symmetry of the wavefunction. Even assuming that the exact ground state wave function can be obtained, step (ii) is difficult because there is no simple choice for coordinate space operators which produce particle–hole excited states when applied to the ground state wave function.

The principal reason for using the ground state solution for the normalized vacuum is that the short-range correlations induced by the singular potential are the same in the ground state and the excited states. But these correlations are not very sensitive to the statistics, so that a symmetric correlation function ψ_0 similar to the boson solution of the ^{3}He Hamiltonian will have the necessary short range correlations. The statistics of the system are then put in by anti-symmetric operators, or model functions, which act upon ψ_0 to produce the trial states of the system. These anti-symmetric functions are the Slater determinants.

This analysis was first suggested by Jastrow, who proposed the trial function for the ground state[85]

$$\psi(x_1 \ldots x_N \mid \{\mathbf{k}_i\sigma_i\}) = \mathscr{S}_{\mathbf{k}_1\sigma_1 \ldots \mathbf{k}_N\sigma_N}(x_1 \ldots x_N)\psi_0(\mathbf{r}_1 \ldots \mathbf{r}_N) \tag{6.4.49}$$

where $\mathscr{S}$ is the ground state Slater determinant of a non-interacting N-body system and ψ_0 is of the Jastrow form (equation 6.1.25). Then the variational principle may be invoked to obtain an upper bound on the ground state energy

$$E = \langle \psi \mid H \mid \psi \rangle / \langle \psi \mid \psi \rangle \geqslant E_0 \tag{6.4.50}$$

and ψ_0 may be varied to obtain an optimum bound on the energy. Of course ψ_0 need not be restricted to the simple Jastrow form, but may also be chosen in the extended Jastrow form (equations 6.2.6 and 6.2.13) for an improved approximation to the energy.[157] No calculation including three-body (or higher) factors has yet been done.

The Slater determinants $\mathscr{S}_{\mathbf{k}_1\sigma_1 \ldots \mathbf{k}_N\sigma_N}$ form a complete set of anti-symmetric N-body states when the set $\{\mathbf{k}_i,\sigma_i\}$ is unrestricted. This completeness property is not changed by a positive definite, symmetric correlation factor ψ_0. Thus the set of states $\psi(x_1 \ldots x_N \mid \{\mathbf{k}_i\sigma_i\})$ forms a complete, non-orthogonal basis in which the problem of the ground state and excited states can be simultaneously treated by orthogonalizing the basis and diagonalizing the Hamiltonian in the orthogonal basis. In the lowest approximation, ignoring the orthogonality problem, the excited states are just the particle–hole excited state Slater determinants multiplying ψ_0.

Unfortunately, the presence of the Slater determinants makes the task of calculating expectation values in this basis enormously more difficult than the similar problem in the boson system. The problem is that $\mathscr{S}$ is an N-body function, in contrast to the symmetrized plane wave $\rho^{(n)}(\mathbf{k}_1 \ldots \mathbf{k}_n)$ which appears in the boson correlated basis and is the sum of n-body functions. There are two distinct procedures which have been developed for dealing with this problem. The first is to cluster expand the Jastrow factor ψ_0 in a manner very similar to the Ursell–Mayer and Van Kampen cluster expansions in classical statistical mechanics.[158] That approach is best suited for low-density systems, and has been applied primarily to nuclear matter calculations, where the hard core takes up a much smaller fraction of the total volume at the density of finite nuclei than the corresponding quantity in liquid ^{3}He at experimental equilibrium density. The other approach, developed by Feenberg, Wu, Woo and Tan involves a cluster expansion of the Slater determinant, leaving ψ_0 intact.[47,159–163] Expectation values are then expressed in terms of correlation functions of ψ_0; the correlation function which appears most often is the liquid structure function of ψ_0, which may be calculated in the same manner as in ^{4}He.

Feenberg and Woo find it most convenient to let ψ_0 be the exact boson liquid ^{3}He ground state, ψ_0^{B}:[161]

$$H\psi_0^{\mathrm{B}} = E_0^{\mathrm{B}}\psi_0^{\mathrm{B}}. \tag{6.4.51}$$

This simplifies some matrix elements, reducing them to commutators with H. It is then only necessary to have certain correlation functions of ψ_0^B, particularly the liquid structure function $S_0^B(k)$, which are obtained from separately determined approximations to ψ_0^B.

This procedure does not leave ψ_0 as a variational function. Feenberg and Woo calculate corrections to the ground state energy by doing second-order non-orthogonal perturbation theory to mix in particle–hole corrections to the Jastrow trial function. A great deal of care must be exercised in this calculation to make certain that anomalous dependences on N are appropriately cancelled by orthogonalization terms. We refer to the literature for details.[47]

The quasiparticle interaction function $f_{\mathbf{k}\sigma\mathbf{l}\sigma'}$ was calculated by Tan and Feenberg in a similar manner, by taking account of second-order perturbation contributions to the two-quasi-particle energy.[163] The calculated Landau–Fermi liquid properties are compared with experiment in Table 6.4.1.

As discussed above, calculations of this sort give access to much more information about the system than the Landau–Fermi liquid parameters through the full quasiparticle interaction function $g_4(\mathbf{k}_1\sigma_1\mathbf{k}_2\sigma_2\mathbf{k}_3\sigma_3\mathbf{k}_4\sigma_4)$. In particular, it should be possible to detect a Cooper instability in the quasi-particles which signals a phase transition to a pair-condensed (BCS superfluid) state. At the time of writing, that analysis has not been carried out.

We now describe a more direct attack on the question of the form of the ground state, including the possibility of pair condensation.

The correlated pairing theory An alternative variational approach to the ground state problem is to consider the trial function

$$\psi = \Phi\psi_0 \tag{6.4.52}$$

where ψ_0 has the same properties as in the preceding section (symmetric positive definite, and containing short range correlations adequate for the strong repulsive part of the potential), and Φ is an antisymmetric function of the N position and spin variables. In the previous section, Φ was taken to be a Slater determinant. Here we would like to include the possibility of a pairing transition, so we choose Φ to minimize the energy expectation value in ψ.

The pair condensed state has off-diagonal long range order (ODLRO) in the two-particle density matrix (see Section 6.1 for a discussion of ODLRO), which must occur because of the form of Φ. Clark and Yang, in their consideration of superfluidity in neutron stars, pointed out that the choice of the BCS pair-condensed ground state for Φ would produce a trial function with ODLRO.[51] The energy of this correlated pairing function can then be compared with the normal ground state energy (where Φ is a Slater determinant or a linear combination of Slater determinants with no ODLRO) to determine which has the lower energy.

The BCS trial ground state is given by

$$|\,\mathrm{BCS}\,\rangle = \prod_{\mathbf{k},\,k_x>0}\ \prod_{\sigma\sigma'} [u_{\mathbf{k}\sigma\sigma'} + v_{\mathbf{k}\sigma\sigma'}\, c^{\dagger}_{\mathbf{k}\sigma} c^{\dagger}_{\mathbf{k}'\sigma'}]\ |\,\mathrm{vac}\,\rangle \tag{6.4.53}$$

where $|\,\mathrm{vac}\,\rangle$ is the vacuum and $c^{\dagger}_{\mathbf{k}\sigma}$ is a bare particle creation operator. The functions $u_{\mathbf{k}\sigma\sigma'}$ and $v_{\mathbf{k}\sigma\sigma'}$ were chosen by BCS to minimize the total energy expectation value of the Hamiltonian. We wish to use this to generate a model function Φ of a fixed number of particles N. This is done in the usual way by projecting out of $|\,\mathrm{BCS}\,\rangle$ the state with N particles and then transforming to the coordinate space representation. The result is that Φ is an anti-symmetrized product of two-body functions

$$\Phi(\mathbf{r}_1\sigma_1 \ldots \mathbf{r}_N\sigma_N) = A[\varphi(\mathbf{r}_1\sigma_1\mathbf{r}_2\sigma_2)\varphi(\mathbf{r}_3\sigma_3\mathbf{r}_4\sigma_4)\ldots\varphi(\mathbf{r}_{N-1}\sigma_{N-1}\mathbf{r}_N\sigma_N)] \tag{6.4.54}$$

where A is the anti-symmetrization operator and φ is given by

$$\varphi(\mathbf{r}_1\sigma_1\mathbf{r}_2\sigma_2)=\frac{1}{\Omega}\sum_{\mathbf{k},\,\mathbf{k}_x>0}\exp(\mathrm{i}\mathbf{k}\cdot\mathbf{r}_{12})\frac{v_{\mathbf{k}\sigma\sigma'}}{u_{\mathbf{k}\sigma\sigma'}}. \tag{6.4.55}$$

This choice includes both the normal and superfluid states. The difference between the two is that $v_{\mathbf{k}\sigma\sigma'}$ has a discontinuity at the Fermi surface for a normal state but is continuous for the superfluid. Note that the ground state Slater determinant has

$$v_{\mathbf{k}\sigma\sigma'}=\delta_{\sigma,-\sigma'}\begin{cases}1 & k\leqslant k_{\mathrm{F}}\\ 0 & k>k_{\mathrm{F}}.\end{cases} \tag{6.4.56}$$

The ratio

$$h_{\mathbf{k}\sigma\sigma'}=v_{\mathbf{k}\sigma\sigma'}/u_{\mathbf{k}\sigma\sigma'} \tag{6.4.57}$$

is determined by minimizing the expectation value of the Hamiltonian with respect to $h_{\mathbf{k}\sigma\sigma'}$. We use the transformation given in equation (6.2.10) on the kinetic energy to obtain

$$\langle\psi\mid H\mid\psi\rangle=E_{\mathrm{B}}+E_{\mathrm{F}}^{(0)}+E_1+E_2 \tag{6.4.58}$$

where

$$E_{\mathrm{B}}=\langle\psi_0\mid H\mid\psi_0\rangle/\langle\psi_0\mid\psi_0\rangle \tag{6.4.59}$$

is a boson energy since ψ_0 satisfies Bose statistics:

$$E_{\mathrm{F}}^{(0)}=\frac{\int\psi_0^2\left\{\Phi^*\sum_{i=1}^{N}(-\hbar^2/4m)\nabla_i^2\Phi+\Phi\sum_{i=1}^{N}(-\hbar^2/4m)\nabla_i^2\Phi^*\right\}\mathrm{d}x_1\ldots\mathrm{d}x_N}{\int\psi_0^2\Phi^*\Phi\,\mathrm{d}x_1\ldots\mathrm{d}x_N} \tag{6.4.60}$$

is closely related to the uncorrelated fermion kinetic energy, and would be

$$E_{\mathrm{F}}^{0}=\sum_{i=1}^{N}\frac{\hbar^2k_i^2}{2m} \tag{6.4.61}$$

if Φ were a Slater determinant;

$$E_1=\frac{\int\psi_0^2\left[V_{\mathrm{B}}^*(\mathbf{r}_1\ldots\mathbf{r}_N)-E_{\mathrm{B}}-\sum_{i=1}^{N}(\hbar^2/8m)\nabla_i^2\right]\mid\Phi\mid^2\mathrm{d}x_1\ldots\mathrm{d}x_N}{\int\psi_0^2\mid\Phi\mid^2\mathrm{d}x_1\ldots\mathrm{d}x_N} \tag{6.4.62}$$

where

$$V_{\mathrm{B}}^*=\sum_{i<j}^{N}V(r_{ij})-\frac{\hbar^2}{8m}\sum_{i=1}^{N}\nabla_i^2\ln\psi_0^2 \tag{6.4.63}$$

and

$$E_2=\frac{\int\psi_0^2(\hbar^2/4m)\sum_{i=1}^{N}\nabla_i^2\mid\Phi\mid^2\mathrm{d}x_1\ldots\mathrm{d}x_N}{\int\psi_0^2\mid\Phi\mid^2\mathrm{d}x_1\ldots\mathrm{d}x_N}. \tag{6.4.64}$$

The advantage of separating out the term E_1 is that it vanishes if ψ_0 is the boson ^{3}He ground state, i.e. if

$$H\psi_0=E_{\mathrm{B}}\psi_0. \tag{6.4.65}$$

(This can be verified by the analysis of Section 6.2).

The cluster expansion methods of Feenberg, Wu and Woo[160–162] are not general enough to evaluate matrix elements such as the normalization integral $\langle \psi | \psi \rangle$ or terms like E_2 for an arbitrary anti-symmetric function Φ. Here we devise a procedure which we call the Effective Bose Expansion which at once handles the general Φ and also takes advantage of previous work done in boson systems. This expansion is motivated by the observation that all expectation values can be written in the form

$$\langle \psi | A | \psi \rangle = \langle \psi_0 | \Phi^* A \Phi | \psi_0 \rangle \tag{6.4.66}$$

where $\Phi^* A \Phi$ is a symmetric operator by virtue of the fact that both Φ and Φ^* are anti-symmetric. As in the case of the energy, most operators of interest can be put into the somewhat simpler form

$$\langle \psi | A | \psi \rangle = \langle \psi_0 | B | \psi_0 \rangle \tag{6.4.67}$$

where B is a symmetric function of the N coordinates and the spins. For E_2,

$$B = \frac{\hbar^2}{4m} \sum_{i=1}^{N} \nabla_i^2 |\Phi|^2 \tag{6.4.68}$$

while for the normalization

$$B = |\Phi|^2. \tag{6.4.69}$$

The function B can be expanded in a complete set of symmetric functions which are just proportional to the symmetrized plane waves $\rho^{(n)}(\mathbf{k}_1 \ldots \mathbf{k}_n)$ discussed in Sections 6.2 and 6.3. To get the normalization of the symmetrized plane waves correct, they should be labelled by the number of equivalent vectors: $\rho^{(n)}(n_{\mathbf{k}_1}, n_{\mathbf{k}_2}, \ldots n_{\mathbf{k}_p})$, $\Sigma_{i=1}^{p} n_{\mathbf{k}i} = n$. Then the properly normalized expansion is

$$\langle \psi_0 | B | \psi_0 \rangle = \frac{1}{N! \Omega^N} \sum_{n=0}^{N} \sum_{\{n_{\mathbf{k}_i} | \sum_i n_{\mathbf{k}_i} = n\}} \frac{(N-n)!}{\Pi n_{\mathbf{k}_i}!} \langle \psi_0 | \rho^{(n)}(n_{\mathbf{k}_1} \ldots n_{\mathbf{k}_p}) | \psi_0 \rangle$$

$$\times \int \rho^{(n)*}(n_{\mathbf{k}_1} \ldots n_{\mathbf{k}_p}) B(x_1 \ldots x_N) \mathrm{d}x_1 \ldots \mathrm{d}x_N \tag{6.4.70}$$

For example, the normalization integral takes the convenient form

$$\langle \psi | \psi \rangle = \frac{1}{N! \Omega^N} \sum_{n=0}^{N} \sum_{\{n_{\mathbf{k}_i} | \sum_i n_{\mathbf{k}_i} = n\}} \frac{(N-n)!}{\Pi n_{\mathbf{k}_i}!} \langle \psi_0 | \rho^{(n)}(n_{\mathbf{k}_1} \ldots n_{\mathbf{k}_p}) | \psi_0 \rangle$$

$$\times \langle \Phi | \rho^{(n)*}(n_{\mathbf{k}_1} \ldots n_{\mathbf{k}_p}) | \Phi \rangle \tag{6.4.71}$$

where $\langle \Phi | \rho^{(n)} | \Phi \rangle$ is the expectation value in the uncorrelated state $|\Phi\rangle$. As was pointed out in Section 6.2, the expectation value of the symmetrized plane wave $\rho^{(n)}$ is the Fourier transform $G_n(\mathbf{k}_1 \ldots \mathbf{k}_n)$ of the n-particle distribution function $g_n(\mathbf{r}_1 \ldots \mathbf{r}_n)$. G_n is related in turn to the set of cumulants $F_n(\mathbf{k}_1 \ldots \mathbf{k}_n)$ by equation (6.2.89). As is seen from the discussion in Section 6.2, the F_n are all $O(N)$. Furthermore, they have some very useful properties for convergence of the above series. For example, the F_n all vanish in the non-interacting Bose ground state, which follows from the fact that the $\rho^{(n)}$ are all orthogonal to a constant. Consequently, we expect that a reasonable approximation scheme is to neglect F_n of ψ_0 beyond some value of n. A similar conclusion can be drawn for the F_n^{F} of the noninteracting Fermi system (i.e. the ground state Slater determinant). Although the F_n^{F} do not vanish, they

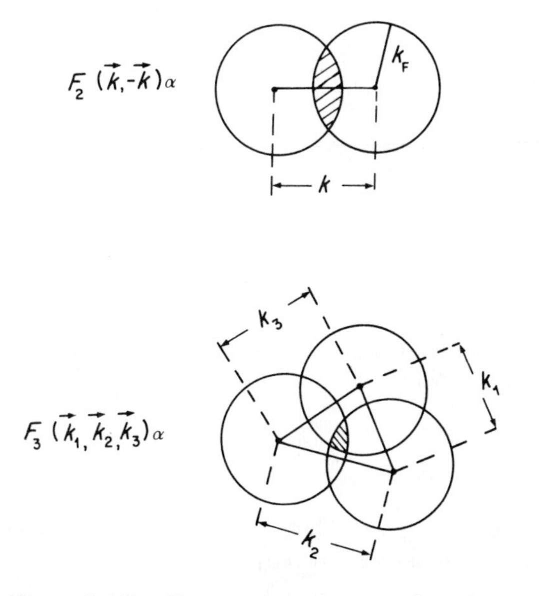

Figure 6.4.2 The cumulant functions $F_2(k,-k)$ and $F_3(k_1, k_2, k_3)$ for a non-interacting Fermi gas expressed as the co-volume of two and three Fermi spheres, respectively

become increasingly less important as n increases. To see this, we recall the well known result that[149]

$$F_2^{\mathrm{F}}(\mathbf{k}, -\mathbf{k}) = \langle \Phi_0 \mid \rho_{\mathbf{k}}\rho_{-\mathbf{k}} - N \mid \Phi_0 \rangle / \langle \Phi_0 \mid \Phi_0 \rangle$$

$$= \quad -2 \sum_{\substack{\mathbf{p} \\ p \leqslant k_{\mathrm{F}},\, |\mathbf{p}+\mathbf{k}| \leqslant k_{\mathrm{F}}}} 1 \quad = N\left[\frac{3}{4}\frac{k}{k_{\mathrm{F}}} - \frac{1}{16}\left(\frac{k}{k_{\mathrm{F}}}\right)^3 - 1\right] \quad k \leqslant 2k_{\mathrm{F}}$$

$$= 0 \qquad k > 2k_{\mathrm{F}} \qquad (6.4.72)$$

which is -2 times the co-volume of two Fermi spheres a distance k apart (Figure 6.4.2).

Similarly, it can be shown that $F_3^{\mathrm{F}}(\mathbf{k}_1, \mathbf{k}_2, \mathbf{k}_3)$ is proportional to the co-volume of three Fermi spheres centred at the corners of the triangle formed by the three momenta $\mathbf{k}_1, \mathbf{k}_2, \mathbf{k}_3$ (Figure 6.4.2):

$$F_3^{\mathrm{F}}(\mathbf{k}_1\, \mathbf{k}_2\, \mathbf{k}_3) = \quad -4 \sum_{\substack{\mathbf{p} \\ p \leqslant k_{\mathrm{F}},\, |\mathbf{p}+\mathbf{k}_1| \leqslant k_{\mathrm{F}},\, |\mathbf{p}-\mathbf{k}_3| \leqslant k_{\mathrm{F}}}} 1 \qquad (6.4.73)$$

This result generalizes so that $F_n^{\mathrm{F}}(\mathbf{k}_1 \ldots \mathbf{k}_n)$ is a measure of the co-volume of n Fermi spheres centred at the closed polygon formed by the momenta $\mathbf{k}_1 \ldots \mathbf{k}_n$. (This is strictly true only when no subset of the momenta add to zero. But F_n^{F} is a continuous function so that those cases where subsets of the momenta add to zero contribute only a negligible amount in sums containing F_n^{F}). This shows that F_n^{F} is an increasingly short range function of its arguments as n increases, and can be reasonably approximated by zero beyond some value of n.

It is clear from this discussion that an obvious systematic approximation scheme is to set F_n to zero beyond some value of n for both the Bose and the Fermi factors. To illustrate, the simplest non-trivial approximation is to retain only F_2:

$$F_2^0(\mathbf{k},-\mathbf{k}) = \langle \psi_0 | \rho_{\mathbf{k}}\rho_{-\mathbf{k}} - N | \Psi_0 \rangle / \langle \psi_0 | \psi_0 \rangle$$
$$= N(S_0(k)-1) \tag{6.4.74}$$

$$F_2^{\mathrm{F}}(\mathbf{k},-\mathbf{k}) = N(S_{\mathrm{F}}^{(0)}(k)-1) \tag{6.4.75}$$

where

$$S_{\mathrm{F}}^{(0)}(k) = \langle \Phi | \rho_{\mathbf{k}}\rho_{-\mathbf{k}} | \Phi \rangle / N \langle \Phi | \Phi \rangle. \tag{6.4.76}$$

Suppose that ψ_0 is the exact Bose ground state and Φ is the ground state Slater determinant. Then the energy expectation value is

$$\frac{\langle \psi | H | \psi \rangle}{\langle \psi | \psi \rangle} = E_{\mathrm{B}} + \sum_{i=1}^{N} \frac{\hbar^2 k_i^2}{2m} - \sum_{\mathbf{k}, k_x > 0} \frac{(S_{\mathrm{B}}(k)-1)(S_{\mathrm{F}}^0(k)-1)}{1-(S_{\mathrm{B}}(k)-1)(S_{\mathrm{F}}^0(k)-1)} \frac{\hbar^2 k^2}{2m} \tag{6.4.77}$$

where E_{B} is the Bose liquid ^{3}He ground state energy E_{B}^0, the second term is the contribution of the Fermi sea, which at equilibrium density (0.0164 Å^{-3}) is

$$\frac{E_{\mathrm{F}}}{Nk_{\mathrm{B}}} = \frac{1}{Nk_{\mathrm{B}}} \sum_{i=1}^{N} \frac{\hbar^2 k_i^2}{2m} = 2.98\,^\circ\mathrm{K} \tag{6.4.78}$$

The final term comes entirely from E_2 (equation 6.4.64). Using the $S_{\mathrm{B}}(k)$ obtained from the paired phonon analysis of Boson liquid ^{3}He (Figure 6.4.1),[70] the contribution of E_2 at equilibrium density is[164]

$$\frac{E_2}{Nk_{\mathrm{B}}} = -1.08\,^\circ\mathrm{K}. \tag{6.4.79}$$

Then

$$\frac{E}{Nk_{\mathrm{B}}} = \frac{E_{\mathrm{B}}^{(0)}}{Nk_{\mathrm{B}}} + 1.90\,^\circ\mathrm{K}. \tag{6.4.80}$$

When the value $-2.64\,^\circ$K obtained in the paired phonon analysis is used for $E_{\mathrm{B}}^0/Nk_{\mathrm{B}}$,[70] the final calculated energy per particle is $-0.74\,^\circ$K, to be compared with the experimental value of $-2.52\,^\circ$K. This result is not as bad as it seems, since the total energy is the sum of a potential energy per particle of about $-14\,^\circ$K and a kinetic energy per particle of about 12 $^\circ$K.

No part of the trial function has been varied in this estimate of the energy; the situation is bound to improve if either Φ or ψ_0 or both are allowed to vary. In fact, it is clear that both Φ and ψ_0 introduce correlations for wave vectors less than k_{F}, and one might expect some overcorrelation. For example, if Φ is held fixed as the Slater determinant, the best ψ_0 will not necessarily be the boson liquid ^{3}He ground state. To check the consequences of varying ψ_0, we considered the trial form

$$\psi_0 = \psi_{\mathrm{B}}^{(0)} \prod_{i<j}^{N} \exp[\tfrac{1}{2}u(r_{ij})] \tag{6.4.81}$$

This choice raises the contribution of E_{B} above E_{B}^0 and changes the values of E_1 and E_2 in equation (6.4.58). The most important change is that E_2 is no longer zero. By minimizing the total energy with respect to u, the value in equation (6.4.80) may be improved. We will not give the algebraic details here, but simply note that the total shift is only $-0.03\,^\circ$K[164] so that

$$\frac{E}{Nk_B} = \frac{E_B^0}{Nk_B} + 1.87\,^\circ\mathrm{K} \tag{6.4.82}$$

where E_B^0 is still the exact Bose liquid ^{3}He ground state energy. So we see that the Bose liquid ^{3}He ground state does provide an excellent correlation function, at least for the purpose of calculating the energy.

The effect of varying Φ by mixing in some additional Slater determinants may be estimated by using the second order perturbation theory result of Feenberg, Woo and Tan.[161–163] They found that at equilibrium density, this second order energy is

$$\frac{\delta E_2}{Nk_B} = -0.36\,^\circ\mathrm{K} \tag{6.4.83}$$

Adding this to the values calculated above gives

$$\frac{E}{Nk_B} = \frac{E_B^0}{Nk_B} + 1.54\,^\circ\mathrm{K} \tag{6.4.84}$$

This gives $-1.1\,^\circ$K when the paired phonon analysis result is used for E_B^0. Most of the remaining discrepancy between this value and the experimental value of $-2.5\,^\circ$K probably lies in the calculated value of E_B^0, although it should be noted that the discrepancy here is about twice as large as in the ^{4}He case.

The effects of the statistics on liquid structure can be studied by calculating the liquid structure function $S(k)$ in the state ψ. Using the same approximations as above, we find that[164]

$$\frac{1}{S(k)} = \frac{1}{S_0(k)} + \frac{1}{S_F^0(k)} - 1 \tag{6.4.85}$$

where S_0 and S_F^0 are defined in equations (6.4.74–76). When we take ψ_0 to be in the form equation (6.4.81) with the $u(r)$ which minimizes the energy, this expression becomes

$$\frac{1}{S(k)} = \left(\frac{1}{S_B^0(k)^2} + \frac{1}{S_F^0(k)^2} - 1\right)^{1/2} \tag{6.4.86}$$

where

$$S_B^0(k) = \frac{1}{N}\frac{\langle \psi_B^0 | \rho_{\mathbf{k}}\rho_{-\mathbf{k}} | \psi_B^0 \rangle}{\langle \psi_B^0 | \psi_B^0 \rangle} \tag{6.4.87}$$

is the liquid structure function for the Bose liquid ^{3}He ground state. Consequently,

$$S(k) = S_B^0(k) \quad k \geqslant 2k_F \tag{6.4.88}$$

and, in the long wavelength limit,

$$S(k) \to \frac{\hbar k}{2m\sqrt{(c_B^2 + c_F^2)}} \equiv \frac{\hbar k}{2mc} \tag{6.4.89}$$

where

$$S_B^0(k) \to \frac{\hbar k}{2mc_B} \tag{6.4.90}$$

$$S_F^0(k) \to \frac{\hbar k}{2mc_F} \tag{6.4.91}$$

and

$$c_F = \frac{2}{3} \frac{\hbar k_F}{m}$$

for the Slater determinant. Since at equilibrium density $c_F = 110$ m/sec is of the same order of magnitude as c_B(168 m/sec from the data of Massey and Woo), the most important effects of statistics come at long wavelengths. Furthermore, the value for c obtained from $\psi_0 = \psi_B^0$ is 278 m/sec, considerably higher than the value of 201 m/sec obtained by including the Jastrow factor in ψ_0. This is still higher than the experimental value of[165]

$$182 \text{ m/sec} < c < 191 \text{ m/sec} \tag{6.4.92}$$

where the limits are the velocities of first sound and zero sound, respectively.

We are presently applying the theory described in this section to the calculation of the pair-condensed ground state of liquid ^{3}He by allowing Φ to have the form of equation (6.4.54) and determining the parameters variationally. The trial function we use must be somewhat more complicated to include the very important effects of paramagnon fluctuations. The results of these calculations will be reported elsewhere.[60]

Acknowledgements

I would like to acknowledge partial support of this work by the National Science Foundation through Grant GH-43836. The Research Corporation and the Graduate School of the University of Minnesota have also supported some of the research reported herein.

I would particularly like to express my gratitude to Eugene Feenberg, who introduced me to the theory of quantum fluids and provided the framework for a powerful theoretical approach to the subject.

I have benefited a great deal from frequent discussions with my colleagues at the University of Minnesota, Chao Ching Chang, J. Woods Halley, and Peter Kleban. Remarks by Halley on the subject of light scattering discussed in Section 6.3 were particularly useful. Finally, special thanks go to Ms. S. Smith for the careful preparation of this manuscript.

References

1. H. Kamerlingh Onnes (1908). Commun. Leiden No. 108, *Proc. Roy. Acad. Amsterdam,* **11**, 169.
2. K. Mendelssohn (1966). *The Quest for Absolute Zero*, World University Library, McGraw-Hill, New York.
3. W. H. Keesom (1926). Commun. Leiden No. 184b, *Proc. Roy. Acad. Amsterdam,* **29**, 1136.
4. J. de Boer (1948). *Physica,* **14**, 139.
   J. de Boer and B. S. Blaisse (1948). *Physica,* **14**, 149.
   J. de Boer and R. J. Lunbeck (1948). *Physica,* **14**, 520.
5. J. de Boer (1963). In G. Careri (Ed.), *Liquid Helium,* Vol. 1. Academic Press, London.
6. L. H. Nosanow, L. J. Parish and F. J. Pinski (1975). *Phys. Rev.,* **B11**, 191.
7. M. D. Miller, L. H. Nosanow and L. J. Parish (1975). *Phys. Rev. Lett.,* **35**, 581.
8. C. E. Campbell and M. Schick (1971). *Phys. Rev.,* **A3**, 691.
9. M. D. Miller, C.-W. Woo and C. E. Campbell (1972). *Phys. Rev.,* **A6**, 1942.
10. A. O. Novaco and C. E. Campbell (1975). *Phys. Rev.,* **B11**, 2525.
11. M. Bretz, J. G. Dash, D. C. Hickernell, E. O. McLean and O. E. Vilches (1973). *Phys. Rev.,* **A8**, 1589.
    M. Schick and J. G. Dash, to be published.

12. M. Wolfke and W. H. Keesom (1928). Commun. Leiden No. 219e, *Proc. Roy. Acad. Amsterdam,* **31**, 81.
13. W. H. Keesom and K. Clausius (1932). Commun. Leiden No. 190a, *Proc. Roy. Acad. Amsterdam,* **35**, 736.
W. H. Keesom and K. Clausius (1931). Commun. Leiden No. 216b, *Proc. Roy. Acad. Amsterdam,* **34**, 605.
14. L. Tisza (1938). *Nature,* **141**, 913.
15. F. London (1938). *Nature,* **141**, 643.
F. London (1938). *Phys. Rev.,* **54**, 947.
16. F. London (1964). *Superfluids,* Vol. II. Dover, New York.
17. N. N. Bogoliubov (1947). *J. Phys. U.S.S.R.,* **11**, 23.
18. O. Penrose and L. Onsager (1956). *Phys. Rev.,* **104**, 576.
19. W. L. McMillan (1965). *Phys. Rev.,* **138**, A442.
20. J. A. Fernandez and H. A. Gersch (1972). *Phys. Rev.,* **A7**, 239 and references cited therein.
21. H. A. Mook, R. Scherm, and M. K. Wilkinson (1972). *Phys. Rev.,* **A6**, 2268.
22. H. W. Jackson (1974). *Phys. Rev.,* **A10**, 278.
23. G. V. Chester (1969). In K. Mahanthappa and W. Brittin (Eds), *Lectures in Theoretical Physics,* Vol. XI B, Gordon and Breach, London.
24. C. N. Yang (1962). *Rev. Mod. Phys.,* **34**, 694.
25. L. Reatto (1969). *Phys. Rev.,* **183**, 334.
26. E. H. Lieb and W. Liniger (1963). *Phys. Rev.,* **130**, 1605.
27. G. V. Chester (1970). *Phys. Rev.,* **A2**, 256 Of course a solid would not be a superfluid in the usual sense because of the nature of its excitation energy.
28. L. Landau (1941). *J. Phys. U.S.S.R.,* **5**, 71.
L. Landau (1941). *Phys. Rev.,* **60**, 356.
29. R. P. Feynman (1954). *Phys. Rev.,* **94**, 262.
30. N. Hugenholtz and D. Pines (1959). *Phys. Rev.,* **116**, 489.
31. S. Beliaev (1958). *Sov. Phys. JETP,* **7**, 289.
32. J. Bardeen, L. N. Cooper and J. R. Schrieffer (1957). *Phys. Rev.,* **108**, 1175.
L. Landau (1944). *J. Phys. U.S.S.R.,* **8**, 1
L. Landau (1947). *J. Phys. U.S.S.R.,* **11**, 91.
33. D. J. Thouless (1960). *Ann. Phys. (N.Y.),* **10**, 553.
34. K. A. Brueckner, T. Soda, P. W. Anderson, and P. Morel (1960). *Phys. Rev.,* **118**, 1442.
35. V. J. Emery and A. M. Sessler (1960). *Phys. Rev.,* **119**, 43.
36. D. D. Osheroff, R. C. Richardson and D. M. Lee (1972). *Phys. Rev. Lett.,* **28**, 885.
D. D. Osheroff, W. J. Gully, R. C. Richardson and D. M. Lee (1972). *Phys. Rev. Lett.,* **29**, 920.
37. J. Wheatley (1975). *Rev. Mod. Phys.,* **47**, 415.
38. A. J. Leggett (1972). *Phys. Rev. Lett.,* **29**, 1227.
39. S. Doniach and S. Engelsberg (1966). *Phys. Rev. Lett.,* **17**, 750.
40. P. W. Anderson and P. Morel (1961). *Phys. Rev.,* **123**, 1911.
41. R. Balian and N. R. Werthamer (1963). *Phys. Rev.,* **131**, 1553.
42. P. W. Anderson and W. F. Brinkman (1973). *Phys. Rev. Lett.,* **30**, 1108.
43. N. D. Mermin and D. Stare (1973). *Phys. Rev. Lett.,* **30**, 1135.
44. A. J. Leggett (1975). *Rev. Mod. Phys.,* **47**, 331.
45. L. W. Bruch and I. J. McGee (1967). *J. Chem. Phys.,* **46**, 2959.
L. W. Bruch and I. J. McGee (1970). *J. Chem. Phys.,* **52**, 5884.
46. B. H. Brandow (1971). *Ann. Phys. (N.Y.),* **64**, 21 and references cited therein.
47. E. Feenberg (1969). *Theory of Quantum Fluids*, Academic Press, New York.
48. E. Feeberg (1970). *Am. J. Phys.,* **38**, 684.
49. See Chapter 8 of Reference 47.
50. See Chapters 4, 5 and 8 of Reference 47.
51. J. W. Clark and C.-H. Yang (1970). *Lett. Nuovo Cim.,* **3**, 272.
J. W. Clark and C.-H. Yang (1971). *Lett. Nuovo Cim.,* **2**, 379.
52. H. W. Jackson and E. Feenberg (1961). *Ann. Phys.,* **15**, 266.
53. H. W. Jackson and E. Feenberg (1962). *Rev. Mod. Phys.,* **34**, 686.
54. R. P. Feynman and M. Cohen (1956). *Phys. Rev.,* **102**, 1189.
55. F. Y. Wu (1971). *J. Math. Phys.,* **12**, 1923.

56. F. Y. Wu (1972). *J. Low Temp. Phys.*, **9**, 177.
57. C.-W. Woo and E. Feenberg (1965). *Phys. Rev.*, **137**, A391.
    C.-W. Woo (1966). *Phys. Rev.*, **151**, 138.
58. H.-T. Tan and E. Feenberg (1968). *Phys. Rev.*, **176**, 370.
59. H.-T. Tan and C.-W. Woo (1970). *J. Low Temp. Phys.*, **2**, 187.
60. C. E. Campbell and T. C. Paulick (1975). *Bull. Am. Phys. Soc.*, **20**, 617.
    T. C. Paulick and C. E. Campbell, to be published.
61. E. K. Achter and L. Meyer (1969). *Phys. Rev.*, **188**, 291.
62. R. B. Hallock (1972). *Phys. Rev.*, **A5**, 320.
63. See Chapter 6 of Reference 47.
64. W. E. Massey (1966). *Phys. Rev.*, **151**, 153; see, however, C. De Michelis and L. Reatto (1974). *Phys. Lett.*, **50A**, 275.
65. W. E. Massey and C.-W. Woo (1967). *Phys. Rev.*, **164**, 256.
66. D. Levesque, Tu Khiet, D. Schiff and L. Verlet (1965). *Orsay Report.*
67. D. Schiff and L. Verlet (1967). *Phys. Rev.*, **160**, 208.
68. R. D. Murphy and R. O. Watts (1970). *J. Low Temp. Phys.*, **2**, 507.
69. V. R. Pandharipande and H. A. Bethe (1973). *Phys. Rev.*, **C7**, 1312.
70. C. E. Campbell and E. Feenberg (1969). *Phys. Rev.*, **188**, 396.
71. C.-W. Woo (1972). *Phys. Rev. Lett.*, **28**, 1442.
    C.-W. Woo (1972). *Phys. Rev.*, **A6**, 2312.
72. C. E. Campbell (1973). *Phys. Lett.*, **44A**, 471.
73. E. Feenberg (1974). *Ann. Phys. (N.Y.)*, **84**, 128; see also E. P. Gross (1962). *Ann. Phys. (N.Y.)*, **20**, 44.
74. C.-W. Woo, H.-T. Tan and W. E. Massey (1969). *Phys. Rev.*, **185**, 287.
    C.-W. Woo, H.-T. Tan and W. E. Massey (1970). *Phys. Rev.*, **A1**, 579.
75. C. E. Campbell (1972). *Ann. Phys. (N.Y.)*, **74**, 43.
76. H.-W. Lai, C.-W. Woo and F. Y. Wu (1971). *J. Low Temp. Phys.*, **5**, 499.
    H.-W. Lai and C.-W. Woo (1972). *Phys. Lett.*, **41A**, 170.
    A. D. Novaco (1973). *Phys. Rev.*, **A7**, 1653.
77. Y.-M. Shih and C.-W. Woo (1973). *Phys. Rev. Lett.*, **30**, 478.
    C. C. Chang and M. Cohen (1973). *Phys. Rev.*, **A8**, 1930, 3131.
    C. C. Chang and M. Cohen (1975). *Phys. Rev.*, **B11**, 1059.
    C. A. Croxton (1973). *J. Phys.* C. **6**, 411.
78. M. W. Cole and M. H. Cohen (1969). *Phys. Rev. Lett.*, **23**, 1238.
79. R. P. Feynman (1955). In C. J. Gorter (Ed.), *Progress in Low Temperature Physics*, Vol. I. North-Holland, Amsterdam. p. 17.
    G. V. Chester, R. Metz and L. Reatto (1968). *Phys. Rev.*, **175**, 275.
80. A. D. B. Woods and R. A. Cowley (1973). *Rep. Prog. Phys.*, **36**, 1135.
81. K. H. Bennemann and J. B. Ketterson (Eds.), *Physics of Liquid and Solid Helium*, to be published, Wiley Interscience, London and New York.
82. J. C. Lee and A. A. Broyles (1964). *Phys. Rev. Lett.*, **17**, 424.
83. G. V. Chester and L. Reatto (1966). *Phys. Lett.*, **22**, 276.
84. P. Berdahl (1974). *Phys. Rev.*, **A10**, 2378; see also E. Feenberg and S. Kilic, to be published.
85. R. Jastrow (1955). *Phys. Rev.*, **98**, 1479.
86. J. K. Percus (1964). In H. L. Frisch and J. L. Lebowitz (Eds.), *The Equilibrium Theory of Classical Fluids*. Benjamin, New York.
87. R. D. Murphy and J. A. Barker (1971). *Phys. Rev.*, **A3**, 1073.
88. R. D. Murphy (1972). *Phys. Rev.*, **A5**, 331.
89. F. Y. Wu and E. Feenberg (1961). *Phys. Rev.*, **122**, 739.
90. D. K. Lee and F. H. Ree (1973). *Phys. Rev.*, **A7**, 730.
91. A. Pokrant (1972). *Phys. Rev.*, **A6**, 1588.
92. C. E. Campbell (1975). *J. Math. Phys.*, **16**, 1076.
93. C. E. Campbell and F. Pinski, to be published.
94. See Reference 69 for an alternative procedure.
95. C. C. Chang and C. E. Campbell, to be published; *Phys. Rev. B*; see also Reference 91.
96. L. S. Ornstein and F. Zernicke (1914). *Proc. Acad. Sci. (Amsterdam)*, **17**, 793.
97. See Reference 83 for a similar analysis. Their long-wavelength behaviour does not come from a variational analysis, but instead is put in by hand.

98. We are indebted to P. Kleban for an illuminating discussion on this point.
99. C. -W. Woo and R. L. Coldwell (1972). *Phys. Rev. Lett.*, **29**, 1062.
100. H. W. Jackson and E. Feenberg (1962). *Phys. Rev.*, **128**, 943.
D. K. Lee (1967). *Phys. Rev.*, **162**, 134.
F. Y. Wu and M. K. Chien (1970). *J. Math. Phys.*, **11**, 1912.
101. C. E. Campbell, C. C. Chang and F. Pinski, to be published.
102. M. Girardeau and R. Arnowitt (1959). *Phys. Rev.*, **113**, 755.
103. R. Hastings and J. W. Halley, to be published.
R. Hastings (1975). *Ph.D. Thesis*, University of Minnesota.
104. R. Hastings and D. Johnson, private communication.
105. H. Wagner (1966). *Z. Physik*, **195**, 273.
106. P. W. Anderson (1958). *Phys. Rev.*, **112**, 1900.
107. R. Hastings and J. W. Halley (1975). *Phys. Rev.*, **B12**, 267.
108. A. Bijl (1940). *Physica*, **7**, 869.
109. N. E. Phillips, C. G. Waterfield and J. K. Hoffer (1970). *Phys. Rev. Lett.*, **25**, 1260.
110. B. M. Abraham, Y. Edlstein, J. B. Ketterson, M. Kuchnir and J. Vignos, (1969). *Phys. Rev.*, **181**, 347.
P. R. Roach, B. M. Abraham, J. B. Ketterson and M. Kuchnir (1972). *Phys. Rev.*, **A5**, 2205.
111. H. J. Maris and W. E. Massey (1970). *Phys. Rev. Lett.*, **25**, 220.
112. I. M. Khalatnikov, (1965). *Introduction to the Theory of Superfluidity*, and references cited therein. W. A. Benjamin, New York.
113. H. J. Maris (1972). *Phys. Rev. Lett.*, **28**, 277.
114. P. R. Roach, B. M. Abraham, J. B. Ketterson and M. Kuchnir (1972). *Phys. Rev. Lett.*, **29**, 32.
115. R. A. Cowley and A. D. B. Woods (1971). *Can. J. Phys.*, **49**, 177.
116. M. Cohen and R. P. Feynman (1957). *Phys. Rev.*, **107**, 13.
117. L. van Hove (1954). *Phys. Rev.*, **95**, 249.
118. H. W. Jackson (1971). *Phys. Rev.*, **A4**, 2386.
119. A. Zawadowski, J. Ruvalds, and J. Solana (1972). *Phys. Rev.*, **A5**, 399.
120. C. A. Murray, R. L. Woerner and T. J. Greytak (1975). *J. Phys.* C., **8**, L90.
121. T. C. Padmore and G. V. Chester (1974). *Phys. Rev.*, **A9**, 1725.
122. C. G. Kuper (1955). *Proc. Roy. Soc. (London)*, **A233**, 223.
123. Y. R. Lin-Liu and C. -W. Woo (1974). *J. Low Temp. Phys.*, **14**, 317.
124. D. K. Lee and F. J. Lee (1975). *Phys. Rev.*, **B11**, 4318.
125. O. W. Dietrich, E. H. Graf, C. H. Huang and L. Passell (1972). *Phys. Rev.*, **A5**, 1377.
126. D. L. Bartley, J. E. Robinson and C. -W. Woo (1974). *J. Low Temp. Phys.*, **15**, 473.
127. H. W. Jackson (1969). *Phys. Rev.*, **185**, 186.
128. H. W. Jackson (1973). *Phys. Rev.*, **A8**, 1529.
129. R. Hastings and J. W. Halley (1974). *Phys. Rev.*, **A10**, 2488.
130. T. Kebukawa, S. Yamasaki and S. Sunakawa (1973). *Prog. Theor. Phys.*, **49**, 1802.
131. N. N. Bogoliubov and D. N. Zubarev (1955). *Sov. Phys.*, **1**, 83.
S. Sunakawa, Y. Yokio and H. Nakatani (1962). *Prog. Theor. Phys.*, **27**, 589, 600.
S. Sunakawa, Y. Yokio and H. Nakatani (1962). *Prog. Theor. Phys.*, **28**, 127.
S. Sunakawa, S. Yamasaki and T. Kebukawa (1969). *Prog. Theor. Phys.*, **41**, 919.
T. Kebukawa, S. Yamasaki and S. Sunakawa (1970). *Prog. Theor. Phys.*, **44**, 565.
A. K. Rajagopal, A. Bagchi and J. Ruvalds (1964). *Phys. Rev.*, **A9**, 2707.
P. Berdahl and F. Bloch, to be published.
132. P. Kleban and J. W. Halley (1975). *Phys. Rev.*, **B11**, 3520.
133. For a different approach see W. B. Gelbart (1974). In I. Prigogine and S. A. Rice (Eds.), *Advances in Chemical Physics*, Vol. 26. Wiley, New York.
134. M. Stephen (1969). *Phys. Rev.*, **187**, 279.
135. L. J. Komarov and I. Z. Fischer (1963). *Sov. Phys. JETP*, **16**, 1358.
136. J. W. Halley (1969). *Phys. Rev.*, **181**, 338.
137. E. R. Pike, J. M. Vaughan and W. F. Vinen (1970). *J. Phys.* C., **3**, L40.
E. R. Pike (1972). *J. de Phys.*, **33**, C1, 25.
138. T. J. Greytak and S. Yan (1969). *Phys. Rev. Lett.*, **22**, 987.
139. T. J. Greytak, R. L. Woerner, J. Yan and R. Benjamin (1970). *Phys. Rev. Lett.*, **25**, 1547.
140. F. Iwamoto (1970). *Prog. Theor. Phys.*, **44**, 1121.

141. R. A. Cowley (1972). *J. Phys.* C.5, L287.
142. J. Ruvalds and A. Zawadowski (1970). *Phys. Rev. Lett.*, **25**, 333; see References 119 and 140 for more details.
143. P. Kleban and R. Hastings (1975). *Phys. Rev.*, **B11**, 1878.
144. See Section II of Reference 92.
145. P. Kleban (1974). *Phys. Lett.*, **A49**, 19.
146. P. O. Löwdin (1950). *J. Chem. Phys.*, **18**, 365.
147. This result was originally obtained by Jackson and Feenberg in the paired phonon analysis.[52] A more general result is obtained by relaxing the assumption that Ψ_0 is the exact ground state.[70] Then (6.3.102) still represents the orthogonalization (nothing was used about Ψ_0 there), but the second quantized Hamiltonian picks up off-diagonal terms proportional to $a_k^\dagger a_{-k}^\dagger$ and $a_{-k}a_k$. A Bogoliubov canonical transformation diagonalizes this Hamiltonian into the form (6.3.103) with an accompanying lowering of the ground state energy and a redefinition of $\epsilon_{BF}(k)$ in terms of the new ground state. Furthermore, this procedure is equivalent to optimizing the Jastrow part of a ground state trial function discussed in Section 6.2; once this optimization is achieved, equation (6.3.103) is obtained with the same set of assumptions.[70]
148. A. L. Fetter and J. D. Walecka (1971). *Quantum Theory of Many-Particle Systems*, McGraw-Hill, New York.
N. H. March, W. H. Young and S. Sampanthar (1967). *The Many-Body Problem in Quantum Mechanics*, Cambridge University Press, London.
149. D. Pines and P. Nozières (1966). *The Theory of Quantum Fluids*, Vol. I. Benjamin, New York.
150. L. D. Landau (1957). *Sov. Phys. JETP*, **3**, 920.
L. D. Landau (1957). *Sov. Phys. JETP*, **5**, 101.
A. A. Abrikosov, L. P. Gorkov and I. E. Dzyaloshinski (1963). *Methods of Quantum Field Theory in Statistical Physics*, Prentice-Hall, Englewood Cliffs, N.J.
151. L. N. Cooper (1956). *Phys. Rev.*, **104**, 1189.
152. For example, if $|\Psi_0\rangle$ is the N particle non-interacting ground state

$$\prod_{\substack{k\sigma \\ k<k_F}} c_{\mathbf{k}\sigma}^\dagger |\text{vac}\rangle \text{ and } a_{\mathbf{k}\sigma}^\dagger = c_{\mathbf{k}\sigma}^\dagger,\ k>k_F \text{ but } a_{\mathbf{k}\sigma}^\dagger = c_{\mathbf{k}\sigma},\ k \leqslant k_F, \text{ then } |\psi_{BCS}\rangle$$

is equivalent to the usual generalized BCS trial function,[41] where the coefficient of $c_{k\sigma}^\dagger c_{-k\sigma'}^\dagger$ for $k<k_F$ is the matrix inverse in spin space, $(h_{\mathbf{k}}^{-1})_{\sigma\sigma'}$.
153. For an excellent review see B. D. Day (1967). *Rev. Mod. Phys.*, **39**, 719.
154. K. A. Brueckner and J. L. Gammel (1969). *Phys. Rev.*, **109**, 1040.
155. L. J. Campbell (1965). *Ph.D. Thesis*, Univ. of California, San Diego.
156. E. Østgaard (1969). *Phys. Rev.*, **187**, 371, and references cited therein.
157. It is not evident that the optimum correlation factor Ψ_0 is real and positive, i.e. that χ in equation (6.2.6) is real. An analysis similar to that of Section 6.2 shows that Ψ_0 should be chosen to be real and positive to minimize the energy if $\mathscr{S}_{k_1\sigma_1\ldots k_N\sigma_N}$ has a constant phase, as is the case for the ground state Slater determinant.
158. J. W. Clark and P. Westhaus (1966). *Phys. Rev.*, **141**, 833.
P. Westhaus and J. W. Clark (1968). *J. Math. Phys.*, **9**, 149, and references cited therein.
159. F. Y. Wu and E. Feenberg (1962). *Phys. Rev.*, **128**, 943.
160. F. Y. Wu (1963). *J. Math. Phys.*, **4**, 1438.
161. E. Feenberg and C.-W. Woo (1965). *Phys. Rev.*, **137**, A391.
162. C.-W. Woo (1966). *Phys. Rev.*, **151**, 138.
163. H.-T. Tan and E. Feenberg (1968). *Phys. Rev.*, **176**, 360.
164. C. E. Campbell, to be published.
165. A. Widom and P. L. Sigel (1970). *Phys. Rev. Lett.*, **24**, 1400.
H.-T. Tan (1971). *Phys. Rev.*, **A4**, 256.

Chapter 7

Microdynamics in Classical Liquids

T. GASKELL

Department of Physics,
The University, Sheffield S3 7RH

7.1 Introduction

The introduction of slow neutron scattering experiments in the 1950s, to probe the density fluctuations in liquids, stimulated a renewed interest in the search for an understanding of the dynamical properties of liquids. The results of such experiments can be expressed in terms of time-dependent autocorrelation functions, or, more precisely, their Fourier transforms with respect to time. More generally the latter type of correlation function has proved to be a powerful and convenient means of discussing the dynamical behaviour of liquids at the microscopic level. Indeed, the establishing of expressions for the transport coefficients in liquids in terms of the time integrals of microscopic correlation functions was a major advance in non-equilibrium statistical mechanics.

There appears to have been three main areas of development in the study of atomic dynamics in liquids over the last fifteen years. The improved use of neutron scattering techniques in experiments on a number of different liquids has of course played an essential role in increasing our understanding of the liquid state. Secondly, the derivation of the generalized Langevin equation for a dynamical variable and from it an equation for the autocorrelation function of that variable, in terms of a memory function, has provided a useful framework from which to discuss time dependent correlations in liquids. Finally, the use of the molecular dynamics method of computer simulation, in the last decade particularly, has been an enormous stimulous to theoretical research. By studying, in this way, a model system of several hundred newtonian particles with a given pairwise interaction, which is believed to accurately reflect that in a real liquid, it is possible to compute for example the density fluctuation spectrum or the transport properties of the model. Not only can some of the results be usefully compared with those from neutron scattering experiments on real liquids, but by carefully analysing the computation it is possible to check the assumptions which may be made in any theoretical attempt to explain the dynamical properties.

The experimental situation, as far as neutron scattering experiments are concerned, has recently been thoroughly discussed by Copley and Lovesey.[1] The latter review makes an especially important contribution in stressing the difficulties of measuring the dynamic structure factor for wavenumber values below 1 Å^{-1}. The interpretation of some of the earlier data from scattering experiments is criticized and the criterion which demonstrates the existence of phonon-like excitations in monatomic classical liquids is clearly stated. Much of the early evidence which was thought to demonstrate the existence of phonons in liquids does not survive this critical survey.

There is little point in covering the same ground here, and we follow this introduction by defining time dependent correlation functions and fairly briefly stating some of their properties. The generalized Langevin equation for a dynamical variable is then derived and the memory function introduced. Schofield[2] has very recently given a comprehensive but rather formal review of existing methods of evaluating time-dependent correlation functions in liquids. In contrast we concentrate here on the longitudinal and transverse current correlations, together with the dynamic structure factor of the liquid, in the particularly interesting wavenumber regime below 1 Å^{-1}. We discuss the theoretical attempts to explain the recent experimental and molecular dynamics data and detailed results are presented. Along with the transverse modes in dense liquids the difficulties involved in a calculation of the shear viscosity coefficient are discussed. In view of the limitations of space the velocity autocorrelation function and the diffusion coefficient are only briefly mentioned. The latter function is probably the most tractable theoretically and has received a great deal of attention over the past few years, most of which has recently been reviewed.[2,3]

7.2 Correlation functions and the generalized Langevin equation

We consider an ensemble of N particles whose evolution is determined by a classical Hamiltonian

$$H_N = \sum_{i=1}^{N} \frac{\mathbf{p}_i^2}{2m_i} + \Phi_N(\{\mathbf{r}_i\})$$

where m_i is the mass of the i-th particle and $\Phi_N(\{\mathbf{r}_i\}) \equiv \Phi_N(\mathbf{r}_1, \mathbf{r}_2, \mathbf{r}_3 \ldots \mathbf{r}_N)$ is a translationally invariant potential energy function.

The time evolution of some dynamical quantity $x_N(\{\mathbf{r}_i(t), \mathbf{p}_i(t)\})$, which is a function of the phase space coordinates, is described by the operator $G_N(t)$ such that

$$x_N(\{\mathbf{r}_i(t), \mathbf{p}_i(t)\}) = G_N(t-t')x_N(\{\mathbf{r}_i(t'), \mathbf{p}_i(t')\}) \tag{7.2.1}$$

where $G_N(t) \equiv \exp(\mathrm{i}t L_N)$ with

$$\mathrm{i}L_N \equiv \sum_i \left(\frac{\partial H}{\partial \mathbf{p}_i} \cdot \frac{\partial}{\partial \mathbf{r}_i} - \frac{\partial H}{\partial \mathbf{r}_i} \cdot \frac{\partial}{\partial \mathbf{p}_i} \right) \tag{7.2.2}$$

the classical Liouville operator.

The time correlation function of two quantities $x_N(\{\mathbf{r}_i(0), \mathbf{p}_i(0)\})$, $y_N(\{\mathbf{r}_i(t), \mathbf{p}_i(t)\})$ will be written as $\langle x(0)\, y(t) \rangle$ where the brackets denote a canonical ensemble average. Hence

$$\phi(t) \equiv \langle x(0) y(t) \rangle = Z_N^{-1} \int \mathrm{d}\mathbf{r}^{3N}\, \mathrm{d}\mathbf{p}^{3N} \exp(-\beta H_N) x(\{\mathbf{r}_i(0), \mathbf{p}_i(0)\}) G_N(t) y(\{\mathbf{r}_i(0), \mathbf{p}_i(0)\} \tag{7.2.3}$$

where Z_N is the canonical partition function for the N particle system and $\beta = (k_\mathrm{B}T)^{-1}$. It is worth noting that the correlation function depends only on the time difference t (since H_N is a constant of the motion), so that we may write $\langle x(0)\, y(t) \rangle$ as $\langle x(t_0)\, y(t + t_0) \rangle$. We shall adopt the convention that variables inside an ensemble average for which the time is unspecified have the values at the instant the thermodynamic average is taken.

The properties of correlation functions are discussed in the recent review by Schofield[2] and we will only summarize those which are employed in this account. We shall be mainly concerned with time-dependent correlation functions which are autocorrelation functions of a single dynamical variable, which we denote as $\langle x\, x^*(t) \rangle$. They have the following properties.

(i) If the autocorrelation function is real then by using the Schwartz inequality it can be shown that its modulus is less than its initial value, i.e.

$$|\langle xx^*(t) \rangle| \leqslant \langle xx^* \rangle. \tag{7.2.4}$$

(ii) If the Hamiltonian is differentiable then there exists a Taylor expansion of the autocorrelation function about $t = 0$. Hence

$$\phi(t) \equiv \frac{\langle xx^*(t) \rangle}{\langle xx^* \rangle} = \sum_{n=0}^{\infty} \frac{1}{n!} a_n t^n \tag{7.2.5}$$

where

$$a_n = \frac{\langle x\, [(\mathrm{i}L_N)^n x]^* \rangle}{\langle xx^* \rangle} = \frac{\langle xx^{*(n)} \rangle}{\langle xx^* \rangle}.$$

Since $\mathrm{i}L_N$ is an odd function of momentum, $a_{2n+1} = 0$ and $\langle x\, x^*(t) \rangle$ is an even function of t. Hence we may write

$$\phi(t) = \sum_{n=0}^{\infty} \frac{1}{(2n)!} a_{2n} t^{2n} \tag{7.2.6}$$

with

$$a_{2n} = \frac{\langle xx^{*(2n)} \rangle}{\langle xx^* \rangle} = (-1)^n \frac{\langle x^{(n)} x^{*(n)} \rangle}{\langle xx^* \rangle}.$$

$x^{(n)}$ represents the n-th time derivative of $x(t)$ ar $t = 0$, the latter form for a_{2n} following from the fact that a correlation function depends only on the time difference.

The moments of the Fourier transform of a correlation function are related to the time derivatives of the correlation function at $t = 0$, leading to the moment sum-rules. Therefore if we define the Fourier transform of the correlation function as

$$\tilde{\phi}(\omega) = \frac{1}{2\pi} \int_{-\infty}^{\infty} \mathrm{d}t\, \phi(t) \exp(-\mathrm{i}\omega t)$$

which is an even function of frequency, then

$$\phi(t) = \int_{-\infty}^{\infty} d\omega\, \tilde{\phi}(\omega) \exp(i\omega t)$$

$$= \int_{-\infty}^{\infty} d\omega \sum_{n=0}^{\infty} \frac{(-1)^n}{(2n)!} \omega^{2n} \tilde{\phi}(\omega)\, t^{2n}$$

and hence

$$a_{2n} = (-1)^n \int_{-\infty}^{\infty} d\omega\, \omega^{2n} \tilde{\phi}(\omega) \equiv (-1)^n\, \overline{\omega^{2n}}\,. \tag{7.2.7}$$

For the correlation functions of interest in the study of liquids, for example the density–density correlation function, a knowledge of the low order moments has proved to be extremely useful. Their values will be stated explicitly when the correlation functions to be discussed are defined.

Before proceeding to do that, however, we shall derive an exact equation which describes the time evolution of a dynamical variable, and from that another formally exact equation for the autocorrelation function of that variable. This latter equation will form the basis for much of the discussion of time dependent correlations in liquids.

We may usefully extend the concept of a dynamical variable at this point by introducing a column vector $\mathbf{a}(t)$ whose components $a_i(t)$ are independent dynamical variables of the phase space coordinates $(\mathbf{r}_1 \ldots \mathbf{r}_N;\ \mathbf{p}_1 \ldots \mathbf{p}_N)$ of the many-body system. A generalized Langevin equation describes the time evolution of the vector $\mathbf{a}(t)$, the latter being such that it has no time invariant part, that is $\langle \mathbf{a}(t) \rangle \equiv 0$. In order to derive the equation a projection operator P is introduced, which is defined by its property that for any arbitrary dynamical variable vector $\mathbf{G}(t)$

$$P\mathbf{G}(t) = \langle \mathbf{G}(t)\mathbf{a}^* \rangle \cdot \langle \mathbf{a}\,\mathbf{a}^* \rangle^{-1} \cdot \mathbf{a}(0) \tag{7.2.8}$$

where $\mathbf{a}^*$ is the Hermitian conjugate of $\mathbf{a}$ and $\langle \mathbf{a}\,\mathbf{a}^* \rangle^{-1}$ is the inverse of the correlation matrix $\langle \mathbf{a}\,\mathbf{a}^* \rangle$. We define $L_0 = PL$ and make use of an operator identity pointed out by Kawasaki[4]

$$\exp(itL)i(L - L_0) - \exp[it(L - L_0)]i(L - L_0)$$

$$= -\int_0^t ds \frac{d}{ds} \{\exp[i(t-s)L] \exp[is(L - L_0)]\, i(L - L_0)\}$$

$$= \int_0^t ds \exp[i(t-s)L]\, iL_0 \exp[is(L - L_0)]\, i(L - L_0). \tag{7.2.9}$$

Hence, from the definition of L_0,

$$\frac{d}{dt} \exp(itL) = \exp(itL) iPL + \exp[it(1-P)L]\, i(1-P)L$$

$$+ \int_0^t ds \exp[i(t-s)L]\, iPL \exp[is(1-P)L]\, i(1-P)L. \tag{7.2.10}$$

We may introduce an additional factor $(1 - P)$ into the integrand without affecting the identity, because of the presence of the term $(1 - P)$ to its right-hand side and the property that $P^2 = P$. Therefore we may write

$$\frac{\mathrm{d}}{\mathrm{d}t}\exp(\mathrm{i}tL) = \exp(\mathrm{i}tL)\mathrm{i}PL + \exp[\mathrm{i}t(1-P)L]\mathrm{i}(1-P)L$$

$$+ \int_0^t \mathrm{d}s \exp[\mathrm{i}(t-s)L]\mathrm{i}PL(1-P)\exp[\mathrm{i}s(1-P)L]\mathrm{i}(1-P)L. \quad (7.2.11)$$

Applying this operator to the vector $\mathbf{a}(0)$ gives

$$\dot{\mathbf{a}}(t) = \langle \dot{\mathbf{a}}\mathbf{a}^* \rangle \cdot \langle \mathbf{a}\mathbf{a}^* \rangle^{-1} \cdot \mathbf{a}(t) + \exp[\mathrm{i}t(1-P)L]\mathrm{i}(1-P)L\mathbf{a}(0)$$

$$+ \int_0^t \mathrm{d}s \exp[\mathrm{i}(t-s)L]P\mathrm{i}L(1-P)\exp[\mathrm{i}s(1-P)L]\mathrm{i}(1-P)L\mathbf{a}(0)$$

$$= \mathrm{i}\boldsymbol{\Omega}\cdot\mathbf{a}(\mathrm{t}) + \mathbf{f}(t) + \int_0^t \mathrm{d}s \exp[\mathrm{i}(t-s)L]\,\langle \mathrm{i}L(1-P)\mathbf{f}(s)\mathbf{a}^* \rangle \cdot \langle \mathbf{a}\mathbf{a}^* \rangle^{-1} \cdot \mathbf{a}(0) \quad (7.2.12)$$

where $\boldsymbol{\Omega}$ is the frequency matrix of the undamped motion and $\mathbf{f}(t)$ is the random force acting upon the vector $\mathbf{a}(t)$. They are given respectively by

$$\mathrm{i}\boldsymbol{\Omega} = \langle \dot{\mathbf{a}}\mathbf{a}^* \rangle \cdot \langle \mathbf{a}\mathbf{a}^* \rangle^{-1} \quad (7.2.13)$$

and

$$\mathbf{f}(t) = \exp[\mathrm{i}t(1-P)L](1-P)\dot{\mathbf{a}}(0). \quad (7.2.14)$$

Now since $\langle \mathrm{i}L(1-P)\mathbf{f}(s)\mathbf{a}^* \rangle = -\langle \mathbf{f}(s)\mathbf{f}^* \rangle$ we may finally write

$$\dot{\mathbf{a}}(t) - \mathrm{i}\boldsymbol{\Omega}\cdot\mathbf{a}(t) + \int_0^t \mathrm{d}s\, \mathbf{M}(s)\cdot\mathbf{a}(t-s) = \mathbf{f}(t) \quad (7.2.15)$$

where $\mathbf{M}(s)$ represents a time-dependent damping term known as the memory function which may be expressed as the autocorrelation function of the random force. That is

$$\mathbf{M}(s) = \langle \mathbf{f}(s)\mathbf{f}^* \rangle \cdot \langle \mathbf{a}\mathbf{a}^* \rangle^{-1}. \quad (7.2.16)$$

This identity is called by Kubo[5] the second fluctuation–dissipation theorem.

Since the random force contains a factor $(1-P)$ it is orthogonal to $\mathbf{a}(0)$, i.e. $\langle \mathbf{f}(t)\,\mathbf{a}^* \rangle = 0$. Therefore by multiplying the above equation from the right by $\mathbf{a}^*(0)\cdot\langle \mathbf{a}\,\mathbf{a}^* \rangle^{-1}$ we have an equation for the autocorrelation matrix $\boldsymbol{\phi}(t) \equiv \langle \mathbf{a}(t)\,\mathbf{a}^* \rangle \cdot \langle \mathbf{a}\,\mathbf{a}^* \rangle^{-1}$ of the form

$$\dot{\boldsymbol{\phi}}(t) - \mathrm{i}\boldsymbol{\Omega}\cdot\boldsymbol{\phi}(t) + \int_0^t \mathrm{d}s\,\mathbf{M}(s)\cdot\boldsymbol{\phi}(t-s) = 0. \quad (7.2.17)$$

In the case of a single variable autocorrelation function, for a system in which the interatomic potential is differentiable,

$$\mathrm{i}\boldsymbol{\Omega} \equiv \frac{\langle \dot{x}x^* \rangle}{\langle xx^* \rangle} = -\dot{\phi}(0) = 0$$

and hence the above equation becomes

$$\dot{\phi}(t) + \int_0^t \mathrm{d}s\, M(s)\phi(t-s) = 0. \quad (7.2.18)$$

Equations (7.2.17) and (7.2.18), which are based on the projection operator formalism developed by Zwanzig[6] and Mori[7], have been extensively investigated and exploited in a number of ways over the past few years. The details will be described in a later section.

It is worth mentioning that there is a useful theorem, pointed out by Singwi,[8] which relates the moments of the spectral function of $\phi(t)$ to the coefficients of t^{2n} in a Taylor expansion of the memory function (which is also an even function of t). By making use of (7.2.18) it may be expressed as

$$\overline{\omega^{2n}} = \sum_{m=1}^{n} (-1)^{m+1} \, \overline{\omega^{2(n-m)}} \, M^{(2m-2)}(0). \tag{7.2.19}$$

All the difficulties involved in describing the time dependence of an autocorrelation function, which in principle of course requires the solution of the many-body problem represented by the liquid, have been re-expressed in terms of the memory function. An important example of the use of the above formalism is represented by the velocity autocorrelation function of an atom in a simple liquid. It is a quantity which is probably the most accessible to theoretical investigation and is important because the physics of the motion of an individual atom in a liquid is reflected in the details of its time decay.[9] Attempts have been made in this case to describe the memory function either by assuming that it has some simple time dependence,[10] or by starting from the equation of motion of an atom in a liquid and adopting a decoupling procedure which eventually leads to a prescription for the memory function.[11,12] Perturbation techniques have also been investigated.[13] In the following section we review the continued fraction expansion of the memory function, which for the velocity autocorrelation function and more complex correlation functions has proved useful in finding approximate solutions.

7.3 Continued fraction representation

Restricting attention again to the single dynamical variable problem, and for the moment to a differentiable interatomic potential, Mori[14] has shown that the Laplace transform of the memory function may be expressed in a continued fraction representation.

Solving equation (7.2.18) by Fourier–Laplace transform, subject to the boundary condition $\phi(0) = 1$, we have for $\tilde{\phi}(z) \equiv \int_0^\infty \mathrm{d}t \, \exp(\mathrm{i}zt)\phi(t)$

$$\tilde{\phi}(z) = \frac{1}{-\mathrm{i}z + \tilde{M}(z)}. \tag{7.3.1}$$

Now $\tilde{M}(z)$ can also be expressed in the form

$$\tilde{M}(z) = \frac{M(0)}{-\mathrm{i}z + \tilde{M}_{(2)}(z)} = \frac{\overline{\omega^2}}{-\mathrm{i}z + \tilde{M}_{(2)}(z)} \tag{7.3.2}$$

where $M_{(2)}(t)$ is the time-dependent part of the memory function associated with the correlation function $\chi(t) \equiv -\ddot{\phi}(t)$. Thus from the property that $\phi(t) \to 0$ as $t \to \infty$ and the results that $\int_0^\infty \mathrm{d}t \, \chi(t) = 0$ and $\int_0^\infty \mathrm{d}t \, t\chi(t) = -1$ we can readily establish the equation

$$\dot{\chi}(t) + \int_0^t \mathrm{d}s \, \{\overline{\omega^2} + M_{(2)}(s)\}\chi(t-s) = 0. \tag{7.3.3}$$

The result in (7.3.2) follows from (7.3.1) and (7.3.3).

We may show also that

$$\tilde{M}_{(2)}(z) = \frac{M_{(2)}(0)}{-\mathrm{i}z + \tilde{M}_{(4)}(z)}$$

or in general that

$$\tilde{M}_{(2n)}(z) = \frac{M_{(2n)}(0)}{-iz + \tilde{M}_{(2n+2)}(z)} \tag{7.3.4}$$

which enables us to obtain the continued fraction representation for $\tilde{M}(z)$ in the form

$$\tilde{M}(z) = \cfrac{M(0)}{-iz + \cfrac{M_{(2)}(0)}{-iz + \cfrac{M_{(4)}(0)}{-iz + \dots}}} \tag{7.3.5}$$

$M_{(2n)}(0)$ is obtained in terms of the moments of the spectral function $\tilde{\phi}(\omega)$ defined earlier, so that for example

$$M(0) = \overline{\omega^2}$$
$$M_{(2)}(0) = \frac{\overline{\omega^4}}{\overline{\omega^2}} - \overline{\omega^2} \tag{7.3.6}$$

It frequently happens that the low order moments of a correlation function are known. In the absence of a theory for the time-dependent correlation function it is sometimes assumed that a useful expression for $\tilde{\phi}(z)$ can be obtained by terminating the continued fraction at some point (which will be determined by the number of available moments) by replacing $\tilde{M}_{(2n)}(z)$ by a constant which is regarded as a variable parameter. Or alternatively some simple time dependence for $M_{(2n)}(t)$ is *assumed*, for example that it has a simple Gaussian form. Such procedures have proved to be useful in correlating computed data, or in predicting some of the qualitative features of $\phi(t)$, although they are difficult to justify. Rigorous derivations of the continued fraction representation and its extension to the multivariable case are given by Mori.[14]

Modifications to the form of the continued fraction representation are necessary, for the situation where the interatomic potential core is that of a rigid sphere, and have been discussed by Schofield.[2]

7.4 Density and longitudinal current correlation functions

The particle number density $\rho(\mathbf{r}, t)$ is defined by

$$\rho(\mathbf{r}, t) = \sum_{j=1}^{N} \delta(\mathbf{r} - \mathbf{r}_j(t)) \tag{7.4.1}$$

from which it follows that the ensemble average $\langle \rho(r, t) \rangle$ is the mean number density of the fluid $\rho = N/V$ The density–density correlation function

$$G(\mathbf{r}, t) = \langle \rho(0, 0)\rho(\mathbf{r}, t) \rangle / \rho \tag{7.4.2}$$

gives the average density of particles at the point $\mathbf{r}$ at time t, given that there is a particle at the origin at $t = 0$. For sufficiently large values of r and t the densities are statistically independent so that in these limits $G(\mathbf{r},t) \to \rho$. It may also be written in the form

$$G(\mathbf{r}, t) = \frac{1}{N} \sum_{j,l=1}^{N} \langle \delta(\mathbf{r} - \mathbf{r}_j(0) + \mathbf{r}_l(t)) \rangle \tag{7.4.3}$$

or expressing the delta function as a Fourier series we may write

$$G(\mathbf{r},t)=\frac{1}{NV}\sum_{\mathbf{q}}\exp(\mathrm{i}\mathbf{q}\cdot\mathbf{r})\langle\rho_{\mathbf{q}}(0)\rho_{\mathbf{q}}^{*}(t)\rangle \tag{7.4.4}$$

where $\rho_{\mathbf{q}}(t)=\Sigma_{j=1}^{N}\exp(-\mathrm{i}\mathbf{q}\cdot\mathbf{r}_j)$ are the density fluctuations.

If we explicitly extract the $\mathbf{q}=0$ term we may write the right-hand side as

$$\frac{N}{V}+\frac{1}{V}\sum_{\mathbf{q}\neq 0}\frac{\exp(\mathrm{i}\mathbf{q}\cdot\mathbf{r})\langle\rho_{\mathbf{q}}(0)\rho_{\mathbf{q}}^{*}(t)\rangle}{N}$$

In the limit $N, V\rightarrow\infty$ this becomes

$$G(\mathbf{r},t)-\rho=\frac{1}{(2\pi)^3}\int \mathrm{d}\mathbf{q}\exp(\mathrm{i}\mathbf{q}\cdot\mathbf{r})F(q,t) \tag{7.4.5}$$

where

$$F(q,t)=\frac{\langle\rho_{\mathbf{q}}(0)\rho_{\mathbf{q}}^{*}(t)\rangle}{N}$$

the density fluctuation autocorrelation function, is sometimes referred to as the intermediate scattering function. The dynamic structure factor, $S(q,\omega)$, which in part determines the coherent scattering cross-section in thermal neutron scattering experiments on liquids, is given by

$$S(q,\omega)=\frac{1}{2\pi}\int_{-\infty}^{\infty}\mathrm{d}t\exp(-\mathrm{i}\omega t)F(q,t) \tag{7.4.6}$$

Two other quantities which figure prominently in discussions of the equilibrium structure of liquids can usefully be introduced at this stage. They are the radial distribution function $g(\mathbf{r})$, which is related to $G(\mathbf{r},0)$ by the equation

$$G(\mathbf{r},0)=\delta(\mathbf{r})+\rho g(r) \tag{7.4.7}$$

and the static structure factor $S(q)$ given by

$$F(q,0)=\int_{-\infty}^{\infty}\mathrm{d}\omega\, S(q,\omega)=S(q) \tag{7.4.8}$$

We may write (7.4.5), with $t=0$, as

$$G(\mathbf{r},0)-\rho=\delta(\mathbf{r})+\frac{1}{(2\pi)^3}\int\mathrm{d}\mathbf{q}\exp(\mathrm{i}\mathbf{q}\cdot\mathbf{r})\{S(q)-1\} \tag{7.4.9}$$

and therefore

$$g(r)-1=\frac{1}{(2\pi)^3\rho}\quad\mathrm{d}\mathbf{q}\exp(\mathrm{i}\mathbf{q}\cdot\mathbf{r})\{S(q)-1\} \tag{7.4.10}$$

It is well known that a knowledge of the low order moment sum-rules has been extremely important in the attempts to establish a theory of the dynamic structure factor in simple liquids. Their derivation is well known and we shall therefore state the results. The zeroth moment has already been given in equation (7.4.8).

The second moment

$$\overline{\omega^2}=\frac{\int_{-\infty}^{\infty}\mathrm{d}\omega\,\omega^2 S(q,\omega)}{\int_{-\infty}^{\infty}\mathrm{d}\omega\, S(q,\omega)}=\frac{q^2}{m\beta S(q)}\equiv\omega_0^2(q) \tag{7.4.11}$$

and the fourth

$$\overline{\omega^4} = \frac{q^2}{m\beta S(q)}\left(\frac{3q^2}{m\beta} + \frac{\rho}{m}\int \mathrm{d}\mathbf{r}\, g(r)(\hat{\mathbf{q}}\cdot\nabla)^2\phi(r)[1-\exp(\mathrm{i}\mathbf{q}\cdot\mathbf{r})]\right)$$
$$\equiv \omega_0^2\,\omega_\mathrm{L}^2 \tag{7.4.12}$$

An expression for the sixth moment has been derived by Forster *et al.*[15] and for the eighth by Bansal and Pathak.[16] The eighth involves the four-particle and the sixth the three-particle distribution function. Because of our ignorance of these distribution functions no reliable estimates of the moments are presently available. We will not therefore quote here the rather extensive expressions for the latter.

The longitudinal current correlation function $C_\mathrm{L}(q,t)$ is defined by

$$C_\mathrm{L}(q,t) = \frac{1}{N}\left\langle \sum_{j=1}^{N} \mathbf{q}\cdot\mathbf{v}_j(0)\exp[-\mathrm{i}\mathbf{q}\cdot\mathbf{r}_j(0)] \sum_{k=0}^{N} \mathbf{q}\cdot\mathbf{v}_k(t)\exp[\mathrm{i}\mathbf{q}\cdot\mathbf{r}_k(t)]\right\rangle \tag{7.4.13}$$

This function is closely related to the density fluctuation function, in fact $C_\mathrm{L}(q,t) = -\ddot{F}(q,t)$. Because of this relationship, it follows from the continued fraction representation discussed earlier, with reference to equation (7.3.3), that $C_\mathrm{L}(q,t)$ satisfies the equation

$$\dot{C}_\mathrm{L}(q,t) + \int_0^t \mathrm{d}s\, M_\mathrm{L}(q,s)C_\mathrm{L}(q,t-s) = 0 \tag{7.4.14}$$

where we may write

$$M_\mathrm{L}(q,t) = \omega_0^2(q) + M_{(2)}(q,t)$$
$$\equiv \omega_0^2(q) + m_\mathrm{L}(q,t). \tag{7.4.15}$$

$M_{(2)}(q,t) \equiv m_\mathrm{L}(q,t)$ is a function which decays to zero as $t\to\infty$ and whose initial value is $\omega_\mathrm{L}^2(q) - \omega_0^2(q)$. The formulation of the density fluctuation function, and hence the dynamic structure factor, in terms of the memory function $m_\mathrm{L}(q,t)$ is a useful and convenient one. For example there is a simple expression for $m_\mathrm{L}(q,t)$ which leads to the hydrodynamic form of the dynamic structure factor. This has been extremely useful in the so-called generalized hydrodynamic theories of the dynamic structure factor where the damping due to the entropy fluctuation term is specifically included. Alternatively, some simple ansatz for the time dependence of $m_\mathrm{L}(q,t)$ has been quite successfully employed. Ailawadi *et al.*[17] were able to describe, by means of a least squares fit, some computer results for $C_\mathrm{L}(q,t)$ in liquid argon by assuming a Gaussian decay for $m_\mathrm{L}(q,t)$ with a wavenumber dependent relaxation time. An exponential decay is another suggested form, the consequences of which have been thoroughly investigated by Schofield[18] and Lovesey.[19] Such theories are frequently referred to as visco-elastic theories.

We can easily derive the relationship between the dynamic structure factor and the memory function $m_\mathrm{L}(q,t)$. It follows from (7.4.14) that the Fourier–Laplace transforms of $C_\mathrm{L}(q,t)$ and $m_\mathrm{L}(q,t)$ are related by the equation

$$\tilde{C}_\mathrm{L}(q,z) = \frac{S(q)\,\omega_0^2(q)}{-\mathrm{i}z - \omega_0^2(q)/\mathrm{i}z + \tilde{m}_\mathrm{L}(q,z)} \tag{7.4.16}$$

Now since

$$\tilde{C}_\mathrm{L}(q,z) = -\mathrm{i}z + z^2\tilde{F}(q,z)$$

it is readily shown that

$$S(q,\omega)=\frac{S(q)\omega_0^2(q)}{\pi}\ \frac{m_c(q,\omega)}{\omega^2 m_c^2(q,\omega)+[\omega^2-\omega_0^2(q)-\omega m_s(q,\omega)]^2} \tag{7.4.17}$$

where

$$m_c(q,\omega)=\int_0^\infty dt\, m_L(q,t)\cos\omega t$$

$$m_s(q,\omega)=\int_0^\infty dt\, m_L(q,t)\sin\omega t$$

The formulation of the neutron scattering cross-section from a liquid in terms of $S(q,\omega)$ by van Hove[20] together with the advent of slow neutron scattering experiments to probe the density fluctuations in liquids has been a major stimulus in the development of the theory of liquids. The essential result is that the differential scattering cross-section per unit solid angle and unit interval of outgoing energy is given by

$$\frac{d^2\sigma}{d\Omega dE_f}=N\frac{k_f}{k_0}\frac{\sigma_c}{4\pi}S(q,\omega)+N\frac{k_f}{k_0}\frac{\sigma_i}{4\pi}S_s(q,\omega) \tag{7.4.18}$$

where E_0 and $\mathbf{k}_0$ are the incident energy and wave vector of the neutrons which are scattered into a state with energy E_f and wave vector $\mathbf{k}_f$. Further, the energy and momentum transfer from the neutron to the system are given by $\hbar\omega=E_0-E_f$ and $\hbar\mathbf{q}=\hbar(\mathbf{k}_0-\mathbf{k}_f)$ respectively. σ_c and σ_i are the single atom coherent and incoherent bound cross-sections and $S_s(q,\omega)$ is the Fourier transform of the self terms in the intermediate scattering function. Hence

$$S_s(q,\omega)=\frac{1}{2\pi}\int_{-\infty}^{\infty} dt\,\exp(-i\omega t)\langle\exp\{-i\mathbf{q}\cdot[\mathbf{r}_i(0)-\mathbf{r}_i(t)]\}\rangle$$

$$=\frac{1}{2\pi}\int_{-\infty}^{\infty} dt\,\exp(-i\omega t)F_s(q,t) \tag{7.4.19}$$

The experimental details and the problems involved in the determination of the dynamic structure factor have recently been extensively reviewed by Copley and Lovesey.[1]

Finally, the connection of $F(q,t)$ and the density–density response function is pointed out. For the classical regime it can be shown (see for example Marshall and Lovesey[21]) that the response function $\chi(q,t)$ may be expressed as

$$\chi(q,t)=-\beta\Theta(t)\dot{F}(q,t) \tag{7.4.20}$$

where $\Theta(t)$ is the step function. The Fourier–Laplace transform $\tilde{\chi}(q,z)$ in the limit $z\to\omega+i\epsilon$ (where ϵ is a positive infinitesimal) defines a generalized susceptibility whose imaginary part will be denoted by $\chi''(q,\omega)$. The latter is related to the dynamic structure factor $S(q,\omega)$ through the equation

$$S(q,\omega)=\frac{1}{\beta\pi\omega}\chi''(q,\omega) \tag{7.4.21}$$

This result, since it relates the fluctuation properties of the system in thermal equilibrium to the response of the system to an external perturbation, is frequently referred to as the fluctuation–dissipation theorem. From (7.4.20) it follows that

$$i z \tilde{\chi}(q,z) = -\beta z^2 \tilde{F}(q,z) + \beta i z\, S(q)$$
$$= -\beta \tilde{C}_L(q,z)$$
$$= \frac{-\beta \omega_0^2 S(q)}{-iz + \tilde{M}_L(q,z)} \tag{7.4.22}$$

Hence

$$\tilde{\chi}(q,z) = \frac{-\beta \omega_0^2 S(q)}{z^2 - \omega_0^2 + iz\, \tilde{m}_L(q,z)} \tag{7.4.23}$$

The analytic properties of the response function and of $\tilde{m}_L(q,z)$ have been extensively discussed,[22,23] and need not be repeated here. Suffice it to say that the elementary excitations of the system will be indicated by the existence of poles in the lower half of the complex z plane. If their coordinates are denoted by $z = \omega + ip$, the real part ω determines the energy of the excitation and the imaginary part the damping. The function $\tilde{m}_L(q,z)$ is frequently referred to as the damping function. Neglecting it completely will lead to the appearance of poles on the real axis at $\omega = \pm\omega_0(q)$. This is reflected in the presence of infinitely sharp side peaks in the dynamic structure factor which would then have the form

$$S(q,\omega) = \frac{1}{2} S(q) \{\delta(\omega - \omega_0(q)) + \delta(\omega + \omega_0(q))\} \tag{7.4.24}$$

giving the dispersion relation $\omega = C_T\, q$, at small q, where C_T is the isothermal sound velocity. We shall be particularly interested in the wavenumber regime $q \lesssim 1$ Å^{-1} where neutron scattering experiments and computer simulation studies indicate the existence of well defined collective modes in the density fluctuations in dense liquids. For these wavenumbers collisional damping is dominant, and any satisfactory theory of $m_L(q,t)$ must adequately include these effects.

7.5 Collective excitations in simple monatomic liquids – generalized hydrodynamic and visco-elastic theories

The existence of high frequency phonon-like excitations in the density fluctuation spectrum of simple monatomic liquids has been the subject of considerable interest and speculation over the past decade.[24] Attempts have been made during this time to infer from neutron scattering data the wavelengths for which such collective excitations are well defined in simple liquids, and even to determine dispersion curves. However, if we use as the criterion for a well defined collective excitation, the existence of a side peak in the frequency dependence of $S(q,\omega)$ for a fixed value of the wavenumber q, then only in the last two years have neutron scattering experiments convincingly demonstrated their existence[1] and accurately determined the wavelength range over which the excitation is reasonably well defined.

Recently, an extensive investigation over a wide wavenumber range in liquid rubidium by Copley and Rowe[25] has shown well defined side peaks up to wavenumbers $\simeq 1$ Å^{-1}. This work, together with some extensive and important molecular dynamics computations on a dense Lennard–Jones fluid by Levesque *et al.*,[26] and more recently by Rahman[27] on a model of liquid rubidium, have provided a great deal of detailed information about the dynamic structure factor in liquids and models of liquids in the particularly interesting regime of wavenumbers $0.1 \lesssim q \lesssim 1$ Å^{-1}.

For sufficiently low frequencies and long wavelengths the density fluctuations are governed by the linearized hydrodynamic equations. The shear and bulk viscosities appear in the latter as factors in the attenuation of sound waves in the liquid. Energy transfer between different parts of the liquid is also important in the damping mechanism and is reflected in the presence of the thermal conductivity. In fact it has been shown[17] that in this regime the memory function $M_L(q, t)$ has the form

$$M_L(q, t) = \frac{q^2}{m\beta S(0)} + \frac{q^2 \eta_L}{\rho m} \delta(t) + \frac{q^2}{m\beta S(0)} (\gamma - 1) \exp[-a(0)q^2 t] \tag{7.5.1}$$

where $\eta_L = \eta_B + 4\eta/3$ is the longitudinal viscosity, $a(0) = K/\rho m C_V$, K being the thermal conductivity and $\gamma = C_P/C_V$. For finite frequencies the slowly decaying final term appears almost as a constant $(\gamma - 1) q^2/m\beta S(0)$. Hence the dispersion relation associated with the propagating sound waves, to order q, becomes

$$\omega = \left(\frac{\gamma}{m\beta S(0)}\right)^{1/2} q = C_s q$$

as distinct from the dispersion obtained via (7.4.24), C_s being the adiabatic sound velocity. If analysed more precisely at small q with this memory function, the dynamic structure factor has the well known three-peak structure,[28] a central line representing non-propagating entropy fluctuations and two side peaks (Brillouin doublet) at $\omega = \pm C_s q$ indicating the damped sound waves.

It is interesting to discuss within the present context the recent computer calculations of the dynamic structure factor by Levesque *et al.*[26] for a model of liquid argon near its triple point, using a Lennard–Jones interatomic potential. At the smallest wavenumber considered ($q \simeq 0.18$ Å^{-1}, corresponding to a wavelength of six to seven times the interatomic separation), the existence of a three-peak structure for $S(q, \omega)$ was established. For smaller wavelengths the excitations are overdamped and there is no well defined collective mode.

The transport coefficients appearing in equation (7.5.1) were also computed, although it should be pointed out that in some cases they differ significantly from the experimentally determined results for liquid argon.[29] In particular, the value for the thermal conductivity is twice the experimental value, but this has since been revised and now is in close agreement with experiment.[30] This is an important correction, because the value of K plays an important role in theoretical attempts to explain the collective mode in argon.

Over the past few years a number of procedures have been put forward to modify the hydrodynamic expression for $S(q, \omega)$ in the hope of extending the range of validity to the region of $q(\sim 1$ Å$^{-1})$ and $\omega(\sim 10^{13}$ s$^{-1})$ values investigated in neutron scattering and computer experiments. Such procedures have become known as generalized hydrodynamic theories. Chung and Yip[31] and Sears[32] suggested a generalization of the above expression for $M_L(q, t)$ which would satisfy the zeroth, second and fourth moments of the dynamic structure factor. The term involving the longitudinal viscosity coefficient is generalized to give a frequency dependence associated with viscous damping such as one might anticipate from visco-elastic theory. Hence the following expression for the memory function:

$$M_L(q, t) = \omega_0^2 + (\omega_L^2 - \gamma\omega_0^2)\exp[-t/\tau(q)] + \omega_0^2(\gamma - 1)\exp[-a(q)q^2 t] \tag{7.5.2}$$

The wavenumber dependence of $\tau(q)$ (which has the role of a Maxwell relaxation time) and of $a(q)$ is chosen so that the hydrodynamic behaviour is recovered at small q whilst at large q the ideal gas form of the dynamic structure factor can be fitted. Thus

$$a(q) = \frac{2(2/\pi m\beta)^{1/2} a(0)}{2(2/\pi m\beta)^{1/2} + a(0)q}$$

and (7.5.3)

$$\tau^{-1}(q) = \left(\tau^{-2}(0) + \frac{8q^2}{\pi m\beta}\right)^{1/2}$$

In order that the hydrodynamic behaviour is reproduced, $\tau(0)$ is chosen so that

$$\tau^{-1}(0) = \frac{4G/3 + B - B_s}{\eta_L} \quad (7.5.4)$$

where $4G/3 + B = \rho m \lim_{q\to 0}\omega_L^2/q^2$, with G the rigidity and B the instantaneous bulk modulus.[33] B_s represents the adiabatic bulk modulus $\gamma/m\beta S(0)$.

Levesque *et al.*[26] succeeded in fitting the computer data with a memory function, $M_L(q, t)$, which was very similar to (7.5.2). They differed because the exponential function in the second term was replaced by a sum of two exponentials with different relaxation times to produce a slowly decreasing decay at large t. Both the thermal conductivity parameter and the parameter γ were functions of the wavenumber. For values of $q \lesssim 0.2$ Å^{-1} however, they are virtually independent of wavenumber; only for large q values is the dependence significant.

Barker and Gaskell[34] also proposed a generalized hydrodynamic scheme, although in this case the memory function $M(t)$ (defined in (7.2.18)) was modified to meet the low order moment requirements rather than $M_{(2)}(t)$. It has been pointed out[35] that a memory function $M(t)$ which is a sum of two exponential functions can reproduce the hydrodynamic structure factor, provided that one makes the appropriate choice of constants and exponents which are combinations of the transport coefficients mentioned in connection with equation (7.5.1). A relaxation time $\tau(q)$ is again introduced in the modification so that each exponential term $\exp[-\alpha(q)t]$ becomes $\exp\{-\alpha(q)[t^2 + \tau^2(q)]^{1/2}\}$, in order that there exists a Taylor expansion about $t = 0$ in even powers of t. Once again in the limit $q \to 0$, $\tau(q)$ satisfies the condition expressed in equation (7.5.4). The thermal conduction contribution is made wavenumber dependent in a way which is guided by the necessity to fit the low order moments.

The latter value of $a(q)$ is also relatively independent of q for small wavenumbers. However the proposed wavenumber dependence of the thermal diffusion coefficient in equation (7.5.3) gives the latter a marked rate of decay as q increases. This appears to be a consequence of insisting that the ideal gas form of $S(q, \omega)$ can be fitted at large enough q. The argument is not very convincing, however, when it has such a marked effect on the wavenumber regime involved here. In Figure 7.5.1 the shape of $a(q)$, according to the theories mentioned above, is indicated along with the form obtained by fitting the computer data. We show in Figure 7.5.2 the results for the dynamic structure factor for $q = 0.18$ Å^{-1} including the molecular dynamics data. Both theories show the existence of a side peak at this wavenumber although its location is apparently not accurately predicted. It should be noted that the heat diffusion term, since it involves the specific heat ratio, is essential in any theory of the dynamic structure factor of argon in this wavenumber regime for two reasons. Firstly, since γ for liquid argon is significantly greater than unity it plays an important role in the determination of the side peak. Secondly, the value of $a(q)$ has a substantial effect on the size of $S(q, 0)$ in the generalized hydrodynamic theories, just as it has in the hydrodynamic regime. It is largely responsible for the sharp peak in $S(q, \omega)$ in the quasi-elastic region around $S(q, 0)$, and because $a(q)$ according to equation (7.5.3) is so small for $q = 0.18$ Å^{-1}, $S(q,0)$ according to the theory of Chung, Yip and Sears greatly exceeds the computed value.

Clearly, therefore, the theoretical schemes described above still do not satisfactorily explain the molecular dynamics data of Levesque *et al.* An important feature is the long time tail associated with the viscous damping which any satisfactory theory should possess. A neutron

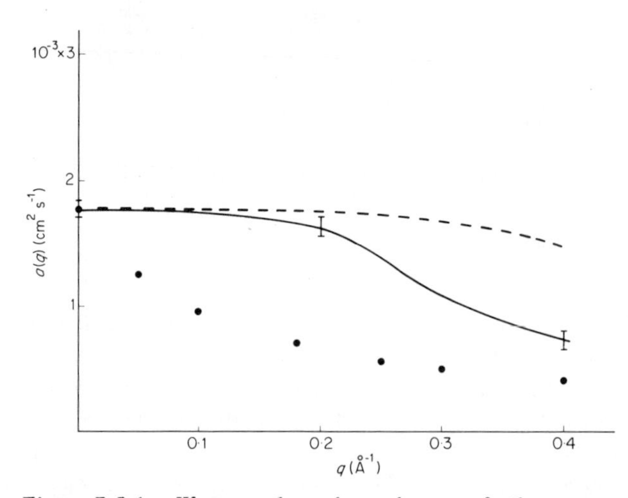

Figure 7.5.1 Wavenumber dependence of the parameter $a(q)$ associated with the thermal diffusion contribution to the damping of collective modes in argon ($a(0) = K/\rho m C_v$). Dots are results obtained from equation (7.5.3), dashed line is from the theory of Barker and Gaskell[34] and the solid line is the computer fit of Levesque *et al.*[26] (Reproduced by permission of the American Institute of Physics)

scattering experiment on liquid argon to determine $S(q, \omega)$ in the wavenumber range $q \lesssim 0.2$ Å^{-1} should be possible with the present availability of high flux reactors, and would be very interesting in view of the molecular dynamics calculations.

There are some results by Bell *et al.*[36] for the dynamic structure factor in fluid neon within the wavenumber ranges $0.06 \leqslant q \leqslant 0.14$ Å^{-1} and $0.27 \leqslant q \leqslant 1.5$ Å^{-1}. The scattering experiments, however, were carried out at comparatively low densities, and for the small values of q a distinct three-peak structure was observed which can apparently be described by hydrodynamic theory. Only at the lower end of the second wavenumber range is there still some indication of the existence of a propagating mode, and within this range a quite successful fit was achieved by a generalization of hydrodynamics corresponding closely to the approach of Chung and Yip.[31]

In contrast to the situation in argon, for the liquid metal rubidium the dynamic structure factor has been determined near the melting point in the extensive neutron scattering experiments of Copley and Rowe,[25] which were mentioned earlier, for wavenumbers where there are well defined collective modes. However, no transport coefficients have been evaluated in the computer calculations on the liquid rubidium model investigated by Rahman.[27] Therefore to apply a generalized hydrodynamic scheme to liquid rubidium we must use computed values of the moments and experimentally determined values for transport coefficients which describe the bulk properties of the liquid. This procedure is probably not inconsistent in view of the good agreement between the neutron scattering experiments and the computed results for the density fluctuation spectrum. Unfortunately, there is no reliable estimate of the bulk viscosity or the ionic contribution to the thermal conductivity. The bulk viscosity term in the attenuation coefficient for ultrasonic waves and the ionic term in the thermal conductivity are swamped in a liquid metal by contributions from the conduction electrons. Fortunately it is possible to estimate both of these quantities in the following way.

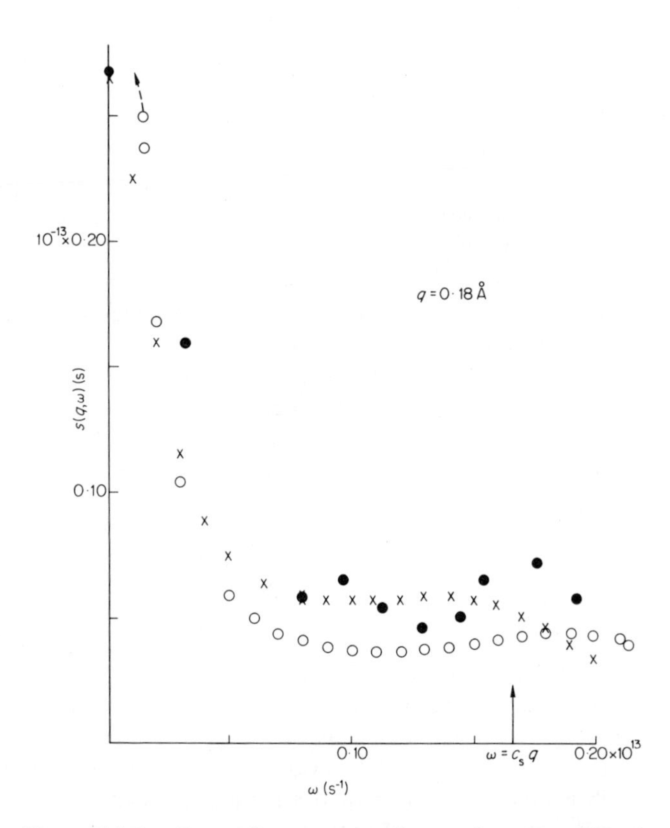

Figure 7.5.2 Dynamic structure factor for a model of liquid argon near its triple point. Open circles are based on equations (7.5.2) and (7.5.3), the crosses show the theory of Barker and Gaskell,[34] and the dots indicate the computer calculation of Levesque *et al.*[26] (Reproduced by permission of the American Institute of Physics)

There is a previous equation which expresses $4G/3 + B$ in terms of the zero wavenumber limit of the quantity $\rho m\omega_L^2(q)/q^2$. Using this and the well known result,

$$B = \frac{5}{3}G + 2\rho k_B T\left(\frac{P}{\rho k_B T} - 1\right) \tag{7.5.5}$$

which is valid for a fluid with two-body central forces (Zwanzig and Mountain[37]), we may estimate B and G provided that both the zero wavenumber limit and the ionic contribution to the pressure are known. This information is indeed available from Rahman's computer calculations, where in fact $P/\rho k_B T = 6.601$, leading to a value of $B = 2.47 \times 10^{10}$ g cm^{-1} s^{-2} with $G = 1.168 \times 10^{10}$ g cm^{-1} s^{-2}.

In the scattering experiments, of course, the conditions of temperature and density were such that $P/\rho k_B T \ll 1$. The discrepancy is due to the fact that in the liquid metal calculation there would be a further contribution to the thermodynamic pressure from the density dependence of the effective inter-ionic potential (produced by the presence of the conduction electrons). That contribution should not be included in the pressure result quoted above, if (7.5.5) is to be employed. Using the experimentally determined shear viscosity a Maxwell relaxation time $\tau(0) = \eta/G = 5.821 \times 10^{-13}$ s, can be estimated. On the assumption that the same relaxation time may be associated with both the nonconserved part of the longitudinal stress correlation function and the transverse stress correlation function, it follows that

$$\frac{4}{3}\eta + \eta_B = \left(\frac{4}{3}G + B - B_s\right)\tau(0) \tag{7.5.6}$$

This assumption, it should be pointed out, is open to some questions (Schofield,[18] Gaskell[35]). Approximately 87% of the contribution to the longitudinal viscosity given above arises from the shear viscosity. The reason for this is the fact that the adiabatic bulk modulus $B_s = 2.24 \times 10^{10}$ g cm^{-1} s^{-2} is within 10% of the instantaneous bulk modulus B and since we have calculated $\eta_B = (B - B_s)\tau(0)$ it is small compared to $4G\tau(0)/3$.

To obtain an estimate for the ionic contribution to the thermal conductivity, some theoretical results for the transport coefficients in a rigid sphere fluid, which have been derived on the basis of kinetic theory arguments (Longuet–Higgins and Pople[38]), may tentatively be used. The theory enables the ratio of transport coefficients to be predicted in terms of the atomic mass. Thus, for the ratio of the viscosity to the thermal conductivity, their result is

$$\frac{\eta}{K} = \frac{2m}{5k_B} \tag{7.5.7}$$

and in the case of argon, for example, is in good agreement with that for the corresponding experimental values at the melting point (1.9 and 2.1×10^{-7} cm^{-2} s^2 °K respectively). It cannot be checked of course for liquid metals because the ionic contribution to the thermal conductivity which the latter equation is intended to describe is insignificant compared with the electronic contribution. It is clear from the previously stated results that the hydrodynamic expression for $S(q, \omega)$ becomes independent of the thermal conductivity in the limit $\gamma = 1$. Fortunately, for rubidium (and a number of other liquid metals) γ is close to unity. Hence the generalized hydrodynamic results for $S(q, \omega)$ are not sensitive to the thermal conductivity (in contrast to the situation in argon where $\gamma \simeq 2.2$) and the use of equation (7.5.7) to estimate K is probably adequate.

A result for the so-called symmetrized scattering function

$$\tilde{S}(q, \omega) = \exp(-\hbar\omega/2k_BT)S(q, \omega)$$

which is based on the approach of Barker and Gaskell is shown in Figure 7.6.2 for one value of the wavenumber. The generalized hydrodynamic approach of Sears consistently exaggerates the height of the side peak, and although both schemes predict the position of the side peak quite well neither of them adequately describes the quasi-elastic region around $\omega = 0$ for any of the wavenumbers for which computer calculations or neutron scattering results are available. A more complete discussion is given by Barker and Gaskell.[39]

An alternative approach to this problem by Akcasu and Daniels[40] should also be mentioned. This too is based on a generalized Langevin equation, but starts from a dynamical variable in the form of a column vector consisting of a number of components which include the fluctuations of the conserved variables – density, momentum and energy. Hence it contains

(*a*) The density fluctuation

$$n_\mathbf{q} = m\rho_\mathbf{q} \tag{7.5.8}$$

(*b*) The momentum density fluctuation

$$J_\mathbf{q} = m\sum_i \mathbf{v}_i \exp(-i\mathbf{q}\cdot\mathbf{r}_i) \tag{7.5.9}$$

(c) The fluctuation in the energy density

$$E_{\mathbf{q}} = \sum_i \left\{ \frac{1}{2} m v_i^2 + \frac{1}{2} \sum_{j \neq i} \phi(r_{ij}) \right\} \exp(-i\mathbf{q} \cdot \mathbf{r}_i) \tag{7.5.10}$$

A result is derived for the memory function $M_L(q, t)$ in the form

$$M_L(q, t) = \omega_0^2(q) + (\omega_L^2(q) - \gamma(q)\omega_0^2(q))n_L(q, t) + \omega_0^2(q)(\gamma(q) - 1)n_K(q, t)$$

with explicit expressions for the wavenumber dependent specific heat ratio $\gamma(q), n_L(q, t)$ and $n_K(q, t)$ in terms of time dependent correlation functions involving the longitudinal component of the momentum density, the nonconserved parts of the longitudinal stress tensor and the energy flux. The theory is formally exact, although evaluation of the correlation functions is too formidable to be carried through, and in practice rather drastic approximations are made. In the original application to argon, for $q > 1$ Å^{-1}, $\gamma(q)$ is put equal to unity. This is probably not too unreasonable at these wavenumbers since the coupling of the density fluctuations to the thermal modes is small in this wavenumber regime, as demonstrated by Levesque *et al.*[26] A simple exponential form was assumed for $n_L(q, t)$ and the wavenumber dependence of the relaxation time chosen on the basis of an interpolation scheme between the correct small and large q limits. Obviously by putting $\gamma = 1$ the damping due to the thermal conduction term is ignored, and it becomes essentially a visco-elastic approach, as distinct from a generalized hydrodynamic theory. In fact, as already mentioned, the exponential choice for the damping function coincides with an approach whose consequences have been analysed independently by Schofield[18] and Lovesey[19]. Copley and Lovesey[1] put forward an argument which scales the time dependence of the memory function with the wavenumber, leading to the result that

$$\tau^{-1}(q) \propto (\omega_L^2(q) - \omega_0^2(q))^{1/2}$$

the constant of proportionality being chosen so that $S(q, 0)$ coincides with the ideal gas result for large q. Neither of the two methods for selecting $\tau(q)$ is particularly convincing. Nevertheless, as pointed out by Barker and Gaskell,[39] the latter choice for the relaxation time gives a better description of the molecular dynamics and neutron scattering results for $S(q, \omega)$, in liquid rubidium at least, than does that of Akcasu and Daniels. The position of the side peak and hence the dispersion curve are quite accurately predicted, although the overall shape of $S(q, \omega)$ is less satisfactory and in particular the region around $\omega = 0$, as demonstrated in Figure 7.6.3.

To summarize, therefore, whilst the generalized hydrodynamic and visco-elastic schemes are capable of describing some features of the frequency spectrum beyond the hydrodynamic regime, where there still exists a well defined collective excitation, such methods are in other respects unsatisfactory. The generalization of the wavenumber dependent parameters which appear in the expressions for the dynamic structure factor rely on some form of interpolation scheme or are chosen so that the low order moments are satisfied, some degree of arbitrariness being a feature of either approach. It was also illustrated by Levesque *et al.* in their attempts to fit the computer results for the argon-like system, that the choice of an exponential decay for that part of the memory function $m_L(q, t)$ associated with viscous damping is not adequate. In the region of q and ω values of interest the relaxation process has essentially two components with quite different rates of decay. The rapidly decaying component due to short range binary collision processes, and another with a long time tail from the collective effects in a dense liquid which the longer-ranged part of the potential probably plays a significant role.

In the following section we discuss attempts to describe the dynamic structure factor from a rather more fundamental point of view, based on the equations of motion of the atoms in the liquid.

7.6 Theories of $\chi(q,t)$ and $m_L(q,t)$ from the equations of motion of the atoms

We discuss two theories, based essentially on the equations of motion of the atoms in the liquid, and which have been carried through to a calculation of the dynamic structure factor. The equations of motion are utilized in two quite different ways. In the first method, due to Hubbard and Beeby,[41] a modified version of linear response theory is developed. This considers the changes in the trajectories of the atoms due to an infinitesimal external perturbation, leading to a theory of the density–density response function. The second approach, developed by Barker and Gaskell,[42] starts essentially from the equation of motion of the longitudinal current component. On the assumption that collisional damping is dominant in dense liquids, for density fluctuations of wavenumbers $\lesssim 1$ Å^{-1}, an expression is derived for the memory function $m_L(q, t)$ in terms of the radial distribution function and another quantity which describes the relative motion of two atoms in the liquid.

Density–density response function $\chi(q, t)$

In order to obtain the response function the change $\langle \delta\rho_{\mathbf{q}} \rangle$ in the Fourier component of the density $\rho_{\mathbf{q}}$ is to be evaluated when some external perturbation of the form $U\exp(\mathrm{i}\mathbf{q}\cdot\mathbf{r} - \mathrm{i}\omega t)$ is applied to the system. If $\mathbf{u}_i(t)$ is the change in the trajectory of the i-th atom, to first order in U one finds

$$m\ddot{\mathbf{u}}_i(t) = -\sum_{j\neq i} \nabla\nabla\,\phi(|\mathbf{r}_i(t) - \mathbf{r}_j(t)|)\cdot(\mathbf{u}_i(t) - \mathbf{u}_j(t)) - \mathrm{i}U\mathbf{q}\exp[\mathrm{i}\mathbf{q}\cdot\mathbf{r}_i(t) - \mathrm{i}\omega t] \tag{7.6.1}$$

Integration of the latter equation gives

$$m\mathbf{u}_i(t) = \int_{-\infty}^{t}\mathrm{d}t'\int_{-\infty}^{t'}\mathrm{d}t''\{\sum_j \mathbf{\Phi}_{ij}(t'')\cdot\mathbf{u}_j(t'') - \mathrm{i}U\mathbf{q}\exp[\mathrm{i}\mathbf{q}\cdot\mathbf{r}_i(t'') - \mathrm{i}\omega t'']\} \tag{7.6.2}$$

with the tensor $\mathbf{\Phi}_{ij}$ given by

$$\begin{aligned}\mathbf{\Phi}_{ij}(t) &= \nabla\nabla\phi(|\mathbf{r}_i(t) - \mathbf{r}_j(t)|) \qquad (i\neq j)\\ \mathbf{\Phi}_{ii}(t) &= -\sum_{j\neq i}\mathbf{\Phi}_{ij}(t)\end{aligned} \tag{7.6.3}$$

Now

$$\begin{aligned}\langle\delta\rho_{\mathbf{q}}(t)\rangle &= \left\langle \sum_i \frac{\partial\rho_{\mathbf{q}}(t)}{\partial\mathbf{r}_i(t)}\cdot\delta\mathbf{r}_i(t)\right\rangle = -\langle \mathrm{i}\mathbf{q}\cdot\sum_i \mathbf{u}_i(t)\exp[-\mathrm{i}\mathbf{q}\cdot\mathbf{r}_i(t)]\rangle\\ &= -\mathrm{i}U\exp[-\mathrm{i}\omega t]\,\langle\mathbf{q}\cdot\sum_i\zeta_i(t)\rangle\end{aligned} \tag{7.6.4}$$

where we define

$$\zeta_i(t) = \frac{\mathbf{u}_i(t)\exp[\mathrm{i}\omega t - \mathrm{i}\mathbf{q}\cdot\mathbf{r}_i(t)]}{U}$$

In terms of the latter quantity (7.6.2) may be written

$$\begin{aligned}m\zeta_i(t) = -\int_{-\infty}^{t}\mathrm{d}t'\int_{-\infty}^{t'}\mathrm{d}t''\{&\mathrm{i}\mathbf{q}\exp[\mathrm{i}\mathbf{q}\cdot(\mathbf{r}_i(t'') - \mathbf{r}_i(t)) - \mathrm{i}\omega(t''-t)]\\ &+ \sum_j \exp[\mathrm{i}\mathbf{q}\cdot(\mathbf{r}_j(t'') - \mathbf{r}_i(t) - \mathrm{i}\omega(t''-t)]\,\mathbf{\Phi}_{ij}(t'')\cdot\zeta_j(t'')\}.\end{aligned} \tag{7.6.5}$$

Hubbard and Beeby solve the above equation for $\zeta_i(t)$ by iteration and to obtain $\langle\zeta_i(t)\rangle$ each term in the resulting series is averaged. The leading term is

$$-\frac{\mathrm{i}}{m}\mathbf{q}\left\langle\int_{-\infty}^{t}\mathrm{d}t'\int_{-\infty}^{t'}\mathrm{d}t''\exp[\mathrm{i}\mathbf{q}\cdot(\mathbf{r}_i(t'')-\mathbf{r}_i(t))-\mathrm{i}\omega(t''-t)]\right\rangle$$

$$=-\frac{\mathrm{i}}{m}\mathbf{q}\int_0^{\infty}\mathrm{d}t\, tF_s(q,t)\exp(\mathrm{i}\omega t)$$

$$\equiv-\frac{\mathrm{i}}{m}\mathbf{q}\,Q(q,\omega). \tag{7.6.6}$$

The second term is decoupled, on averaging, so that the latter term appears as a factor and it becomes

$$-\frac{\mathrm{i}}{m}Q(q,\omega)\boldsymbol{\Psi}(q,\omega)\cdot\mathbf{q}$$

where

$$\boldsymbol{\Psi}(q,\omega)=\frac{1}{m}\left\langle\sum_j\int_{-\infty}^{t}\mathrm{d}t'\int_{-\infty}^{t'}\mathrm{d}t''\boldsymbol{\Phi}_{ij}(t'')\exp[\mathrm{i}\mathbf{q}\cdot(\mathbf{r}_j(t'')-\mathbf{r}_i(t))-\mathrm{i}\omega(t''-t)]\right\rangle. \tag{7.6.7}$$

The decoupling procedure is repeated in each successive term, each factor, together with its associated exponentials being averaged separately, so that the n-th term is $-(\mathrm{i}/m)Q(q,\omega)[\boldsymbol{\Psi}(q,\omega)]^{n-1}\cdot\mathbf{q}$. The series is easily re-summed to give the approximate result

$$\langle\zeta_i(t)\rangle=-\frac{\mathrm{i}}{m}Q(q,\omega)[1-\boldsymbol{\Psi}(q,\omega)]^{-1}\cdot\mathbf{q}. \tag{7.6.8}$$

Using this in equation (7.6.4), from the definition of the response function the Fourier–Laplace transform of the latter is given as

$$\chi(q,\omega)=\frac{1}{m}Q(q,\omega)\,\mathbf{q}\cdot[1-\boldsymbol{\Psi}(q,\omega)]^{-1}\cdot\mathbf{q} \tag{7.6.9}$$

At this stage there is a further approximation. $\boldsymbol{\Psi}(q,\omega)$ may be expressed in the form

$$\boldsymbol{\Psi}(q,\omega)=\frac{1}{m}\left\langle\sum_{j\neq i}\int_{-\infty}^{t}\mathrm{d}t'\int_{-\infty}^{t'}\mathrm{d}t''\boldsymbol{\Phi}_{ij}(t'')\{\exp[\mathrm{i}\mathbf{q}\cdot\mathbf{r}_{ji}(t'')]-1\}\right.$$
$$\left.\times\exp[-\mathrm{i}\mathbf{q}\cdot(\mathbf{r}_i(t)-\mathbf{r}_i(t''))-\mathrm{i}\omega(t''-t)]\right\rangle$$

which is further decoupled, on the assumption that the motion of atom i between t'' and t is independent of the configuration at t'', to obtain

$$\boldsymbol{\Psi}(q,\omega)\simeq\frac{1}{m}\sum_{j\neq i}\int_{-\infty}^{t}\mathrm{d}t'\int_{-\infty}^{t'}\mathrm{d}t''\langle\boldsymbol{\Phi}_{ij}(t'')\{\exp[\mathrm{i}\mathbf{q}\cdot\mathbf{r}_{ji}(t'')]-1\}\rangle$$
$$\times\langle\exp[-\mathrm{i}\mathbf{q}\cdot(\mathbf{r}_i(t)-\mathbf{r}_i(t''))-\mathrm{i}\omega(t''-t)]\rangle. \tag{7.6.10}$$

The first factor within the integral becomes independent of time after carrying out the ensemble average and it is then easy to see that we may write the latter as $Q(q,\omega)\boldsymbol{\Psi}(q)/m$, with

$$\boldsymbol{\Psi}(q)=\frac{\rho}{m}\int\mathrm{d}\mathbf{r}g(r)\boldsymbol{\nabla}\boldsymbol{\nabla}\phi(r)[\exp(\mathrm{i}\mathbf{q}\cdot\mathbf{r})-1]. \tag{7.6.11}$$

Finally, therefore

$$\chi(q,\omega) = \frac{1}{m} Q(q,\omega)\mathbf{q} \cdot \left(\mathbf{1} - \frac{1}{m} Q(q,\omega)\mathbf{\Psi}(q)\right)^{-1} \cdot \mathbf{q}. \tag{7.6.12}$$

In an isotropic liquid the matrix $[\mathbf{1} - (1/m)Q(q,\omega)\mathbf{\Psi}(q)]$ can be immediately expressed in diagonal form, and the response function becomes

$$\chi(q,\omega) = \frac{q^2}{m} \frac{Q(q,\omega)}{1 + \omega_q^2 Q(q,\omega)} \tag{7.6.13}$$

with

$$\omega_q^2 = \frac{\rho}{m} \int d\mathbf{r}\, g(r)(\hat{\mathbf{q}} \cdot \nabla)^2 \phi(r)[1 - \exp(i\mathbf{q} \cdot \mathbf{r})].$$

Referring back to (7.4.21), the dynamic structure factor is now readily obtained from the above result.

The method is more ambitious than those based on the memory function approach which have thus far been discussed. Because the response function is related to the structure factor through the equation $S(q) = \chi(q, 0)/\beta$, equation (7.6.13) also contains a theory of the structure of the liquid. Hence there is the possibility of a self-consistent theory being developed in which $S(q)$ is also determined in terms of the interatomic potential. Unfortunately, the decoupling scheme outlined above does not give a very accurate representation of the structure factor, especially in the region of the principal peak. The calculated dynamic structure factor therefore, although it may be shown to satisfy the second and fourth moments exactly, does not meet the extremely important zeroth moment sum rule at all well.

Schofield[2] has criticized the approximations involved in the theory, as far as its application to dense liquids is concerned, on two counts: (i) no distinction is made in the decoupling scheme between the self terms (where $i = j$) and the distinct terms; and (ii) the final assumption leading to (7.6.10) is not valid for the repulsive core of the interaction, which largely determines the shape of the structure factor. The decoupling technique which enables the series of terms in the iterative solution of (7.6.5) to be re-summed is correct when applied to a perfect lattice, but its validity is doubtful in the liquid. Applications of the method to lead[41] and argon[43] clearly indicate that its treatment of the damping of collective modes is very inadequate. Although it is difficult to see how any improved approximations are going to be made, attempts to further develop the theory seem well worthwhile.

It is clear from the latter approach, and those discussed in the previous section, that an accurate incorporation of the damping effects is crucially important if a good description of the density fluctuation spectrum is to be achieved. A theory of the damping function $m_L(q, t)$ will now be discussed which attempts to describe more accurately the damping of collective modes and to elucidate the connection of the damping mechanism with the characteristics of the interatomic potential.

The damping function $m_L(q, t)$

From the definition of the longitudinal current correlation function in equation (7.4.13)

$$\dot{C}_L(q,t) = -\frac{1}{N}\left\langle \sum_{j=1}^{N} \{\mathbf{q} \cdot \dot{\mathbf{v}}_j(0) - i(\mathbf{q} \cdot \mathbf{v}_j(0))^2\} \exp[-i\mathbf{q} \cdot \mathbf{r}_j(0)] \right.$$
$$\left. \sum_{k=1}^{N} \mathbf{q} \cdot \mathbf{v}_k(t) \times \exp[i\mathbf{q} \cdot \mathbf{r}_k(t)] \right\rangle. \tag{7.6.14}$$

Assuming that we may Fourier transform the effective interatomic potential, then we write

$$\phi(r) = \frac{1}{N} \sum_{\mathbf{n}} \tilde{\phi}(n) \exp(\mathrm{i}\mathbf{n} \cdot \mathbf{r}). \tag{7.6.15}$$

The first term in the curly brackets of (7.6.14) may now be expressed as follows

$$\sum_{j=1}^{N} \mathbf{q} \cdot \dot{\mathbf{v}}_j(0) \exp[-\mathrm{i}\mathbf{q} \cdot \mathbf{r}_j(0)] = -\frac{\mathrm{i}}{Nm} \sum_j \sum_{\mathbf{n}} \mathbf{q} \cdot \mathbf{n}\, \tilde{\phi}(n) \exp[\mathrm{i}(\mathbf{n} - \mathbf{q}) \cdot \mathbf{r}_j(0)]\, \rho_{\mathbf{n}}(0).$$

The term with $\mathbf{n} = \mathbf{q}$ is extracted in the wavenumber sum, because in this case the sum over the atom label j produces a factor N. Hence

$$\sum_{j=1}^{N} \mathbf{q} \cdot \dot{\mathbf{v}}_j(0) \exp[-\mathrm{i}\mathbf{q} \cdot \mathbf{r}_j(0)] = -\frac{\mathrm{i}}{m} \sum_j q^2 \tilde{\phi}(q) \exp[-\mathrm{i}\mathbf{q} \cdot \mathbf{r}_j(0)]$$

$$-\frac{1}{m} \sum_j \sum_{l \neq j} \mathbf{q} \cdot \nabla \phi'(r_{jl}(0)) \exp[-\mathrm{i}\mathbf{q} \cdot \mathbf{r}_j(0)] \tag{7.6.16}$$

where $\phi'(r)$ represents the interaction without the Fourier component $\tilde{\phi}(q)$. In the development of a theory of the damping function in a dense liquid, for wavenumbers $q \lesssim 1$ Å^{-1}, the kinetic terms will be considerably less important than the potential contribution. In equation (7.6.14), therefore, the product of the particle velocities, in the much less significant second term in the curly brackets, will be replaced by its thermally averaged value. The latter equation thus becomes

$$\dot{C}_{\mathrm{L}}(q,t) = -\frac{1}{N} \Bigg\langle \Bigg\{ -\mathrm{i} \left(\frac{q^2 \tilde{\phi}(q)}{m} + \frac{q^2}{m\beta} \right) \sum_j \exp[-\mathrm{i}\mathbf{q} \cdot \mathbf{r}_j(0)]$$

$$-\frac{1}{m} \sum_j \sum_{l \neq j} \mathbf{q} \cdot \nabla \phi'(r_{jl}(0)) \exp[-\mathrm{i}\mathbf{q} \cdot \mathbf{r}_j(0)] \Bigg\} \sum_k \mathbf{q} \cdot \mathbf{v}_k(t) \exp[\mathrm{i}\mathbf{q} \cdot \mathbf{r}_k(t)] \Bigg\rangle$$

and in order to cast this in the memory function form of (7.4.14) we can express it as

$$\dot{C}_{\mathrm{L}}(q,t) = \frac{1}{N} \int_0^t \mathrm{d}s \Bigg\langle \frac{\mathrm{d}}{\mathrm{d}s} \Bigg\{ -\mathrm{i} \left(\frac{q^2 \tilde{\phi}(q)}{m} + \frac{q^2}{m\beta} \right) \sum_j \exp[-\mathrm{i}\mathbf{q} \cdot \mathbf{r}_j(s)]$$

$$-\frac{1}{m} \sum_j \sum_{l \neq j} \mathbf{q} \cdot \nabla \phi'(r_{jl}(s)) \exp[-\mathrm{i}\mathbf{q} \cdot \mathbf{r}_j(s)] \Bigg\} \sum_k \mathbf{q} \cdot \mathbf{v}_k(t) \exp[\mathrm{i}\mathbf{q} \cdot \mathbf{r}_k(t)] \Bigg\rangle$$

since the contribution from the upper limit of the integral, where the ensemble average is taken with all the terms at the same instant of time, is quickly seen to be zero. The frequent atomic collisions will produce rapidly fluctuating changes in the accelerations of the atoms. Provided therefore that q is sufficiently small, the time variation of the exponential factor may be neglected, when the derivative of the second term is taken, compared with that of the potential. Hence to a good approximation

$$\dot{C}_{\mathrm{L}}(q,t) = -\frac{1}{N} \int_0^t \mathrm{d}s \Bigg\langle \Bigg\{ \left(\frac{q^2 \tilde{\phi}(q)}{m} + \frac{q^2}{m\beta} \right) \sum_j \mathbf{q} \cdot \mathbf{v}_j(s) \exp[-\mathrm{i}\mathbf{q} \cdot \mathbf{r}_j(s)]$$

$$+\frac{1}{m} \sum_j \sum_{l \neq j} (\mathbf{v}_j - \mathbf{v}_l) \cdot \nabla(\mathbf{q} \cdot \nabla) \phi'(r_{jl}(s)) \exp[-\mathrm{i}\mathbf{q} \cdot \mathbf{r}_j(s)] \Bigg\}$$

$$\times \sum_k \mathbf{q} \cdot \mathbf{v}_k(t) \exp[\mathrm{i}\mathbf{q} \cdot \mathbf{r}_k(t)] \Bigg\rangle \tag{7.6.17}$$

If in the last term in the curly bracket it is assumed that the component of atomic velocities in the $\mathbf{q}$ direction makes the dominant contribution to the ensemble average, the above equation, with some rearrangement of the terms, becomes

$$\dot{C}_{\mathrm{L}}(q,t) = -\frac{1}{N}\int_0^t \mathrm{d}s \Big\langle \sum_j \mathbf{q}\cdot\mathbf{v}_j(s)\exp[-\mathrm{i}\mathbf{q}\cdot\mathbf{r}_j(s)] \Big\{ \Big(\frac{q^2\tilde{\phi}(q)}{m} + \frac{q^2}{m\beta} \Big)$$

$$+ \frac{1}{m}\sum_{l\neq j}(\hat{\mathbf{q}}\cdot\boldsymbol{\nabla})^2\phi'(r_{jl}(s))(1-\exp[\mathrm{i}\mathbf{q}\cdot\mathbf{r}_{jl}(s)])\Big\} \sum_k \mathbf{q}\cdot\mathbf{v}_k(t)\exp[\mathrm{i}\mathbf{q}\cdot\mathbf{r}_k(t)]\Big\rangle. \tag{7.6.18}$$

The longitudinal and transverse components of velocity are only statistically independent for equal times, and there may be some significant correlation between them over a finite period. The importance of such effects is, however, difficult to assess in the present context. The presence of the final term in the curly brackets of equation (7.6.18) provides the mechanism for the damping of collective modes in the liquid through the interatomic collisions. Without it the memory function would consist only of a constant term, and both $C_{\mathrm{L}}(q,t)$ and $F(q,t)$ would display an undamped oscillatory behaviour. The crucial step in the derivation of an expression for $M_{\mathrm{L}}(q,t)$ which contains the effects of collisional damping is based on the following argument. The dynamical correlation produced by the constant component in the memory function is destroyed in a dense liquid largely through interatomic collisions. One therefore intuitively expects the time scale of the decaying part of the memory function to be intimately related to the duration of a collision. With this idea in mind the terms containing the interatomic potential are decoupled from the components of the current correlation function and replaced by conditionally averaged terms which describe the decay of the interaction between two given atoms over the period $t = 0$ to $t = s$.

The ensemble average is over the phase space coordinates at $t = 0$. For a given atom j and any other atom l from the rest of the ensemble the quantity

$$G^*(x,t) \equiv \langle\delta(\mathbf{x}-\mathbf{r}_{jl}(t)+\mathbf{r}_{jl}(0))\rangle \equiv \frac{1}{(2\pi)^3}\int \mathrm{d}\mathbf{n}\,\exp(-\mathrm{i}\mathbf{n}\cdot\mathbf{x})F^*(n,t) \tag{7.6.19}$$

is introduced, which describes their relative motion. It is essentially the probability density that the change in the separation of two atoms in time t is $\mathbf{x}$. Now each term in (7.6.18) under the first summation is a single particle contribution to the longitudinal current multiplied by the quantity

$$\sum_{l\neq j}(\hat{\mathbf{q}}\cdot\boldsymbol{\nabla})^2\phi'(r_{jl}(s))(1-\exp[\mathrm{i}\mathbf{q}\cdot\mathbf{r}_{jl}(s)]).$$

In the latter the most important consequence of the coupling between the Fourier components of the potential and the exponential term in the brackets has already been taken into account earlier, leading to the extraction of the component $\tilde{\phi}(q)$. The potential terms, it is argued, may therefore be conditionally averaged separately. Hence the mean value of potential term at time $t = s$, for a given initial separation of the atoms at $t = 0$, is

$$\sum_{l\neq j}\int \mathrm{d}\mathbf{x}\,(\hat{\mathbf{q}}\cdot\boldsymbol{\nabla})^2\phi'(|\mathbf{r}_{jl}(0)+\mathbf{x}|)\langle\delta(\mathbf{x}-\mathbf{r}_{jl}(s)+\mathbf{r}_{jl}(0))\rangle$$

$$= \sum_{l\neq j}\int \mathrm{d}\mathbf{x}\,(\hat{\mathbf{q}}\cdot\boldsymbol{\nabla})^2\phi'(|\mathbf{r}_{jl}(0)+\mathbf{x}|)G^*(x,s) \tag{7.6.20}$$

whilst that of the exponential term $\exp[\mathrm{i}\mathbf{q}\cdot\mathbf{r}_{jl}(s)]$ is $\exp[\mathrm{i}\mathbf{q}\cdot\mathbf{r}_{jl}(0)]F^*(q,s)$. There still remains the averaging over the initial configurations of the atoms labelled l. On carrying this out, the mean value of

$$\sum_{l \neq j} (\hat{\mathbf{q}} \cdot \nabla)^2 \phi'(r_{jl}(s)) \, (1 - \exp[\mathrm{i}\mathbf{q} \cdot \mathbf{r}_{jl}(s)])$$

at $t = s$ becomes independent of the index j and is given by

$$m_{\mathrm{L}}(q, s) = \rho \iint \mathrm{d}\mathbf{r} \, \mathrm{d}\mathbf{x} \, g(r)(\hat{\mathbf{q}} \cdot \nabla)^2 \phi'(\,|\mathbf{r} + \mathbf{x}|\,) G^*(x, s)(1 - \exp[\mathrm{i}\mathbf{q} \cdot \mathbf{r}] F^*(q, s)). \tag{7.6.21}$$

By means of the preceding arguments, therefore, the ensemble average in (7.6.18) is intuitively decoupled so that it becomes

$$\dot{C}_{\mathrm{L}}(q, t) = -\frac{1}{N} \int_0^t \mathrm{d}s \, M_{\mathrm{L}}(q, s) C_{\mathrm{L}}(q, t - s)$$

with

$$M_{\mathrm{L}}(q, t) = \frac{q^2}{m\beta}(1 + \tilde{\phi}(q)) + m_{\mathrm{L}}(q, t) \tag{7.6.22}$$

which, with (7.6.21), defines an approximate result for the damping function. In practice $G^*(r, t)$, in the absence of a more satisfactory theory, is approximated by a self-correlation function but containing an effective mass $m^*/2$, the reduced mass in the relative motion of the atoms. This procedure allows the atoms to approach each other arbitrarily closely and (7.6.21) has to be modified to describe adequately the core contribution. The modification is based on the argument that at $t = 0$ only those atoms whose cores are in contact contribute to the memory function. At later times they will move apart under the mutual repulsion. Hence the contribution to the damping function from the potential core $\phi_{\mathrm{c}}(r)$ is expressed as

$$m_{\mathrm{L}}^{\mathrm{c}}(q, t) = \frac{\rho_r}{m} \iint \mathrm{d}\mathbf{r} \, \mathrm{d}\mathbf{x} \, g(r)(\hat{\mathbf{q}} \cdot \nabla)^2 \phi_{\mathrm{c}}(r + x) G_{\mathrm{s}}^*(x, t)(1 - \exp[\mathrm{i}\mathbf{q} \cdot \mathbf{r}] F^*(q, t)). \tag{7.6.23}$$

This means that for $t > 0$ the atoms separate in the $\mathbf{r}$ direction as they would following a direct collision. In an application to liquid rubidium for a number of values of $q \leqslant 1$ Å^{-1} (Barker and Gaskell[44]), the core diameter, r_{c}, is defined as the value of r at which $\phi(r)$ is first zero, and (7.6.23) was applied to that part of the potential. The contribution to the damping function of the potential well and the tail of the potential, denoted by $m_{\mathrm{L}}^{\mathrm{T}}(q, t)$, was evaluated using (7.6.21). The form of the damping function obtained is demonstrated for $q = 0.174$ Å^{-1} in Figure 7.6.1. Note the long tail of the component $m_{\mathrm{L}}^{\mathrm{T}}(q, t)$, and the very rapid decay of the core distribution. These were consistent features of the results for all the wavenumbers used. The exponential choice for $m_{\mathrm{L}}(q, t)$, with the relaxation time discussed in Section 7.5 shows a much more rapid decay at large t for $0.4 \leqslant q \lesssim 1$ Å^{-1} than is displayed by this calculation of the damping function.

The constant term in the total memory function $M_{\mathrm{L}}(q, t)$ is given exactly by $q^2/m\beta S(q)$, as pointed out in (7.4.15), which guarantees that the dynamic structure factor calculated via (7.4.17) satisfies the zero and second moments. The expression derived in this approach is $(q^2/m\beta)(1 + \tilde{\phi}(q))$. It is well known, however, that the random phase argument by which the time independent term was obtained can also be used to determine the static structure factor, although only approximately. To reproduce $S(q)$ exactly within the random phase approximation one must choose $\beta\tilde{\phi}(q) = -c(q) = 1 - S(q)^{-1}$, where $c(q)$ is the Fourier transform of the

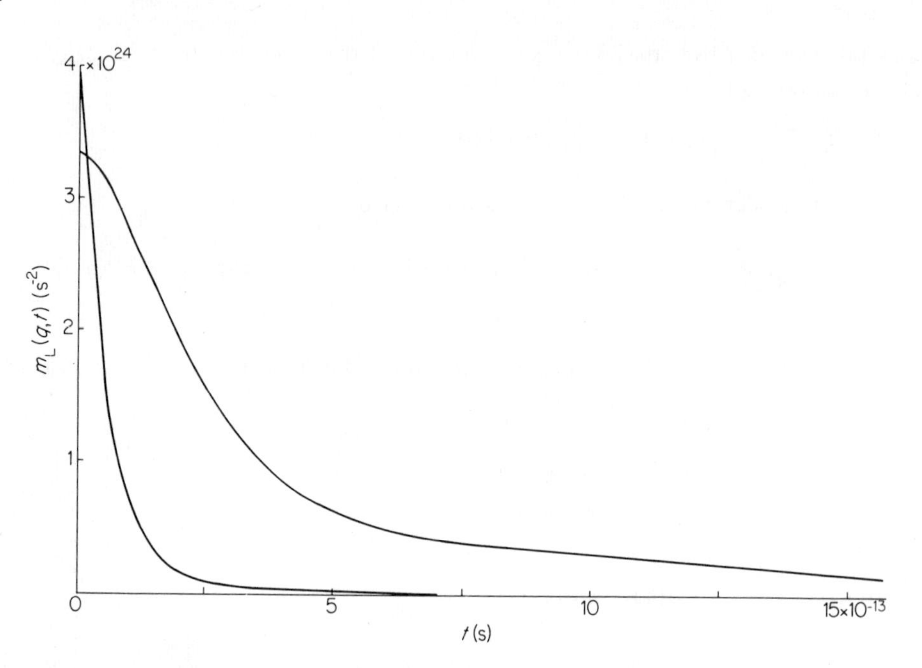

Figure 7.6.1 The two contributions to the memory function $m_L(q, t)$ for a model of liquid rubidium. Upper curves shows $m_L^T(q, t)$ and the lower one $m_L^C(q, t)$ for $q = 0.174$ Å^{-1}

direct correlation function. It may be argued that the random phase approximation to the time independent term should be corrected by making the above replacement in order to meet the extremely important zero order moment condition, maintaining, however, the correct interatomic potential in the remainder of the expression whose time dependence is clearly related to the rate of decay of interatomic collisions. The so-called primary sum rules can therefore be satisfied in this approach. The higher order sum rules involve the potential and atomic distribution functions explicitly. Now the exact initial value of the memory or damping function $m_L(q, t)$ is given by $m_L(q, 0) = \omega_L^2 - \omega_0^2$. Compare this with the initial value of the approximate result for $m_L(q, t)$ in (7.6.21). Apart from the relatively unimportant kinetic terms it has the same form as ω_L^2, defined in equation (7.4.12), except for the missing Fourier component in the potential which represents the subtraction of ω_0^2 within the random phase approximation. As discussed by Barker and Gaskell,[44] because of the modification to the derived expression for the damping function to account adequately for the core contribution, the random phase argument can no longer be applied to that part of the potential and the initial condition on the memory function is not met.

Nevertheless, when the dynamic structure factor for rubidium is calculated via the formula expressed in (7.4.17) the comparison with the molecular dynamics results is encouraging, as demonstrated in Figures 7.6.2 and 7.6.3. None of the generalized hydrodynamic or visco-elastic methods described previously gives as good a description of the quasi-elastic region around $S(q, 0)$ as the present approach, nor is the overall fit so successful.

Some comment about the failure to meet the fourth moment condition is necessary. In the evaluation to the short range collision contribution to the damping function, the condition was incorporated that if two atoms are in contact at $t = 0$ they must subsequently *immediately* move apart, thus treating the potential in a sense as that of a rigid sphere. It is this treatment of the core which is responsible for the failure to satisfy the fourth frequency moment. On the

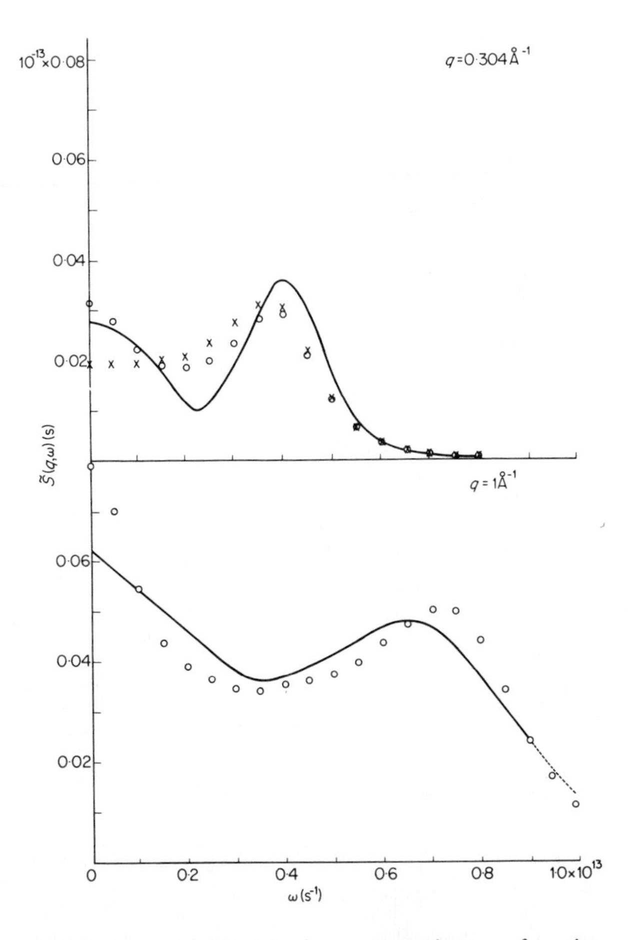

Figure 7.6.2 Symmetrized scattering function $\tilde{S}(q, \omega) = \exp(-h\omega/2k_B T)S(q, \omega)$ for rubidium. Solid curves are Rahman's molecular dynamics results.[27] Open circles are based on equations (7.6.21) and (7.6.23) and the crosses show a result based on the generalized hydrodynamic scheme of Barker and Gaskell.[44] (Reproduced by permission of the American Institute of Physics)

other hand it does enable the hard core limit to be taken in the theory so that the binary encounters are included in an approximate way. The fourth frequency moment has played an essential role in the generalized hydrodynamic or visco-elastic theories of $S(q, \omega)$, largely one suspects because the parameters involved cannot be determined except by relating them in some approximate way to the low order moments. However, the fourth moment, because it involves the second derivative of the potential, is infinite for a rigid sphere fluid. Hence the rigid core limit cannot easily be recovered from such theories even though the rigid sphere fluid is frequently used as a model for real liquids.

In conclusion, therefore, the comparison of the calculated $S(q, \omega)$ and the molecular dynamics results over a range of wavenumbers < 1 Å^{-1} is encouraging and gives some confidence in the ideas on which the memory function is based. It suggests too that a theory of the density fluctuation spectrum should accurately describe the short range collision processes

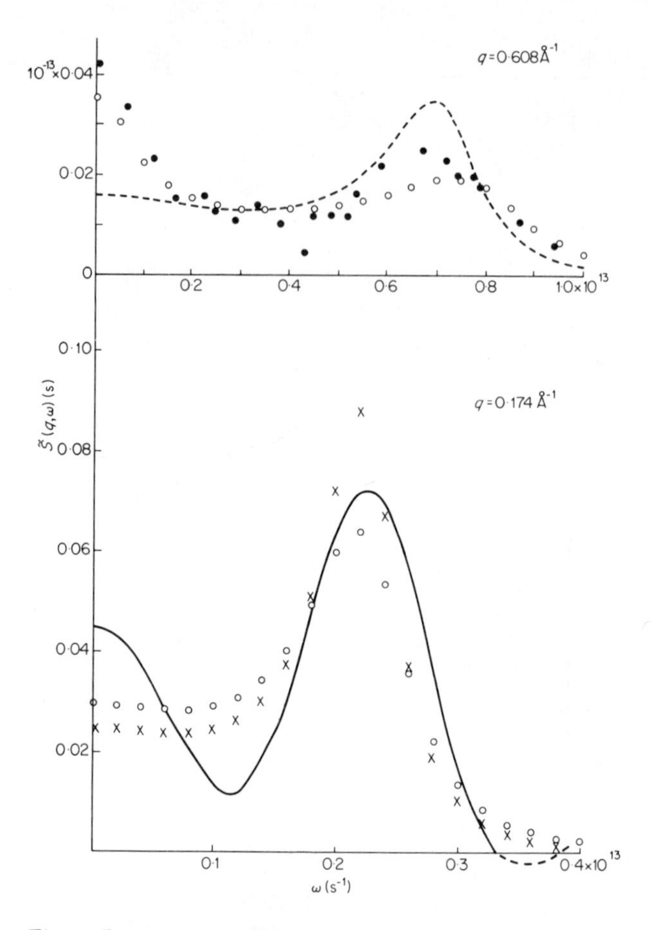

Figure 7.6.3 Symmetrized scattering function $\tilde{S}(q, \omega) = \exp(-h\omega/2k_BT)S(q, \omega)$ for rubidium. Solid curve is Rahman's molecular dynamics result[27] and the dots represent the neutron scattering results of Copley and Rowe.[25] Open circles are based on equations (7.6.21) and (7.6.23), the crosses for $q = 0.174\ \text{Å}^{-1}$ show the effect of omitting the core contribution to the damping. The dashed curve is derived from the choice of a simple exponential time decay for the memory function $m_L(q, t)$ with the wavenumber dependence of the relaxation time given by $\tau(q) = 2(\omega_L^2(q) - \omega_0^2(q))^{1/2}/\pi^{1/2}$. (Reproduced by permission of the American Institute of Physics)

so that the rigid core limit can be recovered. On the other hand, the approach can be criticized because it is based on an intuitive de-coupling of the ensemble averages and in its present form lacks any systematic means of improvement – the last comment being generally true about the existing theories of time dependent correlations in liquids. Neither of the two methods discussed in this section will adequately describe liquid argon in the wavenumber regime below 0.2Å^{-1} where there is a reasonably well defined collective mode, because in their present form they do not contain the hydrodynamic limit. It seems essential in that case to have some form of generalized hydrodynamic scheme, with the heat diffusion term specifically included.

However, it should be pointed out that some calculations of the dynamic structure factor for the liquid argon model, which are based upon a form of kinetic theory, have very recently been presented by Jhon and Forster.[45] A number of authors have been involved in the introduction and development of this type of approach (see Sykes[46] for references), but the theory of Jhon and Forster appears to be the first to be carried through to give satisfactory results for $S(q, \omega)$ in a dense liquid, within the range of q and ω values investigated in molecular dynamics or neutron scattering experiments. A full discussion requires a different framework from the one employed here and for this reason the method will be outlined only very briefly.

The route to the dynamic structure factor involves writing down a kinetic equation for the phase space density–density correlation function

$$S(\mathbf{r}-\mathbf{r}', t-t', \mathbf{p}\mathbf{p}') = \langle (\mathrm{f}(\mathbf{r}, \mathbf{p}, t) - \langle \mathrm{f}(\mathbf{r}, \mathbf{p}, t)\rangle)(\mathrm{f}(\mathbf{r}', \mathbf{p}', t') - \langle \mathrm{f}(\mathbf{r}', \mathbf{p}', t')\rangle)\rangle$$

where

$$\mathrm{f}(\mathbf{r}, \mathbf{p}, t) = \sum_{i=1}^{N} \delta(\mathbf{r} - \mathbf{r}_i(t))\delta(\mathbf{p} - \mathbf{p}_i(t))$$

The density–density correlation function $G(\mathbf{r}, t)$ and hence $S(q, \omega)$ may be obtained from the latter by integration over the momentum variables in the following way

$$\rho G(\mathbf{r}-\mathbf{r}', t-t') - \rho^2 = \int d\mathbf{p} \int d\mathbf{p}'\, S(\mathbf{r}-\mathbf{r}', t-t', \mathbf{p}, \mathbf{p}')$$

The kinetic equation contains a memory function, the collisional part of which is approximated by Jhon and Forster in such a way that the correct hydrodynamic behaviour of the correlation function is guaranteed. The equation is solved analytically but to carry through the calculation the wavenumber dependence of certain coefficients is assumed. The theory describes the location and persistence of the side peak in $S(q, \omega)$ rather well for the argon model, although the results around $\omega \simeq 0$ are less successful. Application of this approach to rubidium will be extremely interesting.

7.7 Transverse current modes and shear viscosity

Transverse modes in simple liquids have been discussed over the past few years from different physical models of a liquid. For example, by Egelstaff[47] on the basis of a visco-elastic approach and Takeno and Goda[48] by means of an extension of a theory of phonons in an amorphous solid. The question immediately at issue is whether transverse modes can be propagated in a liquid, and if so for what frequencies and range of wavenumbers. The generalized Langevin equation once again provides a convenient framework from which to discuss this problem and this has been exploited by Akcasu and Daniels,[40] Murase,[49] Götze and Lücke[50] and Chiakwelu *et al.*[51] in an application to liquid argon. These investigations support the general conclusion that a transverse mode will be propagated in a liquid for wavenumbers greater than some critical value q_c, although there is some discrepancy between the various approaches about its magnitude.

The problem has recently been discussed by Levesque *et al.*[26] by means of some molecular dynamics calculations of the transverse current correlation function, $C_T(q, t)$, for a model of liquid argon under conditions of temperature and density close to those at the triple point of the real liquid. Results were obtained for the frequency spectrum of the correlation function itself to show its dependence on wavenumber. The existence of a peak in the frequency spectrum at a finite value of the frequency indicates oscillatory behaviour of the current correlation function and hence that the liquid can sustain the propagation of shear waves. The transverse current correlation function will be defined as

$$C_T(q,t) = \frac{q^2}{N}\left\langle \sum_{j=1}^{N} v_j^x(0)\exp[-iqz_j(0)] \sum_{k=1}^{N} v_k^x(t)\exp[iqz_k(t)] \right\rangle. \tag{7.7.1}$$

For sufficiently long times and small wavenumbers the form of this correlation function can be deduced from hydrodynamic theory leading to the well known result

$$C_T(q,t) = \frac{q^2}{m\beta}\exp\left(-\frac{\eta}{\rho m}q^2|t|\right). \tag{7.7.2}$$

As $q \to 0$, therefore, a liquid is unable to propagate shear waves. For short wavelengths the situation is not immediately obvious, and molecular dynamics is the only method at present which can throw some light on this problem.

If we denote the memory function in this case by $m_T(q, t)$, the Fourier–Laplace transform of $C_T(q, t)$ is given by

$$\tilde{C}_T(q,z) = \frac{q^2/m\beta}{-iz + m_T(q,z)}. \tag{7.7.3}$$

The hydrodynamic limit can be recovered when the memory function takes the form

$$m_T(q,t) = \frac{\eta q^2}{\rho m}\,\delta(t). \tag{7.7.4}$$

The correct initial value of the $m_T(q, t)$ is

$$\begin{aligned}\omega_T^2(q) &= \frac{q^2}{m\beta} + \frac{\rho}{m}\int \mathrm{d}\mathbf{r}\, g(r)\frac{\partial^2\phi(r)}{\partial x^2}(1-\exp[iqz]) \\ &\equiv \frac{q^2}{\rho m}G(q)\end{aligned} \tag{7.7.5}$$

where $G(q)$ defines a q-dependent shear modulus or rigidity. $G(0)$ is the rigidity, G, introduced earlier in equation (7.5.4). Assuming a simple exponential dependence for the memory function so that

$$m_T(q,t) = \frac{q^2}{m\rho}G(q)\exp(-t/\tau_T(q))$$

it is easy to show that the frequency spectrum $C_T(q, \omega) = (1/\pi)\,\mathrm{Re}\,\tilde{C}_T(q, \omega)$ will have a maximum value for $\omega > 0$ provided that $q > q_c$ where

$$q_c = \left(\frac{\rho m}{2G(q)}\right)^{1/2}\frac{1}{\tau_T(q)}. \tag{7.7.6}$$

Neglecting the wavenumber dependence of the right hand side, and evaluating $\tau_T(0)$ as a Maxwell relaxation time equal to η/G, the critical value of the wavenumber turns out to be $\simeq 0.23$ Å^{-1}. This result for q_c is identical with a visco-elastic estimate, using the above criterion.

Once again in order to use this simple scheme there is the difficult problem of choosing the relaxation time for finite values of q. Even when $\tau_T(q)$ is determined by means of a least squares fit of the calculated frequency spectrum to the computer data, the results for $C_T(q, \omega)$ from this exponential memory function are not especially good. In particular, the side peaks are considerably flatter for the longer wavelengths than those obtained from the molecular dynamics calculations. The source of the discrepancy lies in the fact that the memory function, at least for values of $q \lesssim 1$ Å^{-1}, shows a rapid initial fall with time followed by a much slower decay thereafter. Hence $m_T(q, t)$ has an extensive tail, due to collective effects in the liquid, in which the attractive tail of the potential probably plays a significant (although not the major) role. It cannot therefore be fitted by a single exponential function. Levesque *et al.*[26] achieved a good fit to the frequency spectrum by means of a memory function consisting of two exponential terms with very different relaxation times.

The critical wavenumber value, q_c, is difficult to determine precisely from the molecular dynamics results. However, using the fitted form for the memory function the critical value turns out to be lower than that previously quoted and is given by $q_c \simeq 0.164$ Å^{-1}. The consequences of the long tail of the memory function are therefore: (i) the lowering of the critical wavenumber; and (ii) the rather pronounced shear wave peaks in the frequency spectrum. It should be pointed out that a side peak was not apparent in the molecular dynamics results below $q \simeq 0.22$ Å^{-1}. The discrepancy between this and the latter value of q_c is apparently within the computational uncertainties.

Very recently a molecular dynamics study, by Rahman and Stillinger,[52] of the propagation of sound in water has been carried out in a system of 216 water molecules. The transverse current correlation function was computed in the process. In contrast to a simple dense monatomic liquid, each molecule has only about four to five nearest neighbours with strong bonds which resist both changes of shape and volume of the local structure. For the lowest wavenumber investigated, $q = 0.3$ Å^{-1} or $\lambda \sim 20$ Å^{-1}, the frequency spectrum shows a pronounced side peak indicating propagating shear waves with a well defined velocity of propagation. This is in contrast to the simulation studies in argon for wavelengths of this order, where the damping is considerably more significant. Presumably the different behaviour is a reflection of the considerable rigidity of the hydrogen bond network in the system, and the indications are that water should be able to transmit shear waves of considerably greater wavelengths than 20 Å. In fact the rigidity G for the water model is 9.4×10^{10} g cm^{-1} s^{-2}, which is an order of magnitude greater than for argon near the triple point ($G = 1.0 \times 10^{10}$ g cm^{-1}s^{-2}).

Shear viscosity

The microscopic expression for the shear viscosity coefficient can easily be derived using the procedure discussed by Schofield[18] and which is outlined here. For a one-component fluid the transverse components of the momentum density are independent of the other conserved variables. The fluctuation in the transverse component perpendicular to **q** will be defined as

$$J_{\mathbf{q}}^{\mathrm{T}} = m \sum_{j=1}^{N} v_j^{\mathrm{T}} \exp(-\mathrm{i}\mathbf{q} \cdot \mathbf{r}_j) \qquad (7.7.7)$$

and it satisfies the equation

$$\dot{J}_{\mathbf{q}}^{\mathrm{T}} + \mathrm{i}q\sigma_{\mathbf{q}}^{\mathrm{T}} = 0 \qquad (7.7.8)$$

where $\sigma_{\mathbf{q}}$ is the stress tensor,

$$\sigma_{\mathbf{q}}^{\alpha\beta} = \sum_{j=1}^{N} \left(m v_j^{\alpha} v_j^{\beta} - \frac{1}{2} \sum_{k \neq j} \frac{r_{kj}^{\alpha} r_{kj}^{\beta}}{r_{kj}^2} P_{\mathbf{q}}(r_{kj}) \right) \exp[-\mathrm{i}\mathbf{q} \cdot \mathbf{r}_j]$$

and

$$P_{\mathbf{q}}(r) = r \frac{\partial \phi}{\partial r} \frac{(1 - \exp[-\mathrm{i}\mathbf{q} \cdot \mathbf{r}])}{\mathrm{i}\mathbf{q} \cdot \mathbf{r}}.$$

Now $\langle J_{\mathbf{q}}^{\mathrm{T}}(0)\, J_{\mathbf{q}}^{*\mathrm{T}}(t)\rangle$ is assumed to equal the correlation function of the corresponding macroscopic variable for sufficiently small q and large t. Thus according to hydrodynamic theory, in this limit

$$\langle J_{\mathbf{q}}^{\mathrm{T}}(0) J_{\mathbf{q}}^{*\mathrm{T}}(t)\rangle = \langle J_{\mathbf{q}}^{\mathrm{T}}(0) J_{\mathbf{q}}^{*\mathrm{T}}(0)\rangle \exp\left(-\frac{\eta}{\rho m} q^2 t\right). \tag{7.7.9}$$

Making use of (7.7.8) we may write

$$\langle \dot{J}_{\mathbf{q}}^{\mathrm{T}}(0) \dot{J}_{\mathbf{q}}^{*\mathrm{T}}(t)\rangle = -\frac{\mathrm{d}^2}{\mathrm{d}t^2}\langle J_{\mathbf{q}}^{\mathrm{T}}(0) J_{\mathbf{q}}^{*\mathrm{T}}(t)\rangle = q^2 \langle \sigma_{\mathbf{q}}^{\mathrm{T}}(0)\sigma_{\mathbf{q}}^{*\mathrm{T}}(t)\rangle \tag{7.7.10}$$

from which

$$\langle J_{\mathbf{q}}^{\mathrm{T}}(0) J_{\mathbf{q}}^{*\mathrm{T}}(t)\rangle = \langle J_{\mathbf{q}}^{\mathrm{T}}(0) J_{\mathbf{q}}^{*\mathrm{T}}(0)\rangle \left(1 - q^2 \int_0^t \mathrm{d}s(t-s)\frac{\langle \sigma_{\mathbf{q}}^{\mathrm{T}}(0)\sigma_{\mathbf{q}}^{*\mathrm{T}}(s)\rangle}{\langle J_{\mathbf{q}}^{\mathrm{T}}(0) J_{\mathbf{q}}^{*\mathrm{T}}(0)\rangle}\right). \tag{7.7.11}$$

By identifying the coefficient of $q^2 t$ in an expansion of the expression within the large brackets in equation (7.7.11), for small q and large t, and comparing with (7.7.9) we may obtain

$$\frac{\eta}{\rho m} = \int_0^\infty \mathrm{d}t \lim_{q\to 0} \frac{\langle \sigma_{\mathbf{q}}^{\mathrm{T}}(0)\sigma_{\mathbf{q}}^{*\mathrm{T}}(t)\rangle}{\langle J_q^{\mathrm{T}}(0) J_q^{*\mathrm{T}}(0)\rangle} \tag{7.7.12}$$

where the limit $q \to 0$ must be taken before the integral is evaluated. It is readily appreciated, by bearing in mind equation (7.7.8), that for finite q the time integral is zero. Hence, taking $\sigma_{\mathbf{q}}^{\mathrm{T}}$ as the off-diagonal element $\sigma_{\mathbf{q}}^{xy}$, say, we have

$$\eta = \rho\beta \int_0^\infty \mathrm{d}t \langle \sigma^{xy}(0)\, \sigma^{xy}(t)\rangle \equiv \int_0^\infty \mathrm{d}t\, \eta_{\mathrm{T}}(t) \tag{7.7.13}$$

where

$$\sigma^{xy} = \sum_{j=1}^{N}\left(m v_j^x v_j^y - \frac{1}{2}\sum_{k\neq j}\frac{r_{kj}^x r_{kj}^y}{r_{kj}}\frac{\partial\phi}{\partial r_{kj}}(r_{kj})\right).$$

The time-dependent correlation function $\eta_{\mathrm{T}}(t)$ may be derived in terms of the memory function for transverse modes $m_{\mathrm{T}}(q, t)$. It can be seen from the hydrodynamic form of the correlation function that the shear viscosity is given by taking the following limits

$$\eta = \rho m \pi \lim_{\omega\to 0} \omega^2 \lim_{q\to 0} q^{-2} \frac{C_{\mathrm{T}}(q,\omega)}{C_{\mathrm{T}}(q, t=0)} \tag{7.7.14}$$

Now,

$$\frac{C_{\mathrm{T}}(q,\omega)}{C_{\mathrm{T}}(q,0)} = \frac{m_{\mathrm{T}}^{\mathrm{c}}(q,\omega)/\pi}{[m_{\mathrm{T}}^{\mathrm{s}}(q,\omega)-\omega]^2 + [m_{\mathrm{T}}^{\mathrm{c}}(q,\omega)]^2}$$

where $m_{\mathrm{T}}^{\mathrm{c}}$ and $m_{\mathrm{T}}^{\mathrm{s}}$ are the cosine and sine transforms respectively of $m_{\mathrm{T}}(q, t)$. Therefore if the above limit exists we must have

$$\eta = \rho m \lim_{\omega\to 0}\lim_{q\to 0}\frac{m_{\mathrm{T}}^{\mathrm{c}}(q,\omega)}{q^2}$$

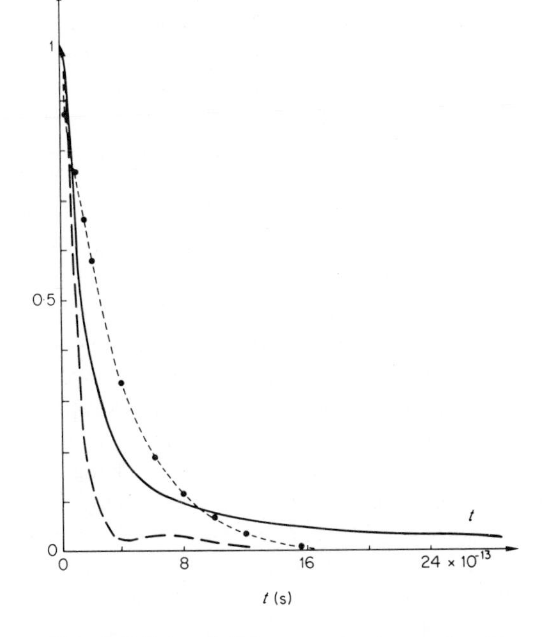

Figure 7.7.1 Solid curve is the molecular dynamics result for the function $\eta_T(t)/\eta_T(0)$. Dashed–dotted line is obtained from the choice of an exponential time dependence for $m_T(q, t)$. The dashed curve is the memory function for the velocity autocorrelation function. All the results are obtained from a Lennard–Jones potential under conditions of temperature and density close to those at the triple point of argon. (From Levesque *et al.*,[26] reproduced by permission of the American Institute of Physics)

or

$$\eta = \rho m \int_0^\infty \mathrm{d}t \lim_{q\to 0} \frac{m_T(q, t)}{q^2}$$

and we may identify

$$\eta_T(t) = \rho m \lim_{q\to 0} \frac{m_T(q, t)}{q^2} \tag{7.7.15}$$

In the computer simulation studies of Levesque *et al.*[26] the function $\eta_T(t)$ was evaluated, along with the shear viscosity coefficient. The computed data are shown in Figure 7.7.1, together with the result obtained from the simple exponential choice for the memory function discussed earlier. The significant feature of the computer calculation is the very long slowly-decaying tail. The memory function for the velocity autocorrelation function, under the same conditions of temperature and density, is shown for comparison. It is extremely interesting that a similar long tail in $\eta_T(t)$ was observed in a computer experiment on a rigid sphere fluid by Alder *et al*[53] near the solidification density. Hence it would seem that the long tail displayed in the shear viscosity function by Lennard–Jones particles is not simply a reflection of the long range part of the potential. The latter authors have also shown that at high densities the rigid sphere shear viscosity is considerably greater than the value obtained from the Enskog formula. The difference must therefore be accounted for by an important contribution from collective effects in the fluid, which makes theoretical attempts to evaluate the viscosity extremely difficult.

It should be pointed out however that the shear viscosity coefficient for the argon model calculated via (7.7.13) is some 30% greater than the experimental value for liquid argon near its triple point ($\eta = 2.7 \times 10^{-3}$ g cm^{-1} s^{-1}).[54] This discrepancy has recently been investigated by Ashurst and Hoover[29] by using a different approach to the computer evaluation of transport coefficients. The computer generates non-equilibrium systems with the desired flux by numerically simulating laboratory methods of measuring transport properties. The results depend upon the width of the system considered, but when extrapolated to infinite width the computed shear viscosity agrees with the experimental value to within a few percent. When the equilibrium molecular dynamics calculation of Levesque *et al.* is similarly adjusted to an infinite width system it too is in excellent agreement with experiment. The apparent discrepancies therefore between the two molecular dynamics methods and the experimental argon data seem to be due to finite flow-field effects. The problem appears to be significant only around the triple point region. Calculations in other regions of the phase diagram indicate no significant width dependence.

References

1. J. R. D. Copley and S. W. Lovesey (1975). *Rep. Prog. Phys.*, **38**, 461.
2. P. Schofield (1975). *Specialist Reports – Stat. Mech. II*, Chemical Society, London.
3. C. A. Croxton (1974). *Liquid State Physics – A Statistical Mechanical Introduction*, Cambridge University Press, London.
4. K. Kawasaki (1973). *J. Phys.* A., **6**, 1289.
5. R. Kubo (1966). *Rep. Prog. Phys.*, **29**, 255.
6. R. Zwanzig (1961). In W. F. Britton, W. B. Downs and J. Downs (Eds.), *Lectures in Theoretical Physics*, Vol. 3 Interscience, New York. p. 135.
7. H. Mori (1965). *Prog. Theor. Phys*, **33**, 423.
8. K. S. Singwi (1968). *Theory of Condensed Matter*, International Atomic Energy Agency, Vienna. p. 603.
9. D. Levesque and L. Verlet (1970). *Phys. Rev.*, **A2**, 2514.
10. B. J. Berne, J. P. Boon and S. A. Rice (1966). *J. Chem. Phys.*, **45**, 1086.
11. K. S. Singwi and A. Sjölander (1968). *Phys. Rev.*, **167**, 152.
12. M. I. Barker and T. Gaskell (1972). *J. Phys.* C., **5** 353.
13. N. Corngold and J. J. Duderstadt (1970). *Phys. Rev.*, **A2**, 836.
14. H. Mori (1965). *Prog. Theor. Phys.*, **34**, 399.
15. D. Forster, P. C. Martin and S. Yip (1968). *Phys. Rev.*, **170**, 155.
16. R. Bansal and K. N. Pathak (1974). *Phys. Rev.*, **A9**, 2773.
17. N. K. Ailawadi, A. Rahman and R. Zwanzig (1971). *Phys. Rev.*, **A4**, 1616.
18. P. Schofield (1968). In H. N. V. Temperley, J. S. Rowlinson and G. S. Rushbrooke (Eds.), *The Physics of Simple Liquids*, North-Holland, Amsterdam. Chap. 13.
19. S. W. Lovesey (1971). *J. Phys.* C., **4**, 3057.
20. L. Van Hove (1954). *Phys. Rev.*, **95**, 249.
21. W. Marshall and S. W. Lovesey (1971). *Theory of Thermal Neutron Scattering*, Oxford University Press, Oxford.
22. L. P. Kadanoff and P. C. Martin (1963). *Ann. Phys. (N.Y.)*., **24**, 419.
23. A. A. Kugler (1973). *J. Stat. Phys.*, **8**, 107.
24. R. Zwanzig (1972). In S. A. Rice, K. F. Freed and J. C. Light (Eds.), *Statistical Mechanics: New Concepts, New Problems, New Applications*, Univ. Chicago Press, Chicago. p. 229.

25. J. R. D. Copley and J. M. Rowe (1974). *Phys. Rev. Lett.*, **32**, 49.
26. D. Levesque, L. Verlet and J. Kurkijarvi (1973). *Phys. Rev.*, **A7**, 1690.
27. A. Rahman (1974). *Phys. Rev. Lett.*, **32**, 53.
28. R. D. Mountain (1966). *Rev. Mod. Phys.*, **38**, 205.
29. W. T. Ashurst and W. G. Hoover (1975). *Phys. Rev.*, **A11**, 658.
30. D. Levesque and M. Rao (1976), (to be published).
31. C. H. Chung and S. Yip (1969). *Phys. Rev.*, **182**, 323.
32. V. F. Sears (1970). *Can. J. Phys.*, **48**, 616.
33. P. Schofield (1966). *Proc. Phys. Soc.*, **88**, 149.
34. M. I. Barker and T. Gaskell (1972). *J. Phys.* C., **5**, 3279.
35. T. Gaskell (1971). *Phys. Chem. Liquids*, **2**, 237.
36. H. Bell, H. Moeller-Wenghoffer, A. Kollmar, R. Stockmeyer, T. Springer and H. Stiller (1975). *Phys. Rev.*, **A11**, 316.
37. R. Zwanzig and R. D. Mountain (1965). *J. Chem. Phys.*, **43**, 4464.
38. H. C. Longuet-Higgins and J. A. Pople (1956). *J. Chem. Phys.*, **25**, 884.
39. M. I. Barker and T. Gaskell (1974). *J. Phys.* C., **7**, L293.
40. A. Z. Akcasu and E. Daniels (1970). *Phys. Rev.*, **A2**, 962.
41. J. Hubbard and J. L. Beeby (1969). *J. Phys.* C., **2**, 556.
42. M. I. Barker and T. Gaskell (1975). *J. Phys.* C., **8**, 89.
43. K. N. Pathak and R. Bansal (1973). *J. Phys.* C., **6**, 1989.
44. M. I. Barker and T. Gaskell (1975). *J. Phys.* C., **8**, 3715.
45. M. S. Jhon and D. Forster (1975). *Phys. Rev.* **A12**, 254.
46. J. Sykes (1976). *J. Phys. A.*, **9**, 1771.
47. P. A. Egelstaff (1967). *An Introduction to the Liquid State*, Academic Press, London. p. 168.
48. S. Takeno and M. Goda (1971). *Prog. Theor. Phys.*, **45**, 331.
49. C. Murase (1972). *J. Phys. Soc. Japan*, **32**, 1205.
50. W. Götze and M. Lücke (1975). *Phys. Rev* , **A11**, 2173.
51. O. Chiakwelu, T. Gaskell and J. W. Tucker (1976). *J. Phys.* C., **9**, 1635.
52. A. Rahman and F. H. Stillinger (1974). *Phys. Rev.*, **A10**, 368.
53. B. J. Alder, D. M. Gass and T. E. Wainwright (1970). *J. Chem. Phys.*, **53**, 3813.
54. D. G. Naugle, J. H. Lunsford and J. R. Singer (1966). *J. Chem. Phys.*, **45**, 4669.

Chapter 8

Polymer Dynamics and the Hydrodynamics of Polymer Solutions

KARL F. FREED

The James Franck Institute and the Department of Chemistry, The University of Chicago, Chicago, Illinois 60637

8.1 Introduction

Polymer systems display a wide and rich variety of physical phenomena which in many cases are only poorly understood, in part, because of the great mathematical complexities present in any realistic theoretical description of these systems. In fact, these difficulties are often found to be of the same generic form as those encountered in other areas of statistical physics, in general, and liquid physics, in particular. It then becomes natural to expect that advances in theoretical methods of statistical mechanics will find fruitful applications to polymer systems. On the other hand, the study of these systems in their own right will also help to produce advances in fundamental statistical mechanics.

Because of the richness and diversity of polymer systems, it is impossible to review recent progress in a single article. The recent monograph on properties of dilute polymer solutions[1] attests to this impossibility. Instead, we wish to review a few selected advances which have arisen from our own research efforts. This involves the study of the dynamics of polymers in solutions at finite concentration as well as the hydrodynamics of these polymer solutions. As an example of the relation of these systems to quite different ones, we can note that the flow of fluid through the polymer solution bears a direct mathematical analogy to the study of electron propagation through a disordered medium.

Unfortunately, polymer systems are generally not discussed in statistical mechanics courses in either physics or chemistry departments, despite their scientific and technological importance. Thus, it is necessary to review first some simple facets of polymers before turning to the description of some recent advancements in the field in a fashion amenable to workers both within and outside the field. This review is intended to aid those familiar with polymer systems in recalling the necessary information. On the other hand, polymer physicists and chemists are often unfamiliar with widely used techniques in the study of liquids and of many-body theory, in general, so it is also useful to introduce some of these methods in a manner requiring no prior knowledge of the latest techniques. Likewise, we digress at various points in the chapter to show how the new and generalized formulation reduces to that of the older theories. This serves to summarize these theories for readers unfamiliar with polymers and to illustrate more fully the generality of the new approach for researchers in the polymer field.

The fundamental outlook involves the consideration of polymers in solution as a set of dynamical objects, obeying equations of motion prescribed by some model of the chain. The solvent introduces random forces that drive the Brownian motion of the chain. Whilst the chain is treated in terms of a microscopic molecular type model, the fluid is treated as a viscous continuum by the use of hydrodynamic equations. Thus, the hydrodynamic boundary conditions couple the dynamics of the fluid and the polymer. By considering the full hydrodynamics of the polymer solution, we are able to describe polymer dynamics and transport properties at finite concentrations, ranging from low to very concentrated ones. This represents a substantial generalization over previous theories which were hitherto restricted to the infinite dilution limit. Our focus on the instantaneous microscopic chain dynamics represents another advancement over other theories as it provides a more detailed understanding of microscopic mechanisms for various observed processes.

8.2 Some basic facts about polymers

A polymer molecule is a repeated series of monomer units which are sequentially bonded to each other. The monomers are labelled 1, 2, . . . , n from one end of the chain to the other. Typical systems can have n ranging up to, say, 10^4. There is an enormous variety of different types of polymers, but it is found that their physical properties can often be adequately represented by just including this chain structure and by endowing the monomer units with simple properties such as size, Stokes friction coefficient, polarizabilities, etc.[1–6] Thus, given these unit characteristics, we seek to describe the properties of polymer systems in terms of them, and a detailed discussion of the chemical nature of the polymer is unnecessary.

Some basic physical properties of polymers that are commonly studied involve light scattering cross-sections, polymer viscosity, and the osmotic pressure of polymer solutions.[1–5] These properties have important bearings on the size, shape, and structure of the individual chains as well as on the characterization of chain interactions. Other important properties involve the study of visco-elastic properties of polymer systems, their neutron scattering cross-sections, and other mechanical relaxation behaviour.[7] These investigations provide information concerning the dynamics of the polymer chain. Certain of the above properties have obvious technological relevance.

Equilibrium properties[1–6]

The equilibrium properties of the polymer systems are completely characterized, in principle, by specifying the interactions between the individual monomers. The positions of the monomers are denoted by $\mathbf{R}_i$, and we begin by consideration of a single polymer chain. The first monomer may be taken to be at the origin of the coordinates,

$$\mathbf{R}_1 = 0 \tag{8.2.1}$$

as factors of the volume, V, in partition functions, etc., are readily included as necessary. The monomer–monomer potential energy, $U(\{\mathbf{R}_i\})$, is decomposed into two qualitatively different parts. The first, u, formally, or explicitly, accounts for the connectivity of the chain, while $V(\{\mathbf{R}_i\})$ contains all other interactions,

$$U = \sum_{j=1}^{n-1} u_j(\mathbf{R}_{j+1} - \mathbf{R}_j) + V(\{\mathbf{R}_j\}) \tag{8.2.2}$$

Since the single polymer chain is only to be found in solution, equation (8.2.2) represents a potential of mean force. Hence, the interactions u and V may be temperature dependent.

All the equilibrium statistical properties of the chain are derivable from the distribution

$$G(\{\mathbf{R}_j\}) = \exp[-\beta U(\{\mathbf{R}_j\})] \tag{8.2.3}$$

where

$$\beta = (k_B T)^{-1}$$

k_B is Boltzmann's constant, and T is the absolute temperature. The partition function, Z, for the chain is obtained by integrating (8.2.3) over all possible positions of monomers 2, 3, . . . , n, while information concerning the size of the polymer chain is contained in the end-to-end vector distribution which gives the probability density that $\mathbf{R}_n$ is some value $\mathbf{R}$,

$$P(\mathbf{R}; n) = Z^{-1} \int G(\{\mathbf{R}_j\}\delta(\mathbf{R}_n - \mathbf{R}) \prod_{j=2}^{n} d\mathbf{R}_j \tag{8.2.4}$$

Since the essential nature of the polymer resides in the fact that the monomer segments are sequentially bonded, it is often convenient to introduce the bond vectors

$$\mathbf{r}_j = \mathbf{R}_j - \mathbf{R}_{j-1}. \tag{8.2.5}$$

The chain connectivity is then expressed in terms of the bond 'probabilities'

$$\tau(\mathbf{r}_j) \equiv \exp[-\beta u_j(\mathbf{r}_j)] \tag{8.2.6}$$

so V in (8.2.2) accounts for all interactions between nonbonded segments. Experimentally, it is found that dilute polymer solutions may possess a special temperature, the Θ-temperature, at which there is a cancellation between polymer–solvent interactions and polymer–polymer interactions between distant monomers *along the chain*, i.e. monomers i and j for $|i - j|$ large.[2] In this case V then involves interactions between next and possibly third nearest neighbouring segments. The resulting evaluation of the distribution function (8.2.4) can then be mathematically represented in terms of a p-fold Markov process with $p \ll n$, or equivalently as a one-dimensional Ising model with spin $p/2$, and therefore it can be exactly evaluated.[5] Because the polymer chains exist in real three-dimensional space, the explicit evaluation of relevant quantities is nontrivial, but much has been accomplished, and the subject is reviewed in Flory's monograph.[5]

Most systems do not involve Θ-conditions, so V must contain interactions between distant monomers along the chain. The simplest approximation takes V to be a sum of pair-wise additive potentials,

$$V = \frac{1}{2} \sum_{i \neq j} v(\mathbf{R}_i - \mathbf{R}_j) \tag{8.2.7}$$

by analogy with the equilibrium theory of ordinary fluids. Employing this analogy further, it is natural to first consider $v(\mathbf{R}_i - \mathbf{R}_j)$ to be a hard core repulsive potential. The equilibrium properties of the hard core fluid have been determined from computer calculations on 100–1000 particles, and various approximate methods, such as the Percus–Yevick and hyper-netted chain theories, often provide useful analytical results for this system. On the other hand, this hard sphere model of a polymer chain has not been amenable to computer solution for, say, 100–1000 monomers because of the essential complication introduced by the chain connectivity. Even when the chain is restricted to a regular lattice, for simplicity, the segments are sequentially added to each other, and the hard core interaction is represented as an exclusion to multiple occupancy of a lattice site.[8] This system has been termed the self-avoiding walk problem or the polymer excluded volume problem. Thus, the description of the system mathematically corresponds more to a dynamical problem involving ordinary particles, e.g. hard spheres, than to an equilibrium one. The monomer indices $1, 2, \ldots, n$ play the role of a time-like variable. Recent advances in the understanding of the polymer excluded volume problem have arisen from the use of self-consistent field formulations[9–12] and renormalization group methods[13–15] which stress the analogies between this problem and that of describing critical phenomena in magnets and fluids. This analogy should not be too surprising, since in the limit of large n, (8.2.4), with (8.2.6) and (8.2.7), becomes an infinite-dimensional Markov process!

Theoretical complexities involved with the excluded volume problem often make it necessary to consider polymer systems at or very near the Θ-temperature. When long polymers, n large, are considered, the short scale interactions become of minor importance as the structure of (8.2.4) is dictated by the central limit theorem,

$$P(\mathbf{R}; n) \longrightarrow \left(\frac{3}{2\pi l^2 n}\right)^{3/2} \exp\left(\frac{-3\mathbf{R}^2}{2nl^2}\right) \qquad n \to \infty \tag{8.2.8}$$

where the mean-square end-to-end vector distance is

$$\begin{aligned} \langle (\mathbf{R}_n - \mathbf{R}_1)^2 \rangle &= \int d\mathbf{R}\, \mathbf{R}^2 P(\mathbf{R}; n) \\ &= \langle \mathbf{R}^2 \rangle = nl^2 \end{aligned} \tag{8.2.9}$$

as appropriate to a random work with the Kuhn effective step length l. Equations (8.2.8) and (8.2.9) are valid only for large enough n, and certain polymers, like DNA, fail to satisfy this condition for experimentally available chain lengths. In these cases, models such as rigid rod chains and random flight chains with some stiffness can be introduced.[1,6] Flexible chains are defined as those for which (8.2.8) and (8.2.9) are obeyed. Provided the polymer physical properties under consideration do not depend in detail upon the chain properties over short distances (or weight heavily highly extended chains), the flexible chains may be replaced by an equivalent random flight chain which exactly reproduces (8.2.8) and (8.2.9). A popular model is the mathematically simplest one of a Gaussian equivalent chain where the bond probabilities

(8.2.6) are taken as

$$\tau_j(\mathbf{r}_j) = \left(\frac{3}{2\pi l^2}\right)^{3/2} \exp\left(\frac{-3\mathbf{r}_j^2}{2l^2}\right) \tag{8.2.10}$$

so the average bond length is l. Because (8.2.2) represents a potential of mean force, l may in fact depend on temperature, but the discussions below refer to it as a constant.

The Gaussian chain, obtained from (8.2.10) and (8.2.6), represents each bond as having a Gaussian distribution of bond lengths *with zero mean*, and this may appear to be somewhat unphysical in view of the nature of real polymer molecules which contain chemical bonds with essentially fixed bond lengths. The effective monomers are, in fact, really taken to be a collection of several actual monomers such that this effective Gaussian monomer displays the assumed distribution (8.2.10). Consequently, the model is only meaningful over distances comparable to or greater than l. Because effective links contain a number of actual monomers, their number n is somewhat at our disposal, and it is customary[1] to define

$$L = nl \tag{8.2.11}$$

as the chain length at maximum extension. When (8.2.11) is combined with the definition

$$\langle (\mathbf{R}_n - \mathbf{R}_1)^2 \rangle \equiv Ll \tag{8.2.12}$$

experimental measurements of $\langle \mathbf{R}^2 \rangle$ provide a unique definition of n and l. For a number of chain models involving fixed bond length and bond angles, but possibly hindered rotation about the chain backbone, expressions are available relating L, l and n to the true chain bond lengths, angles, etc.[1-6]

Alternatively, the freedom of choice of the number of effective monomers enables the use of a continuum limit for the chain, wherein (8.2.10) is replaced by

$$\tau_\epsilon(\mathbf{r}_j) = \left(\frac{3}{2\pi l\epsilon}\right)^{3/2} \exp\left(\frac{-3\mathbf{r}_j^2}{2l\epsilon}\right) \tag{8.2.13}$$

and the limits

$$n \to \infty \qquad \epsilon \to 0 \qquad n\epsilon = L \tag{8.2.14}$$

are applied as they leave (8.2.12) and the overall distribution $P(\mathbf{R}, n)$ invariant.[6] In this limit, integrals like those in (8.2.4) become Wiener integrals and enable the use of functional integrals. In this chapter, the discrete monomer notation is employed at the beginning as it is the most widely used in the polymer field. Later the straightforward transformation to the continuum limit is introduced as a matter of mathematical convenience. In summary, a mathematically simple, but physically realistic, model of the structure of a polymer chain is taken to be provided by the Gaussian chain. The parameters L and l are assumed to be given, and excluded volume interactions $v(\mathbf{R}_i - \mathbf{R}_j)$ are to be incorporated as necessary.

Polymer dynamics

Having introduced simple models for the structural properties of polymers, we now turn to a discussion of dynamical models. Given the underlying Gaussian model, the potential of mean force (8.2.2) becomes, after employing the definitions (8.2.6) and (8.2.10),

$$U = -3\left(\frac{k_B T}{2l^2}\right) \sum_{j=2}^{n} \mathbf{r}_j^2 + V(\{\mathbf{R}_j\}) \tag{8.2.15}$$

Thus, the approximate treatment of chain connectivity is represented in terms of a series of n

beads which are coupled with harmonic forces and force constants $3k_BT/l^2$. This is the origin of the Rouse bead-spring[16] model of polymer chains.

Using Newton's laws the dynamics of individual beads in solution is determined by

$$m\frac{\partial^2}{\partial t^2}\mathbf{R}_i(t)+\left(\frac{3k_BT}{l^2}\right)[2\mathbf{R}_i(t)-\mathbf{R}_{i-1}(t)-\mathbf{R}_{i+1}(t)]+\frac{\partial V}{\partial \mathbf{R}_i(t)}=\text{Forces from fluid} \tag{8.2.16}$$

where m is the mass of the monomer unit and the forces generated by the fluid are to be specified. Each monomer experiences a hydrodynamical friction force $-\zeta(\partial\mathbf{R}_i/\partial t)$ where ζ may be approximated from Stokes' law as

$$\zeta = 6\pi\eta_0 a. \tag{8.2.17}$$

Here η_0 is the viscosity of the pure fluid, and a is an effective hydrodynamical radius to make ζ agree with experiment. Part of the remaining fluid generated forces that are random in nature and drive the Brownian motion of the chain. The generalized Kirkwood diffusion equation[1,17–19] is obtained by postulating that the net effect of these random forces is to lead to a thermodynamic driving force

$$\mathbf{F}_B^j = -k_BT\,\nabla_{\mathbf{R}_j}\ln P(\{\mathbf{R}_j\},t) \tag{8.2.18}$$

where $P(\{\mathbf{R}_j\},t)$ is the time-dependent distribution function for the whole polymer chain (see Section 8.3.1). In this work, we depart from this conventional approach to explicitly include the microscopic origins of the random forces driving the Brownian motion of the chain.

For a variety of applications, it may be assumed that the mass of the chain is enormous, so inertial forces in (8.2.16) may be ignored. Consequently, for the Gaussian chain, equation (8.2.16) reduces to the Rouse model

$$\zeta\frac{\partial}{\partial t}\mathbf{R}_i(t)+(3k_BT/l^2)\sum_{j=1}^{n}A_{ij}\mathbf{R}_j=\text{Forces} \tag{8.2.19}$$

where the nearest neighbour matrix A_{ij} is

$$A_{ij} = 2\delta_{ij}-\delta_{i,j+1}-\delta_{i,j-1} \tag{8.2.20}$$

Although the use of the Gaussian model (8.2.10) for equilibrium properties can rigorously be justified, its use in (8.2.15) may appear to be somewhat *ad hoc* as this approximate potential energy corresponds to an effective one for a subset of the polymer's degrees of freedom. From the Zwanzig–Mori theory of generalized Langevin equations[20,21] this implies the existence of additional friction forces in (8.2.19) on the effective beads as well as memory terms in the chain dynamics of the form

$$\sum_j\int_0^t d\tau M_{ij}(t-\tau)\mathbf{R}_j(\tau) \tag{8.2.21}$$

because of these neglected degrees of freedom. Bixon[22] and Zwanzig[23] have theoretically considered the problem to show how the Rouse model (and the Zimm generalization[24] discussed below) naturally emerges for *arbitrary* bond potentials $u_j(\mathbf{r}_j)$ if the memory terms are ignored. They consider a hypothetical model of n connected units which have no internal degrees of freedom. This may, at first, appear to be in conflict with the structure of real polymers where the backbone atoms are generally attached to other atoms or groups thereof. Bixon and Zwanzig's analysis would still yield the same results in the more realistic case of units with additional degrees of freedom, a Rouse–Zimm model with memory terms which depend on the presence of these extra degrees of freedom. The consideration of this real nature of polymers lends impetus to the belief that the memory effects, arising from these extra degrees of freedom, and/or the difference between true interbead potentials and the harmonic bead spring potential, may lead to additional frictional forces on the Rouse beads.

An additional internal frictional force has been postulated on the basis of somewhat questionable arguments.[19,25–26] However, the resultant theory leads to an improved agreement with experiments of high frequency properties of polymers,[27,28] so the theory of 'internal viscosity' can be taken as an empirical model providing good agreement with experiment. This theory effectively (i.e., apart from a contribution including the fluid velocity) leads to the introduction of a term on the left in (8.2.19) which is of the form

$$\varphi\left(\frac{\partial \mathbf{R}_i}{\partial t}-\frac{\partial \mathbf{R}_{i-1}}{\partial t}\right)\equiv\varphi\sum_j A'_{ij}\frac{\partial}{\partial t}\mathbf{R}_j(t) \tag{8.2.22}$$

where φ is a constant. Equation (8.2.22) would imply a friction force on bead i due to the difference in velocity between bead i and its predecessor $i-1$, *despite the fact that our choice of bead numbering is totally arbitrary and bead i + 1 might just as well have been chosen as the predecessor of bead i.* A symmetry breaking force like (8.2.22) would normally not be considered if it did not have the weight of experimental verification. Recent work[29] indicates that a form similar to (8.2.22) results from (8.2.21) within a long wavelength Markovian approximation from a memory function which correctly treats beads $i-1$ and $i+1$ symmetrically relative to bead i. Thus, if desired, we may append (8.2.19) with some phenomenological internal friction forces like (8.2.22), provided the forces in (8.2.19) also contain random components as required to satisfy the fluctuation–dissipation theorem (see Section 8.6).

The polymers are still in solution, and some of the forces in (8.2.19) arise from the properties of the solvent. Furthermore, in order to discuss transport properties like the viscosity of the polymer solutions, it is desirable to consider the hydrodynamics of the polymer solutions. In simplest terms the fluid hydrodynamics leads to forces of hydrodynamical origin in (8.2.19). When the fluid flows past a given bead i, the bead alters the velocity field of a fluid. But then this perturbed velocity field is incident on bead j, leading to an effective hydrodynamical force between beads i and j which is discussed more fully in Section 8.3. Suffice it to say that in the steady state limit these hydrodynamic forces fall off as $|\mathbf{R}_i-\mathbf{R}_j|^{-1}$ and, therefore, have profound effects on the polymer dynamics. When these hydrodynamic forces are incorporated into (8.2.19), the resultant theory of polymer dynamics is the Rouse–Zimm model, while the use of (8.2.18) and an equation of continuity for the polymer distribution $P(\{\mathbf{R}_j\}, t)$ leads to the diffusion equation method which can, in principle, include additional forces from $V(\{\mathbf{R}_j\})$.

Difficulties with previous theories

The Rouse–Zimm and diffusion equation methods have been very successful for polymer solutions at infinite dilution, low frequencies, and vanishing shear rate. As noted above, the phenomenological use of an internal polymer friction extends the theory's success to higher frequencies. However, at finite concentrations these traditional theories encounter serious difficulties.[30–33] To illustrate these difficulties, it is useful to recall a few facts about the case of infinite dilution. Intrinsic viscosities, $[\eta]$, of polymer solutions are defined by

$$[\eta]=\lim_{c\to 0}(\eta-\eta_0)/c\eta_0 \tag{8.2.23}$$

where η is the viscosity of the polymer solution and c is the polymer concentration in gm/cm^3.

Under Θ-conditions, if hydrodynamic interactions are ignored, the Rouse model describes each bead of the chain as incoherently (i.e. uncorrelated) scattering the fluid. Thus, the disturbance of the fluid is proportional to the number of beads, n, and consequently

$$[\eta] \propto M, \qquad \text{Rouse,} \tag{8.2.24}$$

where M is the molecular weight of the chain,

$$M = LM_A/l \tag{8.2.25}$$

with M_A the unit molecular weight. The long range nature of the hydrodynamic interactions between beads, varying as $|\mathbf{R}_i - \mathbf{R}_j|^{-1}$, has a profound effect on $[\eta]$ when the Rouse–Zimm model is considered. As is discussed in more detail in Section 8.5, the scattering of the fluid by individual beads becomes correlated because of the correlations between the positions of the beads implicit in (8.2.3) even under equilibrium conditions. A solution of this problem leads to

$$[\eta] \propto M^{1/2} \qquad \text{Rouse–Zimm} \tag{8.2.26}$$

and the chain can be thought to behave as an impenetrable sphere of radius proportional to $(Ll)^{1/2}$.

At finite, but still low, concentrations the solution viscosity is empirically found to develop a concentration dependence,

$$\eta = \eta_0(1 + [\eta]c + k_H[\eta]^2 c^2 + \ldots) \tag{8.2.27}$$

thereby defining the dimensionless Huggins coefficient k_H. Just as theories of $[\eta]$ require consideration of a typical chain, those for k_H focus upon a pair of chains in analogy with the treatment of second virial coefficients. Because of the long ranged nature of the hydrodynamic interactions, the average stress tensor acting on a chain, due to the remaining chains, is proportional to r^{-3} where r is the distance from the central chain to a typical partner.[1] This leads to some improper integrals for k_H, depending on the surface of the system, so the theories are of questionable validity.[30–32] Saito[30] has suggested that these improper integrals result from the use of the Stokes approximation to hydrodynamics, implying the need for a nonlinear theory to remedy the deficiencies. However, the fact that (8.2.27) is observed for vanishing shear rates implies that a linear (Stokes) approximation should suffice.

Peterson and Fixman[33] have argued, on the other hand, that the situation is somewhat analogous to that encountered in the theory of dielectrics. They have therefore evaluated both the average stress tensor and the average velocity gradient and have shown that the viscosity, as obtained from the ratio, does not involve improper integrals and is not dependent on the surface. Because of the complicated hydrodynamics involved in the computations, Peterson and Fixman only applied the theory to models of hard and of penetrable spheres. Thus, the standard methods provided no completely satisfactory theory for k_H for chainlike molecules of the random coil, Gaussian chain, variety. Considering the fact that the measurement of intrinsic viscosities, $[\eta]$, as well as a number of infinite dilution properties of polymer solutions, involve extrapolations to zero concentration, extensive data exist for the slopes of these plots. It is, hence, of great importance to provide adequate theories of these slopes, so this experimental data can also be utilized along with the zero concentration intercepts, to provide additional information concerning the properties of the polymer chains.

The profound influence of the hydrodynamic interactions in converting (8.2.24) to (8.2.26) implies that the effects of hydrodynamic interactions should continue to increase with polymer concentration because of the added interchain hydrodynamic interactions like those yielding k_H. However, at high concentrations it is experimentally[34] found that a Rouse model provides a better description of the polymer dynamics than does the Rouse–Zimm model, even for chains which require the Rouse–Zimm model for the infinite dilution $[\eta]$. A resolution of this apparent paradox requires the consideration of the full hydrodynamics of the polymer solution in conjunction with the dynamics of the chains in solution. The formulation of a satisfactory general theory to explain these observed phenomena and to enable further generalizations represents the basic subject matter of this chapter.[35–40]

Before considering the solution hydrodynamics in Section 8.3, the problem of chain entanglements should be noted. Polymer chains begin to become entangled when they become sufficiently long and when the concentration is high enough. These temporary entanglements introduce measurable effects upon the solution viscosity, yielding a behaviour of the form[34]

$$\eta \propto M^{3.4} \qquad \text{entangled} \qquad (8.2.28)$$

in marked contrast to (8.2.24) and (8.2.25). Experiments on polymer melts, pure liquid polymers, indicate that an abrupt transition from Rouse-like behaviour to (8.2.27) occurs at a molecular weight M^* of about 300–700 actual monomer units, depending on the particular polymer.[34] In polymer solutions at high enough concentrations, c, the transition to (8.2.28) occurs at a molecular weight $M_c \propto M^*/c$. It is, therefore, meaningful to first investigate the theory of polymer viscosity for chain molecular weights below M_c, as the complicated problem of entanglements can be ignored, and because the non-entangled regime is pertinent to many experiments.

If, for argument's sake, an entanglement is considered to involve a temporary encounter between a pair of chains, during this encounter the chains behave somewhat as a single polymer chain, a star polymer. Polymer entanglements thereby introduce temporary correlations between chains. Hydrodynamic interactions, between chains which were uncorrelated before entanglement, cause them to become correlated during the encounter. Also, the entanglement hinders the otherwise free motion of the individual chains. Combining these two effects provides a qualitative explanation of the dramatic change represented by (8.2.28). The neglect of entanglements in the remainder of this chapter therefore enables the mathematical simplifications generated from the neglect of temporary correlations between chains. The full hydrodynamic theory of polymer solutions does, however, enable the derivation of effective dynamical equations for a pair of correlated chains, and this should enable the development of a systematic theory of polymer entanglements.

8.3 Hydrodynamics of polymer solutions and chain dynamics

In the absence of polymers the pure fluid is taken to obey the linearized Navier–Stokes equation when the shear rate is low enough. Letting $\mathbf{u}(\mathbf{r}, t)$ denote the fluid velocity at the point $\mathbf{r}$ at time t, ρ_0 the fluid mass density and $P(\mathbf{r}, t)$ the pressure, this equation is

$$\rho_0 \frac{\partial}{\partial t}\mathbf{u}(\mathbf{r}, t) - \eta_0 \nabla^2 \mathbf{u}(\mathbf{r}, t) + \boldsymbol{\nabla} P(\mathbf{r}, t) = \mathbf{F}(\mathbf{r}, t) \qquad (8.3.1)$$

where $\mathbf{F}(\mathbf{r}, t)$ describes the external forces driving the fluid flow. For an incompressible fluid the pressure $P(\mathbf{r}, t)$ is determined by applying the constraint

$$\boldsymbol{\nabla} \cdot \mathbf{u}(\mathbf{r}, t) = 0. \qquad (8.3.2)$$

When polymers are present, bead i of the α-th chain exerts a force of $\boldsymbol{\sigma}_{\alpha i}(t)$ on the fluid at time t. Given that bead αi is at the position $\mathbf{R}_{\alpha i}(t)$, the force $\boldsymbol{\sigma}_{\alpha i}(t)$ represents a point source at the point $\mathbf{r} = \mathbf{R}_{\alpha i}(t)$ at time t. Hence the collection of all polymer chains yields a force density,

$$\sum_{\alpha=1}^{N} \sum_{i=1}^{n} \delta[\mathbf{r} - \mathbf{R}_{\alpha i}(t)]\boldsymbol{\sigma}_{\alpha i}(t) \qquad (8.3.3)$$

acting on the fluid at $\mathbf{r}$ and t. It is convenient to introduce the operator

$$\phi_{\alpha i}(\mathbf{r}, t) = \delta[\mathbf{r} - \mathbf{R}_{\alpha i}(t)] \tag{8.3.4}$$

which transforms between polymer, αi, and fluid, $\mathbf{r}$, spaces, so upon introduction of the polymers, equation (8.3.1) is transformed to

$$\rho_0 \frac{\partial}{\partial t}\mathbf{u}(\mathbf{r}, t) - \eta_0 \nabla^2 \mathbf{u}(\mathbf{r}, t) + \boldsymbol{\nabla} P(\mathbf{r}, t) = \sum_{\alpha=1}^{N} \sum_{i=1}^{n} \phi_{\alpha i}(\mathbf{r}, t)\boldsymbol{\sigma}_{\alpha i}(t) + \mathbf{F}(\mathbf{r}, t) \tag{8.3.5}$$

Newton's laws imply that the solvent exerts the force $-\boldsymbol{\sigma}_{\alpha i}(t)$ on the polymer bead at $R_{\alpha i}(t)$. Hence, this hydrodynamic force must be included in the dynamical equations of Section 8.2. For generality, the dynamical equations for the polymer in solution are then written as

$$\Delta_{\alpha i}\mathbf{R}_{\alpha i} = -\boldsymbol{\sigma}_{\alpha i}(t) + \mathbf{f}_{\alpha i}(t) \tag{8.3.6}$$

where $\Delta_{\alpha i}$ is a linear operator which is independent of $\{\mathbf{R}_{\beta j}\}$, e.g. for the bead-spring model with internal friction it is

$$\Delta_{\alpha i}\mathbf{R}_{\alpha i} \longrightarrow \sum_{j} \left[\left(\frac{3k_B T}{l^2}\right) A_{ij} + \varphi(\delta_{ij} - \delta_{j,i-1}) \frac{\partial}{\partial t} \right] \mathbf{R}_{\alpha j} \tag{8.3.7}$$

Chain stiffness[6,41] would require the introduction of terms involving $j = i \pm 2$ inside the brackets in (8.3.7). The forces $\mathbf{f}_{\alpha i}(t)$ include all external forces $\mathbf{f}_{\alpha i}^{\text{ext}}(t)$, all random forces $\mathbf{f}_{\alpha i}^{*}(t)$ necessary to produce the internal friction, and all remaining monomer–monomer interactions,

$$\mathbf{f}_{\alpha i} = \mathbf{f}_{\alpha i}^{\text{ext}} + \mathbf{f}_{\alpha i}^{*} + \mathbf{f}_{\alpha i}^{\text{int}} \tag{8.3.8}$$

Introducing the potential energy as in (8.2.7) and representing v as a short range repulsion, for simplicity, we get

$$\mathbf{f}_{\alpha i}^{\text{int}}(t) = -\boldsymbol{\nabla}_{\mathbf{R}_{\alpha i}(t)} \sum_{\beta j} v\delta[\mathbf{R}_{\alpha i}(t) - \mathbf{R}_{\beta j}(t)] \tag{8.3.9}$$

with v the monomer–monomer excluded volume.

Equations (8.3.5), (8.3.2) and (8.3.6) represent a set of coupled equations for the polymer and the fluid which still require a hydrodynamical boundary condition to enable the determination of the forces $\boldsymbol{\sigma}_{\alpha i}(t)$. No-slip boundary conditions are mathematically the simplest, and they imply that

$$\dot{\mathbf{R}}_{\alpha i}(t) \equiv \frac{\partial}{\partial t}\mathbf{R}_{\alpha i}(t) = \mathbf{u}[\mathbf{R}_{\alpha i}(t), t] \equiv \int d\mathbf{r}\,\phi_{\alpha i}(\mathbf{r}, t)\mathbf{u}(\mathbf{r}, t) \tag{8.3.10}$$

Variational principles were used previously[35] to derive the set of equations (8.3.5), (8.3.2), (8.3.6) and (8.3.10) which are heuristically introduced here. These general variational principles are very useful in cases with constraints where heuristic arguments encounter difficulties.

We proceed to obtain the solutions to equations (8.3.5) and (8.3.6). First, we use condition (8.3.2) to eliminate the pressure. Taking the divergence of (8.3.5) and using (8.3.2) yields

$$\nabla^2 P(\mathbf{r}, t) = \sum_{\alpha i} \boldsymbol{\nabla}\phi_{\alpha i}(\mathbf{r}, t) \cdot \boldsymbol{\sigma}_{\alpha i}(t) + \boldsymbol{\nabla} \cdot \mathbf{F}(\mathbf{r}, t) \tag{8.3.11}$$

which is readily solved to give

$$P(\mathbf{r}, t) = (4\pi)^{-1} \int d\mathbf{r}' \, |\mathbf{r} - \mathbf{r}'|^{-1} \left[\sum_{\alpha i} \nabla_{\mathbf{r}'} \phi_{\alpha i}(\mathbf{r}', t) \cdot \boldsymbol{\sigma}_{\alpha i}(t) + \nabla_{\mathbf{r}'} \cdot \mathbf{F}(\mathbf{r}', t) \right] \tag{8.3.12}$$

Substitution of (8.3.12) into (8.3.5) gives

$$\rho_0 \frac{\partial}{\partial t} \mathbf{u}(\mathbf{r}, t) - \eta_0 \nabla^2 \mathbf{u}(\mathbf{r}, t) = \sum_{\alpha i} \phi_{\alpha i}(\mathbf{r}, t) \boldsymbol{\sigma}_{\alpha i}(t) + \mathbf{F}(\mathbf{r}, t)$$

$$-(4\pi)^{-1} \nabla_{\mathbf{r}} \int d\mathbf{r}' \, |\mathbf{r} - \mathbf{r}'|^{-1} \nabla_{\mathbf{r}'} \cdot \left[\sum_{\alpha i} \phi_{\alpha i}(\mathbf{r}', t) \boldsymbol{\sigma}_{\alpha i}(t) + F(\mathbf{r}', t) \right] \tag{8.3.13}$$

Equation (8.3.13) is readily solved by introducing space–time Fourier transforms with $\mathbf{k}$ and ω the variables conjugate to $\mathbf{r}$ and t, respectively. (The results are well known.) The formal solution for $\mathbf{u}(\mathbf{r}, t)$ can be represented in terms of the fundamental solution, $\mathbf{G}$, to (8.3.13) when the forces represent a single point source, $\delta(\mathbf{r} - \mathbf{r}')\delta(t - t')$. $\mathbf{G}$ can be shown[35] to satisfy the equation

$$\rho_0 \frac{\partial}{\partial t} G_{ij}(\mathbf{r}\mathbf{r}'; tt') - \eta_0 \nabla_{\mathbf{r}}^2 G_{ij}(\mathbf{r}\mathbf{r}'; tt')$$

$$= [\delta_{ij}\delta(\mathbf{r} - \mathbf{r}') - (\nabla_{\mathbf{r}})_i (4\pi \, |\mathbf{r} - \mathbf{r}'|)^{-1} (\nabla_{\mathbf{r}'})_j] \, \delta(t - t') \tag{8.3.14}$$

or in terms of Fourier transforms

$$(i\rho_0 \omega + \eta_0 k^2) G_{ij}(k, \omega) = [\delta_{ij} - (k_i k_j / k^2)] \tag{8.3.15}$$

where i and j represent the Cartesian components x, y, or z, so $\mathbf{G}$ is given by

$$\mathbf{G}(\mathbf{r} - \mathbf{r}'; t - t') = \int_{-\infty}^{\infty} \frac{d\omega}{2\pi} \exp(i\omega t) \int \frac{d^3 k}{(2\pi)^3} \exp[i\mathbf{k} \cdot (\mathbf{r} - \mathbf{r}')] \frac{\mathbf{1} - k^{-2}\mathbf{k}\mathbf{k}}{i\rho_0 \omega + \eta_0 k^2} \tag{8.3.16}$$

with $\mathbf{1}$ the unit tensor. Using (8.3.14)–(8.3.16) the formal solution for $\mathbf{u}(\mathbf{r}, t)$ is

$$\mathbf{u}(\mathbf{r}, t) = \sum_{\alpha i} \int d\mathbf{r}' \, dt' \, \mathbf{G}(\mathbf{r}\mathbf{r}'; tt') \phi_{\alpha i}(\mathbf{r}', t') \cdot \boldsymbol{\sigma}_{\alpha i}(t')$$

$$+ \int d\mathbf{r}' \, dt' \, \mathbf{G}(\mathbf{r}\mathbf{r}'; tt') \cdot \mathbf{F}(\mathbf{r}', t'). \tag{8.3.17}$$

Equation (8.3.17) represents $\mathbf{u}$ in terms of $\boldsymbol{\sigma}$, and the second step in the derivation involves the use of the constraint (8.3.10) to eliminate $\boldsymbol{\sigma}$ from the polymer and fluid equations. Use of (8.3.17) and (8.3.10) gives

$$\dot{\mathbf{R}}_{\alpha i}(t) = \int d\mathbf{r} \, d\mathbf{r}' \, dt' \phi_{\alpha i}(\mathbf{r}, t) \mathbf{G}(\mathbf{r}\mathbf{r}'; tt') \cdot \mathbf{F}(\mathbf{r}', t')$$

$$+ \sum_{\beta j} \int dt' \, \mathbf{K}_{\alpha i, \beta j}(t, t') \cdot \boldsymbol{\sigma}_{\beta j}(t') \tag{8.3.18}$$

where

$$\mathbf{K}_{\alpha i, \beta j}(t, t') \equiv \mathbf{G}[\mathbf{R}_{\alpha i}(t) - \mathbf{R}_{\beta j}(t'); t - t'] + \zeta^{-1} \mathbf{1} \delta_{\alpha\beta} \delta_{ij} \delta(t - t'). \tag{8.3.19}$$

ζ is the usual bead friction coefficient which has been introduced to account for a deficiency in the analysis. Each bead is of finite extension and must experience a Stokes friction (8.2.17). However, the dynamical model (8.3.6) represents the chain as one with zero thickness, so (8.2.17) would vanish as $a \rightarrow 0$. We have two possible avenues of approach: either employ polymer models of chains with finite thickness[42-44] or phenomenologically append ζ into (8.3.19) as is, in effect, customary.

Reduction to Kirkwood–Riseman theory and to diffusion model

Equation (8.3.18) contains information involving only the fluid dynamics (8.3.5) and the hydrodynamic coupling (8.3.10) between fluid and chains; the polymer dynamics is entirely absent. It is instructive to consider (8.3.18) in the steady state ($\omega \to 0$) limit where the steady state limit of **G** becomes the ordinary Oseen tensor

$$\mathsf{K}_{\alpha i,\beta j} \xrightarrow{\omega \to 0} (8\pi\eta_0 \,|\mathbf{R}_{\alpha i} - \mathbf{R}_{\beta j}|^{-1})[\mathbf{1} + (\mathbf{R}_{\alpha i} - \mathbf{R}_{\beta j})(\mathbf{R}_{\alpha i} - \mathbf{R}_{\beta j})\,|\mathbf{R}_{\alpha i} - \mathbf{R}_{\beta j}|^{2}]$$

$$\times (1 - \delta_{\alpha\beta}\delta_{ij}) + \delta_{\alpha\beta}\delta_{ij}\zeta^{-1}\,\mathbf{1} \equiv \mathsf{T}(\mathbf{R}_{\alpha i} - \mathbf{R}_{\beta j}) + \delta_{\alpha\beta}\delta_{ij}\zeta^{-1}\,\mathbf{1}. \tag{8.3.20}$$

Consider a single chain, for simplicity, and let the steady state velocity of bead αi be $\mathbf{u}_i$ since the index α can be dropped. The steady state limit of (8.3.18) is

$$u_i = \int d\mathbf{r}\, d\mathbf{r}' \phi_i(\mathbf{r}) \mathsf{T}(\mathbf{r} - \mathbf{r}') \cdot \mathbf{F}(\mathbf{r}') + \sum_j \mathsf{K}_{ij} \cdot \sigma_j \tag{8.3.21}$$

where

$$\phi_i(\mathbf{r}) = \delta(\mathbf{r} - \mathbf{R}_i). \tag{8.3.21a}$$

If the polymers were absent, the unperturbed fluid velocity at the position of bead i, $\mathbf{v}_i^0$, would be given by the first term on the right-hand side in (8.3.21), so we get

$$\mathbf{u}_i = \mathbf{v}_i^0 + \zeta^{-1}\boldsymbol{\sigma}_i + \sum_{j \neq i} \mathsf{T}_{ij} \cdot \boldsymbol{\sigma}_j \tag{8.3.22}$$

by using the second line in (8.3.20). If bead i were absent, the fluid velocity at the position of bead i, $\mathbf{v}_i$, would be obtained from (8.3.20) by setting $\mathbf{u}_i$ equal to $\mathbf{v}_i$ and taking $\boldsymbol{\sigma}_i$ to be zero,

$$\mathbf{v}_i = \mathbf{v}_i^0 + \mathbf{v}_i' \tag{8.3.23}$$

where v_i' is defined by

$$v_i' \equiv \sum_{j \neq i} \mathsf{T}_{ij} \cdot \boldsymbol{\sigma}_j. \tag{8.3.24}$$

Combining (8.3.22)–(8.3.24) yields the well known result

$$\boldsymbol{\sigma}_i = \zeta(\mathbf{u}_i - \mathbf{v}_i) \tag{8.3.25}$$

which is generally the starting point in traditional analyses.[1,16,19] Equations (8.3.23)–(8.3.25) form the basis of the Kirkwood–Riseman theory[45] of steady-state transport in polymer systems, so the original equation (8.3.18) is merely a dynamical generalization.

The diffusion equation method builds upon the Kirkwood–Riseman theory (8.2.23)–(8.3.25) and (8.3.20) to incorporate the polymer dynamics by recourse to concepts of irreversible thermodynamics. It is assumed that the average dynamics of the polymer chain can be obtained by equating the friction forces (8.3.25) to the sum of the nonrandom part of the forces (8.2.8), the entropic forces, formally accounting for chain connectivity, and the thermodynamic driving force. The force balance then reads

$$\zeta(\mathbf{u}_i - \mathbf{v}_i) = -\boldsymbol{\nabla}_{\mathbf{R}_i} U - k_{\mathrm{B}} T \boldsymbol{\nabla}_{\mathbf{R}_i} \ln P(\{\mathbf{R}_i\}, t) \tag{8.3.26}$$

where U is the full potential energy (8.2.2), as well as that due to external fields, which is given as (8.2.15) for the bead-spring model. A diffusion equation for the polymer chain results from the use of (8.3.26) and the Kirkwood–Riseman theory (8.3.23)–(8.3.25) along with the equation of continuity for the polymer distribution function,[18,19]

$$\frac{\partial}{\partial t} P(\{\mathbf{R}_j\}, t) + \sum_{i=1}^{n} \nabla_i \cdot [\mathbf{u}_i P(\{\mathbf{R}_j\}, t)] = 0 \tag{8.3.27}$$

which is just a statement of conservation of polymer. The full expression for the diffusion equation is then

$$\frac{\partial}{\partial t} P(\{\mathbf{R}_j\}, t) + \sum_{i=1}^{n} \nabla_i \cdot [\mathbf{v}_i^0 P(\{\mathbf{R}_j\}, t)]$$

$$= \sum_{i,j=1}^{n} \nabla_i \cdot \mathsf{D}_{ij} \cdot [\nabla_j P(\{\mathbf{R}_j\}, t) + P(\{\mathbf{R}_j\}, t) \nabla_j U / k_B T] \tag{8.3.28}$$

where the diffusion tensor is

$$\mathsf{D}_{ij} = (k_B T/\zeta)[\mathbf{1}\delta_{ij} + \zeta \mathsf{T}_{ij}(1 - \delta_{ij})]\,. \tag{8.3.29}$$

The diffustion equation theory is also based upon the equations of steady state hydrodynamics (8.3.21) for the fluid only at the position of the polymer chain. The full fluid dynamics (8.3.17) is ignored and the detailed microscopic chain dynamics (8.3.6) is incorporated only in some averaged sense. Zwanzig[18] has investigated the nature of this effective dynamics by deriving a Langevin equation which can be taken to be a precursor of the diffusion equation (8.3.28). His dynamical equations are

$$\frac{\partial}{\partial t} \mathbf{R}_j(t) = -(k_B T)^{-1} \sum_k \mathsf{D}_{jk} \cdot \nabla_{\mathbf{R}_k(t)} U + \mathbf{v}_i^0$$

$$- \sum_{mp} \Upsilon_{jm} \cdot \nabla_{\mathbf{R}_p(t)} \Upsilon_{pm} + \sum_p \Upsilon_p f_p(t) \tag{8.3.30}$$

where Υ_{jm} are given by

$$\mathsf{D}_{jk} \equiv \sum_m \Upsilon_{jm} \Upsilon_{km} \tag{8.3.31}$$

and $f_m(t)$ are Gaussian random variables with zero mean

$$\langle f_m(t) \rangle = 0 \qquad \text{all } m \tag{8.3.32}$$

and a variance of

$$\langle f_m(t) f_m(t') \rangle = 2\delta_{mn}(t - t'). \tag{8.3.33}$$

Upon introduction of the generalized friction tensor

$$(\zeta^{-1})_{jk} = (k_B T)^{-1} \mathsf{D}_{jk} \tag{8.3.34}$$

(8.3.30) is converted to

$$\sum_j \zeta_{kj} \cdot \left[\frac{\partial}{\partial t} \mathbf{R}_j(t) - \mathbf{v}_j^0 \right] = -\nabla_{\mathbf{R}_k(t)} U - \sum_{j,m,p} \zeta_{kj} \cdot \Upsilon_{jp} \Delta_{\mathbf{R}_p(t)} \cdot \Upsilon_{pm}$$

$$+ \sum_{j,p} \zeta_{kj} \cdot \Upsilon_{jp} f_p(t) \tag{8.3.35}$$

which is of the form of a typical Langevin equation when inertial forces are negligible. The terms in $\dot{\mathbf{R}}_i(t) - \mathbf{v}_j^0$ and $-\nabla_{\mathbf{R}_k(t)}U$ are ones that are anticipated to be present in a Langevin equation for a polymer chain, but the last two terms in (8.3.35) are somewhat strange, involving Υ_{jp}, the 'square root' of the diffusion tensor. It is not at all clear how terms of this form could enter into the actual microscopic dynamics of the chain.

The microscopic dynamics of polymers in solution

The validity of the assumptions implicit in the diffusion equation method and the meaning of terms in (3.3.35) can be assessed by consideration of the dynamical equations (8.3.6) and (8.3.18) that govern the instantaneous chain motion. The former is ignored in the diffusion equation method in favour of statistical and force balance hypotheses, while the latter is retained in the steady state limit. Thus, we now return to a consideration of the true chain dynamics, retaining all the chains and their interactions to enable a treatment of the full concentration dependence of the polymer dynamics.

Multiplying (8.3.6) for bead j of chain β by $\int dt' \Sigma_{\beta j} \mathbf{K}_{\alpha i, \beta j}(t, t') \cdot$ gives

$$\int dt' \sum_{\beta j} \mathbf{K}_{\alpha i, \beta j}(t, t')\Delta_{\beta j}(t') \cdot \mathbf{R}_{\beta j}(t')$$
$$= - \int dt' \sum_{\beta j} \mathbf{K}_{\alpha i, \beta j}(t, t') \cdot \boldsymbol{\sigma}_{\beta j}(t') + \int dt' \sum_{\beta j} \mathbf{K}_{\alpha i, \beta j}(t, t') \cdot \mathbf{f}_{\beta j}(t'). \tag{8.3.36}$$

Comparing (8.3.36) with (8.3.18) gives a pair of equations which readily enable the elimination of the terms involving $\boldsymbol{\sigma}_{\beta j}(t')$ to yield

$$\dot{\mathbf{R}}_{\alpha i}(t) + \sum_{\beta j} \int dt' \mathbf{K}_{\alpha i, \beta j}(t, t')\Delta_{\beta j}(t')\mathbf{R}_{\beta j}(t')$$
$$= \int dt' \, d\mathbf{r} \, d\mathbf{r}' \, \phi_{\alpha i}(\mathbf{r}, t)\mathbf{G}(\mathbf{r} - \mathbf{r}'; t - t') \cdot \mathbf{F}(\mathbf{r}', t')$$
$$+ \sum_{\beta j} \int dt' \mathbf{K}_{\alpha i, \beta j}(t, t') \cdot \mathbf{f}_{\beta j}(t'). \tag{8.3.37}$$

For comparison with equation (8.3.35) we now specialize (8.3.37) to a single chain. The steady-state limits of $\mathbf{K}$ and $\mathbf{G}$ are also used for simplicity, whereupon $\boldsymbol{\zeta}^{-1}$ of (8.3.34) is found to be the steady state limit of $\mathbf{K}$. If internal friction forces are ignored and the equation is multiplied through by $\boldsymbol{\zeta}$, the result is found to be

$$\sum_j \boldsymbol{\zeta}_{kj} \cdot [\dot{\mathbf{R}}_j - \mathbf{v}_j^0] = - \nabla_{\mathbf{R}_k} U + \sum_j \boldsymbol{\zeta}_{kj} \cdot \mathbf{v}_j^{0,*} \tag{8.3.38}$$

where all nonhydrodynamic forces on the polymer in (8.3.37) have been re-expressed in terms of the potential energy U. $\mathbf{v}_j^{0,*}$ is the random thermal velocity fluctuation that is present in the pure fluid at the position of the j-th bead. Since U is taken here to represent the total potential energy of the hypothetical backbone-only chain, there is no residual random force $\mathbf{f}_{\alpha i}^*$ of intrachain origin for the simple limit (8.3.38). Comparison of (8.3.35) and (8.3.38) displays the similarity of the terms involving $\dot{\mathbf{R}}_j - \mathbf{v}_j^0$ and $-\nabla_{\mathbf{R}_k} U$. By the fluctuation–dissipation theorem[46] the random force $\Sigma_j \boldsymbol{\zeta}_{kj} \cdot \mathbf{v}_j^{0,*}$ in (8.3.8) must have an equilibrium autocorrelation function that is proportional to the friction coefficient $\boldsymbol{\zeta}_{kj}$ appearing on the left-hand side of that equation. As demonstrated below,[35] this implies that

$$\langle \mathbf{v}_j^{0,*}(t) \cdot \mathbf{v}_k^{0,*}(t') \rangle = 6k_B T(\boldsymbol{\zeta}^{-1})_{jk}\delta(t - t') \tag{8.3.39}$$

which is consistent with

$$\langle \mathbf{F}^*(\mathbf{r}, t) \cdot \mathbf{F}^*(\mathbf{r}', t') \rangle = 6k_B T \delta(\mathbf{r} - \mathbf{r}') \delta(t - t'). \tag{8.3.40}$$

Equation (8.3.40) shows that $\mathbf{v}_j^{0,*}$ are thermal velocity fluctuations as implicitly assumed above. These randomly occurring fluctuations at the position of, say, bead j induce fluctuations in the velocity of that bead resulting in the Brownian motion of the polymer chain. Since a continuum model has been employed for the fluid, the random forces, arising from collisions between fluid molecules and the polymer, are replaced by the thermal velocity fluctuations in the fluid that are transmitted to the chain by the hydrodynamic boundary conditions.

The random terms in (8.3.38) then correspond to the random terms in (8.3.35), but the autocorrelation function (8.3.39) makes the introduction of the strange $\Upsilon_{jp} f_p$ in (8.3.30) unnecessary. The remaining term, involving $\nabla_{\mathbf{R}_p} \Upsilon_{pm}$, in (8.3.30) and (8.3.35) either vanishes because of the incompressibility condition

$$\nabla_{\mathbf{r}} \cdot \mathsf{T}(\mathbf{r} - \mathbf{r}') = 0 \tag{8.3.41}$$

or it represents an error due to the approximations of the diffusion equation method. The full equations (8.3.37), however, contain all the exact chain dynamics, do not involve steady state limits of the hydrodynamics, allow for internal polymer friction, and include the full concentration dependence. Elsewhere[35] it is shown that the dynamical Oseen tensor K must be used instead of the static limit for frequencies above 10^9 Hz. Thus, in the infrared, optical, etc., regions the static hydrodynamical equations can no longer be used. Chang and Mazo[47] have shown the importance of using the dynamical K in their study of linewidths in the vibration spectra of polymers. Hence, because of the full generality of (8.3.37), it is now analysed in more detail.

8.4 Concentration-dependent effective dynamics for a chain

Equation (8.3.37) is far from simple. Hydrodynamic interactions between different chains, the $\mathsf{K}_{\alpha i, \beta j}$, couple their motions, and direct interchain forces occur at short distances. In the following we derive exact effective dynamical equations for the motion of a single chain in the nonhydrodynamic force field of all other chains where the intrachain hydrodynamics interactions are governed by the exact hydrodynamics of the full polymer solution.[40] This exact hydrodynamics is studied in Section 8.5 to provide self-consistent equations describing polymer dynamics in solutions at finite concentrations. It should be noted that these effective polymer equations are the ones whose existence is implicitly invoked in the introduction of models for chain motion in concentrated solutions, such as the repetation model.[48–50] Hence, the exact equations obtained in this section should aid in the development of a systematic theory of entanglement effects on chain dynamics.

In order to convert (8.3.37) into a more useful form, it is convenient to introduce the auxiliary quantity $\mathcal{G}_{\alpha,ij}$ associated with the dynamics of an isolated chain α,

$$\sum_{k=1}^{n} \int dt' \left(\frac{\partial}{\partial t} \delta_{ik} \delta(t - t') \mathbf{1} + \mathsf{K}_{\alpha i, \alpha k}(t, t') \Delta_{\alpha k}(t') \right) \cdot \mathcal{G}_{\alpha, kj}(t', t'')$$
$$= \mathbf{1} \delta_{ij} \delta(t - t''). \tag{8.4.1}$$

Note that the dependence of $\mathsf{K}_{\alpha i, \alpha k}$ of (8.3.19) upon the $\{\mathbf{R}_{\alpha j}(t)\}$ stresses the formal nature of

(8.4.1). Using (8.4.1), equation (8.3.37) can be converted to the equivalent 'integral' equation

$$\mathbf{R}_{\alpha i}(t) = \sum_k \int \mathrm{d}t' \, \mathscr{G}_{\alpha,ik} \cdot \Big[\int \mathrm{d}\mathbf{r}\, \mathrm{d}\mathbf{r}'\, \mathrm{d}t''\phi_{\alpha k}(\mathbf{r}, t')\mathbf{G}(\mathbf{r}\,\mathbf{r}'; t't'') \cdot \mathbf{F}(\mathbf{r}', t'')$$
$$+ \sum_{\beta j} \int \mathrm{d}t'' \, \mathsf{K}_{\alpha k, \beta j}(t', t'') \cdot \mathbf{f}_{\beta j}(t'')$$
$$- \sum_{\beta \neq \alpha, j} \int \mathrm{d}t'' \, \mathsf{K}_{\alpha k, \beta j}(t', t'')\Delta_{\beta j}(t'') \cdot \mathbf{R}_{\beta j}(t'') \Big]. \tag{8.4.2}$$

We may iterate equation (8.4.2) once to yield

$$\mathbf{R}_{\alpha i}(t) = \sum_k \int \mathrm{d}t'\, \mathrm{d}\mathbf{r}\, \mathrm{d}\mathbf{r}'\, \mathrm{d}t''\, \mathscr{G}_{\alpha, ik}(t, t')\phi_{\alpha k}(\mathbf{r}, t') \cdot \mathbf{G}(\mathbf{r}\,\mathbf{r}'; t\,t') \cdot \mathbf{F}(\mathbf{r}'\,t')$$
$$- \sum_{\beta \neq \alpha} \sum_{j, k, m} \int \mathrm{d}t'\, \mathrm{d}t''\, \mathrm{d}t'''\, \mathrm{d}t^{\mathrm{iv}}\, \mathrm{d}\mathbf{r}\, \mathrm{d}\mathbf{r}'\, \mathscr{G}_{\alpha, ik}(t, t') \cdot \mathsf{K}_{\alpha k, \beta j}(t', t'')\Delta_{\beta j}(t'')$$
$$\times \cdot \mathscr{G}_{\beta, jm}(t'', t''')\phi_{\beta m}(\mathbf{r}, t''') \cdot \mathbf{G}(\mathbf{r}\mathbf{r}'; t'''\, t^{\mathrm{iv}}) \cdot \mathbf{F}(\mathbf{r}', t^{\mathrm{iv}})$$
$$+ \sum_{\beta, k, j} \int \mathrm{d}t'\, \mathrm{d}t''\, \mathscr{G}_{\alpha, ik}(t, t') \cdot \mathsf{K}_{\alpha k, \beta j}(t', t'') \cdot \mathbf{f}_{\beta j}(t'')$$
$$- \sum_{\gamma, \beta \neq \alpha} \sum_{j, k, m, p} \int \mathrm{d}t'\, \mathrm{d}t''\, \mathrm{d}t'''\, \mathrm{d}^{\mathrm{iv}}\, \mathscr{G}_{\alpha, ik}(t, t') \cdot \mathsf{K}_{\alpha k, \beta j}(t', t'')\Delta_{\beta j}(t'')$$
$$\times \cdot \mathscr{G}_{\beta, jm}(t'', t''') \cdot \mathsf{K}_{\beta m\ \gamma p}(t''', t^{\mathrm{iv}}) \cdot \mathbf{f}_{\gamma p}(t^{\mathrm{iv}})$$
$$+ \sum_{\gamma \neq \beta\, \beta \neq \alpha} \sum_{j, k, m} \int \mathrm{d}t'\, \mathrm{d}t''\, \mathrm{d}t'''\, \mathrm{d}t^{\mathrm{iv}}\, \mathscr{G}_{\alpha, ik}(t, t') \cdot \mathsf{K}_{\alpha k, \beta j}(t', t'')\Delta_{\beta j}(t'')$$
$$\times \cdot \mathsf{K}_{\beta j, \gamma m}(t''', t^{\mathrm{iv}}) \cdot \Delta_{\gamma m}(t^{\mathrm{iv}})\mathbf{R}_{\gamma m}(t^{\mathrm{iv}}). \tag{8.4.3}$$

In order to compress otherwise unwieldy equations, it becomes necessary at this juncture to simplify equations substantially by resort to a symbolic supermatrix notation wherein (8.4.3) is rewritten as

$$\mathbf{R}_\alpha = \mathscr{G}_\alpha\phi_\alpha \cdot \mathbf{G} \cdot \mathbf{F} - \sum_{\beta \neq \alpha} \mathscr{G}_\alpha \cdot \mathsf{K}_{\alpha\beta}\Delta_\beta \cdot \mathscr{G}_\beta\phi_\beta \cdot \mathbf{G} \cdot \mathbf{F}$$
$$+ \sum_\beta \mathscr{G}_\alpha \cdot \mathsf{K}_{\alpha\beta} \cdot \mathbf{f}_\beta - \sum\sum_{\gamma, \beta \neq \alpha} \mathscr{G}_\alpha \cdot \mathsf{K}_{\alpha\beta}\Delta_\beta \cdot \mathscr{G}_\beta \cdot \mathsf{K}_{\beta\gamma} \cdot \mathbf{f}_\gamma$$
$$+ \sum_{\gamma \neq \beta} \sum_{\beta \neq \alpha} \mathscr{G}_\alpha \cdot \mathsf{K}_{\alpha\beta}\Delta_\beta \cdot \mathscr{G}_\beta \cdot \mathsf{K}_{\beta\gamma}\, \Delta_\gamma \cdot \mathbf{R}_\gamma \tag{8.4.4}$$

with all the integrations and summations in (8.4.3) implicit. First (8.3.19) is substituted into (8.4.4) and the $\zeta^{-1}\mathbf{1}$ parts are not explicitly displayed but are implicitly taken to be included in **G** and are only reintroduced at the end. Then the dynamical polymer T-matrices

$$\mathscr{T}_\alpha = \phi_\alpha\Delta_\alpha\, \mathscr{G}_\alpha\phi_\alpha$$
$$\equiv \sum_{i, j} \phi_{\alpha i}(\mathbf{r}, t)\Delta_{\alpha i}(t)\, \mathscr{G}_{\alpha, ij}(t, t')\, \phi_{\alpha j}(\mathbf{r}', t') \tag{8.4.5}$$

which are functions of $\mathbf{r}\mathbf{r}', tt'$ and implicitly dependent on the $\{\mathbf{R}_{\alpha j}(t)\}$ through (8.4.1), are substituted to yield the equations

$$\mathbf{R}_\alpha = \mathscr{G}_\alpha\phi_\alpha \cdot \Big[\mathbf{G} \cdot \mathbf{F} - \sum_{\beta \neq \alpha} \mathbf{G} \cdot \mathscr{T}_\beta \cdot \mathbf{G} \cdot \mathbf{F} + \sum_\beta \mathbf{G}\phi_\beta \cdot \mathbf{f}_\beta - \sum_\gamma \sum_{\beta \neq \alpha} \mathbf{G} \cdot \mathscr{T}_\beta \cdot \mathbf{G}\phi_\gamma \cdot \mathbf{f}_\gamma$$
$$+ \sum_{\beta \neq \alpha} \sum_{\gamma \neq \beta} \mathbf{G} \cdot \mathscr{T}_\beta \cdot \mathbf{G}\phi_\gamma\Delta_\gamma \cdot \mathbf{R}_\gamma \Big]. \tag{8.4.6}$$

Equation (8.4.6) can be iterated indefinitely to yield the 'solutions'

$$\begin{aligned}\mathbf{R}_\alpha = {} & \mathscr{G}_\alpha\phi_\alpha \cdot \mathbf{G}\cdot\mathbf{F} + \mathscr{G}\phi_\alpha\mathbf{G}\cdot \mathscr{T}^{(\alpha f)}\cdot\mathbf{G}\cdot\mathbf{F} \\ & + \sum_\beta \mathscr{G}_\alpha\phi_\alpha\cdot\mathbf{G}\cdot\phi_\beta\mathbf{f}_\beta \\ & + \sum_\beta \mathscr{G}_\alpha\phi_\alpha\cdot\mathbf{G}\cdot\mathscr{T}^{(\alpha f)}\cdot\mathbf{G}\phi_\beta\cdot\mathbf{f}_\beta \end{aligned} \tag{8.4.7}$$

where $\mathbf{G}\cdot\mathscr{T}^{(\alpha f)}\cdot\mathbf{G}\cdot\mathbf{F}$ describes the exact 'perturbed' fluid velocity field incident on chain α,

$$\begin{aligned}\mathscr{T}^{(\alpha f)} \equiv {} & -\sum_{\beta\neq\alpha}\mathscr{T}_\beta + \sum_{\beta\neq\alpha}\sum_{\gamma\neq\beta}\mathscr{T}_\beta\cdot\mathbf{G}\cdot\mathscr{T}_\gamma - \sum_{\beta\neq\alpha}\sum_{\gamma\neq\beta}\sum_{\delta\neq\gamma}\mathscr{T}_\beta\cdot\mathbf{G}\cdot\mathscr{T}_\gamma\cdot\mathbf{G}\cdot\mathscr{T}_\delta \\ & + \sum_{\beta\neq\alpha}\sum_{\gamma\neq\beta}\sum_{\delta\neq\gamma}\sum_{\epsilon\neq\delta}\mathscr{T}_\beta\cdot\mathbf{G}\cdot\mathscr{T}_\gamma\cdot\mathbf{G}\cdot\mathscr{T}_\delta\cdot\mathbf{G}\cdot\mathscr{T}_\epsilon + \ldots \end{aligned} \tag{8.4.8}$$

Notice that in the second to fourth terms in (8.4.8), the chain α may reappear in the summations over γ, δ, ϵ, etc. $\mathscr{T}_\alpha$ converts the incident fluid velocity field $\mathbf{G}\cdot\mathbf{F}$ into the force exerted by chain α on the fluid. Hence, $\mathbf{G}\cdot\mathscr{T}_\alpha\cdot\mathbf{G}\cdot\mathbf{F}$ represents the fluid velocity field 'scattered' by a single chain α.

Equation (8.4.8) is of the form encountered in multiple scattering theories of, say, electronic structure in disordered systems.[51,52] Consider the terms $\phi_\alpha\mathbf{G}\cdot\mathbf{F} + \phi_\alpha\mathbf{G}\cdot\mathscr{T}^{(\alpha f)}\cdot\mathbf{G}\cdot\mathbf{F}$ in (8.4.7). $\phi_\alpha\mathbf{G}\cdot\mathbf{F}$ describes that fluid incident on chain α which has not been disturbed by other polymers. Substituting (8.4.8) for $\mathscr{T}^{(\alpha f)}$ shows that $\phi_\alpha\mathbf{G}\cdot\mathscr{T}^{(\alpha f)}\cdot\mathbf{G}\cdot\mathbf{F}$ contains a sum of all possible sequences of scattering by polymers (including possibly chain α) before reaching chain α, e.g. the first term in (8.4.8) involves those contributions to the velocity field incident on chain α which have been perturbed by a single chain (other than α), the second involving disturbance by a pair of polymers, etc. A similar type of multiple scattering series[1] can be obtained by iterating the basic equations of the Kirkwood–Riseman theory (8.3.22)–(8.3.25), where each term involves the scattering of the fluid by an individual bead on the chain. Although equations (8.4.7) and (8.4.8) include all of these scattering processes from all the chains, they are represented in terms of the composite scattering from individual chains which have internal structure and dynamics. We could recover a form involving fluid scattering by individual beads by expanding $\mathscr{T}_\alpha$ from (8.4.1) in an infinite series in $\mathbf{K}_{\alpha i,\alpha k}(t,t')$. However, our main purpose is not in deriving messy expansions; we now introduce exact resummations of (8.4.7) to provide closed, compact, and physically transparent expressions for the dynamics of chains in polymer solutions at finite concentrations.

Since the dynamics of chain α is under consideration, this chain is special. It is useful to isolate those terms in the multiple scattering series (8.4.8) wherein chain α reappears. For this purpose we introduce the full scattering matrix, $\mathscr{T}^{(\alpha)}$, for the full polymer solution *in the absence of chain* α,

$$\begin{aligned}\mathscr{T}^{(\alpha)} = {} & -\sum_{\beta\neq\alpha}\mathscr{T}_\beta + \sum_{\beta\neq\alpha}\sum_{\gamma\neq\beta,\alpha}\mathscr{T}_\beta\cdot\mathbf{G}\cdot\mathscr{T}_\gamma \\ & - \sum_{\beta\neq\alpha}\sum_{\gamma\neq\beta,\alpha}\sum_{\delta\neq\gamma,\alpha}\mathscr{T}_\beta\cdot\mathbf{G}\cdot\mathscr{T}_\gamma\cdot\mathbf{G}\cdot\mathscr{T}_\delta + \ldots \end{aligned} \tag{8.4.9}$$

Equation (8.4.8) may be represented in terms of (8.4.9) and $\mathscr{T}_\alpha$. First there are sequences in (8.4.8) in which chain α never appears, giving a contribution $\mathscr{T}^{(\alpha)}$. Then there are others where α appears first (on the right), but never again, yielding $\mathscr{T}^{(\alpha)}\cdot\mathbf{G}\cdot\mathscr{T}_\alpha$. Next are those where chain α occurs only once at some intermediate position in the terms yielding

$\mathscr{T}^{(\alpha)} \cdot \mathbf{G} \cdot \mathscr{T} \cdot \mathbf{G} \cdot \mathscr{T}^{(\alpha)}$. By induction, the series (8.4.8) can be rewritten as

$$\begin{aligned}\mathscr{T}^{(\alpha_f)} = \mathscr{T}^{(\alpha)} &- \mathscr{T}^{(\alpha)} \cdot \mathbf{G} \cdot \mathscr{T}_\alpha - \mathscr{T}^{(\alpha)} \cdot \mathbf{G} \cdot \mathscr{T}_\alpha \cdot \mathbf{G} \cdot \mathscr{T}^{(\alpha)} \\ &+ \mathscr{T}^{(\alpha)} \cdot \mathbf{G} \cdot \mathscr{T}_\alpha \cdot \mathbf{G} \cdot \mathscr{T}^{(\alpha)} \cdot \mathbf{G} \cdot \mathscr{T}_\alpha \\ &+ \mathscr{T}^{(\alpha)} \cdot \mathbf{G} \cdot \mathscr{T}_\alpha \cdot \mathbf{G} \cdot \mathscr{T}^{(\alpha)} \cdot \mathbf{G} \cdot \mathscr{T}_\alpha \cdot \mathbf{G} \cdot \mathscr{T}^{(\alpha)} \\ &+ \ldots . \end{aligned} \tag{8.4.10}$$

Substituting (8.4.10) into (8.4.7), retaining, for now, those terms beginning with $\mathscr{T}_\alpha$ (on the right) from (8.4.10) and containing $\mathbf{F}$, and using the explicit representation (8.4.5) for $\mathscr{T}_\alpha$, the partial series can be represented as

$$[\mathscr{G}_\alpha - \mathscr{G}_\alpha \cdot \mathbf{X} \cdot \mathscr{G}_\alpha + \mathscr{G}_\alpha \cdot \mathbf{X} \cdot \mathscr{G}_\alpha \cdot \mathbf{X} \cdot \mathscr{G}_\alpha - \ldots]\phi_\alpha \cdot \mathbf{G} \cdot \mathbf{F} \tag{8.4.11}$$

where

$$\mathbf{X} = \phi_\alpha \mathbf{G} \cdot \mathscr{T}^{(\alpha)} \cdot \mathbf{G}\phi_\alpha \Delta_\alpha \tag{8.4.12}$$

The series in brackets in (8.4.11) is just the expansion of the screened polymer Green's function, $\hat{\mathscr{G}}_\alpha$, in powers of $\mathbf{X}$, where $\hat{\mathscr{G}}_\alpha$ obeys the equation

$$\left[\frac{\partial}{\partial t}\mathbf{1} + \zeta^{-1}\mathbf{1}\Delta_\alpha + \phi_\alpha \mathbf{G}\phi_\alpha \Delta_\alpha + \mathbf{X}\right] \cdot \hat{\mathscr{G}}_\alpha = \mathbf{1} \tag{8.4.13}$$

and the $\zeta^{-1}\mathbf{1}\Delta_\alpha$ term has been reintroduced. Taking those terms from (8.4.10) for which $\mathscr{T}_\alpha$ appears first along with those in (8.4.7) involving $\Sigma_\beta \phi_\beta \mathbf{f}_\beta$ again yields the same series in brackets in (8.4.11) which is then multiplied by $\Sigma_\beta \phi_\alpha \cdot \mathbf{G} \cdot \phi_\beta \mathbf{f}_\beta$. These two contributions to (8.4.7) can be represented symbolically as

$$\hat{\mathscr{G}}_\alpha \phi_\alpha \cdot \mathbf{G} \cdot \mathbf{F} + \sum_\beta \hat{\mathscr{G}}_\alpha \phi_\alpha \cdot \mathbf{G}\phi_\beta \cdot \mathbf{f}_\beta \tag{8.4.14}$$

It now remains to incorporate those terms from (8.4.10) and (8.4.7) for which $\mathscr{T}_\alpha$ does not appear first on the right. The remaining terms, say containing $\mathbf{F}$, yield the series

$$\begin{aligned}[\mathscr{G}_\alpha &- \mathscr{G}_\alpha \phi_\alpha \cdot \mathbf{G} \cdot \mathscr{T}^{(\alpha)} \cdot \mathbf{G}\phi_\alpha \Delta_\alpha \cdot \mathscr{G}_\alpha \\ &+ \mathscr{G}_\alpha \phi_\alpha \cdot \mathbf{G} \cdot \mathscr{T}^{(\alpha)} \cdot \mathbf{G}\phi_\alpha \Delta_\alpha \cdot \mathscr{G}_\alpha \phi_\alpha \cdot \mathbf{G} \cdot \\ &\times \mathscr{T}^{(\alpha)} \cdot \mathbf{G}\phi_\alpha \Delta_\alpha \cdot \mathscr{G}_\alpha + \ldots]\,\phi_\alpha \cdot \mathbf{G} \cdot \mathscr{T}^{(\alpha)} \cdot \mathbf{G} \cdot \mathbf{F}\end{aligned} \tag{8.4.15}$$

and the term in brackets is just the previously considered series involving contributions with $\mathscr{T}_\alpha$ coming first (on the right). Introducing these into (8.4.7) gives the additional parts

$$\hat{\mathscr{G}}_\alpha \phi_\alpha \cdot \mathbf{G} \cdot \mathscr{T}^{(\alpha)} \cdot \mathbf{G} \cdot \mathbf{F} + \sum_\beta \hat{\mathscr{G}}_\alpha \phi_\alpha \cdot \mathbf{G} \cdot \mathscr{T}^{(\alpha)} \cdot \mathbf{G} \cdot \phi_\beta \mathbf{f}_\beta \tag{8.4.16}$$

From multiple scattering theory,[51,53] given $\mathbf{G}$, the unperturbed fluid Green's function and $\mathscr{T}^{(\alpha)}$, the full T-matrix in the absence of chain α, the equation

$$\hat{\mathbf{G}}^{(\alpha)} \equiv \mathbf{G} + \mathbf{G} \cdot \mathscr{T}^{(\alpha)} \cdot \mathbf{G} \tag{8.4.17}$$

defines the fluid Green's function for the whole polymer solution *if chain α were absent*. The determination of (8.4.17) requires a consideration of the full hydrodynamics of the polymer solution (minus a single chain), and this problem is discussed in Section 8.5. For the present, (8.4.15) and (8.4.16) are substituted into (8.4.7), and upon use of (8.4.17) we obtain the formal solution

$$\mathbf{R}_\alpha = \hat{\mathscr{G}}_\alpha \phi_\alpha \cdot \hat{\mathbf{G}}^{(\alpha)} \cdot \mathbf{F} + \sum_\beta \hat{\mathscr{G}}_\alpha \cdot \phi_\alpha \hat{\mathbf{G}}^{(\alpha)} \cdot \phi_\beta \mathbf{f}_\beta \tag{8.4.18}$$

Since $\hat{\mathscr{G}}_\alpha$ from (8.4.13) and (8.4.1) still contains the polymer dynamics, i.e. the $\mathbf{R}_\alpha(t')$, equation (8.4.18) is just a useful intermediary. Employing the definitions (8.4.13) and (8.4.17), equation (8.4.18) can be converted to the effective equation of motion

$$\left[\frac{\partial}{\partial t}\mathbf{1} + \hat{\mathbf{K}}_{\alpha\alpha}\Delta_\alpha\right] \cdot \mathbf{R}_\alpha = \phi_\alpha \hat{\mathbf{G}}^{(\alpha)} \cdot \mathbf{F} + \sum_\beta \phi_\alpha \hat{\mathbf{G}}^{(\alpha)} \phi_\beta \cdot \mathbf{f}_\beta \tag{8.4.19}$$

for the effective dynamics of chain α in the exact field of all the other chains. The screened hydrodynamic interactions $\hat{\mathbf{K}}_{\alpha\alpha}$ are given by

$$\hat{\mathbf{K}}_{\alpha\alpha} = \zeta^{-1}\mathbf{1} + \phi_\alpha \hat{\mathbf{G}}^{(\alpha)} \phi_\alpha \tag{8.4.20}$$

and involve the hydrodynamics of the full polymer solution without α. The $\phi_\alpha \hat{\mathbf{G}}^{(\alpha)} \cdot \mathbf{F}$ in (8.4.19) is just the fluid velocity at the position of chain α if α were not present in the solution. The remaining term in (8.4.19) arises because the forces, $\mathbf{f}_\beta$, acting on all chains β are transmitted to the fluid by the hydrodynamic boundary conditions (8.3.10), and then they are propagated by hydrodynamics through the entire polymer solution (with α removed again). Equation (8.4.19) is, therefore, the generalization of the steady state limit, infinite dilution, single chain dynamical equations considered in Section 8.3. The essential feature of (8.4.19) is that it provides the correct definitions for the screened hydrodynamic interactions, $\hat{\mathbf{K}}_{\alpha\alpha}$, and the effective incident velocity field, $\phi_\alpha \hat{\mathbf{G}}^{(\alpha)} \cdot \mathbf{F}$ (for $\mathbf{f}_\beta = 0$).

When entanglement effects are negligible, correlations between chain α and any other chains can be ignored. Consequently, (8.4.19) can be replaced by an approximate equation wherein $\hat{\mathbf{K}}_{\alpha\alpha}$ and $\hat{\mathbf{G}}^{(\alpha)}$ are replaced by their averages over all other chains, $\langle \hat{\mathbf{K}}_{\alpha\alpha} \rangle$ and $\langle \hat{\mathbf{G}}^{(\alpha)} \rangle$, respectively. Since $\langle \hat{\mathbf{G}}^{(\alpha)} \rangle$ describes the effective hydrodynamics of the polymer solution at finite concentration, but without chain α, a negligible error, of order $1/N$, is incurred by ignoring the single chain exclusion and by replacing $\langle \hat{\mathbf{G}}^{(\alpha)} \rangle$ by the full averaged $\langle \hat{\mathbf{G}} \rangle$ for the whole polymer solution. Thus, equation (8.4.19) is reduced to

$$\left[\frac{\partial}{\partial t}\mathbf{1} + \langle \hat{\mathbf{K}}_{\alpha\alpha} \rangle \Delta_\alpha\right] \cdot \mathbf{R}_\alpha = \phi_\alpha \langle \hat{\mathbf{G}} \rangle \cdot \mathbf{F} + \sum_\beta \phi_\alpha \langle \hat{\mathbf{G}} \rangle \phi_\beta \cdot \mathbf{f}_\beta \tag{8.4.21}$$

Consistent with the average over all other chains in $\hat{\mathbf{G}}^{(\alpha)}$, an average over all other chains should also be introduced into the last term in (8.4.21) to give

$$\phi_\alpha \langle \hat{\mathbf{G}}^{(\alpha)} \phi_\alpha \mathbf{f}_\alpha \rangle + \sum_{\beta \neq \alpha} \phi_\alpha \langle \hat{\mathbf{G}}^{(\alpha)} \phi_\beta \cdot \mathbf{f}_\beta \rangle \tag{8.4.22}$$

where $\mathbf{f}_\alpha^{\text{int}}$ of (8.3.9) depends on the position of other chains ($\beta \neq \alpha$). Breaking the averages in (8.4.22) gives

$$\phi_\alpha \langle \hat{\mathbf{G}} \rangle \phi_\alpha \cdot \langle \mathbf{f}_\alpha \rangle^\alpha + \sum_{\beta \neq \alpha} \phi_\alpha \langle \hat{\mathbf{G}} \rangle \langle \phi_\beta \cdot \mathbf{f}_\beta \rangle^\alpha \tag{8.4.23}$$

where the superscript α implies an average over all chains but α. Equation (8.4.22) represents a mean field approximation describing the average force exerted by all other chains on a given chain which are then propagated to chain α by hydrodynamics. This average field is just of the type that confines the motion of a chain to a region surrounding its initial position in concentrated polymer solutions, the basic picture of reptation-like models.[48–50]

When entanglements occur between a pair of chains, say α and γ, the chains develop correlations during the period of entanglement. Hence, it is not permissible to average over chain $\beta = \gamma$ in (8.4.21)–(8.4.23). Both chains α and γ can be isolated from the rest with a derivation, more lengthy than the above, to yield a pair of coupled equations for the two entangled chains. It is necessary to define the Green's function, $\hat{\mathbf{G}}^{(\alpha,\gamma)}$, for the whole polymer solution in the absence of chains α *and* γ, so intra- and inter-chain hydrodynamic interactions are

$$\hat{\mathbf{K}}'_{\alpha\alpha} = \zeta^{-1}\mathbf{1} + \phi_\alpha \hat{\mathbf{G}}^{(\alpha,\gamma)}\phi_\alpha \tag{8.4.24a}$$

$$\hat{\mathbf{K}}'_{\gamma\gamma} = \zeta^{-1}\mathbf{1} + \phi_\gamma \hat{\mathbf{G}}^{(\alpha,\gamma)}\phi_\gamma \tag{8.4.24b}$$

$$\hat{\mathbf{K}}_{\alpha\gamma} = \phi_\alpha \hat{\mathbf{G}}^{(\alpha,\gamma)}\phi_\gamma \tag{8.4.24c}$$

The pair of dynamical equations are then found to be

$$\left(\frac{\partial}{\partial t}\mathbf{1} + \hat{\mathbf{K}}'_{\alpha\alpha}\Delta_\alpha\right)\cdot \mathbf{R}_\alpha + \hat{\mathbf{K}}'_{\alpha\gamma}\Delta_\gamma \cdot \mathbf{R}_\gamma = \phi_\alpha \hat{\mathbf{G}}^{(\alpha,\gamma)}\cdot \mathbf{F} + \sum_\beta \phi_\alpha \hat{\mathbf{G}}^{(\alpha,\gamma)}\phi_\beta \cdot \mathbf{f}_\beta \tag{8.4.25a}$$

and

$$\left(\frac{\partial}{\partial t}\mathbf{1} + \hat{\mathbf{K}}'_{\gamma\gamma}\Delta_\gamma\right)\cdot \mathbf{R}_\gamma + \hat{\mathbf{K}}'_{\gamma\alpha}\Delta_\alpha \cdot \mathbf{R}_\alpha = \phi_\gamma \hat{\mathbf{G}}^{(\alpha,\gamma)}\cdot \mathbf{F} + \sum_\beta \phi_\gamma \hat{\mathbf{G}}^{(\alpha,\gamma)}\phi_\beta \cdot \mathbf{f}_\beta \tag{8.4.25b}$$

Averages over all chains but α and γ can likewise be introduced into (8.4.24) and (8.4.25) as above.

In polymer solutions at infinite dilution, equations (8.4.19) and (8.4.25) could approximately be represented as involving the pure solvent Green's function $\mathbf{G}$ instead of $\hat{\mathbf{G}}^{(\alpha)}$ and $\mathbf{G}^{(\alpha,\gamma)}$, respectively. However, the treatment of polymer dynamics at finite concentrations requires that the hydrodynamics of the full polymer solution be considered. Recalling the fact that it represents an enormous task to solve even steady state (linearized) hydrodynamics for fluid flow past a body of irregular shape, the description of the hydrodynamics of a polymer solution, with irregularly shaped chains and their dynamical motion, might appear to be a horrendous task. The evaluation of $\hat{\mathbf{G}}^{(\alpha)}$ would, in fact, be impossible were it not for the fact that the instantaneous dynamics of a single chain is not of utmost interest. Rather, we usually are interested in an ensemble average over some initial distribution of polymer chain configurations, generally an equilibrium average since all external forces can be considered to be slowly turned on at some convenient prior time. These averages are implicitly incorporated into (8.4.21)–(8.4.23) although their precise nature is yet to be specified.

8.5 Hydrodynamics of the polymer solution

An analysis of the hydrodynamics of the polymer solution serves additional important functions as it enables the discussion of the transport properties of these solutions. Properties like viscosity, translational friction coefficients, rotary friction coefficients, and sedimentation coefficients are of interest.[1] As their treatments are somewhat analogous, only the case of viscosity is considered here in detail.

The traditional analysis of polymer solution viscosity considers a single chain at infinite dilution in a fluid with a constant shear rate. This type of approach becomes quite difficult to generalize to solutions at finite polymer concentrations, but a consideration of the full solution hydrodynamics enables a definition of the polymer viscosity by a modern generalization of the formulation Einstein employed to describe the viscosity of a suspension of spheres.[54] It should be noted that concurrent with our analysis, general correlation function formulas have been developed for the intrinsic polymer viscosity.[55] These have yet to be generalized to polymer solutions at finite concentrations, but it is suspected that, in order to utilize any generalized correlation function expression, it will become necessary to consider polymer dynamics analogous to that in Section 8.4, possibly in terms of diffusion (Fokker–Planck) equations rather than Langevin equations, and to incorporate the solution hydrodynamics described below.

Average solution hydrodynamics

Equation (8.3.17) provides the formal solution for the velocity field in the polymer solution in terms of the forces $\{\boldsymbol{\sigma}_{\alpha i}\}$ exerted by the polymers on the fluid. By evaluating $\{\boldsymbol{\sigma}_{\alpha i}\}$ from the dynamical equations for the polymer, it is possible to convert (8.3.17) to an equation of the form

$$\mathbf{u}(\mathbf{r}, t) \doteq \int \mathrm{d}\mathbf{r}'\, \mathrm{d}t'\, \hat{\mathbf{G}}(\mathbf{r}\mathbf{r}', tt') \cdot \mathbf{F}(\mathbf{r}', t') \tag{8.5.1}$$

with $\hat{\mathbf{G}}$ the full solution Green's function. The actual measured fluid velocity corresponds to the velocity (8.5.1) as averaged over some initial distribution of polymer configurations,

$$\langle \mathbf{u}(\mathbf{r}, t) \rangle = \int \mathrm{d}\mathbf{r}'\, \mathrm{d}t'\, \langle \hat{\mathbf{G}}(\mathbf{r}\mathbf{r}'; tt') \rangle \cdot \mathbf{F}(\mathbf{r}', t') \tag{8.5.2}$$

where the equilibrium distribution (8.2.3) is the one generally employed.

The viscometer, etc., only measures the average properties of the polymer solution, e.g. viscosity, etc. It therefore takes the solution to obey some effective Navier–Stokes equation of the form

$$\left(\rho \frac{\partial}{\partial t} - \eta \nabla^2\right) \langle \mathbf{u}(\mathbf{r}, t) \rangle + \boldsymbol{\nabla} P(\mathbf{r}, t) \doteq \mathbf{F}(\mathbf{r}, t) \tag{8.5.3}$$

where the solution viscosity is

$$\eta = \eta_0 + \delta\eta \tag{8.5.4}$$

defining the excess viscosity, $\delta\eta$, due to the presence of the polymers, and nonlinear terms continue to be ignored. Actually, the presence of polymers with structure and dynamics yields solution hydrodynamics with more complicated and interesting visco-elastic properties. This is conveniently summarized by the effective hydrodynamic equations

$$\left(\rho_0 \frac{\partial}{\partial t} - \eta_0 \nabla^2\right) \langle \mathbf{u}(\mathbf{r}, t) \rangle - \int \mathrm{d}\mathbf{r}'\, \mathrm{d}t'\; \boldsymbol{\Sigma}(\mathbf{r} - \mathbf{r}'; t - t') \cdot \langle \mathbf{u}(\mathbf{r}', t') \rangle$$

$$+ \boldsymbol{\nabla} P(\mathbf{r}, t) = \mathbf{F}(\mathbf{r}, t) \tag{8.5.5}$$

with $\boldsymbol{\Sigma}(\mathbf{r} - \mathbf{r}'; t - t')$ the fluid 'self-energy' or 'mass-operator'. The $\boldsymbol{\Sigma} \cdot \langle \mathbf{u} \rangle$ term in (8.5.5) is the divergence of the polymer contribution to the stress tensor. Solution translational invariance in space and time implies that $\boldsymbol{\Sigma}$ depends only on $\mathbf{r} - \mathbf{r}$, and $t - t'$, so (8.5.5) is readily converted to Fourier variables,

$$[\mathrm{i}\omega\rho_0 + \eta_0 k^2 - \boldsymbol{\Sigma}(\mathbf{k}, \omega) \cdot]\, \langle \mathbf{u}(\mathbf{k}, \omega) \rangle + \mathrm{i}\mathbf{k} P(\mathbf{k}, \omega) = F(\mathbf{k}, \omega) \tag{8.5.6}$$

where identical symbols are used for quantities and their Fourier transforms, for convenience.

Comparing (8.5.2) and (8.5.6) implies that $\langle \hat{G}(\mathbf{k}, \omega) \rangle$ obeys the equation

$$\sum_j [(\mathrm{i}\omega\rho_0 + \eta_0 k^2)\delta_{ij} - \Sigma_{ij}(\mathbf{k}, \omega)] \langle \hat{G}_{jm}(k, \omega) \rangle = \delta_{im} - (k_i k_m / k^2) \tag{8.5.7}$$

as the generalization of (8.3.15).

The dynamical polymer viscosity $\delta\eta(\omega)$ is obtained from (8.5.6) or (8.5.7) from the coefficient of k^2 in the Taylor series expansion of $\mathbf{\Sigma}(k, \omega)$. Isotropic visco-elasticity implies the definition

$$-\lim_{k\to 0} \frac{1}{k^2} [\mathbf{\Sigma}(\mathbf{k}, \omega) - \mathbf{\Sigma}(0, \omega)] = \delta\eta(\omega)\mathbf{1} \equiv \eta_0 c \eta_{\mathrm{sp}}(\omega, c)\mathbf{1} \tag{8.5.8}$$

enabling the determination of the polymer viscosity independent of the imposed velocity field $\mathbf{G} \cdot \mathbf{F}$ subject only to the condition of vanishing shear rate.

Having outlined the basic method, the derivations can now be given. Substitution of the formal solution (8.4.7) for the polymer dynamics into the dynamical equation (8.3.6) leads to a formal solution for $\boldsymbol{\sigma}_{\alpha i}(t)$ in terms of the $\mathscr{T}^{(\alpha f)}$,

$$\begin{aligned}\boldsymbol{\sigma}_\alpha = \mathbf{f}_\alpha &- \Delta_\alpha \mathscr{G}_\alpha \phi_\alpha \cdot \mathbf{G} \cdot \mathbf{F} - \Delta_\alpha \mathscr{G}_\alpha \cdot \phi_\alpha \mathbf{G} \cdot \mathscr{T}^{(\alpha f)} \cdot \mathbf{G} \cdot \mathbf{F} \\ &- \sum_\beta \Delta_\alpha \mathscr{G}_\alpha \cdot \phi_\alpha \mathbf{G} \phi_\beta \cdot \mathbf{f}_\beta - \sum_\beta \Delta_\alpha \mathscr{G}_\alpha \cdot \phi_\alpha \mathbf{G} \cdot \mathscr{T}^{(\alpha f)} \cdot \mathbf{G}\phi_\beta \cdot \mathbf{f}_\beta\end{aligned} \tag{8.5.9}$$

whereupon introduction of (8.5.9) into (8.3.17) and use of the definition (8.4.5) yields

$$\begin{aligned}\mathbf{u} = \mathbf{G} \cdot \mathbf{F} &+ \sum_\alpha \mathbf{G} \cdot [-\mathscr{T}_\alpha \cdot \mathbf{G} \cdot \mathbf{F} - \mathscr{T}_\alpha \cdot \mathbf{G} \cdot \mathscr{T}^{(\alpha f)} \cdot \mathbf{G} \cdot \mathbf{F}] \\ &+ \sum_\alpha \mathbf{G} \cdot [\phi_\alpha \mathbf{f}_\alpha - \sum_\beta \mathscr{T}_\alpha \cdot \mathbf{G} \cdot \phi_\beta \mathbf{f}_\beta - \sum_\beta \mathscr{T}_\alpha \cdot \mathbf{G} \cdot \mathscr{T}^{(\alpha f)} \cdot \mathbf{G} \cdot \phi_\beta \cdot \mathbf{f}_\beta]\end{aligned} \tag{8.5.10}$$

By utilizing the expansion (8.4.8) for $\mathscr{T}^{(\alpha f)}$ it is found that

$$\begin{aligned}\mathscr{T} &\equiv -\sum_\alpha \mathscr{T}_\alpha - \sum_\alpha \mathscr{T}_\alpha \cdot \mathbf{G} \cdot \mathscr{T}^{(\alpha f)} \\ &= -\sum_\alpha \mathscr{T}_\alpha + \sum_\alpha \sum_{\beta \neq \alpha} \mathscr{T}_\alpha \cdot \mathbf{G} \cdot \mathscr{T}_\beta - \sum_\alpha \sum_{\beta\neq\alpha} \sum_{\gamma\neq\beta} \mathscr{T}_\alpha \cdot \mathbf{G} \cdot \mathscr{T}_\beta \cdot \mathbf{G} \cdot \mathscr{T}_\gamma + \ldots\end{aligned} \tag{8.5.11}$$

is the T-matrix for the full polymer solution, so (8.5.10) reduces to

$$\begin{aligned}\mathbf{u} = &[\mathbf{G} + \mathbf{G} \cdot \mathscr{T} \cdot \mathbf{G}] \cdot \mathbf{F} \\ &+ \sum_\beta [\mathbf{G} + \mathbf{G} \cdot \mathscr{T} \cdot \mathbf{G}]\phi_\beta \cdot \mathbf{f}_\beta\end{aligned} \tag{8.5.12}$$

whereupon the definition of the full polymer solution Green's function

$$\hat{\mathbf{G}} \equiv \mathbf{G} + \mathbf{G} \cdot \mathscr{T} \cdot \mathbf{G} \tag{8.5.13}$$

then converts (8.5.12) into

$$\mathbf{u} = \hat{\mathbf{G}} \cdot \mathbf{F} + \hat{\mathbf{G}} \sum_\beta \phi_\beta \mathbf{f}_\beta \tag{8.5.14}$$

which differs from the anticipated form (8.5.1) by the appearance of the term involving the force density $\Sigma_\beta \phi_\beta \mathbf{f}_\beta$. To understand its presence further, it is useful to reintroduce all the indices to obtain

$$\sum_{\beta i} \int \mathrm{d}t' \, \hat{\mathbf{G}}(\mathbf{r} - \mathbf{r}'; t - t')\phi_{\beta i}(\mathbf{r}', t') \cdot \mathbf{f}_{\beta i}(t') \tag{8.5.15}$$

The contribution to (8.5.15) from any force $\mathbf{f}_{\beta i}$ that is derivable from a potential energy, i.e.

$$\mathbf{f}_{\beta i}^{(1)}(t') = -\nabla_{\mathbf{R}_{\beta i}(t')} U \tag{8.5.16}$$

can readily be shown to vanish by using the definition (8.3.4) and the properties of delta functions to convert (8.5.15) to

$$\sum_{\beta i} \int dt' [\nabla_{\mathbf{r}'} \cdot \hat{\mathbf{G}}^{\mathrm{T}}(\mathbf{rr}'; tt')] \delta [\mathbf{r}' - \mathbf{R}_{\beta i}(t')] U \tag{8.5.17}$$

where the superscript T denotes the transpose of a tensor. Equation (8.5.17) vanishes because of the fluid incompressibility (8.3.2) which is preserved in the presence of the polymers, as can be demonstrated by applying the gradient operator to (8.5.13) or its transpose,

$$\nabla_{\mathbf{r}} \cdot \mathbf{G}(\mathbf{rr}'; tt') = \nabla_{\mathbf{r}'} \cdot \mathbf{G}^{\mathrm{T}}(\mathbf{r} - \mathbf{r}'; t - t') = 0$$

$$\Rightarrow \nabla_{\mathbf{r}} \cdot \hat{\mathbf{G}} = \nabla_{\mathbf{r}'} \cdot \hat{\mathbf{G}}^{\mathrm{T}} = 0 \tag{8.5.18}$$

Thus, the only nonvanishing parts of (8.5.15) involve those contributions from random forces $\mathbf{f}_{\beta i}^{*}$ which are not derivable from a potential. However, from the Mori theory of generalized Brownian motion,[20,21] $\mathbf{f}_{\beta i}^{*}$ is most likely expressible in the general from of (8.5.16), but, if not, it is a random function of the degrees of freedom which have been integrated out, so its average vanishes. As we are really interested in the averages, equation (8.5.2) is the final result as expected.

Transport coefficients other than viscosity

Although the polymer viscosity is being utilized as the example of a polymer transport property, we should digress to mention briefly some other transport properties. When a polymer solution is made to flow with constant velocity $\mathbf{u}_0$, the extra averaged drag force, $\mathbf{F}_{\mathrm{d}}$, acting on the polymers is

$$\mathbf{F}_{\mathrm{d}} = Nf\mathbf{u}_0 \tag{8.5.19}$$

where f is the concentration-dependent translational friction coefficient per polymer chain. Elementary analysis implies that $\mathbf{F}_{\mathrm{d}}$ can be represented in terms of $\{\boldsymbol{\sigma}_{\alpha i}\}$via

$$\mathbf{F}_{\mathrm{d}} = -\sum_{\alpha i} \langle \boldsymbol{\sigma}_{\alpha i} \rangle \tag{8.5.20}$$

Using an analysis similar to that given above it can be shown that

$$\mathbf{F}_{\mathrm{d}} = -\int d\mathbf{r}\, d\mathbf{r}'\, dt' \langle \mathscr{T}(\mathbf{rr}'; tt') \rangle \cdot \mathbf{u}_0 \tag{8.5.21}$$

thereby enabling the calculation of f from the same effective polymer dynamics and hydrodynamics as in the case of the viscosity. The sedimentation coefficient for high enough speeds of ultracentrifugation is inversely proportional to f, while the rotary diffusion coefficient can be treated in an analogous fashion.[1]

Solution self-energy at infinite polymer dilution

Given equations (8.5.2), (8.5.6), (8.5.7), and (8.5.13), it remains to evaluate the fluid self-energy $\boldsymbol{\Sigma}$. Equation (8.5.1), with the multiple scattering expansion generated by (8.5.11) and (8.5.13), bears a direct mathematical relation to many other problems associated with the propagation of waves through disordered materials, be they waves, 'electron waves,' in disordered systems, lattice vibrations in substitutionally disordered crystals, or electromagnetic waves in systems with randomly varying refractive indices. Some readymade approximations exist in these areas,[51,52] but the greater complexity of the polymers cautions against their uncritical use since the polymers have additional internal structure and dynamics which are the centre of interest. The polymer problem is readily solved for solutions at infinite dilution. The derivation proves useful in again providing contact with older theories and in suggesting methods for the treatment of solutions at finite concentration.

The n-th term in the series (8.5.11) provides the contribution to the overall velocity field (using (8.5.13) and (8.5.1)) for those elements of fluid which have encountered polymer chains n times, possibly with certain of the chains repeated, so long as the same chain does not disturb the fluid consecutively. At infinite dilution different chains are separated by arbitrarily large distances, so if a chain, α, disturbs the flow of a particular fluid element, it is very unlikely that after subsequent disturbance of the fluid element by one or more other chains, $\beta \neq \alpha$, the fluid element would ever return to be disturbed again by chain α. Thus, at infinite dilution successive scattering events are taken to occur by different uncorrelated chains.[35,36] Since $\mathbf{G}$ is independent of polymers, equation (8.5.13) implies

$$\langle \hat{\mathbf{G}} \rangle = \mathbf{G} + \mathbf{G} \cdot \langle \mathscr{T} \rangle \cdot \mathbf{G} \tag{8.5.22}$$

and at infinite dilution we obtain

$$\begin{aligned} \langle \mathscr{T} \rangle = & -\sum_{\alpha} \langle \mathscr{T}_\alpha \rangle + \sum_{\alpha} \sum_{\beta \neq \alpha} \langle \mathscr{T}_\alpha \rangle \cdot \mathbf{G} \cdot \langle \mathscr{T}_\beta \rangle \\ & - \sum_{\alpha} \sum_{\beta \neq \alpha} \sum_{\alpha \neq \beta} \langle \mathscr{T}_\alpha \rangle \cdot \mathbf{G} \cdot \langle \mathscr{T}_\beta \rangle \cdot \mathbf{G} \cdot \langle \mathscr{T}_\alpha \rangle + \ldots \quad c \to 0. \end{aligned} \tag{8.5.23}$$

It is convenient to designate the terms involving m factors of $\mathscr{T}$ operators as the m-th order contributions. Since the averaging removes the individual identity of a given chain, it is permissible to write

$$\langle \mathscr{T}_\alpha \rangle \equiv \langle \mathscr{T}_0 \rangle \tag{8.5.24}$$

where 0 labels a typical chain in solution. Hence, equation (8.5.23) can be rewritten as

$$\begin{aligned} \langle \mathscr{T} \rangle = & -N \langle \mathscr{T}_0 \rangle + N(N-1) \langle \mathscr{T}_0 \rangle \cdot \mathbf{G} \cdot \langle \mathscr{T}_0 \rangle \\ & - N(N-1)^2 \langle \mathscr{T}_0 \rangle \cdot \mathbf{G} \cdot \langle \mathscr{T}_0 \rangle \cdot \mathbf{G} \cdot \langle \mathscr{T}_0 \rangle + \ldots \quad c \to 0 \end{aligned} \tag{8.5.25}$$

and $N-1$ can be replaced by N in the thermodynamic limit of $N \to \infty$, $V \to \infty$, $N/V \to$ infinitesimal, where V is the volume of the system. Equations (8.5.1), (8.5.22), and (8.5.25) form an operator Born series which can be shown to be the series expansion of (8.5.5) in powers of $\langle \mathscr{T}_0 \rangle$ where we have the approximate association

$$\mathbf{\Sigma} \to -N \langle \mathscr{T}_0 \rangle \quad c \to 0. \tag{8.5.26}$$

Reintroducing all the indices and taking Fourier transforms gives

$$\begin{aligned} \mathbf{\Sigma}(\mathbf{k}, \omega) \to -N \int \mathrm{d}(\mathbf{r}-\mathbf{r}') \exp[-\mathrm{i}\mathbf{k} \cdot (\mathbf{r}-\mathbf{r}')] \int_{-\infty}^{\infty} \mathrm{d}(t-t') \exp[-\mathrm{i}\omega(t-t')] \\ \times \sum_{i,j=1}^{n} \langle \phi_i(\mathbf{r}, t) \Delta_i(t) \mathscr{G}_{ij}(t, t') \phi_j(\mathbf{r}', t') \rangle \quad c \to 0. \end{aligned} \tag{8.5.27}$$

Translational invariance of the averaged polymer solution implies that the average in (8.5.27) depends only on $\mathbf{r}-\mathbf{r}'$ and $t-t'$. Consequently, the integration over $\mathbf{r}-\mathbf{r}'$ can be performed, utilizing the definition (8.3.10), yielding

$$\boldsymbol{\Sigma}(k,\omega)\to-(N/V)\int_{-\infty}^{\infty}\mathrm{d}(t-t')\exp[-\mathrm{i}\omega(t-t')]$$

$$\times\sum_{i,j=1}^{n}\langle\exp\{-\mathrm{i}\mathbf{k}\cdot[\mathbf{R}_i(t)-\mathbf{R}_j(t')]\}\Delta_i(t)\mathcal{G}_{ij}(t,t')\rangle\quad c\to 0. \tag{8.5.28}$$

The polymer intrinsic viscosity

From the association (8.5.8) and the definition of the polymer intrinsic viscosity (8.2.23), equation (8.5.28) leads to

$$[\eta(\omega)]=(N_\mathrm{A}/\eta_0 nM_\mathrm{A})\lim_{k\to 0}k^{-2}\int_{-\infty}^{\infty}\mathrm{d}(t-t')\exp[-\mathrm{i}\omega(t-t')]$$

$$\times\sum_{i,j=1}^{n}\langle\exp\{-\mathrm{i}\mathbf{k}\cdot[\mathbf{R}_i(t)-\mathbf{R}_j(t')]\}\Delta_i(t)\mathcal{G}_{ij}(t,t')\rangle \tag{8.5.29}$$

where N_A is Avagadro's number and c is

$$c=nNM_\mathrm{A}/N_\mathrm{A}V. \tag{8.5.30}$$

Equation (8.5.29) further simplifies in the steady state limit $\omega\to 0$. From the definition (8.4.1) we can write

$$\Delta_i(t)\mathcal{G}_{ij}(t,t')=(\mathsf{K}^{-1})_{ij}(t,t')-\sum_{p=1}^{n}\int\mathrm{d}t''(\mathsf{K}^{-1})_{ip}(t,t'')\cdot\frac{\partial}{\partial t}\mathcal{G}_{pj}(t'',t') \tag{8.5.31}$$

where K^{-1} is the formal inverse of K,

$$\sum_{p=1}^{n}\int(\mathsf{K}^{-1})_{ip}(t,t'')\cdot\mathsf{K}_{pj}(t'',t')\,\mathrm{d}t''=\mathbf{1}\delta_{ip}\delta(t-t') \tag{8.5.32}$$

which is temporarily introduced to accelerate the derivation. In the steady-state limit, $(\partial\mathcal{G}/\partial t)$ can be dropped, and (8.5.29) reduces to the formal expression

$$[\eta]=(N_\mathrm{A}/\eta_0 nM_\mathrm{A})\lim_{k\to 0}k^{-2}\sum_{i,j=1}^{n}\langle\exp(-\mathrm{i}\mathbf{k}\cdot(\mathbf{R}_i-\mathbf{R}_j)]\;\boldsymbol{\zeta}_{ij}\rangle \tag{8.5.33}$$

where the spherical average of the generalized friction tensor (8.3.34) is understood to be taken. K^{-1} or its steady-state limit, $\boldsymbol{\zeta}$ of (8.3.34), is not readily evaluated.

The quantity in brackets can be represented in terms of the solution of a set of linear equations. Let the matrix J_{ij} in bead space be defined by

$$J_{ij}=\zeta^{-1}\sum_{p=1}^{n}\zeta_{ip}\exp[-\mathrm{i}\mathbf{k}\cdot(\mathbf{R}_p-\mathbf{R}_j)](18/nl^2) \tag{8.5.34}$$

so we have the set of algebraic equations

$$J_{ij}+\zeta\sum_{p(\neq i)=1}^{n}T_{ip}J_{pj}=(18/nl^2)\exp[-\mathrm{i}\mathbf{k}\cdot(\mathbf{R}_p-\mathbf{R}_j)] \tag{8.5.35}$$

where T_{ij} is the spherical average of $\mathbf{T}_{ij}$, and, hence, (8.5.33) is transformed into

$$[\eta] = \frac{N_A \zeta l^2}{18\eta_0 M_A} \lim_{k\to 0} k^{-2} \sum_{i=1}^{n} \langle J_{ii} \rangle. \tag{8.5.36}$$

Equation (8.5.35) cannot be solved, as J_{ij} depends explicitly on the $\{\mathbf{R}_p\}$ because of the right-hand side and the definition of T_{ij} in (8.3.20). The customary approximation involves averaging (8.5.35) and approximating $\langle T_{ip} J_{pj} \rangle$ by the product of the averages

$$\langle J_{ij} \rangle + \zeta \sum_{p(\neq i)=1}^{n} \langle T_{ip} \rangle \langle J_{pj} \rangle \approx (18/nl^2) \langle \exp[-\mathrm{i}\mathbf{k} \cdot (\mathbf{R}_i - \mathbf{R}_j)] \rangle \tag{8.5.37}$$

If $\langle J^0 \rangle$ is taken to be the solution to (8.5.37), the average of the desired solution to (8.5.35) can be written as the series

$$\langle J_{ij} \rangle = \langle J_{ij}^0 \rangle + \sum_{m,k,p,q} \zeta^{-1} \zeta_{ip}^0 \langle (-T_{pm} + \langle T_{pm} \rangle) \zeta_{mk}^0 (-T_{kq} + \langle T_{kq} \rangle) \rangle \langle J_{qj}^0 \rangle + \ldots \tag{8.5.38}$$

where

$$\sum_{p=1}^{n} [\delta_{ip} + \zeta \langle T_{ip} \rangle (1 - \delta_{ip})] \, \delta_{pj}^0 \equiv \zeta \delta_{ij} \tag{8.5.39}$$

giving a systematic method for improving the 'pre-averaging' approximation (8.5.37).

The solution of equations like (8.5.37) has been studied in detail by a number of authors,[1] but only an approximate solution is required here. Thus, we may follow the scheme reviewed by Yamakawa.[1] Introduce the change in variables[1]

$$x = (2i/n) - 1 \quad y = (2j/n) - 1 \quad \langle J_{ij} \rangle = J(x, y). \tag{8.5.40}$$

For the Gaussian chain it is found that[1]

$$\langle T_{ij} \rangle = (6\pi \,|i - j|)^{-1/2} (\pi\eta_0 l)^{-1} = (3n\pi \,|x - y|)^{-1/2} (\pi\eta_0 l)^{-1} \tag{8.5.41}$$

Substituting (8.5.40) and (8.5.41) into (8.5.37) and converting the sums to integrals in the resulting equation leads to the integral equation[1]

$$J(x, y) + h \int_{-1}^{1} |x - t|^{-1/2} J(t, y) \mathrm{d}t = f(x, y) \tag{8.5.42}$$

where for Gaussian chains the analogue of (8.2.8) for i and j yields

$$f(x, y) = \frac{18}{nl^2} \exp[-k^2 l^2 n \,|x - y|/12] \tag{8.5.43}$$

and h is the draining parameter

$$h = (\zeta/\eta_0 l)(n/12\pi^3)^{1/2} \tag{8.5.44}$$

The intrinsic viscosity is then

$$[\eta] = \lim_{k\to 0} \frac{N_A \zeta l^2 n}{36\eta_0 M_A k^2} \int_{-1}^{1} J(x, x) \mathrm{d}x. \tag{8.5.45}$$

The integral equation (8.5.42) can be converted to a series of algebraic equations by the use of Fourier expansions for $J(x, y)$ and $f(x, y)$,

$$J(x, y) = \sum_{p=-\infty}^{\infty} J_p(y)\exp(i\pi px) \tag{8.5.46a}$$

$$f(x, y) = \sum_{p=-\infty}^{\infty} f_p(y)\exp(i\pi px) \tag{8.5.46b}$$

The coefficient of $hJ_p(y)$ in the resultant equations is a sum of Fresnel integrals,[1] and there is coupling between different $J_p(y)$. As a simplification, only the diagonal terms are retained, and the asymptotic limit of the Fresnel integrals for p large is employed.[1] The final result can be shown to be

$$J(x, y) = \sum_{p(\neq 0)=-\infty}^{\infty} \frac{f_p(y)\exp(i\pi px)}{1 + h(2|p|^{-1})^{1/2}} \tag{8.5.47}$$

and

$$\int_{-1}^{1} dx J(x, x) = \frac{1}{2}\sum_{p=1}^{\infty} \frac{\int_{-1}^{1} dx \int_{-1}^{1} dy f(x, y)\exp[i\pi p(y - x)]}{1 + h(2|p|^{-1})^{1/2}} \tag{8.5.48}$$

$$= \frac{6k^2}{\pi^2}\sum_{p=1}^{\infty} p^{-2}[1 + h(2|p|^{-1})^{1/2}]^{-1} \tag{8.5.49}$$

where finite chain corrections (terms in $\exp(-dk^2)$) have been ignored in the integral in (8.5.48). Hence, the viscosity (8.5.45) with (8.5.49) is identical to that given by Yamakawa[1] within the same approximation scheme despite the fact that $f(x, y)$ in (8.5.43) differs slightly from the form used in his expression for the intrinsic viscosity.

The derivation (8.5.33)–(8.5.49), although standard, is quite laborious, and the identical approximate results can more simply be obtained by the following scheme.[35–38] By renumbering the beads on the chain, the summation in (8.5.39) can be taken to run from $p = -n/2 + 1$ to $n/2$. If (8.5.39) is approximated by extending the limits to infinity, converting the summation to an integration, and using the chain contour length variables $s = pl$, this equation becomes a convolution integral

$$\int_{-\infty}^{\infty} ds[\zeta^{-1}\delta(s - s'') + T(s - s'')]\zeta^0(s'' - s') \equiv \delta(s - s') \tag{8.5.50}$$

since T and ζ^0 can depend only upon the distance between beads along the chain. Now equation (8.5.50) can be solved exactly by use of Fourier transforms

$$\tilde{T}(q) = \int_{-\infty}^{\infty} d(s - s')\exp[iq(s - s')]T(s - s') \tag{8.5.51a}$$

and

$$\tilde{\zeta}^0(q) = \int_{-\infty}^{\infty} d(s - s')\exp[iq(s - s')]\zeta^0(s - s') \tag{8.5.51b}$$

to give the simple algebraic result

$$\tilde{\zeta}^0(q) = [\zeta^{-1} + \tilde{T}(q)]^{-1} \tag{8.5.52}$$

with the condition that the minimum and maximum wavenumbers are $q_0 = \pi/nl$ and $q_c = \pi/l$, respectively. (For a ring polymer q_0 and q_c are twice as large.) These represent cut-offs on integrations in q-space because of the minimum length l associated with a bead, and the maximum length scale nl, associated with the chain.

In order to employ (8.5.52) in (8.5.33), it is also necessary to convert the summations to integrals over infinite limits; however, as $n \to \infty$, equation (8.5.33) is proportional to n. Since the integrand is only a function of the variable $i - j$, this dependence can be incorporated by the use of the limiting association

$$\sum_{i,j=1}^{n} g(i-j) \longrightarrow n \int_{-n}^{n} \mathrm{d}(i-j) g(i-j) \longrightarrow \frac{n}{l} \int_{-\infty}^{\infty} \mathrm{d}(s-s') g[(s-s')/l] \tag{8.5.53}$$

Replacing the average of the product in (8.5.33) by the product of the averages, just as in (8.5.37), and introducing Fourier transforms, this equation becomes

$$[\eta] = \frac{l^2 N_{\mathrm{A}}}{3\pi\eta_0 M_{\mathrm{A}}} \int_{q_0}^{\infty} \frac{\mathrm{d}q}{q^2} \tilde{\zeta}^0(q) \tag{8.5.54}$$

Notice that the upper integration limit has been extended to infinity as it converges there and as the integrand is negligible between q_c and ∞. (Also the generalized Fourier transform of $|s - s'|$ has been employed.) In order to recover the results (8.5.45) and (8.5.49), the integration in (8.5.54) is reconverted to a summation over the (Fourier) Rouse modes p,

$$[\eta] = \frac{N_{\mathrm{A}} n l^3}{6\pi^2 \eta_0 M_{\mathrm{A}}} \sum_{p=1}^{\infty} p^{-2} \tilde{\zeta}^0(\pi p/nl) \tag{8.5.55}$$

The above procedure is equivalent to the use of a continuous chain model from the outset,[6,35–38] and it simplifies the algebra in treating the polymer viscosity; however, the more tedious type analysis (8.5.33)–(8.5.49) is sometimes necessary in certain cases such as the translational friction coefficient at low polymer concentrations.[40]

Solution self-energy at finite polymer concentrations

The above two sections discuss only the case of infinite dilution in order to regain contact with the older theories. Now we turn to the case of solutions at finite concentrations where fundamentally new results are obtained from our general formulation.[36–40] The dynamical ($\omega \neq 0$) viscosity and a more detailed description of the polymer dynamics follow in Section 8.6.

Dependence on higher order powers of the concentration must be generated by those terms in the average of (8.5.11) for which there is repeated occurrence of a given chain in the sequence of scattering events in any m-th order contribution to the series since all the terms involving no repetition in any m-th order contribution yield the infinite dilution limit (8.5.23). In order to visualize more easily those terms involving the multiple appearance of a given chain, it is useful to represent the averaged multiple scattering series diagrammatically. We begin to motivate the use of diagrams by displaying the series expansion for $\langle \mathbf{u} \rangle$ by introducing the average of (8.5.11) into (8.5.22) and the results into (8.5.2), giving

$$\begin{aligned}\langle \mathbf{u} \rangle = \mathbf{G} \cdot \mathbf{F} &- \sum_{\alpha=1}^{N} \mathbf{G} \cdot \langle \mathcal{T}_\alpha \rangle \cdot \mathbf{G} \cdot \mathbf{F} + \sum_{\alpha} \sum_{\beta \neq \alpha} \mathbf{G} \cdot \langle \mathcal{T}_\alpha \rangle \cdot \mathbf{G} \cdot \langle \mathcal{T}_\beta \rangle \cdot \mathbf{G} \cdot \mathbf{F} \\ &- \sum_{\alpha} \sum_{\beta \neq \alpha} \sum_{\gamma \neq \beta} \mathbf{G} \cdot \langle \mathcal{T}_\alpha \cdot \mathbf{G} \cdot \mathcal{T}_\beta \cdot \mathbf{G} \cdot \mathcal{T}_\gamma \rangle \cdot \mathbf{G} \cdot \mathbf{F} \\ &+ \sum_{\alpha} \sum_{\beta \neq \alpha} \sum_{\gamma \neq \beta} \sum_{\delta \neq \gamma} \mathbf{G} \cdot \langle \mathcal{T}_\alpha \cdot \mathbf{G} \cdot \mathcal{T}_\beta \cdot \mathbf{G} \cdot \mathcal{T}_\gamma \cdot \mathbf{G} \cdot \mathcal{T}_\delta \rangle \cdot \mathbf{G} \cdot \mathbf{F} + \ldots\end{aligned} \tag{8.5.56}$$

Because entanglements are still neglected, chains α and β are uncorrelated in the second order term in (8.5.56), so the factors of $\mathscr{T}_\alpha$ and $\mathscr{T}_\beta$ may be averaged separately. In third order the summation over γ may include α, a multiple occurrence term, or may range over all chains other than α (and β, of course), yielding the two pieces

$$-\sum_{\alpha}\sum_{\beta\neq\alpha} \mathbf{G}\cdot\langle\, \mathscr{T}_\alpha\cdot\mathbf{G}\cdot\langle\, \mathscr{T}_\beta\,\rangle\cdot\mathbf{G}\cdot\mathscr{T}_\alpha\,\rangle\cdot\mathbf{G}\cdot\mathbf{F}$$

$$-\sum_{\alpha}\sum_{\beta\neq\alpha}\sum_{\gamma\neq\alpha,\beta} \langle\, \mathscr{T}_\alpha\,\rangle\cdot\mathbf{G}\cdot\langle\, \mathscr{T}_\beta\,\rangle\cdot\mathbf{G}\cdot\langle\, \mathscr{T}_\gamma\,\rangle\cdot\mathbf{G}\cdot\mathbf{F} \tag{8.5.57}$$

respectively. Contributions of order c^2 to $\mathbf{\Sigma}$ come from the first term in (8.5.57), leading to a portion of the c-dependent part of (8.5.8), i.e. part of the Huggins coefficient k_{H} of (8.2.26). Higher order terms in (8.5.56) also contribute to the c^2 part of $\mathbf{\Sigma}$, and all such terms should be taken together for consistency, etc., for terms leading to the c^n part of $\mathbf{\Sigma}$.

Diagrammatic methods enable the isolation of particular terms, and they are useful to summarize the details of complicated analyses. The diagrammatic representation of (8.5.56) is generated as follows. Each factor of $\mathbf{G}$ is depicted by a single horizontal line, each $-\mathscr{T}_\alpha$ is represented by a vertical wavy line, and the force $\mathbf{F}$ is displayed as a large dot. Whenever a chain reappears more than once in a given order, the wavy lines for all the $-\mathscr{T}_\alpha$ lines for a given chain are connected together. The appropriate integrations over space–time points $\mathbf{r}$, t and summations over all polymer chains are understood to be taken. Using these definitions, (8.5.56) is represented pictorially in Figure 8.5.1. The first term in Figure 8.5.1 provides the contribution to the average fluid velocity field from fluid elements which have not been disturbed by polymers, $\mathbf{G}\cdot\mathbf{F}$. The second term arises from fluid elements that are perturbed by a single chain, while the third involves scattering by a pair of chains. The fourth and fifth diagrams in Figure 8.5.1 are associated with the second, uncorrelated, and the first, repeated, portions of the third order contributions in (8.5.57). Likewise, the next three diagrams in Figure 8.5.1 depict some of the fourth order parts of (8.5.56). The fifth, seventh, and eighth diagrams in Figure 8.5.1 contribute at finite polymer concentrations.

Individual diagrams are termed reducible if, after removing the initial dot, corresponding to F, they can be decomposed into two separate unconnected diagrams, contributing to $\langle\, \hat{\mathbf{G}}\,\rangle$, by breaking one of the internal $\mathbf{G}$ lines. Hence, the second order diagram in Figure 8.5.1 is

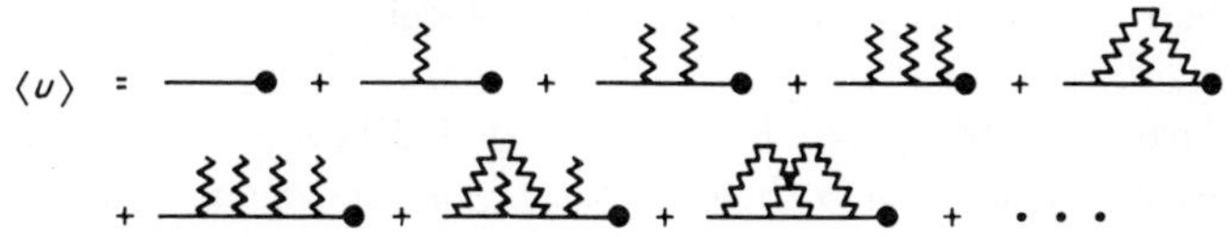

$\langle v \rangle$ = [diagram series] + ...

Figure 8.5.1 Diagrammatic representation of the multiple scattering series (8.5.56) for the average fluid velocity field. The dot on the right corresponds to a factor of $\mathbf{F}$, each solid horizontal line represents a factor of $\mathbf{G}$, while each wavy line indicates a factor of $-\mathscr{T}_\alpha$. Exclusion conditions operate requiring that no two successive wavy lines are associated with the same chain. Wavy lines, which are connected at the top, correspond to terms involving fluid scattering by the same chain. Time runs from right to left; integration over all intermediate times and spatial positions is understood to be taken; and summations over all chains are to be taken for each wave line subject to exclusion conditions and constraints generated by connections between wavy lines

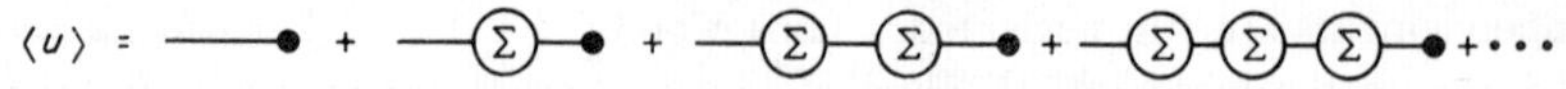

$\langle u \rangle$ = —● + —(Σ)—● + —(Σ)—(Σ)—● + —(Σ)—(Σ)—(Σ)—● + ...

Figure 8.5.2 The expansion of the average fluid velocity field in terms of the fluid 'self-energy' operator. Exclusion principles are operative requiring that individual **Σ** elements have no chain indices in common with each other

reducible into a pair of unconnected first order diagrams, but the fifth diagram is irreducible with respect to the cutting of any interior **G** line (other than the first or last **G** line), because the resultant pieces are connected by the attached wavy $-\mathscr{T}_\alpha$ lines. The irreducible diagrams, ones which are not reducible, are defined as the 'self-energy' operator **Σ**. Hence, the series (8.5.56) can be represented in terms of multiple appearances of the self-energy operator **Σ** as in Figure 8.5.2. However, in contrast to usual diagrammatic series, each repeated occurrence of **Σ** operators cannot contain the same polymers as are present in any other **Σ**-'blobs', since, if they did, the two pieces would have connected wavy lines, and they would not be reducible into a pair of **Σ** inserts as assumed. This multiple occupancy exclusion condition represents a serious restriction in the analogous case of vibrations or electronic structure in disordered systems[51,52,56] where the analog of **G** in these systems usually involves nearest neighbour couplings on a lattice. In the polymer case **G** is very long range, so the neglect of these exclusion constraints leads to errors of order $1/N$ just as described in the infinite dilution limit in (8.5.25).

The diagrammatic series for **Σ** is presented in Figure 8.5.3. The concentration dependence of terms in Figure 8.5.3 is easily shown to be c^ν where ν is the number of different chains appearing in the diagram. Thus, the first term in Figure 8.5.3 is the only one which is of order c, and by (8.5.8) and (8.2.26) this diagram yields the polymer intrinsic viscosity from the Taylor series expansion of **Σ** in powers of k^2. Likewise, the second, third, sixth, . . . diagrams in Figure 8.5.3 are of order c^2 and yield contributions to k_H of (8.2.26).

Σ diagrams come in two varieties. Some, like the first, second, fourth, . . . in Figure 8.5.3 do not contain wavy lines which cross each other, and these are termed the uncorrelated **Σ**-diagram. The correlated **Σ**-diagrams are, hence, the third, fifth, sixth, . . . in Figure 8.5.3. There is an essential mathematical and physical distinction between these two classes of diagrams, and this contrast can be understood by consideration of other situations in which similar types of diagrams arise. The analysis by analogy saves a good deal of the effort associated with performing similar calculations in a new guise.

Diagrams like those in Figures 8.5.1–8.5.3 also appear in the study of electronic states in disordered systems as noted above. The uncorrelated diagrams contribute to the so-called coherent potential approximation, a sophisticated mean field theory in which the average net scattering of electrons by the remaining perturbing potential fluctuations, due to single sites, is taken to vanish.[51,56] The correlated diagrams then describe the effects of the correlated

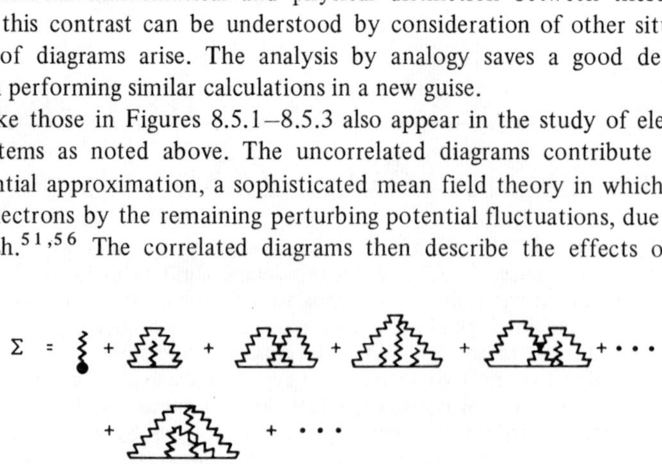

Σ = [diagram] + [diagram] + [diagram] + [diagram] + [diagram] + . . .

+ [diagram] + . . .

Figure 8.5.3 Multiple scattering expansion for the fluid self-energy operator. Diagram rules are those given under Figure 8.5.1

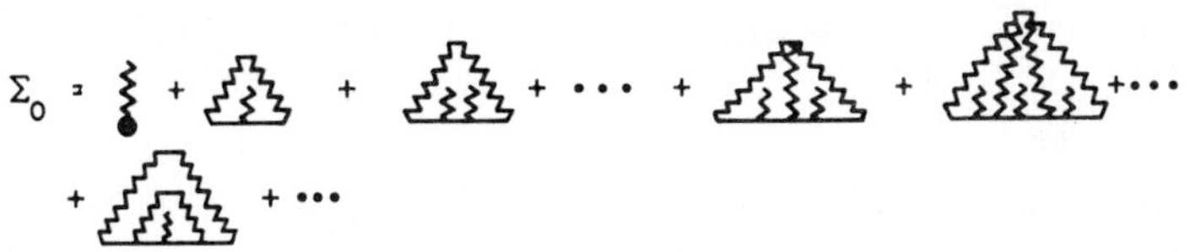

$\Sigma_0 = \cdots + \cdots + \cdots + \cdots + \cdots + \cdots + \cdots$

Figure 8.5.4 Uncorrelated parts of the fluid self-energy operator. Diagram rules are those given under Figure 8.5.1

scattering of electrons by clusters of lattice sites, and such cluster theories have yet to be completely satisfactorily formulated. Hence, the analogy with the polymer problem implies that correlated diagrams are associated with correlated scattering of the fluid by clusters of chains, a phenomenon indicative of the existence of temporary or permanent entanglements. Hence, in the regime where entanglement effects can be ignored, only the uncorrelated $\mathbf{\Sigma}$ diagrams need be included.

Alternatively, the same conclusion follows by consideration of the excluded volume problem generated by (8.2.2)–(8.2.7). Here an analogous, but similar, diagram expansion exists[6] for the unnormalized end-to-end vector distribution $ZP(\mathbf{R}, n)$ of (8.2.4) wherein only the analogues of those $\mathbf{\Sigma}$ diagrams in Figure 8.5.3 appear which have pairs of connected wavy lines, such as the third $\mathbf{\Sigma}$ diagram in Figure 8.5.3. The multiple scattering exclusion, that successive interaction lines be associated with different beads, is not operative for the excluded volume problem. Here, the connected pair of wavy lines is associated with a factor of $v\,(\mathbf{R}_i - \mathbf{R}_j)$, while the horizontal lines correspond to distribution functions for unperturbed random flight chains. The correlated diagrams in this case are responsible for the introduction of the long range correlations within a chain that tend to spread it out in space more than the degree characteristic of a random walk.

With correlated diagrams ignored, the remaining uncorrelated diagrams are presented in Figure 8.5.4 to display some of the typical topological structures involved. Inspection of these diagrams shows that the constraint of no crossing wavy lines requires that the same chain be the first and last one to scatter the fluid. It is, thus, convenient to separate scatterings by this chain from all the rest by representing $\mathbf{\Sigma}_0$ as

$$\mathbf{\Sigma}_0 = \sum_{\alpha=1}^{N} \langle -\mathscr{T}_\alpha + \mathscr{T}_\alpha \cdot \mathbf{G} \cdot \mathscr{T}^{(\alpha)} \cdot \mathbf{G} \cdot \mathscr{T}_\alpha - \mathscr{T}_\alpha \cdot \mathbf{G} \cdot \mathscr{T}^{(\alpha)} \cdot \mathbf{G} \cdot \mathscr{T}_\alpha \cdot \mathbf{G} \cdot \mathscr{T}^{(\alpha)} \cdot \mathbf{G} \cdot \mathscr{T}_\alpha + \ldots \rangle \tag{8.5.58}$$

In writing (8.5.58), we have used the fact that any of the chains α may be the first and last to scatter the fluid, so a sum over all α appears. The series (8.5.58) and the definition of $\mathscr{T}^{(\alpha)}$ in (8.4.9) indicates that scatterings occur by all chains but α between the repeated scatterings by α. The first term in Figure 8.5.4 is just the first in the series (8.5.58), the second, third, and last diagrams are parts of the second term in the series, etc. In principle, as (8.5.58) is written, some correlated scatterings between chains other than α are included because the $\mathscr{T}^{(\alpha)}$ operators can contain internal correlated parts which do not involve α. It is convenient for the analysis to include these terms formally, but the final self-consistent approximations below assume that correlated scatterings for chains other than α likewise are negligible.

By introducing the definition (8.4.5) of $\mathscr{T}_\alpha$ the series (8.5.58) is converted to

$$\mathbf{\Sigma}_0 = \sum_{\alpha=1}^{N} \langle -\phi_\alpha \Delta_\alpha [\, \mathscr{G}_\alpha - \mathscr{G}_\alpha \phi_\alpha \cdot \mathbf{G} \cdot \mathscr{T}^{(\alpha)} \cdot \mathbf{G} \phi_\alpha \Delta_\alpha \cdot \mathscr{G}_\alpha + \mathscr{G}_\alpha \phi_\alpha \cdot \mathbf{G} \cdot \mathscr{T}^{(\alpha)} \cdot \mathbf{G} \phi_\alpha \cdot \mathbf{G} \cdot \mathscr{T}^{(\alpha)} \cdot \mathbf{G} \cdot \phi_\alpha \Delta_\alpha \mathscr{G}_\alpha + \ldots] \phi_\alpha \rangle \tag{8.5.59}$$

The series in brackets in (8.5.59) is identical to that in brackets in (8.4.11) and (8.4.15), so $\mathbf{\Sigma}_0$ simplifies to

$$\mathbf{\Sigma}_0 = - \sum_{\alpha=1}^{N} \langle \phi_\alpha \Delta_\alpha \hat{\mathscr{G}}_\alpha \phi_\alpha \rangle \tag{8.5.60}$$

where the screened polymer Green's function $\hat{\mathscr{G}}_\alpha$ is defined by (8.4.13), which can more simply be represented as

$$\left(\frac{\partial}{\partial t} \mathbf{1} + \zeta^{-1} \mathbf{1} \Delta_\alpha + \phi_\alpha \hat{\mathbf{G}}^{(\alpha)} \phi_\alpha \Delta_\alpha \right) \cdot \hat{\mathscr{G}}_\alpha = \mathbf{1} \tag{8.5.61}$$

The Green's function for the solution without chain α is then replaced by its average value $\langle \hat{\mathbf{G}}^{(\alpha)} \rangle$ as in (8.4.21), and then the omission of α can be neglected to yield the approximation,

$$\left(\frac{\partial}{\partial t} \mathbf{1} + \zeta^{-1} \mathbf{1} \Delta_\alpha + \phi_\alpha \langle \hat{\mathbf{G}} \rangle \phi_\alpha \Delta_\alpha \right) \cdot \hat{\mathscr{G}}_\alpha = \mathbf{1} \tag{8.5.62}$$

to (8.5.61) which is applicable to the absence of correlated scatterings. The set of equations (8.5.60), (8.5.62), and (8.5.7) are nonlinear self-consistent integral equations for $\mathbf{\Sigma}_0$. These equations are most conveniently presented in terms of coupled equations between $\mathbf{\Sigma}_0$ and the effective concentration-dependent hydrodynamic interaction for chain

$$\hat{\mathbf{K}}_{\alpha\alpha} = \zeta^{-1} \mathbf{1} + \phi_\alpha \langle \hat{\mathbf{G}} \rangle \phi_\alpha . \tag{8.5.63}$$

Equations (8.5.62) and (8.5.63) represent the dynamical equations for a single chain where hydrodynamic interactions propagate through the effective medium $\langle \hat{\mathbf{G}} \rangle$ which is determined self-consistently. Equations (8.5.60) and (8.5.8) then represent the solution viscosity as a sum (incoherent) of contributions from these concentration dependent 'dressed' chains. While this mean field approach is familiar in other fields, the essential nonlinearities of this problem make its justification and derivation quite difficult.

Hydrodynamic screening in concentrated solutions

For simplicity, the analysis is further limited to a consideration of the steady state limit, whereupon the coupled equations can be represented more simply. From the definitions (8.5.62) and (8.5.63), the analogue of (8.5.31) is

$$\Delta_\alpha \mathscr{G}_\alpha = (\hat{\mathbf{K}}_{\alpha\alpha})^{-1} - (\hat{\mathbf{K}}_{\alpha\alpha})^{-1} \cdot \frac{\partial}{\partial t} \mathscr{G}_\alpha \tag{8.5.64}$$

and in the static limit the term in $\partial \hat{\mathscr{G}}_\alpha / \partial t$ can be dropped. Hence reintroducing the explicit indices, equation (8.5.60) becomes

$$\mathbf{\Sigma}_0(\mathbf{r} - \mathbf{r}', \omega = 0, c) = -N \sum_{i,j=1}^{n} \langle \phi_i(\mathbf{r}) [\hat{\mathbf{K}}^{-1}]_{ij} \phi_j(\mathbf{r}') \rangle \tag{8.5.65}$$

where $\phi_i(\mathbf{r})$ is the steady state limit operator (8.3.21a), and the factor of N arises because each chain gives an identical contribution (so the index α has been dropped). Fourier transforms are introduced just as in (8.3.15)–(8.3.16), and the average over the centre of mass distribution for the polymer results in a factor of V^{-1}, so (8.5.65) is reduced to

$$\mathbf{\Sigma}_0(\mathbf{k}, c) = -(N/V) \sum_{i,j=1}^{n} \langle \exp[i\mathbf{k} \cdot (\mathbf{R}_i - \mathbf{R}_j)] \, [\hat{\mathbf{K}}^{-1}]_{ij} \rangle \tag{8.5.66}$$

where $\hat{\mathbf{K}}$ is, of course, the static limit of the fully screened (8.5.63), and $\omega = 0$ has been omitted from the arguments of $\mathbf{\Sigma}_0$. Arbitrary models for the configurational statistics (8.2.2)–(8.2.7) may be used in (8.5.66). The only approximations inherent in the equation stem from the use of the uncorrelated approximation and the restriction to the static ($\omega = 0$) limit.

Define the function

$$\mathbf{J}_{ij}(k, c) = \sum_{p=1}^{n} [\hat{\mathbf{K}}^{-1}]_{ip} \exp[i\mathbf{k} \cdot (\mathbf{R}_p - \mathbf{R}_j)] \tag{8.5.67}$$

so it obeys the equation

$$\sum_{p=1}^{n} \hat{\mathbf{K}}_{ip} \cdot \mathbf{J}_{pj}(k, c) = \mathbf{1} \exp[i\mathbf{k} \cdot (\mathbf{R}_i - \mathbf{R}_j)] \tag{8.5.68}$$

eliminating the unwieldy $\hat{\mathbf{K}}^{-1}$ operator. Parelleling the transition from (8.5.33) to (8.5.36), equations (8.5.66)–(8.5.68) imply that

$$\mathbf{\Sigma}_0 = -(N/V) \sum_{i=1}^{n} \langle \mathbf{J}_{ii}(k, c) \rangle \tag{8.5.69}$$

Just as in (8.5.37), equation (8.5.68) is averaged over chain configurations, and the average of the product of $\hat{\mathbf{K}}$:$\mathbf{J}$ is replaced by the product of averages. Corrections for the inadequacy of this approximation can be obtained as in (8.5.38). When $\hat{\mathbf{K}}$ is averaged over chain conformations, the results must be proportional to a unit tensor, (likewise for $\mathbf{J}$) so the definitions

$$\langle \hat{\mathbf{K}}_{ij} \rangle \equiv \mathbf{1}\, \hat{K}_{ij}(c) \tag{8.5.70a}$$

$$\langle \mathbf{J}_{ij}(k, c) \rangle = \mathbf{1}\, J_{ij}(k, c) \tag{8.5.70b}$$

can be used to simplify the equations to

$$\mathbf{\Sigma}_0(k, c) = \Sigma_0(k, c)\mathbf{1} = -(N/V)\mathbf{1} \sum_{i=1}^{n} J_{ii}(k, c) \tag{8.5.71}$$

$$\sum_{p=1}^{n} \hat{K}_{ip}(c) J_{pj}(k, c) = \langle \exp[i\mathbf{k} \cdot (\mathbf{R}_i - \mathbf{R}_j)] \rangle. \tag{8.5.72}$$

The equations become closed upon introduction of the definition of $\hat{\mathbf{K}}$ to manifest its dependence on Σ_0. Equations (8.5.63) and (8.5.70a) enable us to write

$$\hat{K}_{ij}(c)\mathbf{1} = \zeta^{-1}\delta_{ij}\mathbf{1} + (1 - \delta_{ij}) \langle \hat{\mathbf{G}}(\mathbf{R}_i - \mathbf{R}_j; \omega = 0) \rangle \tag{8.5.73}$$

After introduction of the Fourier representation for $\langle \hat{\mathbf{G}} \rangle$ generated by (8.5.7), equation (8.5.73) gives

$$\hat{K}_{ij}(c)\mathbf{1} = \zeta^{-1}\delta_{ij}\mathbf{1} + (1 - \delta_{ij}) \int \frac{d^3 k}{(2\pi)^3} [\mathbf{1} - k^{-2}\mathbf{k}\mathbf{k}]$$

$$\times [\eta_0 k^2 - \Sigma_0(k, c)]^{-1} \langle \exp[i\mathbf{k} \cdot (\mathbf{R}_i - \mathbf{R}_j)] \rangle \tag{8.5.74}$$

Recognizing that the static correlation function

$$S_{ij}(k) = \langle \exp[i\mathbf{k} \cdot (\mathbf{R}_i - \mathbf{R}_j)] \rangle \tag{8.5.75}$$

can only be a function of $k = |\mathbf{k}|$, the angular average over directions of $\mathbf{k}$ can be evaluated to yield the final equation,

$$\hat{K}_{ij}(c) = \zeta^{-1}\delta_{ij} + (1-\delta_{ij})\frac{2}{3}\frac{4\pi}{(2\pi)^3}\int_0^\infty k^2\,\mathrm{d}k S_{ij}(k)[\eta_0 k^2 - \Sigma_0(k,c)]^{-1} \tag{8.5.76}$$

complementing (8.5.71) and (8.5.72).

Flexible chains have Gaussian distributions for $(\mathbf{R}_i - \mathbf{R}_j)$, so $S_{ij}(k)$ then simplifies to

$$S_{ij}(k) = \exp[-k^2\langle(\mathbf{R}_i - \mathbf{R}_j)^2\rangle/6] \tag{8.5.75a}$$

It is important to notice that $\langle(\mathbf{R}_i - \mathbf{R}_j)^2\rangle$ may, in fact, depend upon polymer concentration because of interchain interactions, and this fact has been incorporated in a calculation of quasi-elastic neutron scattering cross-sections from concentrated polymer solutions[39] as well as in a theory of the excluded volume dependence of the Huggins coefficient for solutions near Θ conditions.

Given this further simplification to flexible chains, the definition (8.5.8) enables (8.5.71)–(8.5.76) to be converted to the concentration-dependent generalization of the Kirkwood–Riseman equations (8.5.34)–(8.5.36),

$$\sum_{p=1}^{n} \hat{K}_{ip}(c)J_{pj}(c) = (18/nl^2)\langle\exp[-\mathrm{i}\mathbf{k}\cdot(\mathbf{R}_i - \mathbf{R}_j)]\rangle \tag{8.5.77a}$$

$$\eta_{\mathrm{sp}}(c) = \lim_{k\to 0}\frac{N_A\zeta l^2}{18\eta_0 M_A k^2}\sum_{i=1}^{n} J_{ii}(c) \tag{8.5.77b}$$

where the average of $\hat{\mathbf{K}}$ has been further employed and the normalization in (8.5.77a) has been altered so the equation corresponds directly to (8.5.36). The traditional theories of polymer dynamics and the transport properties of polymer solutions,[1] described in Section 8.2 and equations (8.5.33ff), all begin by introducing the pre-averaged Oseen tensor $\langle T_{ip}\rangle$ from (8.3.20) as in (8.5.37), and then possibly introduce corrections for this pre-averaging by perturbation theory as in (8.5.38). For finite concentrations (and ignoring entanglements) this procedure cannot simply yield the concentration-dependent $\hat{K}_{ip}(c)$ in (8.5.72) and (8.5.77a) which incorporates the full hydrodynamics of the whole solution that governs the true hydrodynamic interactions in polymer solutions at finite concentrations. After deriving the correct $\hat{\mathbf{K}}$, the use of its average in (8.5.72) and (8.5.77a) is just one of numerical convenience.

The equations (8.5.77) can be obtained from the k^2 coefficient of (8.5.71), (8.5.72), and (8.5.76) (with altered normalization), so the latter more general equations are considered first. We can analyse these equations by pursuing an analysis similar to that used at infinite dilution in (8.5.40)–(8.5.49). Instead, the same final results are generated by using the Fourier transformation method[35–38] outlined in (8.5.50)–(8.5.55). In this case we can employ a continuous chain notation $\mathbf{R}(s)$, and equations like (8.5.72) are readily solved by Fourier transformation. Using transformations like those in (8.5.53), the pre-averaged form of (8.5.66) reduces (8.5.66) to

$$\Sigma_0(k,c) = -(N/V)(n/l)\int_{-\infty}^{\infty}\mathrm{d}(s-s')S(s-s',k)\hat{K}^{-1}(s-s',c) \tag{8.5.78}$$

where $S(s-s',k)$ and $\hat{K}^{-1}(s-s',c)$ are the continuum limits of $S_{ij}(k)$ and $[K^{-1}(c)]_{ij}$. Employing the Fourier transforms,

$$\Gamma(k,q) = \int_{-\infty}^{\infty}\mathrm{d}(s-s')\exp[\mathrm{i}q(s-s')]\,S(s-s',k) \tag{8.5.79a}$$

and

$$K(q,c) = \int_{-\infty}^{\infty} \mathrm{d}(s-s')\exp[\mathrm{i}q(s-s')]\,k(s-s',c) \tag{8.5.79b}$$

equation (8.5.78) can be transformed to

$$\Sigma_0(k,c) = -\left(\frac{nNl}{V}\right)\int_0^{\infty} \mathrm{d}q\,\frac{\Gamma(k,q)}{\pi K(q,c)} \tag{8.5.80}$$

which follows from the simple relation between the Fourier transforms of $\hat{K}^{-1}$ and $\hat{K}$,

$$K^{-1}(q,c) = \int_{-\infty}^{\infty} \mathrm{d}(s-s')\exp[\mathrm{i}q(s-s')]\hat{K}^{-1}(s-s',c) = \frac{1}{K(q,c)} \tag{8.5.81}$$

Similarly, the Fourier transform of (8.5.76) yields

$$K(q,c) = (\zeta/l)^{-1} + (3\pi^2)^{-1}\int_0^{\infty} k'^2\,\mathrm{d}k'\,\Gamma(k',q)[\eta_0 k'^2\,\Sigma_0(k',c)]^{-1} \tag{8.5.82}$$

so (8.5.80) and (8.5.82) are the basic pair of coupled integral equations.[36,38] Again, the simplest case of a Gaussian random coil is considered further, and this leads to

$$\Gamma(k,q) = (k^2 l/3)[(k^2 l/6)^2 + q^2]^{-1} \tag{8.5.83}$$

Equations (8.5.80) and (8.5.82) can be solved in the limiting cases of $k \to 0$ or $k \to \infty$. The general nature of these solutions can be deduced in a bootstrap fashion. For $k \to 0$ the $\Gamma(k,q)$ term in (8.5.80) has dominant contributions for $q \propto (k^2 l/6)$, if $K(q,c)$ is then slowly enough varying with q over this region. For $q \propto (k^2 l/6)$ the $\Gamma(k',q)$ part of (8.5.82) has its major contributions for $q \propto (k'^2 l/6)$, so the important values of k' are $k' \approx k$. When $\Sigma_0(0,c)$ is nonvanishing, the second factor in (8.5.82) gives contributions for $k' \propto [-\Sigma_0(0,c)/\eta_0]^{1/2}$, providing an additional length scale delineating when $\Sigma_0(0,c)$ is nonzero and when $K(q,c)$ is slowly varying for small q. Thus, for $k \to 0$ the $\Sigma_0(k',c)$ in the denominator of (8.5.82) can be expanded through terms of order k'^2,

$$\Sigma_0(k,c) = \Sigma_0(0,c) - k^2 c\eta_0\eta_{\mathrm{sp}}(c) \tag{8.5.84}$$

and the slow variation of $K(q,c)$ with q means that the $q \to 0$ limit can be taken in (8.5.82), provided that $\Sigma_0(0,c)$ is nonzero. Hence, (8.5.82) is reduced to

$$K(0,c) = (\zeta/l)^{-1} + (3\pi^2)^{-1}\int_0^{\infty} k'^2\,\mathrm{d}k'(12/k'^2 l)[\eta k'^2 - \Sigma_0(0,c)]^{-1} \tag{8.5.85}$$

where η is the viscosity of the full polymer solution. Similarly, employing the small q limit of $K(q,c)$ in (8.5.80) transforms this equation to

$$\Sigma_0(0,c) = -\left(\frac{clN_{\mathrm{A}}}{M_{\mathrm{A}}K(0,c)}\right)\int_0^{\infty} \mathrm{d}q\left(\frac{k^2 l}{3\pi}\right)\left[\left(\frac{k^2 l}{6}\right)^2 + q^2\right]^{-1} \tag{8.5.86}$$

Now (8.5.85) and (8.5.86) are a pair of coupled equations for $\Sigma_0(0,c)$ and $K(0,c)$ which are readily solved since the integrals are elementary, so we obtain

$$K(0,c) = (\zeta/l)^{-1} + 2/l\pi[-\eta\Sigma_0(0,c)]^{-1/2} \tag{8.5.87}$$

and

$$- \Sigma_0(0, c) = clN_A/M_A K(0, c). \tag{8.5.88}$$

This pair of equations has the general solutions

$$- \Sigma_0(0, c) = \eta^{-1}(\zeta/\pi l^2)^2\{[1 + cN_A\eta l^4\pi^2/M_A\zeta]^{1/2} - 1\}^2 \tag{8.5.89}$$

and

$$K(0, c) = (l/\zeta)(1 + 2\{[1 + cN_A\eta l^4\pi^2/M_A\zeta]^{1/2} - 1\}^{-1}). \tag{8.5.90}$$

The finite $q \to 0$ limit $K(0, c)$ is to be contrasted with its value at infinite dilution. As $c \to 0$, the $\Sigma_0(k', c)$ can be dropped in (8.5.82) as Σ_0 in (8.5.80) has an overall factor of c. Then $K(q, 0)$ is readily evaluated as

$$K(q, 0) = (\zeta/l)^{-1} + [\pi\eta_0(3l\,|\,q\,|\,)^{1/2}]^{-1} \tag{8.5.91}$$

which is singular in the limit $q \to 0$. (Note q still has a minimum value of π/nl.) The q-dependence of $K(q, 0)$ results from the fact that the long range nature of hydrodynamics interactions and the correlations between bead positions on a chain, induced by (8.2.2)–(8.2.8), combine to produce correlations in the fluid scattering from different beads on a given chain. Thus, as hydrodynamic interactions are turned on, the intrinsic polymer viscosity undergoes a transition from the incoherent limit Rouse behaviour (8.2.23) to the coherent fluid scattering Rouse–Zimm limit (8.2.25). The contrasting constancy of $K(q, c)$ for small enough q and large enough c(limits to be determined below) implies that the long wavelength modes somehow have their hydrodynamic interactions screened, so they provide an incoherent fluid scattering contribution to the polymer viscosity. Using the leading coefficient of k^2 in (8.5.80) and reintroducing the lower cutoff q_0, yields the concentration-dependent polymer viscosity

$$\eta_{sp}(c) = \frac{N_A n\zeta l^2}{36\eta_0 M_A}\{1 + 2[(1 + cN_A\eta l^4\pi^2/M_A\zeta)^{1/2} - 1]^{-1}\}^{-1} \tag{8.5.92}$$

The coefficient in equation (8.5.92) is just the Rouse intrinsic viscosity

$$[\eta(0)]^R = N_A n\zeta l^2/36\eta_0 M_A \tag{8.5.93}$$

and in the screened regime, equation (8.5.92) obeys the relation

$$\eta_{sp}(c) \leqslant [\eta(0)]^R \tag{8.5.94}$$

with equality ensuing only in the high concentration limit where the parameter

$$x = cN_A\eta l^4\pi^2/\zeta M_A \tag{8.5.95}$$

is much greater than unity.

The presence of screening of hydrodynamic interactions at high enough concentrations leads to a qualitative difference between the properties of high and low concentration solutions. The mechanism for the introduction of this screening is best understood by determining the concentration at which screening commences. When screening is present, $\Sigma_0(0, c)$ is nonzero. Substituting the above solution for $\Sigma_0(0, c)$ as well as the simple coefficient of k^2 in its Taylor series expansion into (8.5.82), leads to an approximation to the full $K(q, c)$ which can be employed to find the transition. Defining the screening constant by the equation

$$\kappa = [-\Sigma_0(0, c)/\eta]^{1/2} \tag{8.5.96}$$

and the inverse screening length by

$$\delta = \kappa^2(l/6) = \xi^{-1}(c) \tag{8.5.97}$$

the integral

$$K(q,c) = (\zeta/l)^{-1} + (3\pi^2\eta)^{-1} \int_0^\infty k^2\, dk\Gamma(k,q)[k^2+\kappa^2]^{-1} \tag{8.5.98a}$$

can be evaluated to give the approximation

$$K(q,c) = (\zeta/l)^{-1} + \{\pi\eta(3l/2)^{1/2}[\delta^2+q^2]^{-1}\}\{\delta^{3/2}+\delta(|q|/2)^{1/2}+(|q|^3/2)^{1/2}\}. \tag{8.5.98b}$$

As the polymer concentration $c \to 0$, $\Sigma_0(0,c)$ and δ tend to zero, so (8.5.98) reduces to the value (8.5.91) as it must. The longest wavelength Rouse mode $q_0 = \pi/nl$ becomes screened for

$$\pi/nl \lesssim \delta \tag{8.5.99}$$

whereupon q_0 may be neglected with respect to δ.

The screening condition (8.5.99) can simply be understood by introducing two dimensions associated with the polymers in solution. The radius of the average volume allotted per chain on a democratic basis is

$$r_0 = (V/N)^{1/3} = (nM_A/N_Ac)^{1/3} \tag{8.5.100}$$

for a solution at concentration c. The average radius of a random coil from (8.29) is written as

$$R_0 = n^{1/2}l. \tag{8.5.101}$$

Equation (8.5.99) then becomes

$$r_0 \lesssim R_0(\pi/24)^{1/6} \tag{8.5.102a}$$

or

$$c \gtrsim M_A/N_Al^3(n\pi/24)^{1/2} \tag{8.5.102b}$$

and screening begins at concentrations when the natural size of a single chain, r_0, leads to significant interpenetration between chains because r_0 now exceeds the average spacing between chains. Since the different chains have been assumed to be uncorrelated with respect to each other in the screening regime governed by (8.5.102), then once a fluid element is scattered by, say, bead i on chain α, it is overwhelmingly probable that the fluid is then incoherently scattered by beads on other chains, $\beta \neq \alpha$ before it again can encounter another bead, say j, on chain α. This then destroys the correlations between successive fluid scatterings among different beads on a single chain which produces the Rouse to Rouse–Zimm transition at infinite dilution.

The hydrodynamic screening in polymer solutions is to be contrasted from the Debye screening arising in electrolyte solutions. In both cases the bare interactions vary as r^{-1}, and they become screened to $r^{-1}\exp(-\delta r)$. However, the Debye screening occurs as a result of a reorganization of charges, an introduction of correlations between them, whereas in the polymer case screening signals a *destruction* in already existing correlations. The Debye screening δ varies as $c^{1/2}$, whereas in the polymer case it has much richer behaviour as given by (8.5.89), (8.5.96), and (8.5.97).

Higher Rouse modes, larger q, become screened at progressively larger values of δ or c. Thus, there is a concentration-dependent correlation length $\xi(c)$ such that hydrodynamic correlations are screened for lengths greater than $\xi(c)$. De Gennes and coworkers[13,57] have shown that this same correlation length $\xi(c)$ also determines the screening of excluded volume interactions between different beads on a given chain, and Edwards[58] has described the mean square end-to-end distance of chains in the high concentration limit when this distance is greater than $\xi(c)$. Again, this excluded volume screening can be understood to arise from a destruction of intrachain correlations by the interpenetration of chains which are uncorrelated with respect to each other. A full theory of the combined hydrodynamic and excluded volume screening in concentrated polymer solutions has been provided[46] in a theory of the quasi-elastic neutron scattering cross-sections from these solutions.[39]

The behaviour of the polymer solution in the screening regime is dependent on the magnitude of the parameter x in (8.5.95), exhibiting a lower concentration screening regime for

$$x \ll 1 \quad \text{or} \quad r_0 \gg R_0^{2/3}(\pi^2 \eta l^2/\zeta)^{1/3} \qquad \text{(low)} \qquad (8.5.103)$$

a high concentration limit for

$$x \gg 1 \quad \text{or} \quad c \gg \zeta M_A / N_A \eta l^4 \pi^2 \qquad \text{(high)} \qquad (8.5.104)$$

and a transition zone about $x \approx 1$. In the low limits, we obtain

$$\kappa = c N_A l^2 \pi / 2M_A \qquad \text{(low)} \qquad (8.5.103a)$$

$$K(0, c) = 4M_A / l^3 \pi^2 \eta c N_A + (\zeta/l)^{-1} + \ldots \qquad \text{(low)} \qquad (8.5.103b)$$

$$\eta_{sp}(c) = \tfrac{1}{4} c k_H^d ([\eta(0)]_d^{RZ})^2 [1 - \pi^2 l^4 \eta N_A c / 4\zeta M_A + \ldots] \qquad \text{(low)} \qquad (8.5.103c)$$

where

$$[\eta(0)]_d^{RZ} = (N_A l^3 / M_A)(n/6\pi)^{1/2} \sum_{p=1}^{\infty} p^{-3/2} \qquad (8.5.105)$$

is the Rouse–Zimm intrinsic viscosity generated by (8.5.45) and (8.5.49) in the nondraining limit when $n(2/|\mathrm{p}|)^{1/2} \gg 1$, and where

$$k_H^d = \pi \sum_{p=1}^{\infty} p^{-2} \Big/ \left(\sum_{p=1}^{\infty} p^{-3/2} \right)^2 = 0.7574. \qquad (8.5.106)$$

Notice the non-Debye-like behaviour of (8.5.103a) and the overall proportionality of $\eta_{sp}(c)$ in (8.5.103c) to cM.

The higher concentration screening regime yields

$$\kappa = [c\zeta N_A / \eta M_A]^{1/2} \qquad \text{(high)} \qquad (8.5.104a)$$

$$K(0, c) = (\zeta/l)^{-1} + 2/l\pi(\eta c N_A \zeta / M_A)^{1/2} + \ldots \qquad \text{(high)} \qquad (8.5.104b)$$

$$\eta_{sp}(c) = [\eta(0)]^R \left[1 - \left(\frac{2}{\pi l^2} \right) (\zeta M_A / \eta c N_A)^{1/2} + \ldots \right] \qquad \text{(high)} \qquad (8.5.104c)$$

Equation (8.5.104a), at first, appears Debye-like, but the Stokes' law estimate (8.2.17) should now involve the full solution viscosity which is dominated by the polymers in the limit (8.5.104). Hence, equation (8.5.103a) has $\kappa \propto c(M)^{1/2}$. The high concentration limit (8.5.104c) is the Rouse result found experimentally.

It is interesting to consider the case of a single very long polymer chain in solution where the local concentration is high enough to satisfy (8.5.102). In this case the polymer viscosity is just the intrinsic viscosity, so if some model, such as a Gaussian chain with or without excluded volume, is employed, we would expect to obtain $\eta_{sp} = [\eta]$. Here $[\eta]$ is just the high molecular weight limit of the intrinsic viscosity for that model. However, a given chain monomer will, in general, find itself in the region of a number of very distant monomers along the chain, so distant, in fact, that the monomer has no way of 'knowing' that they belong to the same chain. This long chain will be so self-entangled that the Gaussian backbone model of the chain is no longer applicable. Near an entanglement junction new intrachain correlations are introduced. Since the entanglements are randomly distributed along the chain, we may picture this long chain as composed of a set of effective Gaussian chains between entanglement junctions (with or without excluded volume). Hydrodynamic and excluded volume interactions within one of these subchains will then be screened by the other subchains. Hence, as $M \to \infty$, $[\eta]$ must approach the fully screened Rouse limit of being proportional to M.

The basic integral equations (8.5.80) and (8.5.82) not only describe the effects of hydrodynamic screening on the properties of polymer solutions in the concentration region governed by (8.5.102). Given that entanglement effects are negligible, they also must describe the solution hydrodynamics for lower concentrations where $\Sigma_0(0, c)$ vanishes. At low, but finite, concentrations it is of interest to compute the Huggins coefficient, k_H, in (8.2.26). From equations (8.2.26) and (8.5.8), it is clear that an evaluation of k_H requires only the knowledge of the coefficient of k^2 in $\Sigma_0(k, c)$ through second order in c. Inspection of (8.5.80) with an overall proportionality to $c \propto N/V$ indicates that a substitution of the zeroth order $K(q, 0)$ of (8.5.91) suffices to generate only a first order value for $\Sigma_0(k, c)$, but this can be employed in (8.8.52) to generate $K(q, c)$ through order c. Subsequent use of this first order $K(q, c)$ from (8.5.80) in (8.5.8) provides $\Sigma_0(k, c)$ through order c^2. The algebra is somewhat simplified by choosing the nondraining limit $\zeta^{-1} \to 0$ in (8.5.91). Changing integration variables to

$$y = (6 \mid q \mid / k^2 l)^{1/2} \tag{8 5.107}$$

converts (8.5.80) with (8.5.91) to

$$\begin{aligned} \Sigma_0^{(1)}(k, c) &= -[4cl^2 k N_A \eta_0 / M_A (2)^{1/2}] \int_0^\infty y^2 \, dy (1 + y^4)^{-1} + O(c^2) \\ &= -(cN_A k l^2 \pi \eta_0 / M_A) + O(c^2). \end{aligned} \tag{8.5.108}$$

Notice the nonhydrodynamic proportionality of the first order $\Sigma_0^{(1)}(k, c)$ to k. This corresponds to the large k behaviour of $\Sigma_0(k, c)$. (In the small k limit required for the intrinsic viscosity $[\eta]$ from (8.5.8) and (8.5.80), the expansion of (8.5.80) to order k^2 requires the introduction of the lower cut-off q_0 for the integration. For k finite the factor of $\Gamma(k, q)$ leads to convergent behaviour as $q \to 0$, so the lower limit can be set to zero, resulting in the large k behaviour (8.5.108).)

Given (8.5.108) and the change of variables

$$z = (l/6 \mid q \mid^{1/2}) k \tag{8.5.109}$$

in (8.5.82) with (8.5.108) and $\zeta^{-1} \to 0$, yields

$$K^{(1)}(q, c) = (2/3\pi^2 \eta_0)(6/l \mid q \mid)^{1/2} \int_0^\infty dz (1 + z^4)^{-1} [z^2 + c\alpha z + O(c^2)]^{-1} \tag{8.5.110}$$

where

$$\alpha \equiv \alpha(q) \equiv \pi N_A l^{5/2} / M_A (6 \mid q \mid)^{1/2}. \tag{8.5.111}$$

The integral (8.5.110) is found to be

$$\begin{aligned} K^{(1)}(q, c) &= [\pi \eta_0 (3l \mid q \mid)^{1/2}]^{-1} [(c\alpha)^2 + 1]^{-1} \{1 + (c\alpha)^2 + c\alpha(2)^{-1/2}\} \\ &= K(q, 0)[1 - c\alpha(2)^{-1/2} + O(c^2)] \end{aligned} \tag{8.5.112}$$

and is less than $K(q, 0)$ to order c, corresponding to a diminution of the correlated scattering by the q-th Rouse mode of a chain because of hydrodynamic interactions with typical uncorrelated other chains. This decrease in intrachain hydrodynamic interactions is converted by (8.5.8) and (8.5.80) with the first order (8.5.112), in the low k-limit to an increased solution viscosity. When the integration is converted back to the more precise summation, as in (8.5.54)–(8.5.55), we obtain

$$\eta_{sp}(c) = [N_A l^3 n^{1/2}/M_A(6\pi)^{1/2}] \sum_{p=1}^{\infty} p^{-3/2} + c(N_A^2 l^6 n/6M_A) \sum_{p=1}^{\infty} p^{-2} + O(c^2), \qquad (8.5.113)$$

and the nondraining limit of the Huggins coefficient is the value quoted in equation (8.5.106).

The mechanism for the concentration dependence of the viscosity of polymer solutions becomes clear from an analysis of the high and low concentration limits for polymers with no interchain correlations. At infinite dilution the hydrodynamic interactions lead to coherent fluid scattering by the individual Rouse modes as given in (8.5.91). As the concentration is increased, the presence of hydrodynamic interactions with other chains leads to a reduction of the magnitude of the effective intrachain hydrodynamic interactions because of the intervention of uncorrelated fluid scattering by other chains between the successive fluid scatterings by a particular chain. This reduction of hydrodynamic interactions produces a concomitant increase in the solution viscosity, and the reduction in the former and increase in the latter continues as the polymer concentration is raised. Throughout this dilute region, a Rouse–Zimm type behaviour is observed (assuming the draining parameter h is large enough). However, when the concentration reaches the semi-dilute region, characterized by an approximate equality in (8.5.102), the previously singular ($|q|^{-1/2}$) long wavelength hydrodynamic interactions now become finite (constant, independent of $|q|$), and the viscosity now takes on a concentration-dependent Rouse-like form. Equation (8.5.94) shows that, as the concentration is further increased, the viscosity monotonically increases to the limiting Rouse value, while equation (8.5.90) displays a monotonic decrease in the fully screened ($|q|$-independent) hydrodynamic interactions to their asymptotic null value. The more detailed equation (8.5.98) indicates that the increasing concentration also yields the full screening of hydrodynamic interactions of higher and higher Rouse modes.

A related analysis of the translational friction coefficient per chain from (8.5.19) and (8.5.21) displays a similar behaviour.[40] At infinite dilution the correlated scatterings, arising from hydrodynamic interactions, produce a Rouse–Zimm value which is smaller than the pure Rouse limit. Increasing the concentration decreases the hydrodynamic interactions, as above, and leads to an increased translational friction coefficient, approaching the Rouse value at high concentration. Again the screening transition (8.5.102) is the demarcation between a concentration-dependent Rouse–Zimm character at lower concentrations and a concentration-dependent Rouse form on the higher side.

8.6 Random forces and dynamical viscosity

Previous sections only consider specific applications to the steady state limit wherein the explicit treatment of the polymer dynamics is unnecessary. This exposition lays the necessary groundwork for a more general discussion of dynamics. Dynamical transport properties of polymer solutions are considerably more interesting as they contain information about this chain dynamics. A theoretical description of these dynamical processes requires the solution of the chain Langevin equations (8.4.19)–(8.4.23). Random forces in these equations are responsible for driving the Brownian motion of the chain, and an explicit analysis of these random forces is necessary before quantities like the dynamical viscosity can be treated.

Random forces and the fluctuation–dissipation theorem[35]

First, consider the case where the chain has no internal structure, so the internal friction effects in (8.2.22) and (8.3.7) can be neglected for the moment. Equation (8.4.19), or the more approximate infinite dilution (8.3.38) with static hydrodynamic interaction, then constitutes the Langevin equations for the chain. For the former, more general, equation the lack of internal chain structure implies the absence of a random component $\mathbf{f}_\rho^*$. Further, it is permissible to consider a fluid which is at rest in the absence of polymers. Hence, **F** involves only random forces **F*** in the pure fluid. It might be assumed that these forces are associated with purely random thermal fluctuations and are governed by having zero mean and the variance (8.3.40), where the average in the latter corresponds only to an ensemble average over these random forces. The use of (8.3.40) and the 'group' property of Green's functions,

$$\int d\mathbf{r}'' \, \hat{\mathbf{G}}^{(\alpha)}(\mathbf{r}-\mathbf{r}''; t-t'') \cdot \hat{\mathbf{G}}^{(\alpha)}(\mathbf{r}''-\mathbf{r}'; t''-t') = \hat{\mathbf{G}}^{(\alpha)}(\mathbf{r}-\mathbf{r}'; t-t') \tag{8.6.1}$$

leads to the conclusion that

$$\langle \phi_\alpha \hat{\mathbf{G}}^{(\alpha)} \cdot \mathbf{F}^* \rangle = 0 \tag{8.6.2}$$

and the result that

$$\int d\mathbf{r}' \, d\mathbf{r} \, d\mathbf{r}_0' \, d\mathbf{r}_0 \, dt' \, dt_0' \, \langle \mathbf{F}^*(\mathbf{r}', t') \cdot \hat{\mathbf{G}}^{(\alpha)}(\mathbf{r}'-\mathbf{r}; t'-t) \phi_{\alpha i}(\mathbf{r}, t) \times \phi_{\alpha j}(\mathbf{r}_0, t_0) \hat{\mathbf{G}}^{(\alpha)}(\mathbf{r}_0-\mathbf{r}_0'; t_0-t_0') \cdot \mathbf{F}^*(\mathbf{r}_0', t_0') \rangle = 6 k_B T \hat{\mathbf{K}}_{\alpha i, \alpha j}(t, t_0) \tag{8.6.3}$$

as might be anticipated from (8.4.19) if this equation were to obey the fluctuation–dissipation theorem.[46] Note that (8.3.39) is but a special limit of (8.6.3).

The fluctuation–dissipation theorem is, in part, just a requirement on the random forces assuring the existence of a proper equilibrium limit for the system.[46] Consequently, it is only necessary to verify that (8.6.3) is consistent with the many-chain generalization of (8.2.3). As the analysis with the general case is quite tedious, for simplicity, we explicitly consider only the Gaussian chain at infinite dilution. A further approximation is convenient, but not necessary, when investigating relatively low frequency phenomena. The time dependence in (8.3.16) becomes relevant as soon as $\rho_0 \omega$ becomes comparable to $\eta_0 k^2$, where k is proportional to an inverse length of interest. If this length is chosen as the mean radius of a chain, (8.5.101), this fluid dynamics can be ignored for

$$\omega < \eta_0 / \rho_0 n^{1/2} l^2 \sim 2 \times 10^{10} / \text{Hz} \tag{8.6.4}$$

where the values $(\eta_0/\rho_0) = 0.01$ cm^2 sec^{-1}, $l \sim 7$ Å, and $n = 10^4$ have been employed to obtain the numerical estimate. As current experiments on dynamical viscosities involve considerably lower frequencies,[27,28] the static limit for the fluid may safely be employed; however, in studies of flow birefringence the full dynamics must be included.

Since the pre-averaging approximation is eventually used, it can be introduced from the outset, whereupon the infinite dilution, low frequency dynamical equations become

$$\frac{\partial}{\partial t} \mathbf{R}_i(t) + \sum_{p, m-1}^{n} [\zeta^{-1} \delta_{ip} + T_{ip}] (3 k_B T / l^2) A_{pm} \mathbf{R}_m(t) = \mathscr{F}_i(t). \tag{8.6.5}$$

$\mathscr{F}_i$ are the random forces $\phi \mathbf{G} \cdot \mathbf{F}^*$, which may be specialized to the static limit, and A_{pm} is defined in (8.2.28). Equation (8.6.5) represents the general Rouse–Zimm equations with a random driving force. Fourier transformation of (8.6.5) with respect to time yields

$$i\omega \, \mathbf{R}_i(\omega) + \sum_{p, m=1}^{n} [\zeta^{-1} \delta_{ip} + T_{ip}] (3 k_B T / l^2) A_{pm} \mathbf{R}_m(\omega) = \mathscr{F}_i(\omega) \tag{8.6.6}$$

and the general solution may be expressed in terms of the eigenvalues $\omega(q)$ of the homogeneous equation with $\mathscr{F}_i(\omega)$ absent. Just as in the case of the Kirkwood–Riseman equations, a good deal of analysis has gone into the solution of this equation.[1] Approximations similar to those used in the static case can be introduced again. The final approximate results are obtained more rapidly by recourse to the continuum limit, as in (8.5.50), with the upper limit n in (8.6.5) tending to infinity so Fourier transforms in bead space may be used.

Expressing the continuum limit of $\mathbf{R}_i(t)$ as $\mathbf{R}(s, t)$, equation (8.6.5) is approximated by

$$\mathrm{i}\omega\,\mathbf{R}(s,\omega) - \int_{-\infty}^{\infty} \mathrm{d}s'\,[(\zeta/l)^{-1}\delta(s-s') + T(s-s')] \times (3k_\mathrm{B}T/l)\frac{\partial^2}{\partial s'^2}\mathbf{R}(s',\omega) = \mathscr{F}(s,\omega) \tag{8.6.7}$$

whereupon Fourier transforms like those in (8.5.51) yield

$$\mathrm{i}\omega\,\mathbf{R}(q,\omega) + [(\zeta/l)^{-1} + \tilde{T}(q)]\,(3k_\mathrm{B}Tq^2/l)\mathbf{R}(q,\omega) = \mathscr{F}(q,\omega) \tag{8.6.8}$$

with the same symbol $\mathbf{R}$ used for $\mathbf{R}(q, \omega)$ for simplicity. The solution to (8.6.8) is simply written as

$$\mathbf{R}(q,\omega) = \mathscr{G}(q,\omega)\;\mathscr{F}(q,\omega), \tag{8.6.9}$$

where the polymer Green's function is

$$\mathscr{G}(q,\omega) = \{\mathrm{i}\omega + [(\zeta/l)^{-1} + \tilde{T}(q)]\,(3k_\mathrm{B}Tq^2/l)\}^{-1}. \tag{8.6.10}$$

The Rouse mode relaxation rates $\omega(q)$ are the imaginary parts of the poles of (8.6.10), so the Rouse relaxation times are

$$\begin{aligned}\tau(q) &= \omega(q)^{-1}\\ &= \{\,[(\zeta/l)^{-1} + \tilde{T}(q)]\,(3k_\mathrm{B}Tq^2/l)\}^{-1}.\end{aligned} \tag{8.6.11}$$

The polymer correlation function

$$\langle \mathbf{R}(q,\omega)\cdot\mathbf{R}(q',\omega')\rangle = \mathscr{G}(q,\omega)\;\mathscr{G}(q',\omega')\langle\,\mathscr{F}(q,\omega)\cdot\;\mathscr{F}(q',\omega')\rangle \tag{8.6.12}$$

can be used to evaluate the dynamical mean square end-to-end distance via

$$\begin{aligned}&\langle\,[\mathbf{R}(s,t) - \mathbf{R}(s',t')]^2\,\rangle\\ &= \left\langle\left\{\int_{-\infty}^{\infty}\frac{\mathrm{d}\omega}{2\pi}\int_0^{\infty}\frac{\mathrm{d}q}{\pi}\,[\exp(\mathrm{i}\omega t - \mathrm{i}qs) - \exp(\mathrm{i}\omega t' - \mathrm{i}qs')]\,\mathbf{R}(q,\omega)\right\}^2\right\rangle.\end{aligned} \tag{8.6.13}$$

By employing the assumption

$$\langle\,\mathscr{F}(q,\omega)\cdot\;\mathscr{F}(q',\omega)\,\rangle = 2\pi^2\theta(q)\delta(\omega+\omega')\delta(q+q') \tag{8.6.14}$$

which is found below to be consistent with the equilibrium limit, equation (8.6.13) is rewritten as

$$\langle\,[\mathbf{R}(s,t) - \mathbf{R}(s',t')]^2\,\rangle = 2\int_{-\infty}^{\infty}\frac{\mathrm{d}\omega}{2\pi}\int_0^{\infty}\frac{\mathrm{d}q}{\pi}\,\mathscr{G}(q,\omega)\,\mathscr{G}(-q,-\omega) \times \theta(q)\,\{1 - \cos[q(s-s') + \omega(t-t')]\}. \tag{8.6.15}$$

Since (8.6.15) involves an equilibrium average, the equal time, $t = t'$, limit must reduce to the equilibrium value (8.2.9) for the Gaussian chain. Performing the ω-integration leads to

$$\langle [\mathbf{R}(s,t)-\mathbf{R}(s',t)]^2\rangle = l\,|\,s-s'\,| \tag{8.6.16a}$$

$$= \int_0^\infty \mathrm{d}q\,\frac{l\theta(q)\{1-\cos[q(s-s')]\}}{3\pi k_B T q^2\,[(\zeta/l)^{-1}+\tilde{\mathrm{T}}(q)]}\,. \tag{8.6.16b}$$

The fact that

$$\int_0^\infty \mathrm{d}q\, q^{-2}\{1-\cos[q(s-s')]\} = |\,s-s'\,|\,\pi/2 \tag{8.6.17}$$

can be used in equation (8.6.16) to give

$$\theta(q) = 6k_B T[(\zeta/l)^{-1}+\tilde{\mathrm{T}}(q)] \tag{8.6.18}$$

the static limit of the Fourier transform of (8.6.3).

Inserting (8.6.18) into (8.6.15) enables the determination of dynamical expectation values like $\langle [\mathbf{R}(s,t)-\mathbf{R}(s,t')]^2\rangle$ which determine the incoherent neutron scattering cross-section

$$\sigma_{\mathrm{incoh}}(\mathbf{k},\omega) \propto \int_0^L \mathrm{d}s \int_{-\infty}^{\infty} \mathrm{d}(t-t')\exp[-\mathrm{i}\omega(t-t')]$$

$$\times \langle \exp\{\mathrm{i}\mathbf{k}\cdot[\mathbf{R}(s,t)-\mathbf{R}(s,t')]^2\}\rangle$$

$$= \int_0^L \mathrm{d}s \int_{-\infty}^{\infty} \mathrm{d}(t-t')\exp[-\mathrm{i}\omega(t-t')]\exp\{-k^2\langle[\mathbf{R}(s,t)-\mathbf{R}(s,t')]^2\rangle/6\}. \tag{8.6.19}$$

When s' is set equal to s in (8.6.15), we obtain

$$\langle [\mathbf{R}(s,t)-\mathbf{R}(s,t')]^2\rangle \propto |\,t-t'\,|^{2/3} \qquad \text{Rouse–Zimm} \tag{8.6.20}$$

leading to the relation for the half-width of (8.6.19) of[59]

$$\Delta\omega_{\mathrm{RZ}} \propto k^3 \tag{8.6.21}$$

in contrast to the Rouse limit of[59]

$$\Delta\omega_{\mathrm{R}} \propto k^4 \tag{8.6.22}$$

which follows from

$$\langle [\mathbf{R}(s,t)-\mathbf{R}(s,t')]^2\rangle \propto |\,t-t'\,|^{1/2} \qquad \text{Rouse.} \tag{8.6.23}$$

A Gaussian chain with internal friction and hydrodynamic interactions yields the Langevin equation

$$\frac{\partial}{\partial t}\mathbf{R}_i(t) + \sum_{p,m}\,[\zeta^{-1}\delta_{im}+T_{im}(1-\delta_{im})]\left[\left(\frac{3k_BT}{l^2}\right)A_{mp}+\varphi A'_{mp}\frac{\partial}{\partial t}\right]\mathbf{R}_p(t)$$

$$= \mathscr{F}_i^{\mathrm{I}}(t) + \mathscr{F}_i(t) \tag{8.6.24}$$

where A'_{ip} is defined in (8.2.22) and $\mathscr{F}_i^{\mathrm{I}}(t)$ are the random forces from $\Sigma_j \mathrm{K}_{\alpha i,\alpha j} f_j^*(t)$. The use of Fourier variables transforms equation (8.6.24) to[35]

$$\mathrm{i}\omega\mathbf{R}(q,\omega) + [(\zeta/l)^{-1}+\tilde{T}(q)]\,[(3k_BT/l)q^2+\varphi\mathrm{i}\omega q]\,\mathbf{R}(q,\omega)$$

$$= \mathscr{F}^{\mathrm{I}}(q,\omega) + \mathscr{F}(q,\omega). \tag{8.6.25}$$

Again the use of (8.6.13) and the condition that the equilibrium value (8.6.16a) be maintained leads to (8.6.14) and (8.6.18) with the additional results that

$$\langle \mathscr{F}^{\mathrm{I}}(q,\omega)\cdot\mathscr{F}^{\mathrm{I}}(q',\omega')\rangle = 12\pi^2 k_{\mathrm{B}}T\varphi q[(\zeta/l)^{-1}+\tilde{T}(q)]^2\delta(q+q')\delta(\omega+\omega') \tag{8.6.26}$$

and

$$\langle [\mathbf{R}(s,t)-\mathbf{R}(s',t')]^2\rangle = \int_{-\infty}^{\infty}\frac{d\omega}{2\pi}\int_0^{\infty}\frac{dq}{\pi}\,12k_{\mathrm{B}}T[(\zeta/l)^{-1}+\tilde{T}(q)]$$

$$\times\{1+\varphi q[(\zeta/l)^{-1}+\tilde{T}(q)]\}\{1-\cos[q(s-s')+\omega(t-t')]\}$$

$$\times\{\omega^2(1+\varphi q[(\zeta/l)^{-1}+\tilde{T}(q)])^2+(3k_{\mathrm{B}}T[(\zeta/l)^{-1}+\tilde{T}(q)]q^2/l)^2\}^{-1}. \tag{8.6.27}$$

Notice that the independent nature of the random forces $\mathscr{F}$ and $\mathscr{F}^{\mathrm{I}}$ make them statistically independent. Use of the definition of $\mathscr{F}^{\mathrm{I}}$ and (8.6.26) exhibit the purely internal nature of the random forces $\mathbf{f}^*(q,\omega)$,

$$\langle \mathbf{f}^*(q,\omega)\cdot\mathbf{f}^*(q',\omega')\rangle = 6k_{\mathrm{B}}T\Psi q\delta(q+q')\delta(\omega+\omega') \tag{8.6.28}$$

The equilibrium limit, $t=t'$, of (8.6.27) can be verified to reduce to (8.6.16a), and the leading behaviour of the $s=s'$ term is found to vary as in (8.6.20) for the nondraining limit as in (8.6.23) for the Rouse limit.

As anticipated in (8.3.39)–(8.3.40), an analysis of the random forces, acting on the polymer, provides a clearer understanding of the mechanism for the Brownian motion of the chain. When the chain lacks internal structure and there is no 'internal friction', the random forces are due to thermal velocity fluctuations in the fluid (8.3.40). These forces are then transmitted to the polymer by the hydrodynamic boundary conditions (8.3.10), providing the random forces correlation functions (8.6.3) and its static limit (8.6.14) with (8.6.18). This provides the detailed microscopic description of chain Brownian motion that is lacking in the statistical force balance (8.3.26) with the thermodynamic driving force (8.2.18). These correlation functions clarify the occurence of the random force term in Zwanzig's Langevin equation (8.3.30) or (8.3.35).

If the polymer has internal structure and/or additional potentials beyond the bead spring terms in (8.2.19), 'internal friction' must be present along with the corresponding random forces with spectra like (8.6.28). Conceptually, this implies that these internal friction terms may be understood on the basis of the Zwanzig–Mori theory of generalized Langevin equations by a consideration of the memory function; however, the analysis is extremely difficult. Recent works by Fixman and Kovac[60] and by others[61,62] have considered the dynamics of a freely hinged chain, a polymer with fixed bond lengths but random bond angles. Their results can be taken to represent an exact solution to the problem, and they display features of the same general form as those introduced in the 'theory' of internal viscosity. Unfortunately, exact solutions do not provide a simple physical description of this internal friction, so even approximate analyses of the memory function should provide very useful physical information. These approximate treatments may also be extremely valuable for more complicated models of polymers when exact solution of the equations is no longer feasible. Because of the physical simplicity, it is useful to continue the treatment of polymer dynamics using the simple phenomenological model in (8.6.25), assuming that further studies will provide an improved description of the memory function.

Dynamical viscosity of polymer solutions

Equation (8.5.29) provides the expression for evaluating the dynamical intrinsic viscosity. This is considered in more detail, as the general concentration-dependent case can be discussed following (8.5.8) and the analysis in Section 8.5. Our purpose here is to illustrate how $[\eta(\omega)]$ can be evaluated with the general theory and to demonstrate that the phenomenological model (8.6.25) for the 'internal friction' leads to almost identical expressions for $[\eta(\omega)]$ as generated by the Cerf–Peterlin theory of internal viscosity. This then shifts the questions posed by the empirical success of their theory to questions concerning the structure of the memory function associated with the dynamics of individual effective beads. The phenomenological model (8.6.25) also corresponds more closely to the Verdier–Stockmayer lattice hopping models of polymer dynamics. Equation (8.6.25) corresponds to an internal friction coefficient which increases with the Rouse mode index q, thereby increasing for shorter scale motions which presumably have higher barriers, etc., to overcome. The model (8.6.25) also indicates how hydrodynamic forces could, in principle, be introduced to also produce bead hopping.

Equation (8.5.29) is simplified as in the zero frequency limit by employing the Kirkwood–Riseman type approximation of replacing the average of a product by the product of averages

$$[\eta(\omega)] = -(N_A/6\eta_0 n M_A)\int_{-\infty}^{\infty} \mathrm{d}t \exp(-\mathrm{i}\omega t) \times \sum_{i,j=1}^{n} \langle [\mathbf{R}_i(t) - \mathbf{R}_j(0)]^2 \rangle \Delta_i(t) \langle \mathcal{G}_{ij}(t,0) \rangle \tag{8.6.29a}$$

where the fact that $\langle \mathcal{G} \rangle$ is a scalar has been used. As noted above, static hydrodynamic interactions may be used for frequencies lower than those in (8.6.4). The general case is considered elsewhere. Suffice it to say that for frequencies in the range of (8.6.4), there should be an additional change in the visco-elastic properties of polymer solutions as the fluid can no longer instantaneously follow the disturbance. Hence, Rouse–Zimm-like hydrodynamic interactions at lower frequencies than (8.6.4) become frequency-dependent Rouse-like behaviour at sufficiently high frequencies ($\omega \gg \eta_0/\rho_0 n^{1/2} l$). On this timescale, the form of the polymer memory function may also be considerably altered, so further experimental study is highly desirable.

Introducing the continuum limit and the approximations (8.5.53), equation (8.6.29a) becomes, upon use of the convolution theorem,

$$[\eta(\omega)] = -(N_A l/6 M_A \eta_0)\int_{-\infty}^{\infty} \frac{\mathrm{d}\omega'}{2\pi}\int_{-\infty}^{\infty} \mathrm{d}(s-s')\Delta^{\dagger}(s,\omega) \times \langle [\mathbf{R}(s,t) - \mathbf{R}(s',0)]^2 \rangle_{\omega-\omega'} \langle \mathcal{G}(s-s',\omega) \rangle \tag{8.6.29b}$$

where the subscript $\omega - \omega'$ indicates the appropriate Fourier transform, and where partial integration has been applied to the original $(s-s')$-integration with $\Delta^{\dagger}(s,t)$ the adjoint of $\Delta(s,t)$. The latter removes the term arising from the unity in (8.6.27), leaving only the cosine term that contributes to $[\eta(\omega)]$. Employing Fourier transforms to the Rouse mode variables q reduces (8.6.29b) to

$$[\eta(\omega)] = (l N_A/3 M_A \eta_0)\int_{-\infty}^{\infty} \frac{\mathrm{d}\omega'}{2\pi}\int_{q_0}^{\infty} \frac{\mathrm{d}q}{\pi}\, 6k_B T K(q)[1+\varphi q K(q)] \times \{(\omega-\omega')^2[1+\varphi q K(q)]^2 + [3k_B T q^2 K(q)/l]^2\}^{-1} \times [\mathrm{i}\omega'\varphi q + (3k_B T q^2/l)]\{\mathrm{i}\omega'[1+\varphi q K(q)] + 3k_B T q^2/l\}^{-1} \tag{8.6.30}$$

where, for convenience, $K(q)$ is defined by

$$K(q) \equiv (\zeta/l)^{-1} + \tilde{T}(q) \tag{8.6.31}$$

The last factor in (8.6.30) arises from the polymer Green's function as can be seen by re-expressing (8.6.25) as

$$[1/\mathcal{G}(q,\omega)]\mathbf{R}(q,\omega) \equiv \mathcal{F}^{\mathrm{I}}(q,\omega) + \mathcal{F}(q,\omega) \tag{8.6.32}$$

The next to last factor arises from Δ, and the remaining q and ω-dependent terms come from the transform of the cosine part of (8.6.27).

After contour integration over ω' and rearrangement of terms, equation (8.6.30) is found to be

$$[\eta(\omega)] = (lN_{\mathrm{A}}k_{\mathrm{B}}T/M_{\mathrm{A}}\eta_0)\int_{q_0}^{q_c}\frac{\mathrm{d}q}{\pi}([(1/\varphi q) + K(q)]^{-1}(3k_{\mathrm{B}}Tq^2/l)^{-1}$$
$$+ [1 + \varphi qK(q)]^{-2}\{\mathrm{i}\omega + 6k_{\mathrm{B}}Tq^2K(q)/l[1 + \varphi qK(q)]\}^{-1}) \tag{8.6.33}$$

Reintroducing the discrete Rouse variables $q = \pi p/nl$, as in the transition from (8.5.54) to (8.5.55), and defining the Rouse–Zimm relaxation times in the absence of internal friction

$$\tau_p = [n^2l^3/6k_{\mathrm{B}}T\pi^2p^2K(\pi p/nl)] \tag{8.6.34}$$

and those involving internal friction

$$\tau_p' = \left[1 + K(\pi p/nl)\varphi\left(\frac{\pi p}{nl}\right)\right]\tau_p \tag{8.6.35}$$

equation (8.6.33) can be rewritten as

$$[\eta(\omega)] = 2[\eta_\infty] + \frac{k_{\mathrm{B}}TN_{\mathrm{A}}}{nM_{\mathrm{A}}\eta_0}\sum_p\frac{\tau_p^2}{\tau_p'(1 + \mathrm{i}\omega\tau_p')} \tag{8.6.36}$$

The Cerf–Peterlin high frequency limiting viscosity is

$$[\eta_\infty] = \frac{k_{\mathrm{B}}TN_{\mathrm{A}}}{nM_{\mathrm{A}}\eta_0}\sum_p\frac{\tau_p(\tau_p' - \tau_p)}{\tau_p'} \tag{8.6.37}$$

Apart from the factor of 2 before $[\eta_\infty]$ in (8.6.36), equation (8.6.37) is identical to that of the Cerf–Peterlin theory. Their result differs from (8.6.36) because of contributions from their deformational viscosity. A microscopic theory of the correct form of such a term would require an analysis of polymer memory functions under steady flow conditions, a subject which is presently under study.

8.7 Acknowledgements

Much of the work reviewed and analysed in this chapter was done in collaboration with Professor S. F. Edwards. The research is supported, in part, by Grant DMR74-11830 from the National Science Foundation, and it has benefited from MRL(NSF) facilities at the University of Chicago. I am grateful to the Camille and Henry Dreyfus Foundation for a Teacher–Scholar grant.

References

1. H. Yamakawa (1971). *Modern Theory of Polymer Solutions*, Harper and Row, New York.
2. P. J. Flory (1953). *Principles of Polymer Chemistry*, Cornell University Press, Ithaca, N.Y.
3. M. V. Volkenstein (1963). *Configurational Statistics of Polymeric Chains*, Interscience, New York.
4. T. M. Birshtein and O. B. Ptitsyn (1966). *Conformations of Macromolecules*, Interscience, New York.
5. P. J. Flory (1969). *Statistical Mechanics of Chain Molecules*, Interscience, New York.
6. K. F. Freed (1972). *Adv. Chem. Phys.*, **22**, 1.
7. J. D. Ferry (1970). *Viscoelastic Properties of Polymers*, 2nd ed. Wiley, New York.
8. C. Domb (1969). *Adv. Chem. Phys.*, **15**, 229.
9. S. F. Edwards (1965) *Proc. Phys. Soc. (London)*, **85**, 613.
   S. F. Edwards (1966). *Misc. Publ. U.S. Natl. Bur. Std.*, **273**, 225.
10. H. Reiss (1967). *J. Chem. Phys.*, **47**, 186.
11. K. F. Freed (1971). *J. Chem. Phys.*, **55**, 3910.
    K. F. Freed and H. P. Gillis (1971). *Chem. Phys. Lett.*, **8**, 384.
    H. P. Gillis and K. F. Freed (1975). *J. Chem. Phys.*, **63**, 852.
12. S. G. Whittington (1970). *J. Phys.* A., **3**, 28.
    S. G. Whittington and J. F. Harris (1972). *J. Phys.* A., **5**, 411.
13. P.-G. de Gennes (1972). *Phys. Lett.*, **A38**, 339.
    E. Brezin, J. C. le Guillou, J. Justin and B. G. Nickel (1973). *Phys. Lett.*, **A44**, 227.
14. J. des Cloizeaux (1975). *J. de Phys. (Paris)*, **36**, 281.
15. M. Daound, J. P. Cotton, B. Farnoux, G. Jannink, G. Sarma, H. Benoit, C. Duplessix, C. Picot and P.-G. de Gennes (1975). *Macromol.*, **8**, 804.
16. P. E. Rouse, Jr. (1953) *J. Chem. Phys.*, **21**, 1272.
17. J. G. Kirkwood (1949). *Rec. Trav. Chim.*, **68**, 649.
    J. J. Erpenbeck and J. G. Kirkwood (1958). *J. Chem. Phys.*, **29**, 909.
    J. J. Erpenbeck and J. G. Kirkwood (1963). *J. Chem. Phys.*, **38**, 1023.
18. R. Zwanzig (1969). *Adv. Chem. Phys.*, **15**, 325.
19. W. H. Stockmayer (1976). *Les Houches Lectures – 1973*, Gordon and Breach, London.
20. R. Zwanzig (1960). *J. Chem. Phys.*, **33**, 1338.
    R. Zwanzig (1961). In W. E. Britten, B. W. Downs and J. Downs (Eds), *Lectures in Theoretical Physics*, Vol. 3. Interscience, New York. p. 106.
    R. Zwanzig (1965). *Ann. Rev. Phys. Chem.*, **16**, 67.
21. H. Mori (1965). *Prog. Theor. Phys.*, **33**, 423.
22. M. Bixon (1973). *J. Chem. Phys.*, **58**, 1459.
23. R. Zwanzig (1974). *J. Chem. Phys.*, **60**, 2717.
24. B. H. Zimm (1956). *J. Chem. Phys.*, **24**, 269.
25. R. Cerf (1958). *J. Phys. Radium*, **19**, 122.
    R. Cerf (1966). *J. Chim. Phys.*, **66**, 479.
    R. Cerf (1973). *Chem. Phys. Lett.*, **15**, 613.
26. A. Peterlin (1967). *J. Polym. Sci.* A–2, **5**, 179.
    G. B. Thurston and A. Peterlin (1967). *J. Chem. Phys.*, **46**, 488.
27. D. J. Massa, J. L. Schrag and J. D. Ferry (1971). *Macromol.*, **4**, 210.
28. K. Osaki and J. L. Schrag (1971). *Polymer J.*, **2**, 541.
29. S. Adelman and K. F. Freed, *J. Chem. Phys.* (submitted).
30. N. Saito (1950). *J. Phys. Soc. (Japan)*, **5**, 4.
    N. Saito (1952). *J. Phys. Soc. (Japan)*, **7**, 447.
31. J. Riseman and R. Ullman (1951). *J. Chem. Phys.*, **19**, 578.
32. H. Yamakawa (1961). *J. Chem. Phys.*, **34**, 1360.
33. J. M. Peterson and M. Fixman (1963). *J. Chem. Phys.*, **39**, 2516.
34. W. W. Graessley (1974). *Adv. Polymer Sci.*, **16**, 1.
35. S. F. Edwards and K. F. Freed (1974). *J. Chem. Phys.*, **61**, 1189.
36. K. F. Freed and S. F. Edwards (1974). *J. Chem. Phys.*, **61**, 3626.
37. K. F. Freed and S. F. Edwards (1975). *J. Chem. Phys.*, **62**, 4032.
38. K. F. Freed and S. F. Edwards (1975). *J.C.S. Faraday Trans.*, **1**, 71, 2025.
39. K. F. Freed, S. F. Edwards and M. Warner (1976). *J. Chem. Phys.*, **64**, 5132.
40. K. F. Freed (1976). *J. Chem. Phys.*, **64**, 5126.

41. R. A. Harris and J. E. Hearst (1966). *J. Chem. Phys.*, **44**, 2595.
42. R. Ullman (1969). *Macromol.*, **2**, 27.
R. Ullman (1970). *J. Chem. Phys.*, **53**, 1734.
43. S. F. Edwards and M. A. Oliver (1971). *J. Phys.* A., **4**, 1.
44. H. Yamakawa and M. Fujii (1974). *Macromol.*, **7**, 128.
45. J. G. Kirkwood and J. Riseman (1948). *J. Chem. Phys.*, **16**, 565.
46. R. Kubo (1966). *Rep. Prog. Phys.*, **29**, 255.
47. E. L. Chang and R. M. Mazo (1976). *J. Chem. Phys.*, **64**, 1389.
48. P.-G. de Gennes (1971). *J. Chem. Phys.*, **55**, 572.
49. S. F. Edwards and J. Grant (1973). *J. Phys.* A., **6**, 1169, 1186.
50. M. Doi (1974). *Chem. Phys. Lett.*, **26**, 269.
51. K. F. Freed and M. H. Cohen (1971). *Phys. Rev.* B., **3**, 3400.
52. E. N. Economou, M. H. Cohen, K. F. Freed and E. S. Kirkpatrick (1976). In J. Tauc (Ed.) *Amorphous and Liquid Semiconductors*, Plenum, New York.
53. M. L. Goldberger and K. M. Watson (1964). *Collision Theory*, Wiley, New York.
54. A. Einstein (1906). *Ann. Phys.*, **19**, 289.
A. Einstein (1911). *Ann. Phys.*, **34**, 591.
55. B. U. Felderhof, J. M. Deutch and U. Titulaer (1975). *J. Chem. Phys.*, **63**, 740, and references therein.
56. F. Yonezawa and S. Homma (1968). *J. Phys. Soc. (Japan)*, **36**, Suppl. 68.
P. L. Leath (1968). *Phys. Rev.*, **171**, 725.
P. L. Leath (1972). *Phys. Rev.*, **B5**, 1643.
57. P.-G. de Gennes (1976). *Macromol.*, **9**, 587, 594.
58. S. F. Edwards (1975). *J. Phys.* A., **8**, 1670.
59. P.-G. de Gennes (1967). *Physics*, **3**, 37.
E. Dubois-Violette and P.-G. de Gennes, (1967) *Physics*, **3**, 181.
60. M. Fixman and J. Kovac (1974). *J. Chem. Phys.*, **61**, 4939, 4950.
61. U. M. Titulaer and J. M. Deutch (1975). *J. Chem. Phys.*, **63**, 4505.
62. M. Doi, H. Nakajima and Y. Wada (to be published).

Chapter 9

Machine Simulation of Water

PAUL BARNES

Department of Crystallography, Birkbeck College, Malet Street, University of London, WC1E 7HX, UK

9.1 Introduction

This subject is barely three-quarters of a decade old and has coincided more with the mushrooming of computer facilities than did the simulation of classically simple liquids with the birth of computing. The problems, unlike the solutions, of water simulation have always been obvious – the non-spherical shape of the water molecule demands an acknowledgement of its rotational as well as translational degrees of freedom, and the strong directional interactions must be reconciled with the present concepts of hydrogen-bonding; also the molecular interactions are notably non-additive, and individual responsiveness of each water molecule to its polarizing environment leads to distinct molecular distinguishability. One by one the major simplifications of classical simulation are lost and the resultant programming becomes inevitably more complex, though not without its own peculiar attractiveness. We have already reached the stage where we can realistically translate all these conceptual expansions into programmable algorithms.

The historical development of water simulation is particularly interesting in that it has of necessity been phenomenologically rather than thermodynamically oriented. In 1969, Barker *et al.*[1] presented a paper at a conference on the Physics of Liquids[2] demonstrating the feasibility of Monte Carlo simulations on water using a four-point-charge model derived from the Rowlinson[3] potential. With further initiative at the hands of Rahman and Stillinger, a four-point-charge model was used for the molecular dynamic studies of bulk water molecule interactions. The initial report of this work in 1971[4] produced what has really turned out to be a classic account of basic water molecular interactions. In the summer of 1972, a Workshop on 'Molecular Dynamics and Monte Carlo Calculations on Water' was held at Orsay under the auspices of CECAM (Centre Européen de Calcul Atomique et Moléculaire) and organized by Professor C. Moser and Professor H. J. C. Berendsen. This brought together some eleven participants and at least as many water models, and the fruits of this highly successful workshop has made it a notable landmark in this fast-developing field. Two further, partly connected, workshops followed and, together with the original workshop, have contributed largely to our present understanding, and indeed to the contents of this chapter.

The author, while recognizing that one must eventually come to terms with the equation of state (viral coefficients) and transport properties of water, concentrates here on the phenomenological description of water molecule interactions (particularly the schemes of hydrogen-bonding) which the various models have portrayed. Also, the author makes no attempt to establish this report as a literature/subject-review, but rather as a vehicle for acquainting others with the types of problem one must face in this topic and their possible solutions. In this spirit, these aspects are illustrated using examples mainly from the work of Rahman and Stillinger and of the author. The topic of water molecule interactions within aqueous systems encompasses an enormous range of biological, physical, and chemical interest, and so with the intent of netting just as wide a range of disciplines, the author has restricted the subject terminology as much as possible.

9.2 Water models

At this point it is convenient to summarize the types of model used in water simulation before proceeding further into their application. A given model describes the basic energetic or force interactions between a pair of molecules, extending from this to the complete interaction of any assembly of such molecules. The total potential energy U_N of any collection of N molecules may be expressed as an expansion of the form

$$U_N(\mathbf{x}_1, \ldots, \mathbf{x}_N) = \frac{1}{2!} \sum_{\substack{i,j=1 \\ i \neq j}}^{N} U^{(2)}(\mathbf{x}_i, \mathbf{x}_j) + \frac{1}{3!} \sum_{\substack{i,j,k=1 \\ i \neq j \neq k \neq i}}^{N} U^{(3)}(\mathbf{x}_i, \mathbf{x}_j, \mathbf{x}_k) + \ldots + U^{(N)}(\mathbf{x}_1, \mathbf{x}_2, \ldots\ldots \mathbf{x}_N) \qquad (9.2.1)$$

This expansion covers interactions of all possible pairs, triplets, quadruplets, etc., of molecules (the factorial divisors are necessary as each pair is summed twice, each triplet six times, etc.) where $\mathbf{x}_i$ refers to the configurational vector of molecule i and $U^{(2)}$, $U^{(3)}$, $U^{(4)}$, etc. are thus a uniquely defined set of interaction functions. In the classically simple liquids, the functions higher than $U^{(2)}$ are generally thought to be negligible, leading to the simplifying approximation of *pair-wise additivity* – that is, one may derive the total energy from a consideration of just the pairs of interacting molecules. It will be seen, however, that this is certainly not the case for water, and one must make an appropriate allowance either by replacing the pair potential by an *effective* pair-potential, thus effectively averaging the higher terms (see, for example, the methods of Rahman *et al.*[4,5,9], or by making more specific allowances as with the polarization models of Barnes and Berendsen (see below).

In the models presented here, it has been found convenient to separate out the interactions into two parts comprising (i) a spherically symmetric portion to account for the London dispersion–attractive and cloud charge–overlap repulsive forces, and (ii) an electrostatic part to which all the orientational and non-additive aspects are ascribed. The symmetric potential portion U_{LJ}, is more or less universally assumed to be of the centrally directed Lennard–Jones '6–12' type (or its exponential equivalent)

$$U_{LJ} = 4\epsilon\left[\left(\frac{\sigma}{R}\right)^{12} - \left(\frac{\sigma}{R}\right)^{6}\right] \tag{9.2.2}$$

where the constants ϵ and σ respectively represent the minimum value (well depth) of U_{LJ} and the separation for zero interaction energy ($U_{LJ}(R = \sigma) = 0$), and the separation R is usually taken as the distance between the oxygen centres of each molecule. The values of the constants ϵ and σ are estimated from either theoretical considerations[10] by comparing with an isoelectronic system, by trial and error, or by a combination of all of these within physically realistic limits (e.g. ϵ = 0.05 to 0.4 kcal/mole, σ = 2.7 to 3.5 Å).

It remains for the electrostatic portion of the interaction to bear the responsibility for the extraordinary characteristics of water and on which we now focus our greater attention. The present wealth of experimental and theoretical evidence[11] shows that the isolated water molecule forms hybridized bond orbitals leading to an HOH angle of about 104½° and with the two lone pair electrons are assumed to maintain a near-tetrahedral arrangement around the central oxygen (see Figure 9.2.1). This charge distribution is reflected in the considerable permanent dipole moment of the isolated water molecule (1.84 Debye) directed along the molecule's two-fold rotation axis of symmetry, and indirectly in the nearly symmetrical but negative quadrupole moment (see Reference 12 for example). The structure of ordinary (Ih-hexagonal) ice shows that essentially the same water molecule can be incorporated into a tetrahedral arrangement of bonding by merely increasing the HOH angle from 104½° up to the tetrahedral angle of nearly 109½° (see Figure 9.2.2).†

Each tetrahedral bond between two molecules then contains the partly-shielded hydrogen of one molecule and lone-pair electron of the other, viz, the 'hydrogen-bond'. The hydrogens and lone pairs more or less maintain their positions relative to the central oxygen, with a resultant bond distance of about 2.76 Å which itself is close to the first radial distribution peak (2.8 Å) in liquid water near 0 °C.

The ice structure is a very open structure with wide channels in the *c*-direction, and it is an indication of the strength of the tetrahedral hydrogen-bond that water crystallizes into this form with its consequent decrease in density. The overall picture that emerges is of an essentially rigid water molecule, whether in isolation or in condensed crystallized or liquid form; the greater changes are to be seen in the function of the hydrogen-bond which is profoundly influenced by subtle changes in the electron cloud distribution and shielding of the hydrogens and lone-pair electrons.

The immediate aim then of any realistic electrostatic model is to mimic the known charge distribution of the isolated water molecule and also its propensity to form strongly cooperative and near-tetrahedrally disposed hydrogen bonds with suitably neighbouring molecules. We therefore look at some typical electrostatic models in use.

(i) *The Stockmayer Potential* consists of a Lennard–Jones function with the electrostatic portion represented by the point dipole formulation. McDonald[15] has shown that although there exist useful perturbation theories for such a potential, the energy convergence is too slow for use with strongly polar molecules such as water.

†The ice bonding angle is not necessarily exactly tetrahedral since the ratio of the unit cell dimensions c/a is about 0.25% less than the ideal 1.633,[13] and the hydrogens may not be exactly collinear with the bond direction.[14]

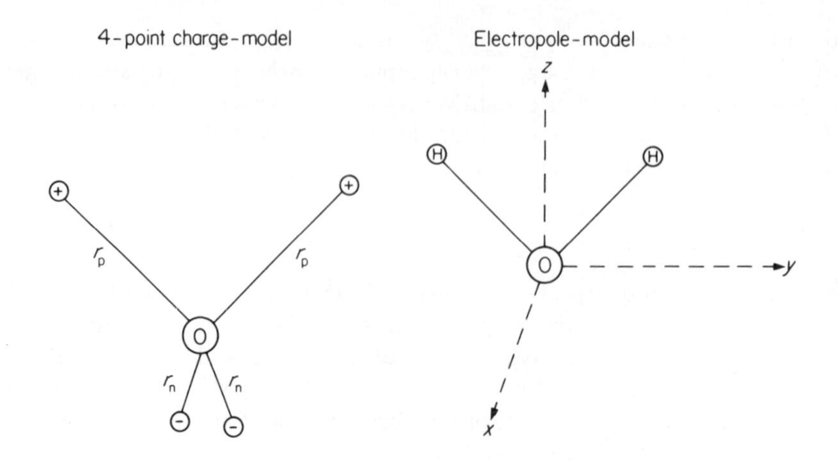

Figure 9.2.1 Geometrical representations of the water models described in the text. In the four-point-charge models, the two positive charges represent the two hydrogen atoms and the two negative charges the two lone-electron pairs, all disposed more or less tetrahedrally around the central oxygen atom. The OH-distances (r_p) are taken to be 1 Å, while the lone-pair–oxygen distances (r_n) vary with the model; BNS (1 Å), ST2 (0.8 Å), Bol-potential (0.7 Å), Berendsen polarizable four-point-charge model (0.5 Å).

The electropole model figure indicates the three axes as defined for the dipole, quadrupole and polarizability tensors. The z-axis bisects the two OH-directions, and the y-axis is contained in the HOH-plane

(ii) *The BNS (Ben-Naim Stillinger) model*[9] consists of a Lennard–Jones function with the constants for neon (the water molecule and the neon atom are iso-electronic closed-shell systems) and a somewhat modulated four-point-charge model (see Figure 9.2.1) in which the two positive charges (+0.19e) represent the hydrogen atoms and the two equally negative charges (−0.19e) represent the lone-pair positions, all distant 1 Å from the central oxygen atom. This charge distribution yields a dipole moment of 2.170 Debye as distinct from the isolated value (1.84) – the increase represents an average allowance for the cooperativity discussed later. An important implication of the charge symmetry is that in any given configuration all charges could be reversed in sign without any energetic differences. A modulation function must be introduced to avoid the singularity that might otherwise occur when two neighbouring molecules are within 2 Å of each other and two point charges (one from each molecule) coincide.

In practice the position and orientation of a given molecule may be specified by three coordinates and three angles (Euler angles) which act as a basis for the Newton–Euler equations of motion.

(iii) *The Bol*[16] *potential* is similar to that of the BNS model apart from a reduction in the negative point charge–oxygen distances r_n from 1 Å to 0.7 Å. This removes the charge symmetry inherent in the BNS model.

(iv) *The ST2 model*[17] is essentially a revised version of the BNS model in which charge asymmetry has been introduced by reducing the negative charge–oxygen distances r_n from 1 to 0.8 Å. This charge distribution yields a dipole moment of 2.353 Debye.

(v) *The Berendsen*[18] *polarizable four-point-charge model* is one of the first models to incorporate the cooperative effects of polarization in a specific analytical fashion.

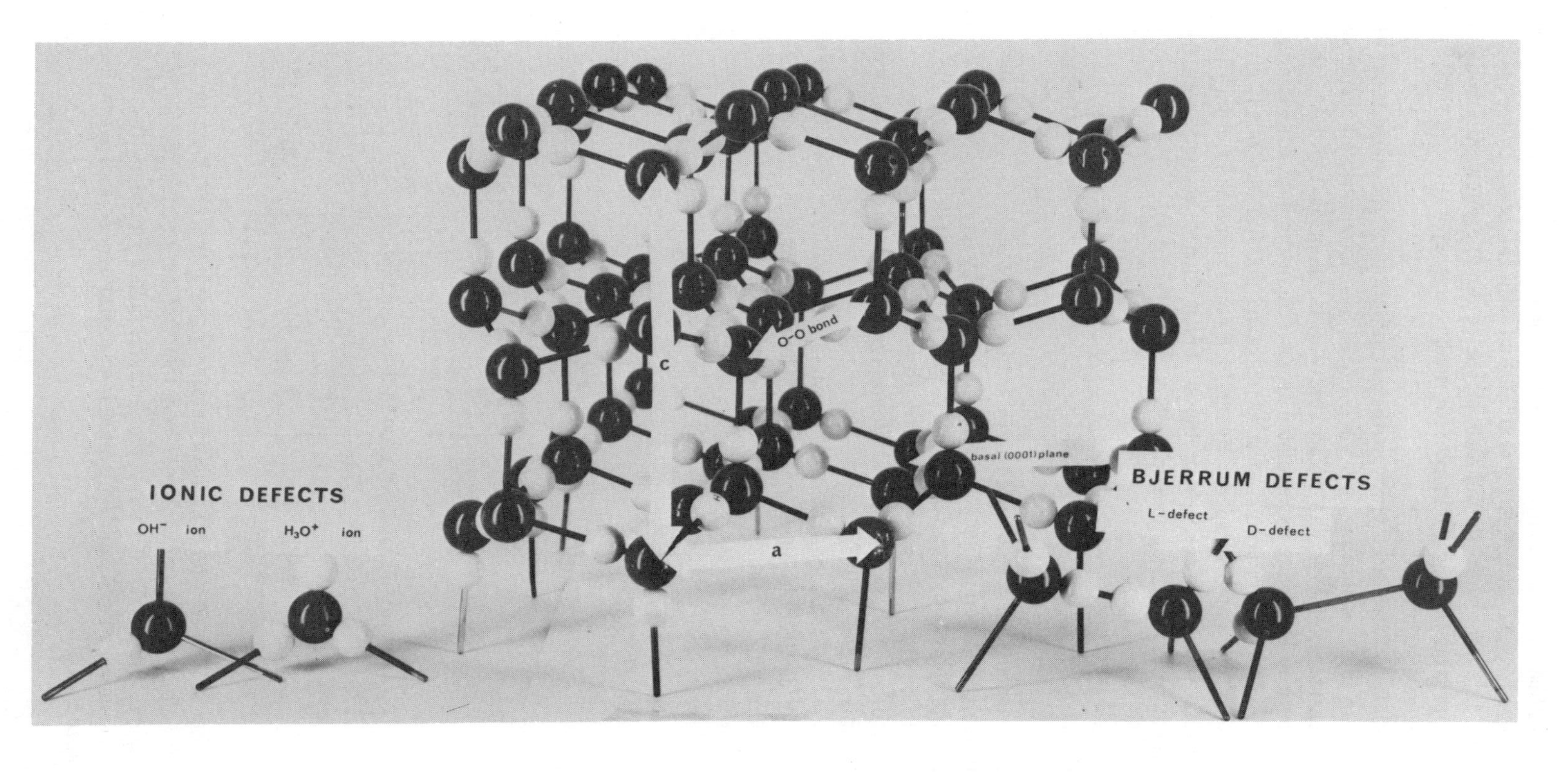

Figure 9.2.2 Model illustrating the structure of Ice I and its point defects. The smaller white balls represent the hydrogen atom positions which are seen to obey the Bernal–Fowler rules (see text and Figure 9.4.1) but nevertheless are non-repeating. Departures from the Bernal–Fowler rules produce ionic and bond defects as illustrated in the figure

In the unperturbed state, the charge assembly is similar to that of the Bol potential but with the negative charge–oxygen distances set at 0.5 Å and with the point charges set to yield the required isolated dipole moment of 1.84 Debye. The coulombic singularities mentioned above are avoided by approximating the interaction for small charge separations to that obtained for a near-Gaussian charge distribution.

The polarizability is introduced through the perturbation of the magnitudes of the point charges rather than their separations, in such a way as to maintain overall charge neutrality of the perturbed charge assembly. The polarizability α is taken as 1.444 $Å^3$, this being the normally accepted isotropic value extrapolated to zero frequency (see, for example, Eisenberg and Kauzmann[19]). The final molecular electric fields and polarizations of any given system of water molecules are calculated in an iterative fashion (see (vi) below) until a 'steady state' is achieved. The total potential energy U of the system is then calculated using the following summation over all molecules i and point charges α, β (α,β = 1 to 4).

$$U = \tfrac{1}{2} \sum_{i\alpha} q_\alpha^i \cdot V_\alpha^i + \tfrac{1}{2} \sum_{i} \sum_{\alpha\beta} \delta q_\alpha^i \cdot k_{\alpha\beta} \cdot \delta q_\beta^i + \tfrac{1}{2} \sum_{\substack{i,j \\ i \neq j}} 4\epsilon \left[\left(\frac{\sigma}{R_{ij}} \right)^{12} - \left(\frac{\sigma}{R_{ij}} \right)^{6} \right] \tag{9.2.3}$$

The first term represents the Coulombic energy of the charge q_α^i in the local potential V_α^i, and the second the internal energy changes or the 'polarization energy' U_P which is expended in changing the point charges by $\delta q_\alpha^i, \delta q_\beta^i$. (The ½-factors account for the double summing of pairs, and the coefficient $k_{\alpha\beta}$ is related to the polarization tensor.)

As with the other models, a Lennard–Jones potential is superimposed with constants ϵ = 0.330 kcal/mole, and σ = 2.985 Å.

(vi) *The Barnes*[20] *Polarizable-Electropole model.* Here the electrostatic interaction is represented by the electropole series expansion. The permanent electropoles, referred to the axes given in Figure 9.2.1 are those given by Glaeser and Coulson[12] as follows:

0th. electropole	charge q	zero	
1st	dipole μ_i	$\begin{vmatrix} 0 \\ 0 \\ 1.84 \end{vmatrix}$	Debye
2nd	quadrupole $\mathbf{Q}_{ij}$	$\begin{vmatrix} -6.56 & 0 & 0 \\ 0 & -5.18 & 0 \\ 0 & 0 & -5.51 \end{vmatrix}$	Debye Ang.
	octupole Ω_{ijk} not used in these calculations		
	dipole polarizability tensor α_{ij}	$\begin{vmatrix} 1.444 & 0 & 0 \\ 0 & 1.444 & 0 \\ 0 & 0 & 1.444 \end{vmatrix}$	$Ang.^3$

Both polarization models (v and vi) use an iterative procedure to evaluate the changes in the dipole moments $\Delta\mu$ of each water molecule from their isolated permanent value μ^0. For a given water molecule in a given molecular array, the resultant field **F** emanating from the surrounding molecules is calculated and the molecule polarized accordingly so that a new dipole moment vector μ is realized:

$$\mu = \mu^0 + \Delta\mu$$

where

$$\Delta\mu = |\alpha| \times \mathbf{F}.$$

This calculation is performed for each molecule in the assembly, all this constituting the 'first iteration'. However, these dipolar increments give rise in turn to corresponding changes to the initial field strengths which must therefore also be appropriately incremented. Similarly, these field increments imply a further (second) set of dipolar-increments, the evaluation of which is termed the 'second iteration'. This procedure is continued until the required accuracy (6%–0.2% in energy) is obtained – normally about 2–5 iterations in molecular dynamics or Monte Carlo simulations. It is worth emphasizing that, with this method described here, no effective averaging process is used – each and every molecular polarization is calculated individually. Once this has been completed, the electropolar energy U_{E} of the assembly can be evaluated over all pairs of molecules i and j using the electropolar energy expansion of Buckingham[21]

$$U_{\mathrm{E}} = \tfrac{1}{2}\sum_{\substack{i,j\\ i\neq j}} \sum_{\alpha,\beta} q_i \cdot \Phi_j + \mu_i^{\alpha} F_j^{\alpha} + \tfrac{1}{2} Q_i^{\alpha\beta} \cdot F_j'^{\alpha\beta} + \text{higher terms} \tag{9.2.4}$$

The first term represents the charge q-potential Φ interaction which in the absence of foreign ions is clearly zero. The $\mu_\alpha^i F_\alpha^i$ term represents the total dipole–electric field F interaction (the *permanent* and *induced* dipole of molecule i in the electric field emanating from molecule j) and similarly the third term $\tfrac{1}{2} Q_i^{\alpha\beta} \cdot F_j'^{\alpha\beta}$ term represents the quadrupole–electric field gradient F' interactions, summed over all direction-axes α and β. If one characterizes the potential, field, field-gradient, etc. at i as due to the dipole, quadrupole, octupole moments etc. of j, one can alternatively subdivide the above electropolar energy into the traditional divisions of dipole–dipole, dipole–quadrupole, quadrupole–quadrupole, etc. interactions. Finally, as with the Berendsen model, we add a Lennard–Jones term E_{LJ} and take into account the polarization energy E_{P} (see Reference 22) in evaluating the total energy of the system:

$$U = U_{\mathrm{E}} + U_{\mathrm{P}} + U_{\mathrm{LJ}}$$

where

$$U_{\mathrm{P}} = \sum_i \sum_\beta \tfrac{1}{2}(\mu_i^\beta - \mu_0^\beta)^2 / \alpha^{\beta\beta} \tag{9.2.5}$$

These models encompass a range of ideas and allow some room for manoeuvre. In general terms the four-point-charge models (ii to v) tend directly towards tetrahedral bonding schemes, as such isolated pairs have an energy minimum when a positive charge of one molecule approaches a negative charge of the other; whereas electropole models (i and vi) have a corresponding minimum some 25° away in the case of an *isolated pair*. Clearly, therefore, the near-tetrahedral bonding networks acquired with these models arise partly from overall structural reasons. The four-point-charge models provide for some flexibility in the charge distribution and symmetry, and experience has shown (Bol[16], Rahman and Stillinger[6]) that improvements can be achieved by careful empirical variation of the oxygen point-charge separations r_{n}. The charge asymmetry is undoubtedly important seeing that there is a wealth of experimental evidence, albeit controversial in conclusion, alluding to electric potentials of a particular direction across specific ice/water interfaces (both free and imposed). One obvious weakness of the four-point-charge models is their inability to reflect either in magnitude or sign the true quadrupole moment of water – this might be important for specialized interface studies where dipole–quadrupole effects may predominate (N. H. Fletcher[23]). The modulation function used in the BNS and ST2 models is of a rather artificial construction, and a physically realistic improvement is to be found in the Gaussian rather than the point-charge representation used by van der Velde and Berendsen.[18,24] The electropole models leave little room for empirical variation – a surprising consequence (of perhaps mixed blessings) of there being little real numerical difference in the electropole values (unlike the polarizability) obtained from quantum mechanical studies on the water molecule. What empiricism remains here is mainly in the exact choice of the Lennard–Jones constants.

The first four models do not account for non-additive events in a specific way, though, as Rahman and Stillinger point out, one may talk of optimized *effective* pair potentials. The two polarizable models (v and vi), however, cater for *local*, non-additive events in an analytical fashion. In Sections 9.3, 9.4 and 9.6 we look into this topic in greater depth.

The reader is referred to the relevant literature[4–8] for further details on these models and their actual application in simulation calculations. The simulation may be of either the molecular dynamic[25] (MD) or Monte Carlo[26] (MC) variety, and both types are discussed here. In MD simulations, the evolution of a system is approximated by applying the classical equations of motion to each represented molecule over time intervals which are sufficiently small to conserve reasonable constancy in the force fields. If the time intervals are made too large, energy is not conserved and configurational catastrophes (superpositioning of molecules) may occur. Typically for water, some thousands of such steps of 10^{-16} to 10^{-15} sec. duration are computed, though the requisite number depends on the initial configuration and the properties being examined. In this way, one can study most thermodynamic and structural features of an assembly, though time-dependent and non-equilibrium properties can also be evaluated in principle.

In contrast, the MC method is devised so that the represented molecules are constrained to undergo a set of trial random displacements in such a way that the new configurations are generated at a frequency proportional to the Boltzmann weighting factor, $\exp(-U/k_B T)$, where U is the potential energy of a given configuration. As no time scale is directly involved, only equilibrium thermodynamic and structural properties can thus be averaged over a MC simulation sequence. Both MD and MC techniques have their respective advantages and collectively represent the term 'machine simulation' used here.

In both MD and MC simulations, bulk conditions can be approximated by using a relatively small number of molecules (50–500) by the use of periodic boundaries. The desired number of molecules is initially constrained within a cubical container of appropriate size (L^3); then all possible interactions are permitted between any given pair of molecules and all their combinations of spatial translations by $\pm nL$ (n integral) along the container edge directions. For each molecule there are 27 such (translated) images for just the first three values of n (–1, 0, +1) but in practice the number of interactions is limited by setting a 'cut-off' distance of interaction between any pair; typically, a cut-off of 2.5–3.5σ (6½ to 10 Å depending on σx) would produce about 50–150 interactions per molecule. McDonald[27] has shown using the Ewald method on the BNS model that such a cut-off results in an average error of only + 1½ to 2½% in energy and very much less than 5% in force calculations. Although the bonding in water may be cooperative and extensive, the large orientational variations in water cause the *angle-averaged* interactions to disappear rapidly with separation, unlike the situation in plasmas and ionic salts where such severe truncation is inadvisable.

9.3 Small water clusters

Before moving on to considerations of bulk water, it is worthwhile to consider first the applicability of non-additive models to specific cases for which we have some additional information concerning the hydrogen-bond, viz the quantum-mechanical calculations for small water clusters, and also the structure of the ice-Ih lattice (see Section 9.4).

We compare here the predictions of the polarizability models of Barnes[20] and Berendsen[18] with the quantum-mechanical predictions for the configuration energetics of small water chains and rings that have been provided by several workers.[29]

The Lennard–Jones portion of the interactions is purely additive and therefore of no consequence in this context, and so the results here concentrate on the electrostatic contribution. The tests on the Berendsen model were performed jointly during a week of the Orsay 'Water Workshop' and so are less extensive than with the electropole model, though the overlap is good. Futher details of both sets of results are to be published elsewhere.[28] It should be pointed out immediately that the essential consideration is to compare the various predictions and general trends as to the relative stabilities and non-additivities of different cluster configurations. The absolute energies are of considerably less consequence since variations of over 100% are to be found among the quantum-mechanical results for even the simple dimer.

The concept of non-additivity requires some formal introduction. As an alternative to equation (9.2.1), we can represent the total binding potential energy U_N of a cluster of N water molecules as a sum of pairwise additive and pairwise non-additive (the residue of equation (9.2.1)) terms:

$$U_N(\mathbf{x}_1, \ldots, \mathbf{x}_N) = \frac{1}{2!} \sum_{\substack{i,j=1 \\ i \neq j}}^{N} U^{(2)}(\mathbf{x}_i, \mathbf{x}_j) + E_{(N-2)}(\mathbf{x}_1, \mathbf{x}_2, \ldots, \mathbf{x}_N). \qquad (9.3.1)$$

In other words, we can refer to the residual energy $E_{(N-2)}$ as the *pairwise non-additivity function*, which is thus a measure of the failure of the pure-pair potential to represent the total energy of the given system as well as a measure of the extent of cooperative molecular interactions. Higher non-additivity functions ($E_{(N-3)}$ – tripletwise non-additivity, $E_{(N-4)}$, etc.) are equally tractable though our concern here is only with the inadequacies of the pair potential.

Table 9.3.1 compares the results of the two polarizability models with a number of quantum-mechanical predictions for water chain series. In the case of the trimer water chain, investigation into some depth has been made to determine which of the three

H H H H H

possible types – sequential (OH–OH–OH), double donor (O–HOH–O) or double acceptor

H H H

(HOH–O–HOH) trimer (see Figure 9.3.1) is the most stable. Both quantum-mechanical treat-

H

ments and both polarizability models all agree that the sequential trimer is the most stable with its negative non-additivity indicating that it is more strongly bound than two 'isolated hydrogen-bonds', thus giving the simplest example of the cooperative hydrogen-bond. The Del Bene and Pople trimers, which are geometrically quite different from the Hankins, Moscowitz and Stillinger trimers, are seen to be the more strongly bound configurations and this is also borne out with the polarizability model evaluations. Geometrically speaking, the DPB trimers are essentially two stable dimers with a common central molecule as a logical predecessor to the larger 'branching network' of clusters, in contrast to the HMS configurations which appear in the ice Ih bonding network. In passing, it is also worth pointing out that contrary to the DBP trimers and most other structures studied here, the HMS trimers exhibit unusually large dipole–quadrupole (compared with the normally larger dipole–dipole) interactions and this might bear interesting implications on the ice Ih structure (see also Fletcher's 'ice-surface' calculations[23]).

Table 9.3.1 Binding energies (and pairwise non-additivities) of the waterchain series illustrated in Figure 9.3.1

Water chain	Type	Quantum-mechanical	Polarizable four-point-charge model	Polarizable electropole pole model	Energy breakdown $\mu - \mu$	$\mu - Q$	$Q - Q$	U_P	Polarization $\bar{\mu}$
Dimers	DBP	−6.09	−5.15	−4.65	90%	15%	9%	−14%	2.05
Trimers									
Double-donor	DBP	−9.33 (+1.87)	−9.39 (+0.23)	−8.17 (+0.07)	89%	17%	10%	−16%	2.09
	HMS	−2.97 (+0.87)		−4.17 (+0.50)	11%	95%	8%	−14%	1.92
Double-acceptor	DBP	−10.14 (+2.05)	−9.28 (+0.89)	−8.18 (+0.79)	91%	14%	10%	−15%	2.09
	HMS	−3.14 (+0.35)		−4.12 (+0.62)	9%	93%	9%	−11%	1.91
Sequential	DBP	−14.66 (−2.06)	−11.91 (−1.21)	−10.24 (−0.61)	94%	18%	8%	−20%	2.16
	HMS	−6.77 (−1.36)		−5.90 (−0.26)	47%	71%	8%	−26%	1.98
Tetramers									
DBP-closed		−24.55	−18.73 (−2.53)	−16.76 (−1.56)	98%	19%	8%	−25%	2.61
DBP-open		−23.45	−18.30 (−2.34)	−15.86 (−1.28)	97%	19%	8%	−24%	2.21
Difference		−1.10	−0.43	−0.90					
Pentamers									
DBP-closed		−41.39		−26.46 (−5.47)	101%	23%	7%	−31%	2.33
DBP-open				−22.18 (−2.23)	99%	20%	8%	−27%	2.25
Difference				−4.28					

The main values refer to the potential energy and the bracketed value the pairwise non-additivity both expressed in kcal/mole. The quantum-mechanical solutions(29) refer to the Del Bene, Pople (DBP) (R = 2.73 Å) and Hankins, Moscowitz, Stillinger (HMS) treatments[8] (R = 2.76 Å), and the polarizable models[28] are those of Berendsen (four-point-charge) and Barnes (electropole). Only the electrostatic portions are included here. The energy breakdown

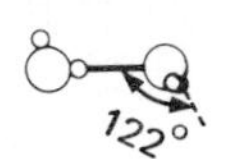

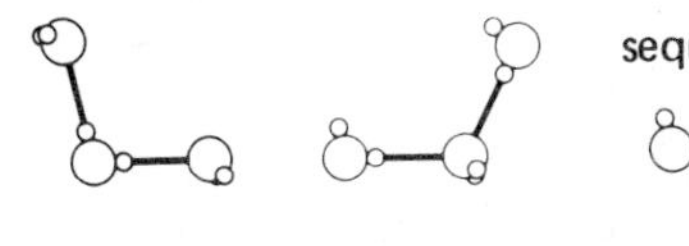
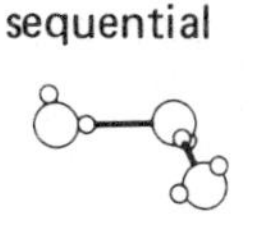

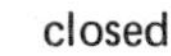
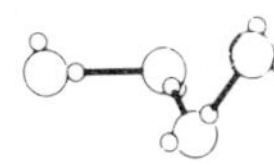
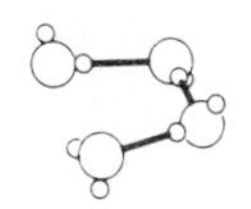

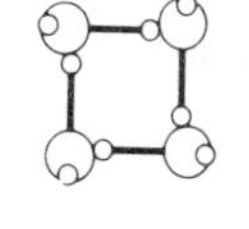
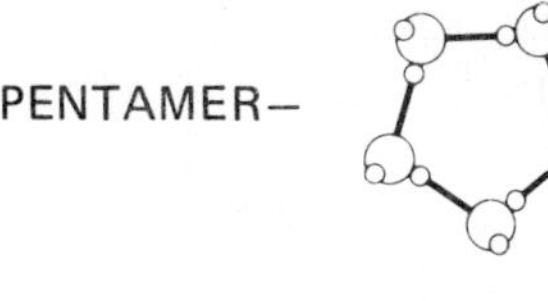
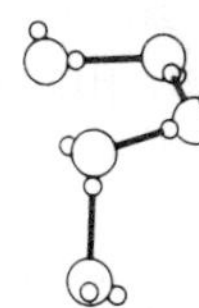
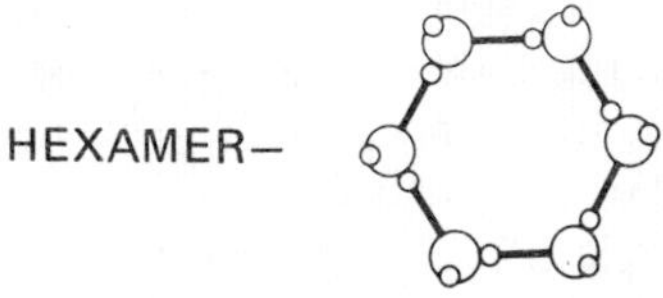

Figure 9.3.1 Representation of the water chain and cyclic cluster series as defined by Del Bene and Pople,[29] and drawn here using a microfilm-molecular drawing program. The larger spheres represent oxygen atoms and smaller spheres the hydrogen atoms. In the chain series the HOH-angle is set at 102° and successive molecules are placed 2.73 Å apart with their dipole directions (HOH-bisector) disposed at relative angles of 109° (= 102°/2 + (180° – 122°)) to each other. In addition to the stable configurations, the less stable non-sequential and part-open alternatives are also shown.

Only the polar (all non-bonding hydrogens on the same side of the cyclic plane) variety of cyclic clusters is shown here, the intermolecular distances being 2.56, 2.47, 2.44, 2.44 Å for the trimer, tetramer, pentamer and hexamer rings respectively

Following Del Bene and Pople's treatment, we now look at the higher-order clusters but consider only those configurations with the sequential-type link (i.e. OH–OH–OH, etc.) which proved to be the most stable link for the trimers. However, in adding the fourth water molecule to the sequential trimer, we find (see Figure 9.3.1) that we can generate either a more closed or more open type of chain. The quantum-mechanical treatment and both polarizability models all agree that the more closed variety is the most stable by amounts in the region of 1 kcal/mole. Similarly with the pentamer, we may add the fifth water molecule to the more closed tetramer in two different ways, thereby producing either a more closed or more open pentamer form. Again, the more closed variety is found unanimously to be the more stable.

Table 9.3.2 Electrostatic binding energies (kcal/mole) of the cyclic water clusters illustrated in Figure 9.3.1

Water cycle	Type	Quantum Mechanical	Polarizable Electropole model	Energy breakdown $\mu-\mu$	$\mu-Q$	$Q-Q$	U_P	Polarization $\bar{\mu}$
Cyclic trimer	DBP-polar	−16.83	+4.47	26%	−8%	16%	+66%	1.75
Cyclic tetramer	DBP-polar	−37.91	−22.24	96%	59%	5%	−60%	2.35
Cyclic pentamer	DBP-polar	−53.15	−44.84	117%	45%	3%	−65%	2.69
Cyclic hexamer	DBP-polar	−64.53	−56.24	135%	38%	1%	−74%	2.79

The quantum-mechanical solutions refer to the Del Bene and Pople (DBP) treatment,[29] and the polarizable electropole model of Barnes[20, 28]. Only the electrostatic portion of the latter is included here. The energy breakdown refers to the dipole–dipole, dipole–quadrupole, quadrupole–quadrupole, and polarization terms expressed in percentages, and the mean dipole moment in Debye, as calculated using the electropole model

It is interesting to look at the chain series from the point of view of the binding strength per bond. Table 9.3.3 shows that according to all four models, the hydrogen-bond strength increases with the length of the chain, with an average rate of over 20% increase per additional bond. At the same time there is a corresponding increase in the polarization of the molecules in accordance with our proposed model of the cooperative hydrogen-bond. The slightly disproportionate increase with $N = 5$ may be due to the almost complete closure of the pentamer ring-like shape, thereby providing effectively a part-extra bond.

The inherent stability of the semi-closed chains led Del Bene and Pople to investigate the cyclic water aggregates. In our study we have so far considered only the polar variety which have all the unbounded hydrogens on the same side of the cyclic cluster plane (see Figure 9.3.1). The results obtained using the electropole model are compared with those of Del Bene and Pople in Table 9.3.2 for the cyclic clusters. Again we see the trend of stronger binding per hydrogen-bond and greater polarization as the cyclic cluster size increases. Also one may note that the cyclic clusters are generally more strongly bound than the corresponding chain, on account, presumably, of the extra cyclic bond. This is borne out by the comparison made in Table 9.3.3 which shows in the overlap region ($N = 3$ to 5) that the actual *strength per bond* in the stable chain and cyclic clusters varies with the number of molecules N in similar fashion,

Table 9.3.3 Variation of hydrogen-bond energy (in kcal/mole) with number N of molecules throughout the cluster series of Figure 9.3.1 and Tables 9.3.1 and 9.3.2

N (number of molecules)	2	3	4	5	6
Chain series	dimer	trimer	tetramer	pentamer	hexamer
N_B (number of bonds)	1	2	3	4	5
Quantum-mechanical – DBP	−6.09	−7.33	−8.18	−10.35	
Polarizable four-point-charge model	−5.15	−5.95	−6.24		
Polarizable electropole model	−4.65	−5.12	−5.59	−6.62	
Cyclic cluster series		trimer	tetramer	pentamer	hexamer
N_B (number of bonds)		3	4	5	6
Quantum-mechanical – DBP		−5.61	−9.48	−10.63	−10.76
Polarizable electropole model		+1.49	−5.56	−8.97	−9.37

The values obtained are those of Tables 9.3.1 and 9.3.2 divided by the appropriate number of bonds N_B, the cyclic series having an additional bond with the notable exception of the cyclic trimer. It is interesting to note the more recent conclusions of Lentz and Scheraga[29] who, using a larger basis set for the molecular orbital calculations, also favour this departure with the cyclic trimer; they consider the cyclic series to be more stable only on account of the consequent additional bond which more than offsets the greater straining of the bonds (from the ideal chain configurations) imposed by the ring geometry. The exception, they also find, is with the cyclic trimer for which the bond straining (19° rotation compared with 7° with the cyclic tetramer) is too excessive for compensation by the extra bond, and in this respect they dispute the DBP trimer results.

Finally, it is fair to say that the polarizability models qualitatively reproduce the right trends in the cases discussed above, and the quantitative aspects might well be said to be on a par with the present accuracy of quantum-mechanical results. In summary, we see throughout the chain and cyclic cluster series an overall trend of greater polarization and stronger hydrogen-bonding with the cluster size, and a definite preference for the sequential link. This summarizes one of the main contentions of this particular study, that the cooperative hydrogen-bonding in water is largely a consequence of the mutual polarization of the water molecules as envisaged by the Barnes and Berendsen models. It is fortunate (and perhaps amusing) that such an important cooperative effect, hitherto evaluated and conceptually disguised in lengthy quantum-mechanical calculations, can be solved using essentially classical techniques and can be evaluated relatively quickly in computer simulation models. This could well be of importance to the future study and understanding of the special role of water in more complex systems – this aspect is briefly considered in the later sections of this chapter.

9.4 The ice I lattice

Ice in its own way is as anomalous as water and provides a somewhat undeveloped area of interest for the simulation models. As explained in Section 9.2, the ice lattice is such that the essential model features of the water molecule are consistent with the near-tetrahedral hydrogen-bonding scheme. However, it is now well known that ice I is not a true crystal structure in that the hydrogens are somewhat disordered, and continually move around the lattice between successive configurations in which the two Bernal–Fowler rules[30] (i.e. two hydrogens associated with each oxygen, and one hydrogen on each oxygen–oxygen bond) are largely obeyed.

This 'quasi-randomness' of the hydrogen positions is in itself an interesting simulation problem. It is well illustrated by following an ingenious method used by Rahman and Stillinger[17] in order to randomly generate Bernal–Fowler configurations for evaluating the Kirkwood dielectric correlation factor for ice I. The method is illustrated two-dimensionally in Figure 9.4.1. An initially hydrogen-ordered structure is disordered (hydrogen-wise) by moving the hydrogens around a chosen closed circuit as shown, so that they read HO–HO–HO etc. in contrast to the initial OH–OH–OH etc. This new configuration, although no longer perfectly repeating, still obeys the Bernal-Fowler rules and has preserved its original electric polarity. These circuits may deviate from the basic lattice into the surrounding lattice of periodic images provided the structural mappings are maintained and the final closure of the circuit occurs in the real lattice. By generating large numbers of such 'randomizing circuits' one can effect the disordering of an assembly of identical unit cells of ice I so that the final configuration is independent of its original ordered state.†

†This is not strictly true as the original lattice has an inbuilt structural parity. The opposite parity, which is however energetically equivalent, can only be attained by introducing ionic and Bjerrum-bond defects into the lattice.

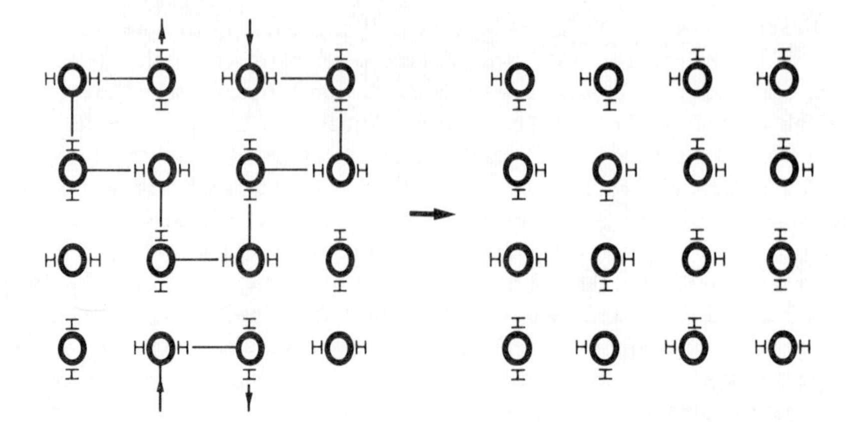

Figure 9.4.1 Schematic illustration of the disordering of a two-dimensional lattice using the technique of closed loop generation. The left-hand perfectly-repeating lattice (repeats every alternate molecule) is made to undergo a disordering process around the closed circuit shown, by moving all the H's on this loop to the opposite extremity of their respective bonds. A new non-repeating lattice (right-hand figure) is thus generated in which the two Bernal–Fowler rules (two H's per O, one H per O–O bond) and the polar neutrality are maintained.

As illustrated, the loop may extend out of the basic lattice, as long as closure finally occurs back in the basic lattice

Ideally such a randomizing procedure ought to be performed with a Monte Carlo weighting of the total energy of each configuration. However, as ultimately the energy differences will be small, this may be justifiably neglected in some applications.

Our immediate interest is to examine the energy-polarization characteristics of the ice I lattice using the polarizable electropole model and the randomizing technique discussed above. In passing, we note that Coulson and Eisenberg[22] had previously obtained an averaged result using a rather clever geometric rather than energetic weighting technique; their averaged results compare well with the specific interactions evaluated here. In order to get reliable results for the small energy differences involved in this problem, the accuracy of the computer calculations was increased considerably from that normally tolerated in the water simulations: 216 molecules were placed on an initially ordered ice Ih lattice and set into an orthorhombic (13.52 Å x 23.42 Å x 22.08 Å) container with the usual system of periodic images. The cut-off radius was increased to about 13 Å (200 nearest neighbours) and about seven polarization iterations were required to obtain an accuracy of 0.1% and 0.2% respectively in the equilibrium polarization and energy.

The results are given in Table 9.4.1 and displayed in Figure 9.4.2 as histograms of dipole-moment and molecular energy populations. It should be noted that the energy values include a Lennard–Jones portion ($\sigma = 2.985$ Å, $\epsilon = 0.330$ kcal/mole) to give an idea of the overall energies involved. These Lennard–Jones terms are unlikely to be the optimum values (see Section 9.5). However, their introduction affects only the absolute position of the energy histograms and not the shape of the distributions.

It is interesting to observe that although the average energy and dipole moment of the repeating lattice does not change much during the randomization, the individuality of the molecules increases enormously, as demonstrated by the width of the distributions: whereas the fully ordered configuration yields just two narrow bands of molecular identity, a well-randomized configuration yields broad, roughly Gaussian distributions of molecular energy and

Table 9.4.1 Breakdown of energy components and polarizations for ice I, water and a small water cluster at 0 °C, using the polarizable electropole model

	Ice Ih	Water	Cluster
Dipole–dipole	–8.897 (109%)	–9.209 (192%)	–4.285 (120%)
Dipole–quadrupole	–3.395 (42%)	–1.857 (39%)	–0.434 (12%)
Quadrupole–quadrupole	–0.556 (7%)	–0.153 (3%)	–0.024 (1%)
Polarization	+3.247 (–40%)	+3.462 (–72%)	+1.081 (–30%)
Total electrostatic	–9.600 (118%)	–7.758 (162%)	–3.662 (103%)
Lennard–Jones	+1.488 (–18%)	+2.957 (–62%)	+0.100 (–3%)
Total energy	–8.112	–4.800	–3.562 kcal/mole
Mean dipole moment $\bar{\mu}$	2.643	2.588	2.218 Debye

All the energies are in units of kcal/mole and the dipole moments in Debye, and are calculated using the polarizable electropole model of Barnes.[20, 28] Some further details regarding the ice I, water, and 50-molecule cluster are given in Sections 9.4, 9.5 and 9.6 respectively. The *total* energy difference (Ice – water) implies a poor value (3.312) for the latent heat (1.440) whereas the electrostatic energy difference yields a more respectable value of 1.842 kcal/mole. This is in keeping with the general notion that these preliminary Lennard–Jones constants (ϵ = 0.33 kcal/mole, σ = 2.985 Å) are far from ideal for this model and over-emphasize their contribution to the total potential energy and virial pressure

dipole moment. At the same time, the average dipole moment $\bar{\mu}$ drops slightly from 2.649 to 2.635 Debye while the average electrostatic binding energy increases from –9.420 to –9.599 kcal/mole, this being the effective contribution of two hydrogen-bonds per molecule. Although these changes are small and may be refined later using a larger set of configurations, on account of the convergence and accuracy of the calculations the author is convinced that the signs of the changes are correct and so discusses their interesting implications. These reorientations have in their own subtle and cooperative way produced a more efficient scheme of binding since, unlike the binding energy, the polarization has actually decreased in addition to its broadening. As mentioned previously, it is now well-accepted that at temperatures near 0 °C, ice I displays this quasi-randomness which itself is reflected in the residual entropy of ice I at the absolute zero.[31] However, the interesting feature here is that this energy difference shows that the quasi-randomized structure is the more stable (by about 0.18 kcal/mole) apart from considerations of its increased entropy – this unusual alliance between entropy and internal energy suggests that there cannot be a low temperature transition from the quasi-random to the ordered state. The reportings of time-dependent piezoelectric (and ferroelectric) effects in ice I[32] must then be a consequence of impurities in the ice samples used. The extraordinary broadness of the binding energy distribution is in itself a curious phenomenon for a single component crystal structure. The 1 to 2 kcal/mole energy width (*half* maximum peak width) is comparable even with the liquid value (about 5 kcal/mole – see Figure 9.4.2) and is clearly not negligible in comparison with the activation energy of many important physical processes, e.g. self-diffusion dislocation climb, ionic/Bjerrum defect creation/migration (6–30 kcal/mole), latent heat (1.44 kcal/mole), etc. A caution is then necessary regarding the dangers of using theories developed for the 'normal' crystalline state to the ice lattice with its extraordinary molecular individuality.

Table 9.4.1 shows that bulk ice, like water, is dominated by the dipole–dipole energy interaction term (109%) which more than offsets the considerable polarization energy (–40%) required to polarize the molecules. Orban and Bellemans[33] have made a preliminary study of cubic ice using a four-point-charge model and report polarizations of more than 70%. There is clearly much scope for deeper investigation into the various ice phases and also its contentious surface (see References 23 and 34) which displays anomalous mechanical and electrical

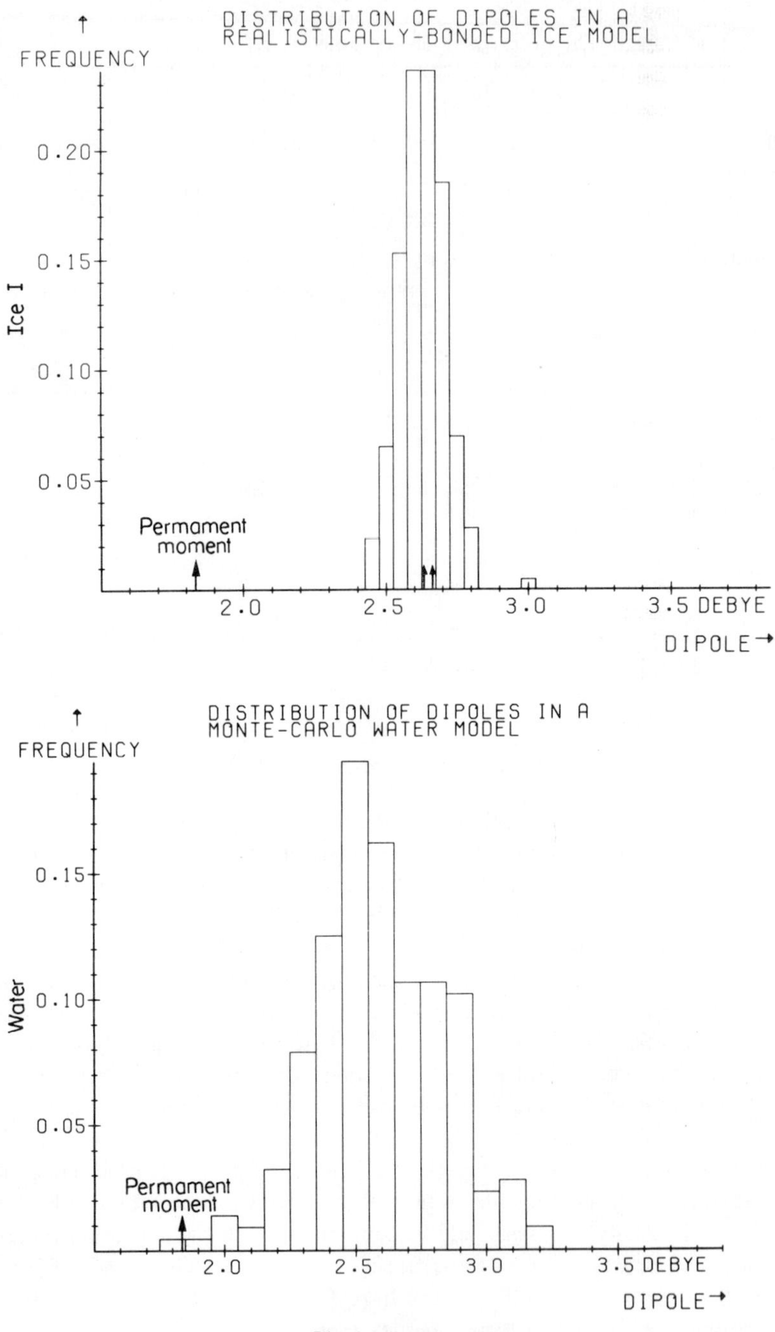

Figure 9.4.2 Distribution of dipoles and molecular potential energies in ice and water at 0 °C, obtained using the polarizable electropole model[20,28] on repeating cells containing 216 H_2O molecules. The ice lattice was disordered using 1000 randomizing loops (see Figure 9.4.1), each molecule suffering an average of nearly 40 disordering events in the process. The resulting molecular (dipole) distinguishability is greatly enhanced in comparison with that of the original ordered state which

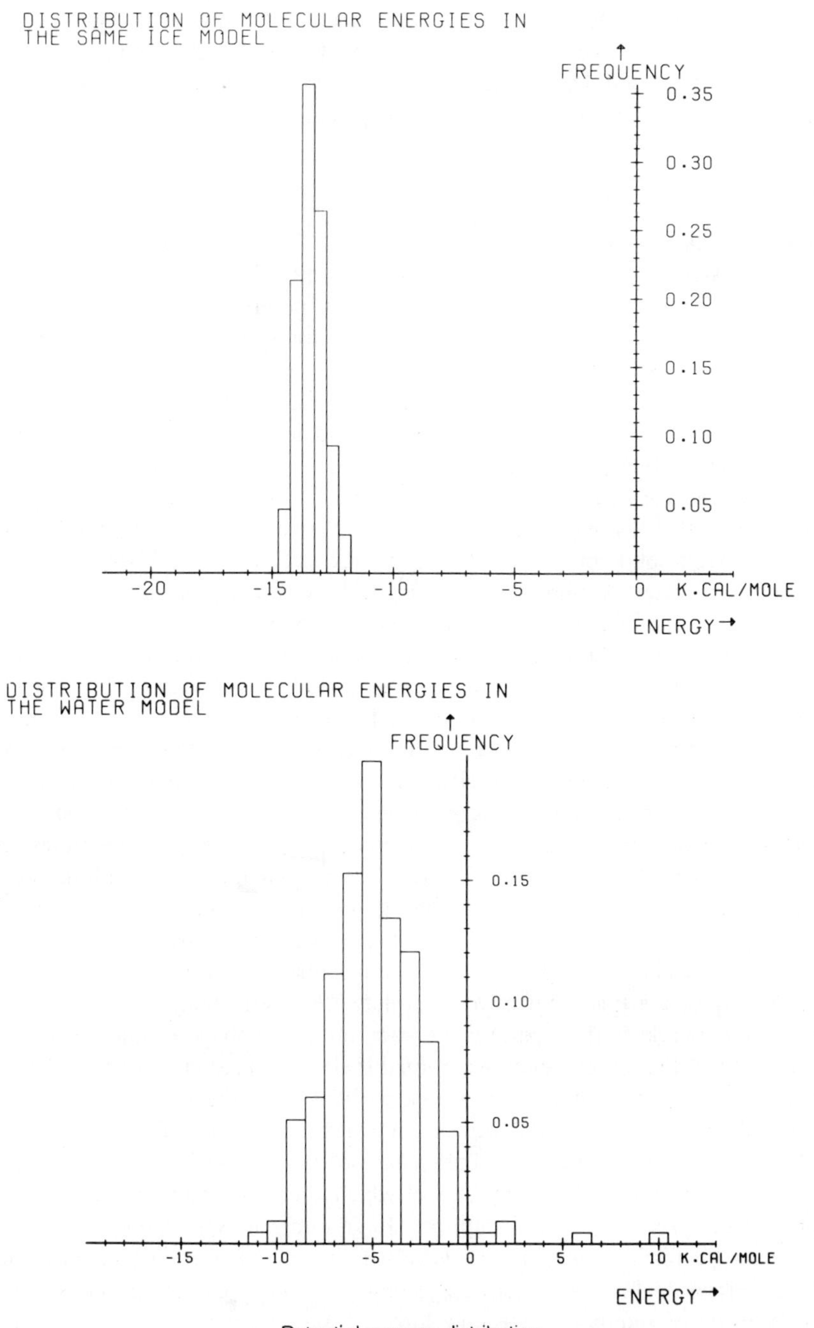

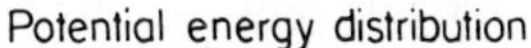
Potential energy distribution

possessed only two (at 2.631 and 2.651 Debye – see arrowed positions) narrow bands of dipolar identity.

The water configuration (see Section 9.5 and Figure 9.5.1 for details) yields a molecular distinguishability which is about three times greater than that of ice.

Both distributions of potential energy derive their Gaussian characters almost entirely from the dipolar variations described above properties near the melting point. Some refinements along the lines of making allowances for the exact Lennard–Jones contribution, defect concentrations, and pure quantum-mechanical contributions to bonding will be necessary before the *absolute* accuracy is achieved to quantify other basic lattice parameters like latent heat and sublimation energy.

9.5 The simulation of bulk water

Inevitably we lean heavily here on the pioneering work of Rahman and Stillinger[4–6] using the BNS and ST2 water models. The simulations so far could well be considered as preliminary in their ability to achieve *accurate* thermodynamic results; nevertheless a wealth of phenomenological detail has been recorded in areas which had previously been clouded with stagnant and protracted theoretical controversy.

The picture drawn by Rahman and Stillinger[4] of bulk water is one in which the water molecules move so as to be continually breaking and remaking highly strained directional hydrogen-bonds which themselves form uniformly space-occupying but essentially disordered networks. These networks are roughly tetrahedral in nature and strongly temperature-dependent as predicted by earlier model work of Bernal,[30,38] but do not promote any appreciable density fluctuations, thus dismissing the self-clathrate, low-density cluster or 'iceberg' theories of water. A number of dangling OH bonds together with a small number of unbonded or 'interstitial' molecules exist, but the overall bond distribution is singly maximized thereby ruling out the two-state liquid water models which postulate simultaneous regions of bonded and unbonded molecules. At low temperatures the structural rigidity dominates the self-diffusion of the water molecule, which is therefore highly oscillatory in nature and proceeds through a gradual restructuring of the bond networks rather than by the so-called 'hopping' process. The dipole libration–relaxation time is found to be about 5.6×10^{-12} sec. at 34.3 °C, which is consistent with the measured dielectric relaxation time and frequency dependence of the dielectric constant. The simulations at four different temperatures[6] exhibit a clear density maximum at about 20 °C (the 'TMD') which compares well with the well-known experimental effect at 4 °C. This anomalous density maximum is presumably a consequence of the unique temperature dependence and straining of the highly directional hydrogen-bond networks in water – at lower temperatures the thermal disruption is insufficient to prevent the substantial forming of the space-expensive bond networks; as the temperature rises the gradual collapse of these networks initially permits the water molecules to pack more freely and closely together until, at still higher temperatures, normal thermal expansion takes over. However, the fine detail in many physical properties of water, particularly in the vicinity of solid surfaces[35] indicates that hydrogen-bonding still persists in a restricted sense up to even 70 °C. The elevated TMD value of 20 °C may be a consequence of the admitted over-directionality of the hydrogen-bonds in the four-point-charge model which therefore requires this higher temperature to provide the necessary bond disruption. Other refinements, including perhaps quantum corrections, may be forthcoming, but nevertheless the overall phenomenological success of this admittedly provisional model is quite staggering and is perhaps a lesson to other disciplines on the timely fruits of simulation. We now look a little deeper into the more structurally-oriented results of bulk water simulation using the BNS, ST2, Bol and polarizable models of Barnes and Berendsen.

The radial distribution function

Central to all liquid studies is the radial distribution function (RDF or radial pair-correlation function) which describes the averaged density of molecules as a function of distance R from

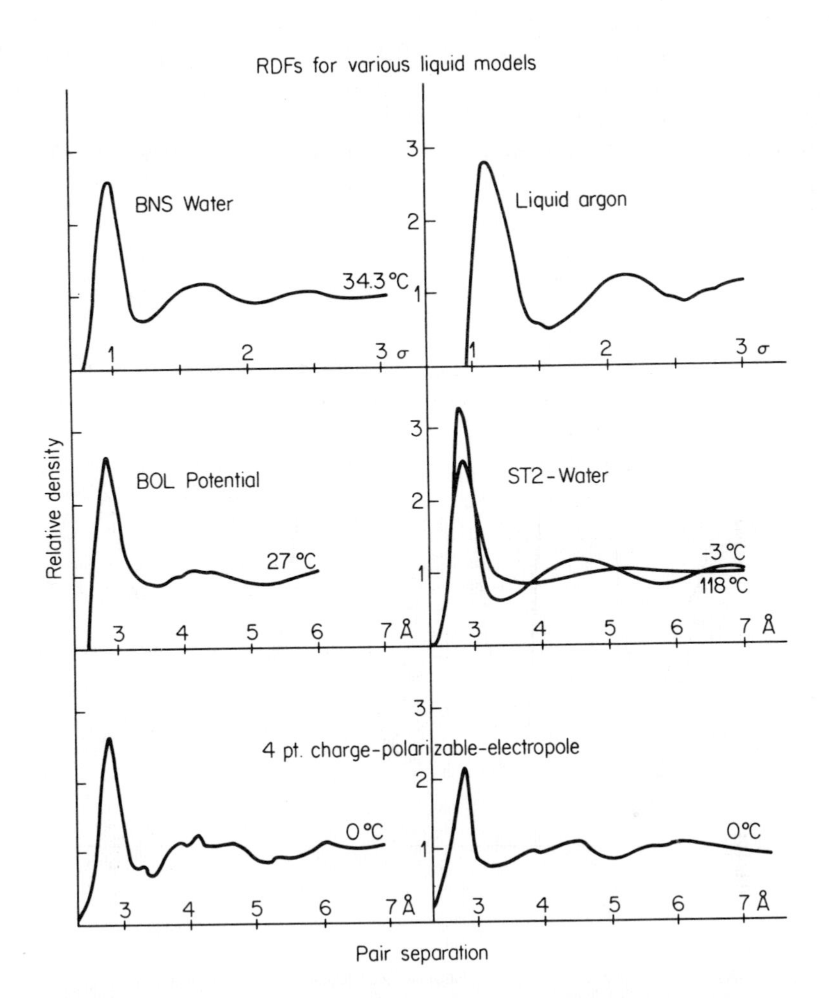

Figure 9.5.1 A collection of RDFs (pair correlation functions) obtained from simulations using various water models described in Section 9.2. The curves have been subjected to varying degrees of averaging and smoothing and so only the broad features (peak positions) are comparable. The simulation details are as follows: BNS – up to 10,000 MD steps of 4.355 x 10^{-16} sec., cut-off of 3.25σ (9.165 Å), $\epsilon = 0.0721$ kcal/mole, $\sigma = 2.82$ Å. BOL – up to 140,000 MC configurations, $\epsilon = 0.2750$–0.1910 kcal/mole, $\sigma = 3.0373$–3.1310 Å.; ST2 – 20,000 to 40,000 MD steps of 4–8 ps, cut-off of 8½ Å, $\epsilon = 0.07575$ kcal/mole, $\sigma = 3.10$ Å.; Polarizable models – cut-offs of 7–7½ Å, $\epsilon = 0.330$ kcal/mole, $\sigma = 2.985$ Å, the four-point-charge simulation result is from a preliminary study of 250 MD steps of 10^{-16} sec. using one of the BNS results as a starting configuration, the electropole simulation similarly consists of 10,000 MC configurations using the four-point-charge result as a starting configuration; Liquid argon – result (see Reference 4) for reduced state parameters of $\rho^* = \rho\sigma^3 = 0.81$, $T^* = k_B T/\epsilon = 0.74$. (Figures reproduced by permission of the American Institute of Physics (top LHS, top RHS, middle RHS) and CECAM (middle LHS, bottom LHS))

any molecule. While opposing liquid state theories all have a knack of producing acceptable RDFs, the RDF remains a necessary if insufficient criterion of success. Figure 9.5.1 shows a collection of normalized RDFs obtained from the oxygen positions using various model simulations (details in the legend) which may be compared with the experimental scattering results of Narten *et al.*,[36] and, for comparison, a computed RDF for liquid Argon.[4]

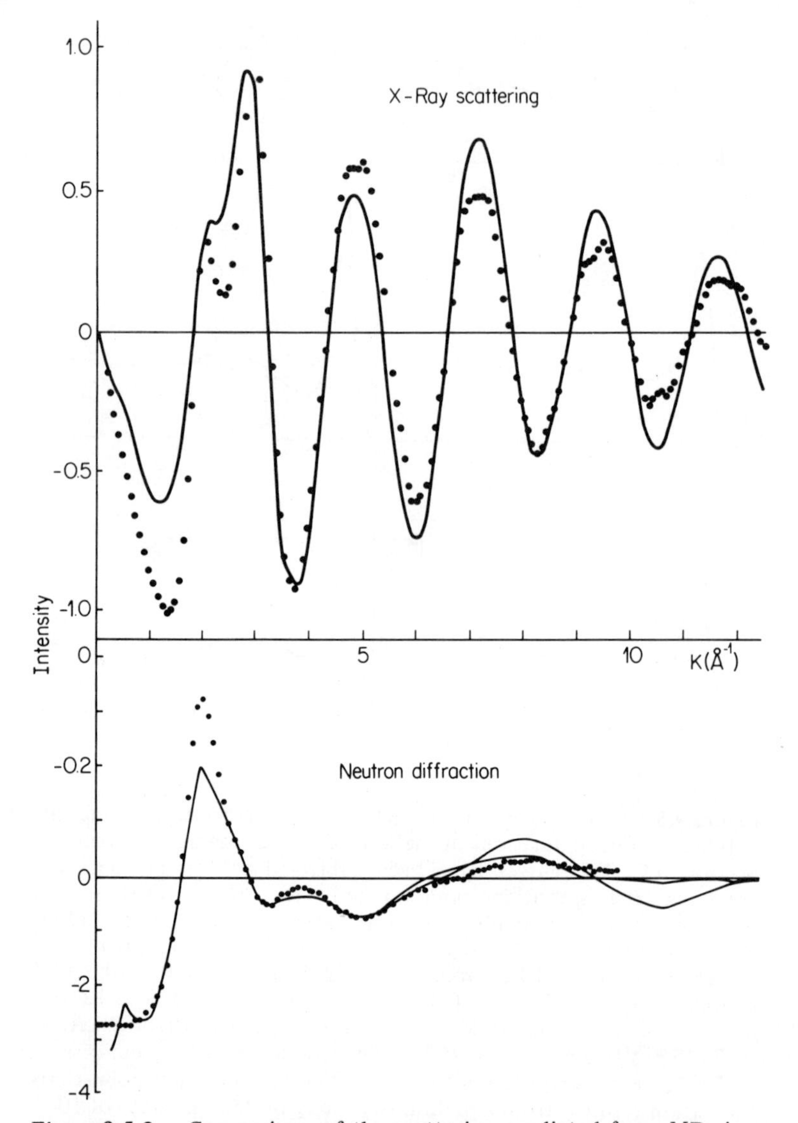

Figure 9.5.2 Comparison of the scattering predicted from MD simulations of water at 10°C (solid lines – Rahman and Stillinger[6]) and experimental (dots) X-ray (water at 0 °C) and neutron (D^2O at 25 °C) scattering intensities from Narten;[36] the additional line in the lower figure indicates the effect of making a vibrational correction. The fits indicate reasonable agreement on the spatial (X-ray scattering) and orientational (mainly neutron scattering) ordering within simulated bulk water. (Reproduced by permission of the American Institute of Physics)

Firstly, the broad conclusions as noted by Rahman and Stillinger[4] concern the relatively sharp first peak at a separation of 2.75 Å (0.957σ) and broad second peak at 4.65 Å (1.65σ) with substantial filling between these two peaks at 34.3 °C. The first peak encompasses an average of 5.5 (nearest) neighbours while the height of the second main peak is strongly temperature–dependent. By comparison the computed RDF for argon (pure Lennard–Jones potential) has a first peak which encompasses over twelve nearest neighbours and the distance ratio of the second to first peaks is 1.9 compared with the 1.69 for water. In fact, this distance ratio for water is close to that for ice (1.63) and indeed to that of the chain trimers of Figure 9.3.1. The first two RDF peaks are therefore clearly dominated by the directional hydrogen-bonds with the filling between, representing a proportion of the strained tetrahedral bonds, increasing with temperature at the expense of the second peak.

More recently, Rahman and Stillinger[6], have reversed the method of comparison between experiment and computation by predicting X-ray or neutron scattering intensities from the Fourier transforms of the various pair correlation functions and the known atomic scattering factors. Figure 9.5.2 shows that qualitative agreement is achieved with the ST2 model; with X-ray scattering both curves show the unique double peak at 2.5 $Å^{-1}$ and the departure at smaller K may be due to distorted atomic scattering factors, and for the neutron scattering of D_2O the departure at larger K can be reduced by correcting for the equivalent hydrogen vibrational modes. The former indicates the agreement on spatial ordering (X-rays are mainly scattered by the oxygen atoms) and the latter on orientational ordering (neutrons are highly sensitive to the deuteron positions) within water.

The RDFs obtained are consistent with the idea of directional hydrogen-bonds dominating the water interactions, and indicate basic agreement with experimental scattering data, though some improvement is still needed. It is worth re-emphasizing that it is still early days and all the models presented have plenty of scope for 'fine-tuning' provided this can be balanced with other structural and kinetic considerations. For this reason relative comparisons between the RDFs from the various models is at present a little premature.

Hydrogen-bonding schemes for water

In this section we examine various hydrogen-bonding schemes for bulk water. The discussion is restricted to those results obtained from the MD simulations by Rahman and Stillinger using the BNS and ST2 models, and the MC runs by the author using the polarizable electropole model.

Clearly a satisfactory bonding scheme should use criteria based on energy rather than just configurational considerations. In this spirit, Rahman and Stillinger examined the distribution of interaction energies U_{ij} between pairs of molecules – the distributions are shown in Figure 9.5.3. Apart from the large central band which represents the majority of non-interacting ($U_{ij} \simeq 0$) molecules, and the small maximum at +2.5 kcal/mole which turned out to be an artifact of the BNS model, the most striking feature is the residual plateau or hump (depending on temperature) on the negative energy side of the central band with an invariant point near −4 kcal/mole. This feature is not found with pure Lennard–Jones liquids (see the curve for argon) and it was concluded that this residual binding energy represents the distribution of hydrogen-bond energies in water. This residual effect is confirmed using the MC results from the polarizable electropole model (see Figure 9.5.4) though the distribution is noticeably without any distinct plateau or hump. In other words, the hydrogen-bonding is less well defined, with a far greater spread of energies than with the BNS model.

Nevertheless, as a first attempt it seems reasonable to define a hydrogen-bond between two

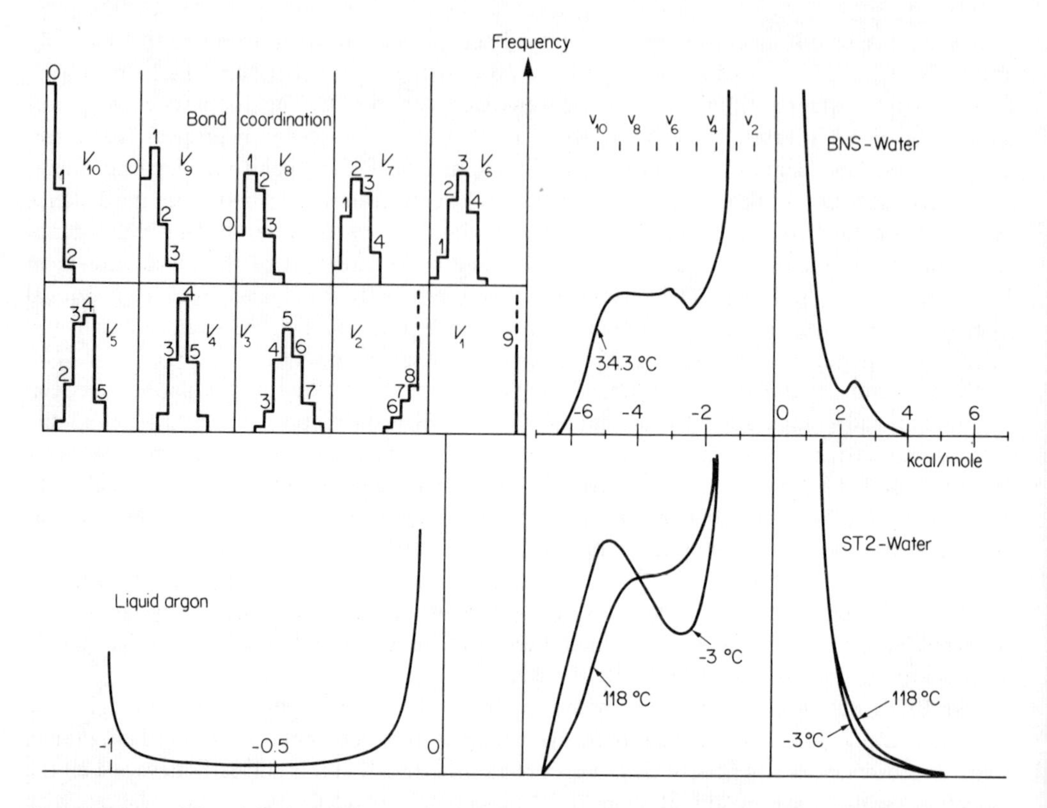

Figure 9.5.3 The right-hand figures represent the pair-energy ditributions obtained with the water models (upper right) of Rahman and Stillinger,[4,6] and for comparison (lower left) the distribution for liquid argon (see Figure 9.5.1). The various bond coordination distributions are derived using a range (V_1 to V_{10}) of bond-energy criteria with the BNS water model at 34.3 °C. (Reproduced by permission of the American Institute of Physics)

molecules *i* and *j* as:

$U_{ij} < V_{\mathrm{HB}}$, *i* and *j* are hydrogen-bonded

$U_{ij} \geqslant V_{\mathrm{HB}}$, *i* and *j* are not hydrogen-bonded.

Rahman and Stillinger used this definition for various values of V_{HB} (see Figure 9.5.3) ranging from a permissive choice (too many bonds) to a stringent choice (too few bonds). The corresponding bond coordination distributions are also shown in Figure 9.5.3 and are all seen to be singly maximized over a range of temperatures, indicating that two-state liquid models are untenable. On the basis of the four-fold coordination acquired using the BNS model for ice and the acceptable bond distribution obtained with essentially nearest neighbours, Rahman and Stillinger suggested the $V_{\mathrm{HB}} = V_6$ (−2.9 kcal/mole) value which produced a most probable hydrogen-bond coordination number of 3 for bulk water at 34.3 °C.

Using the same technique for the polarizable electropole MC simulations requires a slightly different procedure, as non-additive energies are involved. To use the general formulation of Rahman and Stillinger[6] would have been cumbersome for this particular application, so the value ($U_{\mathrm{E}} + U_{\mathrm{LJ}}$) of equations (9.2.2), (9.2.4) and (9.2.5) was used; this effectively attributes the various *polarized* electropolar and Lennard–Jones interactions to the required hydrogen-bond components, and deems the polarization energy to be a more communal property. This

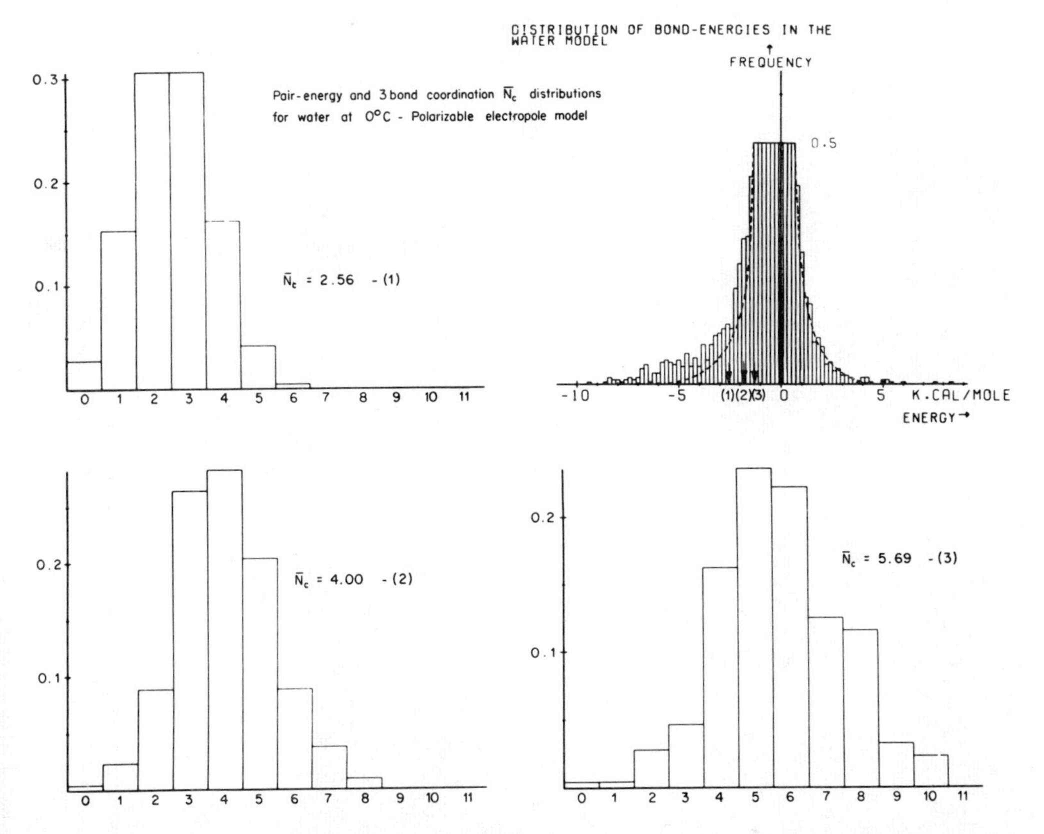

Figure 9.5.4 The upper right figure represents the pair-energy distribution obtained using the polarizable electropole model.[20,28] Three bond-energy criteria are chosen (see arrowed positions) leading to three-bond coordination distributions (upper left and lower figures) which are characterized by their average coordination number $\bar{N}_c$

choice may not be exactly rigorous, but then the use of a singly valued V_{HB} is equally questionable.† Using this definition of interaction energy one obtains singly maximized coordination distributions for 0 °C which are similar, but somewhat broader, than those obtained by Rahman and Stillinger at 34.3 °C. Figure 9.5.4 shows just three of these distributions which yielded characteristic average bond coordination numbers $\bar{N}_c$ of 2.56, 4.00 and 5.69 at 0 °C.

Figure 9.5.5 visually illustrates the sort of bonding networks obtained using the three different energy criteria. Even without the periodic images the profusion from 216 water molecules is considerable, though in practice stereo-pairs are used to assimilate the three-dimensional character. The 'half-slice' pictures (108 molecules) showing also the hydrogen positions, give an insight into the localized bonding, and a simulation sequence of these pictures (not shown here) confirms Rahman and Stillinger's description of rapidly changing random networks of bonds. The author was particularly struck by the rapidity of change in bonding associated with barely perceptible molecular movements or rotations, reflecting perhaps the arbitrary nature of the bond criterion used. Various polygonal networks were evident, including

†Ideally, $(\partial U/\partial \mathbf{R})_{\mathbf{R}=\mathbf{R}_{ij}}$ should be calculated for each appropriate pair (i,j) in the presence of all the molecules, but these are time-consuming calculations. Ignoring the polarizing energy variation $(\partial U_P/\partial \mathbf{R})$ is a good first approximation and is also time-saving.

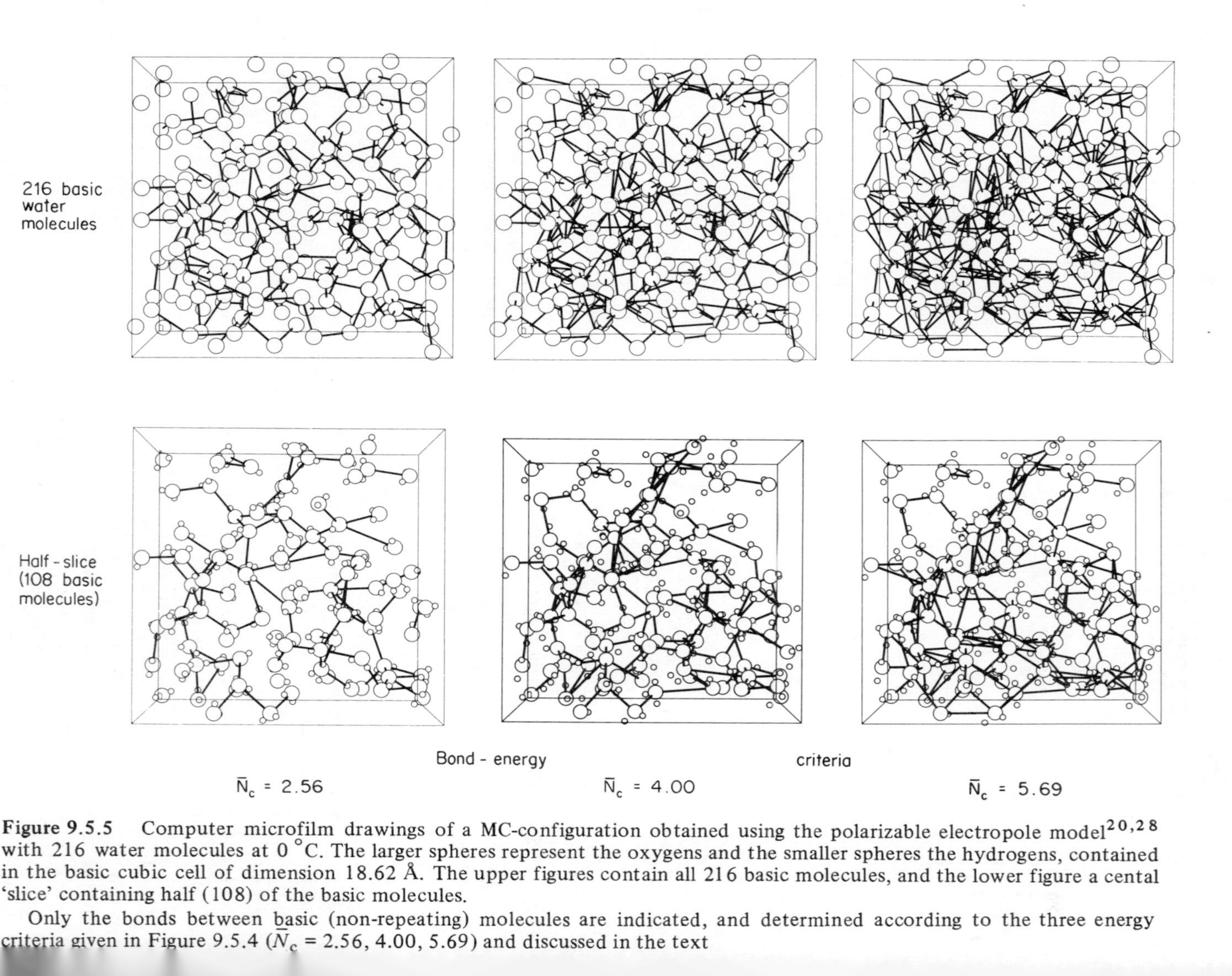

Figure 9.5.5 Computer microfilm drawings of a MC-configuration obtained using the polarizable electropole model[20,28] with 216 water molecules at 0 °C. The larger spheres represent the oxygens and the smaller spheres the hydrogens, contained in the basic cubic cell of dimension 18.62 Å. The upper figures contain all 216 basic molecules, and the lower figure a cental 'slice' containing half (108) of the basic molecules.

Only the bonds between basic (non-repeating) molecules are indicated, and determined according to the three energy criteria given in Figure 9.5.4 ($\bar{N}_c$ = 2.56, 4.00, 5.69) and discussed in the text

Figure 9.5.6 A building rig designed for making bonded liquid models fairly rapidly. Polystyrene balls are located in the three-dimensional space using the positioner, and then incorporated into the model by affixing with bond-sticks and fast-setting glue to the appropriate neighbours. In this way a 216-molecule liquid structure can be built with intense application in a few days

even the occasional cyclic trimer which we have previously noted to be unstable in isolation (see Section 9.3). In Dr Finney's liquid laboratory, we are trying to develop methods of rapid model building[37] (see Figure 9.5.6) to investigate the characterization of various bond networks obtained using different bond criteria. The statistical geometry of these models will be determined in terms of the equivalent network of Voronoi polyhedra used previously on simple liquids by Bernal and Finney.[38]

In surveying the various bonding criteria presented so far, it is easy to dismiss the extremes of permissiveness or rigidity, but less clear how to choose between the middle candidates (e.g. V_4 to V_8 in Figure 9.5.3, and the three given in Figure 9.5.4). In the author's opinion the assumptions do not permit one to attach overt physical significance to the numerical value of

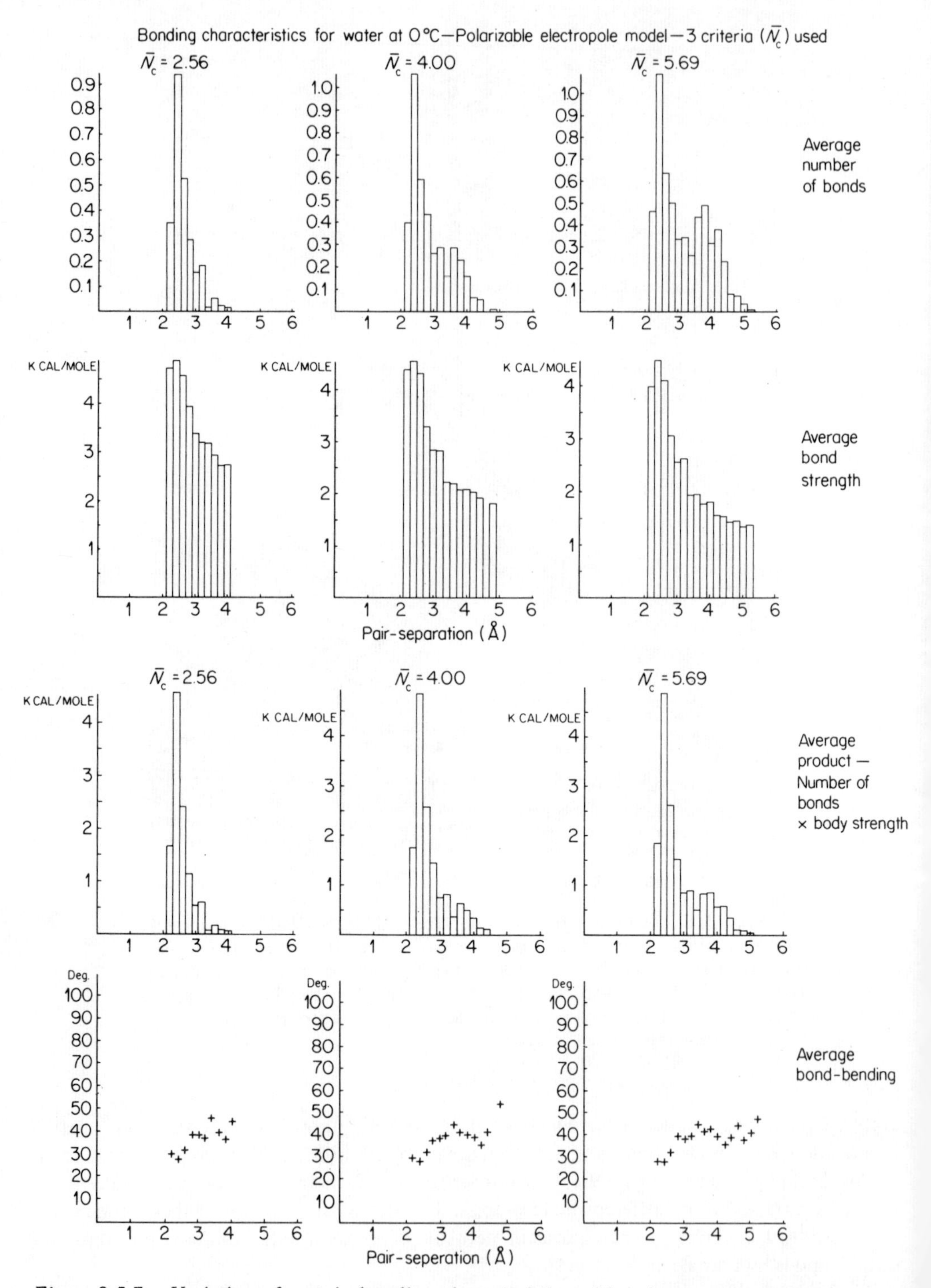

Figure 9.5.7 Variation of certain bonding characteristics with pair separation using the three bond-energy criteria ($\overline{N_c}$ = 2.56, 4.00, 5.69) of Figure 9.5.4. The average bond number, bond energy, and their product, display sharpened versions of the first RDF peak with a finite bonding contribution from the second RDF peak, particularly with the more permissive bond criterion ($\overline{N}_c$ = 5.69). The average bond-bending is seen to increase with pair separation

V_{HB}, and the application to ice I (presumably $\bar{N}_c = 4$) is pertinent but not without the complications of bridging two different structures with one model; an extension of the ideas discussed earlier (Section 10.4) shows[28] that even in ice the bond coordination, like the polarization, becomes statistical and not uniquely four-fold if we use this idealized (singly valued V_{HB}) approach. Likewise, an examination of the simplest triplet interaction (the sequential trimer studied earlier) indicates all is not straightforward: a division of the non-additive energy by three would imply that all three potential bonds (1–2, 2–3, 1–3) contribute equally to the non-additivity, whereas it is easily shown with the polarizable model that the non-additivity mostly derives from the two closest (U_{12} and U_{23}) interactions. Also, in the light of the variation of the hydrogen-bond energy with the number of molecules N for the cluster series (Table 9.3.3), should one necessarily associate hydrogen-bonding with a single-valued 'on–off' energy criterion? Clearly there are problems in trying to rationalize a concept of bonding, which is a pairwise property, to cases where non-pairwise effects are large – more work and thought are needed along these lines.

Bearing in mind the ambiguities outlined above, we proceed with the single-valued energy criterion and examine further its implications. The criterion chosen by Rahman and Stillinger with the BNS model ($V_{HB} = V_6 = -2.9$ kcal/mole) leads to a variation of coordination number $\bar{N}_c$ within the range 4 to 1 over a temperature range of –8 to +315 °C, and the criterion preferred for the polarizable model ($V_{HB} = -1.75$ kcal/mole) gives an average coordination of 4.00 at 0 °C.

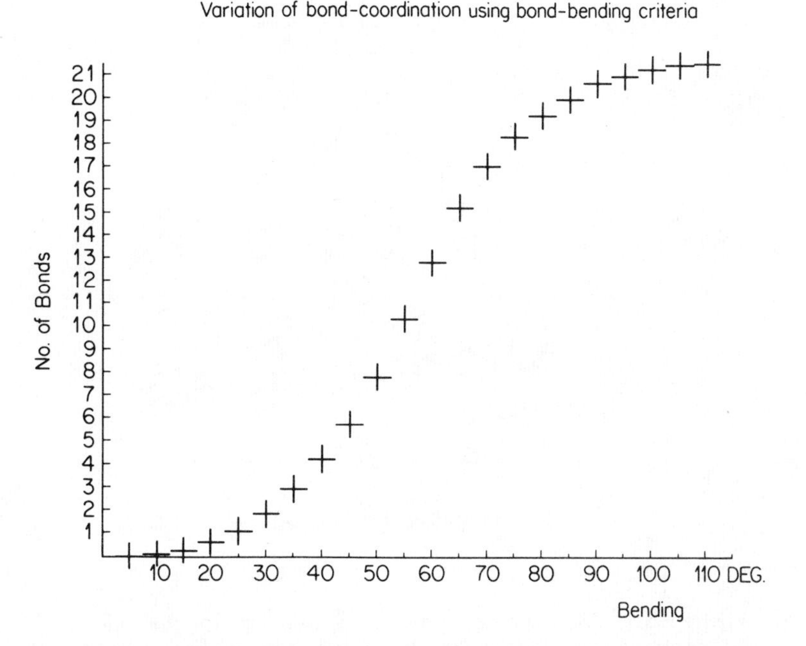

Figure 9.5.8 Variation of average bond coordination $\bar{N}_c$ with permitted bond-bending for pair separations up to 5.4 Å. As with the various energy criteria (Figure 9.5.4), different bond criteria may be selected on a basis of permitted bond-bending. For example, a maximum permitted bending of 40° yields an average bond coordination $\bar{N}_c$ of 4, and bond networks similar to those of Figure 9.5.9

216 basic water molecules

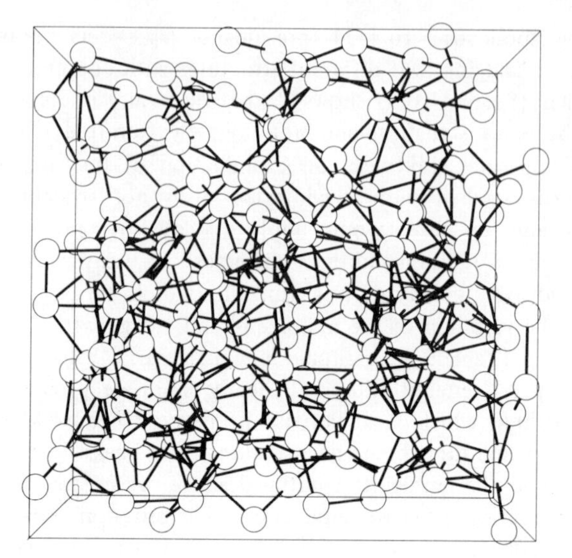

Half-slice (108 basic molecules)

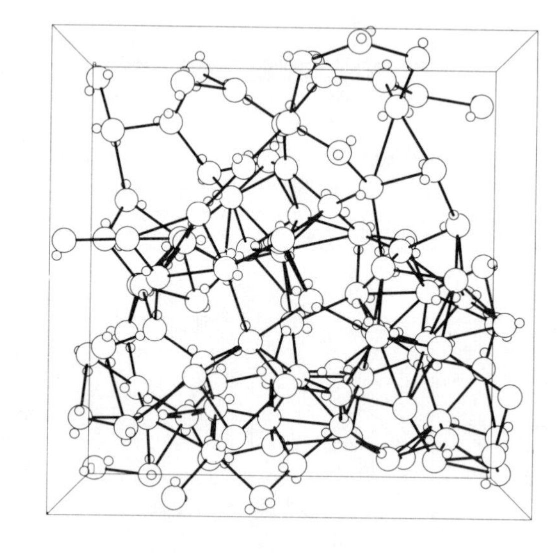

Bond-bending criterion with $\overline{N}_c = 4.02$

Figure 9.5.9 A computer microfilm drawing of the same MC-configuration of Figure 9.5.5, but with the bonding selected using a bond-bending criterion. The permitted bond-bending variation with separation is proportional to that found in Figure 9.5.7, yielding an average bond coordination $\overline{N}_c$ of 4.02 (compare with Figure 9.5.5 for energy criterion $\overline{N}_c = 4.00$); an agreement of about 85% in bond identity is found between the bending and energy criteria

Bond-bending in water

The bending of the hydrogen-bonds between neighbouring water molecules induces a strain into the overall bond network, and it is the exact degree of strain permitted by the thermal movements that is the essence of the peculiar water-like interactions. Rahman and Stillinger[4,5] examined this effective bending using a technique of constructing icosahedra around the water molecules. The long-range effect of the surrounding structure is to quench the dipole moment of a given molecule, while the close-range bonding structure can be determined from considering neighbouring pairs of icosahedra. The three first sets of neighbouring molecules can be classified according to the types of icosahedron faces presented *vis-à-vis.* The statistical departure from idealism gives a measure of the bond-bending which Rahman and Stillinger deduced to be 'substantial' at even 34.3 °C.

The author has computed the distribution of bond types in a structure obtained from the MC simulations with the polarizable electropole model. Figure 9.5.7 shows, using a given bond criterion, the variation of bond coordination, bond energy and their product with radial pair separation. These variations give a good idea of the spatial extent of the bonding and show that some second neighbours are bonded too, the proportion being dependent on the stringency of the bond criterion used. Of particular interest is the average distribution of bond-bending as a function of radial separation: the bond-bending is defined simply as the average angular deviation between the closest OH-direction, or OP-direction (P representing the lone-pair electron) as the case may be, from the linear bond direction defined by the two molecular centres. This definition is not ideal but gives an easily appreciated measure of bending which should be noted as clearly excessive above 55° ($\frac{1}{2}\theta_T$) when it becomes possible for both a hydrogen and lone electron pair of *one* molecule to be simultaneously identified with the bond. Figure 9.5.7 indicates an average bond-bending in the range of 25–45° with a small but noticeable increase with separation, though in some individual cases bending over 70°. This variation presumably reflects the expected sharper mismatching at close separations, though bonding at larger separations is less frequent, if more mismatched, on account of the overall decrease of U_{ij} with separation.

Some trial and error attempts were made to see if an angular bonding criterion could be devised which approximated the results obtained with the energy criteria. The simplest geometrical criterion used was to define the existence of a bond between two molecules when their separation is less than a critical value R_{HB} (about 3.5 Å) and the bending less than some fixed critical angle θ_{HB} (e.g. 35–49°). Figure 9.5.8 shows how a range of average bond coordination may be selected from the choice of θ_{HB} (analogous to the V_{HB} used previously), though the fit is improved further by allowing for some variation of θ_{HB} with separation (as in Figure 9.5.7). The bond coordination distributions and bond networks (Figure 9.5.9) so obtained were qualitatively similar to those obtained with energy criteria, and an agreement of 85% was achieved regarding the actual bond identities; this is quite reasonable in view of the considerable 'grey areas' of bonding and partial arbitrariness with even the energetic criteria. These geometric schemes, however, cannot cater for non-additive interactions, though they have proved useful in the preparation of instructional films. They do give some limited conceptual insight into the bonding patterns in *bulk* water.

9.6 Studies of heterogeneous aqueous systems

In this present context, we are concerned with systems in which the local ordering has been perturbed from that of bulk water by the influence of spatial confinement or introduction of other components. In the broadest sense we see this as a potentially fruitful field which will enlarge to embrace a host of disciplines in which water plays a central part. For the moment we concentrate on some preliminary investigations into water droplets and also the role of limited neutral or ionic components in bulk water.

Spatially confined water

The subject of confined water has generated controversy, particularly over the last twenty years, from the extremes of invoking another stable water phase (polywater[39]) to the accepted peculiar freezing characteristics of water when confined to dimensions of less than 1 μm.[40] The possibility of a frozen glassy state at low temperatures was investigated by Rahman[41] using a droplet of 475 BNS-molecules. During two MD runs of 3750 steps ($\Delta t \sim 10^{-16}$ sec.) at 59 °K and 2200 steps ($\Delta t \sim 2 \times 10^{-16}$ sec.) at 263 °K no such unambiguous 'freezing' was recorded although the interior of the droplet was quite dense (1.2 g cm^{-3}) and at 59 °K did exhibit some temporal rigidity.

Although inconclusive in one aspect, the results of Rahman's droplet simulation leave us, not surprisingly, in no doubt as to the perturbing role of the free surface. The author was particularly interested to learn whether a pairwise non-additive model increased the disparity between bulk and surface-induced interactions, and so in this spirit made a preliminary MC-simulation at 0 °C using the polarizable electropolar model. The results are illustrated in the various plots of Figure 9.6.1. The initial cluster of 50 water molecules was chosen to be non-spherical in order to get a large surface:volume ratio, and was in fact taken from an initially larger (216 molecules) repeating configuration. Leaving aside the problems of defining a cluster,[42] the main interest here was to exaggerate the differences between bulk and confinement; the results are at the time of writing being repeated with a more spherical cluster and a longer chain of configurations.

Figure 9.6.1 shows that there is a dramatic change in the bonding (previous energy criterion, $\bar{N}_c = 4.00$ used) on initially isolating the cluster from its original bulk environment as a starting configuration. While this is to be expected, the extent of influence of the non-additive effects on hydrogen bonding is such that the cluster potential energy goes *positive*, the polarization drops, and nearly all the bonds are broken by the isolation. This dramatic change should be placed in its appropriate perspective – that an equivalent experiment using additive models would have left the bonding network initially unaltered after isolation. The cooperative nature of hydrogen-bonding should not be visualized as a mere strengthening of the bond networks, but as part and parcel of their existence.

Equally remarkable is the subsequent subtle readjustment of the water molecules to their new spatial environment: in the space of just 23,000 MC-configurations, after which time the molecular positions in the cluster still strongly resemble those of its genesis (see Figure 9.6.1), another hydrogen-bond network is re-established with the potential energy of the system relatively equilibriated. This overall procession may well be analogous to the initial orientational readjustments obtained in Rahman's solvation studies (see next section) to be followed by a more gradual spatial readjustment to include, presumably, ultimate sphericalization.

The chain of configurations generated shows some striking additional features. The re-established bonding network differs significantly from that of the bulk. The RDF, although not properly defined on account of the smallness and shape of the cluster, develops more clearly defined second- and third-neighbour peaks (though not at ice-like separations). At the same time a curious double-peaking in the polarization distribution (reminiscent of ordered ice, see earlier) is found as the potential energy settles down. (These results may be compared with those obtained using additive models.[42])

The final cluster is more highly bonded than its counterpart in the bulk and this raises some

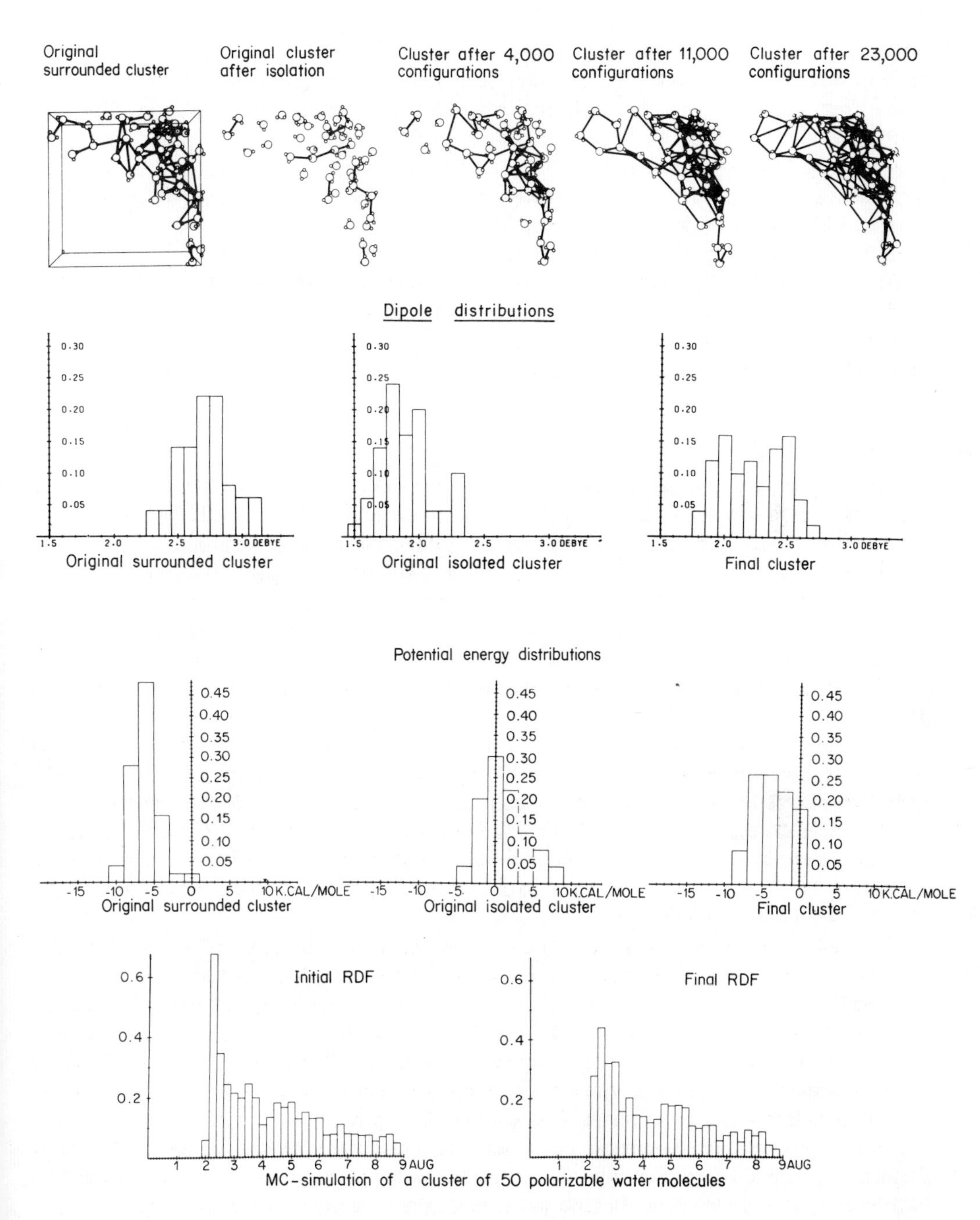

Figure 9.6.1 An illustrated MC-simulation of a cluster of 50 water molecules. The original cluster is taken from a larger, near-equilibrated 216-molecule repeating configuration. The surrounding molecules are not shown, but their considerable effect can be seen from the bonding within the cluster (first frame) which is almost completely destroyed on isolation (second frame) of the cluster from the original bulk. However, bonding is quickly re-established and strengthened during the subsequent MC simulation. The RDF and dipolar/potential energy populations develop characteristics markedly different from the bulk. The sequence underlines the importance of non-additive polarization effects in localized events atypical of the bulk, and is discussed further in the text

interesting conjecture, though the strength of the conclusions must be tempered with the caution that the initial bulk may not have been fully equilibrated. Although the polarizable model may account for the different non-additive energies experienced in bulk and confinement, are we right to employ the same simple bonding criteria in both cases? If the increased rigidity of the structure is then a genuine feature, the effect must be critically dependent on the size of the cluster. The extensive cooperativity of hydrogen-bonding in bulk is undeniable, but at the same time the constraints encountered by fitting this ordering into the overall structure must profoundly modify the final result. With the removal of these long-range constraints and the introduction of another – the emergence of the surface – it is perhaps hardly surprising that we have witnessed such dramatic changes. The unknown long-term future of this cluster cannot detract from the unavoidable conclusion that non-additive events in bulk and confinement are profoundly different with water. More work and thought are needed here.

Inert and ionic components

Notwithstanding the complex nature of the bonding networks in bulk water, aqueous ionic solutions have already emerged to the forefront of simulation study. Although the experimental concentration range of below, say, 0.01 mole is beyond the simulation range described here, a significant start[8B] has been made using systems with about one ion per 100 solvent molecules, and even at the time of writing more ambitious proportions have been reported.[43] Nevertheless, we have been able to glean meaningful information on the energetics of the solvation process and the basic structure of simple hydration shells. The following discussion is mainly a brief resumé of the simulation studies carried out at the two CECAM Workshops of 1974.[8B,8C]

As a preliminary to discussing the role of ions in water, one notes that the presence of an inert (pure Lennard–Jones) particle also strongly influences the solvent order through the combined effects of its space exclusion and interruption of the hydrogen-bonding networks, so that a molecular 'cage' is formed surrounding the inert particle. Rahman's study[44] has shown that two inert particles in 214 water molecules display, after an initial somewhat sluggish oscillatory 'captive' phase, a tendency to move apart to an equilibrium separation of 1.5σ (4.2 Å) after 4,000 steps (about 1 ps).

The behaviour of two oppositely-charged ions in a fluid of polar molecules[45] is also complex, and somewhat unpredictable from simplistic considerations. The MD simulations by Rahman[44] of two structureless (Lennard–Jones constants as for water) and oppositely-charged ($\pm 0.9428e$) ions in 214 ST2 water molecules show two distinct phases of interionic motion (Figure 9.6.2). After an initial oscillatory period of 2000 to 7000 steps (½ to 1½ ps) in which the local water molecules quickly readjust to their new ionic environment, the ions move apart from their neighbouring positions until an equilibrium separation is reached. Hence during the separation process the systematic couloumb attraction between the opposite charges is dramatically over-compensated by the stochastic force on the ions from the surrounding environment, thus yielding an effective net repulsion between their respective shells. The first hydration shell of each ion is consistently comprised of six water molecules, the greater interaction around the negative ion reflecting the inherent charge asymmetry of the ST2 model.

In fact, a number of runs were performed under slightly different initial conditions (285 °K, 271 °K, 278 °K) during which two distinct equilibrium separations of 1.3σ (3.7 Å) and 1.8σ (5.1 Å) were found. Rahman interpreted these two equilibria along the lines of two energetically and geometrically similar structures in which the positions of the water molecules attached to the two ions describe two octahedra, one centred on each ion. The two equilibrium separations then correspond to the formation of these octahedra so as to share either a common

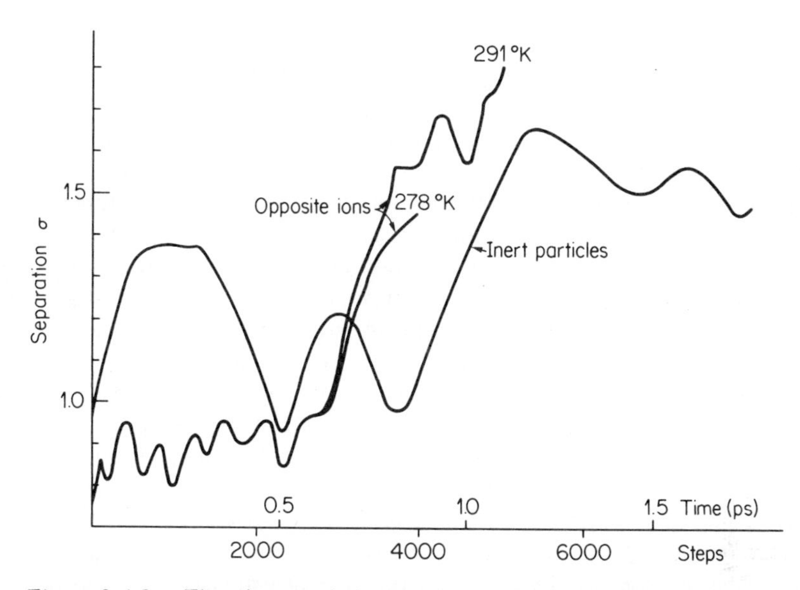

Figure 9.6.2 The characteristic separation of two foreign particles in 214 ST2 host water molecules (Rahman[44]). After an initial period of 'oscillatory readjustment' (2,000–4,000 MD steps), the oppositely charged ions or uncharged inert particles move apart to an equilibrium separation found to be between 1.3σ and 1.8σ. (Reproduced by permission of CECAM)

vertex (1.8σ) or a common face (1.3σ) respectively. Presumably, longer runs would exhibit transitions between these two complementary stable configurations.

9.7 Conclusions

The account presented here, although necessarily sketchy, should give some idea of the considerable interest and energy that has been invested in *'l'enfant terrible'* of liquids. The pace has hardly abated, so that during the present writing the relevant literature is growing alarmingly. In the author's opinion the returns have been very worthwhile, especially in view of the intellectual impasse which existed prior to the advent of machine simulation. The purists may wince at the lack of precision in predicting even the basic thermodynamic results (equation of state, specific heat, etc.) which have been particularly excluded in this present report. On the credit side remains the phenomenological success of admittedly incomplete models. Exemplary here is the outstanding work of Rahman and Stillinger who have shown how effective pair potentials can be used in simulating the complex subject of hydrogen-bonding in bulk water. These contrasting fortunes require some comment, however conjectural. In the author's opinion the answer lies in the peculiarity of water itself, that its anomalous behaviour necessitates features in the models which can clearly dominate the molecular interactions. The electrostatic models have this ability to propagate the required hydrogen-bond networks and to relate these networks to the configurational aspects of the dipolar water molecules. This also explains why relatively short cut-offs can be used; the isolated interaction between two distant water molecules is relatively unimportant in the normal sense, the long-range effects being transmitted via the bond linkage (through the intervening molecules) over distances in excess of the cut-off radius. Given the predilection for such bond networks, the dynamic behaviour of these networks reflects the delicate balance between the disruptive power of thermal motions and the amount of strain that these bonds can tolerate without losing their essential identity. The balance is then critically affected on one hand by the actual bond strength and elasticity, and on the other by such terms as the repulsive Lennard–Jones component. Put more crudely, this reasserts the ability of these models to predict qualitatively a range of phenomenological characteristics of water, but in doing so are likely to get the actual temperature variations wrong – perhaps a fair summary description of the present art! The author feels that this picture gives a phenomenologically reasonable description of basic bulk water interactions in spite of the over-simplified hydrogen-bond concept inherent in the use of effective pair potentials. The non-additive models strongly suggest that both geometrical and singly-valued energy criteria do not give a full description of hydrogen-bonding, which must really be considered as a more statistical phenomenon in both ice and bulk water. The cooperative effects described are more consistent with a wider continuous range of hydrogen-bond energies. Rather than imagine processes involving the breaking and reforming of discrete hydrogen-bonds, the author feels that one should visualize a continuous progression of weakening and strengthening of many cooperative hydrogen-bonds between the interacting molecules.

In guessing the future trends, the author feels that two obvious avenues present themselves:

(i) Further refinement of the present models aimed at improving basic thermodynamic agreement, or possibly the introduction of conceptually new models;

(ii) A wider application of the models to more complex aqueous systems.

Regarding the first direction, the need is clear and the necessary scope should be found – the manipulation of the charge distributions in the point-charge models, the Lennard–Jones constants, possibly quantum corrections, and it should be noted that the polarizable models depend critically on an accurate polarizability tensor. Rahman and Stilliger[6] wisely point out that: 'it would be pointless . . . to achieve near-perfect agreement with measured pair correlation functions if at the same time kinetic aspects . . . became grossly erroneous.'[8] There is no need yet for pessimism, although the agreement is not immediately forthcoming.

The latter endeavours are unlikely to wait for the completion of the former aims. Wider applications will require a more judicious choice of models to suit the need, and the time question (the present computer time:real time ratio is around 10^{16}) remains a problem for any modest institution. The 1974 CECAM Workshop[8C] concentrated on time-saving techniques and found some limited success. Where long-term simulation is being held up by essentially irrelevant faster motions (e.g. bond vibrations, rotational motion), specially tailored methods can aid convergence. For example, the stretching vibrations of a polymer chain can be separated out by appropriately modifying the intramolecular forces, the artificial increasing of the inertial moments of water can aid with the time convergence of translation-dominated processes (diffusion, viscosity). Other short-cuts rely on some sort of physical intuition or specific programming skill, as for example, the subdivision of the pair interaction sphere into concentric shells with ranging effective time-steps. Various useful time-savings of up to 10 were reported, though it remains to be seen if this can be extended or indeed proves to be sufficient.

The penetration into other aqueous systems will doubless raise new problems. It will be interesting to see whether non-additive effects increase in importance; spatial confinement or foreign components locally modify the structure so that local polarization and bonding is perturbed from the bulk average. If these effects are as large as predicted in Sections 9.3 and 9.6, then effective pair models may have to be supplanted by the more time-consuming polarizable models – though this may be prejudiced conjecture on the part of the author. If one's imagination is allowed to wander a little, one can visualize the application of water simulation to small clusters and droplets and their role in the nucleation of ice phases, to the confinement of water in narrow capillaries (artificial and biological), to surfaces (membranes?), not to mention the contentious free surfaces of both ice and water. The study of ionic solutions has already given limited insight into the solvation process and the structure of the hydration shell. Extending the scale of these methods leads ultimately to the role of water in the crystallization and conformation of biological macromolecules. Some seven or eight years ago, the attempted simulation of bulk water was thought equally ambitious, and so certainly a further equal space of time will prove this program also to be realistic. Whether this advance will be achieved through improved simulation techniques or with a new breed of more powerful computers or both is not at present clear, yet surely just one reasonably large and modern computer internationally devoted solely to the study of water would 'break the back' of this engaging and important study.

Acknowledgements

I would like to thank all the innumerable friends and colleagues who have given me inspirational and technical help throughout this work. Among those I must find space to mention here are Professor C. A. Croxton, who introduced me, albeit by circuitous routes, to liquid physics, Dr J. L. Finney, who has all along been closely associated with this work as have the other members, past and present, of our 'liquid group' (Mr I. Cherry, Mrs S. Bailey, Dr P. Timmins, Mr J. Nicholas, Mr S. Heathman) and Professor J. B. Hasted and Dr A. Appleton of the Physics Department. Special mention must be made of the University's Computing resources (ULCC) and the technical assistance of Birkbeck College's computer link and department, especially Mr M. Farmer and Mr L. Moore, and our departmental programmer, Dr C. A. Reynolds.

As a direct consequence of the inspiration and cooperation gained from participation in the CECAM Workshops, I remain indebted to Professors C. Moser, H. J. C. Berendsen, K. Singer and Dr A. Rahman, and all the other participants, for their interest and advice.

Finally, to the preparation of this manuscript, I pay tribute to the typing of Mrs G. Dryer, the photography of Mr D. Parry and to my wife Desirée for preparing the diagrams.

References

1. Barker, J. A. and Watts, R. O. (1969). *Chem. Phys. Lett.*, **3**, 144.
2. *Physics of Liquids*, Institute of Physics Conference, University of East Anglia, Norwich, 15–18 April 1969. (unpublished proceedings)
3. Rowlinson, J. S. (1951). *Trans. Faraday Soc.*, **47**, 120.
4. Rahman, A. and Stillinger, F. H. (1971). *J. Chem. Phys.*, **55**, 3336.
   Rahman, A. and Stillinger, F. H. (1973). *J. Amer. Chem. Soc.*, **95**, 7943.
5. Rahman, A. and Stillinger, F. H. (1972). *J. Chem. Phys.*, **57**, 1281.
6. Rahman, A. and Stillinger, F. H. (1974). *J. Chem. Phys.*, **60**, 1545.
7. Rahman, A. and Stillinger, F. H. (1972). *J. Chem. Phys.*, **57**, 4009.

8A. CECAM (Centre Européen de Calcul Atomique et Moléculaire, Bâtiment 506, Université de Paris XI, 91405 Orsay, France) *Report of Workshop on Molecular Dynamics and Monte Carlo Calculations on Water, 19th June–11th August 1972.*

8B. CECAM *Report of Workshop on Ionic Liquids, 20th May–13th July 1974.*

8C. CECAM *Report of Workshop on Molecular Dynamics – Long Time Scale Events, 19th August–13th September 1974.*

9. Ben-Naim, A. and Stillinger, F. H. (1972). Aspects of the Statistical-Mechanical Theory of Water. In R. A. Horne (Ed.), *Structure and Transport Processes in Water in Aqueous Solutions*, Wiley–Interscience, New York.
   Stillinger, F. H. (1970). *J. Chem. Phys.*, **74**, 3677.
10. Eisenberg, D. and Kauzmann, W. (1969). *The Structure and Properties of Water*, Oxford University Press, New York. p. 44.

11. See, for example, Eisenberg, D. and Kauzmann, W. (1969). *The Structure and Properties of Water*, Oxford University Press, New York. p. 44. See also F. Franks (Ed.) (1973–5). *Water – A Comparative Treatise*, Vols. 1–5. Plenum Press, New York and London.
12. Eisenberg, D. and Kauzmann, W. (1969). *The Structure and Properties of Water*, Oxford University Press, New York. p. 12. Glaeser, R. M. and Coulson, C. A. (1965). *Trans. Faraday Soc.*, **61**, 389.
13. Lonsdale, K. (1958). *Proc. Roy. Soc.*, **A247**, 424.
14. Chidambaram, R. (1961). *Acta Cryst.*, **14**, 467.
15. See, for example, McDonald, I. R. Monte Carlo Calculations for the Stockmayer Potential. In CECAM *Report of Workshop on Molecular Dynamics and Monte Carlo Calculations on Water, 19th June–11th August 1972.* p. 26.
16. Bol, W. Monte Carlo Calculations with Some Effective Pair Potentials. In CECAM *Report of Workshop on Molecular Dynamics and Monte Carlo Calculations on Water, 19th June–11th August 1972.* p. 46.
17. See, for example, Rahman, A. and Stillinger, F. H. (1974). *J. Chem. Phys.*, **60**, 1545.
18. Berendsen, H. J. C. and van der Velde, G. A. Molecular Dynamics on 216 Polarisable Water Molecules. In CECAM *Report of Workshop on Molecular Dynamics and Monte Carlo Calculations on Water, 19th June–11th August 1972.* p. 63.
19. Eisenberg, D. and Kauzmann, W. (1969). *The Structure and Properties of Water*, Oxford University Press, New York. p. 16.
20. Barnes, P. Polarisability and Quantum Calculations. In CECAM *Report of Workshop on Molecular Dynamics and Monte Carlo Calculations on Water, 19th June–11th August 1972.* p. 77.
21. Buckingham, A. D. (1959). *Quart. Rev. Chem. Soc.*, **13**, 183.
22. Coulson, C. A. and Eisenberg, D. (1966). *Proc. Roy. Soc.*, **A291**, 445, 454.
23. Fletcher, N. H. (1962). *Phil. Mag.*, **7**, 255.
Fletcher, N. H. (1968). *Phil. Mag.*, **18**, 1287.
24. Van der Velde, G. A. A Realistic Coulomb Potential. In CECAM *Report of Workshop on Molecular Dynamics and Monte Carlo Calculations on Water, 19th June–11th August 1972.* p. 38.
25. Alder, B. J. and Wainwright, T. E. (1960). *J. Chem. Phys.*, **33**, 1439.
26. Metropolis, N., Rosenbluth, A. W., Rosenbluth, M. N., Teller, A. H. and Teller, E. (1953). *J. Chem. Phys.*, **21**, 1087.
27. McDonald, I. R. Ewald Sums in Calculations on Water. In CECAM *Report of Workshop on Molecular Dynamics and Monte Carlo Calculations on Water, 19th June–11th August 1972.* p. 36.
28. Barnes, P. and Berendsen, H. J. C. *Hydrogen Bonding and the Cooperative Nature of Water Models using Polarisable Electropoles.* In preparation.
29. Del Bene, J. E. and Pople, J. A. (1970). *J. Chem. Phys.*, **52**, 4858.
Del Bene, J. E. and Pople, J. A. (1973). *J. Chem. Phys.*, **58**, 3605.
Hankins, D., Moscowitz, J. W., and Stillinger, F. H. (1970). *J. Chem. Phys. Letters*, **4**, 527.
Hankins, D., Moscowitz, J. W., and Stillinger, F. H. (1970). *J. Chem. Phys.*, **53**, 4544.
Kollman, P. A., and Allen, L. C. (1969). *J. Chem. Phys.*, **51**, 3286.
Morokuma, K., and Pedersen, L. (1968). *J. Chem. Phys.*, **48**, 3275.
Morokuma, K., and Winick, J. R. (1970). *J. Chem. Phys.*, **52**, 1301.
Lentz, B. R., and Scheraga, H. A. (1973). *J. Chem. Phys.*, **58**, 5296.
30. Bernal, J. D., and Fowler, R. H. (1933). *J. Chem. Phys.*, **1**, 515.
31. Pauling, L. (1935). *J. Amer. Chem. Soc.*, **57**, 2680.
Nagle, J. F. (1966). *J. Math. Phys.*, **7**, 1484.
Suzuki, Y. (1966). *Cont. Inst. Low Temp. Sci.*, Series **A21**, 1.
Giaque, W. F., and Stout, J. W. (1936). *J. Amer. Chem. Soc.*, **58** 1144.
Kamb, B. (1964). *Acta Cryst.*, **17**, 1437.
32. Teichmann, I., and Schmidt, G. (1945). *Physica Status Solidi*, **8**, 145.
Kopp, M. (1962). Progress Report No. C, "A Transitional Film on the Surface of Ice", Melbon Institute, 4400 Fifth Avenue, Pittsburgh, Pa. 15213, USA.
Tippe, A. (1967). *Naturwissenschaften*, **54**, 95.
Deubner, A., Heise, R., and Wenzel, K. (1960). *Naturwissenschaften*, **47**, 600.
Dengel, O., Eckener, U., Plitz, H., and Riehl, N. (1964). *Phys. Letters*, **9**, 291.

33. Orban, J. and Bellemans, A. Polarisability Effects in Ice. In CECAM *Report of Workshop on Molecular Dynamics and Monte Carlo Calculations on Water, 19th June–11th August 1972.* p. 90.
34. Jellinek, H. H. G. (Chairman) (1967). In the 'Ice Symposium', 1966, *J. Colloid Sci.*, **25**, 131–294.
See also, for example, the following references:
Telford, J. W. and Turner, J. S. (1963). *Phil. Mag.*, **8**, 527.
Weyl, W. A. (1951). *J. Colloid Sci.*, **6**, 389.
Kopp, M. (1967). *Surface Sci.*, **7**, 82.
Bullemer, B. and Riehl, N. (1966). *Solid State Comms.*, **4**, 447.
Kopp, M. (1962). *Z. Angew. Math. Phys.*, **13**, 431.
Jaccard, C. (1966). *Int. Conf. Low Temp. Sci.*, Hokkaido University, Sapporo, Japan, (unpublished).
Bradley, R. S. (1957). *Trans. Faraday Soc.*, **53**, 687.
Latham, J. (1963). *Nature*, **200**, 1087.
Kuroiwa, D. (1960). *Low Temp. Sci.*, **A**, **19**, 1.
Hobbs, P. V. and Mason, B. J. (1964). *Phil. Mag.*, **9**, 181.
Jellinek, H. H. G., Snow, Ice and Permafrost Establishment (SIPRE), Corps of Engineers, US Army, Wilmette, Illinois, USA, *Res. Rep.* Nos. **23**, (1957), **38** (1957), **62** (1960).
Hoekstra, P., and Miller, R. D. (1965). Cold Regions Research and Engineering Laboratory (CRREL), Hanover, New Hampshire, USA, *Res. Rep.* No. **153**, (1965).
Maeno, N. (1973). Measurements of Surface and Volume Conductivities of Single Ice Crystals. In E. Whalley, S. J. Jones and L. W. Gold (Eds), *Physics and Chemistry of Ice*, Royal Society of Canada, Ottawa. p. 173.
Maeno, N. (1966). Air Bubble Formation in Ice Crystals. In *Int. Conf. Low Temp. Sci.*, Hokkaido University, Sapporo, Japan. (Parts of this paper to be found in *Low Temp. Sci.*, Series A, **24**, English summary p. 109.)
Nakaya, U., Hananjima, M. and Muguruma, J. (1958). *J. Fac. Sci.*, Hokkaido University Series II, V, **3**, 97–106.
Hallet, J. (1961). *Phil. Mag.*, **6**, 1073.
Jellinek, H. H. G. (1959). *J. Colloid Sci.*, **14**, 268.
35. See, for example, Drost-Hansen, W. (1968). *Chem. Phys. Lett.*, **2**, 647.
36. Narten, A. H. and Levy, H. A. (1969). *Science*, **165**, 447.
Narten, A. H. (1971). *J. Chem. Phys.*, **55**, 2263.
Narten, A. H. (1972). *J. Chem. Phys.*, **55**, 5681.
37. Heathman, S. P., Nicholas, J. D., Timmins, P. A. and Finney, J. L. (1976). *J. Phys.*, E. **9**, 143.
38. Bernal, J. D. (1964). *Proc. Roy. Soc., Lond.*, **A280**, 299.
Finney, J. L. (1970). *Proc. Roy. Soc., Lond.*, **A319**, 479, 495.
39. See, for example:
Derjaguin, B. V. (1966). *Disc. Faraday Soc.*, **42**, 109.
Derjaguin, B. V. (1970). *Sci. Amer.*, **223**, 52.
Bellamy, L. J., Osborn, A. R., Lippincott, E. R. and Bandy, A. R. (1969). *Chem. Ind.*, 686.
Lippincott, E. R., Stromberg, R. R., Grant, W. H. and Cessac, G. L. (1969). *Science*, **164**, 1482.
Barnes, P., Cherry, I., Finney, J. L. and Petersen, S. (1971) *Nature*, **230**, 31.
40. Franks, F. (Ed.), Water, A Comprehensive Treatise, Vols **1–5** (up to 1975). Plenum Press, New York. Volume 5 deals specifically with water in disperse systems. Other references include Reference 39, and Derjaguin, B. V., Fedjakin, N. N. and Talayev, M. V. (1967). *J. Colloid and Interface Sci.*, **24**, 132.
Hori, T. (1960). *SIPRE transl. No. 62.*
Schufle, J. A. and Venngopalan, M. (1967). *J. Geophys. Res.*, **72**, 3271.
41. Rahman, A. Molecular Dynamics of a Droplet of 475 Molecules. In CECAM *Report of Workshop on Molecular Dynamics and Monte Carlo Calculations on Water, 19th June–11th August 1972.* p. 40.
42. Binder, K. (1975). *J. Chem. Phys.*, **63**, 2265.
Binder, K. and Stauffer, D. (1972). *J. Stat. Phys.*, **6**, 49.
Lee, J. K., Barker, J. A. and Abraham, F. F. (1973). *J. Chem. Phys.* **58**, 3166.

Abraham, F. F. (1974). *J. Chem. Phys.*, **61**, 1221.
Briant, C. L. and Burton, J. J. (1975). *J. Chem. Phys.* **63**, 2045, 3327.
43. Vogel, P. C. In CECAM *Report of Workshop on Ionic Liquids, 20th May–13th July 1974.*
Heinzinger, K. and Vogel, P. C. (1974). *Z. f. Naturf.*, **29**, 1164.
44. Rahman, A. Molecular Dynamics of One Ion Pair in ST2-Water. In CECAM *Report of Workshop on Ionic Liquids, 20th May–13th July 1974.* p. 263.
45. McDonald, I. R. and Rasaiah, J. C. Monte Carlo Simulation of the Average Force Between Two Ions in a Stockmayer Solvent. In CECAM *Report of Workshop on Ionic Liquids, 20th May–13th July 1974.* p. 230.
Gosling, E. M. and Singer, K. Molecular Dynamics of a Solution of One Ion in 255 Polar Molecules. In CECAM *Report of Workshop on Ionic Liquids, 20th May–13th July 1974.* p. 242.

Chapter 10

Statistical Mechanics of Aqueous Fluids

A. BEN-NAIM

Department of Physical Chemistry,
The Hebrew University, Jerusalem, Israel

10.1 Introduction

The ultimate goal of a statistical mechanical theory of matter is to provide methods of predicting and explaining the experimentally measurable quantities of a given substance in terms of the properties of its elementary constituent particles. The question that arises immediately is: what are the elementary particles that one should start with when devising a molecular theory of a complicated, but important, liquid such as water?

The answer to this question is never a unique one, and is certainly not so for water. In general there seems to exist a hierarchy of different 'levels' of approaches, from which we make our choice. In the case of water, some possible choices are the following:

(1) The fundamental particles are electrons and nuclei of two kinds (the hydrogen's and the oxygen's nucleus). Here the interaction among the particles is presumed to be essentially coulombic, and pairwise additive (of course one can start with more fundamental particles to describe the liquid, but this approach is never used in chemistry).

(2) The fundamental particles are hydrogen and oxygen atoms, H and O, or the corresponding ions H^+ and O^{--}.

(3) The fundamental particles are single water molecules. Here we view the system as a one-component system; the interaction among the particles is yet to be described.

(4) Recognizing the fact that hydrogen-bonds exist, and that their precise characterization is difficult, one can view water as a multicomponent system, which may be defined in various ways. For instance, one can systematically define clusters of hydrogen-bonded molecules;

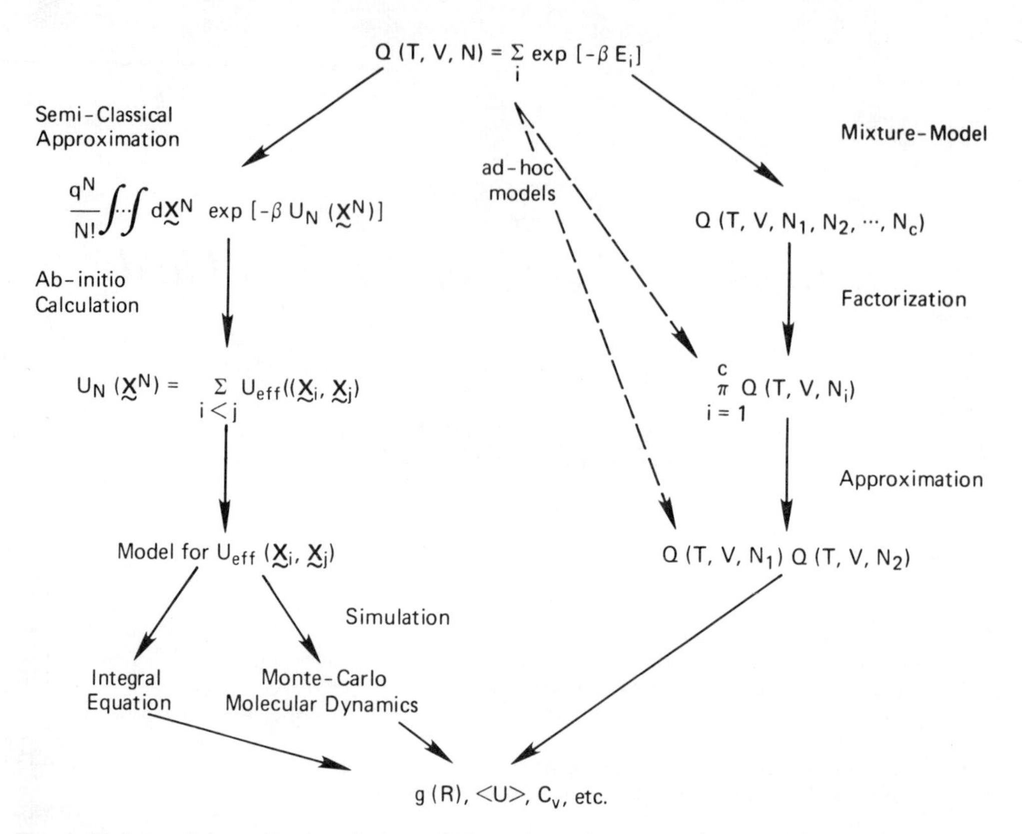

Figure 10.1.1 Schematic description of the two main routes along which theories of liquid water have been developed. The right hand route is referred to as the mixture-model approach, whereas the left hand one is the *ab-initio* approach

alternatively the species could be molecules participating in different numbers of hydrogen-bonds (see Section 10.5 for more details).

It is important to stress that each of the above listed possibilities can be used as a basis for an exact theory of water. In each case, however, the mathematical description of the system will be different, and, more important, the inevitable approximations that we are compelled to use in any approach will have different physical significance.

In making our choice among the various approaches we must maintain a balance between the more fundamental approach, usually more elegant but more difficult to pursue, and the cruder, yet more feasible approach that leads to some useful information on the nature of the system.

In the case of water, there have been essentially two routes along which theories were developed. The two routes are schematically depicted in Figure 10.1.1, and are referred to as the mixture-model and the *ab-initio* approaches.

Clearly the full quantum mechanical partition function, $Q(T, V, N)$, is a fundamental starting point. However, soon one realizes that nothing can be achieved without resorting to some drastic approximations.

In the so-called *ab-initio* approach, depicted at the left hand side of Figure 10.1.1, one starts with the replacement of the quantum mechanical partition function by the semiclassical partition function. This kind of approximation has been traditionally used in the theory of classical fluids.

The basic idea underlying this approximation is that the internal partition function of a *single* molecule is, on the average, not affected by the presence of neighbouring molecules. Consider for instance a single argon atom, the internal partition function of which is designated by q_A. We may classify all possible configurations of the surrounding atoms into two classes. In one, we include all configurations for which no repulsive forces are exerted on the particular atom under observation, with all other configurations included in the second class. Clearly, because of the weak attractive intermolecular forces, we expect that for all the configurations in the first class q_A remains virtually unchanged. On the other hand, configurations in the second class, for which we expect strong environmental effects on the internal properties of the atom, occur with very low probability. Hence, on the average it is safe, in this case, to assign a quantity q_A for each atom in the system, whether the system is a gas or a liquid.

This very elementary step is already questionable for a liquid with strong attractive forces, such as water. Here the situation is markedly different, since in many configurations, which occur with large probability, a single molecule experiences strong influence by its neighbouring molecules. Therefore the very usage of the semiclassical partition function is already a serious drawback in the *ab-initio* approach.

The next difficulty that we have to face in applying traditional methods in the theories of simple fluids to water is the content of the potential function $U_N(\mathbf{X}^N)$, where $\mathbf{X}^N = \mathbf{X}_1 \dots \mathbf{X}_N$ stands for a specific configuration of the N molecules.

Most of the progress in the theory of simple fluids has been based on one simplifying assumption, namely the pairwise additivity of the total potential energy, i.e.

$$U_N(\mathbf{X}^N) = \sum_{i<j} U(\mathbf{X}_i, \mathbf{X}_j). \tag{10.1.1}$$

It is believed that this approximation is a reasonably good one for simple fluids, but today it is well known that it is not a good approximation for complex liquids such as water (for more details see Section 10.4). At this juncture we again face a dilemma of whether to include higher order potentials (in the characterization of $U_N(\mathbf{X}^N)$), vastly complicating the structure of the formal theory, or adopt the pairwise assumption as in (10.1.1) but replace the true pair potential (i.e. the potential function for two water molecules in vacuum) by an *effective* pair potential, and write

$$U_N(\mathbf{X}^N) = \sum_{i<j} U_{\text{eff}}(\mathbf{X}_i, \mathbf{X}_j). \tag{10.1.2}$$

In a way, by using an effective pair potential we commit ourselves to a study of a system of model-particles that obey the 'law of force' contained in U_{eff}. We propose to call these particles *waterlike* molecules, to stress the fact that we are uncertain as to the extent of relevance of such a study to the problem of real liquid water.

Once we have chosen an effective pair potential, the formal road for applying standard statistical mechanical methods for computing average quantities is opened, yet there are still many obstacles impeding a specific computation. The difficulties here are essentially technical and arise from the fact that each pairwise property of the system is a function of six coordinates, not just one as in the case of simple fluids. In spite of these difficulties, many attempts to compute some average quantities for waterlike particles have been carried out by various methods such as Monte Carlo,[1–7] integral equation,[8] molecular dynamics[9–14] and cluster expansions.[15,16]

The essence of the right hand route of Figure 10.1.1, the so-called mixture model approach, is a clever way of 'absorbing' the difficulties mentioned above by redefining species that contain, as part of their 'internal' structure, the strong attractive forces. This approach has been long recognized to be useful for interpreting a vast amount of experimental facts on water and aqueous solutions.

The basic idea of the mixture model approach is similar to the theory of associated liquids (see for example the treatment by Prigogine[17]). For simplicity, suppose that we are concerned with molecules that can participate only in one hydrogen-bond, and that this is the only significantly strong attractive interaction between a pair of molecules. In such a case, instead of viewing the system as composed of 'monomers' interacting via strong attractive forces, one adopts a relatively simpler point of view; every pair engaged in a hydrogen-bond will be called a 'dimer'. In this way we have defined a two-component system, in such a way that all the strong attractive forces are 'absorbed' in the definition of the dimer. The interaction between the various species is considered to be conceptually simpler, at least in the sense of inviting simple and reasonable approximation.

Clearly in the case of water there is a multitude of ways of classifying species containing varying numbers of monomers. It is no wonder, therefore, that a plethora of mixture-models have been developed ever since it was first conceived (there apparently is no precise knowledge of the origin of the mixture model approach. It is commonly known that the first written document on this is by Roentgen[18] but Roentgen himself admits that he is not the originator of this idea).

Although one can start with an exact classification of different species in the construction of a mixture model approach, one is soon forced to resort to some drastic approximations. The most common one is the factorization of the total partition function, each factor representing the partition function of one of the species. This procedure is indicated on the right hand side of Figure 10.1.1. It is also a common practice to take one more simplifying step, i.e. replacing the product over all the possible species by a product of two or three factors representing those species which are believed to be the most important. Finally one assigns some properties to each of the species, thereby opening the way for actual computations of the properties of the liquid.

At the present stage it is very difficult to establish an unequivocal preference for one route over the other. Each approach involves drastic approximations, the assessment of which is almost impossible at present. It is the conviction of the author that both approaches will continue to contribute to our understanding of the properties of aqueous solutions in the foreseeable future.

We have presented above what we believe are the fundamental difficulties that arise in developing a theory of water. Some more specific topics will be described in a similar spirit in the following sections. Our survey of topics will be far from being extensive, as several reviews are already available covering a large spectrum of experimental and theoretical aspects of these systems.[19-24]

10.2 Survey of properties of water

The outstanding properties of aqueous fluids have been compiled and reviewed extensively in the past two decades.[19-23] We shall present here only a short list of properties the explanation of which, on a molecular level, seems to be the most challenging aspect of a statistical mechanical theory of these systems.

(1) Perhaps the most unusual, and very well known, phenomenon exhibited by water, at one atmosphere, is the temperature dependence of the density between 0° to 4 °C. All known liquids expand on heating; water, however, has a range of temperatures at which it contracts upon raising the temperature. Similar behaviour is observed for the heavier isotopes D_2O and T_2O. The overall change in the volume of water in the temperature range of 0–4 °C is admittedly very small relative to the molar volume of water, but nevertheless this property seems to be a unique one among liquids. We shall see in Section 10.5 that this property does

reveal some important information on the mode of packing of water molecules in the liquid state.

A related phenomenon is the contraction of ice upon melting. Most solids expand on melting, though there are some exceptions to this rule besides ice.

(2) The coefficient of isothermal compressibility of water, at one atmosphere, decreases with temperature in the range of 0–46 °C. Normal liquids exhibit a monotonic increase of compressibility as a function of temperature. This phenomenon as well as the previous one gradually disappears as the pressure increases, leading to the notion that water tends to behave as a 'normal liquid' at high pressures.

(3) The heat capacity (both at constant volume and constant pressure) is much higher than the value expected from classical considerations. Assuming that liquid water is a collection of noninteracting classical particles one estimates that the heat capacity would be about $3R$ or 6 cal/mole deg, due to translational and rotational degrees of freedom (neglecting the contribution due to vibrational excitations). The experimental value is about 18 cal/mole deg. Although such a high heat capacity is not a unique property of water (e.g. liquid ammonia also has a high heat capacity), it is still quite unusual amongst 'normal' liquids.

(4) The large dielectric constant is another property which is outstanding though not a unique property of water. This property is central to the understanding of the behaviour of aqueous solutions of ionic solutes.

(5) Solutions of the simplest solutes such as the noble gases exhibit some unusual and remarkable thermodynamic properties. For instance, the process of dissolution of an argon atom in water involves a large and negative heat and entropy of solution compared with the same process in nonaqueous solvents. These phenomena have been attributed to structural changes in the solvent, which in turn is a manifestation of the unique structure of liquid water.

(6) The phenomenon of hydrophobic interaction, believed to be of crucial importance in biochemistry, is another property which is outstanding if not unique among liquids. Here we are concerned with the tendency of nonpolar solutes to attract each other in aqueous solutions, more so than they do in nonaqueous solvents. Also the enthalpy and the entropy associated with the process of hydrophobic interaction seems to be very characteristic to the aqueous environment. We shall see in Section 10.7 that this behaviour may also be accounted for by using the notion of structural changes in the solvent.

(7) Finally we note the existence of a large body of experimental data on the properties of aqueous solutions of ionic solutes, which is obviously of fundamental importance in achieving an understanding of the physical chemistry of biological solutions. Unfortunately, these systems are too complicated for a molecular theoretical approach, and await further progress at a preliminary level.

10.3 The radial distribution function of water

The radial distribution function (RDF), considered as one of the properties of liquid water, could well be discussed under the headline of the previous section. Yet because of its special importance as a source of information on the mode of packing of water molecules, it deserves a more detailed consideration.

The pair correlation function for water, when viewed as a collection of rigid undissociable molecules, is defined in the canonical ensemble as

$$g(\mathbf{R}_1, \mathbf{\Omega}_1, \mathbf{R}_2, \mathbf{\Omega}_2) = \frac{N(N-1) \int \dots \int d\mathbf{X}_3 \dots d\mathbf{X}_N \exp[-\beta U_N(\mathbf{X}^N)]}{\rho^2 \int \dots \int d\mathbf{X}_1 \dots d\mathbf{X}_N \exp[-\beta U_N(\mathbf{X}^N)]} \tag{10.3.1}$$

where $\mathbf{R}_i$ is the locational, and $\mathbf{\Omega}_i$ the orientational, vector describing the configuration of the

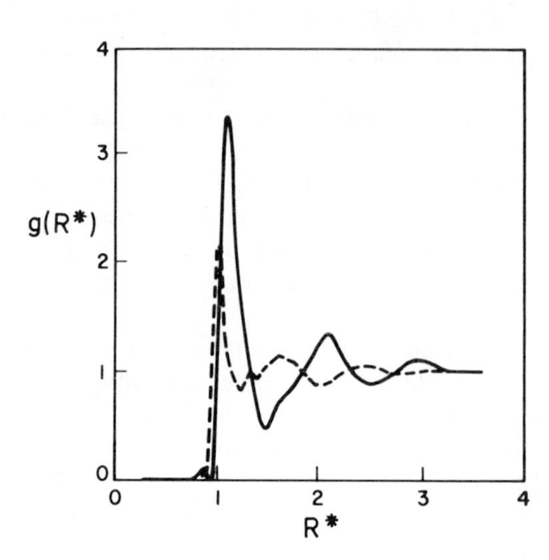

Figure 10.3.1 Radial distribution function $g(R^*)$ for water (dashed line) at 4 °C and 1 atm, and for argon (solid line) at 84.25 °K and 0.71 atm, as a function of the reduced distance $R^* = R/\sigma$. Data for argon were provided by N. S. Gingrich, and for water by A. H. Narten, for which the author is very grateful

i-th molecule. ρ is the number density, $\beta = (k_B T)^{-1}$ and $U_N(\mathbf{X}^N)$ is the total potential energy of interaction among N water molecules being at a specific configuration $\mathbf{X}^N = \mathbf{X}_1, \ldots \mathbf{X}_N$; $\mathbf{X}_i$ stands for $(\mathbf{R}_i, \mathbf{\Omega}_i)$.

The spatial pair correlation function is defined by

$$g(\mathbf{R}_1, \mathbf{R}_2) = (8\pi^2)^{-2} \iint d\mathbf{\Omega}_1\, d\mathbf{\Omega}_2\, g(\mathbf{R}_1, \mathbf{\Omega}_1, \mathbf{R}_2, \mathbf{\Omega}_2). \tag{10.3.2}$$

In an homogeneous liquid, $g(\mathbf{R}_1, \mathbf{R}_2)$ is a function of the scalar distance $R = |\mathbf{R}_2 - \mathbf{R}_1|$ only. Henceforth we shall always refer to $g(R)$ as the spatial pair correlation function, or alternatively, the radial distribution function.

From here on we shall also refer to the $g_{OO}(R)$ pair correlation function and assume that this function has been reasonably well resolved from the combined diffraction pattern obtained from the O–O, O–H and H–H correlations.[25,26]

Figure 10.3.1 shows the RDF of water and of argon as a function of the reduced distance $R^* = R/\sigma$ where σ, the 'effective' diameter, has been taken as 2.82 Å for water and 3.4 Å for argon. (The effective diameter may be conveniently defined here as the location of the first peak of the RDF.)

There are two features of the RDF of water that should be noted. Consider first the 'coordination number' defined by

$$n_{CN} = \rho \int_0^{R_M} g(R) 4\pi R^2\, dR. \tag{10.3.3}$$

If we choose R_M to be the location of the first minimum that follows the first peak, then n_{CN} is a measure of the average number of water molecules occupying the first 'coordination shell', i.e. the average number of molecules that are found in a spherical region of radius R_M around the centre of a selected molecule (excluding the molecule at the centre from this counting). The value of n_{CN} for water at 4 °C and 1 atm is about 4.4, whereas for argon at 84.25 °K and 0.71 atm the value is about 10. Clearly the difference in the coordination number reflects the relatively 'open' mode of packing of water as compared with the 'close' packing of argon, in the liquid state. The coordination number 4.4 of liquid water is strongly reminiscent of the exact value, 4, in the solid ice structure (here we refer to ordinary ice Ih). Therefore this piece of information already indicates some degree of structural preservation upon melting of ice.

The second important feature is the location of the second peak of the RDF. In argon the

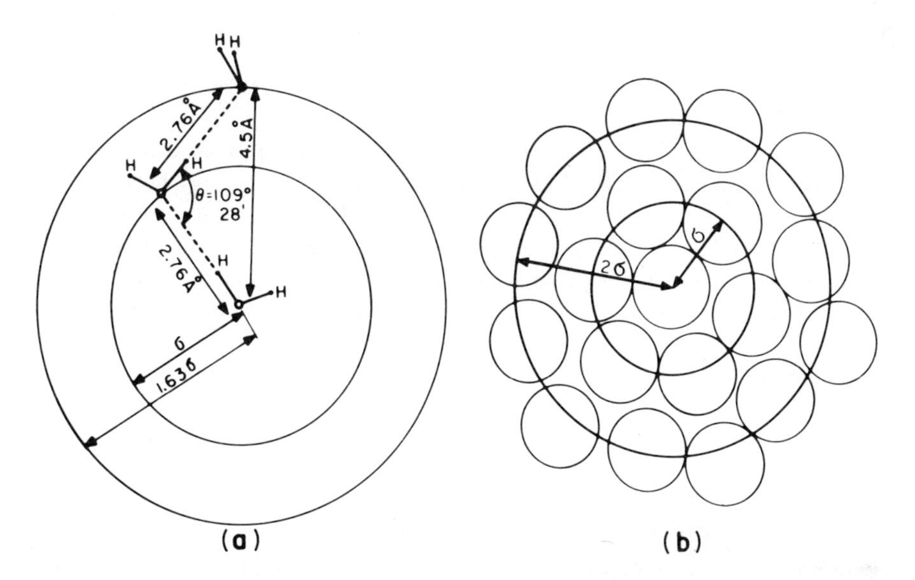

Figure 10.3.2 Schematic description of the distribution of first- and second-nearest neighbours (a) in water and (b) in a simple fluid. The tetrahedral orientation of the hydrogen bond induces a radial distribution of first- and second-nearest neighbours at σ and 1.63σ, respectively, $\sigma = 2.76$ Å being the O-O distance in ice Ih. The almost equidistant and concentric nature of the packing of particles in a simple fluid produces corresponding neighbours at σ and 2σ

second peak is found at about $R^* = 2$. This reflects the tendency of spherical molecules to pack in roughly concentric and equidistant spheres around a given particle. This is shown schematically in Figure 10.3.2. In contrast, the second peak of the RDF of water is found at $R^* \approx 1.6$, corresponding to $R \approx 4.5$ Å. A simple calculation shows that this distance coincides with the exact distance of the second nearest neighbours in the ice lattice of ice Ih. Taking the O–O nearest neighbour distance to be 2.76 Å we get for the second nearest neighbour distance

$$2 \times 2.76 \times \sin(\theta_T/2) = 4.5$$

where θ_T is the tetrahedral angle 109.46°.

The essential difference in the mode of packing between argon and water is schematically demonstrated in Figure 10.3.2. In water, because of the strong directional forces (hydrogen-bonds) the second nearest neighbour is likely to be found at the distance of about 4.5 Å from a given molecule.

The two features mentioned above lead to the following important conclusion: the basic *local* geometry of packing of water molecules around a given molecule has on the average a great similarity to that found in solid ice. This is to say that on average each molecule has about four nearest neighbours, and that the angle between triplets of bonded molecule is on the average very close to the tetrahedral angle θ_T. It must be emphasized that the kind of retention of the structure of ice in the liquid that we consider here refers only to the local, or immediate environment of the molecule. Liquid water, certainly, does not possess the long range order that is typical to the solid state.

The above structural resemblance between water and ice has been the source of many models for water. In the past the very form of the RDF has often been referred to as a measure of the 'structure of water'; however, we shall reserve this term for a different concept on which we shall further elaborate in Section 10.5.

10.4 Basic features of an effective pair potential for water

As we have noted in Section 10.1, a basic ingredient in a molecular theory of water is an effective pair potential. The immediate question arises: from where does one get the pertinent information for such a function? There is no experimental method that provides information on the full six-dimensional function that describes the interaction energy between two water molecules. Therefore the only sound source for such information is a quantum mechanical calculation of the total energy of a pair of water molecules at various configurations. Indeed, much progress on this line has been achieved in recent years,[4,24,27,28,29] but still it seems that we are a long way from having a detailed analytical description of the full pair potential function.

Thus, at present, we must resort to a more empirical attitude towards constructing such a pair potential, in doing so we combine information from experiments, computations and physical intuition. (In this section we shall use the term 'pair potential' instead of the more lengthy term 'effective pair potential'.)

We know for instance the dipole moment of a water molecule, and it is clear that the long range interaction between two water molecules will be dominated by the dipole–dipole interaction. We known also that a water molecule is iso-electronic with neon, and that its effective van der Waals diameter is about the same as that of neon, namely about 2.82 Å.

With these two pieces of information one can immediately suggest that a pair potential for water is likely to contain dipole–dipole, U_{DD}, and Lennard–Jones, U_{LJ}, components:

$$U_{DD}(\mathbf{X}_1, \mathbf{X}_2) = R_{12}^{-3}\ [\mathbf{u}_1 \cdot \mathbf{u}_2 - 3\ (\mathbf{u}_1 \cdot \mathbf{u}_{12})(\mathbf{u}_2 \cdot \mathbf{u}_{12})] \tag{10.4.1}$$

$$U_{LJ}(R_{12}) = 4\epsilon\left[\left(\frac{\sigma}{R_{12}}\right)^{12} - \left(\frac{\sigma}{R}\right)^{6}\right] \tag{10.4.2}$$

where the first part takes care of the long range interaction, whereas the second part accounts for the very short range interaction, say $R_{12} < 2.5$ Å. As for the parameters appearing in (10.4.1) and (10.4.2), a reasonable choice would be to take for μ the dipole moment of a free water molecule, i.e. $\mu = 1.84 \times 10^{-18}$ esu, and for the Lennard–Jones parameters ϵ and σ to take the corresponding values as for neon, namely

$$\epsilon = 7.21 \times 10^{-2}\ \text{Kcal/mole} \qquad \sigma = 2.82\ \text{Å}. \tag{10.4.3}$$

The unit vector $\mathbf{u}_{12}$ in (10.4.1) is along the direction $\mathbf{R}_{12} = \mathbf{R}_2 - \mathbf{R}_1$.

The main, difficult, and so far unsettled, question is how to describe the analytical form of a pair potential in an intermediate range of distances, say $2\ \text{Å} \lesssim R_{12} \lesssim 5$ Å. All we know, both from experiments and from quantum mechanical computation, is that a hydrogen-bond (HB) is formed between pairs of water molecules, with a characteristic energy of a few kilocalories per mole, and that this interaction is highly directional.

We shall henceforth refer to this part of the potential as the HB part, and denote it by $U_{HB}(\mathbf{X}_1, \mathbf{X}_2)$. Thus the simplest description of a pair potential for water would have the form

$$U(\mathbf{X}_1, \mathbf{X}_2) = U_{LJ}(\mathbf{R}_{12}) + U_{DD}(\mathbf{X}_1, \mathbf{X}_2) + U_{HB}(\mathbf{X}_1, \mathbf{X}_2) \tag{10.4.4}$$

pending further characterization of the HB part.

We now turn to a brief description of three possible ways of formulating a pair potential for water. (For more details see References 3 and 24.)

The most recent work on this problem is due to Popkie *et al.*[4] which is based on an extensive Hartree–Fock calculation for water dimers. (The assumption is made, as in any other work of this kind, that the geometry of the single water molecule is fixed.) They fitted their results to an expression which contains coulombic and exponential terms which represent interaction between four centres of force, defined in each molecule. This potential function was then used by the authors in their Monte Carlo simulation of liquid water.

A different, and somewhat simpler form of a pair potential was suggested by Ben-Naim and Stillinger[30] and was later used for various computations such as the approximate solution of Percus–Yevick equation,[8] molecular dynamic calculations,[10–14] and in a cell model theory.[31]

The essence of this function is a four-point-charge model, originally suggested as a representation of a water molecule by Bjerrum. Embedding the four point charges in the van-der-Waals sphere of a water molecule one writes the pair potential as

$$U(\mathbf{X}_1,\mathbf{X}_2) = U_{LJ}(R_{12}) + S(R_{12})U_{HB}(\mathbf{X}_1,\mathbf{X}_2) \tag{10.4.5}$$

where $U_{HB}(\mathbf{X}_1,\mathbf{X}_2)$ is essentially a sum over all coulombic interactions between the different pairs of point charges, each of which belong to one molecule. The function $S(R_{12})$, referred to as the 'switching function', serves to suppress the possibility that charges of opposite signs come too close to each other, leading to a divergence of the potential.

Finally, we describe another form for the HB potential in (10.4.4) which we believe conveys the major properties of this part of the pair potential for water,[23] and it has also the advantage of being useful in the construction of a definition of the structure of water (Section 10.5) and structural changes in the solvent induced by processes like solubility (Section 10.6), hydrophobic interaction (Section 10.7), etc.

We start by writing

$$U_{HB}(\mathbf{X}_1,\mathbf{X}_2) = \epsilon_{HB}G(\mathbf{X}_1,\mathbf{X}_2) \tag{10.4.6}$$

where ϵ_{HB} is referred to as the HB energy and ϵ_{HB}/k_BT is presumed to be of the order of -10 at room temperature. The function $G(\mathbf{X}_1,\mathbf{X}_2)$ is essentially a stipulation on the relative geometry of the pair of water molecules. The most important property of this function is that it attains a maximum value of unity whenever the two molecules are hydrogen-bonded to each other, and the value declines sharply to zero when the configuration deviates considerably from the one required to form a HB.

Of course, with the above description of $U_{HB}(\mathbf{X}_1,\mathbf{X}_2)$ we have left a great latitude for an analytical form for this function. We have recently[23] suggested one possible form that we shall now describe with the help of Figure 10.4.1.

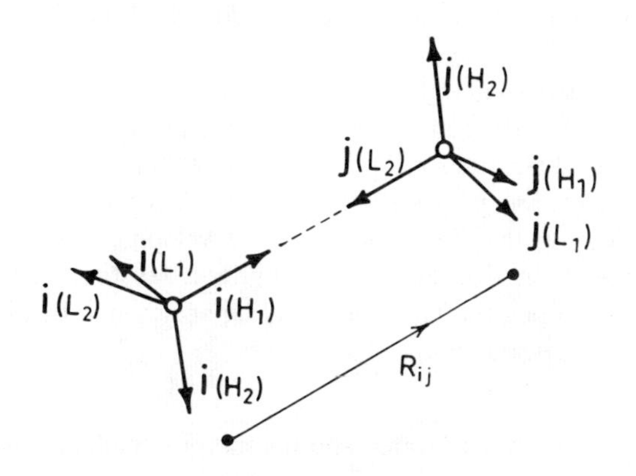

Figure 10.4.1 Schematic description of two water-like particles in a favourable orientation to form a hydrogen-bond

For each water molecule we define four unit vectors emanating from the centre (say the nucleus of the oxygen) and pointing towards the four vertices of a regular tetrahedron. The vectors belonging to the i-th molecule are denoted by $\mathbf{h}_{i1}$, $\mathbf{h}_{i2}$, $\mathbf{l}_{i1}$ and $\mathbf{l}_{i2}$, and they roughly correspond to the direction of the two OH bonds and the two lone-pair electrons (for more details see Reference 23). With the choice of $R_{HB} = 2.76$ Å, we now write

$$U_{HB}(\mathbf{X}_1, \mathbf{X}_2) = \epsilon_{HB}\bar{G}(R_{12} - R_{HB})\{\sum_{\alpha,\beta=1}^{2} \bar{G}(\mathbf{h}_{1\alpha} \cdot \mathbf{u}_{12} - 1)\bar{G}(\mathbf{l}_{2\beta} \cdot \mathbf{u}_{12} + 1)$$

$$+ \sum_{\alpha,\beta=1}^{2} \bar{G}(\mathbf{l}_{1\alpha} \cdot \mathbf{u}_{12} - 1)\bar{G}(\mathbf{h}_{2\beta} \cdot \mathbf{u}_{12} + 1)\} \qquad (10.4.7)$$

Clearly a bond is formed whenever the separation R_{12} is close to R_{HB} and when one of the OH directions of one molecule (the donor) is directed towards the centre of the second molecule, and simultaneously the lone-pair vector of the second molecule (or the acceptor) is directed toward the centre of the first molecule.

The functions $\bar{G}(x)$ in (10.4.7) may be chosen as either an unnormalized Gaussian

$$\bar{G}(x) = \exp(-x^2/2\delta^2) \qquad (10.4.8)$$

with variance δ^2, as a double step function

$$\bar{G}(x) = \begin{cases} 1 & \text{for } |x| < \delta \\ 0 & \text{for } |x| \geqslant \delta \end{cases} \qquad (10.4.9)$$

or any other function that has similar properties to (10.4.8) and (10.4.9). Of course this is a very crude description, and one can easily refine it by adding repulsive terms, further restrictions on the rotation about the O–H . . . O axis, and removing the symmetrical behaviour around $x = 0$ in (10.4.8) and (10.4.9). There is no point in elaborating on these refinements now, since all we need at present is the qualitative features of $G(\mathbf{X}_1, \mathbf{X}_2)$ as discussed after equation (10.4.6).

It is clear from the above discussion that we are still quite far from having a 'good' pair potential for water. This does not mean that our problems will be solved once we acquire one. In fact the contrary is true. If we have an exact pair potential that correctly describes the interaction between two water molecules, it is not clear to what extent this function will be useful for the study of liquid water. On the other hand, suppose we have devised a good effective pair potential, the employment of which in the theory of liquids leads to a good agreement with experimental results; does this function have any physical significance, as a property of water?

The above troublesome situation in fact exists to a certain extent in the field of simple fluids also, but it becomes far more acute in the case of a complex liquid such as water. Therefore, in the author's opinion these difficulties will probably stay with us for quite a long time. It would be difficult to foresee significant progress without either incorporating higher order potentials in the framework of existing theories of liquids, or developing new tools to handle complex liquids that do not depend on the additivity assumption for the total potential energy. We believe that some progress along the latter line may be achieved with the help of the mixture model approach, which we describe in the next section.

10.5 The concept of the 'structure of water' and the mixture model approach to the theory of water

Perhaps one of the most ubiquitous concepts appearing in the literature on aqueous fluids is the 'structure of water' and structural changes in the solvent induced by processes taking place in water. Because of lack of a precise definition, there has been a considerable controversy on the very meaning of these concepts.

In the theory of simple fluids, one often refers to the radial distribution function as a quantity that conveys structural information on the liquid. This is certainly true for water. However, the majority of workers in the field of aqueous solution have a notion other than the radial distribution function when they speak of the structure of water: there seems to be a general opinion that this concept should reflect, in a quantitative way, the extent of similarity between liquid water and ice. In fact, one can encounter many examples where authors refer to the 'degree of ice-likeness' as a measure of the structure of water.

In this section we elaborate on a possible definition of the concept of the structure of water, which, we believe, conforms with the common meaning assigned to this notion. Furthermore, in this definition the structure of water is a scalar quantity, not an entire function. This makes it a more convenient quantity for the discussion of various structural changes induced by processes in aqueous solutions (see the next two sections).

The older way of defining the structure of water was to start with some kind of a mixture-model for water. Suppose one assumes that water is a mixture of ice-like molecules and monomers, i.e. nonhydrogen-bonded molecules. If X_{ice} stands for the mole fraction of ice-like molecules, then, intuitively, it was also assigned the meaning of a measure of the structure of water.

The above definition is not satisfactory, however, since the definition of the two components presumed to exist in water is incomplete, and certainly this quantity cannot be either measured or computed by any of the available simulation techniques. Its only relevance is within the *ad-hoc* model that one has adopted for liquid water.

We now turn to a different definition of the structure of water which is valid within the context of the definition of the pair potential described in the previous section. We shall later see that the same definition can be reinterpreted from the point of view of the mixture model approach to water.

We start from the description of the pair potential (10.4.4) and (10.4.6) which we write in a more compact form as

$$U(\mathbf{X}_1, \mathbf{X}_2) = U'(\mathbf{X}_1, \mathbf{X}_2) + \epsilon_{\text{HB}} G(\mathbf{X}_1, \mathbf{X}_2) \tag{10.5.1}$$

where we have lumped together in U' all the ingredients of the pair potential except for the HB-part. We further assume, for simplicity, that $G(\mathbf{X}_1, \mathbf{X}_2)$ is a discontinuous function, attaining the value of unity whenever the two particles are in the proper range of configurations to form a HB, and zero for any other configuration. (This may be done, for instance, by using (10.4.7) and (10.4.9) in the definition of the function $G(\mathbf{X}_1, \mathbf{X}_2)$.)

Next we define the quantity

$$\psi_i(\mathbf{X}^N) = \sum_{\substack{j=1 \\ j \neq i}}^{N} G(\mathbf{X}_i, \mathbf{X}_j) \tag{10.5.2}$$

which, for each configuration of the system $\mathbf{X}^N$, measures the number of HBs in which the i-th molecule is participating.

Clearly, because of the discrete nature of the function $G(\mathbf{X}_1, \mathbf{X}_2)$, the function $\psi_i(\mathbf{X}^N)$ may attain only one of the five possible integral values 0, 1, 2, 3, 4. (One assumes that δ in (10.4.9) is chosen sufficiently small so that two water molecules can form at most *one* HB between themselves at any given time.) On the other hand if we relax the requirement of discontinuity of $G(\mathbf{X}_1, \mathbf{X}_2)$, say by using (10.4.8) instead of (10.4.9), then the function $\psi_i(\mathbf{X}^N)$ will change continuously on roughly the same range of values (0,4).

Next we define the ensemble average of $\psi_i(\mathbf{X}^N)$, say, in the T, V, N ensemble, by

$$\bar{\psi} = \int \dots \int d\mathbf{X}^N P(\mathbf{X}^N) \psi_1(\mathbf{X}^N) \tag{10.5.3}$$

where

$$P(\mathbf{X}^N) = \frac{\exp[-\beta U_N(\mathbf{X}^N)]}{\int \dots \int d\mathbf{X}^N \exp[-\beta U_N(\mathbf{X}^N)]}. \tag{10.5.4}$$

The quantity $\bar{\psi}$ measures the average number of HBs in which any selected water molecule participates when the system is characterized by the variables T, V, N. Alternatively the quantity $\bar{\psi}/2$ measures the average number of HBs in the system.

It is intuitively clear that the quantity $\bar{\psi}$ may serve as a measure of the 'structure of water'. The more structured the system, i.e. the larger is its 'degree of ice-likeness', the larger would be the number $\bar{\psi}$. Of course, this definition is valid only within the context of the model of water-like particles that are subjected to the 'law of force' given by (10.5.1). Once we have adopted this model, one can certainly compute the value of $\bar{\psi}$ by either the Monte Carlo or the molecular dynamics simulation methods. At present there is no way of measuring this quantity, but as we shall see in the next section, one can devise an approximate way of measuring changes in $\bar{\psi}$ induced by processes that take place in water.

The definition of the structure of water as presented here is based on purely geometrical considerations. A different definition based on the interaction energy between pairs of water molecules was suggested by Ben-Naim and Stillinger.[30] We now show that the same quantity $\bar{\psi}$ may be derived by using the mixture model approach to liquid water, and at least in this respect it conforms with the more conventional ideas about the structure of water.

The mixture model approach starts by classifying molecules into classes. In the present case we use the function $\psi_i(\mathbf{X}^N)$, defined in (10.5.2), to define the average number of molecules in a T, V, N ensemble that participate in exactly n–HB, that is

$$\bar{N}_n = N \int \dots \int d\mathbf{X}^N P(\mathbf{X}^N) \delta[\psi_1(\mathbf{X}^N) - n] \tag{10.5.5}$$

where δ may be either a Kronecker delta or a Dirac delta function depending on whether the function $\psi_i(\mathbf{X}^N)$ attains discrete or continuous values. We shall adopt the former case. The mole fraction of such molecules is

$$x_n = \bar{N}_n / N. \tag{10.5.6}$$

Once we have defined our 'components' we may adopt the point of view that our system is a mixture of, say, five components, the composition of which is given by the vector $(x_0, x_1, x_2, x_3, x_4)$. The traditional mixture model approach may be viewed as an approximation of the preceding procedure: namely one views the system as say, a two-component system whose mole fractions are x_0, x_4.

The average number of HB in the mixture model approach is

$$\frac{N}{2} \sum_{n=0}^{4} n x_n = \frac{N}{2} \int \dots \int d\mathbf{X}^N P(\mathbf{X}^N) \sum_{n=0}^{4} n \delta[\psi_1(\mathbf{X}^N) - n]$$

$$= \frac{N}{2} \int \dots \int d\mathbf{X}^N P(\mathbf{X}^N) \psi_1(\mathbf{X}^N) = \frac{N}{2} \bar{\psi} \tag{10.5.7}$$

which leads to the same quantity $\bar{\psi}$ as in our previous definition. In the approximate, two-structure model, one would have simply

$$\frac{N}{2}(0x_0 + 4x_4) = 2Nx_4 \tag{10.5.8}$$

i.e. only the mole fraction of the 'ice-like' component serves as a measure of the structure.

It should be noted that the so called two-structure model, where one neglects all intermediate components (with mole fractions x_1, x_2, x_3) may not be a good approximation for liquid water, yet it may serve a useful purpose by providing a simple way of interpreting a large number of unusual properties of liquid water as well as aqueous solutions. We shall present below a simple illustration of this kind of application, but first we describe a generalization of the idea of a mixture model approach which we believe is of potential interest in the study of liquid water.

Within the context of a molecular model for water we define the 'binding energy' of the i-th molecule to the rest of the system, being at any specific configuration $\mathbf{X}^N$, by

$$B_i(\mathbf{X}^N) = \sum_{\substack{j=1 \\ j \neq i}}^{N} U(\mathbf{X}_i, \mathbf{X}_j) \tag{10.5.9}$$

and as in (10.5.5) we define the corresponding distribution function

$$x(\nu) = \int \ldots \int d\mathbf{X}^N P(\mathbf{X}^N)\delta\,[B_1(\mathbf{X}^N) - \nu] \tag{10.5.10}$$

where here δ is the Dirac delta function. Clearly $x(\nu)d\nu$ is the average 'mole fraction' of water molecules whose binding energy lies between ν and $\nu + d\nu$. Some interesting features of this function for a system of water-like particles in two dimensions have been discussed recently.[23] A study of this function in three dimensions has not been carried out so far, but it surely deserves serious attention:† in the first place because the form of this function reveals some important information on the mode of packing of water molecules in the liquid state, and in the second place, knowledge of the form of the function $x(\nu)$ for some model particles may suggest useful approximate mixture models for water. As an illustrative example, suppose that for a system of water-like particles we compute the function $x(\nu)$ which turns out to have the form as in Figure 10.5.1. In such a case one would be induced to use the 'natural' cut-off point ν^* to define an *exact* two-structure model, namely by defining the two mole fractions:

$$x_A = \int_{-\infty}^{\nu^*} x(\nu)d\nu \qquad x_B = \int_{\nu^*}^{\infty} x(\nu)d\nu. \tag{10.5.11}$$

One can now view the system as a mixture of two components A and B, the properties of which may be approximated by two pure fluids for which the distribution functions $x(\nu)$ roughly coincide (after proper normalization) with the two peaks of $x(\nu)$ in Figure 10.5.1.

In connection with the application of the two-structure model for water, it is instructive to present what may be viewed as the traditional interpretation of the negative temperature dependence of the molar volume of water between 0–4 °C. The coefficient of thermal expansion of any system may be expressed as the cross fluctuation of volume and enthalpy, that is

$$\alpha = \frac{1}{\langle V \rangle}\frac{\partial \langle V \rangle}{\partial T} = \frac{\langle (V - \langle V \rangle)(H - \langle H \rangle) \rangle}{kT^2 \langle V \rangle} \tag{10.5.12}$$

where the symbol $\langle \ldots \rangle$ stands for an average in the T, P, N ensemble. A negative value of α means that on the average a positive fluctuation from $\langle V \rangle$ is associated with a negative

†Recently the function $x(\nu)$ has been computed for several models of water in three dimensions. The results should appear in the literature in the near future.

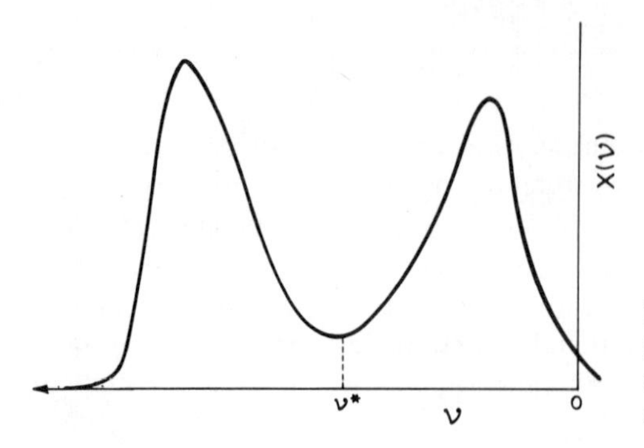

Figure 10.5.1 A schematic form of the function $x(\nu)$ for water-like particles. ν^* may be used as a cut-off point for defining a two-structure model

fluctuation from $\langle H \rangle$ and vice versa. This is an interesting situation since in most cases an expansion of the system means that on average the intermolar distances increase, and as a result of this the total interaction energy of the system is expected to decrease (in absolute magnitude). Water is the only liquid where this correlation between volume and energy is reversed (in the temperature range 0–4 °C).

The traditional interpretation of the negative temperature dependence of the volume, within the context of the two-structure model, is as follows. One writes the volume as

$$V = N_A \bar{V}_A + N_B \bar{V}_B \tag{10.5.13}$$

with $\bar{V}_A$ and $\bar{V}_B$ being the partial molar volumes of components A and B respectively. The temperature derivative is

$$\left(\frac{\partial V}{\partial T}\right)_{P,N} = N_A \frac{\partial \bar{V}_A}{\partial T} + N_B \frac{\partial \bar{V}_B}{\partial T} + (\bar{V}_A - \bar{V}_B)\frac{\partial N_A}{\partial T}\,. \tag{10.5.14}$$

Now, one assumes that the two components A and B behave in a 'normal fashion' in the sense that both of the first two terms on the RHS of (10.5.14) are positive. The third term is therefore supposed to explain the negative value of the whole term on the RHS of (10.5.14). This term may be rewritten as[23]

$$(\bar{V}_A - \bar{V}_B)\frac{\partial N_A}{\partial T} = \frac{(\bar{V}_A - \bar{V}_B)(\bar{H}_A - \bar{H}_B)}{T(\mu_{AA} - 2\mu_{AB} + \mu_{BB})}\,. \tag{10.5.15}$$

where $(\mu_{ij} = \partial^2 G/\partial N_i \partial N_j)$. Since the denominator on the RHS of (10.5.15) is always positive[17,23] one must conclude that

$$(\bar{V}_A - \bar{V}_B)(\bar{H}_A - \bar{H}_B) < 0 \tag{10.5.16}$$

which is similar to the condition that the cross fluctuation between volume and enthalpy in (10.5.12) is negative. Here, however, we have adopted the mixture model approach, and assumed that the two components A and B behave normally, in the sense that their temperature derivatives are positive. With this assumption one must attribute the negative temperature derivative of the volume of water to the third term on the RHS of (10.5.14) which, in turn, leads to (10.5.16).

Although we have no straightforward relation between the partial molar enthalpy and volume on one hand and local molecular properties on the other, we believe that relation (10.5.16) is a manifestation of a general and important aspect of the mode of packing of water molecules in the liquid state: namely, it is possible to define two major species, say A and B, in such a way that the component which is found in an environment having a large local density (roughly related to a small partial molar volume) will, on the average, have a small binding energy (roughly related to the partial molar enthalpy). This particular coupling of local density and binding energy seems to be the major molecular reason for many anomalous properties of water. A more detailed discussion of this topic is found in Reference (23).

10.6 Very dilute solutions of simple solutes

The simplest solutes that can be dissolved in water are the noble gases and some low molecular weight hydrocarbons such as methane, ethane, etc. Any molecular theory of pure water can be readily extended to apply to these solutions, without introducing fundamentally new assumptions. From the molecular point of view, the properties of very dilute solutions of simple solutes may be studied in a system of pure water with a single solute.

Thus the standard free energy of solution of a simple solute, s, is given by

$$\Delta\mu_s^0 = -k_B T \ln \frac{\int \ldots \int d\mathbf{X}^N \exp[-\beta U_N(\mathbf{X}^N) - \beta B_s(\mathbf{X}^N)]}{\int \ldots \int d\mathbf{X}^N \exp[-\beta U_N(\mathbf{X}^N)]}$$

$$= -k_B T \ln \langle \exp[-\beta B_s] \rangle_0 \qquad (10.6.1)$$

where $B_s(\mathbf{X}^N)$ stands for the total 'binding energy' of the solute to the rest of the system, being at a particular configuration $\mathbf{X}^N$.

$$B_s(\mathbf{X}^N) = \sum_{i=1}^{N} U(\mathbf{X}_i, \mathbf{R}_s) \qquad (10.6.2)$$

with $U(\mathbf{X}_i, \mathbf{R}_s)$ the solute-solvent pair potential. The symbol $\langle \ldots \rangle_0$ stands for an average, here in the T, V, N ensemble, over all the configurations of the solvent molecules.

An important theoretical simplification is obtained by replacing the real solute s by a hard-sphere solute, which may be viewed as an idealized solute, for which the solute-solvent pair potential is

$$U(\mathbf{X}_i, \mathbf{R}_s) = \begin{cases} 0 & \text{for } R_{is} > \sigma_{ws} \\ \infty & \text{for } R_{is} \leqslant \sigma_{ws} \end{cases} \qquad (10.6.3)$$

where $R_{is} = |\mathbf{R}_i - \mathbf{R}_s|$ and σ_{ws} is the effective distance of closest approach between the solute and the solvent molecules.

Using the pair potential (10.6.3) in (10.6.1) we get an important relation between the standard free energy of solution of a hard-sphere solute and the probability of finding a cavity, at $\mathbf{R}_s$, the radius of which is at least σ_{ws}.

$$\Delta\mu_{HS}^0 = -k_B T \ln P_{cav}(\mathbf{R}_s, \sigma_{ws}). \qquad (10.6.4)$$

The last relation is the basis for the application of the 'scaled particle theory'[32,33] to aqueous solutions of simple solutes.[34–43]

Perhaps one of the most fascinating aspects of the behaviour of aqueous solutions of noble gases is the relatively large and negative entropy and enthalpy of solution of these solutes in water as compared with other solvents. Table 10.6.1 presents some illustrative experimental values for $\Delta\mu_s^0$, ΔS_s^0 and ΔH_s^0 for methane in water and in some other solvents. It is very clear that water is outstanding as a solvent for this solute, at least as far as it is compared with those solvents for which the relevant data are available.

Almost thirty years ago it was suggested by Frank and Evans[45] that the large negative entropy of solution of simple solutes in water may be accounted for by the assumption that these solutes 'stabilize the structure of water'. Whatever that statement meant at that time, it had a dramatic impact on the field of aqueous solutions. Two fundamental questions have been

Table 10.6.1 Values of the standard free energy, entropy, enthalpy, and heat capacity of methane in water and in some nonaqueous solvents at two temperatures†

Solvent	t °C	$\Delta\mu_s^0$ cal/mole	ΔS_s^0 e.u.	ΔH_s^0 cal/mole	ΔC_s^0 cal/mol deg
Water	10	1747	−18.3	−3430	53
	25	2000	−15.5	−2610	
Heavy water (D_2O)	10	1703	−19.2	−3740	52
	25	1971	−16.5	−2940	
Methanol	10	343	−2.6	−390	−21
	25	390	−3.7	−710	
Ethanol	10	330	−3.2	−570	−5
	25	380	−3.5	−650	
1-Propanol	10	345	−4.3	−880	25
	25	400	−3.0	−500	
1-Butanol	10	369	−2.8	−420	−33
	25	430	−4.5	−910	
1-Pentanol	10	399	−3.3	−530	−7
	25	450	−3.6	−630	
1,4-Dioxane	10	538	−0.8	+310	−6
	25	553	−1.1	+220	
Cyclohexane	10	154	−1.9	−390	11
	25	179	−1.4	−230	

†All values refer to the process of transferring one solute from a fixed position in the gas to a fixed position in the liquid at constant T and P. (Data from Yaacobi and Ben-Naim[44])

debated ever since; the first concerns the type of structure that the solute is presumed to enhance, while the second inquires into the molecular reasons for such a 'stabilization effect' and how this effect is related to the peculiarities of liquid water.

Even today we have no definite answers to either of these questions. We shall elaborate on some theoretical aspects of the second one below. Treatments of the first question have been confined mainly to within the context of *ad hoc* models for water; a review of various approaches may be found elsewhere.[22,23,46]

We shall first present here what may be referred to as the traditional thermodynamic treatment of the problem, and later we shall turn to the more formal statistical mechanical formulation of the same problem.

Consider a two-structure model, which may be either an *ad hoc* or an exact one, as discussed in Secion 10.5. Let N_L and N_H be the average number of L and H molecules respectively. We assume that the component L is the more structured one (by whatever criterion we choose to define the structure). The standard enthalpy and entropy of solution of a solute s are written as[23]

$$\Delta H_s^0 = \Delta H_s^* + (\bar{H}_L - \bar{H}_H)\left(\frac{\partial N_L}{\partial S_s}\right)_{P,T} \tag{10.6.5}$$

$$\Delta S_s^0 = \Delta S_s^* + (\bar{S}_L - \bar{S}_H)\left(\frac{\partial N_L}{\partial N_s}\right)_{P,T} \tag{10.6.6}$$

where ΔH_s^* and ΔS_s^* refer to the standard enthalpy and entropy of solution of s into a system in which the conversion 'reaction' L ⇄ H is 'frozen in', and the second terms on the RHS of

(10.6.5) and (10.6.6) are the contributions to ΔH_s^0 and to ΔS_s^0 due to 'structural changes in the solvent'.

The traditional explanation of the anomalously large and negative values of ΔH_s^0 and ΔS_s^0 is based on the assumption that the solute stabilizes the structure of water, in the sense that $\partial N_L/\partial N_s$ is positive. If this is true then the increase in structure, presumably by creating more hydrogen bonds, provides a convenient interpretation for the negative values of ΔH_s^0 and ΔS_s^0. However, the crucial question is still left unanswered, i.e. why does an inert solute, like argon, stabilize the structure of water in the first place?

Indeed, there exists a large body of experimental facts that are consistent with the above assumption, although no direct measurement of the structural changes in the solvent has ever been suggested. We shall return to this point below, but before doing that it is instructive to write down the statistical mechanical analogues of relations (10.6.5) and (10.6.6). This is obtained by differentiation of (10.6.1) with respect to temperature. The result, after some rearrangement, is

$$\Delta E_s^0 = \langle B_s \rangle_s + [\langle U_N \rangle_s - \langle U \rangle_0] \tag{10.6.7}$$

$$T\Delta S_s^0 = k_B T \ln \langle \exp[-\beta B_s] \rangle_0 + \langle B_s \rangle_s + [\langle U_N \rangle_s - \langle U_N \rangle_0]. \tag{10.6.8}$$

(For our purposes we may disregard the small difference between enthalpy and energy of solution.) The symbol $\langle \ldots \rangle_s$ in (10.6.7) and (10.6.8) stands for the conditional average over all configurations of the solvent molecules, given a solute s at a fixed position $\mathbf{R}_s$.

The comparison between (10.6.5) and (10.6.6) on one hand and (10.6.7) and (10.6.8) on the other is instructive. With a simple reinterpretation one can assign to the term $[\langle U_N \rangle_s - \langle U \rangle_0]$ the meaning of structural change in the solvent, induced by the addition of the solute s. To do that we use the distribution function $x(\nu)$ defined in (10.5.10) and rewrite $\langle U_N \rangle_0$ as

$$\begin{aligned}
\langle U_N \rangle_0 &= \frac{1}{2} \int \ldots \int \mathrm{d}\mathbf{X}^N P(\mathbf{X}^N) \sum_{i=1}^{N} \sum_{\substack{j=1 \\ j \neq i}}^{N} U(\mathbf{X}_i, \mathbf{X}_j) \\
&= \frac{1}{2} \int \ldots \int \mathrm{d}\mathbf{X}^N P(\mathbf{X}^N) \sum_{i=1}^{N} B_i(\mathbf{X}^N) \\
&= \frac{N}{2} \int \ldots \int \mathrm{d}\mathbf{X}^N P(\mathbf{X}^N) \int_{-\infty}^{\infty} \nu \mathrm{d}\nu \delta [B_1(\mathbf{X}^N) - \nu] \\
&= \frac{N}{2} \int_{-\infty}^{\infty} \nu x(\nu) \, \mathrm{d}\nu
\end{aligned} \tag{20.6.9}$$

and similarly for $\langle U_N \rangle_s$ we have

$$\langle U_N \rangle_s = \frac{N}{2} \int_{-\infty}^{\infty} \nu x \left(\frac{\nu}{\mathbf{R}_s} \right) \mathrm{d}\nu \tag{10.6.10}$$

where $x(\nu/R_s)$ is the conditional distribution function for binding energy of the solvent molecules, given a solute s at $\mathbf{R}_s$.

Clearly, the quantity $N[x(\nu/\mathbf{R}_s) - x(\nu)] \mathrm{d}\nu$ measures the change of the average number of the ν-components induced by the addition of s. Hence the whole difference $[\langle U_N \rangle_s - \rangle U_N \rangle_0]$ may be viewed as a generalization of the concept of the contribution to ΔE_s^0 due to structural change in the solvent. From (10.6.7) and (10.6.8) it is also clear, as it was from (10.6.5) and (10.6.6), that the contribution to $T\Delta S_s^0$ from structural changes is exactly the same as to ΔE_s^0.

This result is often referred to as the enthalpy–entropy compensation law.[47] The formulation of this law through equation (10.6.7) and (10.6.8) may be viewed as the more fundamental one compared with the one derived from (10.6.5) and (10.6.6), which is often based on *ad hoc* models for water.

We now turn to the more conventional notion of structural changes in the solvent. We use equations (10.5.1) and (10.5.2) and rewrite the total potential energy of the system of N water molecules as

$$U_N(\mathbf{X}^N) = U'_N(\mathbf{X}^N) + \epsilon_{HB} \sum_{i<j} G(\mathbf{X}_i, \mathbf{X}_j). \tag{10.6.11}$$

Hence the quantity $\langle U_N \rangle_s - \langle U_N \rangle_0$ may be split into two terms

$$\langle U_N \rangle_s - \langle U_N \rangle_0 = [\langle U'_N \rangle_s - \langle U'_N \rangle_0] + \frac{N\epsilon_{HB}}{2} [\langle \psi \rangle_s - \langle \psi \rangle_0] \tag{10.6.12}$$

where the second term is now identified as the more 'conventional' structural change in the water, induced by the solute. The first term on the RHS of (10.6.12) is probably small and has a similar magnitude to the corresponding term in nonaqueous solvents.

Next we present an approximate estimate of the structural change in water due to the processes of solubility. This, in a way, provides additional evidence in favour of the contention that simple solutes stabilize the structure of water.

Consider the standard free energies of solution of a solute s in H_2O and D_2O. The assumption is made that both H_2O and D_2O may be represented by a system of water-like particles, interacting through pair potentials of the form (10.5.1), and that the main difference between the two solvents may be ascribed to the difference in the parameter ϵ_{HB}, which we denote by ϵ_H and ϵ_D for H_2O and D_2O respectively. We also assume that the solute-solvent interaction is the same for the two solvents. (This is certainly true for a hard sphere solute, and approximately so for simple solutes.)

Differentiating $\Delta\mu_s^0$ in (10.6.1) (but now taking the average in the T, P, N ensemble) with respect to ϵ_{HB} we get the exact relation

$$\left(\frac{\partial \Delta\mu_s^0}{\partial \epsilon_{HB}}\right)_{T,P,N} = \langle \sum_{i<j} G(\mathbf{X}_i, \mathbf{X}_j) \rangle_s - \langle \sum_{i<j} G(\mathbf{X}_i, \mathbf{X}_j) \rangle_0$$

$$= \frac{N}{2} [\langle \psi \rangle_s - \langle \psi \rangle_0]. \tag{10.6.13}$$

We further assume that $\epsilon_D - \epsilon_H$ is small compared with ϵ_{HB}, so that we can write the approximate relation

$$\frac{\Delta\mu_s^0(D_2O) - \Delta\mu_s^0(H_2O)}{\epsilon_D - \epsilon_H} = \frac{N}{2} [\langle \psi \rangle_s - \langle \psi \rangle_0]. \tag{10.6.14}$$

The isotope effect in $\Delta\mu_s^0$ in H_2O and D_2O is known for a large number of solutes.[48–50] The quantity $\epsilon_D - \epsilon_H$ is not directly measurable, but it is widely agreed that this is negative (i.e. HB is stronger in D_2O compared with H_2O), and a reasonable estimate suggested by Nemethy and Scheraga[51] is

$$\epsilon_D - \epsilon_H \approx -0.23 \text{ kcal/mole of bonds}. \tag{10.6.15}$$

Using this value we can easily estimate the structural changes in water due to addition of various solutes. Table 10.6.2 presents some illustrative examples, which, though approximate, are in agreement with the conjecture on the effect of simple solutes on the structure of water.

Table 10.6.2 Values of $\langle G \rangle_s - \langle G \rangle_0$ for some simple solutes at different temperatures[50]

t °C	5	10	15	20
Argon	0.28	0.25	0.23	0.20
Methane	0.21	0.19	0.17	0.15
Ethane	0.19	0.14	0.12	0.12
Propane	0.14	—	—	—
Butane	0.17	0.13	0.1	0.03

Next we turn briefly to discuss the molecular reasons for the existence of a stabilization effect.[52] Suppose we classify water molecules in two classes: those with low-local-density, denoted by L, and those with relatively high-local-density, denoted by H. Let N_L and N_H be the average number of molecules in the two classes respectively (in a system characterized by the temperature T, pressure P and total number of water molecules ($N_W = N_L + N_H$). Using the Kirkwood–Buff[53] theory of solution one can derive an exact expression for the derivative $\partial N_L/\partial N_s$ (in 10.6.5 and 10.6.6) in terms of molecular distribution functions,[52] which reads

$$\lim_{\rho_s \to 0} \left(\frac{\partial N_L}{\partial N_s}\right)_{T,P,N_W} = \rho_W x_L x_H [(G_{WH} - G_{WL}) + (G_{Ls} - G_{Hs})] \tag{10.6.16}$$

where ρ_W is the total number density of water, x_L and x_H are the mole fractions of the two components, L and H respectively, and G_{ij} are integrals over the corresponding pair correlation functions for the two species i and j,

$$G_{ij} = \int_0^\infty [g_{ij}(R) - 1]\, 4\pi R^2 \, dR. \tag{10.6.17}$$

The quantity $\rho_W [G_{WH} - G_{WL}]$ measures the average excess of water molecules around H relative to L. This quantity may, by definition of the two species, be made positive. Thus we have one term on the RHS of (10.6.16) which is positive, and which is independent of the properties of the solute s.

The second term $G_{Ls} - G_{Hs}$ is related to the relative 'affinity' between s and L and between s and H. Since L was chosen to be the component of lower local density one can argue that the solute s would tend to concentrate around L more than around H. Hence the second term $G_{Ls} - G_{Hs}$ would also be positive. (More details on this topic may be found in Reference 23.)

The above argument is very general and can be applied to any solvent; it merely depends on the classification of the solvent molecules into two classes according to their local densities. The important feature that makes water different is that the L component is also the more structured one, and by stabilizing this component one gets a considerable lowering in energy and entropy of the system. For instance in (10.6.5) the product of a positive derivative $\partial N_L/\partial N_s$ and a negative quantity $\bar{H}_L - \bar{H}_H$ is probably a unique feature of aqueous solutions. This is similar to the primary reason for a negative $\partial V/\partial T$ of water as discussed in Section 10.5.

We conclude this section by suggesting that one of the most important aspects of the study of aqueous solution is to have an estimate of the relative contribution to, say, ΔE_s^0 from 'binding energy' on the one hand, and structural changes in the solvent on the other. The latter part is likely to turn out to be very characteristic of liquid water. One way of studying this aspect is suggested by equations (10.6.7)–(10.6.10), i.e. by studying the properties of the function $x(\nu)$ and its response to the addition of solutes.

10.7 Moderately dilute solutions of simple solutes; the problem of hydrophobic interaction

The behaviour of the very dilute solutions of simple solutes, discussed in the previous section, is characterized by the absence of solute–solute interaction. From the formal point of view it is sufficient to study the properties of the one-solute-in-a-solvent systems as representatives of the very dilute solutions. However, most aqueous solutions of interest, and in particular those that occur in biological systems, are not 'very dilute', and therefore one must consider also the solute–solute interactions. The first order deviation from the very dilute solution may be studied, for instance, by considering the virial expansion of the osmotic pressure Π, namely

$$\Pi/k_BT = \rho_s + B_2^*\rho_s^2 + \dots \tag{10.7.1}$$

where ρ_s is the (simple) solute density, and B_2^* is the so-called second virial coefficient of the osmotic pressure. This quantity is related to the solute–solute pair correlation function through

$$B_2^* = -\frac{1}{2}\int_0^\infty [g_{ss}^0(R) - 1]\, 4\pi R^2 \mathrm{d}R \tag{10.7.2}$$

where

$$g_{ss}^0(R) = \lim_{\rho_s \to 0} g_{ss}(R). \tag{10.7.3}$$

Again, from the formal point of view it is sufficient to study the properties of the two-solute-in-a-solvent-systems as representatives of the slightly concentrated solutions (i.e. solutions that are adequately represented by the two terms in the density expansion in (10.7.1)).

There exists a considerable amount of theoretical knowledge on $g_{ij}(R)$ for mixtures of simple fluids. (For some specific references see Reference 23.) However, almost nothing is known about the behaviour of these functions for a mixture of, say, argon in water. Yet this kind of information is crucial for understanding the so-called hydrophobic interaction (HI). This term has been used for many years, especially by biochemists and biologists, in reference to a kind of excess attraction between two nonpolar solutes in aqueous solution.[54–63]

These forces are presumed to be operative between side-chain nonpolar groups in various biopolymers, and probably are partially responsible for maintaining a very specific conformation of the biopolymer in aqueous media. Similar forces are likely to play an important role in the process of self assembling of many subunits to form large multisubunit biological molecules.[64]

Of course, theoretical tools are still far from being sufficiently advanced for handling problems of such immense complexity, and one must resort to the simplest systems for which the term 'hydrophobic interaction' is still relevant.

Consider a system of N_W water molecules at a given temperature T and pressure P, and containing two simple solutes, such as argon, which we denote by s.

An appropriate measure of the *strength* of the HI may be defined as follows; suppose one starts with the system as described above but in which the two solutes are held at two fixed positions $\mathbf{R}_1$ and $\mathbf{R}_2$ at very large separation, $R = |\mathbf{R}_2 - \mathbf{R}_1|$. We now define the strength of the HI as the work $\Delta G(R)$ (in the T, P, N ensemble) required to bring the two solutes from $R = \infty$ to some close distance $R = \sigma$, where σ is of the order of magnitude of the molecular diameter of the solute s. This work, i.e. the free energy change, is related to the solute–solute pair correlation function $g_{ss}^0(R = \sigma)$ by

$$g_{ss}^0(R = \sigma) = \exp\left(\frac{-\Delta G(R = \sigma)}{k_BT}\right). \tag{10.7.4}$$

Clearly, information on either $g_{ss}^0(R)$ or on $\Delta G(R)$ is equivalent. The main question concerning the properties of $g_{ss}^0(R)$ are the following:

(*a*) Is there any major difference between the functions $g_{ss}^0(R)$ for a given solute s, in water and in other nonaqueous solvents? In particular, one is interested to know whether there exists a pronounced maximum in $g_{ss}^0(R)$ at $R \approx \sigma$ which will indicate that the two-solute particles have a relatively large probability of being found at close distance $R \approx \sigma$ when they are in water as compared with other solvents.

(*b*) What is the value of R at which such a maximum exists? For simple liquids, as well as for mixtures of simple liquids, the first maximum of $g_{ij}(R)$ is usually found at $R \approx \sigma_{ij}$, where σ_{ij} is the effective distance of closest approach between the two species i and j. In water, we do not have, so far, any information on the precise location of the maximum of $g_{ss}^0(R)$. It is conceivable, because of the peculiarities of the structure of water, that a first maximum would occur at a distance which substantially differs from σ_{ss}. Some suggestions have been made that this distance could be $\sigma_{ss} + \sigma_{ww}$ where σ_{ww} is the effective diameter of water,[65] but as far as the author is aware there exists no support for this conjecture on either theoretical or experimental grounds.

(*c*) Suppose we have already established that $g_{ss}^0(R)$ in water has an unusually high peak at some distance $R \approx \sigma_{ss}$. What is the temperature dependence of $g_{ss}^0(R \approx \sigma_{ss})$? There exists some information (see below) that the strength of the HI *increases* with temperature,[54,55,58] and that this behaviour is probably unique to water as a solvent. Of course, we have no direct information on the temperature dependence of $g_{ss}^0(R)$ at $R \approx \sigma$. The available information that will be discussed below indicates that $g_{ss}^0(R)$ at $R \approx \sigma$ increases with temperature only in water, whereas in other solvents it decreases with temperature (as is also expected on qualitative grounds). This aspect of the behaviour of solutes in water is very interesting, and is certainly related to the phenomenon of structural changes in the solvent, induced by the process of HI.[23]

Before turning to an approximate estimate of the HI and its temperature dependence, we note that for each R we can write

$$\Delta S(R) = -\frac{\partial \Delta G(R)}{\partial T} = k_B \ln g_{ss}^0(R) + k_B T \frac{\partial \ln g_{ss}^0(R)}{\partial T}$$

$$\Delta H(R) = \Delta G(R) + T\Delta S(R) = k_B T^2 \frac{\partial \ln g_{ss}^0(R)}{\partial T}. \tag{10.7.5}$$

Thus in principle the temperature dependence of $g_{ss}^0(R)$ at any R may be extracted from the corresponding enthalpy change $\Delta H(R)$. At present both theory and experiment are incapable of providing information on $g_{ss}^0(R)$ for simple solutes in water.

Because of these difficulties one must resort to some fragmental information in order to establish the properties of the HI. From now on we shall be focusing our attention on one source of information, namely on an approximate way of measuring $\Delta G(R)$ at $R = \sigma_1$ (with $\sigma_1 < \sigma_{ss}$) and its temperature dependence.

For any distance R one may write the statistical mechanical expression for $\Delta G(R)$ as follows

$$\Delta G(R) = U(R) - k_B T \ln \frac{\langle \exp[-U(\mathbf{X}^N/R)/k_B T] \rangle_0}{\langle \exp[-U(\mathbf{X}^N/R = \infty)/k_B T] \rangle_0} \tag{10.7.6}$$

where $U(\mathbf{X}^N/R)$ stands for the total interaction energy between all the solvent molecules at a specific configuration $\mathbf{X}^N$ and the pair of solutes s at a given separation R. $U(R)$ is the direct pair potential operating between the two solutes, and the symbol $\langle \ldots \rangle_0$ stands for an average, in the T, P, N ensemble, over the configurations of the solvent molecules.

Since $U(R)$ is presumed to be approximately unaffected by the type of solvent, we write (10.7.6) as

$$\Delta G(R) = U(R) + \delta G^{HI}(R) \tag{10.7.7}$$

and focus attention only to the indirect part of the work required for bringing the two solutes from infinite separation to the final distance R. Clearly, any peculiarities of the hydrophobic interaction must be contained within the function $\delta G^{HI}(R)$, to which we may refer in the context of aqueous solution as the hydrophobic interaction part of the total work $\Delta G(R)$.

We now notice that the function $\delta G^{HI}(R)$ in (10.7.7) does not have the divergence behaviour at $R \lesssim \sigma_{ss}$, as in $\Delta G(R)$, which is imposed by the *direct* strong repulsive part, $U(R)$. In fact, all the information that is presently available indicates that this function attains finite values even at $R = 0$. Exploiting this fact we have recently suggested an approximate relation between $\delta G^{HI}(R = \sigma_1 = 1.533 \text{ Å})$ and experimental quantities, that reads[23]

$$\delta G^{HI}(\sigma_1) = \Delta\mu^0_{Et} - 2\Delta\mu^0_{Me} \tag{10.7.8}$$

where $\Delta\mu^0_{Et}$ and $\Delta\mu^0_{Me}$ are the standard free energy of solution of ethane and methane respectively.

Using this particular approximation we can now compare water with some other solvents. A sample of values of $\delta G^{HI}(x_1)$ as well as the corresponding entropies and enthalpies are reported in Table 10.7.1. It is clearly seen that the values of water are markedly different from the corresponding values in all other solvents for which the relevant experimental data are available.

Relation (10.7.8), though approximate, is instructive in stressing the fact that HI is not measured by a *single* standard free energy of solution but depends on the difference between the free energy of solution of an elongated molecule (e.g. ethane) and two spherical molecules (methane).

In the past, standard free energies of solution of single solutes were used as a measure of the strength of the HI.[54] In fact, some of the conclusions based on standard free energies were consistent with conclusions arrived at from quantities like $\delta G^{HI}(\sigma_1)$. However, there are examples for which the two quantities lead to contradictory conclusions.[23] This is no wonder, since $\Delta\mu^0_s$ basically involves the properties of a single solute which is entirely surrounded by solvent molecules, whereas $\delta G^{HI}(R)$ measures the *difference* in the solvation of the two solutes at two different separations.

Table 10.7.1 Values of $\delta G^{HI}(\sigma_1)$, $\delta S^{HI}(\sigma_1)$, and $\delta V^{HI}(\sigma_1)$ for bringing two methane molecules to the distance $\sigma_1 = 1.533$ Å at 10 °C†

Solvent	δG^{HI} kcal/mole	δS^{HI} e.u.	δH^{HI} kcal/mole	δV^{HI} cm^3 mole
Water	−1.99	12	1.60	−23††
Heavy water	−1.94	13	1.70	—
Methanol	−1.28	0	−1.4	—
Ethanol	−1.34	0	−1.3	—
1-Propanol	−1.39	2	−0.6	—
1-Butanol	−1.44	−1	−1.9	—
1-Pentanol	−1.49	0	−1.3	—
1-Mexanol	−1.51	2	0.7	—
1,4-Dioxane	−1.61	3	−0.8	—
Cyclohexane	−1.36	1	−1.2	—

†Based on data from Yaacobi and Ben-Naim[44].
††In benzene, the value is −31 cm^3/mole.

From the theoretical point of view there exists one interesting example for which the two quantities δG^{HI} and $\Delta\mu_s^0$ furnish the same information. This occurs for a hard sphere solute for which the standard free energy of solution will be denoted by $\Delta\mu_{\mathrm{HS}}^0$. On the other hand, consider the hydrophobic interaction between two hard spheres at zero separation, namely $\delta G^{\mathrm{HI}}(R=0)$. This is equivalent to the work required to bring two cavities of suitable diameter from infinite to zero separation. The relation between the two quantities can easily be shown to be[23]

$$\delta G^{\mathrm{HI}}(R=0) = \Delta\mu_{\mathrm{HS}}^0 - 2\Delta\mu_{\mathrm{HS}}^0 = -\Delta\mu_{\mathrm{HS}}^0$$

which is almost obvious on intuitive grounds as well, i.e. the work required to bring two cavities from infinite to zero separation (within the liquid at constant P, T, N) is the same as the work required to eliminate a single cavity from the liquid. Another common way of saying the same thing is that the HI is a partial reversal of the dissolution process. As it is clear from the above example, this statement is valid exactly only for two cavities brought to *zero* separation. For real solutes and for any $R \neq 0$ the concept of *partial* reversal of solubility is also used in connection with the HI process,[55,65] but, as we have noted before, the analogy between the two processes should not be pursued too seriously, since the two involve quite different properties of the system.

There are some other problems related to hydrophobic interaction which theory could help to clarify:

(1) The process of hydrophobic interaction may, in principle, induce some structural changes in the solvent. This problem may be formulated by a simple extension of the corresponding problem in the solubility process, discussed in the previous section. One can easily show[66] that structural changes in the solvent, of the type discussed in Section 10.6, would not affect the *strength* of the hydrophobic interaction, but may well affect the entropy and the enthalpy change of this process.

(2) The hydrophobic interaction between large molecules, such as subunits forming a multisubunit biopolymer, is very important in biochemistry. There exist examples where two or more subunits form aggregates which do not involve any covalent bonds. The driving forces for these processes must be due either to the direct interaction between the subunits or the indirect, i.e. hydrophobic, interaction which originates from the peculiarities of the solvent.

(3) Processes in biological systems usually involve more than two simultaneously interacting particles. The natural question that immediately arises is to what extent the hydrophobic interaction among many particles is additive. There is very little information on this problem from the theoretical point of view. Experimental information, mainly from the process of micells formation, indicates that the simultaneous interaction among a large number of molecules is highly non-additive. Elucidation of this aspect will be very important to the understanding of many biochemical processes.

Acknowledgement

This chapter has been written while the author spent a sabbatical year at the Laboratory of Molecular Biology, National Institutes of Health, Bethesda, Maryland, for which the author is very grateful.

References

1. J. A. Barker and R. O. Watts (1969). *Chem. Phys. Lett.*, **3**, 144.
2. J. A. Barker and R. O. Watts (1973). *Molecular Phys.*, **26**, 789.

3. R. O. Watts (1974). *Molecular Phys.*, **28**, 1069.
4. H. Popkie, H. Kistenmacher and E. Clementi (1973). *J. Chem. Phys.*, **59**, 1325.
5. H. Kistenmacher, H. Popkie, E. Clementi and R. O. Watts (1974). *J. Chem. Phys.*, **60**, 4455.
6. G. C. Lie and E. Clementi (1975). *J. Chem. Phys.*, **62**, 2195.
7. G. N. Sarkisov, V. G. Dashevsky and G. G. Malenkov (1974). *Molecular Phys.*, **27**, 1249.
8. A. Ben-Naim (1970). *J. Chem. Phys.*, **52**, 5531.
9. A. Rahman and F. H. Stillinger (1971). *J. Chem. Phys.*, **55**, 3336.
10. F. H. Stillinger and A. Rahman (1972). *J. Chem. Phys.*, **57**, 1281.
11. A. Rahman and F. H. Stillinger (1973). *J.A.C.S.*, **95**, 7943.
12. F. H. Stillinger and A. Rahman (1974). *J. Chem. Phys.*, **60**, 1545.
13. F. H. Stillinger and A. Rahman (1974). *J. Chem. Phys.*, **61**, 4973.
14. K. Heinzinger and P. C. Vogel (1974). *Z. Naturforsch.*, **29a**, 1164.
15. H. C. Andersen (1973). *J. Chem. Phys.*, **59**, 4714.
16. H. C. Andersen (1974). *J. Chem. Phys.*, **61**, 4985.
17. I. Prigogine and R. Defay (1954). In D. H. Everett (Translator). *Chemical Thermodynamics*, Longman Green, New York.
18. W. C. Roentgen (1892). *Ann. Phys.*, **45**, 91.
19. D. Eisenberg and W. Kauzmann (1969). *The Structure and Properties of Water*, Oxford University Press.
20. N. H. Fletcher (1971). *Rep. Prog. Phys.*, **34**, 913.
21. R. A. Horne (Ed.) (1972). *Water and Aqueous Solutions, Structure, Thermodynamics and Transport Processes*, Wiley-Interscience, New York.
22. F. Franks (Ed.) (1972). *Water, a Comprehensive Treatise*, Vol. I. Plenum Press, New York.
23. A. Ben-Naim (1974). *Water and Aqueous Solutions, Introduction to a Molecular Theory*, Plenum Press, New York.
24. F. H. Stillinger. In I. Prigogine and S. A. Rice (Eds), (1975) *Advances in Chemical Physics*, Vol. XXV.
25. A. H. Narten (July 1970). *ONRL Report* No. 4578.
26. A. H. Narten and H. A. Levy (1972). In F. Franks (Ed.), *Water, A Comprehensive Treatise*, Vol. I. Plenum Press, New York.
27. D. J. Evans and R. O. Watts (1974). *Molecular Phys.*, **28**, 1233.
28. B. R. Lentz and H. A. Scheraga (1973). *J. Chem. Phys.*, **58**, 5296.
29. J. G. Lane and H. W. Leidecker (1974). *J. Chem. Phys.*, **60**, 705.
30. A. Ben-Naim and F. H. Stillinger (1972). In R. A. Horne (Ed.), *Water and Aqueous Solutions*, Wiley-Interscience, New York.
31. O. Weres and S. A. Rice (1972). *J. Am. Chem. Soc.*, **94**, 8983.
32. H. Reiss, H. L. Frisch and J. L. Lebowitz (1959). *J. Chem. Phys.*, **31**, 369.
33. H. Reiss (1966). *Adv. Chem. Phys.*, **9**, 1.
34. R. A. Pierotti (1963). *J. Phys. Chem.*, **67**, 1840.
35. R. A. Pierotti (1965). *J. Phys. Chem.*, **69**, 281.
36. R. A. Pierotti (1967). *J. Phys. Chem.*, **71**, 2366.
37. A. Ben-Naim and H. A. Friedman (1967). *J. Phys. Chem.*, **71**, 448.
38. E. Wilhelm and R. Battino (1971). *J. Chem. Phys.*, **55**, 4012.
39. E. Wilhelm and R. Battino (1972). *J. Chem. Phys.*, **56**, 563.
40. M. Lucas (1973). *J. Phys. Chem.*, **77**, 2479.
41. F. H. Stillinger (1973). *J. Sol. Chem.*, **2**, 141.
42. C. Jolicoeur and G. Lacroix (1973). *Can. J. Chem.*, **51**, 3051.
43. P. R. Philip and C. Jolicoeur (1975). *J. Sol. Chem.*, **4**, 105.
44. M. Yaacobi and A. Ben-Naim (1974). *J. Phys. Chem.*, **78**, 170, 175.
45. H. S. Frank and N. W. Evans (1945). *J. Chem. Phys.*, **13**, 507.
46. A. Ben-Naim (1972). In R. A. Horne (Ed.), *Water and Aqueous Solutions*, Wiley-Interscience, New York.
47. R. Lumry and S. Rajender (1970). *Biopolymers*, **9**, 1125.
48. E. M. Arnett and D. R. McKelvey (1969). In J. F. Coetzee and C. D. Ritchie (Eds), *Solute–Solvent Interaction*, M. Dekker, New York.
49. H. L. Friedman and C. V. Krishman (1973). In F. Franks (Ed.), *Water, A Comprehensive Treatise*, Vol. III. Plenum Press, New York.
50. A. Ben-Naim (1975). *J. Phys. Chem.*, **75**, 1268.

51. G. Nemethy and H. A. Scheraga (1962). *J. Chem. Phys.*, **36**, 3382, 3401.
52. A. Ben-Naim (1973). *J. Statist. Phys.*, **7**, 3.
53. J. G. Kirkwood and F. P. Buff (1951). *J. Chem. Phys.*, **19**, 774.
54. W. Kauzmann (1959). *Adv. Protein Chem.*, **14**, 1.
55. G. Nemethy and H. A. Scheraga (1962). *J. Phys. Chem.*, **66**, 1773.
56. A. Ben-Naim (1971). *J. Chem. Phys.*, **54**, 1387.
57. J. J. Kozak, W. S. Knight and W. Kauzmann (1968). *J. Chem. Phys.*, **48**, 675.
58. H. A. Scheraga, G. Nemethy and I. Z. Steinberg (1962). *J. Biol. Chem.*, **237**, 2506.
59. R. B. Hermann (1971). *J. Phys. Chem.*, **75**, 363.
60. R. B. Hermann (1972). *J. Phys. Chem.*, **76**, 2754.
61. R. B. Hermann (1975). *J. Phys. Chem.*, **79**, 163.
62. M. H. Klapper (1973). In E. T. Kaiser and F. J. Kezdy (Eds), *Progress in Bioorganic Chem.*, Vol. 2, p. 55.
63. T. S. Sarma and J. C. Ahluwalia (1973). *Chem. Soc. Rev.*, **2**, 203.
64. A. L. Lehninger (1973). *Biochemistry*, Worth, New York.
65. F. Franks (Ed.) (1975). *Water, A, Comprehensive Treatise*, Vol. IV. Plenum, New York.
66. A. Ben-Naim (1975). *Biopolymers*, **14** 1337.

Chapter 11

Phase Transitions in Classical Liquids

H. N. V. TEMPERLEY

Department of Applied Mathematics,
University College of Wales, Swansea, Wales

11.1 Introduction

We shall be concerned with liquids whose molecules can be treated as spherically symmetrical interacting with central forces that depend simply on the distances apart of the centres of the molecules. This does *not* restrict us to the rare gases, because many substances like benzene and carbon tetrachloride have molecules that are reasonably symmetrical. In the liquid phase they are probably rotating, so the field of a molecule 'seen' by a neighbour is spherically symmetrical. We exclude polar molecules, the forces between which are directional, molecules such as water and polymers which cannot be thought of as effective spheres, and 'rigid-rod-like' molecules such as hydrocarbons. The properties of all these assemblies are being intensively studied and the effects of these complications on the finer details of the evaporation and solidification transitions may be expected to become clearer in due course.

It can be said that the properties of a 'simple' substance like argon are reasonably well understood except in the immediate neighbourhood of the triple point at which vapour, liquid and solid are in equilibrium. In Figure 11.1.1 we have the diagram of state of a simple substance, the line OT representing equilibrium between solid and vapour, the line TC (ending at the critical point C) representing equilibrium between liquid and vapour and the line TA that determining equilibrium between solid and liquid. Experimentally, this line TA has been traced up to temperatures many times the critical temperature T_c above which it is meaningless to distinguish between liquid and gas. No indication of a 'solid–fluid critical temperature' has ever been found experimentally and the entropy difference between the ordered solid phase and the disordered fluid phase remains finite, and can even increase, as we travel up the curve TA.

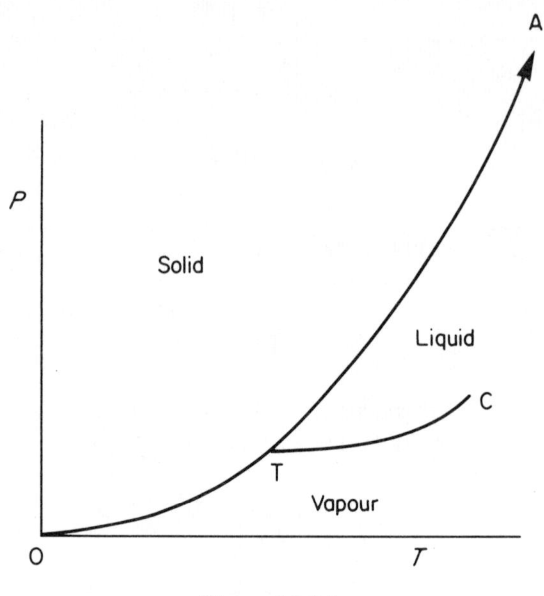

Figure 11.1.1

Admittedly, the melting curves of certain substances do show 'breaks' and local maxima. These have invariably been traced to the occurrence of solid–solid phase transitions, so that the observed melting curve really consists of two or more portions each determining equilibrium between the fluid phase and one particular solid phase. These solid–solid transitions are extremely frequent, and occur even in the inert gases. They are being intensively studied, but cannot be regarded as part of liquid state theory.

Although, strictly speaking, it is not part of liquid state theory either, it is helpful to summarize our understanding of the curve OT. Surprisingly, our understanding of it becomes better the *lower* the temperature. According to standard thermodynamics, the equilibrium between two phases is determined by the condition that their partial potentials (Gibbs functions per unit mass) should be equal. Statistical mechanics enables us to calculate the Helmholtz free energy and hence the Gibbs function for certain ideal assemblies, in particular for the perfect gas and the 'ideal' crystalline solid. If we can do this, we can predict not only the vapour-pressure curve OT, but also the latent heat and discontinuity in density at all points along the curve.

For a justification of the above statements the reader is referred to standard works on statistical mechanics, such as those by Fowler, Mayer and Mayer, Rushbrooke or Terrell Hill. In fact it turns out that we get a good first approximation to the curve OT if we treat the vapour phase as perfect and then treat the solid as an ideal crystal following the work of Debye, Born and van Karman and Blackman. It will be helpful to our subsequent discussions to trace briefly why the prediction of the curve OT becomes better as we go to *lower* temperatures. If we follow the vapour-pressure curve, the vapour phase becomes *more* perfect as we approach absolute zero. At constant density any gas becomes less perfect as the temperature falls, but this is more than counterbalanced by the extremely rapid fall in density along the vapour-pressure curve. For low temperatures the curve is described roughly by

$$P \approx P_0 \exp(-l/kT) \tag{11.1.1}$$

where l is the latent heat per molecule. For T small compared with l/k, the exponential decrease in pressure implies an equally rapid decrease in the density of the vapour, and a corresponding progress towards perfection, which more than offsets the effect of lowering T.

To understand the situation in the solid phase we return to Debye's old, but extremely fruitful, idea that the normal modes of vibration of a solid can be roughly mapped by treating the solid as a continuum and determining the possible frequencies by acoustic theory. If we assign to each mode an energy according to Planck's formula we arrive at the following expression for the energy content

$$E(T) = E_0 + \int \frac{h\nu g(\nu)\,\mathrm{d}\nu}{\exp(h\nu/kT) - 1} \tag{11.1.2}$$

(where $g(\nu)\mathrm{d}\nu$ is the number of modes between ν and $\nu + \mathrm{d}\nu$ and E_0 the zero-point energy $\frac{1}{2}h\nu$ for each mode).

At low temperatures only the low frequency modes, that is those with long wavelength, are excited appreciably. The continuum approximation becomes better as the wavelength becomes longer and longer relative to the intermolecular spacing. Thus, the solid as well as the gas approaches ideal behaviour more and more closely as the temperature falls.

11.2 What do we ask of a theory?

In the remainder of this chapter, we examine how far we can account for the parameters of the solidification and evaporation transitions if the intermolecular interaction function is supposed to be known. There is a complication here because, over the years, conclusive evidence has been obtained that the effect of three-molecular interactions is significant even in so simple an assembly as liquid argon. Although we can account for the equation of state of the liquid quite well in terms of a two-molecular interaction of the Lennard–Jones type, we cannot, try as we will, account for all the available information (equations of state of the solid and gas, direct measurement of the interaction by molecular beam scattering) by any two-molecular interaction. It is at present beyond the power of theoretical methods to predict the effect of the multi-molecular interactions on the critical parameters. This situation may change soon: we shall see below that multi-molecular interactions may affect the critical exponents of *lattice* models very significantly. Until this work can be extended to continuum models the best that we can hope for is to define an 'effective two-body interaction' and to examine how closely this can represent all the observed properties of a real substance. Surprisingly, this enables us to extend our ideas to metallic liquids, following the work of March[1] and others. Although quantum effects are always significant for electrons in a metal, it is nevertheless possible to think of the ions as surrounded by a 'sea' of electrons and as interacting classically with an 'effective two-ion interaction' of 'screened Coulomb' type. At large distance, however, the effective attraction, produced by the sea of electrons surrounding the ions., becomes oscillatory. Again it is difficult to predict what effect this should have on the critical exponents (which are difficult to measure for metals in any case).

For these reasons and because of possible mutual cancellation of errors, close agreement between 'calculated' and 'observed' curves is not quite the occasion for rejoicing that it is in other branches of physics! One exercise that has become popular recently has been to examine the consequences of an assumed purely two-body interaction first according to a particular theory being tested, secondly according to computer calculations of molecular dynamics or Monte Carlo type. These exercises are often described as 'computer experiments'.

From experience with lattice models we know that firm prediction of critical exponents needs very precise calculations indeed. We should need at least 10–12 virial coefficients of the gas to make a satisfactory extrapolation of the equation of state into the critical region. This is far more than we can ever hope to measure experimentally and no more than seven are available for any theoretical model at present. This rather unsatisfactory situation has prompted a fresh start. This is known as 'scaling theory'. Instead of beginning from gaseous densities and laboriously trying to extrapolate into the critical region, can we not find approximations that will be valid in the immediate neighbourhood of the critical point? This is indeed possible and good predictions of the critical exponents can be made, but at present only at the expense of fairly drastic approximations and assumptions whose effects are very hard to assess.

The solidification transition is certainly more sensitive to the finer details of the interaction, but again little is known theoretically about the possible effect of multi-molecular interactions. If we disregard complications due to this, then to the proximity of the triple point and to solid–solid transitions it seems that we can account for the solidification curve fairly well in terms of the repulsive part of the inter-molecular interaction.

11.3 The critical region. Metastable effects

The basic idea of van der Waals remains perfectly valid today. The critical phenomenon is caused by a balancing-out of the effects of short-range intermolecular repulsion and longer-range attraction. In fact, the famous a/V^2 term in the van der Waals equation has been shown by quite a number of workers to be the rigorous consequence of an attractive force that is both very weak and of extremely long range.[2] This result was suspected for a very long time before it was rigorously established.

One can make a simple theory of a lattice gas by transcription from the Weiss theory of ferromagnetism. An occupied cell in the vessel corresponds to an atomic magnet pointing along the magnetic field, an empty cell to a magnet pointing against the field. To make his theory of ferromagnetism Weiss simply described the interaction between magnets by adding to the true magnetic field an 'effective field' proportional to the intensity of magnetization, that is, to the excess of positive over negative magnets. Whether they are near to or distant from the point in question makes no difference: *all* the magnets in the specimen have the same chance of contributing to the 'effective field' at a given point. Thus any 'effective field' type of theory is the consequence of assuming a long-range interaction. This result is elementary for a lattice model, but is much more difficult to establish rigorously for a continuum-type model.[2]

It is easy to calculate the critical exponents for the van der Waals model. The critical temperature is located by the condition that $\partial P/\partial V$ and $\partial^2 P/\partial V^2$ both vanish, and it is elementary to deduce that, for example, $|T_c - T|$ is proportional to the square of $|V - V_c|$ whether we are talking about the liquid or vapour branches of the equation. The assumption of van der Waals that the equation of state of a gas of non-attracting rigid spheres would be simply $P = NkT/(V - b)$ could now be improved upon considerably, since the rigid-sphere equation of state is now known very accurately (e.g. from computer studies using molecular dynamics[11]). At high densities the variation of pressure is more like $(V - b)^{-3}$ than $(V - b)^{-1}$. However, it can easily be shown that making this change would not affect the critical *exponents* at all as long as we describe the intermolecular attractions by the long-range term a/V^2.

We repeat that van der Waals' basic idea is entirely correct. But how would the theory be modified if we replaced the long-range attraction by something more realistic physically? In particular, we have to try to answer two crucial questions:

(*a*) What would be the predicted critical exponents?

(*b*) What interpretation can we give to the metastable states (superheated liquid and supercooled vapour) which seem to be predicted quite naturally by van der Waals and which *are* experimentally observed? (They are respectively the basic mechanisms of the bubble-chamber and the cloud-chamber!) There has been considerable discussion and controversy about this, because a completely accurate statistical mechanical treatment of *any* assembly at equilibrium should predict no metastability of any kind! If we refer to the standard derivation of pressure in statistical mechanics, we arrive at it by listing all the micro-configurations of the assembly, calculating for each the quantity $\partial E_r/\partial V$ where E_r is the total energy (kinetic and potential) of the r-th micro-configuration and V is the volume of the vessel and then weighting each micro-configuration with the probability of its occurrence, proportional to the Boltzmann factor $\exp(-E_r/kT)$. This reasoning leads to the standard formula for the pressure as the average of $\partial Er/\partial V$ over all the micro-configurations open to the assembly, that is

$$P = -\left(\frac{\partial E_r}{\partial V}\right)_{\text{Av}} = \frac{\sum_r -(\partial E_r/\partial V)\exp(-E_r/kT)}{\sum_r \exp(-E_r/kT)} = -\frac{kT\partial \log f}{\partial V}. \tag{11.3.1}$$

Plainly, the pressure *must* be a one-valued function of volume and temperature if the averaging is properly carried out. The paradox has been resolved for a van der Waals type attraction, but has to be considered anew for attractions of finite range.

In the van der Waals case, we are assuming interactions that are small, of strength inversely proportional to the size of the assembly, and of long range. What happens when we go to the limiting case of a large assembly – the only one for which mathematical infinities and discontinuities can occur? Several limiting processes are involved in discussing such an assembly and it is not immediately obvious in what order we are performing them when we make the calculation. Which comes first, making the range of the interaction long, making its strength small, or making the assembly large? In fact, careful investigations have shown[2] that a correct calculation *would* lead to a one-valued pressure without metastable portions, that is to liquid and vapour branches connected by a horizontal 'tie-line'. But what happens for a real interaction? How do we interpret the metastable effects that are undoubtedly observed?

We meet the same problems with the solidification transition. Very considerable super-cooling of liquids is very often observed, for example water can remain liquid down to –40 °C under proper conditions and many substances can remain in the glassy state indefinitely. It is less easy to superheat a solid (or, having done so, to obtain positive experimental evidence that superheating has in fact occurred). Again, we are faced with the difficulty that a true 'equilibrium' statistical mechanical calculation, that is a true average over all possible micro-configurations of the assembly as given by equation (11.3.1), *must* necessarily produce a single value for the pressure under all conditions rather than a van der Waals loop. We shall return to this question below.

11.4 Towards better critical exponents

Our objective is to show that better critical exponents are a correct consequence of a more realistic assumption for the attractions between molecules. For example a Lennard–Jones-type interaction conforms with the very general quantum-mechanical prediction that the interaction energy of real molecules (without net charge or permanent dipole moment) should be inverse-sixth power at large distances. For a wide variety of real substances, the observation is $|V_c - V| \propto (T_c - T)^{1/3}$ whereas van der Waals predicts an index of ½. There is very little doubt that the correct consequences of an inverse-sixth power law of attraction are critical indices much nearer to the observed ones, but at present the evidence for this is rather indirect. It is already possible to make Monte Carlo or molecular dynamics studies of assemblies with realistic interactions (such as Lennard–Jones or attracting rigid spheres),[11] but neither these calculations nor treatments using integral equations have yet reached the degree of refinement that would enable firm predictions of critical indices to be made.

The indirect evidence is of two distinct types, from lattice models and from scaling theory. The *modus operandi* of scaling, or renormalization group theory to give it its alternative name, is to look for relations between the partition functions of two assemblies that resemble one another closely enough for us to be reasonably confident that, in the neighbourhood of the critical point, the density and entropy vary in the two assemblies with pressure and temperature in similar ways. Since we are mainly interested in the immediate neighbourhood of the critical region, it is customary to express these variations by means of power-law relationships of the type $|V - V_c| \propto$ some power of $|P - P_c|$.

The relations sought for have taken various forms; for example, in a continiuum model we might change an elementary system from a single molecule to an aggregate of several molecules at the same time averaging over an appropriate region of phase space in order to make the two problems physically comparable. In a lattice model we might average over all permissible configurations of the atoms on alternate lattice sites, thus obtaining a related problem for atoms on the remaining sites with new, 'effective interactions' mathematically related to the old ones. To be more specific, suppose that atoms A and B and C are each interacting directly, through pairwise forces, with another atom D, and with one another. Now group together in the partition function all these terms corresponding to the same positions, velocities and quantum states of the atoms A, B, and C, but sum over all possible values of these variables relating to D. We shall arrive at an 'effective partition function' for atoms A, B and C only, with 'effective interactions' different from their actual interactions to allow for the indirect effects of atom D. The fundamental difficulty facing this sort of approach is that the effective interaction in a case like the above may involve the positions, velocities and quantum states of atoms A, B and C jointly, that is *the original assembly may have only pair-wise interactions but the transformed assembly may involve multi-atom interactions.* In only a few instances, involving lattice type models, can an assembly be *exactly* related to another assembly of exactly the same kind. The relations obtained are therefore approximate.

We now look for the so-called 'fixed points' of our transformation, that is to say we ask in what circumstances the original and transformed assemblies have the same interaction parameters. If the pressure, temperature etc. are reasonably close to the observed critical values, we may assume further that the assembly that transforms approximately into itself is a reasonable model of the real critical assembly. With the further assumption, justified by experiment, that small departures from critical conditions can be described by power laws, we can obtain relations between the exponents by using the condition that the critical assembly is transformed approximately into itself.

This comparatively recent approach to the problem of the critical region already has a very extensive literature, to which we must refer the reader.[3] A very early example of this type of reasoning may be helpful. In their study of the two-dimensional ferromagnetic Ising problem Kramers and Wannier[4] found an *exact* relationship between the partition functions at a high and at a related low temperature – the famous dual transformation. Making the natural assumption (afterwards verified by Onsager's exact solution) that there was only one transition for such a simple model, they identified the critical temperature as the fixed point of their transformation. They were then able to use the fact that the critical assembly transforms into itself to make various rigorous deductions about the analytic form of the transition. (For example, they were able to exclude the possibility that the assembly has, as functions of temperature, a continuous energy and a jump in the specific heat, which is the behaviour of the Weiss model of ferromagnetism.)

This instance was exceptional because the partition function could be exactly transformed into another one. Why can we, nevertheless, have reasonable confidence in this type of argument even when the transformations used are only approximate? Earlier work on phase transitions had seemed to prove, over and over again, that the consequences of quite plausible approximations could be grossly misleading if we tried to push them to the point of actually predicting critical temperatures and critical exponents. For example, all the early treatments of the nearest-neighbour ferromagnetic problem predicted jump discontinuities in the specific heat, whereas the correct behaviour in two dimensions is certainly a logarithmic infinity[5] and in three dimensions is probably a $|T - T_c|^{1/8}$ type of law. To put it crudely, an *infinite* exponent was predicted, when the true one is small or zero! Why should the new type of theory, which certainly involves drastic approximations, be expected to fare any better?

This criticism is of crucial importance and we must try to deal with it properly. It has long been known, for example because of the phenomenon of critical opalescence, that 'domains' must exist in a system near critical, of effective size at least comparable with the wave-length of light, that is, of the order of hundreds of interatomic distances. If we represent a liquid by means of a lattice model, exact calculations on the latter show indeed that correlations between atoms extend far beyond those in neighbouring cells, out to distances of the order of hundreds of cell diameters. Indeed, one definition of the appearance of a two-phase region is that a suitably defined 'correlation length' is approaching infinity.

If the atoms are so strongly correlated that, for example, local density remains practically uniform out to distances of this order, we can further argue that 'local' properties like lattice structure, shape of molecule and the like should have little influence on the critical behaviour, because the effect of local variations will be very largely smoothed out. This is the postulate of 'universality' in scaling theory. This leaves parameters like the number of spatial dimensions, the number of low-lying quantum states of each atom and the type of interaction function (is it strictly pairwise or is it many-body?) as those that *are* expected to have major influence on the critical exponents.

This argument also helps us to understand why the critical properties of a liquid are apparently represented quite well by lattice models. It may seem crude to try to represent the effect of repulsive forces by dividing our vessel into cells and distinguishing only between occupied and empty cells. Recall the van der Waals result $V_c = 3b$, where b is four times the actual volume of the molecules. This means that, at critical, the mean intermolecular spacing is around twice the gas-kinetic diameter, so close encounters are comparatively rare and the finer details of the repulsive part of the interatomic field are of little importance. Since, as we have just said, the correlation length is large compared with the interatomic spacing, we are concerned with averages over the attractive part of the interatomic field, rather than with its finer details.

11.5 Lattice models of the critical region

Lattice models throw further light on several important points. In the first place, as was shown by Yang and Lee,[6] it is possible to transcribe any magnetic model involving $S = ½$ into a cell model of liquid. If we identify down spin with an empty cell and up spin with an occupied cell we can relate intensity of magnetization to relative numbers of occupied and empty cells, that is to density of the gas or liquid, and analogous relations exist for all other thermodynamic quantities: magnetic field is related to pressure, susceptibility to compressibility and so on. The number of micro-configurations, and therefore the entropy, will be the same in both problems. In two dimensions with nearest-neighbour interactions, precise calculations can be made and we can, for example, calculate correlations[7] between spins in the magnetic problem, and hence between cells in the liquid problem, as a function of distance. We obtain a qualitative picture of the phase-transition of a real liquid and confirmation of the statements made above that in the critical region correlations extend over distances much greater than the cell dimensions.

We can make similar calculations for three-dimensional models. No closed results are available yet, but series expansion methods, in the hands of Domb and others,[8] have now been carried to a stage at which, starting from low densities, quite reliable extrapolations into the critical region can be made. We now summarize briefly the very interesting and important conclusions that can be drawn from this work:

(*a*) The critical exponents obtained for two- and three-dimensional models are quite different from the van der Waals or mean-field values. The three-dimensional values are closer to the mean field values than are the two-dimensional ones. The reader is referred to the review literature for numerical details.[3] Scaling theory suggests that in four or more dimensions we should be back again to mean-field values.

(*b*) The predicted exponents are close to the values observed[3] for liquids (and for suitable magnetic materials for which the assumption of interactions between neighbouring spins only is probably a good one).

(*c*) The data now available are equivalent to knowledge of around 15–20 virial coefficients of a liquid. It is often found that the first few terms of such an expansion are untypical, and that attempts to extrapolate on the basis of too short a series can be misleading. Therefore, we cannot expect to make realistic extrapolations into the critical region on the basis of the five or six coefficients which are all that are available for any realistic continuum liquid model.

(*d*) Very recent work on lattice models, both exact and using series expansions, has shown quite clearly that the presence of multi-body interactions can alter critical exponents quite significantly.[9] It is also known that it is impossible to represent all the observed properties of argon (solid, liquid and gas) by a two-atom interaction function. There is fairly conclusive evidence that appreciable three-atom interaction is involved and attempts have been made to estimate it quantum mechanically. It does seem likely that the small differences observed between the critical exponents of magnetic materials and liquids will ultimately be accounted for in this fashion.

To sum up, we can claim a rather happy situation, in spite of the fact that no realistic model showing critical behaviour in three dimensions has yet been solved analytically! The experimentally observed critical exponents of liquids and magnetic materials agree reasonably well, and such agreement is expected as a consequence of 'universality', that is because correlation distances are large in the critical region. The three distinct approaches: analytically soluble models, series expansions and scaling theory give predictions that agree reasonably closely in cases that can be treated by two or three of the methods. Finally we have, at least in the present state of knowledge, the possibility of explaining away some of the remaining small discrepancies by invoking multi-body interactions!

11.6 The solidification transition

This seems to be a decidedly more difficult theoretical problem than evaporation. The solidification transition is between a disordered and a partly ordered assembly and it has always been a difficult problem to find a parameter that satisfactorily measures the 'degree of order' of a continuum type assembly. The lattice structure of the solid obviously has to be considered explicitly and no 'universality' hypothesis (at least in the form that it occurs in evaporation theory) is permissible. Moreover, although we can obtain considerable insight into solidification by appropriate transcriptions of magnetic models (this time those that show antiferromagnetism), we know of no simple analogue of van der Waals' equation that produces a simple 'working model' of solidification.

Despite these difficulties, we can trace a progressive improvement in our understanding of solidification as the years have gone by. We cannot yet make satisfactory predictions in the neighbourhood of the triple point, where individual terms in the virial expansion become numerically large and of mixed signs, nor can we predict (after assuming a given interaction function) which particular solid structure will *first* appear when a liquid freezes (though we can calculate the energies, and so predict which is the most stable lattice structure at very low temperatures). We can, however, claim a reasonably accurate understanding of the solidification curve at high temperatures and pressures.

The first step was due to Lindemann. Although it is little more than a dimensional argument, that a solid may be expected to melt when the amplitude of oscillation of a typical molecule about its equilibrium position in the crystal lattice reached a certain fraction of the distance between neighbouring molecules, it is certainly correct in principle. We should not expect this fraction to be sensitive to the particular lattice structure. On this basis, we might expect a scaling law for melting. Many further attempts were made to gain insight into the exact mechanism by which the solid structure breaks down: for example, it was suggested that this could be related to the number of unoccupied sites in the crystal lattice or to the vanishing of the shear modulus of the lattice. (Strictly speaking, the *limit of stability* of the solid need not be the same as the *melting point.*)

The problem was approached from the liquid side by Kirkwood and Monroe in 1941.[10] They found that their, admittedly approximate, integral equation for the two-molecule distribution function of a liquid showed clear indications of a transition at a certain density, probably by the liquid-type distribution function solution being replaced by a spatially periodic solution. Born and Green later reached similar conclusions on the basis of slightly different approximations. The very interesting prediction was that this transition would occur even for a gas of rigid spheres with no mutual attractions at all. At that time almost the only available theories of phase transitions were of 'mean-field' type like van der Waals and the Weiss theory of ferromagnetism, in all of which it *seemed* that to get a sharp phase transition one needed an interaction of 'cooperative' type, whose effect became progressively stronger as the transition progressed. For example the a/V^2 term in van der Waals' theory is of this type, the correction to the pressure becoming progressively more important as the density increases. There was, therefore, considerable difficulty about accepting this conclusion until it was finally confirmed in 1957[11] by machine calculations. (Until then it was not possible to exclude the possibility that the predicted transition was a spurious consequence of the approximations that have to be made in the integral equation approach.)

Actually, a great deal of experimental work has been done on the behaviour of powders and of aggregates of spheres. It has invariably been found that if, with suitable precautions, a vessel is randomly filled with spheres, the end result is *not* a regular close-packed aggregate but a random structure of about 15% lower density, very similar to the observed structure of a liquid.[11] Under the combined influence of shaking and external pressure, an ordered solid-like structure does appear. In thermodynamic language if, say, 30 neighbouring spheres form themselves into an ordered close-packed structure, there is an increase in order and a local loss of entropy. However, there is also a saving of space, which becomes available to the remaining spheres in the assembly, leading to a gain in entropy. The transition is associated with a balance between these two effects.

The machine calculations were performed independently on aggregates of about 100 spheres, by Wood and Jacobson[11] using a Monte Carlo approach to the calculation of the equation of state, the spheres being moved in turn in a random fashion, and by Alder and Wainwright[11] using a molecular dynamics method, that is to say actually solving the equations of motion of the spheres and allowing for collisions. Both of these approaches predicted a first-order type of transition, involving a finite jump in the density of about 10%, occurring at about 0.6 of close-packed density. This entirely confirmed the original conclusion that a transition is to be expected for the geometrical reasons discussed above. The first seven virial coefficients are now known for the 'rigid sphere gas', but these are not enough by themselves to enable a firm inference about the transition that some of the higher ones may become negative. Some of the higher virial coefficients in lattice models are definitely known to be negative.

If we accept the above conclusion that the main reason for the rigid sphere gas transition is geometrical, we can quite easily deduce an approximate scaling law for melting, relating the melting curve to the law of repulsion between the molecules. We neglect altogether the effect of the attractive forces, which means that our law will only be a limiting one (approached at temperatures large compared with $1/k$ times the depth of the 'bowl' of the intermolecular interaction function).

Suppose the law of force between the molecules is a pure repulsion of the form Ar^{-n}. Then, as collisions become more vigorous with rising temperature, the centres of the two molecules in a typical collision can approach more closely. We might expect such an assembly to behave like rigid spheres of diameter equal to the distance of closest approach, which will be proportional to $T^{-1/n}$, that is, we have an 'effective volume' proportional to $T^{-3/n}$. Now, the result of the machine calculations is that the transition of the rigid sphere gas occurs at a pressure given by $PV_0 \approx 6NkT$ where V_0 is the close-packed volume. Thus, we infer for soft spheres whose effective volume varies with temperature a law of the type

$$P + P_0 = CT^{1+3/n} \tag{11.6.1}$$

where the term P_0 is similar to the van der Waals term a/V^2 and is introduced to allow, to a first approximation, for the effect of the attractive forces. This is exactly the type of law found empirically by Simon. We can go further:for rare gases the effective value of n is of the order of 12–14, leading to a temperature exponent of the order of 1.2–1.25 as is observed. If we use March's work[1] that a liquid metal can, for many purposes, be regarded as ions interacting with an effective interaction of 'screened Coulomb' type, we infer that we might get an effective temperature exponent as high as 4. (The actual determination of the experimental values of the temperature exponents can be complicated by solid–solid transitions which occur even in the rare-gas solids.)

It is remarkable that we have obtained such insight even in the absence of a precise analytic model of melting. This circumstance has prompted the study of lattice models. In the magnetic case, it has been realized by many workers that any interaction that favoured unlike magnets as nearest neighbours would, at low enough temperatures, produce an alternating arrangement of spins. The antiferromagnetic structure can be destroyed by the effect of thermal agitation (point A in Figure 11.6.1) since a random arrangement has a higher entropy, *or* by the effect of a magnetic field, which, if strong enough would line up all the magnets in one direction (point B in the Figure 11.6.1), or by some combination of these two. In other words, the antiferromagnetic state can only exist in a 'corner' of the H–T plane (Figure 11.6.1). We can transcribe this to a model of the liquid and solid by identifying, as before, down spins with empty lattice sites and up spins with occupied sites, the antiferromagnetic arrangement then corresponding to the ordered solid.

Considerable care is necessary in using series expansion methods in the antiferromagnetic problem because the series very often diverge before we reach the phase boundary shown in Figure 11.6.1, so the 'physically obvious' variables such as density or fugacity are not always the best ones to use. From the point of view of liquid state theory (in view of the result mentioned above that a gas of rigid spheres would have a transition) the most interesting case to study is that in which two molecules are completely prevented from being in neighbouring cells. This analogue of the rigid sphere case corresponds to the point B in Figure 11.6.1, in the limiting case in which both the interaction energy and the magnetic field become infinite.

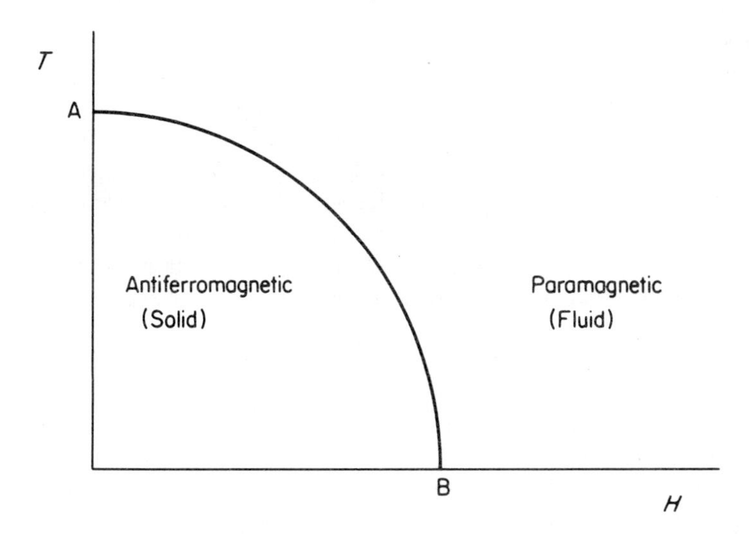

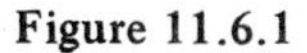
Figure 11.6.1

This model turned out to be analytically simpler than the general antiferromagnetic model, but it is still of great mathematical delicacy and several approximate approaches gave inconsistent and misleading results. Finally, Gaunt and Fisher[12] showed that the correct consequence of the model is an extremely weak transition, possibly just a point of inflection of P (expressed as a function of density), possibly a jump in some fairly high derivative of P with respect to ρ, but certainly nothing resembling the rigid sphere transition with its finite jump in density at the transition. Attempts have been made to study 'rigid sphere' models in which the interaction extends over several times the lattice distance[17] (which should resemble the continuum model more closely), but results are rather inconclusive. Temperley[13] has, however, shown that a certain rather pathological model, on a lattice of 'cactus' type, *does* show a first order transition rather like that for rigid spheres.

A clue to possible reasons for this mathematical delicacy of lattice models of solidification (quite unlike the universality that we get in the liquid–gas critical region) is provided by an important paper by Lieb and Heilmann.[14] They showed quite generally that no model of 'rigid-dimer' type could show a transition of any order (except possibly in the limiting case when the lattice is completely covered with dimers). In these models, a 'dimer' is an object that occupies just *two* neighbouring sites on the lattice, and dimers are not allowed to overlap. One might hope that such an assembly would itself provide a simple model of melting, but the result is that it cannot have a transition whatever the number of dimensions.

This model is important in the present context because it gives a clue to the unexpected weakness of the transition of the 'rigid sphere' model on the square lattice in which occupation of a site prevents occupation of any of the four neighbouring sites. Consider now the dimer model on the plane honeycomb lattice. If any bond of this is occupied by a dimer, no dimer may occupy any of the four bonds that intersect it. Consequently, we may convert this problem *exactly* into the 'nearest neighbours forbidden' model on the kagomé latticé (the covering lattice of the plane hexagonal lattice) and we must conclude that this model can have no transition either. Yet in the kagomé lattice each site has four nearest neighbours, just like the square lattice, so the weakness of the transition in the latter is understandable.

We conclude by mentioning the Gaussian model.[15] This is a continuum-type model in which the actual Boltzmann factor corresponding to interaction between two molecules at a distance r

apart is replaced by

$$1 - A\exp(-r^2/a^2)$$

where A is a constant between 0 and 1 and a is a distance measuring the effective range of the repulsion. In the limiting case $A = 1$, two molecules are prevented from coinciding, but have a finite probability of being at any finite distance apart. Ford and Uhlenbeck[13] studied this model in detail, because any assigned term in the virial series can be evaluated exactly, each of the cluster integrals of Mayer type that occur being expressible in terms of the 'complexity' of the corresponding graph (the complexity being defined as the number of Cayley trees that the graph contains).

Ford and Uhlenbeck were not able to settle whether or not this model has a transition, though they did find that for the case $A = 1$ and in three dimensions, the fourth and seventh virial coefficients are negative. The present author has found that another limiting case of this model *does* show an extremely weak transition. This limiting case is one in which A is small and a large, in such a way that Aa^d is finite where d is the number of dimensions. We use the theory given by Ford and Uhlenbeck. To calculate the virial series we only need irreducible graphs (i.e. those without articulation points), and we group them according to their cyclomatic number. We find, in terms of $x(=N/V$, the density) the result for the pressure, putting $A[\sqrt{(\pi)}a]^d = b$:

$$\frac{P}{kT} = x \text{ (perfect gas term)}$$

$$+\frac{bx^2}{2}\text{(graph }\bullet\!\!-\!\!-\!\!\bullet)$$

$$+\frac{A}{2}\left(\frac{2b^2x^3}{3.3^{d/2}} - \frac{3b^3x^4}{4.4^{d/2}} + \frac{4b^4x^5}{5.5^{d/2}}\cdots\right)\text{ (ring graphs)}$$

$$+O(A^2).$$

$$+\text{ (theta graphs of cyclomatic number two)} \tag{11.6.2}$$

For the limiting case of A small but a large so that b is finite, that is to say an extremely weak repulsion of very long range, the virial series reduces to two positive terms and can show *no* transition. For a repulsion that is weak, but now of finite range so that b remains finite, the next approximation is obtained from the coefficient of A. For $d = 2$ we can sum the expression for $\partial P/\partial x$ in finite terms to give

$$\frac{1}{kT}\left(\frac{\partial P}{\partial x}\right) = 1 + bx + \frac{A}{2b}\left(\frac{bx}{2} + \frac{1}{1+bx} - \frac{\ln(1+bx)}{bx}\right) + O(A^2).$$

Since the coefficient of A has no singularity on the positive real axis, we can continue it analytically for all positive values of x, and obtain an expansion in inverse powers of bx for $bx > 1$. It is easily shown that the third derivative of P vanishes for bx of the order of 4.5. This means that, to the first approximation in A, the pressure has a zero third derivative here. Thus we infer that the Gaussian model *has* a transition, if we grant that making A larger and keeping b constant, that is shortening the range of the interaction but strengthening its amplitude, is most unlikely to wipe out this transition but may be expected instead to make it stronger. (The higher powers of A involve 'lattice sums' that have not yet been evaluated in closed form.)

We have had to continue the virial series analytically beyond its radius of convergence to reach the transition density. Exactly similar behaviour is found with some antiferromagnetic lattice models; the closest singularity of the virial series is often on the negative real axis, but does not produce a transition.

11.7 Metastable effects

We must not assume that the van der Waals loop is just an artifact resulting from the difficulty of carrying out the limiting processes of making the range of interaction large, making its amplitude small and making the assembly large in the physically proper order. Other approximate equations of state also have van der Waals loops. For example, if we truncate a virial series after its third term, and use either the calculated or the observed variations of the second and third virial coefficients B and C with temperature, we can repeat van der Waals' argument, obtain a cubic equation for the volume, and get a theory that is qualitatively the same as that of van der Waals. Also the machine calculations on the rigid sphere model do show very pronounced hysteresis effects in the solidification transition, in the sense that it may take a time corresponding to hundreds of thousands of collisions before we get a transition from fluid to solid or *vice versa.* Therefore we want a theory that gives the correct result of a one-valued function for equilibrium pressure, but is nevertheless consistent with metastable effects with quite long lifetimes. This must involve an excursion into non-equilibrium statistical mechanics. In the absence of firm information about transition probabilities we shall make some 'informed guesses'.

If the true equilibrium isotherm for a gas just below critical, or for a liquid near the solidification density, is of the form of two disconnected portions connected by a horizontal tie-line, it is obviously not surprising that approximate expressions for it show van der Waals loops. In fact, there is a striking analogy with the problem of expressing a function with a discontinuous derivative as a Fourier series. If we truncate such a series after a few terms we nearly always get an overshoot, the celebrated Gibbs phenomenon. As we add more and more terms to the series, we introduce more and more maxima and minima, but it may take a great many terms to bring the deviations down to (say) 1%.

Something rather similar almost certainly happens for the virial series. The second and third terms each correspond to a single Mayer diagram so their variations with temperature are usually fairly simple, though, in general, not monotonic. Already with the fourth virial coefficient we have contributions from three distinct types of Mayer diagram, whose variations with temperature may well differ and whose signs are not the same even if we have a purely repulsive interaction. Indeed, many authors have called attention to this tendency for the Mayer diagrams to cancel one another. The fourth virial coefficient may, therefore, have quite a complicated variation with temperature.

We cannot at present hope to evaluate very high order virial coefficients for real assemblies, but it is well known that the contributions of *certain types* of diagram (in particular those which can be represented by planar graphs) can be evaluated and summed to all orders. A striking example of the type of thing that we are suggesting may occur is given by some recent work by Croxton.[16] He studies the solid–fluid transition for rigid spheres. In first approximation the Percus–Yevick virial expansion gives a very good representation of the equation of state for the fluid region, but it fails to predict the transition.

Croxton shows that the introduction of a class of diagrams not included in the Percus–Yevick virial expansion changes the situation dramatically. The predicted equation of state now consists of two branches connected by a van der Waals loop, the corresponding metastable regions agreeing well with those obtained from the machine calculations. If we were able to set up a *consistent* set of successive approximations, involving more and more types of diagram, we could expect the successive predictions to come closer and closer to a one-valued expression for pressure. Unfortunately, we do not yet have a prescription leading (in terms of diagrams) to a *consistent* process of successive approximation, except the grouping of diagrams that we have just used for the simple Gaussian model.

However, we can make the following reasonable assumption about the effect of taking in more and more terms of the virial series, that is of including diagrams involving more and more lines and points. This assumption is that, in non-equilibrium statistical mechanics, the diagrams associated with more and more points and lines are associated with close encounters between more and more atoms, that is to say with rarer and rarer processes involving longer and longer time constants. Such assumption would imply the desired conclusion, that a high degree of overshoot is associated with a shorter metastable lifetime that is a low degree of overshoot and we, by analogy with nucleation theory, might expect the time-constants associated with the high-order diagrams to increase very rapidly with the number of molecules involved. In this way we can hope to explain the existence of metastable effects associated with extremely long time constants, which are in fact, observed in association with both transitions. An instance of a process with a time-constant of the order of years is the 'devitrification' of glass, the metastable glassy state forms crystals after a lapse of years or centuries.

References

1. See, for example, Temperley, H. N. V., Rowlinson, J. S. and Rushbrooke, G. S. (Eds) (1968). *Physics of Simple Liquids*, North-Holland, Amsterdam.
2. Kac, M., Uhlenbeck, G. E. and Hemmer, P. C. (1963). *J. Math. Phys.*, **4**, 216, 229.
3. See, for example, Fisher, M. E. (1967). *Rep. Prog. Phys.*, **30** (II), 615.
   Kadanoff, L. P. *et al.* (1967) *Rev. Mod. Phys.*, **39**, 395.
4. Kramers, H. A. and Wannier, G. H. (1941). *Phys. Rev.*, **60**, 252.
5. Onsager, L. (1944). *Phys. Rev.*, **65**, 117.
6. Yang, C. N. and Lee, T. D. (1952). *Phys. Rev.*, **87**, 404, 410.
7. See, for example, Kaufmann, B. and Onsager, L. (1949). *Phys. Rev.*, **76**, 1232, 1244.
   Temperley, H. N. V. (1971). In C. Domb and M. S. Green (Eds), *Phase Transitions and Critical Phenomena*, Vol. I. Academic Press, New York. Chapter 6.
8. Domb, C. *et al.* (1973). In C. Domb and M. S. Green (Eds), *Phase Transitions and Critical Phenomena*, Vol. III. Academic Press, New York.
9. See, for example, C. Domb and M. S. Green (Eds) (1971). *Phase Transitions and Critical Phenomena*, Vol. I. Academic Press, New York. Chapter 8.
10. Kirkwood, J. G. and Monroe, E. (1941). *J. Chem. Phys.*, **9**, 514.
11. See Temperley, H. N. V., Rowlinson, J. S. and Rushbrooke, G. S. (Eds) (1968). *Physics of Simple Liquids,* North-Holland, Amsterdam. Chapters 4, 5 and 6.
12. Gaunt, O. G. and Fisher, M. E. (1965). *J. Chem. Phys.*, **43**, 2840.
13. Temperley, H. N.V. (1965). *Proc. Phys. Soc.*, **86**, 185.
14. Lieb, E. H. and Heilmann, O. J. (1972). *Comm. Math. Phys.*, **25**, 190.
15. Ford, G. W. and Uhlenbeck, G. E. (1962). *Studies in Statistical Mechanics,* Vol. I. North-Holland, Amsterdam. p. 123.
16. Croxton, C. A. (1974). *J. Phys.* C., **7**, 3723.
17. Runnels, K. (1972). In Domb, C. and M. S. Green (Eds), *Phase Transitions and Critical Phenomena,* Vol. II. Academic Press, New York. Chapter 8.

Chapter 12

Triplet Correlations

HAROLD J. RAVECHÉ and RAYMOND D. MOUNTAIN

National Bureau of Standards, Washington, D.C. 20234

12.1 Structure in fluids

(a) Origins

Experiments on the scattering of neutrons or X-rays with wavelengths of the order of 10 nm by a fluid reveal the existence of local molecular order. It is this local order which is referred to as the structure of fluids. Analysis of the interference in the radiation which is elastically scattered by individual atoms in a monatomic fluid shows that the positions of two arbitrary atoms are radially correlated in space, the observed diffraction patterns cannot be explained if the positions of the atoms in the fluid are taken to be completely random. Near the melting curve, there are large deviations from the completely random value which extend over distances comparable to several atomic dimensions. The variation of the correlation over this distance in the fluid has some features which can be directly related to the diffraction pattern obtained from a single crystal of the same chemical substance.

Diffraction measurements in polyatomic fluids also reveal a definite correlation between the positions of two molecules. The scattering data can, in these systems, be analysed either in terms of the distance between atoms in different molecules, or in terms of a radial centre of mass distance and angular variables.

Measurements of the elastically scattered radiation from fluids have been made for more than three decades (Schmidt and Tompson 1968; Larsson *et al.* 1968). These measurements form part of the core of what is known about the structure of fluids. All the data can be conveniently represented in terms of the function $g_2(\mathbf{r}_1, \mathbf{r}_2; \rho, T)$, which depends on the molecular positions, $\mathbf{r}_1$ and $\mathbf{r}_2$, and also on the density, ρ, and temperature, T, of the fluid. The function is well known as the pair correlation function. For a uniform monatomic fluid, the spatial dependence of g_2 is determined by the scalar distance $r_{12} = |\mathbf{r}_1 - \mathbf{r}_2|$. In a fluid of diatomic molecules the spatial dependence is specified by r_{12} and three angular variables.

The pair correlation function, however, tells us about only one aspect of the structure present in fluids at equilibrium. One can also ask about the structure manifested by the simultaneous correlation of triples, quadruples, pentuples, etc. of molecules. In particular, one seeks to understand whether the local order present in fluids is accurately approximated by assuming that, in an arbitrary set of molecules, only the positions between all possible pairs of molecules are correlated. That is, one asks whether the view of molecules being only pairwise correlated is a good model for the structure of fluids, or whether one must also consider the simultaneous correlation of three, of four, or of more molecules. This problem is known as the statistical many-body problem in the molecular theory of fluids. The problem is to determine the contribution of the higher order correlation functions to the thermal properties.

The phenomenon of the simultaneous correlation of the position of three arbitrary molecules in a fluid is the subject of this chapter. We will consider the structural information contained in the triplet correlation function, $g_3(r_{12}, r_{13}, r_{23}; \rho, T)$, and the contribution of this structure to the bulk equilibrium properties. Emphasis will be placed on classical monatomic fluids, but we will discuss correlations in the quantum fluid, ^{4}He. The phenomena we consider are almost exclusively high density phenomena. As we will later observe, the simultaneous correlation of three atoms is only appreciable at densities near to and higher than the triple point density. We will use the term fluid to describe a fluid system at these densities whether or not the temperature is above the critical temperature.

Analogous to the measurement of the pair correlation function from the scattering of radiation from single molecules, one can in principle also consider determining $g_3(r_{12}, r_{13}, r_{23}; \rho, T)$, and even higher order correlation functions, from multiple elastic scattering. However, these experiments are not, to date, feasible. A major obstacle is the problem of obtaining a beam of radiation with sufficient intensity so that the multiply scattered radiation is appreciable compared with the radiation which is scattered from only one molecule. Even if this were possible, one must then devise an algorithm to analyse binary, ternary and higher order scattering separately and accurately. Such experiments are a challenge of the future.

In spite of the difficulties with direct measurements of triplet correlations, it is in fact possible to study these phenomena experimentally from the density dependence of the pair correlation function. We will report later in Section 12.3 on results of these measurements.

In addition to this experimental probe, it is also possible to compute the $g_3(r_{12}, r_{13}, r_{23}; \rho, T)$ directly, and exactly, from the computer simulation of a fluid with a given potential energy function. Both the Monte Carlo and molecular dynamics techniques have been successfully applied to this problem. In Section 12.3 we will discuss the results of these computations.

It is also possible to study triplet effects theoretically by solving equations for the correlation functions. The equations which have been derived to date are cumbersome and the incorporation of triplet correlations requires elaborate numerical computations, but nevertheless results are available for the hard sphere and Lennard–Jones potentials. We will discuss these computations and also the incorporation of triplet correlations in the molecular theory of fluids in Section 12.2.

(b) Significance of the correlation functions

The importance of the correlation functions follows from the fact that, together with the intermolecular potential energy function, they determine all the equilibrium macroscopic properties of a classical fluid.

In the idealized case that the intermolecular potential energy of the fluid is composed of a sum of the potential energy between only pairs of molecules, the pair correlation function and the pair potential energy determine all the instantaneously measurable thermodynamic properties. The formulae for the pressure and energy will be given in the next section. Derivative properties like, the constant pressure and constant volume heat capacities, require knowledge of more complicated structural information. Even in the idealized case of a pairwise additive potential energy function, the constant volume heat capacity involves triplet and quadruplet correlations. As we will show, the entropy involves even higher order correlation functions.

If, in addition to pairwise additive terms, triplet and more complicated interactions are present in the intermolecular potential energy function, then the instantaneously measurable thermodynamic properties involve more detailed structural information than is contained in the pair function. A useful guide is that if n-body terms exist in the intermolecular potential energy, then the equilibrium thermal properties involve at least n-body correlation functions. The pressure and energy will involve only terms up to the n-body correlation function, but derivative properties will involve higher order correlation functions.

All of the comments we have made in this subsection apply strictly to classical fluids. When quantum mechanical effects are important, the kinetic energy introduces added complications. The correlation functions are in this case matrices and the off-diagonal elements can make important contributions to the properties of the fluid, especially at very low temperatures. It is only under a certain approximation for the wave function of the fluid at the absolute zero of temperature that the radial pair correlation function determines the average kinetic and potential energy.

In this chapter we will only briefly consider quantum fluids. The relation between the pair and triplet correlation functions which we use is an exact relation. We will report on the diagonal elements of the triplet correlation function in ^4He.

The chapter proceeds as follows. In Section 12.2 we introduce the reduced probability densities through the defining relation in terms of the intermolecular potential energy function. The correlation functions follow naturally from properties of the reduced probability densities. We give results for thermodynamic properties emphasizing the role of the triplet correlation function.

From the definition of the correlation functions in terms of certain spatial averages involving the potential energy function, it is possible to derive integral equations for the correlation functions. In Section 12.2 we will consider the inclusion of triplet correlations in known theories for the pair correlation function. The improvement over results obtained from considering only pair correlations will be discussed. The explicit analysis of the triplet correlation function itself is dealt with in Section 12.3. In this section we investigate three sources of information about the function: the configurational entropy, the density derivative of the pair correlation function and computer simulations. The latter two techniques are then used to assess the validity of closure approximations for the triplet correlation function.

12.2 Theory

(a) Definitions and thermodynamic properties

Consider a system of a single chemical species whose thermodynamic state is specified by the temperature, the chemical potential, μ, and the volume, V. For convenience we use the activity variable $Z = \exp(\beta\mu)/[h/(2\pi m k_B T)^{1/2}]^3$, where $\beta = (k_B T)^{-1}$ and k_B is the Boltzmann constant, instead of the chemical potential directly. Let the potential energy function for the set of N molecules in V be denoted by $U_N(\mathbf{r}_1, \ldots, \mathbf{r}_N)$, then the classical probability density for finding molecules at the positions $\mathbf{r}_1, \ldots, \mathbf{r}_N$ is,

$$P_N(\mathbf{r}_1, \ldots, \mathbf{r}_N) = \frac{Z^N}{N!} \frac{\exp(-\beta U_N)}{\Xi} \tag{12.2.1}$$

where $\ln \Xi = \beta PV$ with P the pressure of the system (for a detailed discussion see Hill 1956). The quantity Ξ is the grand canonical partition function and from the normalization,

$$\frac{1}{\Xi} \sum_{N \geqslant 0} \int \cdots \int \mathrm{d}^3 r_1 \ldots \mathrm{d}^3 r_N P_N = 1 \tag{12.2.2}$$

one has a relation for obtaining the pressure in terms of the potential energy. The integrals in (12.2.2) extend over all space and we assume, as is almost always done in practice, the existence of the thermodynamic limit.

Now (12.2.1) and (12.2.2) contain, in principle, all the equilibrium thermodynamic and structural information for a system of molecules whose motion in space is determined by classical mechanics. The problem lies in the implementation of a direct computation. For a macroscopic system at liquid densities, the average number,

$$\langle N \rangle = \sum_{N \geqslant 0} \int \cdots \int \mathrm{d}^3 r_1 \ldots \mathrm{d}^3 r_N P_N N \tag{12.2.3}$$

is typically of the order of 10^{20} or larger. At low densities, below the liquid–vapour condensation, the well known virial series, i.e. the activity expansion of (12.2.2), is useful in calculating bulk properties of the system (Hill 1956). At liquid and higher densities, this is no longer feasible. Therefore, it seems appropriate to consider simpler structural information and relate this to thermodynamic properties.

The reduced probability densities, $\rho_n(\mathbf{r}_1, \ldots, \mathbf{r}_n; Z, T)$, which determine the probability of finding molecules at the positions $\mathbf{r}_1, \ldots, \mathbf{r}_n$, independent of the positions of the remaining molecules in the system, are a logical alternative to (12.2.1). As their physical meaning suggests, the reduced probability are densities defined by averaging (12.2.1) over all but n of the particles in the system,

$$\rho_n(\mathbf{r}_1, \ldots, \mathbf{r}_n; Z, T) = \frac{1}{\Xi} \sum_{N \geqslant 0} \frac{Z^N}{(N-n)!} \int \cdots \int \mathrm{d}^3 r_{n+1} \ldots \mathrm{d}^3 r_N \tag{12.2.4}$$

$$\exp[-\beta U_N(\mathbf{r}_1, \ldots, \mathbf{r}_N)] .$$

Since the number density, $\rho = \langle N \rangle / V$, is a function of Z and T it is convenient to take the reduced probability densities as depending on $\rho = \rho(Z,T)$ and T. For a fluid at equilibrium, $\rho_1(\mathbf{r}_1) = \rho$. Since in the limit of infinite dilution

$$\rho_n \to \rho^n$$

it is natural to define the correlation functions, $G_n(\mathbf{r}_1, \ldots, \mathbf{r}_n)$, by

$$G_n(\mathbf{r}_1, \ldots, \mathbf{r}_n) = \frac{\rho_n(\mathbf{r}_1, \ldots, \mathbf{r}_n)}{\rho^n} . \tag{12.2.5}$$

Since we must also have, in a single phase, that in the limit that the sets n and m become infinitely far apart,

$$\rho_{m+n} \to \rho_n \rho_m$$

it is logical to decompose G_n into a product of functions. That is, we write,

$$G_n(\mathbf{r}_1, \ldots, \mathbf{r}_n) = \prod_{\text{all pairs}}\prod g_2 \prod_{\text{all triples}}\prod\prod g_3 \times \ldots \times g_n \tag{12.2.6}$$

where for any $\mathbf{r}_i$ and $\mathbf{r}_j$ in the set of positions $\mathbf{r}_1, \ldots, \mathbf{r}_n$,

$$\lim_{\mathbf{r}_i - \mathbf{r}_j \to \infty} g_n = 1. \tag{12.2.7}$$

All the correlation functions depend on the density and temperature, and their position dependence in a uniform monatomic fluid is specified by $|\mathbf{r}_i - \mathbf{r}_j| = r_{ij}$ for all different pairs of vectors in their argument. For convenience we use g_2 instead of G_2; the two functions are identical in a uniform fluid.

It is the set of functions $g_2, \ldots, g_n$ in (12.2.6) which nonredundantly contains the structural information of the system. Our concern in this chapter is the function $g_3(r_{12}, r_{13}, r_{23}; \rho, T)$; in particular, we are interested in the contribution of this function to bulk properties of the fluid and in determining the structural information which it contains.

Writing the potential energy function $U_N(\mathbf{r}_1, \ldots, \mathbf{r}_N)$ as a sum over the interactions between all different pairs in N, plus all different triples in N etc,

$$U_N(\mathbf{r}_1, \ldots, \mathbf{r}_N) = \sum_{1 \leqslant i < j \leqslant N}\sum u_2(r_{ij}) + \sum_{1 \leqslant i < j < k \leqslant N}\sum\sum u_3(r_{ij}, r_{ik}, r_{jk})$$
$$+ \ldots + u_N(r_{ij}, 1 \leqslant i < j \leqslant N) \tag{12.2.8}$$

enables one to represent the average potential energy as,

$$\langle U \rangle = \sum_{N \geqslant 0} \int \cdots \int \mathrm{d}^3r_1 \ldots \mathrm{d}^3r_N P_N U_N$$
$$= \frac{\rho^2}{2} \iint \mathrm{d}^3r_1\, \mathrm{d}^3r_2 g_2(r_{12}) u_2(r_{12}) \tag{12.2.9}$$
$$+ \frac{\rho^3}{6} \iiint \mathrm{d}^3r_1\, \mathrm{d}^3r_2\, \mathrm{d}^3r_3 G_3(r_{12}, r_{13}, r_{23}) u_3(r_{12}, r_{13}, r_{23}) + \ldots .$$

Now, for a pairwise additive system, $u_q = 0, q \geqslant 3$, and

$$\langle U \rangle = \frac{\rho^2}{2} \iint \mathrm{d}^3r_1\, \mathrm{d}^3r_2\, g_2(r_{12}) u_2(r_{12}). \tag{12.2.10}$$

The pressure can be calculated from the virial theorem of Clausius (Hill 1956) which relates the average value of the virial, $\mathbf{r} \cdot \nabla_r U_N$, to the pressure. The result is known as the virial equation of state, which in terms of the correlation functions is,

$$P - \rho\beta^{-1} = -\frac{\rho^2}{6} \int \mathrm{d}^3r_2\, r_{12} \frac{\mathrm{d}u_2}{\mathrm{d}r_{12}} (r_{12}) g_2(r_{12})$$
$$- \frac{\rho^3}{18} \iint \mathrm{d}^3r_2\, \mathrm{d}^3r_3 \left(r_{12} \frac{\mathrm{d}u_3}{\mathrm{d}r_{12}} + r_{13} \frac{\mathrm{d}u_3}{\mathrm{d}r_{13}} + r_{23} \frac{\mathrm{d}u_3}{\mathrm{d}r_{13}} \right)$$
$$\times G_3(r_{12}, r_{13}, r_{23}) - \ldots . \tag{12.2.11}$$

For a pairwise additive system only the first term on the right exists.

The pressure can also be computed from

$$\beta^{-1}\left(\frac{\partial \rho}{\partial P}\right)_T = 1 + \rho \quad d^3r_2\, h(r_{12}) \tag{12.2.12}$$

where

$$h(r_{12}) = g_2(r_{12}) - 1. \tag{12.2.13}$$

The relation (12.2.12) is called the compressibility equation of state, and to compute it one must know h, and therefore g_2 as a function of density. The compressibility equation of state does not directly involve the potential energy function.

Analogous to this result it is also possible to write the entropy directly in terms of the correlation functions. This is achieved by inverting (12.2.4) so that the potential energy can be written as a sum over functionals involving only the correlation functions. The entropy expression was derived by Nettleton and Green (1958) using graphical techniques and by Raveché (1971) using a generating function involving isothermal activity derivatives of correlation functionals. The result for the configurational entropy to terms up to and including the triplet correlation function is

$$\begin{aligned}\frac{\Delta S}{\langle N\rangle k_B} = &-\frac{1}{2}\rho \int d^3r_2 g_2(r_{12}) \ln g_2(r_{12}) + \frac{\rho}{2}\int d^3r_2[g_2(r_{12}) - 1]\\ &-\frac{1}{6}\rho^2 \iint d^3r_2\, d^3r_3\, G_3(r_{12}, r_{13}, r_{23}) \ln g_3(r_{12}, r_{13}, r_{23})\\ &+\frac{\rho}{2}\iint d^3r_2 [g_2(r_{12}) - 1] + \frac{\rho^2}{6}\iint d^3r_2\, d^3r_3 [G_3(r_{12}, r_{13}, r_{23})\\ &- g_2(r_{12})g_2(r_{13}) - g_2(r_{12})g_2(r_{23})\\ &- g_2(r_{13})g_2(r_{23}) + g_2(r_{12}) + g_2(r_{13}) + g_2(r_{23}) - 1].\end{aligned} \tag{12.2.14}$$

Terms involving four, five, . . . etc., particle functions have a similar structure. Since this expression does not involve the potential energy function explicitly, it is possible to make a direct assessment of the contribution of pair and triplet correlation functions separately without resort to any approximation for the potential energy function. To do this one needs the pair and triplet terms from an experimental measurement. Using the fact that one of the triplet terms in (12.2.14) is related to an activity derivative (Raveché 1971), it is possible to compute all the terms (12.2.14) except the term in ln g_3 from the measurement of g_2 and its density derivative. We will report on entropy computations for liquid neon, hard spheres and liquid rubidium in Section 12.3.

The constant volume heat capacity, C_V, is given in a monatomic pairwise additive system by Mayer (1962) and by Schofield (1966),

$$\begin{aligned}C_V - \frac{3}{2}k_B = \frac{\beta}{T}\Bigg(&\frac{1}{2}\rho \int d^3r_2 g_2(r_{12})[u_2(r_{12})]^2\\ &+ \rho^2 \iint d^3r_2\, d^3r_3\, G_3(r_{12}, r_{13}, r_{23}) u_2(r_{12}) u_2(r_{13})\\ +\frac{\rho^3}{4}\iiint &d^3r_2\, d^3r_3\, d^3r_4 [(G_4(r_{12}, r_{13}, r_{14}, r_{23}, r_{24}, r_{34}) - g_2(r_{14})g_2(r_{24}) u_2(r_{14}) u_2(r_{24})]\end{aligned} \tag{12.2.15}$$

$$\left. \frac{\{\rho\int d^3r_2 g_2(r_{12})u_2(r_{12}) + \tfrac{1}{2}\rho^2\int\int d^3r_2\, d^3r_3\, [G_3(r_{12}, r_{13}, r_{23}) - g_2(r_{12})]\, u_2(r_{12})\}^2}{1 + \rho\int d^3r_2\, [g_2(r_{12}) - 1]} \right)$$

The origin of the higher order correlation functions in the heat capacity can be appreciated if one recalls that C_V is the isochoric temperature derivative of the average energy. Now from (12.2.10) and the fact that potential energy functions must be independent of density and temperature, we note that $(\partial g_2(r_{12})/\partial T)_\rho$ determines C_V. For pairwise additive potentials one has (Schofield 1966),

$$\begin{aligned}\frac{T}{\beta}\left(\frac{\partial g_2(r_{12})}{\partial T}\right)_\rho &= u_2(r_{12})g_2(r_{12}) + \rho\int d^3r_3\{G_3(r_{12}, r_{13}, r_{23})[u_2(r_{13}) + u_2(r_{23})]\} \\ &+ \frac{\rho^2}{2}\int\int d^3r_3\, d^3r_4\, [G_4(r_{12}, r_{13}, r_{14}, r_{23}, r_{24}, r_{34}) - g_2(r_{12})g_2(r_{34})]\, u_2(r_{34}) \\ &- \left(\frac{\partial\rho^2 g_2(r_{12})}{\partial\rho}\right)_T \left(\int d^3r_2 g_2(r_{12})u_2(r_{12})\right. \\ &+ \left.\frac{\rho}{2}\int\int d^3r_2\, d^3r_3\, [G_3(r_{12}, r_{23}, r_{13}) - g_2(r_{12})\right) u_2(r_{12}).\end{aligned} \tag{12.2.16}$$

Because of our initial choice of activity and temperature as the independent intensive variables, this result involves the transformation from derivatives at constant activity to derivatives at constant density. The isothermal density derivative of the pair correlation function also involves a higher order correlation function (Yvon 1937, Buff and Brout 1955), with $\kappa_T = \rho^{-1}(\partial\rho/\partial P)_\beta$ denoting the isothermal compressibility,

$$\begin{aligned}\left(\frac{\partial\rho^2 g_2(r_{12})}{\partial\rho}\right)_T &= \frac{\beta}{\rho^2\kappa_T}\left(2\rho^2 g_2(r_{12}) + \rho^3 g_2(r_{12})\right. \\ &\times \left.\int d^3r_3\, [g_2(r_{13})g_2(r_{23})g_3(r_{12}, r_{13}, r_{23}) - 1]\right).\end{aligned} \tag{12.2.17}$$

These results illustrate that the specification of the density and temperature dependence of the pair correlation function involves the triplet and quadruplet correlation functions. Since the pair function can be measured accurately, this suggests the possibility of probing higher order correlation functions through measurements of thermodynamic derivatives of g_2. Because of the complexity of (12.2.16) and the fact that the pair potential energy enters, the logical investigation is a probe of the triplet correlation function through the isothermal density derivative of the pair correlation function. One should note that (12.2.17) cannot be used as an integral equation to obtain g_3 because, as an equation for g_3, it is underdetermined (Raveché and Mountain 1970). However, it is possible to use (12.2.17) to show that in real fluids $g_3(r_{12}, r_{13}, r_{23})$ has a nonconstant amplitude which oscillates about unity. We will give these results in Section 12.3.

From the above relations we see that the correlation functions play a key role in determining the thermodynamic properties. The formulae establish the connection between local molecular structure and bulk properties.

We will in this chapter focus almost exclusively on systems with only pairwise additive potential energy functions, the exception being our results for neon and helium. It is known (for references see Present 1971) that pairwise additivity is only an idealization even for simple fluids of rare gas atoms. Barker *et al.* (1968, 1969, 1971) have pioneered in showing that when the triple dipole dispersion forces obtained by Axilrod and Teller (1943) are included, it is possible to find a pair potential energy such that with these and g_2 and g_3, the properties of liquid argon are well determined. It is important to note that if triplet terms were present in the potential energy, $\ln g_3$ would contain a density and temperature independent component which would be precisely the triplet potential energy in units of $k_B T$.

Our goal is to study general features of triplet correlation functions. We are not directly concerned with those features which arise solely from triplet terms in the potential energy function.

(b) Integral equations

It is convenient to introduce the so-called direct correlation function, $c(r_{12})$, which is related to $h_2(r_{12})$ by

$$h(r_{12}) = c(r_{12}) + \rho \int d^3 r_3 c(r_{13}) h(r_{23}) . \tag{12.2.18}$$

From the defining relation for the pair correlation, (12.2.4), it is possible to obtain an expansion of $g_2(r_{12})$ in terms of $c(r_{12})$ and the potential energy function (Percus 1964, Stell 1964). For the case of pairwise additive potentials only the factor $\exp(-\beta u_2)$ enters the relation. In general, it is not possible to sum all the graphs which enter the relation between $h(r_{12})$ and $c(r_{12})$, but there are two approximate summations which have led to useful results in the molecular theory of fluids. These are the Percus–Yevick (PY) (Percus and Yevick 1958) approximation and the hypernetted chain approximation (HNC) (van Leeuwen *et al.* 1959, Morita and Hiroike 1960, Green 1960, Rushbrooke 1960, Verlet 1960). In terms of

$$g_2^*(r_{12}) = \exp[\beta u_2(r_{12})] g_2(r_{12}) \tag{12.2.19}$$

and the Mayer function

$$f(r_{12}) = \exp[-\beta u_2(r_{12})] - 1 \tag{12.2.20}$$

the PY result is

$$c(r_{12}) = f(r_{12}) g_2^*(r_{12}) \tag{12.2.21}$$

and the HNC result is

$$c(r_{12}) = f(r_{12}) g_2^*(r_{12}) + g_2^*(r_{12}) - 1 - \ln g_2^*(r_{12}) . \tag{12.2.22}$$

These results, together with the original definition (12.2.18) of $c(r_{12})$, lead to integral equations for the pair correlation function in terms of the pair potential energy. Both approximations have been extensively studied for a variety of model potential energy functions. For hard spheres the PY equation is superior to the HNC equation, although the latter is graphically more complete. For Lennard–Jones type potentials, the PY equation generally is more accurate at higher temperatures, but there is evidence that below the critical temperature the HNC equation is more accurate. The numerical results for both the PY and HNC equations have been reviewed in detail (for a recent bibliography see Munster 1974).

We are concerned with the consequences of including triplet correlations in the PY and HNC theories. Extensions of these theories have been given by Verlet (1964, 1965) which we denote by PY II(V) and HNC II(V), and by Wertheim (1967), which we denote by PY II(W) and HNC II(W). Baxter (1968) has also proposed a generalization of the PY and HNC methods but he has obtained numerical results for only the nearest-neighbour lattice gas.

Wertheim's procedure is more general than Verlet's in that it allows one in principle to incorporate contributions from not only triplet but also higher order correlations into the PY and HNC equations. However, both methods lead to cumbersome equations. Although we will report on results of both techniques, we will for the sake of brevity give only the formulae for Verlet's method and we refer to the original literature for Wertheim's formulae. The PY II(V) approximation differs from (12.2.19) in an additive linear functional of g_3, that is,

$$c(r_{12}) = f(r_{12})g_2^*(r_{12}) + \Phi(r_{12}) \tag{12.2.23}$$

where, using δ for the Dirac delta function,

$$\begin{aligned}\Phi(r_{12}) = \frac{\rho^2}{2}\iint d^3r_3\, d^3r_4\, d^3r_5\, [c(r_{13})c(r_{14})\{G_3(r_{34}, r_{35}, r_{45}) - g_2(r_{34})\} \\ \times [\delta(r_{25}) - \rho c(r_{25})] \\ - \frac{\rho^2}{2}\iint d^3r_3\, d^3r_4 c(r_{13})c(r_{14})g_2(r_{34})[c(r_{23}) + c(r_{24})]\end{aligned} \tag{12.2.24}$$

and the HNC II(V) differs from (13.2.22) by

$$\begin{aligned}c(r_{12}) = f(r_{12})g_2^*(r_{12}) + g_2^*(r_{12}) - 1 - \ln g_2^*(r_{12}) + \Phi(r_{12}) \\ - \tfrac{1}{2}(h(r_{12}) - c(r_{12}))^2\end{aligned} \tag{12.2.25}$$

For a more detailed discussion see Rushbrooke (1965, 1968). Since the triplet correlation function appears in $\Phi(r_{12})$ the theory is not self-contained. A closure approximation was proposed which assumes,

$$g_3(r_{12}, r_{23}, r_{13}) = 1 + \rho\int d^3r_4 c(r_{34})\left(\frac{G_3(r_{12}, r_{23}, r_{13})}{g_2(r_{12})} - g_2(r_{14}) - g_2(r_{24}) + 1\right) \tag{12.2.26}$$

and this gives

$$\Phi(r_{12}) = \frac{\rho^2}{2}\iint d^3r_3\, d^3r_4 c(r_{23})c(r_{24})g_2(r_{34})h(r_{14})h(r_{34}). \tag{12.2.27}$$

With this, (12.2.23) can be solved numerically, and computations for both hard sphere systems (Verlet 1965) and the Lennard–Jones potential (Levesque 1966, Verlet 1966) have been reported.

Wertheim's procedure introduces the direct triplet correlation function, $c_3(r_{12}, r_{13}, r_{23})$ and effects a closure which involves the isothermal derivative,

$$\frac{\partial c}{\partial \rho}(r_{12}) = \int d^3r_3 c_3(r_{12}, r_{13}, r_{23}).$$

The resulting equations for PY II(W) and HNC II(W) therefore involve the density derivative, and the solution of the equation requires an integration along an isotherm.

The HNC and PY theories give the second and third virial coefficients exactly, but not the fourth; the PY II and HNC II equations give the fourth exactly but not the fifth. For hard spheres the following results are obtained for the fifth virial coefficient (in units of $2\pi\sigma^3/3$

where σ is the hard sphere diameter) from the compressibility equation of state,

$$\begin{aligned} \text{PY II(V)} &= 0.107 \\ \text{HNC II(V)} &= 0.123 \\ \text{PY II(W)} &= 0.111 \\ \text{HNC II(W)} &= 0.107 \\ \text{PY} &= 0.121 \\ \text{HNC} &= 0.049\,. \end{aligned}$$

The exact result obtained by Ree and Hoover (1964) is 0.1103 ± 0.0003 and by Katsura and Abe (1963) is 0.1097 ± 0.0008. The results for PY II(V) and HNC II(V) agree with results reported by Rushbrooke (1965). Oden *et al.* (1966) computed values for the fifth virial coefficient from PY II(V) and HNC II(V) using the virial equation of state, (12.2.11). Their results are,

$$\begin{aligned} \text{PY II(V)} &= 0.124 \\ \text{HNC II(V)} &= 0.066\,. \end{aligned}$$

Rowlinson (1968) reports the same result for PY II(W) as

$$\text{PY II(W)} = 0.110$$

which indicates that Wertheim's procedure is more selfconsistent than Verlet's. The Lennard–Jones results reported by Rowlinson (1968) show that when compared to the exact results of Barker *et al.* (1966), PY II(V) is more accurate than PY II(W) at lower temperatures.

It is difficult to assess the validity of a theory for dense fluids by considering only the virial coefficients in the density expansion for the equation of state. Unfortunately, numerical results for Wertheim's method are limited, but there are additional numerical results for Verlet's extension of the PY approximation. The results show that the critical constants for the Lennard–Jones system (Levesque 1966) are predicted with an accuracy which is substantially improved over the original PY results. The pair correlation function obtained from PY II(V) for both the hard spheres and Lennard–Jones models (Verlet and Levesque 1967) at moderate densities also indicates an improvement over the original PY approximation. The principal improvement appears to be in the vicinity of the first maximum.

The numerical solution of both the Wertheim and Verlet extensions to the PY and HNC approximations requires substantial computing. Furthermore, it should be noted that Verlet's approximation is not symmetrical in the exchange of particle positions; this would cause serious problems in multicomponent fluids. In order to solve Wertheim's equations it is necessary to perform an isothermal density integration, and this may not be possible for fluids below their critical temperature. Nevertheless, the fact remains that the extensions which incorporate triplet correlations lead to considerable improvement over the original PY and HNC approximations.

In contrast to the diagrammatic expansions, one can compute the gradient of the original definition, (12.2.4), of the reduced probabilities at the vector position $\mathbf{r}_1$; for a pairwise additive potential this gives

$$\begin{aligned} -\boldsymbol{\nabla}_{r_1} \ln G_n(\mathbf{r}_1, \ldots, \mathbf{r}_n) &= \beta \boldsymbol{\nabla}_{r_1} \sum_{1 \leqslant i < j \leqslant n}\sum u_2(r_{ij}) \\ &\quad + \rho\beta \int \mathrm{d}^3 r_{n+1} \, \frac{G_{n+1}(\mathbf{r}_1, \ldots, \mathbf{r}_{n+1})}{G_n(\mathbf{r}_1, \ldots, \mathbf{r}_n)} \boldsymbol{\nabla}_{r_1} u_2(|\mathbf{r}_1 - \mathbf{r}_{n+1}|). \end{aligned} \tag{12.2.28}$$

This hierarchy of equations was derived independently by Bogoliubov, Born and Green, Kirkwood and Yvon; they are known as the equilibrium BBGKY equations. For uniform fluids the first nontrivial member is $n = 2$, but this involves G_3 and therefore a closure is required. The first closure was proposed by Kirkwood (1935). It is referred to as the superposition approximation because it was originally defined in terms of the potentials of mean force which are the natural logarithms of the correlation functions. The Kirkwood closure assumes

$$G_3(r_{12},r_{23},r_{13}) = g_2(r_{12})g_2(r_{23})g_2(r_{13}) \tag{12.2.29}$$

which states that the triplet probability density is composed of a product of pair probabilities, that is the molecules are pairwise independently correlated. Extensive computations have been performed with this closure, and a recent discussion is given elsewhere (Munster 1974). The BBGKY equation for $n = 2$ and (12.2.29) give the second and third virial coefficient exactly, but beyond this it is generally a poorer approximation than either the PY or the HNC equations. We will later show the error inherent in the approximation for various configurations of three atoms. The error can be quite sizeable (±40%) for certain configurations and high density, but the function g_3 is comparatively short ranged. It appears that the BBGKY equation for $n = 2$ with the superposition approximation magnifies the error inherent in (12.2.29). It should be noted that the resulting equations for a binary fluid mixture of say species a and b, are not symmetrical in the sense that one should have $g_2^{\mathrm{ab}}(r_{12}) = g_2^{\mathrm{ba}}(r_{21})$. This gives serious errors even at moderate densities (Alder 1955).

It is important to consider whether the inclusion of explicit triplet correlations improves the results obtained from the BBGKY hierarchy. In particular, it is natural to ask whether the generalization of the Kirkwood approximation to the next level, namely at $n = 3$, results in an improvement over the results obtained from (12.2.28). The closure at this level is

$$\begin{aligned} &G_4(r_{12},r_{13},r_{14},r_{23},r_{24},r_{34}) \\ &\quad = \frac{G_3(r_{12},r_{23},r_{13})G_3(r_{12},r_{24},r_{14})G_3(r_{13},r_{14},r_{34})G_3(r_{23},r_{24},r_{34})}{g_2(r_{12})g_2(r_{13})g_2(r_{14})g_2(r_{23})g_2(r_{24})g_2(r_{34})} \\ &\quad = g_2(r_{12})g_2(r_{13})g_2(r_{14})g_2(r_{23})g_2(r_{24})g_2(r_{34}) \\ &\qquad \times g_3(r_{12},r_{13},r_{23})g_3(r_{12},r_{14},r_{24})g_3(r_{13},r_{14},r_{34})g_3(r_{23},r_{24},r_{34}) \end{aligned} \tag{12.2.30}$$

and was discussed initially by I. Z. Fisher and Kapeliovich (1960). The following coupled set of equations, known as the BBGKY II, is obtained from (12.2.26) for $n = 2$ and $n = 3$,

$$\begin{aligned} -\nabla_{r_1} \ln g_2(r_{12}) &= \beta \nabla r_1 u(r_{12}) \\ &\quad + \rho\beta \int \mathrm{d}^3 r_3 g_2(r_{13})[g_2(r_{23})g_3(r_{12},r_{13},r_{23}) - 1]\nabla_{r_1} u_2(r_{13}) \end{aligned} \tag{12.2.31}$$

$$\begin{aligned} -\nabla_{r_1} \ln g_3(r_{12},r_{23},r_{13}) &= \rho\beta \int \mathrm{d}^3 r_4 \\ &[g_2(r_{14})g_2(r_{24})g_2(r_{34})g_3(r_{12},r_{14},r_{24})g_3(r_{13},r_{14},r_{34})g_3(r_{23},r_{24},r_{34}) \\ &-g_2(r_{14})g_2(r_{24})g_3(r_{12},r_{14},r_{24}) - g_2(r_{14}) \\ &-g_2(r_{24})g_2(r_{34})g_3(r_{23},r_{24},r_{34}) - g_2(r_{24})]\nabla r_1 u_2(|\mathbf{r}_1 - \mathbf{r}_4|). \end{aligned} \tag{12.2.32}$$

The equations were first investigated by Lee, Ree and Ree (1968) and by Ree, Lee and Ree (1971). Since the solution is a formidable numerical task, these authors concentrated first on the hard sphere case. The fourth virial coefficient is given exactly by the BBGKY II approximation as it is with the PY II and HNC II approximations. The fifth virial coefficient from the virial equation of state, (12.2.11), gives

BBGKY II = 0.1090 ± 0.0008

and from the compressibility equation, (12.2.12), the result is

BBGKY II = 0.1112 ± 0.0005.

Both results are in close agreement with the exact values quoted earlier and the selfconsistency is impressive. The improvement can be seen by comparing the virial and compressibility equation of state results from the original BBGKY equation; these are, respectively,

BBGKY = 0.0475

and

BBGKY = 0.1335.

Since it is known that the BBGKY equation with the superposition approximation is inferior to the PY equation, even at moderate densities, it is of interest to compare the pair correlation function obtained from the BBGKY II approximation with the PY method. Results are available for hard spheres at densities $\rho = 0.298\rho_0$ and $\rho = 0.372\rho_0$ where ρ_0 is the closest packing density, $\sqrt{2}$, of the crystalline phase. Ree, Lee and Ree (1971) find that in terms of the intermolecular distance r/σ, where σ is the hard sphere diameter, the BBGKY II and computer simulation results are in very good agreement; this is shown in Table 12.2.1.

The agreement between the Monte Carlo computer simulation and the BBGKY II equations is within the numerical uncertainty of the simulation, except at $r/\sigma = 1.6$ where the difference is still less than 1%. At higher densities $\rho = 0.372\rho_0$ (but still moderate, the freezing density is $\rho = 0.667\rho_0$) while the relative agreement with the Monte Carlo results is, for $r/\sigma < 1.4$, still within the numerical uncertainty of the simulation and therefore still superior to the PY approximation. There is evidence that at large separations the PY equation is closer to the computer simulation result, but because of a technical problem which is discussed by Ree *et al.* (1971) an unambiguous comparison is not possible. It is sufficient to remark that the BBGKY II equations represent a substantial improvement over the original BBGKY equation obtained with the Kirkwood closure. It would be valuable to have more results to intercompare the BBGKY II, PY II and HNC II approximations.

Table 12.2.1 BBGKY II results for hard spheres at $\rho = 0.298\rho_0$, from Ree *et al.* (1971)

r/σ	g_2(computer simulation)	g_2(BBGKY II)	g_2(PY)
1.0	1.906	1.886	1.828
1.2	1.390	1.390	1.393
1.4	1.107	1.103	1.111
1.6	0.967	0.959	0.962
1.8	0.920	0.921	0.924
2.0	0.978	0.976	0.974
2.2	1.022	1.023	1.020
2.4	1.018	1.018	1.017
2.6	1.005	1.002	1.002
2.8	0.995	0.993	0.994
3.0	0.996	0.995	0.996
3.2	1.000	1.000	1.000
3.4	1.002	1.002	1.002
3.6	1.001	1.001	1.001
3.8	1.000	1.000	1.000

The success of the extensions of the integral equations shows that the inclusion of triplet correlations gives better results for both thermodynamic properties and the pair correlation functions. Although the PY II, HNC II and BBGKY II equations are difficult to solve numerically, they invariably result in a considerable improvement over the original theory.

(c) Functional expansions

The fact that the correlation functions can be expanded in terms of the Mayer function, $f(r) = \exp[-\beta u_2(r)] - 1$, and the fact that the definition at (12.2.4) can be inverted to express the U_N in terms of the ρ_n (Mayer 1942) suggests that it is possible to write the correlation function g_n as a functional in $g_2 \ldots g_{n-1}$ is such a way that the potential energy does not appear explicitly.

We are interested in the triplet correlation function as a functional of the lower order correlation function; for a uniform fluid this leaves only the pair correlation function. Representations of this type have been investigated by Abe (1959) and also by Stell (1963) and Lee *et al.* (1967). Abe (1959) found that,

$$\begin{aligned}\ln g_3(r_{12}, r_{13}, r_{23}) = {} & \rho \int \mathrm{d}^3 r_4 h(r_{14}) h(r_{24}) h(r_{34}) \\ & + \rho^2 \iint \mathrm{d}^3 r_4 \, \mathrm{d}^3 r_5 \, [h(r_{14}) h(r_{24}) h(r_{25}) h(r_{35}) h(r_{45}) \\ & + h(r_{14}) h(r_{15}) h(r_{24}) h(r_{35}) h(r_{45}) \\ & + h(r_{14}) h(r_{25}) h(r_{34}) h(r_{35}) h(r_{45}) \\ & + h(r_{14}) h(r_{15}) h(r_{24}) h(r_{25}) h(r_{34}) h(r_{45}) \\ & + h(r_{14}) h(r_{15}) h(r_{24}) h(r_{34}) h(r_{35}) h(r_{45}) \\ & + h(r_{14}) h(r_{24}) h(r_{25}) h(r_{34}) h(r_{35}) h(r_{45}) \\ & + \tfrac{1}{2} h(r_{14}) h(r_{15}) h(r_{24}) h(r_{25}) h(r_{34}) h(r_{35}) h(r_{45})] + \ldots \end{aligned} \tag{12.2.33}$$

This series is not one in density since the integrands depend on density as well as on temperature. The first two sets of terms are sufficient to give the fourth and fifth virial coefficients exactly.

Not all of the integrals are easily evaluated numerically. It is possible to evaluate the first term which, discarding the higher order terms, is equivalent to,

$$g_3(r_{12}, r_{13}, r_{23}) \simeq 1 + \rho \int \mathrm{d}^3 r_4 h(r_{14}) h(r_{24}) h(r_{34}) . \tag{12.2.34}$$

We will in Section 12.3 compare this closure with exact results obtained from computer simulations.

On the basis of an analysis of a cluster type representation for the triplet probability density,

$$\rho_3(r_{12}, r_{23}, r_{13}) = \rho^3 g_2(r_{12}) g_2(r_{13}) g_2(r_{23}) + \rho^3 h(r_{12}) h(r_{23}) h(r_{13}) + \rho^3 \psi(r_{12}, r_{13}, r_{23}) \tag{12.2.35}$$

Feenberg (1969) proposed

$$\psi(r_{12}, r_{13}, r_{23}) = \rho \int \mathrm{d}^3 r_4 h(r_{14}) h(r_{24}) h(r_{34}) \tag{12.2.36}$$

which leads to the so-called convolution approximation

$$g_3(r_{12},r_{23},r_{13}) = 1 - \frac{h(r_{12})h(r_{23})h(r_{13}) - \rho\int \mathrm{d}^3 r_4 h(r_{14})h(r_{24})h(r_{34})}{g_2(r_{12})g_2(r_{13})g_2(r_{23})}.$$

(d) Determination from computer simulations

Direct experimental measurement of the triplet correlation function has not to date been realized. In the absence of a laboratory experiment and in the absence of an exact solution of a realistic theory, it is necessary to find an alternative determination of $g_3(r_{12},r_{13},r_{23})$. Without this it is not possible to advance our understanding of the structure of fluids beyond that expressed by the pair correlation function.

The principal advantage of computer simulations is that they can provide exact, albeit numerical, solutions of a given model. The two techniques presently available for computer simulation in statistical mechanics are the Monte Carlo method, initially proposed by Metropolis *et al.* (1953) and the molecular dynamics method, investigated for hard spheres by Alder and Wainwright (1958), and by Rahman (1964a,b) for the Lennard–Jones system. There are several recent reviews (McDonald and Singer 1970, Ree 1971, Wood 1974), and we refer the reader to these for detailed discussions of the techniques. The molecular dynamics method consists of an algorithm for integrating the equations of motion in both the position and the conjugate momentum space of a number, say N, of particles. The force acting on a molecule at the position $\mathbf{r}_i$ is $\nabla_{r_i} U_N(\mathbf{r}_1, \ldots, \mathbf{r}_N)$. The Monte Carlo technique numerically executes a biased random walk by randomly moving particles, one at a time, according to criteria which give a series of configurations, each having a probability proportional to $\exp[-\beta U_N(\mathbf{r}_1, \ldots, \mathbf{r}_N)]$, where U_N is the energy of the given configuration. Periodic boundary conditions are used in practice, and current computing capabilities limits the size of the system to approximately $N \sim 1000$.

The absolute size of the system is not as important as the relative amount of computing for a given size. It is possible with systems of a few hundred particles to achieve an accuracy of within one or two percent for both microscopic and bulk properties with good precision. For equilibrium properties, the principal effect of the size of the system is to limit the distance in configuration space over which one can calculate correlation functions. For the systems with which we will be concerned in this chapter, it is possible to make accurate calculations for distances as large as 3 atomic diameters. This distance is adequate for fluid structure studies away from the liquid–vapour critical point.

The determination of the pair correlation function in computer simulations is discussed in Barker and Henderson (1972), Ree (1971) and Wood (1969). These articles also contain references to the extensive amount of computations of $g_2(r_{12})$ for a variety of potentials.

Computations of the triplet correlation function, $g_3(r_{12},r_{13},r_{23})$, are comparatively few in number. The first computation for hard spheres was by Alder (1964) and for the Lennard–Jones system by Rahman (1964b). Since these investigations there were no results until Krumhansl and Wang (1972a,b) and by Raveché, Mountain and Streett (1972, 1974); both sets of computations were for the Lennard–Jones system by the Monte Carlo method. Recently Tanaka and Fukui (1975) have reported molecular dynamics results with the model potential of liquid sodium by Paskin and Rahman (1966).

The computations involve the probability of observing three particles separated by the distances r_{12}, r_{13} and r_{23} conditional to the probability of finding two particles separated by the distance r_{12}. That is, the ratio $G_3(r_{12},r_{13},r_{23})/g_2(r_{12})$ is computed for fixed r_{12} and then one divides by $g_2(r_{13})g_2(r_{23})$ to obtain $g_3(r_{12},r_{13},r_{23})$. A procedure for performing these computations has been devised by Streett and is described in detail by Raveché, Mountain and Streett (1974) in an appendix. This procedure eliminates certain redundancies which exist in Krumhansl and Wang and also in Tanaka and Fukui.

12.3 Direct analysis of triplet correlations

(a) Configurational entropy

We return to (12.2.14), which is the configurational entropy per molecule up to and including terms in the triplet correlation function. The expression does not explicitly involve the potential energy function and therefore it can be used whether or not the true potential energy is pairwise additive. Using known values for the correlation functions it is then possible to compare the computed entropy with the known value of either a model system or a real fluid. The expression can also be used to assess the relative contribution of the pair and triplet terms. We will address both issues.

As we discussed in Section 12.2, in the absence of $g_3(r_{12}, r_{13}, r_{23})$ for all values of its argument, it is not possible to calculate the term in $\ln g_3$ in (12.2.14). It is, however, possible to compute the other triplet term from the equation of state. Therefore, with known value of $g_2(r_{12})$ it is possible to compute all of the terms in the entropy expression in (12.2.14), except the term in $\ln g_3$. The fact that the non-logarithmic terms can be expressed in terms of isothermal activity derivatives of compressiblity-like terms (Raveché 1971) leads to,

$$S[G_3] = \frac{\rho^2}{6} \iint d^3r_2\, d^3r_3 \,[G_3(r_{12}, r_{13}, r_{23}) - g_2(r_{12})g_2(r_{23})$$

$$- g_2(r_{12})g_2(r_{13}) - g_2(r_{12})g_2(r_{23}) + g_2(r_{12}) + g_2(r_{13}) + g_2(r_{23}) - 1]$$

$$= \frac{1}{6} + \frac{\gamma}{2} - \frac{\gamma^2}{3} + \frac{1}{6}\rho\gamma\left(\frac{\partial \gamma}{\partial \rho}\right)_\beta \tag{12.3.1}$$

where $\gamma = \beta^{-1}(\partial\rho/\partial P)_\beta$. In the Kirkwood superposition approximation this becomes,

$$S[g_2g_2g_2] = \frac{\rho^2}{6} \iint d^3r_2 d^3r_3 h(r_{12})h(r_{13})h(r_{23}). \tag{12.3.2}$$

These terms and the two-body terms in the entropy expression have been computed by Mountain and Raveché (1971) using diffraction data of DeGraaf and Mozer (1971) for liquid neon at $T = 35.05$ K. The results are given in Table 12.3.1 where we use $\langle \ln g_2 \rangle$ and $S[g_2]$ to denote respectively the first and second terms in (12.2.14) and Σ to denote the sum of these and (12.3.1). The results are plotted in Figure 12.3.1, while the Percus–Yevick and hard sphere results are shown in Figure 12.3.2.

Using Rosenbaum's data (1972) for liquid rubidium and Rahman's (1975) molecular dynamics data for the pair correlation function, the entropy terms have been calculated for liquid rubidium and the results are shown in Table 12.3.2.

Table 12.3.1 Entropy results for liquid neon at 35.05 K

$\rho(10^{22}/\text{cm}^3)$	$\langle \ln g_2 \rangle$	$-S[g_2]$	$-S[G_3]$	$-\Sigma$	$-\Delta S/\langle N \rangle k_B$ expt	$-S[g_2g_2g_2]$
3.170	1.519	0.420	0.116	2.055	2.126	−0.064
3.339	1.653	0.442	0.129	2.224	2.317	−0.089
3.471	1.778	0.457	0.138	2.373	2.463	−0.149

Table 12.3.2 Entropy results for liquid rubidium at $T = 315$ K

$\rho(10^{22}/\text{cm}^3)$	$\langle \ln g_2 \rangle$	$-S[g_2]$	$-S[G_3]$	$-\Sigma$	$-\Delta S/\langle N \rangle k_B$ expt	$-S[g_2g_2g_2]$	Σ(supp)
1.058	2.87	0.49	0.16	3.52	3.87	−1.46	1.90

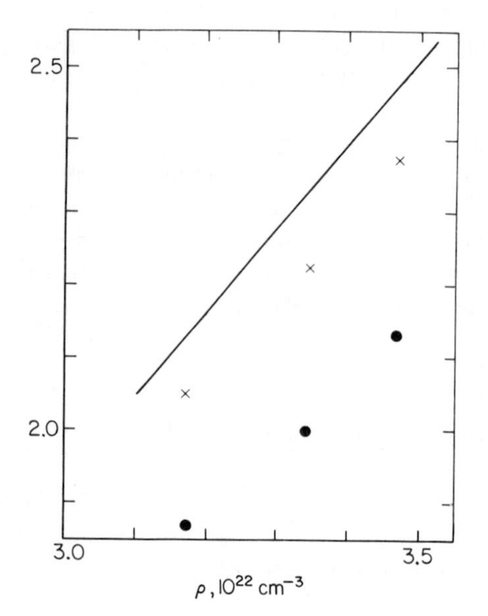

Figure 12.3.1 Configurational entropy of liquid neon times (−1) as a function of density at $T = 35.05$ K, from Mountain and Raveché (1971). Solid curve is experimental result, the x points denote Σ and the solid circles denote the superposition value of Σ

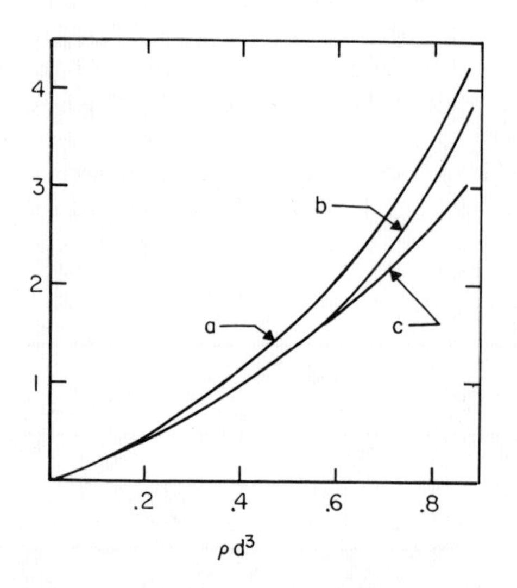

Figure 12.3.2 Configurational entropy of hard spheres times (−1) as a function of density from Mountain and Raveché (1971). Curve (a) is obtained from the analytic formula for the Percus–Yevick compressibility equation of state, curve (b) is obtained by applying the Percus–Yevick equation for g_2 and its density derivative, curve (c) is obtained from applying the Percus–Yevick theory for g_2 and the superposition approximation for $S[G_3]$

The results for liquid neon, hard spheres and liquid rubidium suggest that the entropy expression is convergent, but at present a proof does not exist. Moreover, the results show that in order to obtain accurate entropy values, triplet correlations must be included. Figures 12.3.1 and 12.3.2 show that the superposition approximation does not do well because it gives both the incorrect sign and magnitude of $S[G_3]$. The approximation appears to be much poorer for liquid rubidium where Σ is within 9% of the measured entropy and the superposition value, $\Sigma(\text{supp})$, is inaccurate by about a factor of two.

(b) Density derivative of the pair correlation function

The entropy results show that triplet correlations can make important contributions. However, a bulk property is, of course, a very coarse indication of the structure in $g_3(r_{12},r_{13},r_{23})$, and it is necessary to probe this structure on a molecular scale. This can be done with 12.2.17, the isothermal density derivative of the pair correlation function. It is important to recall that the relation (12.2.17) is independent of whether or not the potential energy function is pairwise additive. Therefore with measured values of $g_2(r_{12})$ and $(\partial g_2/\partial\rho)_T$ one can probe the simultaneous correlation of three atoms in a real fluid without resort to any approximation about the interactions in the fluid.

The relation at (12.2.17) can be written in a form which focuses on the departure of g_3 from unity (Raveché and Mountain 1970);

$$\left(\frac{\partial g_2(r_{12})}{\partial\rho}\right)_\rho = \frac{1}{\gamma}\left(\int d^3r_3 h(r_{13})h(r_{23}) + \int d^3r_3 g_2(r_{13})g_2(r_{23})[g_3(r_{12},r_{13},r_{23})-1]\right). \tag{12.3.3}$$

It is convenient to define

$$F(r_{12}\,|\,g_3) = \int d^3r_3 g_2(r_{13})g_2(r_{23})[g_3(r_{12},r_{13},r_{23})-1] \tag{12.3.4}$$

and since the coefficient of $g_3 - 1$ is positive, a nonzero value of F necessarily requires $g_3 \neq 1$. Egelstaff *et al.* (1969, 1971) preferred to use

$$H(r_{12}) = -g_2(r_{12})F(r_{12}\,|\,g_3). \tag{12.3.5}$$

The neutron diffraction measurements of deGraaf and Mozer (1971) on liquid neon at $T = 35.05$ were designed, in part, to construct the derivative. The results for $F(r_{12}\,|\,g_3)$ obtained by Raveché and Mountain (1972) are given by the solid curve in Figure 12.3.3, the result for $H(r_{12})$ is given by the long dashed curve and $g_2(r_{12})$ (at $\rho = 3.339 \times 10^{22}/\text{cm}^3$) is given by the short dashed curve. The form of $F(r_{12}\,|\,g_3)$ clearly shows that $g_3(r_{12},r_{13},r_{23})$ must oscillate about unity in value; however, as we have mentioned, the amplitudes of the oscillations cannot be obtained from (12.3.3). Similar conclusions are reached from the form of $H(r_{12})$, but this function somewhat masks the contribution of g_3 which is the quantity of interest.

The main results obtained to date for the density derivative of $g_2(r_{12})$ are for argon (Raveché and Mountain 1970), neon (Raveché and Mountain 1972) and helium (Raveché and Mountain 1974). The Fourier transform of (12.3.3) gives the density derivative of the structure factor $S(k,\rho,T)$ as a function of the magnitude of the wave vector k. This has been investigated for argon, rubidium and carbon tetrachloride by Egelstaff *et al.* (1969, 1971), and Winfield and Egelstaff (1973) have applied this analysis to krypton.

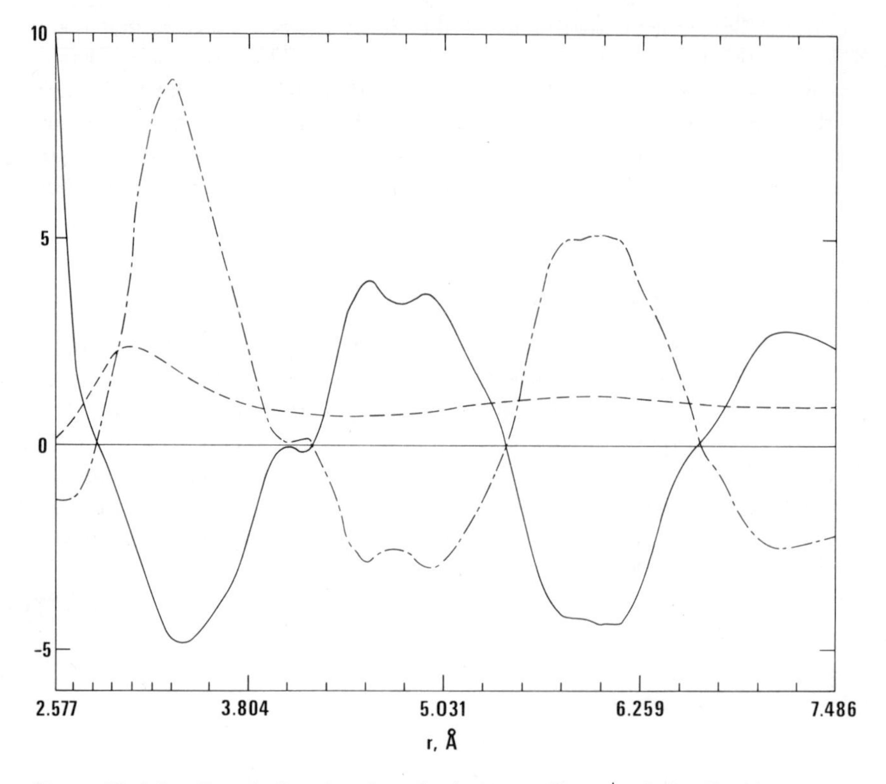

Figure 12.3.3 Result for the three body term $F(r_{12} \mid g_3)$ for liquid neon at $T = 35.05$ K and $\rho = 3.339 \times 10^{22}/\text{cm}^3$, from Raveché and Mountain (1972). The solid curve is (12.3.4), the long-short dashed curve is (12.3.5) and the short dashed curve is $g_2(r_{12})$

The important point of all these investigations is that the simultaneous correlation of the positions of three molecules has been detected in a variety of real fluids. That is, triplet correlations are a real phenomena. Since the density derivative gives only a functional of g_3 we must now investigate the detailed structure in the function. However, since this is not at present possible in the laboratory, we pursue the information contained in computer experiments with model potentials.

(c) Results of computer simulations

We now investigate characteristics of the actual structure present in g_3 for the Lennard–Jones fluid and for a model of liquid sodium. At first we examine detailed aspects of g_3 from various isosceles configurations. Then we consider the general structure present in the function.

As we have already mentioned in Section 12.2*d*, computer simulation results for g_3 have been only recently available. The earlier work of Rahman (1964b) and Alder (1964) pertained to a very limited number of configurations. The recent investigations of Krumhansl and Wang (1972a,b) (see also Wang and Krumhansl 1972) and of Raveché, Mountain and Streett (1972, 1974) include more configurations for several thermodynamic states of the Lennard–Jones system. The computations of Tanaka and Fukui (1975) are for the Paskin–Rahman model (1966) of liquid sodium.

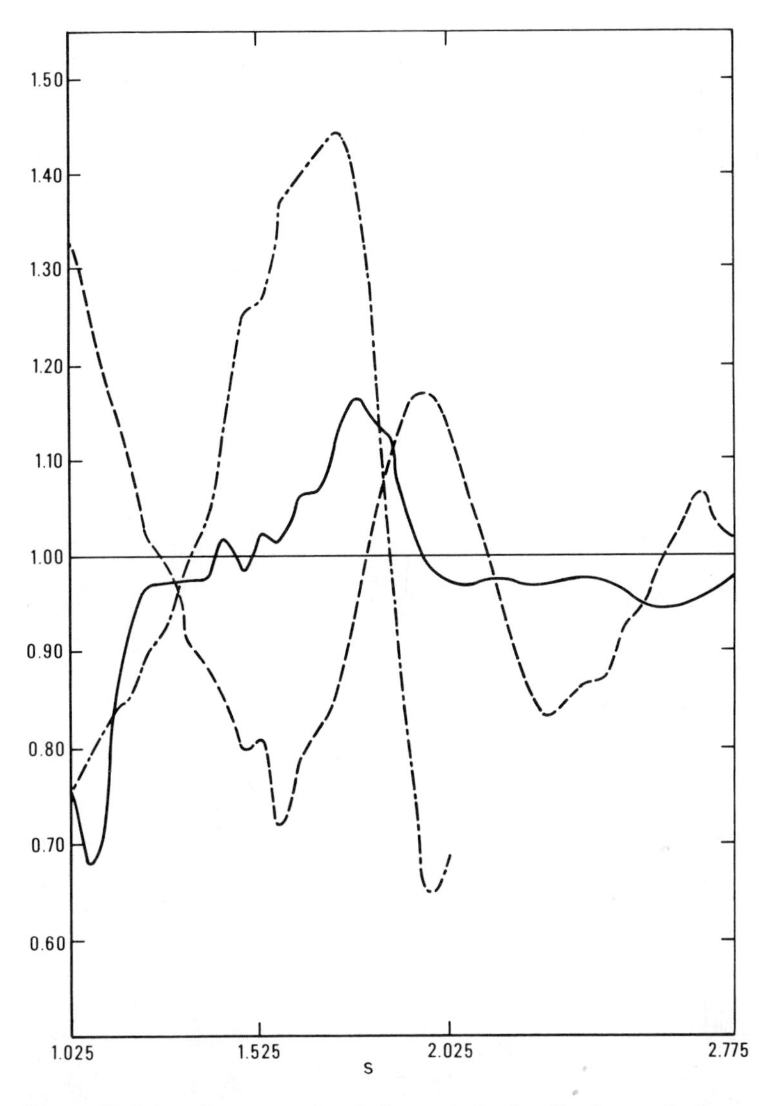

Figure 12.3.4 Computer simulations results for the Lennard–Jones system at $\rho = 0.850$ and $T = 0.719$ from Raveché, Mountain and Streett (1972). The dashed–dotted curve is $g_3(1.025, 1.025, s)$, the dashed curve is $g_3(1.525, s, s)$ and the solid curve is $g_3(2.025, s, s)$

Since the simulations of Raveché *et al.* have better statistics, and have been done for more thermodynamic states than those of Krumhansl and Wang, we focus on the former investigations. Egelstaff (1973) has also commented on the relative accuracy of these simulations.

The results for $g_3(1.025, 1.025, s)$, $g_3(1.525, s, s)$ and $g_3(2.025, s, s)$ at $\rho = 0.850$ and $T = 0.719$ are shown in Figure 12.3.4. This state is near the triple point. All quantities are reported in units of the Lennard–Jones potential parameters. For reference, the triple point of the model system is $\rho = 0.85$, $T = 0.68$; the critical temperature is 1.36.

The results for $g_3(1.025, 1.025, s)$, $g_3(1.525, s, s)$ and $g_3(2.025, s, s)$ at $\rho = 1.066$ and $T = 2.74$ are given in Figure 12.3.5; this thermodynamic state is near the freezing curve. The general

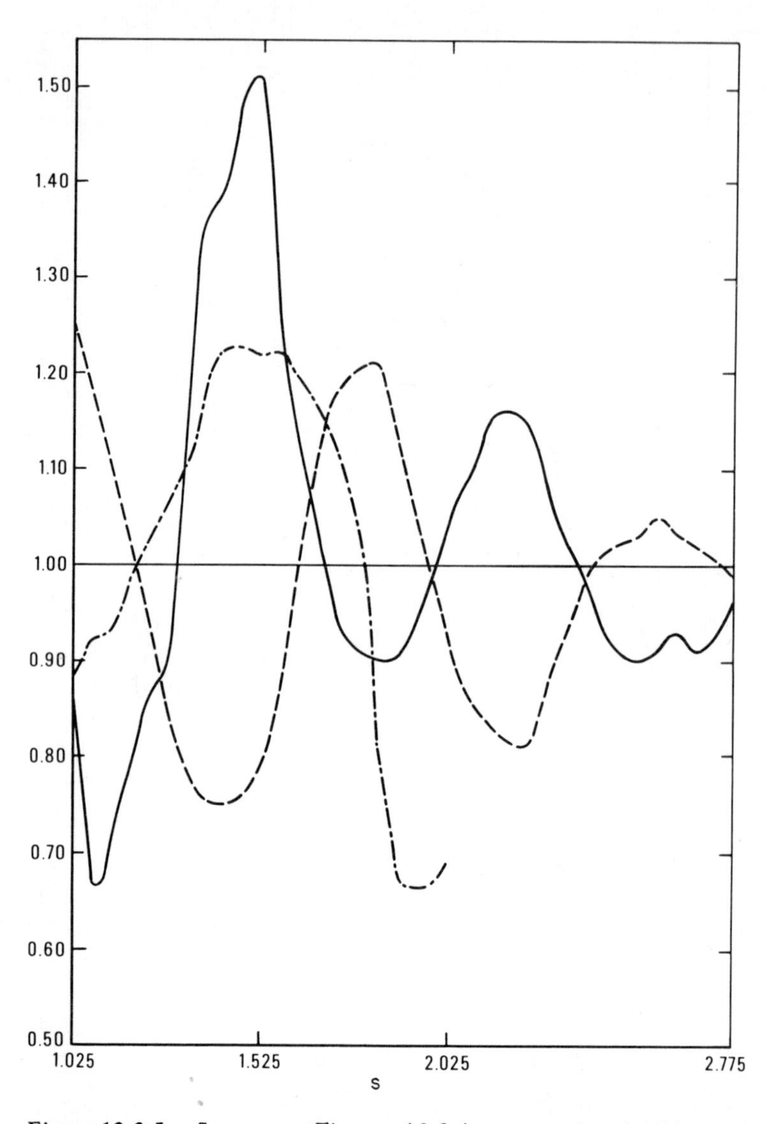

Figure 12.3.5 Same as Figure 12.3.4 except $\rho = 1.066$ and $T = 2.74$, from Ravecheé, Mountain and Streett (1974)

trends in Figures 12.3.4 and 12.3.5 are similar but the positions of the maxima in Figure 12.3.4 are shifted to smaller internuclear distances reflecting that the system is more dense. The amplitudes of the denser state are also larger, with the exception of the maximum in $g_3(1.025, 1.025, s)$: the amplitude of this peak is smaller than that shown by the dashed-dotted curve in Figure 12.3.4. This result is consistent with computations at other states which show that the peak is stronger at lower temperatures.

It is apparent that g_3 has appreciable structure out to internuclear separations as large as three diameters. The states corresponding to the results in Figures 12.3.4 and 12.3.5 are both near the fluid-solid boundary and, as the state of the system moves away from the boundary, the amplitude of g_3 decreases. This is shown in Figure 12.3.6 where we have plotted g_3 for the same configurations as in Figures 12.3.4 and 12.3.5, but at $\rho = 0.750$ and $T = 1.070$. We note

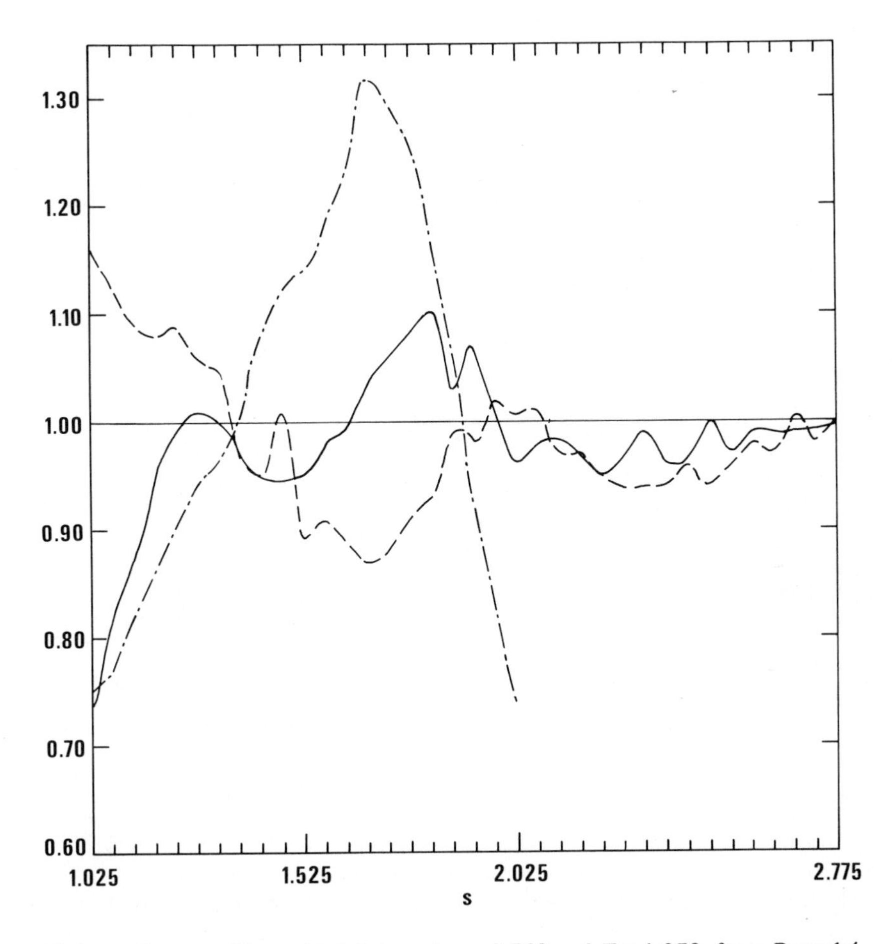

12.3.6 Same as Figure 12.3.4 except $\rho = 0.750$ and $T = 1.070$, from Raveché *et al.* (1972)

that the structure in $g_3(1.525,s,s)$ and $g_3(2.025,s,s)$ has diminished and that, for $s \geqslant 2.025$, the function is quite close to unity. This is consistent with results on other states away from the fluid-solid boundary where we have found that, except for configurations in which all internuclear distances are within two atomic diameters, the triplet function is close to unity.

The uncertainty in the computations is estimated to be about ±4% and we believe not greater than this, except for very improbable configurations: for example, the configurations where the positions of the atoms determine a straight line (these are given by the first point in the solid curve and the last point in the dashed-dotted curves in Figures 12.3.4, 12.3.5 and 12.3.6). We note that the disagreement between these two identical configurations is larger at the higher densities. But at $\rho = 0.750$ and $T = 1.070$, $g_3(1.025, 1.025, 2.025) = 0.736$ and $g_3(2.025, 1.025, 1.025) = 0.744$. Therefore, the dip between the first and second points on the solid curves and the rise at the end of the dashed-dotted curves in Figures 12.3.4 and 12.3.5 are artifacts of the numerical precision. We have done computations which are 35% longer than those for which the data in Figure 12.3.5 were reported. The resulting curves are quite close to those in Figure 12.3.5 except for improved precision of the 'straight line' configurations; we found $g_3(1.025, 1.025, 2.025) = 0.666$ and $g_3(2.025, 1.025, 1.025); = 0.664$.

We can make other checks on the uncertainty, for example the point $g_3(1.025,$

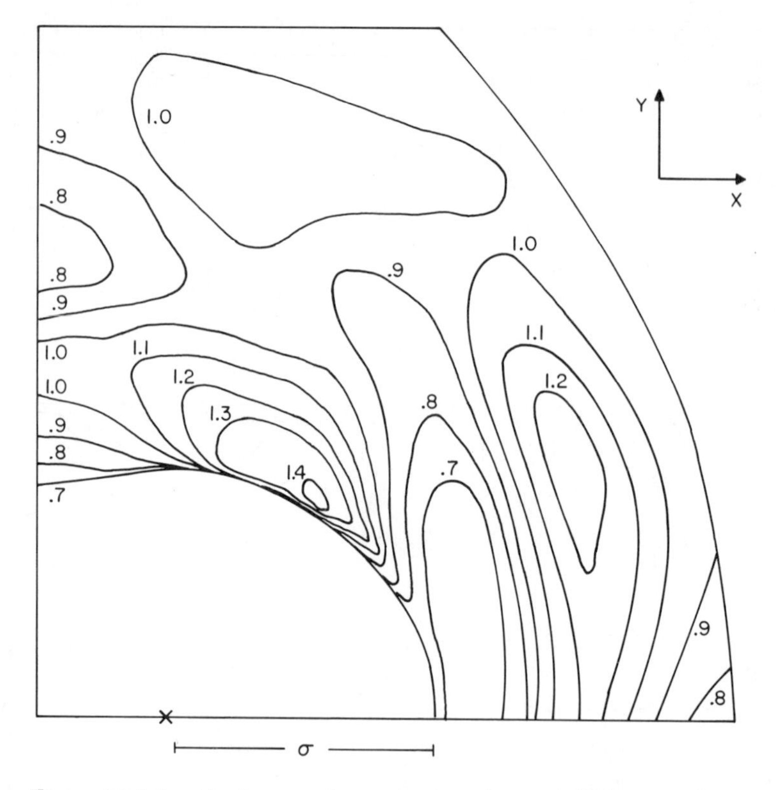

Figure 12.3.7 Contours of constant $g_3(r_{12} = 1.025, r_{13}, r_{23})$ at $\rho = 0.850$ and $T = 0.719$ from Raveché *et al.* (1972). The criss-cross denotes the position of one atom and the the second atom is fixed at an equal distance from the y axis. The position of the atom is located in the $x - y$ plane

1.025, 1.525) on the dashed-dotted curve and the first point on the dashed curve. In Figures 12.3.4, 12.3.5 and 12.3.6 the differences between these identical configurations is within ±4%. For example at $\rho = 1.066$ and $T = 2.74$, $g_3(1.025, 1.025, 1.525) = 1.223$ and $g_3(1.525, 1.025, 1.025) = 1.252$.

The results we have considered thus far pertain to special configurations and it is desirable to have a more global view of the triplet correlation function. Envisage an x,y coordinate system with two atoms fixed along the horizontal x-axis at equal distances from the y-axis. Let the position of a third atom be confined to the xy plane. The projections in the plane of contours of constant $g_3(r_{12} = 1.025, r_{13}, r_{23})$ for $\rho = 0.850$ and $T = 0.719$ are shown in Figure 12.3.7. We note that the values range from 0.7 to 1.4 and that for large r_{13} and r_{23} the function is unity. Points along the y-axis correspond to isosceles configurations, $g_3(r_{12} = 1.025, r_{13} = r_{23} = s)$, the contour at 1.4 corresponds, approximately, to the maximum in the dashed-dotted curve in Figure 12.3.3.

An even broader prospective of the structure in g_3 can be visualized from the surface generated by $g_3(r_{12}, r_{13}, r_{23})$ for fixed values of r_{12}. For this we envisage an x,y,z coordinate system with the z-axis vertical and we fix two atoms along the y-axis at equal distance from the z-axis. The locus of points in the xy plane denotes positions of the third atom and the vertical axis denotes the magnitude of g_3. The surface generated by $g_3(1.025, r_{13}, r_{23})$ at $\rho = 1.066$ and $T = 2.74$ is shown in Figure 12.3.8. The cross corresponds to the centre of one atom and the

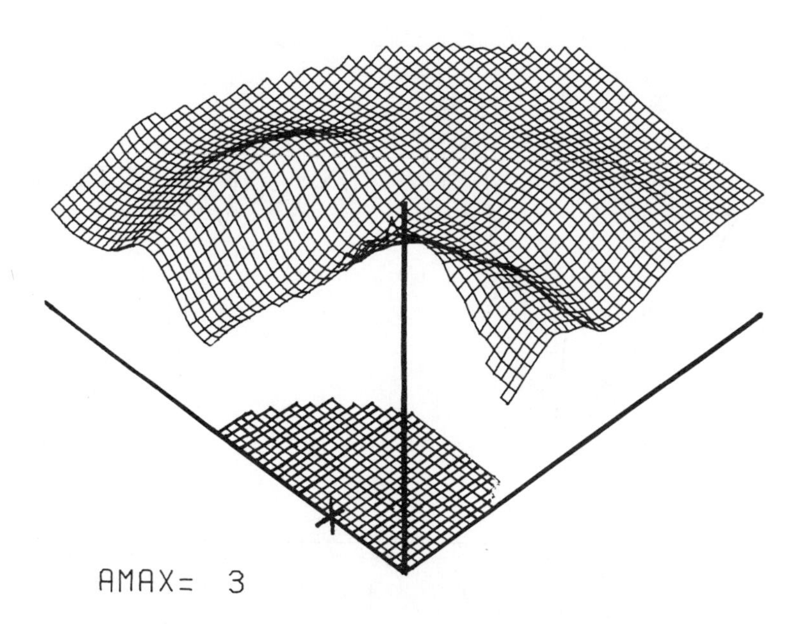

Figure 12.3.8 The surface generated by $g_3(r_{12} = 1.025, r_{13}, r_{23})$ at $\rho = 1.066$ and $T = 2.74$ from Raveché *et al.* (1974). The cross denotes the position of one atom and the second is located on the same axis at an equal distance from the vertical axis. The spacing of the grid is 0.05 in units of the diameter σ and vertical scale is determined by dimensionless number AMAX. Points inside the grid correspond to small interatomic separations and these are not shown because the numerical uncertainty rapidly increases in this region

spacing of the grid in the xy plane is 0.05 in units of the atomic diameter, σ. The vertical scale is denoted by the dimensionless number AMAX = 3, and therefore a plane which is everywhere co-planar to the xy plane and at a height of one third of the vertical line would correspond to g_3 in the Kirkwood superposition approximation.

The structure of g_3 is evidently complex if for no other reason than the fact that it is a function of five variables, three internuclear separations, the density and temperature. However, the function does appear to have one simplifying aspect (Raveché *et al.* 1974). Consider a face-centred cubic lattice with a cell parameter that is determined by the density of the fluid. The triplet correlation function is greater than unity for all configurations which 'fit' on the lattice. If one, or more than one, of the internuclear separations does not correspond to a lattice separation the triplet function is less than unity. Therefore the maxima and minima in the function appear to reflect the structure of the stable solid phase. Now there is no proof that this is necessarily the case for g_3 and one should consider the structural rule as provisional until it is either confirmed or denied by further investigations. The exception we have noted corresponds to the configuration where three atoms are just at contact, but here g_3 is close to unity.

Tanaka and Fukui (1975) investigated g_3 from a molecular dynamics simulation of liquid sodium using the semi-empirical potential function of Paskin and Rahman (1966). The results pertain to equilateral and isosceles configurations. Unfortunately, these authors smoothed their results in a way which emphasizes physically excluded configurations. For instance, the equilateral configurations given in Table 2 of their article are not in agreement with their Figure 2 which displays the smoothed results for those configurations. We show in

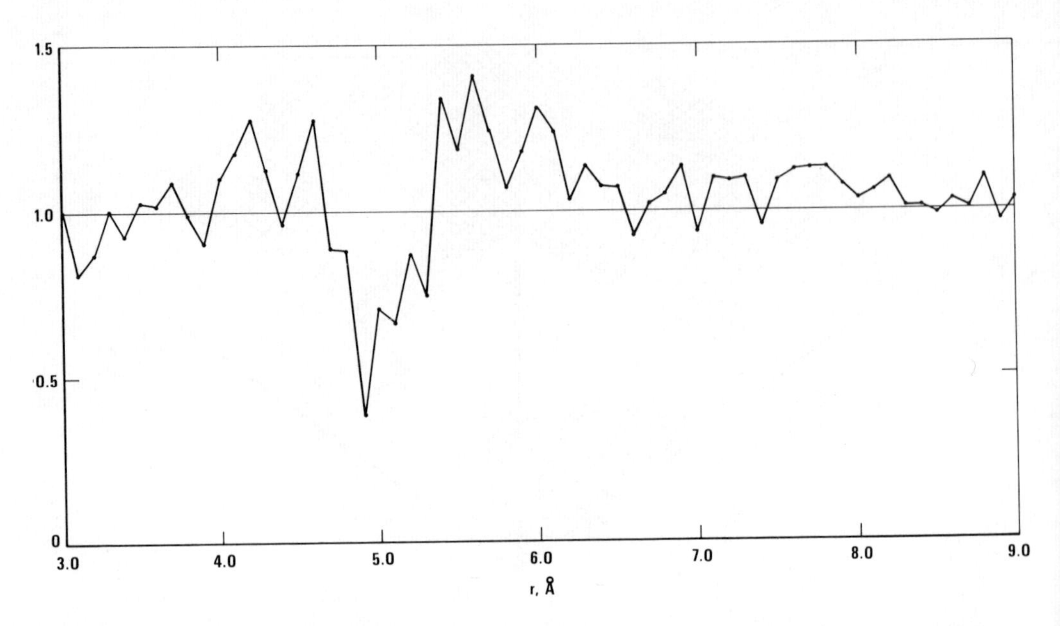

Figure 12.3.9 Unsmoothed data for $g_3(r, r, r)$ from the molecular dynamics computations of Tanaka and Fukui (1975) for the Paskin–Rahman (1966) model of liquid sodium at $T = 377.01$ K and $\rho = 2.41 \times 10^{22}/\text{cm}^3$. The horizontal axis is in Angstrom units

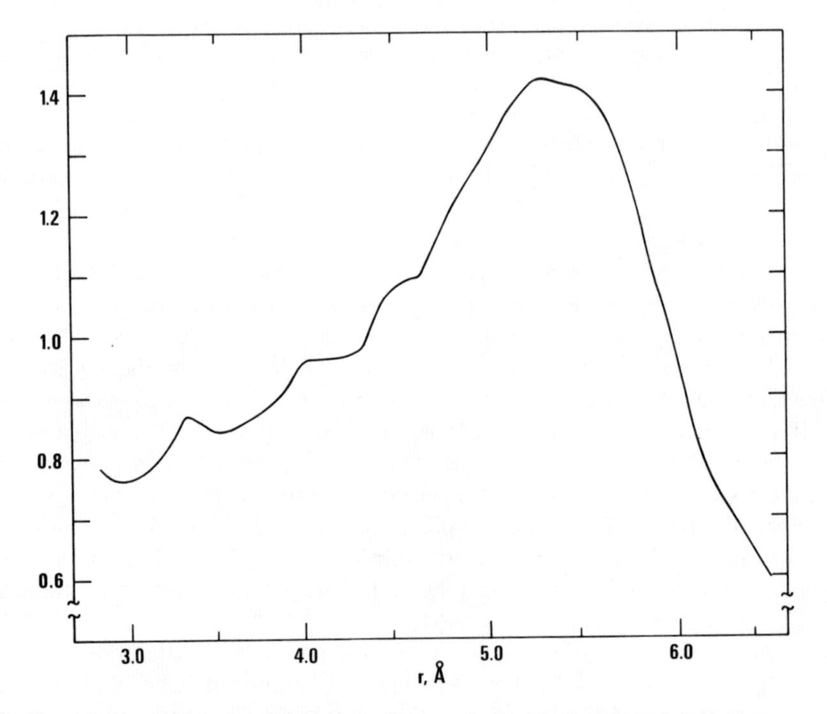

Figure 12.3.10 Monte Carlo results for $g_3(s, s, r)$ for the Paskin–Rahman (1966) model of liquid sodium at $T = 377$ K and $\rho = 2.41 \times 10^{22}/\text{cm}^3$, $s = 3.32$ Å; unpublished. These should be compared with the results of Tanaka and Fukui (1975) in Figure (12.3.9)

Figure 12.3.9 $g_3(r,r,r)$ as constructed from Tanaka and Fukui's Table 2. No smoothing is involved in Figure 12.3.9. The discrepancy between our Figure 12.3.9 and their Figure 2 is due to their smoothing technique.

Since the unsmoothed data of Tanaka and Fukui indicate inadequate statistics, we repeated the same calculation using our Monte Carlo program. Our results are shown in Figure 12.3.10, and these should be compared to the data of Tanaka and Fukui shown in Figure 12.3.9. We note that the maximum in Figure 12.3.9 between 5 and 6Å is well resolved in Figure 12.3.10, and that the local minimum at 4.9 Å in Figure 12.3.9 is apparently due to inadequate statistics. Even so, the data of Tanaka and Fukui reveal an interesting result which is clearly manifested in our results in Figure 12.3.10. The maximum in our data corresponds to a configuration which 'fits' on the bcc lattice at the specified density. This is quite consistent with our earlier observations that the maxima and minima in g_3 are reflections of the lattice structure of the stable solid phase. We remark parenthetically that the observations made earlier were based on results for the Lennard–Jones system and this has a different equilibrium lattice structure, namely the fcc pattern.

(d) Tests of closures

The concept of a closure for the triplet correlation function emerged in Section 12.2*b* of our discussion. There we noted that when one tries to extend the validity of an equation for a given correlation function, or of equations for a given set of correlation functions, a higher order correlation function enters the development. We have seen examples of this with the PY, HNC and BBGKY equations. In these cases the extension of the equation for the pair correlation function to a more general, and apparently more accurate, form required knowledge of the triplet correlation function, g_3.

We have also noted that the logical first assumption, namely to assume $g_3 = 1$, can be a poor approximation, being in error by as much as ±40% near the fluid–solid phase boundary. The problem is to find a better approximation to the triplet correlation function. In Section 12.2*c* we discussed two of these, the so-called convolution approximation (Feenberg 1969) and the approximation obtained by assuming that the first term in the functional expansion of g_3 in terms of g_2 (Abe 1959) is adequate. These are,

$$g_3 \simeq 1 - \frac{h(r_{12})h(r_{13})h(r_{23}) - \rho\int d^3r_4 h(r_{14})h(r_{24})h(r_{34})}{g_2(r_{12})g_2(r_{13})g_2(r_{23})} \tag{12.3.6}$$

and

$$g_3 \simeq 1 + \rho\int d^3r_4 h(r_{14})h(r_{24})h(r_{34}). \tag{12.3.7}$$

We refer to (13.3.6) as closure I and (13.3.7) as closure II. For purely heuristic reasons we also consider a nonsingular version of closure I,

$$g_3 \simeq 1 - \frac{h(r_{12})h(r_{13})h(r_{23}) - \rho\int d^3r_4 h(r_{14})h(r_{24})h(r_{34})}{g_2(r_{12})g_2(r_{13})g_2(r_{23}) + 1} \tag{12.3.8}$$

and we refer to this as closure III.

Until recently it has been difficult to make direct checks of the validity of these, or indeed any other, approximations for the triplet correlation function. For example with the PY II and HNC II equations, there are other assumptions involved which make unambiguous testing difficult. A similar situation exists with the BBGKY hierarchy of equations. With the density derivative of the pair correlation function, one can compute $F(r_{12} \mid g_3)$ at (12.3.4) with a given

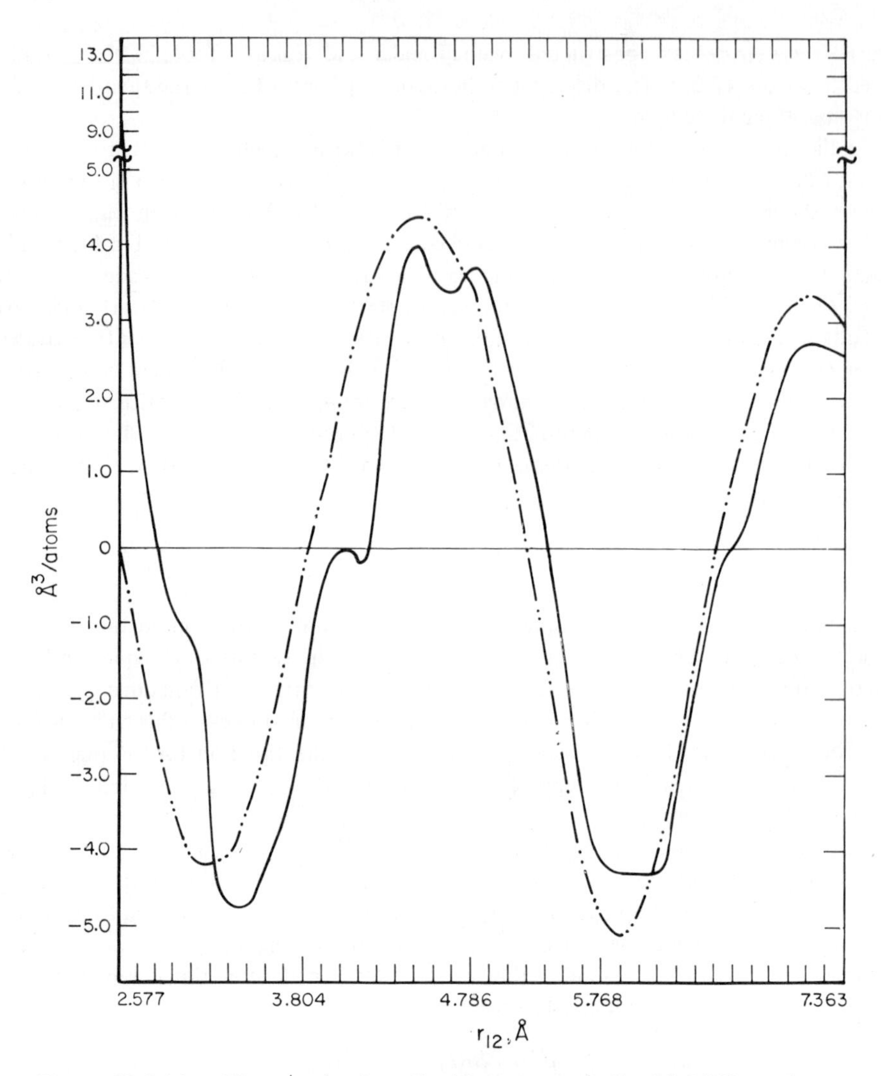

Figure 12.3.11 $F(r_{12} | g_3)$ for liquid neon at $T = 35.05$ K and $\rho = 3.339 \times 10^{22}/\text{cm}^3$ from Raveché and Mountain (1972). The solid curve is determined from the diffraction data of de Graaf and Mozer (1972) and the dashed-dotted curve is closure I, equation (12.3.6)

closure and then compare it with the known value of $F(r_{12} | g_3)$. Since the potential energy function does not enter explicitly, there are no problems with the assumption of pairwise additivity and there are also no problems associated with assumptions about higher order correlation functions than the third. But as we will later see, an accurate prediction of $F(r_{12} | g_3)$ is a necessary but not sufficient condition in establishing the validity of a closure.

The result of applying closure I to compute $F(r_{12} | g_3)$ for liquid neon at $\rho = 3.339 \times 10^{22}/\text{cm}^3$ and $T = 35.05$ K (Raveché and Mountain 1972) is shown by the dashed-dotted curve in Figure 12.3.11. The experimental result obtained from (12.3.3) and the known values of $(\partial g_2(r_{12})/\partial \rho)_T$ and $g_2(r_{12})$ is shown by the solid curve in Figure 12.3.11. Closure I is in good agreement with the data except at small values of r_{12}. The close agreement is due in part to the fact that the diffraction data show that the convolution term in (12.3.3)

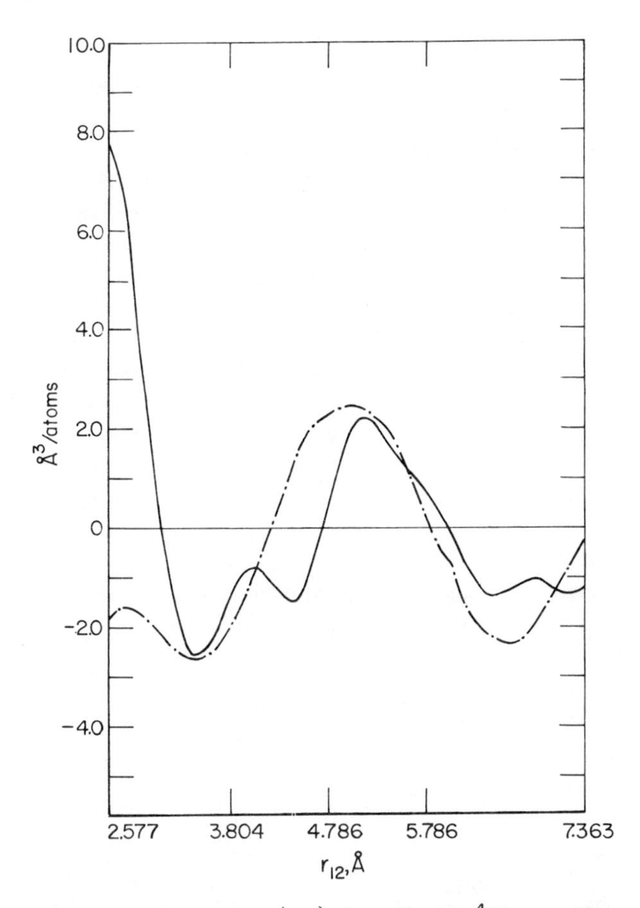

Figure 12.3.12 $F(r_{12} \mid g_3)$ for liquid ^{4}He at $T = 2.86$ K and $\rho = 2.63 \times 10^{22}/\text{cm}^3$ from Raveché and Mountain (1974). The solid curve is determined from the diffraction data of Mozer *et al.* (1974) and the dashed curve from closure I, equation (12.3.6)

and the three body term are comparable in magnitude, as a function of r_{12}, but out of phase. Applying (12.3.6) to (12.3.4) gives,

$$F(r_{12} \mid g_3) = \left(\frac{\gamma}{g_2(r_{12})} - 1 \right) \int d^3 r_3 h(r_{13}) h(r_{23})$$

and since the first bracketed term is small, except near the origin, the convolution term is the dominant contribution.

Winfield and Egelstaff (1973) report that from density derivative measurements closure I is qualitatively correct for krypton near its critical point. We remark that the necessary condition $G_3(r_{12} = 0, r_{13}, r_{23}) = 0$, is not satisfied with closure I.

Because of extensive numerical problems in constructing $F(r_{12} \mid g_3)$ from closures II and III we did not make a comparison of these with the diffraction data. Instead we compared the g_3 predicted by the closures for certain configurations of triplets. The results (Raveché and Mountain 1972) indicate that closure III would be better than II. We will shortly illustrate the actual accuracy of each closure from the computer simulation results.

Closure I was originally proposed for liquid ^{4}He and it has been investigated for this system by Raveché and Mountain (1974) in the course of studies on the structural aspects of the superfluid transition. The same relation between $(\partial g_2(r_{12})/\partial\rho)_T$ and g_3, namely (12.3.3), holds for quantum systems as well as classical systems. The result for $F(r_{12} \mid g_3)$ obtained from closure I is indicated by the dashed curve in Figure 12.3.12 and the result for $F(r_{12} \mid g_3)$ from the measured value of $(\partial g_2(r_{12})/\partial\rho)_T$ (Mozer *et al.* 1974) for helium at $\rho = 2.63 \times 10^{22}/\text{cm}^3$ and $T = 2.86$ K is shown by the solid curve in Figure 12.3.12. As with liquid neon, closure I also is in good agreement with the measured value of $F(r_{12} \mid g_3)$ for liquid helium and probably for the same reasons as was discussed in the classical case.

It is interesting to note, as general information, that the correlation functions in liquid ^{4}He behave anomalously as a function of temperature when compared with classical fluids. In a classical fluid the amplitudes of the correlation functions increase as the temperature is decreased at fixed density. This simply reflects the fact that the average local order in the fluid increases as the temperature is decreased at fixed density. But as the temperature of superfluid ^{4}He is decreased at fixed density, the amplitudes of $g_2(r_{12})$, $g_3(r_{12}, r_{13}, r_{23})$ etc. decrease. This is because of the relative increase in the macroscopic occupation of the ground momentum state of the fluid and the Heisenberg uncertainty principle. The occupation of the momentum state primarily reflects itself in the off-diagonal elements of the reduced density matrices and not the diagonal elements which correspond to the correlation functions g_n.

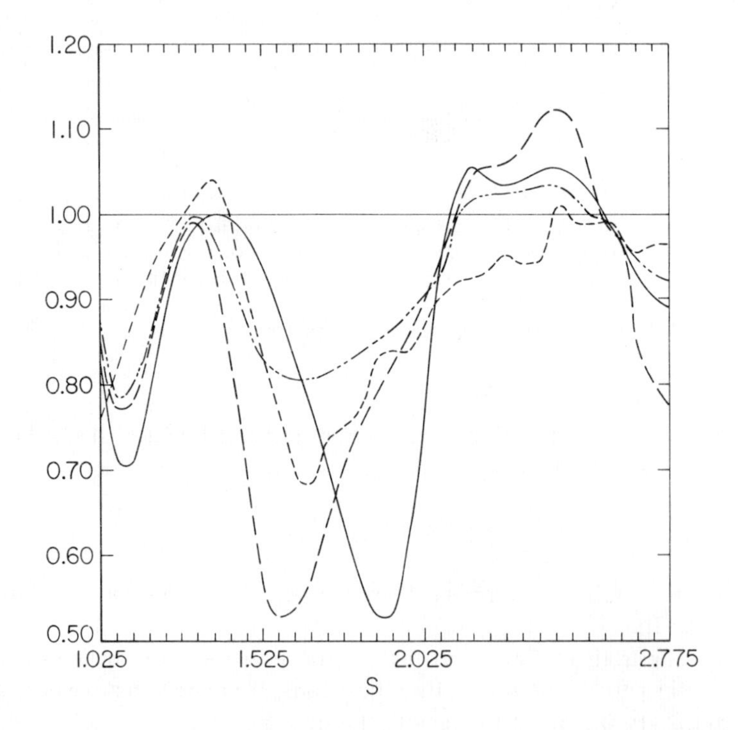

Figure 12.3.13 Results for $g_3(1.025, s, s)$ at $\rho = 0.850$ and $T = 0.719$ from Raveché *et al.* (1972). Short dashed curve is Monte Carlo result, the long dashed curve is closure I (equation 12.3.6), the solid curve is closure II (equation 12.3.7) and the long-short dashed curve is closure III (equation 12.3.8)

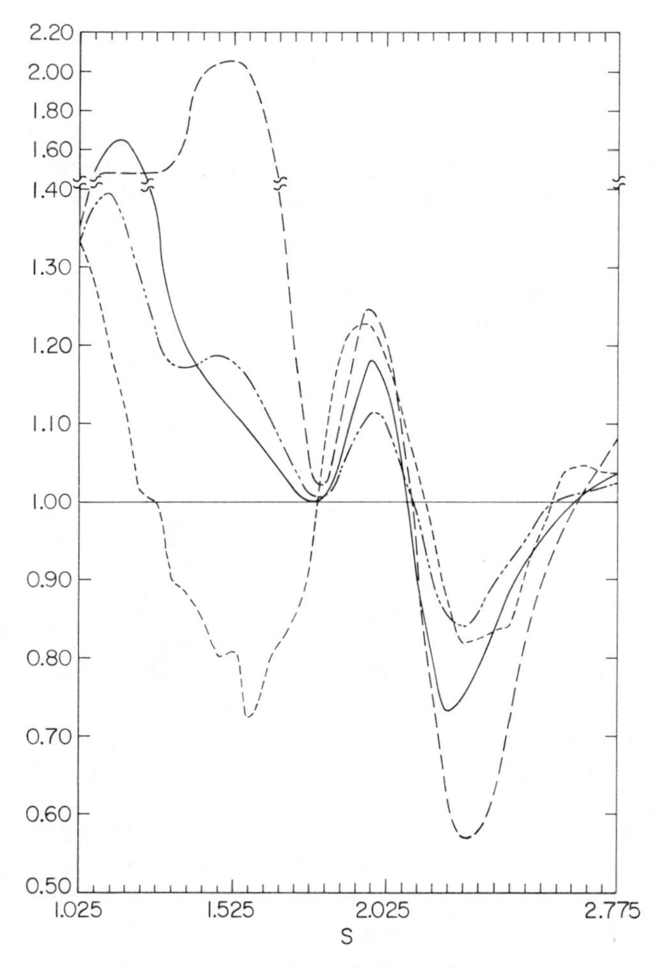

Figure 12.3.14 Results for $g_3(1.525,s,s)$ at $\rho = 0.850$ and $T = 0.719$ from Raveché *et al.* (1972). Same notation as in Figure (12.3.13)

We must now look in greater detail at the accuracy of the closures. To do this we must have exact results on $g_3(r_{12},r_{13},r_{23})$ for a given model or from experiment, and, since the latter is not currently available, we focus on the computer simulations.

Closures I (12.3.6), II (12.3.7) and III (12.3.8) have been investigated by Raveché *et al.* (1972) for the Lennard–Jones system at several densities and temperatures. We consider here the states near the triple point and the configurations corresponding to $g_3(1.025,s,s)$ and $g_3(1.525,s,s)$. The results at $\rho = 0.850$ and $T = 0.719$ are shown in Figure 12.3.13. All the closures are similar near $s \simeq 1.025$, but at intermediate separations, $s \simeq 1.525$, closures I and II are poor and closure III does moderately well. At somewhat large separations, $s \simeq 2.525$, closures I and II continue to be poor but closure II approaches unity faster than closure I which does not agree with exact results. On the basis of these results one would infer that closure III is the better approximation. However, as we examine other configurations we find that none of the closures are quantitative. The results for the same thermodynamic state but for $g_3(1.525,s,s)$ are shown in Figure 12.3.14 where the notation of each curve is exactly the same as in Figure 12.3.13. The results for $g_3(1.525,s,s)$ show that at intermediate configurations,

$s \simeq 1.525$, all the closures are very poor in that they seriously overestimate the triplet correlation function. The overestimation persists to even larger internuclear separations. The same trends obtain at lower densities and higher temperatures (Raveché *et al.* 1972), and on the basis of this we must conclude that none of the closures accurately reproduces the known structure in the triplet correlation function for the isosceles configurations. Now, it is clear that the isosceles configurations constitute only a small class of configurations; nevertheless an accurate closure must correctly predict all the structure present in g_3.

Abramo and Tosi (1972) have used the simulation results of Krumhansl and Wang (1972a,b) for the Lennard–Jones system to test closure II. They also find that this closure does not quantitatively reproduce the data. Similarly, Tanaka and Fukui (1975) have used their molecular dynamics results for the Paskin–Rahman potential in liquid sodium to test closures I and II. Although they are limited by somewhat poor statistics for g_3, they are also able to show that these closures do not reproduce the simulation results. Similar calculations were done by Block and Schommers (1975) using a special model potential for liquid rubidium; they also find that closures I and II are not quantitative.

12.4 Summary

Results for the triplet correlation function, $g_3(r_{12},r_{13},r_{23};\rho,T)$, in both real and model fluids have been reviewed. The diffraction data for the three body term $F(r_{12} \mid g_3)$, (12.3.4), given by the solid curves in Figures 12.3.3 and 12.3.12 established that in real fluids g_3 has a nonconstant amplitude with oscillates about unity over distances comparable to several molecular diameters. It is important to recall that the computation of $F(r_{12} \mid g_3)$ from the diffraction data does not involve any assumption about the intermolecular potential energy function. These results do not determine the magnitude of the deviation of g_3 from unity, but they do show that the view that the positions of only pairs of molecules are correlated cannot accurately reproduce the local order in real fluids. Triplet correlations must be considered.

The expression (12.2.14) for the configurational entropy is also independent of any assumption about the form of the intermolecular potential energy function. While the convergence of the entropy series remains to be mathematically proven, the results in Figures 12.3.1 and 12.3.2 imply that it is convergent. Assuming that this is true, we see that two and three body correlations determine more than 90% of the known configurational entropy of a fluid. Moreover, three body correlations must be included in order to reproduce accurately the entropy of a fluid. This has been shown for liquid neon, hard spheres and liquid rubidium. Tables 12.3.1 and 12.3.2 show that the Kirkwood approximation, which ignores triplet correlations, gives both the incorrect sign and magnitude for $S[G_3]$ defined by (12.3.1).

The importance of triplet correlations in determining thermal properties is emphasized by the fact that in real fluids the potential energy function is not only pairwise additive. Three body interactions exist, and when they are included in molecular expressions for the pressure and energy the triplet correlation function must also be included. This is illustrated in the pioneering work of Barker *et al.* (1968, 1969, 1971) where it is shown that assuming the triple dipole dispersion term is the dominant nonadditive interaction, the properties of real liquid argon can be reproduced with high accuracy over wide ranges of density and temperature.

We have seen in Section 12.2*b* that the triplet correlation function also plays an important role in theories of the pair correlation function. In the PY II, HNC II and BBGKY II theories, triplet correlations cause a substantial improvement over the original PY, HNC and BBGKY theories.

The actual structure present in $g_3(r_{12},r_{13},r_{23};\rho,T)$ has been investigated for the Lennard–Jones system at several thermodynamic states by computer simulations. The results for certain isosceles configurations are shown in Figures 12.3.4, 12.3.5 and 12.3.6. These show that near the fluid–solid phase boundary the triplet function can deviate by more than ±40% from unity, and that deviations as large as ±10% persist over large internuclear separations. As the thermodynamic state moves away from the fluid–solid phase boundary, the structure in g_3 rapidly decreases except for certain high symmetry points.

A broader view of the structure present in g_3 is shown by the contours of constant $g_3(r_{12} = 1.025, r_{13}, r_{23})$ in Figure 12.3.7 and also by the actual surface of this function which is shown at a different thermodynamic state in Figure 12.3.8.

The seemingly complicated structure of g_3 appears to have a simplifying aspect. Imagine the lattice structure of the crystalline phase with a lattice parameter that is determined by the density of the fluid. For configurations of triplets which fit on this lattice, g_3 is greater than unity, while for configurations which are not commensurate with the lattice, g_3 is less than unity. Therefore the local maxima and minima in the triplet function reflect the structure of the solid phase. As was discussed in Section 12.3*c*, the validity of this structural rule is subject to further investigation.

Since g_3 is a function of five variables, there is substantial motivation to find a closure for the function: that is, a relation which accurately expresses the triplet function in terms of the pair correlation function. We have, in Section 12.3*d*, reviewed the most widely used closures. The computer simulation results show that none of the closures are quantitative for all of the configurations which were investigated.

The present understanding of the structure of fluids is based primarily on studies of the pair correlation function. There is little doubt that the triplet correlations are significant in determining both microscopic and macroscopic properties of fluids at high densities. The direct laboratory measurement and the theoretical determination of the triplet correlation function are important unsolved problems.

References

Abe, R. (1959). *Progr. Theor. Phys. (Tokyo)*, **21** 421.
Abramo, M. C. and M. P. Tosi (1972). *Nuovo Cimento Lett.*, **5**, 1044.
Alder, B. J. (1955). *J. Chem. Phys.*, **23**, 263.
Alder, B. J. and T. E. Wainwright (1958). In I. Prigogine (Ed.), *International Symposium on Statistical Mechanical Theory of Transport Process,* Interscience, New York.
Alder, B. J. (1964). *Phys. Rev. Lett.*, **12**, 317.
Axilrod, B. M. and Teller, E. (1943). *J. Chem. Phys.*, **11**, 299.
Barker, J. A., P. J. Lenard and A. Pompe (1966). *J. Chem. Phys.*, **44**, 4206.
Barker, J. A., D. Henderson and W. R. Smith (1968). *Phys. Rev. Lett.*, **21**, 134.
Barker, J. A., D. Henderson and W. R. Smith (1969). *Mol. Phys.*, **17**, 579.
Barker, J. A., R. A. Fisher and R. O. Watts (1971). *Mol. Phys.*, **21**, 657.
Barker, J. A. and D. Henderson (1972). *Ann. Rev. Phys. Chem.*, **23**, 439.
Baxter, R. J. (1968). *Annls. Phys.*, **46**, 509.
Block, R. and U. Schommers (1975). *J. Phys.* C., **8**, 1997.
Buff, F. P. and R. Brout (1955). *J. Chem. Phys.*, **23**, 458.
Egelstaff, P. A., D. I. Page and C. R. T. Heard (1969). *Phys. Lett.*, **30A**, 376.
Egelstaff, P. A., D. I. Page and C. R. T. Heard (1971). *J. Phys.* C., **4**, 1453.
Egelstaff, P. A. (1973). *Ann. Rev. Phys. Chem.*, **24**, 159.
Fisher, I. Z. and B. L. Kopeliovich (1960). *Soviet Phys.–Doklady*, **5**, 761.
Feenberg, E. (1969). *Theory of Quantum Fluids*, Academic Press, New York. pp 28–36.
deGraaf, L. A. and B. Mozer (1971). *J. Chem. Phys.*, **55**, 4967.
Green, M. S. (1960). *J. Chem. Phys.*, **33**, 1403.
Hill, T. L. (1956). *Statistical Mechanics*, McGraw-Hill, New York, Chap. 5 and 6.
Katsura, S. and Y. Abe (1963). *J. Chem. Phys.* **39**, 2068.
Kirkwood, J. G. (1935). *J. Chem. Phys.*, **3**, 300.

Krumhansl, J. A. and Wang, S. (1972a). *J. Chem. Phys.*, **56**, 2034.
Krumhansl, J. A. and Wang, S. (1972b). *J. Chem. Phys.*, **56**, 2179.
Larsson, K. E., U. Dahlborg and K. Skold (1965). In H. L. Frisch and Z. W. Salsburg (Eds), *Simple Dense Fluids.* Academic Press, New York.
Lee, D. K., H. W. Jackson and E. Feenberg (1967). *Annls. Phys.*,**44**, 84.
Lee, Yon-Teh, F. H. Ree and T. Ree (1968). *J. Chem. Phys.*, **50**, 1581.
van Leeuwen, J. M. J., J. Groeneveld and J. deBoer (1959). *Physica,* **25**, 792.
Levesque, D. (1966). *Physica,* **32**, 1985.
Mayer, J. E. (1942). *J. Chem. Phys.*, **10**, 629.
Mayer, J. E. (1962). *J. Phys. Chem.*, **66**, 591.
McDonald, I. R. and K. Singer (1970). *Quarterly Reviews,* **XXIV**, 238.
Metropolis, N. A., A. W. Rosenbluth, M. N. Rosenbluth, A. H. Teller and F. Teller (1953). *J. Chem. Phys.*, **21**, 1087.
Morita, T. and K. Hiroike (1960). *Prog. Theoret. Phys. (Kyoto),* **23**, 1003.
Mountain, R. D. and H. J. Raveché (1971). *J. Chem. Phys.*, **55**, 2250.
Mozer, B., L. A. deGraaf and B. LeNeindre (1974). *Phys. Rev.*, **9**, 448.
Munster, A. (1974). *Statistical Thermodynamics,* Vol. II. Springer-Verlag, Berlin, Chap. XVI.
Nettleton, R. E. and M. S. Green (1958). *J. Chem. Phys.*, **29**, 1365.
Oden, L. D., D. Henderson and R. Chen (1966). *Phys. Lett.*, **21**, 420.
Paskin, A. and A. Rahman (1966). *Phys. Rev. Lett.*, **16**, 300.
Percus, J. K. and G. J. Yevick (1958). *Phys. Rev.*, **110**, 1.
Percus, J. K. (1964). In H. L. Frisch and J. L. Lebowitz (Eds), *The Equilibrium Theory of Classical Fluids,* Benjamin, New York.
Present, R. D. (1971). *Contemp. Phys.*, **12**, 595.
Rahman, A. (1964a). *Phys. Rev.* A, **136**, 405.
Rahman, A. (1964b). *Phys. Rev. Lett.*, **12**, 575.
Rahman, A. (1975). (Private communication).
Raveché, H. J. and R. D. Mountain (1970). *J. Chem. Phys.*, **53**, 3101.
Raveché, H. J. (1971). *J. Chem. Phys.*, **55**, 2242.
Raveché, H. J. and R. D. Mountain (1972). *J. Chem. Phys.*, **57**, 3987.
Raveché, H. J., R. D. Mountain and W. B. Streett (1972). *J. Chem. Phys.*, **57**, 4999.
Raveché, H. J. and R. D. Mountain (1974). *Phys. Rev.* A., **9**, 435.
Raveché, H. J., R. D. Mountain and W. B. Streett (1974). *J. Chem. Phys.*, **61**, 1970.
Ree, F. H. and W. G. Hoover (1964). *J. Chem. Phys.*, **4**, 1635.
Ree, F. H., Yong-Teh Lee and T. Ree (1971). *J. Chem. Phys.*, **55**, 234.
Ree F. H. (1971). In *Physical Chemistry: An Advanced Treatise,* Vol. VIII A. Academic Press, New York.
Rosenbaum, I. R. (1972). NOLTR 72–107 (National Technical Information Service, 5285 Port Royal Road, Springfield, Va.)
Rowlinson, J. S. (1968). In H. N. V. Temperley, J. S. Rowlinson and G. S. Rushbrooke (Eds), *Physics of Simple Liquids.* North Holland, Amsterdam.
Rushbrooke, G. S. (1960). *Physica,* **26**, 259.
Rushbrooke, G. S. (1965). In J. Meixner (Ed.), *Statistical Mechanics of Equilibrium and Non-Equilibrium.* North-Holland, Amsterdam.
Rushbrooke, G. S. (1968). In H. N. V. Temperley, J. S. Rowlinson and G. S. Rushbrooke (Eds), *Physics of Simple Liquids,* North Holland, Amsterdam.
Schmidt, P. W. and C. W. Tompson (1968). In M. L. Frisch and Z. W. Salsburg (Eds), *Simple Dense Fluids,* Academic Press, New York.
Schofield, P. (1966). *Proc. Phys. Soc.*, **88**, 149.
Stell, G. (1963). *Physica,* **29**, 517.
Stell, G. (1964). In H. L. Frisch and J. L. Lebowitz (Eds), *The Equilibrium Theory of Classical Fluids,* Benjamin, New York.
Tanaka, M. and Y. Fukui (1975). *Prog. Theor. Phys.*, **53**, 1547.
Verlet, L. (1960). *Nuovo Cimento,* **18**, 77.
Verlet, L. (1964). *Physica,* **30**, 95.
Verlet, L. (1965). *Physica,* **31**, 959.
Verlet, L. (1966). *Physica,* **32**, 304.
Verlet, L. and D. Levesque (1967). *Physica,* **63**, 254.
Wang, S. and Krumhansl, J. A. (1972). *J. Chem. Phys.*, **56**, 4287.

Wertheim, M. S. (1967). *J. Math. Phys.*, 8, 927.
Winfield, D. J. and P. A. Egelstaff (1973). *Can. J. Phys.*, **51**, 1965.
Wood, W. W. (1968). In H. N. V. Temperley, J. S. Rowlinson and G. S. Rushbrooke (Eds), *Physics of Simple Liquids*, Wiley, New York.
Wood, W. W. (1974). In E. G. D. Cohen (Ed.), *Fundamental Problems in Statistical Mechanics*, III, North-Holland, Amsterdam.
Yvon, J. (1937). *Fluctuations En Densitê*, Actualités Scientifiques et Industrielles, Vol. 542, Hermann et Cie, Paris.

Chapter 13

Recent Developments in Neutron Scattering

R. A. HOWE

Department of Physics,
University of Leicester, LE1 7RH, UK

13.1 Introduction

The liquid state is characterized by the absence of any permanent structure and by an overall isotropy resulting from the time-averaged motion of the constituent ions, atoms or molecules. In many ways it is useful to regard a liquid as the intermediate phase between a solid and a gas in which the long range order of the former has been greatly decreased but the complete thermal disorder of the latter has not been attained. It is the residual amount of comparatively short range order that we seek to quantify when describing the structure of a liquid.

The simplest description is based upon the distribution function approach and is concerned with the probability of certain structural configurations occurring in the liquid. In particular, the majority of work so far has centered upon the pair distribution function, $g(\mathbf{r})$, which for a given atom or molecule at the origin expresses the probability of another atom or molecule being a distance r away. In a liquid, it is naturally assumed that $g(\mathbf{r})$ is independent of direction on account of the isotropy referred to above.

For a simple, monatomic liquid $g(\mathbf{r})$ can be related[1,2] to the structure factor, $S(\mathbf{k})$, by the equation

$$S(\mathbf{k}) = 1 + (N/V) \int [g(\mathbf{r}) - 1] \exp(i\mathbf{k} \cdot \mathbf{r}) d\mathbf{r}$$

or, by Fourier inversion,

$$g(\mathbf{r}) = 1 + (V/8\pi^3 N) \int [S(\mathbf{k}) - 1] \exp(i\mathbf{k} \cdot \mathbf{r}) d\mathbf{k}.$$

Thus a knowledge of $S(\mathbf{k})$ over the full range of momentum exchange is equivalent to a complete knowledge of $g(\mathbf{r})$. Neutron scattering affords a method of measuring $S(\mathbf{k})$ although, contrary to the impression often given, $S(\mathbf{k})$ is not *directly* available from experiment and suitable approximations must be made.

The most commonly used method of measuring $S(\mathbf{k})$ by neutrons is by means of a monochromatic, two-axis spectrometer. A specimen is placed in a monochromatic beam of thermal neutrons and the intensity of scattered neutrons is measured as a function of angle by means of some movable counter of fixed counter array. All scattered radiation is detected, irrespective of any energy transfer that might occur. The conversion of these measured intensities to $S(\mathbf{k})$ is discussed in detail and the nature and extent of the approximations used are emphasized in the basic theory of neutron scattering as it appears in Sections 13.2 and 13.4.

Although descriptions of liquid structure based upon the pair distribution function are known to be inadequate for many purposes,[3] the experimental inaccessability of the three-body and higher order functions has prevented progress. Accordingly, the main emphasis in this chapter centres upon the experimental determination of the pair distribution functions in a wide variety of liquid systems using neutron diffraction techniques. A discussion of work on deriving the three-body functions is reserved until Section 13.10.

The idea throughout is to highlight the outstanding problems that currently face the experimentalist. For this reason, some of the data presented will necessarily be provisional, in some cases controversial and for the most part incomplete.

13.2 The theoretical framework of neutron scattering

The theory of neutron scattering in condensed matter is discussed extensively in the literature. In particular, Marshall and Lovesey[4] provide a very comprehensive treatment whilst the article by Lomer and Low[5] gives a more condensed review. The present approach will be to abstract the relevant equations from such articles and to leave out the rigorous justifications contained in the original works.

The generally accepted starting point is to define a double differential scattering cross-section, $d^2\sigma/d\Omega d\omega$ which represents the fraction of an incident neutron beam that is scattered into a solid angle $d\Omega$ within an energy range $\hbar d\omega$. It is usually normalized to the scattering per nucleus per unit area. If the scattering strength of a nucleus is measured in terms of a scattering length b then, adopting the first Born approximation, we can write

$$\left.\frac{d^2\sigma}{d\Omega d\omega}\right|_{\text{ideal}} = \frac{1}{2\pi} \int \exp(-i\omega t) dt \sum_{ll'} \langle t_l t_{l'} \exp\{i\mathbf{k} \cdot \mathbf{r}_l(t)\} \exp\{i\mathbf{k} \cdot \mathbf{r}_{l'}(0)\} \rangle \qquad (13.2.1)$$

where $r_l(t)$ is the position coordinate of the l-th nucleus at a time $t = t$, and $\hbar\mathbf{k}$ is the momentum imparted to the nucleus by the neutron. In practice, the scattered neutrons are counted by a detector whose sensitivity D depends upon the velocity of the neutron through the detector. Further, the count *rate* depends upon the neutron velocity, which for a free neutron is proportional to its momentum. Hence, a factor k_f/k_i is introduced to allow for the momentum change between the incident and scattered neutron. Thus, experimentally,

$$\left.\frac{d^2\sigma}{d\Omega d\omega}\right|_{\exp} = D(k_f) \frac{k_f}{k_i} \left.\frac{d^2\sigma}{d\Omega d\omega}\right|_{\text{ideal}}. \qquad (13.2.2)$$

The double differential cross-section can be divided into two components by separating the 'self' terms in which $l = l'$ from the 'interference' terms in which $l \neq l'$. Then we can write

$$\left.\frac{d^2\sigma}{d\Omega d\omega}\right|_{\text{ideal}} = \left.\frac{d^2\sigma^{\text{self}}}{d\Omega d\omega}\right|_{\text{ideal}} + \left.\frac{d^2\sigma^{\text{int}}}{d\Omega d\omega}\right|_{\text{ideal}} \tag{13.2.3}$$

where

$$\left.\frac{d^2\sigma^{\text{self}}}{d\Omega d\omega}\right|_{\text{ideal}} = \frac{1}{2\pi}\int \exp(-i\omega t)\,dt\,\langle b^2\rangle \sum_{l=l'} \langle \exp\{i\mathbf{k}\cdot\mathbf{r}_l(t)\}\exp\{i\mathbf{k}\cdot\mathbf{r}_{l'}(0)\}\rangle \tag{13.2.4}$$

$$\left.\frac{d^2\sigma^{\text{int}}}{d\Omega d\omega}\right|_{\text{ideal}} = \frac{1}{2\pi}\int \exp(-i\omega t)\,dt\,\langle b\rangle^2 \sum_{l=l'} \langle \exp\{i\mathbf{k}\cdot\mathbf{r}_l(t)\}\exp\{i\mathbf{k}\cdot\mathbf{r}_{l'}(0)\}\rangle. \tag{13.2.5}$$

Alternatively, we can divide the double differential cross-section into 'coherent' and 'incoherent' parts by writing

$$\left.\frac{d^2\sigma}{d\Omega d\omega}\right|_{\text{ideal}} = \left.\frac{d^2\sigma^{\text{coh}}}{d\Omega d\omega}\right|_{\text{ideal}} + \left.\frac{d^2\sigma^{\text{inc}}}{d\Omega d\omega}\right|_{\text{ideal}} \tag{13.2.6}$$

where

$$\left.\frac{d^2\sigma^{\text{coh}}}{d\Omega d\omega}\right|_{\text{ideal}} = \frac{1}{2\pi}\int \exp(-i\omega t)\,dt\,\langle b\rangle^2 \sum_{l=l'} \langle \exp\{i\mathbf{k}\cdot\mathbf{r}_l(t)\}\exp\{i\mathbf{k}\cdot\mathbf{r}_{l'}(0)\}\rangle \tag{13.2.7}$$

$$\left.\frac{d^2\sigma^{\text{inc}}}{d\Omega d\omega}\right|_{\text{ideal}} = \frac{1}{2\pi}\int \exp(-i\omega t)\,dt\{\langle b^2\rangle - \langle b\rangle^2\} \sum_{l=l'} \langle \exp\{i\mathbf{k}\cdot\mathbf{r}_l(t)\}\exp\{i\mathbf{k}\cdot\mathbf{r}_{l'}(0)\}\rangle. \tag{13.2.8}$$

Let us now introduce the van Hove[6] space-time correlation functions $G(\mathbf{r},t)$, $G_d(\mathbf{r},t)$ and $G_s(\mathbf{r},t)$. Given that an atom is situated at the origin at $t = 0$, the total correlation function $G(\mathbf{r},t)$ represents the probability of *any* atom being found at the position $\mathbf{r}$ at $t = t$. Similarly, the self correlation function $G_s(\mathbf{r},t)$ and the distinct correlation function $G_d(\mathbf{r},t)$ represent, respectively, the probability of the *same* atom and a *different* atom being at $\mathbf{r}$ at $t = t$. Then, rewriting equations (13.2.4) and (13.2.5), we obtain

$$\left.\frac{d^2\sigma^{\text{self}}}{d\Omega d\omega}\right|_{\text{ideal}} = \frac{N\langle b^2\rangle}{2\pi}\iint d\mathbf{r}\,dt\,\exp i(\mathbf{k}\cdot\mathbf{r} - \omega t)G_s(\mathbf{r},t) \tag{13.2.9}$$

$$\left.\frac{d^2\sigma^{\text{int}}}{d\Omega d\omega}\right|_{\text{ideal}} = \frac{N\langle b\rangle^2}{2\pi}\iint d\mathbf{r}\,dt\,\exp i(\mathbf{k}\cdot\mathbf{r} - \omega t)G_d(\mathbf{r},t) \tag{13.2.10}$$

or, in place of (13.2.7) and (13.2.8),

$$\left.\frac{d^2\sigma^{\text{coh}}}{d\Omega d\omega}\right|_{\text{ideal}} = \frac{N\langle b\rangle^2}{2\pi}\iint d\mathbf{r}\,dt\,\exp i(\mathbf{k}\cdot\mathbf{r} - \omega t)G(\mathbf{r},t) \tag{13.2.11}$$

$$\left.\frac{d^2\sigma^{\text{inc}}}{d\Omega d\omega}\right|_{\text{ideal}} = \frac{N\{\langle b^2\rangle - \langle b\rangle^2\}}{2\pi}\iint d\mathbf{r}\,dt\,\exp i(\mathbf{k}\cdot\mathbf{r} - \omega t)G_s(\mathbf{r},t) \tag{13.2.12}$$

where

$$G(\mathbf{r},t) = G_s(\mathbf{r},t) + G_d(\mathbf{r},t). \tag{13.2.13}$$

In many cases, the scattering of neutrons from condensed matter is most conveniently described in terms of the so-called scattering law, $S(\mathbf{k},\omega)$. This quantity measures the probability that a neutron, in collision, imparts a momentum $\hbar\mathbf{k}$ to the system and gains an amount of energy, $\hbar\omega$. $S(\mathbf{k},\omega)$ can be divided into self and interference components or coherent and incoherent components. We can relate these quite simply to the van Hove correlation functions. For example,

$$G_s(\mathbf{r}, t) = \frac{1}{(2\pi)^3} \iint S^{\mathrm{inc}}(\mathbf{k}, \omega) \exp[-(\mathbf{k} \cdot \mathbf{r} - \omega t)]\, \mathrm{d}\mathbf{k}\, \mathrm{d}\omega \tag{13.2.14}$$

$$G(\mathbf{r}, t) = \frac{1}{(2\pi)^3} \iint S^{\mathrm{coh}}(\mathbf{k}, \omega) \exp[-(\mathbf{k} \cdot \mathbf{r} - \omega t)]\, \mathrm{d}\mathbf{k}\, \mathrm{d}\omega. \tag{13.2.15}$$

It then follows that the double differential scattering cross-sections can be written as

$$\left.\frac{\mathrm{d}^2 \sigma^{\mathrm{coh}}}{\mathrm{d}\Omega \mathrm{d}\omega}\right|_{\mathrm{ideal}} = N \langle b \rangle^2 S^{\mathrm{coh}}(\mathbf{k}, \omega) \tag{13.2.16}$$

$$\left.\frac{\mathrm{d}^2 \sigma^{\mathrm{inc}}}{\mathrm{d}\Omega \mathrm{d}\omega}\right|_{\mathrm{ideal}} = N\{\langle b^2 \rangle - \langle b \rangle^2\} S^{\mathrm{inc}}(\mathbf{k}, \omega). \tag{13.2.17}$$

The treatment outlined above has illustrated the fact that by measuring $\mathrm{d}^2\sigma/\mathrm{d}\Omega\mathrm{d}\omega$ over all ranges of $\mathbf{k}$ and ω we could determine the van Hove correlation functions which, in turn, contain the structural and dynamical information of the system. In practice, such measurements can only be made over a limited range of $\mathbf{k}$ and ω. Even this requires extensive experimentation on a triple axis or time of flight spectrometer, and proves to be a formidable task for most systems. In the present context, we are interested only in the structural properties of the system, and $S(\mathbf{k},\omega)$ contains a surplus of information. As Powles[7] points out, our aim is to abstract the dynamics information from $S(\mathbf{k},\omega)$ and, hopefully, to be left with the task of performing a much simplified experiment to obtain the purely structural information.

With this in mind, we can attempt to remove the time-dependent features of $\mathrm{d}^2\sigma/\mathrm{d}\Omega\mathrm{d}\omega$ and $S(\mathbf{k},\omega)$ by integrating over all values of ω at constant $\mathbf{k}$. This integration is readily achieved by adopting what is usually referred to as the *static approximation* in which one assumes that the exchange of energy between the nucleus and the neutron is negligible (a condition that is well satisfied for X-rays but not, in general, for neutrons).

Consider a neutron of wave vector $\mathbf{k}_i$ scattered from a free nucleus of mass m through an angle 2θ and ending up with a final wave vector $\mathbf{k}_f$. In such a process, the neutron transfers a momentum $\hbar(\mathbf{k}_f - \mathbf{k}_i) = \hbar\mathbf{k}$ to the nucleus and acquires an additional energy $h\omega$ where, by simple physical considerations,

$$k^2 = 2k_i^2 \left[1 - \frac{m\omega}{\hbar k_i^2} - \left(1 - \frac{2m\omega}{\hbar k_i^2}\right)^{1/2} \cos 2\theta\right] \tag{13.2.18}$$

If we adopt the static approximation, then $\omega = 0$, $k_i = k_f$ and the above equation reduces to

$$k = 2k_i \sin\theta = (4\pi/\lambda_i) \sin\theta \tag{13.2.19}$$

making constant k equivalent to constant θ.

Then, from equation (13.2.1), we obtain the differential scattering cross-section

$$\left.\frac{\mathrm{d}\sigma}{\mathrm{d}\Omega}\right|_{\mathrm{ideal}}^{\mathrm{static}} = \int_{\mathrm{const}\,k} \left.\frac{\mathrm{d}^2\sigma}{\mathrm{d}\Omega \mathrm{d}\omega}\right|_{\mathrm{ideal}} \mathrm{d}\omega = \sum_{ll'} \langle b_l b_{l'} \exp\{i\mathbf{k} \cdot (\mathbf{r}_l - \mathbf{r}_{l'})\}\rangle. \tag{13.2.20}$$

We can now proceed in a manner analogous to that adopted for the double differential scattering cross-section and divide $d\sigma/d\Omega$ in to either 'self' and 'interference' or 'coherent' and 'incoherent' terms.

$$\left.\frac{d\sigma}{d\Omega}\right|_{\text{ideal}}^{\text{static}} = \left.\frac{d\sigma^{\text{self}}}{d\Omega}\right|_{\text{ideal}}^{\text{static}} + \left.\frac{d\sigma^{\text{int}}}{d\Omega}\right|_{\text{ideal}}^{\text{static}}$$

$$= N\langle b^2\rangle + \sum_{l\neq l'} \langle b\rangle^2 \langle \exp\{i\mathbf{k}\cdot(\mathbf{r}_l - \mathbf{r}_{l'})\}\rangle \tag{13.2.21}$$

or

$$\left.\frac{d\sigma}{d\Omega}\right|_{\text{ideal}}^{\text{static}} = \left.\frac{d\sigma^{\text{inc}}}{d\Omega}\right|_{\text{ideal}}^{\text{static}} + \left.\frac{d\sigma^{\text{coh}}}{d\Omega}\right|_{\text{ideal}}^{\text{static}}$$

$$= N\{\langle b^2\rangle - \langle b\rangle^2\} + \sum_{ll'} \langle b\rangle^2 \langle \exp\{i\mathbf{k}\cdot(\mathbf{r}_l - \mathbf{r}_{l'})\}\rangle. \tag{13.2.22}$$

For an isotropic system, the average value expression contained in the summation can be written in terms of an integral over r, involving the radial distribution function $g(r)$:

$$\langle \exp\{i\mathbf{k}\cdot(\mathbf{r}_l - \mathbf{r}_{l'})\}\rangle = V^{-1}\int \exp(i\mathbf{k}\cdot\mathbf{r})g(r)\,d\mathbf{r}. \tag{13.2.23}$$

The integral, however, can be shown to contain a delta function at the origin (see, for example, March[1]) and this can be removed to give

$$\langle \exp\{i\mathbf{k}\cdot(\mathbf{r}_l - \mathbf{r}_{l'})\}\rangle = V^{-1}\int \exp(i\mathbf{k}\cdot\mathbf{r})[g(r) - 1]\,d\mathbf{r}. \tag{13.2.24}$$

Equation (13.2.21) now can be written

$$\left.\frac{d\sigma}{d\Omega}\right|_{\text{ideal}}^{\text{static}} = N\langle b^2\rangle + \frac{N^2}{V}\langle b\rangle^2 \int \exp(i\mathbf{k}\cdot\mathbf{r})[g(r) - 1]\,d\mathbf{r} \tag{13.2.25}$$

and equation (13.2.22) as

$$\left.\frac{d\sigma}{d\Omega}\right|_{\text{ideal}}^{\text{static}} = N\{\langle b^2\rangle - \langle b\rangle^2\} + N\langle b\rangle^2 \left(1 + \frac{N}{V}\int \exp(i\mathbf{k}\cdot\mathbf{r})[g(r) - 1]\,d\mathbf{r}\right) \tag{13.2.26}$$

Further, by integration of (13.2.16),

$$\left.\frac{d\sigma^{\text{coh}}}{d\Omega}\right|_{\text{ideal}}^{\text{static}} = N\langle b\rangle^2 S(k) \tag{13.2.27}$$

where we have introduced the *static structure factor* $S(k)$ defined as

$$S(k) = \int_{\text{const } k} S^{\text{coh}}(\mathbf{k}, \omega)\,d\omega. \tag{13.2.28}$$

For consistency between equations (13.2.27) and (13.2.26) we now require

$$S(k) = 1 + (N/V)\int \exp(i\mathbf{k}\cdot\mathbf{r})[g(r) - 1]\,d\mathbf{r} \tag{13.2.29}$$

which gives

$$\left.\frac{d\sigma}{d\Omega}\right|_{\text{ideal}}^{\text{static}} = N\langle b^2\rangle + N\langle b\rangle^2\{S(k) - 1\} = N\{\langle b^2\rangle - \langle b\rangle^2\} + N\langle b\rangle^2 S(k). \tag{13.2.30}$$

Under the conditions of the static approximation, the detector efficiency becomes a simple, angular-dependent normalization constant, $\alpha(\theta)$, and from (13.2.2) by integration

$$\left.\frac{d\sigma}{d\Omega}^{coh}\right|_{exp}^{static} = \alpha(\theta)\left.\frac{d\sigma}{d\Omega}^{coh}\right|_{ideal}^{static} = \alpha(k)N\langle b\rangle^2 S(k) \tag{13.2.31}$$

$$\left.\frac{d\sigma}{d\Omega}^{inc}\right|_{exp}^{static} = \alpha(\theta)\left.\frac{d\sigma}{d\Omega}^{inc}\right|_{ideal}^{static} = \alpha(k)N\{\langle b^2\rangle - \langle b\rangle^2\}. \tag{13.2.32}$$

For neutrons, the conditions of the static approximation cannot be realized in practice. Experimentally, one usually measures the count rate of a detector which collects neutrons at a particular scattering angle, 2θ, and counts them with an efficiency that depends upon their velocity and hence their energy loss within the scattering medium. The first order differential scattering cross-section that is measured experimentally is therefore given by

$$\left.\frac{d\sigma}{d\Omega}^{coh}\right|_{exp}^{meas} = N\langle b\rangle^2 \int_{0 \atop \text{const } \theta}^{\infty} D(k)\frac{k_f}{k_i} S(\mathbf{k},\omega)\,d\omega \tag{13.2.33}$$

which means that $S(k)$ is not readily accessible by experiment unless one assumes that the static approximation can be used.

(Strictly, the above integral is bounded at the upper limit to a value determined by the energy of the incident neutrons since the free neutron cannot have a negative energy value. In general (and in the following treatment) this limitation is disregarded for thermal neutrons.)

Placzek[8] has, however, determined a relationship between the integral of equation (13.2.33) and the static structure factor, $S(k)$, by expanding the integral in energy moments of $S(\mathbf{k},\omega)$ where the n-th moment is given by

$$M_n(k_0) = \int_{0 \atop \text{at } k=k_0}^{\infty} \omega^n S(\mathbf{k},\omega)\,d\omega. \tag{13.2.34}$$

Egelstaff and Poole[9] and Enderby[10] both give a detailed description of the way in which this leads to an expression for the experimentally measured cross-section of equation (13.2.33) in terms of the energy moments and their derivatives with respect to k. They show that, for a detector of constant efficiency ($D(k) = 1$),

$$\left.\frac{d\sigma}{d\Omega}^{coh}\right|_{exp}^{meas}{}_{D=1} = N\langle b\rangle^2\Bigg(M_0^{coh}(k_0) - \frac{\hbar}{2E_0}\{M_1^{coh}(k_0) + k_0^2\dot{M}_1^{coh}(k_0)\} - \frac{\hbar^2}{16E_0^2}\{2M_2^{coh}(k_0) - [2k_0^2 + 4k_0^2]\dot{M}_2^{coh}(k_0) - 2k_0^4\ddot{M}_2^{coh}(k_0)\} + \ldots\Bigg). \tag{13.2.35}$$

For a detector of efficiency $D(k) = 1/k_f$, the corresponding equation is

$$\left.\frac{d\sigma}{d\Omega}^{coh}\right|_{exp}^{meas}{}_{D=1/k_f} = N\langle b\rangle^2\Bigg(M_0^{coh}(k_0) - \frac{\hbar k_0^2}{2E_0}\dot{M}_1^{coh}(k_0) + \frac{\hbar^2}{16E_0^2}\{[4k_i^2 - 2k_0^2]\dot{M}_2^{coh}(k_0) + 2k_0^4\ddot{M}_2^{coh}(k_0)\} + \ldots\Bigg) \tag{13.2.36}$$

where, with quantum mechanical corrections,

$$M_0^{coh} = S(k) \quad M_1^{coh} = \frac{\hbar^2 k^2}{2m} \quad M_2^{coh} = \frac{2\bar{K}k^2}{3m} + \frac{\hbar^2 k^4}{4m^2} + \frac{\hbar^2 k^2}{4m\bar{K}}\sum_j \langle S_{0j}\rangle + \ldots \tag{13.2.37}$$

In the above,

$$\langle S_{0j} \rangle = \left\langle \exp[\mathrm{i}\mathbf{k} \cdot (\mathbf{r}_0 - \mathbf{r}_j)] \times \frac{\partial^2 U}{\partial z_0 \partial z_j} \right\rangle \text{ is a thermal average}$$

$\bar{K}$ is the mean kinetic energy of the atoms, of mass m

U is the potential energy of the system

z_j is the component of r_j in the direction of $\mathbf{k}$.

In addition,

$$\dot{M} = \frac{\partial M}{\partial k} \quad \text{and} \quad \ddot{M} = \frac{\partial^2 M}{\partial k^2}.$$

If all the coherent suffixes of equations (13.2.35) and (13.2.36) are replaced by incoherent ones and $\{\langle b^2 \rangle - \langle b \rangle^2$ is substituted for $\langle b \rangle^2$ then expressions for

$$\left.\frac{\mathrm{d}\sigma}{\mathrm{d}\Omega}^{\mathrm{inc}}\right|_{\mathrm{exp}}^{\mathrm{meas}}$$

are obtained for the two detector frequencies. In the incoherent expressions,

$$M_0^{\mathrm{inc}} = 1 \quad M_1^{\mathrm{inc}} = M_1^{\mathrm{coh}} = \frac{\hbar^2 k^2}{2m} \quad M_2^{\mathrm{inc}} = \frac{2k^2 \bar{K}}{3m} + \frac{\hbar^2 k^4}{4m^2} + O(\hbar)^4. \tag{13.2.38}$$

In principle, if the moments of $S(\mathbf{k},\omega)$ can be calculated, the measured coherent and incoherent cross-sections can be fully corrected for the effects of departures from the static approximation.

It is apparent from these equations, however, that although the zeroth and first moments can be evaluated precisely, higher moments than these require a knowledge of the pair potential function, U. In the absence of such knowledge, only approximate values can be obtained for the second and higher moments. In principle, an approximate value of $S(k)$ allows one to calculate an approximate value for U using one of the models available (HNC, PY or BG). This value of U could then be used to correct the previous $S(k)$. Such iterative techniques do not appear to have been attempted and whether or not they would lead to a converging solution is in doubt since the pair potential is notoriously sensitive to the precise form of the structure factor and also to the model used.[11] Alternatively, one might attempt to calculate U and hence the derivative from a model system of the type available from computer simulation studies. Such an approach is currently being studied for the case of water,[12] where Placzek effects are large and need to be estimated accurately in order to obtain meaningful data over a wide range of k-values.

In the majority of work published on neutron scattering experiments, the Placzek corrections are small and equations (13.2.35) and (13.2.36) can be approximated by the expression given by Enderby,[10]

$$\left.\frac{\mathrm{d}\sigma}{\mathrm{d}\Omega}^{\mathrm{coh}}\right|_{\mathrm{exp}}^{\mathrm{meas}} = N\langle b \rangle^2 \left(S(k) + \frac{\bar{K}M_n}{3E_0 m} - \frac{\hbar^2 k^2}{2mE_0}\frac{1}{D(k)} \right) \tag{13.2.39}$$

where M_n is the neutron mass. This approximation is useful except for samples in which the nuclear mass, m, is small, or in which more than one atom type is present (see next section).

In conclusion, the quantity $S(k)$ has been shown to contain the complete structural information of the system as described in $g(r)$. It is, therefore, the required end product of our experimentation, and our aim will be to determine it as precisely as possible over as large a range of k as possible. This will minimize the truncation and cumulative errors involved in the Fourier transform of equation (13.2.29).

By consideration of the relationship between $S(k)$ and $g(r)$ expressed in equation (13.2.29), we can deduce certain important properties of the static structure factor that will assist us by providing a check on experimental values.

Firstly, the long wavelength limit has been shown to be determined uniquely from macroscopic thermodynamics and is given by

$$S(0) = Nk_{\mathrm{B}}T\chi_T V^{-1} \tag{13.2.40}$$

where χ_T is the isothermal bulk compressibility.

Secondly, if we assume that for real molecules $g(0) = 0$ then

$$\int_0^\infty [1 - S(k)]\,k^2\,\mathrm{d}k = 2\pi^2 N/V. \tag{13.2.41}$$

Finally, as shown by Faber[2] and others, as k tends to infinity, $S(k)$ tends to unity in the absence of any singularities in $g(r)$,

$$S(\infty) = 1. \tag{13.2.42}$$

13.3 Multicomponent systems

In the preceding discussion it has been assumed on a number of occasions that only one atom type was present in the system, although the element could well occur in a variety of isotopic forms. As a result of this, the derived equations can, in general, only be applied to simple, monatomic systems. The majority of current interest centres upon liquid systems containing more than one atom type, the so-called multicomponent systems of which the simplest example is a binary liquid. By this we mean an atomic mixture of two atom types rather than a diatomic molecular system. It is, therefore, necessary to generalize the preceding theoretical framework to accommodate systems with more than one type of component atom.

Let us first of all introduce the partial radial distribution function $g_{\alpha\beta}(\mathrm{r})$ to measure the average distribution of nuclei of type β observed from an α-type nucleus at the origin. For a liquid containing m components, α and β can assume any integer value from 1 to m, giving a total of $\frac{1}{2}m(m+1)$ distinct partial distribution functions for the system. By analogy with (13.2.29) we can follow Faber[2] and Enderby[10] and define a partial structure factor $S_{\alpha\beta}(k)$ by the equation

$$S_{\alpha\beta}(k)_{\mathrm{F-E}} = 1 + (N/V)\int [g_{\alpha\beta}(r) - 1]\exp(\mathrm{i}\mathbf{k}\cdot\mathbf{r})\,\mathrm{d}\mathbf{r}. \tag{13.3.1a}$$

Other authors, however, including Powles[7] and Page and Mika[13] prefer to define the partial structure factor by the equation

$$S_{\alpha\beta}(k)_{\mathrm{P-P}} = \delta_{\alpha\beta} + \frac{2Nc_\alpha c_\beta}{V(c_\alpha + c_\beta)}\int [g_{\alpha\beta}(r) - 1]\exp(\mathrm{i}\mathbf{k}\cdot\mathbf{r})\,\mathrm{d}\mathbf{r} \tag{13.3.1b}$$

where $c\alpha$ is the fraction of atoms of type α. They correctly argue that the unity occurring in (13.2.29) arises from the 'self' term of the interference functions and can only arise if α and β refer to the same atom type. The disadvantage of the Powles–Page structure factor is that it is necessarily concentration-dependent and, unlike the Faber–Enderby factor, does not consistently approach unity in the asymptotic limit.

In view of this disagreement of authors let us therefore define the partial structure integral, $\mathcal{I}g_{\alpha\beta}$, by the equation

$$\mathcal{I}g_{\alpha\beta} = (N/V)\int [g_{\alpha\beta}(r) - 1]\exp(i\mathbf{k}\cdot\mathbf{r})d\mathbf{r}. \tag{13.3.2}$$

The various partial structure factors are then simply different algebraic functions involving $\mathcal{I}g_{\alpha\beta}$ and the scattering cross-section for a multicomponent system can, by a simple generalization of equation (13.2.25), be written as

$$\left.\frac{d\sigma}{d\Omega}\right|_{\text{ideal}}^{\text{static}} = \underset{\text{(self)}}{N\sum_{\alpha=1}^{m} c_\alpha\langle b_\alpha^2\rangle} + \underset{\text{(interference)}}{N\sum_{\alpha=1}^{m}\sum_{\beta=1}^{m} c_\alpha c_\beta\langle b_\alpha\rangle\langle b_\beta\rangle\mathcal{I}g_{\alpha\beta}} \tag{13.3.3}$$

or from (13.2.26)

$$\left.\frac{d\sigma}{d\Omega}\right|_{\text{ideal}}^{\text{static}} = \underset{\text{(incoherent)}}{N\sum_{\alpha=1}^{m} c_\alpha\{\langle b_\alpha^2\rangle - \langle b_\alpha\rangle^2\}} + \underset{\text{(coherent)}}{N\sum_{\alpha=1}^{m} c_\alpha\langle b_\alpha\rangle^2 + N\sum_{\alpha=1}^{m}\sum_{\beta=1}^{m} c_\alpha c_\beta\langle b_\alpha\rangle\langle b_\beta\rangle\mathcal{I}g_{\alpha\beta}} \tag{13.3.4}$$

and for a simple binary system

$$\left.\frac{d\sigma^{\text{coh}}}{d\Omega}\right|_{\text{ideal}}^{\text{static}} = N\{c_1\langle b_1\rangle^2[1 + c_1\mathcal{I}g_{11}] + c_2\langle b_2\rangle^2[1 + c_2\mathcal{I}g_{22}] + 2c_1c_2\langle b_1\rangle\langle b_2\rangle\mathcal{I}g_{12}\} \tag{13.3.5}$$

The quantity $\mathcal{I}g_{\alpha\beta}$ is the desired end product since it contains the structural information of the system as defined by $g_{\alpha\beta}(r)$.

Although the partial structure integrals used in the above formulation offer a complete description of the structure of any multicomponent system, various authors have found it convenient to use an alternative approach when there is evidence of molecular behaviour. Egelstaff, Page and Powles[14,15] have attempted to abstract the molecular structure from the experimental distribution functions and discuss the residual liquid structure in terms of molecular orientational correlation coupled with a molecular centres distribution function. The underlying assumption is that the structure of the molecule is sufficiently well known to enable one to calculate the contribution to the neutron scattering from the intramolecular correlations. In order to apply the formulation to a multicomponent liquid system one must attempt to allow for the molecular distortion within the liquid. In the first instance, the molecular structure is normally derived from rotational spectra in the gaseous state and must be adapted to the high density liquid state and, secondly, the molecular structure within the liquid may well be distorted in accord with the molecular environment. Further, it is necessary to define the degree of orientational correlation as a function of molecular separation. Some authors have assumed this to be independent of distance but as Egelstaff *et al.*[14] point out, this is a highly implausible assumption for a liquid.

For a molecular system, we can write equation (13.2.20) as the sum of intramolecular and intermolecular components. Thus, in accord with Egelstaff *et al.*.[14]

$$\left.\frac{d\sigma}{d\Omega}\right|_{\text{ideal}}^{\text{static}} = N_m\langle|\sum_n\langle b_n\rangle\exp(i\mathbf{k}\cdot\mathbf{r}_{cn})|^2\rangle + \langle\sum_{i\neq j}\exp\{i\mathbf{k}\cdot(\mathbf{r}_i - \mathbf{r}_j)\}\sum_{n_in_j}\langle b_{n_i}\rangle\langle b_{n_j}\rangle\exp\{i\mathbf{k}\cdot(\mathbf{r}_{cn_i} - \mathbf{r}_{cn_j})\}\rangle \tag{13.3.6}$$

where the sum over n is a sum over all the nuclei within the molecule and i and j label different molecules such that n_i refers to the n-th nucleus of the i-th molecule, r_{cn} is the intramolecular distance of the n-th nucleus from the molecular centre c, and r_i and r_j are position coordinates of the molecular centres.

It is convenient to define a molecular form factor

$$f_1(k) = [\sum_n \langle b_n \rangle]^{-2} \langle | \sum_n \langle b_n \rangle \exp(\mathrm{i}\mathbf{k} \cdot \mathbf{r}_{cn}) |^2 \rangle \tag{13.3.7}$$

which would represent the scattering from an isolated molecule, and can be calculated for a given molecular structure.

The second term in equation (13.3.6) is dependent upon the molecular centres separation, r_{cn}, and the orientational correlation contained in $(r_{cn_i} - r_{cn_j})$, and to proceed further it is necessary to make one of a variety of assumptions. The simplest assumption is that the orientational correlation between molecules is completely independent of their separation (as mentioned above). The two summations contained in the second term are then statistically independent and can be separated. The first of these is now handled by defining a structure factor for molecular centres, $S_c(k)$, by analogy with the case of monatomic liquids, where

$$S_c(k) = 1 + N_m^{-1} \langle \sum_{i \neq j} \exp\{\mathrm{i}\mathbf{k} \cdot (\mathbf{r}_i - \mathbf{r}_j)\} \rangle. \tag{13.3.8}$$

Equation (13.3.6) can then be written as

$$\left. \frac{\mathrm{d}\sigma}{\mathrm{d}\Omega} \right|_{\text{ideal}}^{\text{static}} = N_m [\sum_n \langle b_n \rangle]^2 [f_1(k) + f_2(k)\{S_0(k) - 1\}] = N_m [\sum_n \langle b_n \rangle]^2 S_m(k) \tag{13.3.9}$$

where $S_m(k)$ is a molecular structure factor and

$$f_2(k) = [\sum_n \langle b_n \rangle]^{-2} \langle \sum_{n_i n_j} \langle b_{n_i} \rangle \langle b_{n_j} \rangle \exp\{\mathrm{i}\mathbf{k} \cdot (\mathbf{r}_{cn_i} - \mathbf{r}_{cn_j})\} \rangle \tag{13.3.10}$$

is the molecular orientation form factor.

Further progress necessitates a method of calculating $f_2(k)$, and this is only possible under certain circumstances. If one assumes that the molecular orientation is completely uncorrelated, then, from the above,

$$f_{2u} = [\sum_n \langle b_n \rangle]^{-2} [\sum_n \langle b_n \rangle \langle \exp(\mathrm{i}\mathbf{k} \cdot \mathbf{r}_{cn}) \rangle]^2 \tag{13.3.11}$$

as used by Zachariason.[16] In this case, if we know the molecular structure, f_1 and f_{2u} are calculable and $S_c(k)$ can be obtained by a single scattering experiment, as attempted by Egelstaff *et al.*[14] for CCl_4 and $GeBr_4$, and Page and Powles[15] for water (see Section 13.6).

Alternatively, if we assume that a particular orientational correlation exists (such as all molecules parallel) throughout the entire sample, then once again f_{2c} is calculable and $S_c(k)$ accessible by direct measurement.

Neither of these cases is realizable in a true liquid system and, in general, the range of orientational correlation can be represented by $C(R)$, which ranges from zero for a completely uncorrelated system to unity for one in which there is total correlation. Egelstaff[14] shows that, in general,

$$f_{2g}(k)\{S_c(k) - 1\} \cong N_m \int \exp(\mathrm{i}\mathbf{k} \cdot \mathbf{R})[g_c(R) - 1]\,[f_{2c}(k) + \{1 - C(R)\}f_{2u}(k)]\,\mathrm{d}\mathbf{R} \tag{13.3.12}$$

where $g_c(R)$ is a molecular centres pair distribution function and the equation is analogous to (13.2.29) for a monatomic liquid.

Let us now consider the corrections to the static approximation for these systems. The Placzek corrections were originally developed for simple atomic liquids and present difficulties when applied to multicomponent systems. In particular, the formulation (as outlined previously) contains only a single mass parameter which in a typical multicomponent system could vary from the lightest atom present to some comparatively large 'molecular' weight. The methods of handling the corrections have tended to vary according to the particular liquid system being studied. In general, the most serious attempts have been confined to systems which contain light atoms and therefore exhibit the most noticeable effects. Strictly speaking, the corrections will involve a summation of terms similar to those presented in (13.2.35) and (13.2.36) over a number of different types of nuclei, and Powles[7] has provided formulations of this for an ideal detector and a $1/k$ detector. Unfortunately, they include terms involving the average kinetic energy, $\bar{K}$, of *each* atom type; information that is not readily accessible.

The Placzek effects are, of course, most apparent at high k-values. At the other end of the spectrum, the long wavelength limit ($k \rightarrow 0$) of the partial structure functions has received considerable theoretical attention within the past few years. In a paper on the resistivity of liquid binary alloys, Bhatia and Thornton[17] have derived expressions for the long wavelength limits of the partial structure factors of equation (13.3.1a) in terms of measureable thermodynamic quantities. This can be applied to any two-component liquid system, as demonstrated by McAlister and Turner.[18] Beeby[19] has further extended this work to cover aqueous solutions.

Furthermore, the partial structure integrals obey general constraints similar to those outlined in Section (13.2) for the single element case. In particular, as pointed out by Enderby *et al.*,[20]

$$\int k^2 [\mathscr{I} g_{\alpha\beta}]\, \mathrm{d}k = -2\pi^2 (N/V). \tag{13.3.13}$$

Unlike the structure factor of a single element, there is no requirement that the *partial* structure factors remain positive. The analogous condition for a multicomponent system arises from the fact that the differential coherent cross-section of equation (13.3.4) must be positive, and this has been shown[2] to give rise to the following conditions for a binary mixture:

$$\begin{gathered} c_1 \mathscr{I} g_{11} \geqslant 1 \qquad c_2 \mathscr{I} g_{22} \geqslant 1 \\ [c_1 \mathscr{I} g_{11} + 1]\,[c_2 \mathscr{I} g_{22} + 1] \geqslant c_1 c_2 [\mathscr{I} g_{12}]^2 . \end{gathered} \tag{13.3.14}$$

13.4 Problems in extracting $S(k)$ from experimental data

In a typical neutron diffraction experiment, the intensity, $I(\theta)$, of neutrons scattered from a sample, which is usually in the form of a thin slab or cylinder, is measured as a function of the scattering angle, 2θ. In order to obtain the required cross-section,

$$\left.\frac{\mathrm{d}\sigma^{\mathrm{coh}}}{\mathrm{d}\Omega}\right|_{\mathrm{exp}}^{\mathrm{meas}}$$

from $I(\theta)$ it is necessary to make allowance for absorption, incoherent and multiple scattering effects and to normalize the intensities accurately. The situation is further complicated in the case of liquid samples by the necessary inclusion of some container. There is no simple generalization of these corrections and they must be calculated for the particular experimental configuration used, taking into account the geometry of the sample, container, incident beam and detector.

In general, we can write

$$I(\theta)=\gamma(\theta)\beta(\theta)\left\{\left[\frac{d\sigma}{d\Omega}^{coh}\Bigg|_{exp}^{meas}+\frac{d\sigma}{d\Omega}^{inc}\Bigg|_{exp}^{meas}\right](1-\delta)+\Delta_{inc}+\Delta_{coh}\right\} \tag{13.4.1}$$

where $\gamma(\theta)$ is the usual absorption correction, $\beta(\theta)$ is a normalization factor which allows for the number of scattering centres in the sample and the counter efficiency, and δ, Δ_{inc} and Δ_{coh} arise from multiple scattering effects. Let us look, briefly, at each of these correction factors.

For samples in the form of a thin slab of thickness, t, the absorption factor, $\gamma(\theta)$, is known analytically and is given by

$$\gamma(\theta)=t\sec\theta\exp(-\sigma_T t) \tag{13.4.2}$$

where σ_T is the total interactive cross-section per unit volume and includes both scattering and absorption terms. In the case of cylindrical and annular geometries, it is normal practice to use the numerical methods of Paalman and Pings[21] which, although originally devised for X-rays, can be applied directly to neutrons. They depend upon the sample dimensions, scattering angle and absorption cross-section; the latter is a function of the incident neutron energy.

The normalization constant, $\beta(\theta)$, was originally obtained from the limiting behaviour of $S(k)$, and hence $d\sigma^{coh}/d\Omega$, at high and low k-values. This method was somewhat inaccurate on account of the experimental inaccessibility of data outside a range of k-values of ~ 0.4 to 12 Å^{-1}. A far better technique is to compare the sample scattering with that of a piece of solid vanadium of corresponding shape and size.[22] Since the total scattering from vanadium is over 99% incoherent, it is essentially isotropic and enables one to make a simple normalization over all θ values. This method is now widely adopted and applies to almost all data published after 1968. The only requirement for accurate normalization is that the numbers of scattering centres in both the sample and the vanadium are accurately known.

The multiple scattering effects can, in principle, be calculated if one has a complete knowledge of the scattering law $S(\mathbf{k},\omega)$ and information on the geometry of the system. The corrections to total scattering data should be simpler than those for inelastic data but do, nevertheless, involve very complex computer calculations. As a consequence of the work by Vineyard,[23] who showed that, in the case of total scattering experiments, the multiple scattering from a thin slab was essentially isotropic, various approximations are commonly made. Let J_n represent the fraction of incident neutrons that are transmitted through the sample per unit solid angle which have experienced n scattering events. Vineyard obtained an analytic expression for J_1, and an integral expression for J_2. Furthermore, he showed that to a good approximation

$$\Delta=\sum_{n=2}^{\infty}J_n=J_2\,[1-(J_2/\bar{J}_1)]^{-1} \tag{13.4.3}$$

where $\bar{J}_1$ is the value obtained for J_1 by replacing $d\sigma/d\Omega$ by Nb^2. Assuming that the sample was of infinite lateral extent, he evaluated J_1 and J_2 and, using the above expression, established the isotropic nature of the total multiple scattering subject only to the condition that $\sigma_T t<1$. Subsequently, Cocking and Heard[24] have refined the work of Vineyard and adopting a more realistic expression for the cross-section and Blech and Averbach[25] have extended the discussion to cylindrical samples. In all cases, the multiple scattering is found to be isotropic to within a few percent. This isotropic nature of the multiple scattering suggests that the corrections to the data can be applied in the following simple manner. Δ_{inc} and Δ_{coh} are independent of θ and can be estimated from one or other of the formulations referred to above and subtracted from $I(\theta)$. These multiple scattered counts can now be put back into the measured cross-sections by a suitable choice of the factor δ, subject to the normalization condition

$$\sum_{\text{all }\theta} \delta \left[\frac{d\sigma}{d\Omega}^{\text{coh}} \Bigg|_{\text{exp}}^{\text{meas}} + \frac{d\sigma}{d\Omega}^{\text{inc}} \Bigg|_{\text{exp}}^{\text{meas}} \right] = \sum_{\text{all }\theta} (\Delta_{\text{inc}} + \Delta_{\text{coh}}). \tag{13.4.4}$$

We observe from this that the overall effect of multiple scattering is to reduce peak heights and fill in minima in $I(\theta)$. For structureless incoherent scattering the effect of multiple scattering tends to zero since, at any particular angle, as many counts are lost due to multiple scattering *to* other angular values as are gained by multiple scattering *from* other directions. On this basis multiple scattering effects in vanadium should be negligible.

The observed scattering from the sample + container must, in addition, be corrected for container effects. It is not sufficient to simply measure the scattering from the empty container and subtract this. Due allowance must be made for the fact that the scattering from the container in the absence of the sample is not the same as the scattering from the container in the presence of the sample. Since there will be some uncertainty in this correction, it is advisable to select a container such that the container scattering is a small fraction of the sample scattering. This is most readily achieved for slab samples. There are, however, practical difficulties attached to maintaining an accurate geometry with slab samples, particularly if the sample is heated, and for this reason annular containers are frequently adopted. Apart from having a small total cross-section, containment materials should, preferably, be incoherent scatterers. For this reason vanadium and null-matrix alloys (consisting of a random mixture of two elements whose coherent scattering lengths are of opposite sign) are the best materials to use. Materials which exhibit appreciable amounts of structure in the form of Bragg peaks etc. should be avoided since these peaks are modified by the presence of the sample, and consequently it is hard to make proper allowance for them. In addition, the presence of such peaks renders the assumption of isotropic multiple scattering somewhat suspect.

The correction procedures outlined above are now well established and, until recently, have proved adequate. At the present time, however, data are becoming available from higher flux neutron facilities in which the statistical uncertainty is very small. It now seems likely that we are approaching the point where a limitation on the final accuracy of the data will be imposed by the inadequacy of the correction techniques employed in the data reduction. In particular, the complete separation of absorption and multiple scattering corrections may lead to considerable errors in highly absorptive samples.

In certain cases the container corrections could possibly be derived experimentally, thus eliminating the errors incurred in the calculated corrections. To do this one must be able to construct a solid sample of the same dimensions as the liquid sample and which scatters neutrons in a similar manner to the liquid sample, i.e. it has a similar structure factor and total interactive cross-section. The ideal material would be some kind of doped glass. Two intensity measurements, one of this solid sample and one of this sample plus the container, will then give, by simple subtraction, the scattering from the container *in the presence of the sample.* Alternatively, the accuracy of the container, absorption and multiple scattering corrections could be tested by taking measurements of the neutron scattering from the liquid in a number of different sized containers. Neither of these procedures appears to have been adopted so far.

13.5 Current studies of molten salts

The simplest model of a molten salt is that based upon localized ionic charges interacting via coulombic forces. As such the pair potential, although long range, is precisely quantified. The main assumptions implicit in such a model are that the salt is completely ionized in the melt, the ions are spherically symmetric and they experience simple pairwise additive forces. The most significant problem currently associated with the structural studies of molten salts must be the extent to which such a model has to be modified to take into account the polarization effects recently discussed by Dixon and Sangster[26] or, in the limit, effects of covalency.

The potential power of neutron diffraction as a technique to be applied to the study of molten salts has been illustrated by Page and Mika.[13] The k-dependent neutron scattering, $F(k)$, from a molten salt MX_n can be written from equation (13.3.5) as

$$\frac{1}{N} F(k) = c_1^2 \langle b_M \rangle^2 \mathscr{I} g_{MM} + n^2 c_1^2 \langle b_X \rangle^2 \mathscr{I} g_{XX} + 2nc_1^2 \langle b_M \rangle \langle b_X \rangle \mathscr{I} g_{MX} \tag{13.5.1}$$

where $c_1 = (n+1)^{-1}$ and $c_2 = n(n+1)^{-1}$ are the fractional concentrations of the positive and negative ions respectively.

The neutron method enables b_M and b_X to be changed by isotopic substitution and hence, in principle, enables one to extract the three partial structure integrals. The success of the method depends upon selecting isotopes whose scattering amplitudes vary by sufficiently large amounts that the solution of the linear set of equations obtained from (13.5.1) is as insensitive

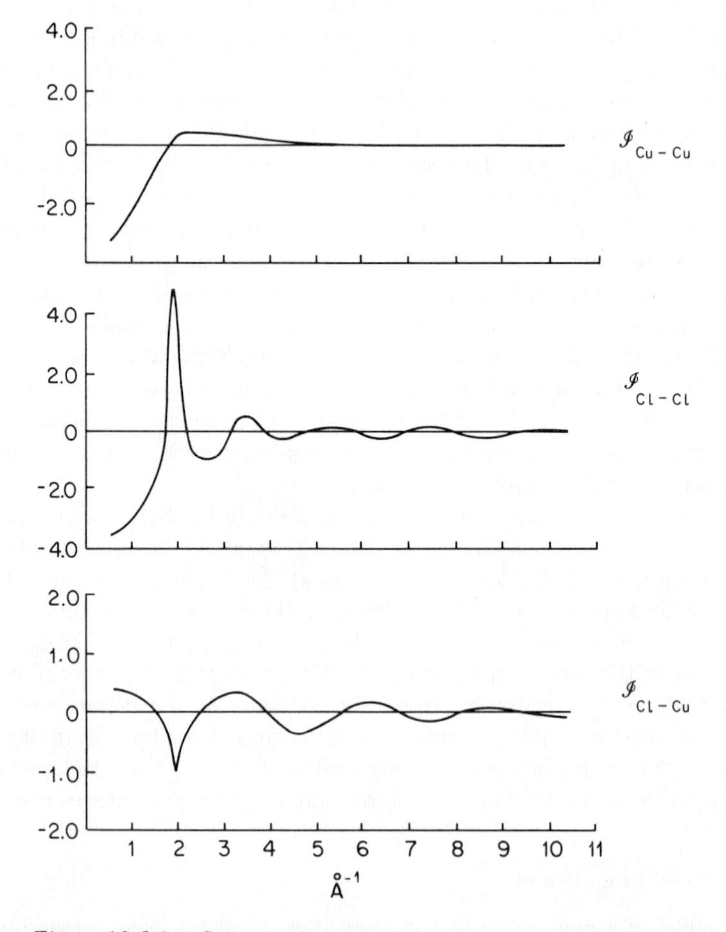

Figure 13.5.1 Smoothed partial structure integrals for molten CuCl (from Page and Mika[13])

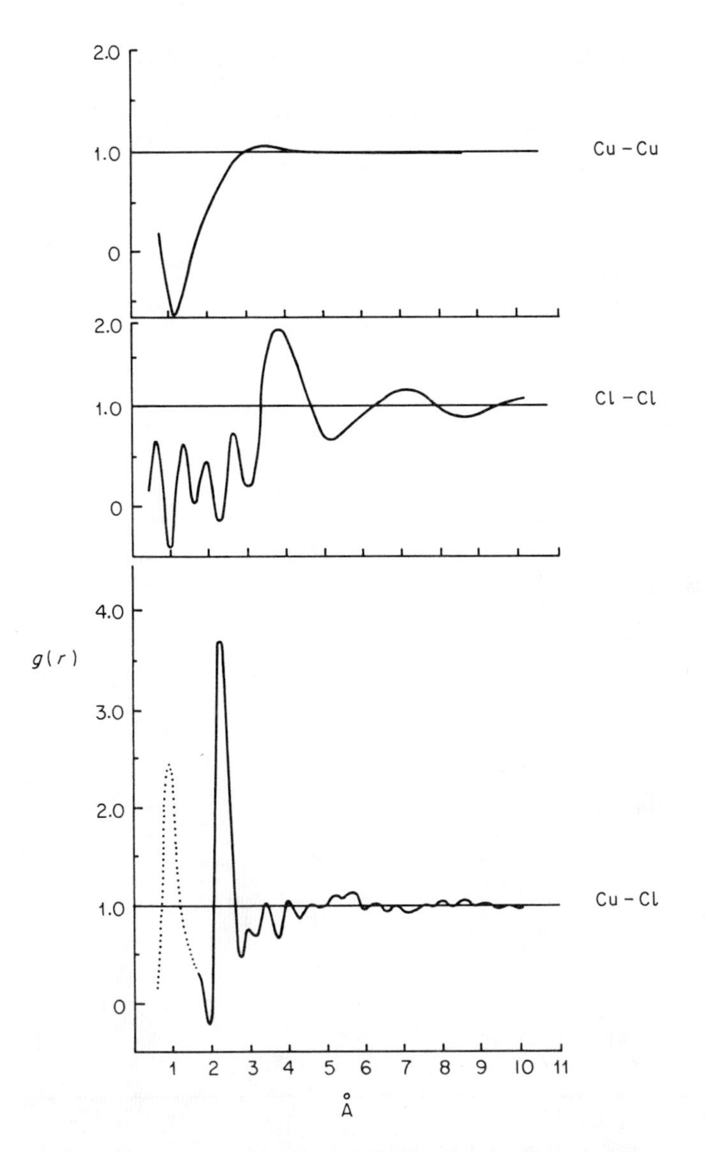

Figure 13.5.2 The partial radial distribution functions for liquid CuCl obtained from the smoothed partial structure factors (from Page and Mika[13])

as possible to experimental uncertainties. An inappropriate choice of isotopes (see Table 13.5.1) can produce ill-conditioned equations where one or more of the three partial structure integrals is indeterminate.[31]

Page and Mika attempted to minimize this effect by selecting liquid copper chloride as their system and using both chlorine (^{35}Cl, ^{37}Cl) and copper (^{63}Cu, ^{65}Cu) isotopes. In this way they were able to perform scattering experiments on four (rather than three) distinct isotopic mixtures and use the redundant information as an independent check. The choice of CuCl as the ideal system from the experimental point of view is somewhat unfortunate from a theoretical point of view. Although the transport properties (in particular the electrical conductance of $4.0\ \Omega^{-1}\ \mathrm{cm}^{-1}$ at 1200 K) and thermodynamic data[27] would suggest a high

Table 13.5.1 List of common, stable isotopes which have a useful variation in coherent scattering length

Element	z	Isotope	Coh. scatt.[a] length (10^{-12} cm)	Coh. cross-section σ^{coh} (barns)	Inc. cross-section[b] σ^{inc} (barns)	Abs. cross section[c, d] σ^{abs} (barns)
Ca	20	Nat Ca	0.466	2.73	0.5	0.25
		^{40}Ca	0.49	3.0	0.1	0.12
		^{44}Ca	0.18	0.4		
Cl	17	Nat Cl	0.958	11.55	3.4	19.1
		^{35}Cl	1.18	17.5	1.9	25.3
		^{37}Cl	0.26	0.85	0.0	0.0
Cu	29	Nat Cu	0.76	7.4	0.7	2.20
		^{63}Cu	0.67	5.6		2.6
		^{65}Cu	1.11	15.5		1.3
H	1	^{1}H	−0.372	1.8	79.8	0.190
		^{2}D	0.670	5.7	2.2	0.0
Fe	26	Nat Fe	0.951	11.4	0.4	1.50
		^{54}Fe	0.42	2.2		1.3
		^{56}Fe	1.01	12.8		1.5
		^{57}Fe	0.23	0.7		1.4
Li	3	Nat Li	−0.214	0.58	0.7	41
		^{6}Li	0.3 (+0.3i)	1.1		(540)
		^{7}Li	−0.233	0.68	0.8	
Ni	28	Nat Ni	1.03	13.3	4.6	2.7
		^{58}Ni	1.44	26.1		2.6
		^{60}Ni	0.282	1.0		1.5
		^{62}Ni	−0.87	9.5		8.6
Ag	47	Nat Ag	0.61	4.48	1.9	36
		^{107}Ag	0.83	8.7	1.3	18
		^{109}Ag	0.43	2.3	3.7	50
Ti	22	Nat Ti	−0.335	1.4	3.0	3.3
		^{46}Ti	0.48	2.9		0.3
		^{48}Ti	−0.58	4.2		4.7

[a] Source – MIT Compilation, C. G. Schull, Feb. 1971.
[b] Derived from total cross-sections of various sources.
[c] Absorption at $\lambda = 1.05$ Å.
[d] Obtained from Handbook of Physics and Chemistry, 1968.

degree of ionization in molten CuCl, measurements of the molar volume[28] give sufficient cause to doubt that CuCl is anything like an ideal, ionic salt in the liquid phase. Indeed, the partial structure integrals obtained by Page and Mika (shown in Figure 13.5.1) and their Fourier transforms (Figure 13.5.2) show substantial departures from simple ionic behaviour and they have attempted to explain these in terms of molecular complexes occuring in the liquid together with a plasma-like gas of itinerant Cu atoms. In a more recent article, Powles[29] has carried this a stage further and has suggested that the results are in line with a liquid model based solely on CuCl molecules with an atomic separation of 2.3 Å. The degree of covalency required in this model is, however, somewhat inconsistent with the observed electrical conductance. Furthermore, molecular dynamics calculations by Dixon and Sangster[26] on NaI

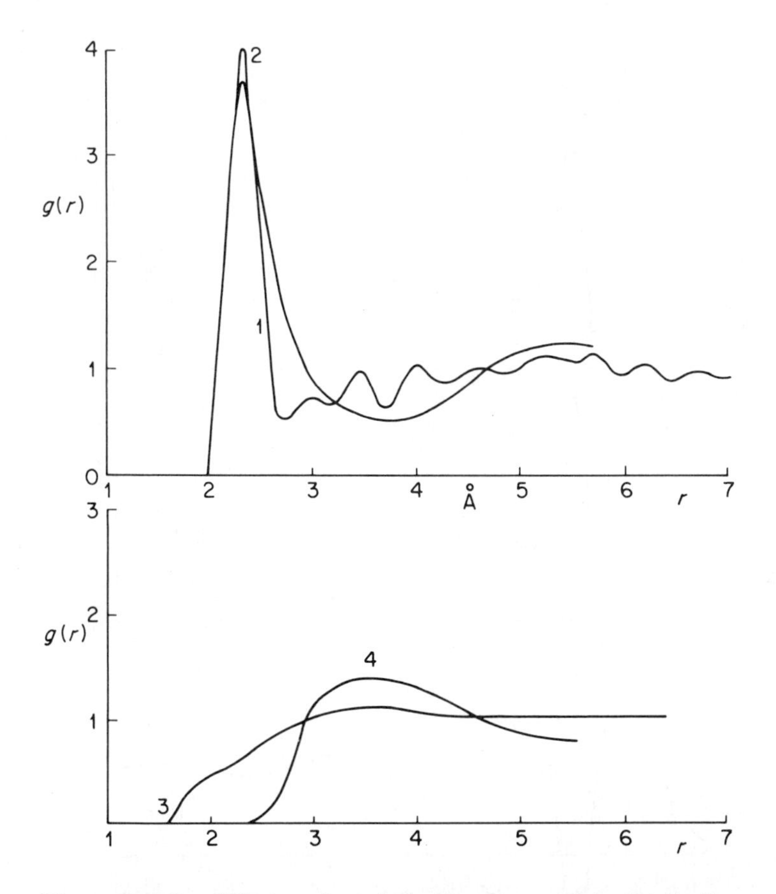

Figure 13.5.3 Effects of polarization on radial distribution functions. A comparison between the experimental radial distribution functions in molten CuCl (from Page and Mika[13]) with the molecular dynamics calculations of Dixon and Sangster[26] on molten NaI. (1) Experimental $g_{\mathrm{Cu-Cl}}$; (2) calculated g_{+-} (scaled) including polarization; (3) experimental $g_{\mathrm{Cu-Cu}}$; (4) calculated g_{++} (scaled) including polarization

show that the inclusion of polarization effects in the ionic model produces a flattening of the principal peak in $g_{++}(r)$ and an enhancement of the differences between $g_{++}(r)$ and $g_{--}(r)$ as compared with a simple rigid ion model. A comparison of their results (after renormalization to the density of CuCl) with those of Page and Mika are shown in Figure 13.5.3 and, whilst obvious differences exist, it is apparent that the polarization effects could go a long way towards explaining the anomalies in the CuCl system.

Attention has recently turned to the more ideal systems of molten alkali halides. Unfortunately, no suitable stable isotopes exist amongst the alkali metals, and the neutron scattering data now available for KCl[30] and NaCl[31] are obtained from chlorine isotope substitution experiments. The precise details of the partial structure factors of KCl as presented by Derrien and Dupuy[30] could be indicative of inadequacies in their data reduction techniques and, similarly, some of the oscillations in the radial distribution functions are not unlike those that arise from truncation errors (see Section 13.9). Nevertheless, the data clearly contribute to our understanding of simple ionic liquids and contrast sharply with the CuCl system. A marked

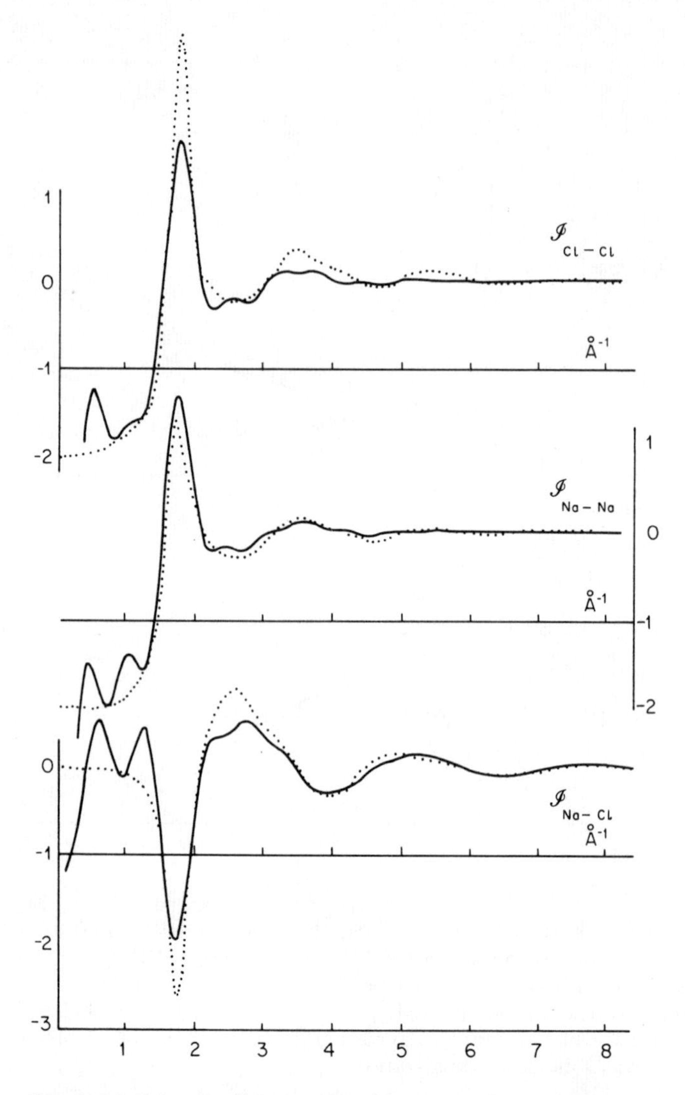

Figure 13.5.4 Partial structure integrals for NaCl (from Edwards *et al.*[31]). Experimental data (. . . . .) are compared with the Fourier transform of the radial distribution functions of Lantelme *et al.*[32] obtained from molecular dynamics

similarity exists between the Cl–Cl and K–K structure factors (and radial distribution functions) which would be expected in a simple ionic system of this kind with two similar-sized ions. Furthermore, the Cl–K radial distribution function tends to be 180° out of phase with the Cl–Cl and K–K radial distribution functions, indicating the distinct possibility of charge oscillations in the liquid.

When all the experimental corrections and theoretical approximations are taken into account, it is highly unlikely that experimental data can ever be obtained with sufficient accuracy to enable a direct calculation of useful pair potentials for molten salts. The only alternative approach is to calculate the pair distribution functions and structure factors from various model potentials and by comparing the results with the experimentally determined

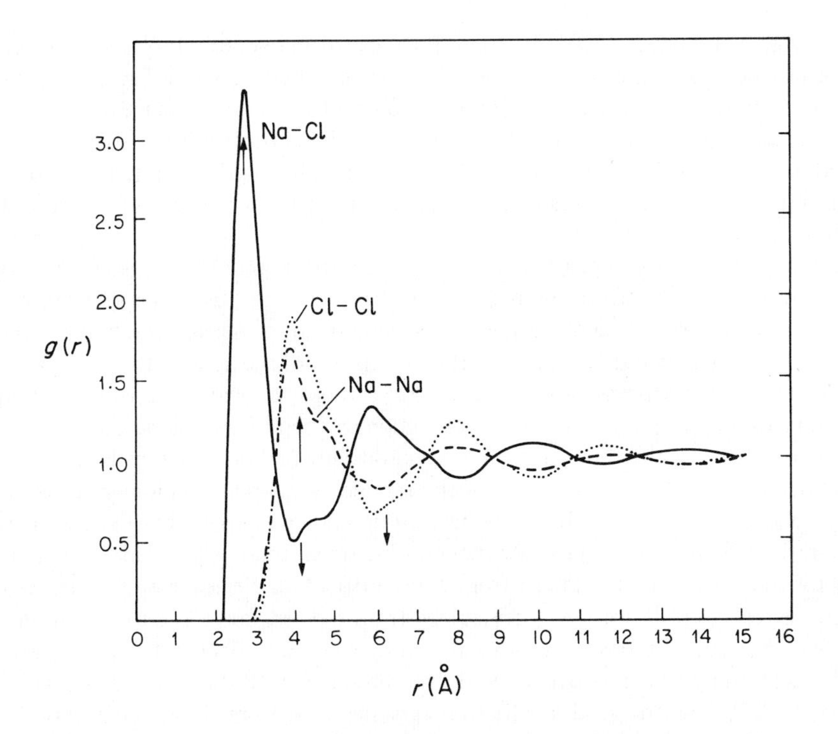

Figure 13.5.5 Partial radial distribution curves for molten sodium chloride. The arrows indicate the positions of the maxima (↑) and minima (↓) in the calculated values of Lantelme *et al.*[32]

quantities arrive at the most representative model potential. In particular, it is of fundamental interest to determine the extent to which a simple rigid ion potential can reproduce the measured distribution functions. Such systems can then be described as 'ideally' ionic. Departures from this ideal behaviour can often be accounted for, at least in a qualitative sense, by introducing polarisation effects, one method being the use of shell model potentials.[26] In the extreme case of covalency, it may be possible to identify certain features in the experimentally determined distribution functions which characterise such a state. In particular, a very sharp peak in the unlike partial distribution function is frequently observed in systems known to be covalently bonded.

Recent measurements on isotopically enriched samples of molten NaCl at 1150K by Edwards *et al.*[3.1] are compared to the theoretical calculations of Lantelme *et al.*[32] in Figures 13.5.4 and 13.5.5. The calculations are based upon simple rigid ion potentials.[33] The excellent agreement demonstrates that NaCl is a good approximation to an 'ideal' 1:1 ionic system. Such a system has a number of characteristic properties: $\mathscr{I}_{11}$ and $\mathscr{I}_{22}$ are very similar and have well defined oscillations, likewise g_{11} and g_{22}; g_{12} has a broad first peak and a subsequent liquid-like structure to high r; charge oscillations ($g_{11}g_{22}$ out of phase with g_{12}) and charge cancellation are evident and there is little or no fine structure.

13.6 The structure of water

The existence of two stable hydrogen isotopes with coherent scattering lengths of −0.372 and +0.67 x 10^{-12} cm suggests that, in principle, one could perform the standard isotopic substitution experiments on water, similar to those discussed in the preceding section on molten salts, and obtain the three partial structure integrals for the system. In practice, however, the existence of a very large incoherent cross-section in light hydrogen (see Table 13.5.1) obscures the structural information contained within the coherent scattering. This means that in order to obtain sufficiently good statistics for the differential coherent scattering cross-section very large numbers of neutrons (of the order of 1 million per step) have to be counted. Nevertheless, the importance of water and the diversity of opinion regarding its structure would suggest that such an experiment is well worth while.

Powles *et al.*[34] have undertaken isotopic substitution of hydrogen in water, but concentrated their efforts upon one particular isotopic composition rather than attempting the three independent measurements required to evaluate the three partial structure integrals. They realized that by taking advantage of the difference in sign of the coherent scattering lengths for deuterons and protons it was possible to devise a sample of water for which the hydrogen contribution to the coherent scattering cross-section was zero. This assumes, of course, that the distribution of hydrogen within the sample is isotopically random. They measured the neutron scattering from such a 'semi-transparent' water sample over a range of momentum exchange up to $\hbar k_{max}$ where $k_{max} = 6\,\text{Å}^{-1}$. In analysing their data, they assumed that there were negligible changes in the structure of water consequent on the substitution of protons for deuterons, and support for this argument is obtained from X-ray diffraction measurements.[35] In their unique mixture, the coherent scattering arises solely from oxygen–oxygen correlation and can be interpreted, crudely, in terms of a molecular centres structure factor if one assumes that the molecular centre in water is associated with the oxygen nucleus. Indeed, the peaks observed at 2, 3 and $5\,\text{Å}^{-1}$ are in good qualitative agreement with model calculations.[15] A full, quantitative discussion of the data is difficult in view of the severe effects of the Placzek corrections for light nuclei such as hydrogen. Springer *et al.*[36] used a number of incident neutron wavelengths in order to determine the importance of Placzek effects and showed that in water they vary from around 3% at $2\,\text{Å}^{-1}$ to 20% at $10\,\text{Å}^{-1}$. Attempts at deriving the higher order Placzek terms for a multicomponent system are frustrated by the inability to define a single effective mass for the nuclei. Conversely, one possible approach is to extend the experimental data to sufficiently high k values to enable an effective mass to be obtained from the Placzek corrections required to flatten out the data. Using results of Sinclair and Day for heavy water up to a k value of $25\,\text{Å}^{-1}$, Page and Powles[15] showed that an effective mass of about 6.5 A.U. produces a flat high k plateau in the data: the figure of 6.5 A.U. is roughly in line with the average mass of the nuclei within the molecule.

It is useful to observe at this point that the leading terms in the Placzek corrections apply only to the *self* term of equation (13.3.3). Since $S(k)$ is contained within the *interference* term we can conclude that the structural information is less affected by the Placzek corrections than is the overall scattering.

Although the structure of water has been investigated by neutron diffraction techniques as early as 1958[36,37] the most extensive study to date is that of Page and Powles.[15] They used heavy water (for the reasons mentioned) and in the absence of any attempt to change the isotopic composition they interpreted their results in terms of the molecular liquids framework, adopting equation (13.3.9). Values for $S_c(k)$ were determined by calculating f_1 and f_2 for a number of molecular structures from ice-like to gas-like and varying the range of orientational correlation from zero to infinity. A plot of the experimentally derived $S_m(k)$ together with the calculated f_1 appears in Figure 13.6.1. Rearrangement of (13.3.9) gives

$$S_0(k) = 1 + \frac{S_m(k) - f_1(k)}{f_2(k)} \tag{13.6.1}$$

and, since $S_c(k)$ is finite, this means that any zeros in $f_2(k)$ must be accompanied by

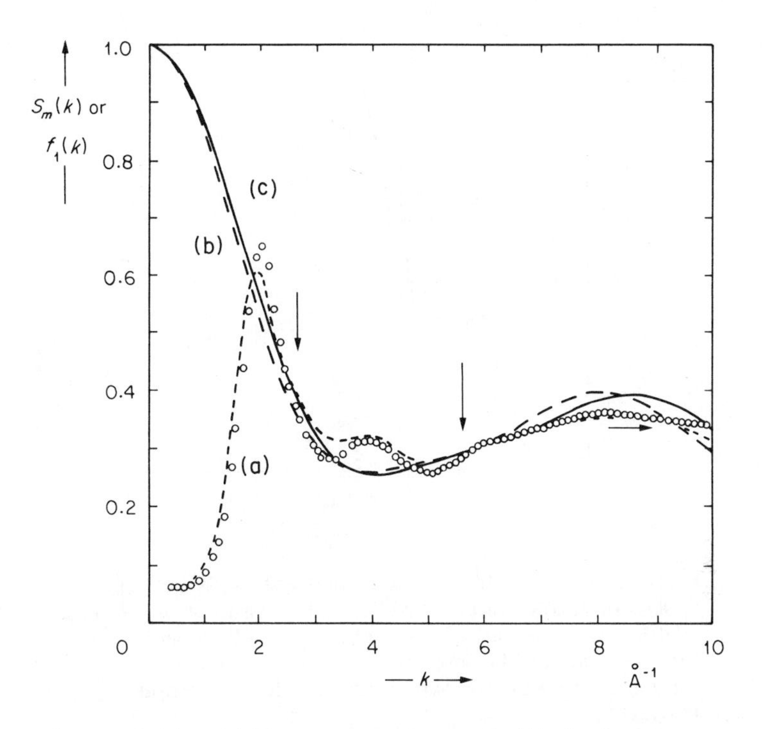

Figure 13.6.1 (a) The structure factor, $S_m(k)$, for heavy water at 22°C together with the molecular form factor, $f_1(k)$, for the molecule; (b) — — — as in ice; and (c) —— as in vapour. Zeros in $[S_m(k) - f_1(k)]$ are indicated by arrows (from Page and Powles[15]). (d) The data of Narten for pure D_2O is shown for comparison

corresponding zeros in $[S_m(k) - f_1(k)]$. Page and Powles suggested that the absence of such a correspondence was sufficient reason to invalidate any particular model structure used for f_2. On this basis, they concluded that both ice-like structures and the so-called clathrate model for water are ruled out. In addition, they calculated $f_{2u}(k)$ for totally uncorrelated orientation and although this function contains no zeros the resulting $S_c(k)$, shown in Figure 13.6.2, exhibits a maximum at ~ 4 Å^{-1}, which is sufficiently implausible to suggest that one cannot adopt such a model for water.

More recently, neutron diffraction data by Narten[38] has thrown further light on this problem. His results, whilst in broad agreement with those cited above, exhibit appreciable differences in the important region of k-space from 2 to 5 Å^{-1} This, he explains, is due in part to the use of a more up-to-date value for the coherent scattering length of deuterium (see Table 13.5.1) but also reflects, in the present author's opinion, the existing degree of uncertainty in such measurements. In combining neutron data with X-ray data, Narten also derives a set of partial distribution functions for the system based on the assumption that the X-ray scattering is predominantly due to oxygen and directly yields the oxygen–oxygen distribution function. A comparison of his results with the molecular dynamics calculations of Rahman and Stillinger[39] suggests that in the liquid state the water molecule undergoes considerable relaxation, and Narten attempts to estimate the distribution of intramolecular distances and angles about the mean. This again underlines the limited usefulness of a molecular approach that is based on a single, recurring molecular structure. Finally, Narten suggests that a

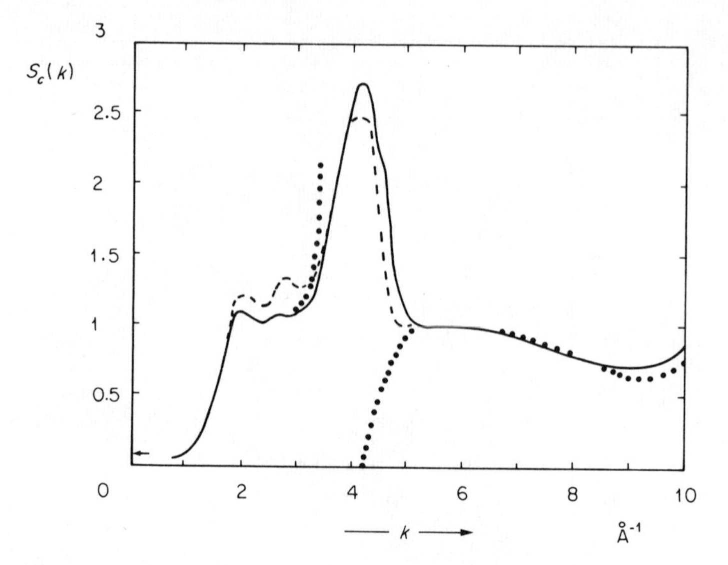

Figure 13.6.2 Deduced molecular-centres structure factor, $S_c(k)$, for heavy water at 22°C on the assumption of no correlation of orientation for (a) --- ice-like molecule, and (b) —— vapour-like molecule. Also shown in (c) · · · · is $S_c(k)$ for ice correlation for vapour-like molecules. (From Page and Powles[15])

molecular model based on only *near neighbour* orientational correlation is sufficient to explain his results, although it is necessary to invoke a significantly larger range of positional correlation of molecular centres.

At first sight, such usage of X-ray data in conjunction with neutron data seems very attractive. There are, however, two associated problems. Firstly, there is no method by which the X-ray data can be normalized to the extent that is possible for neutrons by the usage of vanadium. Secondly, the results are confused by a lack of information regarding the molecular-electron-density distribution; the most common assumption is that this can be approximately represented by a simple superposition of atomic-like distributions. This topic is discussed again in Section 13.9.

In summary, it appears that further investigations are necessary of high statistical quality, particularly in the mid- and high-k regions, before a complete picture of the structure of water can be assembled. At the present time it seems that water can be described in terms of relaxed molecules, with a molecular centres separation of around 2.8 Å. There is evidence of a limited amount of orientational correlation combined with a longer range positional correlation.

13.7 Aqueous solutions

The potential power of neutron diffraction combined with isotopic substitution has long been appreciated,[40] but not until recently[41,42] has it been exploited as a technique to unravel the complex systems of aqueous solutions. This is more easy to understand if we recognize that, in order to describe the structure of an aqueous solution in terms of partial structure factors, we require *ten* such factors. The differential neutron scattering cross-section for a system consisting of a salt MX_n dissolved in heavy water can, by expansion of equation (13.3.3), be written as

$$\frac{1}{N}\frac{d\sigma}{d\Omega}\bigg|_{\text{ideal}}^{\text{static}} = (1-c)[\langle b_O^2\rangle + 2\langle b_D^2\rangle] + c[\langle b_M^2\rangle + n\langle b_X^2\rangle]$$
$$+ (1-c)^2[\langle b_O\rangle^2 \mathscr{I}g_{OO} + 4\langle b_D\rangle^2 \mathscr{I}g_{DD} + 4\langle b_O\rangle\langle b_D\rangle \mathscr{I}g_{OD}]$$
$$+ c^2[\langle b_M\rangle^2 \mathscr{I}g_{MM} + n^2\langle b_X\rangle^2 \mathscr{I}g_{XX} + 2n\langle b_M\rangle\langle b_X\rangle \mathscr{I}g_{MX}]$$
$$+ 2c(1-c)[\langle b_O\rangle\langle b_M\rangle \mathscr{I}g_{OM} + 2\langle b_D\rangle\langle b_M\rangle \mathscr{I}g_{DM}$$
$$+ n\langle b_O\rangle\langle b_X\rangle \mathscr{I}g_{OX} + 2n\langle b_D\rangle\langle b_X\rangle \mathscr{I}g_{DX}]. \qquad (13.7.1)$$

Here $\langle b_0 \rangle$ represents the isotopically averaged value for the oxygen coherent scattering length and $\langle b_0^2 \rangle$ the isotopically averaged value of the squares; c is the molecular concentration of the solute and the subscripted notation is self-evident. The neutron method now enables us to make a systematic variation of the scattering lengths by changing the isotopic concentrations. As in the case of water it is assumed that the structure of the system is invariant under such changes; an assumption that is more easily justified for the heavy solute nuclei than for hydrogen and deuterium. Then, by successive measurements of the scattered intensity, the partial structure integrals can, in principle, be determined.

Now, at first sight, equation (13.7.1) looks hopelessly complicated and it is natural to ask whether, in spite of its fundamental nature, it can lead to a practical solution. It is possible, however, to concentrate upon certain aspects of the system. For example, simple algebra reveals that only three isotopic substitutions are required to obtain any one of the 'self' partial structure integrals and that only six substitutions are needed to yield the three solute partials – $\mathscr{I}g_{MM}$, $\mathscr{I}g_{MX}$ and $\mathscr{I}g_{XX}$. Any one of these fundamental quantities would contribute immensely to our knowledge of the physical processes governing the degree of solubility and would help resolve the discrepancies in the variety of models of aqueous solutions.[43]

Structural studies of aqueous solutions can be broadly classified into two areas: those concerned with the changes that occur in the structure of water on the introduction of a solute, and often described loosely in terms of the so-called hydration phenomena, and those concerned with the positional correlation of the solute ions themselves.

Until comparatively recent times, neutron diffraction had not been applied to aqueous solutions and the only information available came from the extensive X-ray work that has gone on from as early as 1929.[44,45,46] The advantage of the X-ray method is that there is no need to apply the difficult Placzek corrections to the data, but the major disadvantage is that the measured coherent cross-section is an admixture of all ten partial structure integrals (defined in equation (13.7.1)) and there is no method of separating them. Accordingly, the standard procedure adopted has been to Fourier transform the k-dependent scattering to yield an *average* radial distribution function, $G(r)$, and then discuss features arising in it. Wertz and Kruh,[47] for example, examined a 3.75 M solution of $CoCl_2$ and evaluated $G(r)$. They interpreted the first feature at 2.1 Å, solely in terms of Co–O correlations, as implying that cobalt is octahedrally co-ordinated by an average of six water molecules. The weakness of this approach (discussed further in Section 13.8) lies in the fact that it is impossible to identify unambiguously the structural combinations present in $G(r)$. In general, the results have been somewhat qualitative and, on occasions, not altogether convincing.

The first published neutron diffraction data on aqueous solutions was by Narten *et al.*[41] on LiCl and Enderby *et al.*[42] on $NiCl_2$, $BaCl_2$ and NaCl. The most obvious feature in both cases was a shift in position of the principal peak of the diffracted intensity to higher k-values with increasing concentration. Now the main peak in the diffraction pattern of D_2O is dominated by the D–D and D–O correlation functions, and so this shift might be interpreted as a reduction

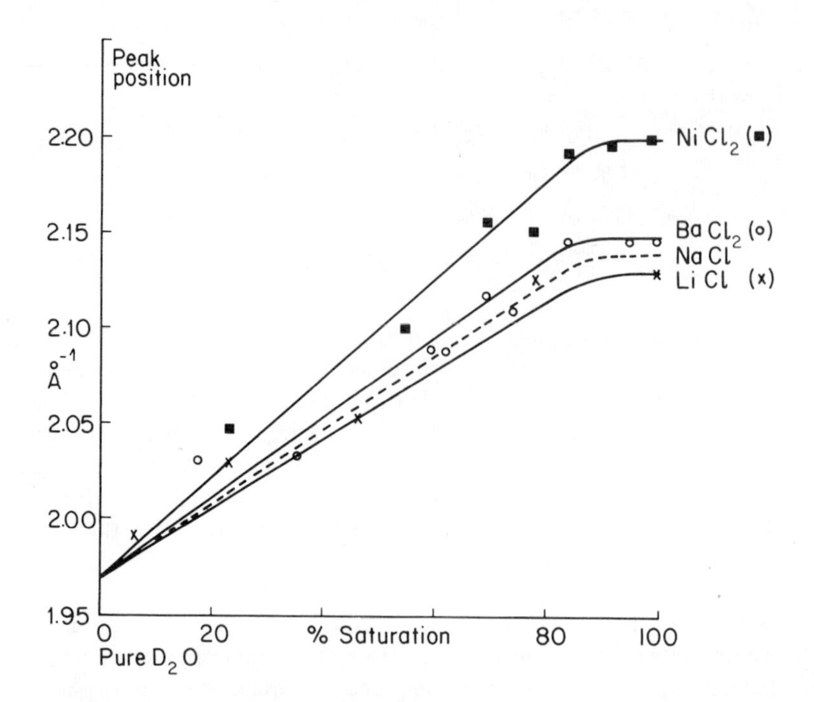

Figure 13.7.1 Principal peak position in a variety of aqueous solutions, as a function of solute concentration. (From Enderby *et al.*,[42] with adapted LiCl data of Narten *et al.*[41])

in the *inter*molecular D–D distance; the *intra*molecular D–O distance being relatively inflexible. Furthermore, with reference to Figure 13.7.1, the observation that (dk_{max}/dc) varies from ~0.04 Å^{-1} per mole % for $NiCl_2$ to only 0.003 Å^{-1} per mole % for NaCl is in line with previous ideas in which transition metal salts are more strongly 'hydrated' than others. It was observed by Enderby *et al.* that the shift described above was independent of the isotopic form of the Ni in $NiCl_2$ or the isotopic form of the Cl in NaCl. This tells us that the shift is not due to the confusing superpositioning of any of the solute–solute or solute–solvent partial structure factors, but is indicative of fundamental changes in the structure of water upon dissolution of a salt. Finally, the linearity of the effect suggests that these changes take place even in very dilute solutions usually described in terms of the so-called 'primitive' models,[48] some of which include the effects of the water only as a simple dielectric constant.

Narten *et al.*[41] combined neutron data for LiCl with X-ray measurements on the same system. Their reasons for adopting this particular system are that the X-ray scattering amplitudes of Li^+, Cl^-, H and O contrast markedly with the neutron scattering amplitudes of ^{7}Li, Cl, D and O. This guarantees a maximum of discrimination between the two measurements, and the result is that the X-ray scattering is determined almost entirely by the intermolecular correlations whilst the neutron intensity is principally a measure of the intramolecular correlation. Features in $G(r)_{X\text{-rays}}$ are compared and contrasted with those in $G(r)_{Neutron}$ and the behaviour of both as a function of concentration is carefully followed. In this way a number of conclusions are made concerning the nature of 'hydration' around the solute ions. It is suggested that the Cl ion in LiCl solutions displays octahedral coordination with 6 ± 1 water molecules, whilst Li shows tetrahedral coordination with 4 ± 1 water molecules. As in the case of pure water, Narten *et al.* suggest that the orientational correlation extends only to near neighbour molecules.

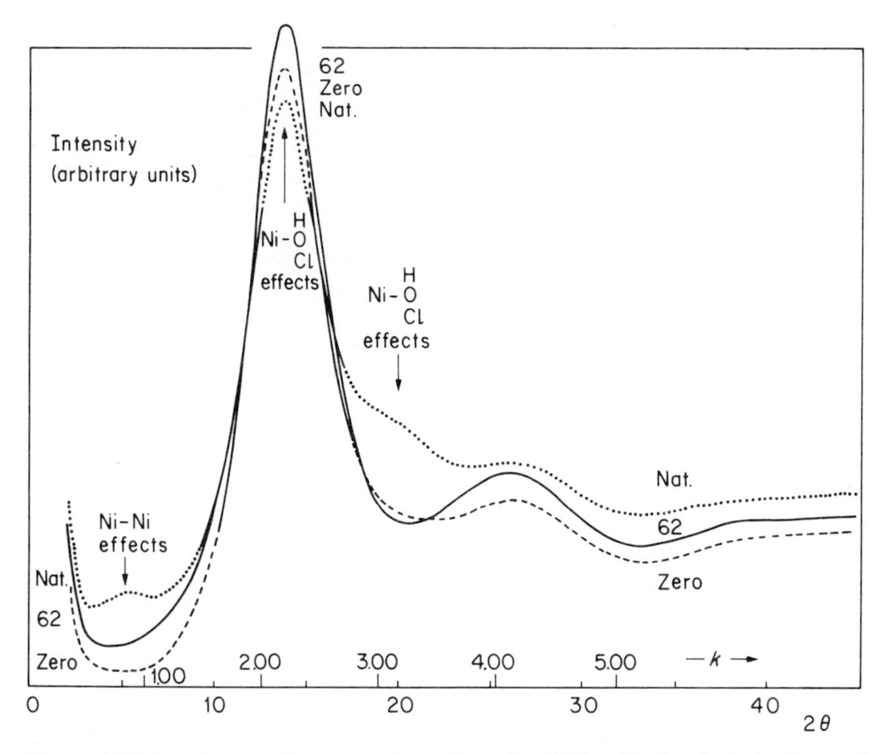

Figure 13.7.2 Scattered neutron intensities for $NiCl_2-D_2O$ solutions showing the variations due to substituting Ni and 'zero' Ni for nat. Ni

More detailed information than this requires isotopic substitution. The present author, with Howells and Enderby,[49] has substituted ^{62}Ni into $NiCl_2$ solutions. Figure 13.7.2 shows the magnitude of the changes that occur in the diffracted intensity. Since ^{62}Ni has a *negative scattering length* it is possible to construct a 'zero nickel' sample (analogous to the 'semi-transparent' water mentioned in the previous section), in which the contribution of the Ni to the coherent scattering cross-section is zero. Furthermore, one can readily differentiate between Ni–Ni effects and Ni–O, Ni–Cl, Ni–D effects. In the former, the ^{62}Ni will produce a change from the 'zero Ni' which is in the same sense as the change produced by natural Ni. In this way we can immediately associate the Ni–Ni partial structure factor with the effects observed around 0.9 $Å^{-1}$, whereas those at 2.1 $Å^{-1}$ and 3.1 $Å^{-1}$ are caused by Ni–O, and/or Ni–Cl, and/or Ni–D partials.

In order to proceed in a more quantitative manner let us introduce the function $F_0^N(k)$ to represent the k-dependent part of the differential coherent cross-section for the natural Ni sample. This zeroth order function can be obtained directly from equation (13.7.1) and is, therefore, a relatively straightforward quantity to measure to a high degree of accuracy. It is, however, extremely difficult to interpret the function since, as we have already observed, it contains ten partial structure factors.

If measurements are now made on a second sample containing a different isotopic concentration of Ni ('zero' Ni) then the first order function, $F_1^{N-O}(k) = F_0^N(k) - F_0^O(k)$ is a weighted admixture of the Ni–Ni, Ni–Cl, Ni–D and Ni–O structure integrals. Furthermore, the weighting factors contain the product of the concentrations of the two species, and consequently, to a first order approximation, F_1^{N-O} is an admixture of the Ni–D and Ni–O partial structure integrals only. The Fourier transform of F_1^{N-O} can therefore be interpreted in terms of the

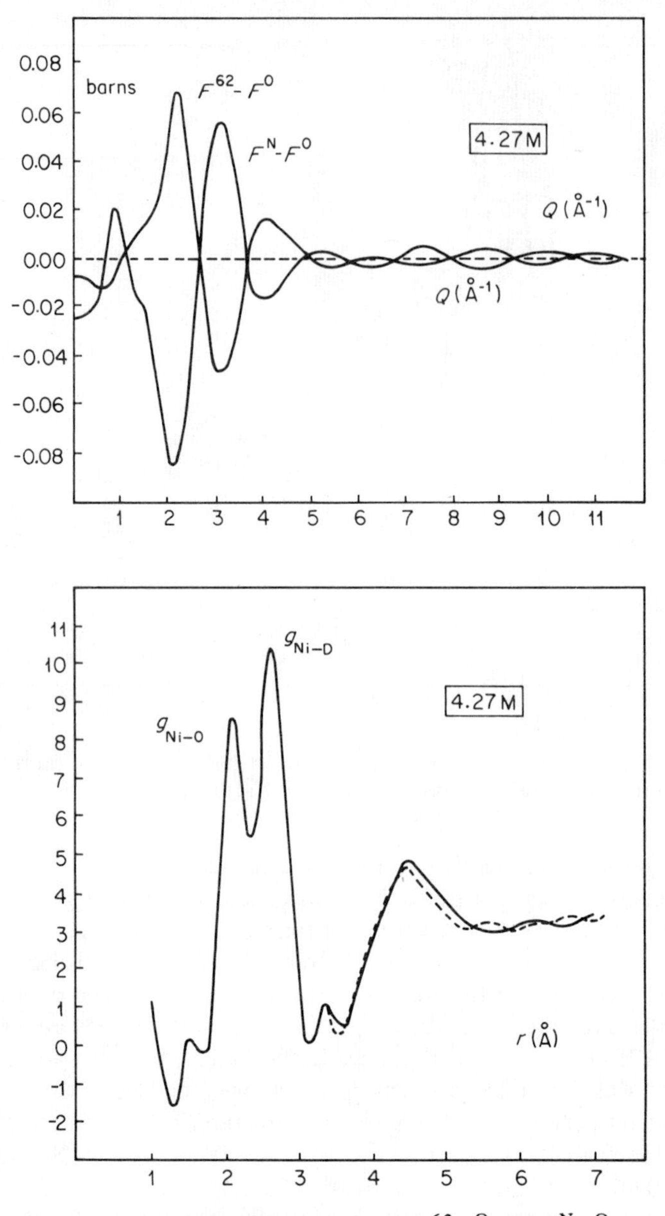

Figure 13.7.3 First order functions, F_1^{62-O} and F_1^{N-O} for saturated solutions of $NiCl_2$ in D_2O, together with their Fourier transforms. (Adapted from the data of G. W. Neilson, private communication)

Ni–O and Ni–D radial distribution functions. Figure 13.7.3 shows the experimental results obtained for these quantities using Natural Ni, 'Zero' Ni and ^{62}Ni samples of $NiCl_2$ in D_2O. The peaks occurring in the regions of 2.0 Å and 2.7 Å in the Fourier transform are interpreted as the principal peaks in g_{Ni-O} and g_{N-D} respectively and this is supported by the fact that their weighted areas are in the ratio of 1:2. These peaks are surprisingly sharp and suggest a strong positional correlation of the D_2O and the nickel ion. Furthermore, analysis of the weighted areas indicates that there are, on average, 5.4 ± 0.2 molecules of D_2O in the first hydration shell.

Analogous to the way in which our first order functions F_1^{A-B} were defined in terms of the difference between two zeroth order functions F_0^A and F_0^B, we can now introduce second order functions $F_2^{(A,B-C,D)}$ defined as a weighted difference of two first order functions. If the appropriate weighting is chosen these second order functions are nothing more than the partial structure integrals of the system.

For example, a straightforward rearrangement of (13.7.1) gives

$$\mathscr{I}g_{\text{Ni-Ni}} = \frac{\langle b_{62}\rangle[F_1^{\text{N-O}}] - \langle b_{\text{Nat}}\rangle[F_1^{62-\text{O}}]}{c^2\langle b_{62}\rangle\langle b_{\text{Nat}}\rangle[\langle b_{\text{Nat}}\rangle - \langle b_{62}\rangle]} \,. \tag{13.7.2}$$

Such second order functions contrast sharply with the zeroth order functions. They are extremely difficult to determine experimentally, but the interpretation is relatively straightforward in that the ambiguities associated with the zeroth order functions are completely removed. If equation (13.7.2) is to offer a feasible experimental solution for the Ni–Ni partial structure integral, the data must have good signal-to-noise ratios for $F_1^{\text{N-O}}$ and $F_1^{62-\text{O}}$, and this in turn requires counts of the order of 10^6 neutrons per step. This is only feasible with a high incident flux of neutrons.

Figure 13.7.4 shows the Ni–Ni partial structure integral, defined as in equation (13.7.2). The principal peak at around 1 $Å^{-1}$ gives rise to a first peak in the Ni–Ni pair distribution function at roughly 6 Å, which is close to the calculated value of the *mean* separation for Ni ions at a concentration of 4.27 *M*, if one assumes that the ions are located on a cubic lattice. This suggests that the Ni ions, in solution, may well be arranged in a highly regular manner. The suggestion is given further credibility by the extraordinary magnitude of the structure factor in comparison with a wide range of liquid systems in which the main maxima are of the order of unity.

Final confirmation for the existence of a lattice-like or cellular arrangement of the Ni ions comes in a recent publication by Neilson *et al.*[50] which shows that the principal peak position in the Ni–Ni partial structure factor is concentration dependent and in agreement with a $c^{1/3}$

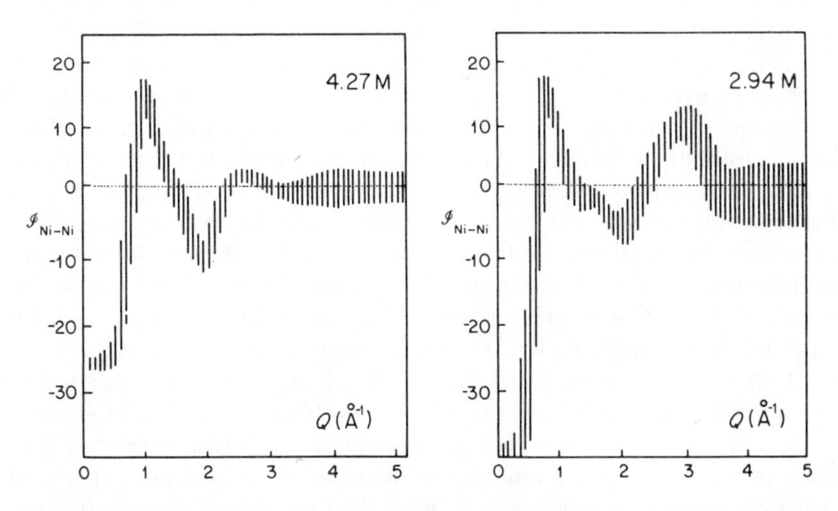

Figure 13.7.4 The Ni–Ni partial structure integrals for solutions of $NiCl_2$ in D_2O at two concentrations. (Adapted from the data of G. W. Neilson, private communication)

law. Since the peak position reflects inversely the closest distance of approach, it follows that a *random* arrangement of Ni ions would result in a value for the peak position that was roughly independent of concentration. Only when the *closest* separation corresponds to the *mean* separation, as in a lattice, would one expect a $c^{1/3}$ behaviour.

These results offer the first clear experimental evidence in favour of a *quasi-lattice* description for concentrated aqueous solutions. Theoretical work exists[51] which suggests the possible criteria for such a lattice-like structure, but the calculations are based on rather simplified model systems and are not readily applicable to $NiCl_2$ solutions.

In conclusion, it is now apparent that further measurements of the first order functions in a variety of solutions of differing concentrations and solute components will play an important role over the next few years in our understanding of the hydration effects in strong electrolytes. At the present time, the only other data available are on the $NaCl-D_2O$ system and in this case the positional correlation of the D_2O and the chlorine ion appears to be much less than that of the nickel ion in $NiCl_2-D_2O$.

The situation regarding second order functions is not so clear. In view of the experimental difficulties involved, relatively few systems can be studied in this way. The availability of suitable isotopes together with present neutron fluxes will undoubtedly be the limiting factors here.

13.8 Liquid semiconductors

Under this heading are included those materials which, in the liquid state, exhibit electronic properties which are similar to those normally associated with semiconductors. Two groups have been identified, namely those in which two metals of widely differing electronegativities are alloyed together and those based on the chalcogens and including pure Te and Se. The unusual electrical properties of these materials have attracted considerable theoretical interest in recent years and attempts have been made to relate these properties to structural features, in particular, the local order.[52] As a consequence of this, various phenomenological descriptions have arisen, such as 'clusters'[53] and 'bands', most of which are simply concerned with the varying degree of covalency or ionicity of the system. The observed electrical conductivity has even been explained in terms of classical percolation theory applied to an inhomogeneous melt composed of 'islands' of nonconducting material dispersed in a metallic medium.[54]

It is obvious that we still lack a fundamental understanding of these materials and, in particular, are hampered by a lack of information concerning their structure. As a first step towards correcting this, neutron diffraction experiments combined with isotopic substitution will yield pair distribution functions. At the present time data are becoming available for a limited number of systems of this type but so far the results tend to be somewhat inconclusive.

Neutron scattering measurements on Te[55] and Se[56] immediately reveal basic differences in comparison with normal liquid metals. In both Te and Se, $S(k)$ exhibits an oscillatory behaviour which would appear to extend well beyond the range of conventional two-axis scattering instruments (>12 Å^{-1}). Now in order to obtain the pair distribution function, $g(r)$, we apply equation (13.1.2) which involves an integration, in principle, to infinity and, in practice, to a sufficiently high k-value to ensure that $[S(k) - 1]$ has decayed to zero. Failure to observe this limitation introduces spurious features in $g(r)$ arising from the truncation error. We must, therefore, regard detailed features in $g(r)$ for these systems with some suspicion. The spurious features can, in part, be identified and removed by adopting various forms of truncation and observing the consequent changes in the Fourier transform. The corrected $g(r)$ can then be back-transformed and compared with the original data. Such repetitive Fourier inversion used in conjunction with experimental data in an iterative manner has been described in detail

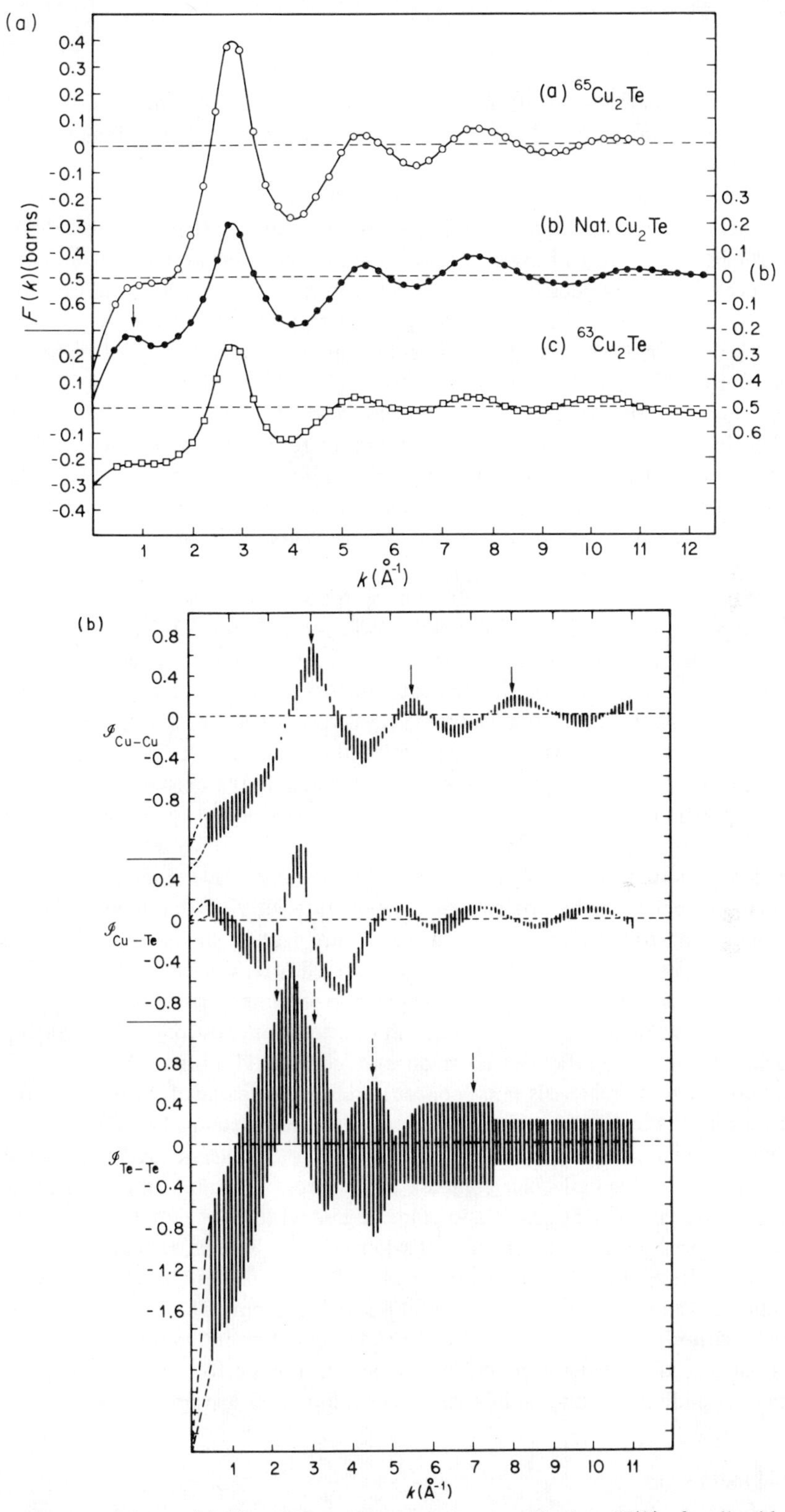

Figure 13.8.1 (a) The k-dependent neutron scattering, $F(k)$, for liquid $CuCl_2$ at 1000 °C. (b) The three partial structure integrals for liquid Cu_2Te at 1000 °C (from Hawker *et al.*[59]). Also shown are the peak positions for the pure Cu (↓) and pure Te (↓) structure factors in the liquid state

elsewhere.[57] Unfortunately, not all authors are so rigorous and, in addition, various publications of $g(r)$ are made in the absence of the original scattering data, thus preventing any subsequent check.

Even allowing for such corrective procedures the ultimate aim must be to extend the range of k-values over which measurements are made by using such systems as the pulsed, time-of-flight spectrometer (LINAC at Harwell).[58] A pulsed electron beam, incident upon a mercury target, produces Bremsstrahlung radiation which can then interact with natural uranium to produce neutrons. These can be subsequently moderated and collimated to a diffractometer consisting of a number of fixed angle counters in conjunction with a time-of-flight analyser. In this way scattering data can be obtained up to k-values of around 35–40 $Å^{-1}$. Information concerning the pair potential is then readily revealed in the following manner. For a hard sphere potential, the leading edge of $g(r)$ rises abruptly from zero at the hard sphere diameter. This leads directly to persistent oscillations in $S(k)$ and a minimum of damping. Now for Se and Te, even less damping of $S(k)$ occurs. Since the damping associated with hard spheres represents a minimum for simple liquids we must conclude that these materials cannot be discussed in terms of simple, central pair-wise interactions.

Studies of multicomponent semiconducting systems using isotopic substitution methods have been hampered by a lack of suitable stable isotopes and only Cu–Te has been extensively studied, using Cu isotopes.[59] In this case, it was conclusively shown that a melt inhomogeneity of the type mentioned above was inconsistent with the experimental data. This was done by postulating that in such a condition Te would need to exist in two forms, the one contained within a nonconducting island and the other within the metallic melt. There is no experimental support for this although the approach used cannot preclude the existence of 'clusters' which vary with composition, i.e. 'clusters' whose structure is dependent upon their size and atomic content.

The partial structure integrals for Cu_2Te obtained by Hawker *et al.*[59] are shown in Figure 13.8.1 together with the original k-dependent intensities of the three isotopic mixtures. An important point to observe is the absence of any feature in the three partial structure integrals corresponding to the pre-peak in the scattered intensities at ~ 0.8 $Å^{-1}$. This, again, clearly illustrates that discussions based solely upon features arising in the total intensity functions (as is often the case in X-ray work) can lead to erroneous conclusions. Such pre-peaks have frequently been cited in the past as evidence of 'compound' formation.

Future work on these materials must concentrate on the degree of short range ordering and on a clarification of the possible role of covalency in these systems. Certainly there is almost always an intermetallic compound formed in the solid phase of these alloy systems, and it is not clear what happens to this molecular complex in the liquid. Moscinski *et al.*[56] suggest that the chain-like structure of solid Se persists to some degree within the liquid phase, although it is gradually lost with increasing temperature. Hawker *et al.*[59] observe a great similarity between the Cu–Cu partial structure factor in Cu_2Te and the pure Cu structure, whereas the Te–Te correlation in Cu_2Te differs markedly from that occurring in pure Te. This can be interpreted as indicative of a disordered array of Cu in Cu_2Te and a reduction in the covalent character of Te. These both suggest that an ionic model might be appropriate for liquid Cu_2Te and that the strong covalent bonding in the solid, intermetallic compound is lost on melting.

13.9 Liquid metals

It is a well-documented fact[60] that the gross features of $S(k)$ in liquid metals arise from the repulsive part of the pair potential, $\phi(r)$. To obtain information from a 'measured' $S(k)$ on structural features other than these requires a high degree of accuracy in the experimental data.

In particular, the extreme sensitivity of the pair potential to the low-k region of $S(k)$ has been discussed in some detail by Gehlen and Enderby,[11] and Ballentine and Jones[61] showed that a 25% error in this region was sufficient to completely remove the attractive well in $\phi(r)$. With the overriding priority being paid to precision of the data, particularly at low-k values, it is pertinent to compare the relative roles of X-rays and neutrons in determining the structure factors of pure metal systems where, unlike the multicomponent system, there is no obvious advantage to be gained from the use of neutrons.

The main problems associated with X-rays are related to the uncertainty attached to the atomic form factor and to normalization. For neutrons, the chief difficulty has, in the past, centred around the comparatively poor statistics and upon an accurate determination of the multiple scattering and Placzek corrections.

Egelstaff[62] criticized the usage of X-rays for three reasons: (i) that the free atom values of the form factor were not known with sufficient accuracy; (ii) that the so-called 'solid state effects' involved in allowing for the environment of the atom within the solid were difficult to apply; and (iii) that temperature dependence of polarization would confuse the data. Greenfield *et al.*[63] have refuted these arguments, claiming that many form factors are known to within one percent and are particularly well defined in the important low-k region. They also argue that the solid state and polarization effects are almost negligible. As a consequence they are able to present data for liquid Na which represent the highest precision obtained to date, with an overall error claimed to be less than 2.5%.

Narten,[64] however, has made a detailed comparison of X-ray and neutron techniques in studying liquid Ga, and concludes that the Placzek corrections are valid to a high degree of accuracy, thereby removing the main anxiety of using neutrons. He established the equivalence of the two techniques in terms of an overall accuracy of around 3% even at low k-values, where the multiple scattering for neutrons corresponds to around four times the normal scattering and is assumed to be isotropic in the liquid in accord with Paalman and Pings.[21] The advent of higher flux neutron sources should enable further improvements, and experimentalists are for the first time in a position to supply structure factors of sufficient accuracy to distinguish not only between liquid gases and liquid metals but to discern subtle differences between various liquid metals. In conjunction with special low angle techniques[65] high precision data is now accessible over a wide range of momentum exchange.

Having established the broad equivalence of X-rays and neutrons for use with liquid metal systems, let us now turn to a class of metals where the neutron technique offers a distinct advantage – the transition metal elements. In particular, the structure of liquid iron has been investigated by neutron diffraction and a pair distribution function evaluated.[66] Now for a system containing unpaired spins, the neutron experiences an additional interaction via its own spin, and equation (13.2.27) becomes

$$\left.\frac{d\sigma}{d\Omega}^{\text{coh}}\right|_{\text{ideal}}^{\text{static}} = N\langle b\rangle^2 S(k) + N'\langle b_m(k)\rangle^2 \tag{13.9.1}$$

where b_m is the k-dependent magnetic scattering length and N' is the number of nuclei which possess a net magnetic moment. The magnetic scattering length can be related to the magnetic form factor, $F(k)$, by

$$\langle b_m(k)\rangle^2 = \frac{1}{3}\left[\frac{\gamma e^2}{mc^2}\right]^2 [gF(k)]^2 J(J+1) \tag{13.9.2}$$

where γ is the magnetic moment of the neutron and g is the Landé splitting factor. By ascribing a particular spectroscopic state to the sample, the magnetic form factor can be calculated and allowance made for the magnetic scattering. This is the procedure adopted by Waseda and

Suzuki in the work cited above.[66] Alternatively, however, the neutron scattering can be used to measure $F(k)$ and hence determine the magnetic state of the material. One way of doing this was reported by Enderby and Nguyen[67] for liquid Ce. They measured the neutron scattering from two liquids of similar structure, Ce and La, where only one of the liquids (Ce) exhibits magnetic scattering. A simple subtraction of the two sets of data yielded the magnetic form factor, $F(k)$, which was subsequently normalized to unity in the long wavelength limit. In this way they confirmed that the spectroscopic state of Ce is as suggested by Russel–Saunders coupling and that, at any one time, roughly 20% of the Ce ions within the liquid are magnetically inactive. This particular experimental method relies upon the assumed similarity of structure of the two metals. A more general approach requires the use of polarized neutrons and a full spin flip analysis and could usefully be applied to a variety of liquid, transition and rare earth metals.

13.10 Pressure dependence studies

So far we have been concerned solely with the application of radiation scattering techniques to the determination of $S(k)$ and hence $g(r)$. It is, however, of great importance that information should be available on the three body and higher order correlation functions if we are to progress beyond simple liquids. These quantities can be obtained, in principle, from experimental studies of the derivatives of the structure factor. In particular, Yvon[68] has shown that the triplet correlation function $g(\mathbf{r},\mathbf{s})$ is related to the isothermal pressure derivative of the pair distribution function and is given by

$$k_B T \left.\frac{\partial (N/V)^2 g(r)}{\partial P}\right|_T = \left(\frac{N}{V}\right)^2 \int \{g_3(\mathbf{r}, \mathbf{s}) - g(r)\}\, \mathrm{d}s + 2\left(\frac{N}{V}\right) g(r). \tag{13.10.1}$$

The pressure derivative of $g(r)$ can be related to that of $S(k)$ by the equation

$$\frac{N}{V} k_B T \left.\frac{\partial S(k)}{\partial P}\right|_T = \left(\frac{N}{V}\right)^2 k_B T \int \left.\frac{\partial g(r)}{\partial P}\right|_T \exp(\mathrm{i}\mathbf{k}\cdot\mathbf{r})\, \mathrm{d}r + S(0)[S(k) - 1]. \tag{13.10.2}$$

The first experiment designed to measure $g_3(\mathbf{r},\mathbf{s})$ along these lines was carried out by Egelstaff *et al.*[3] on liquid Rb. In particular, their measurements were an attempt to test the validity of the superposition theorem, used in many liquid state theories, in which the triplet correlation function is built up as the product of the pair function, namely

$$g_3(\mathbf{r},\mathbf{s}) = g(r) g(s) g(\mathbf{r} - \mathbf{s}). \tag{13.10.3}$$

They defined a function $H(\mathbf{r},\mathbf{s})$ which was a measure of the departure from the superposition equation, such that

$$g_3(\mathbf{r},\mathbf{s}) = g(r) g(s) g(\mathbf{r} - \mathbf{s}) - H(\mathbf{r},\mathbf{s}) \tag{13.10.4}$$

and then showed that the Fourier transform of this quantity could be related to experimentally observable quantities. Using (13.10.1) and (13.10.2),

$$\left[\left(\frac{N}{V}\right)(2\pi)^3\right]^{-1} \int [S(\mathbf{k} - \mathbf{k}') - 1]\,[S(k') - 1]^2\, \mathrm{d}\mathbf{k}' + [S(k) - 1][S(0) + S(k) - 1]$$
$$- \frac{N}{V} k_B T \left.\frac{\partial S(k)}{\partial P}\right|_T = \left(\frac{N}{V}\right)^2 \int \exp(\mathrm{i}\mathbf{k}\cdot\mathbf{r})\, \mathrm{d}\mathbf{r} \int H(\mathbf{r}, \mathbf{s})\, \mathrm{d}\mathbf{s} = \tilde{H}(k). \tag{13.10.5}$$

If the superposition theorem were valid, $\tilde{H}(k)$ would be identically zero for all k-values. This was not found to be the case for Rb, and Egelstaff *et al.* concluded that, at least in the vicinity of the triple point, the superposition theorem was a poor representation of the liquid system.

Although experiments of this nature are fundamental to our understanding of the liquid state, very little experimental data are yet available, reflecting the inherent, practical difficulties. In addition to the work mentioned above on liquid Rb, a brief, but somewhat inconclusive, study has been made of the alkali metals Na, K and Cs[69] and some data are available for liquid Ne.[70] It is obvious that there is a large amount of work yet to be done on these lines.

13.11 Conclusions

We have explored a number of aspects of the current investigations into the structure of liquids that employ neutron diffraction techniques. There are, necessarily, some serious omissions and in no sense should this be taken as a complete review of the field. Apart from the deliberate decision to leave out any discussion of inelastic neutron scattering there are, for example, a number of interesting avenues of research into liquid gases and non-aqueous solutions which are not included here. What it is hoped has emerged is some impression of what has been achieved, the present limitations of the work and where, in the immediate future, progress is likely to be made.

In particular, we have emphasized that the availability of higher flux sources has provided us not only with new opportunities and horizons but also with new problems in that some of the commonly adopted methods of extracting $S(k)$ from experimental data may well need to be reformulated if we are to obtain the full benefit of the improved statistics. In the past, experimentalists were content to identify the gross features that distinguished such systems as liquid gases from liquid metals. Now we are in a position to explore the subtleties of systems which at first sight appear similar. As a consequence of this, much of our current interest centres upon determining the extent to which a system can be described as molecular or, related to this, the degree of ionic or covalent behaviour exhibited by the system. It is by no means obvious that structural investigations alone will resolve this problem, but undoubtedly the observation of charge oscillation effects gives support to ionicity in the same way that the presence of certain molecular structure can indicate covalency.

Undoubtedly the immediate future will see a wider application of the isotopic substitution methods, in which case neutrons are indisputably superior to other forms of radiation. We have seen the dangers in attempting to obtain information on multicomponent systems from total Fourier transforms and there is plenty of evidence to suggest that methods based upon the concentration invariance of partial structure integrals are ill-founded. The use of X-rays alone must be limited to monatomic systems, but used in conjunction with neutron techniques they are likely to provide useful supplementary information, as the studies on water and aqueous solutions have shown.

References

1. N. H. March (1968). *Liquid Metals,* Pergamon, Oxford.
2. T. E. Faber (1972). *An Introduction to the Theory of Liquid Metals,* Cambridge University Press, London.
3. P. A. Egelstaff, D. I. Page and C. R. T. Heard (1971). *J. Phys.* C., **4**, 1453.
4. W. C. Marshall and S. W. Lovesey (1971). *Theory of Thermal Neutron Scattering*, Clarendon, Oxford.
5. W. M. Lomer and G. G. Low (1965). *Thermal Neutron Scattering*, Academic Press, New York.

6. L. van Hove (1954). *Phys. Rev.*, **95**, 249.
7. J. G. Powles (1973). *Adv. in Phys.*, **22**, 1.
8. G. Placzek (1952). *Phys. Rev.*, **86**, 337.
9. P. A. Egelstaff and M. J. Poole (1969). *Experimental Neutron Thermalisation*, Pergamon Press, Oxford.
10. J. E. Enderby (1968). *Physics of Simple Liquids*, North-Holland, Amsterdam. Chap. 14.
11. P. C. Gehlen and J. E. Enderby (1969). *J. Chem. Phys.*, **51**, 547.
12. N. Quirke, private communication.
13. D. I. Page and K. Mika (1971). *J. Phys.* C., **4**, 3034.
14. P. A. Egelstaff, D. I. Page and J. G. Powles (1971). *Molec. Phys.*, **20**, 881.
15. D. I. Page and J. G. Powles (1971). *Molec. Phys.*, **21**, 901.
16. W. H. Zachariason (1935). *Phys. Rev.*, **47**, 277.
17. A. B. Bhatia and D. E. Thornton (1970). *Phys. Rev.* B., **2**, 3004.
18. S. P. McAlister and R. Turner (1972). *J. Phys.* F., **2**, L51.
19. J. L. Beeby (1973). *J. Phys.* C., **6**, 2262.
20. J. E. Enderby, D. M. North and P. A. Egelstaff (1966). *Phil. Mag.*, **14**, 961.
21. H. H. Paalman and C. Pings (1962). *J. Appl. Phys.*, **33**, 2635.
22. D. M. North, J. E. Enderby and P. A. Egelstaff (1968). *J. Phys.* C., **1**, 784.
23. G. H. Vineyard (1954). *Phys. Rev.*, **95**, 93.
24. S. J. Cocking and C. Heard (1965). AERE Rep. No. R5016.
25. I. A. Blech and B. L. Averbach (1965). *Phys. Rev.*, **137**, 1113.
26. M. Dixon and M. J. L. Sangster (1975). *J. Phys.* C., **8**, L8.
27. W. C. Nieuwpoort and G. Blasse (1968). *J. Inorg. Nucl. Chem.*, **30**, 1635.
28. I. G. Murgulescu (1969). *Rev. Roum. Chim.*, **14**, 965.
29. J. G. Powles (1975). *J. Phys.* C., **8**, 895.
30. J. Y. Derrien and J. Dupuy (1975). *Jour de Physique,* **36** 191.
31. F. G. Edwards, J. E. Enderby, R. A. Howe and D. I. Page (1975). *J. Phys.* C., **8**, 3483.
32. F. Lantelme, P. Turq, B. Quentrac and J. W. E. Lewis (1974). *Mol. Phys.*, **28**, 1537.
33. M. P. Tosi and F. G. Fumi (1964). *J. Phys. Chem. Sol.*, **25**, 31.
34. J. G. Powles, J. C. Dore and D. I. Page (1972). *Mol. Phys.*, **24**, 1025.
35. A. H. Narten (1970). Oak Ridge Nat. Lab. Rep. 4578.
36. T. Springer, C. Hofmeyr, S. Kornblicher and H. D. Lemmel (1964). *3rd Int. Conf. on Peaceful Uses of Atomic Energy (Geneva),* I.A.E.A. **2**, 351.
37. B. N. Brockhouse (1958). *Nuovo Cim. (suppl.)*, **9**, 45.
38. A. H. Narten (1972). *J. Chem. Phys.*,**56**, 5681.
39. A. Rahman and F. H. Stillinger (1971). *J. Chem. Phys.*, **55**, 3336.
40. D. T. Keating (1963). *J. Appl. Phys.*, **34**, 923.
41. A. H. Narten, F. Vaslow and H. A. Levy (1973). *J. Chem. Phys.*, **58**, 5017.
42. J. E. Enderby, R. A. Howe and W. S. Howells (1973). *Chem. Phys. Lett.*, **21**, 109.
43. P. S. Ramathan and H. L. Friedman (1971). *J. Chem. Phys.*, **54**, 1086.
44. J. A. Prinz (1929). *Z. Phys.*, **56**, 617.
45. G. W. Brady (1958). *J. Chem. Phys.*, **28**, 464.
46. R. M. Lawrence and R. F. Kruh (1967). *J. Chem. Phys.*, **47**, 4758.
47. D. L. Wertz and R. F. Kruh (1969). *J. Chem. Phys.*, **50**, 4313.
48. D. N. Card and J. P. Valleau (1970). *J. Chem. Phys.*, **52**, 6232.
49. R. A. Howe, W. S. Howells and J. E. Enderby (1974). *J. Phys.* C., **7**, L111.
50. G. W. Neilson, R. A. Howe and J. E. Enderby (1975). *Chem. Phys. Lett.*, **33**, 284.
51. F. H. Stillinger (1968). *Proc. Nat. Acad. Sci.*, **60**, 1138.
52. A. F. Joffe and A. R. Regel (1960). In A. F. Gibson (Ed.), *Progress in Semiconductors*, Wiley, New York.
53. R. J. Hodgkinson (1970). *Phil. Mag.*, **22**, 1187.
54. M. H. Cohen and J. Sak (1972). *J. Non-Cryst. Sol.*, **8–10**, 696.
55. G. Tourand and M. Breuil (1971). *Journ. de Physique*, **32**, 813.
56. J. Moscinski, A. Renninger and B. L. Averbach (1973). *Phys. Lett.*, **42A**, 453.
57. R. Kaplow, S. L. Strong and B. L. Averbach (1965). *Phys. Rev.*, **138A**, 1336.
58. R. N. Sinclair, D. A. G. Johnson, J. C. Dore, J. H. Clarke and A. C. Wright (1974). *J. Nucl. Inst. Meth.*, **117**, 445.
59. I. Hawker, R. A. Howe and J. E. Enderby (1974). In J. Stuke and W. Brenig (Eds), *Amorphous and Liquid Semiconductors*, Taylor and Francis, London. p. 85.

60. N. W. Ashcroft and J. Lekner (1966). *Phys. Rev.*, **145**, 83.
61. L. E. Ballentine and J. C. Jones (1973). In S. Takeuchi (Ed.), *The Properties of Liquid Metals*, Taylor and Francis, London.
62. P. A. Egelstaff (1967). *Adv. in Phys.*, **16**, 147.
63. A. J. Greenfield, J. Wellendorf and N. Wiser (1971). *Phys. Rev.* A., **4**, 1607.
64. A. H. Narten (1972). *J. Chem. Phys.*, **56**, 1185.
65. J. E. Enderby (1973). In S. Takeuchi (Ed.), *The Properties of Liquid Metals,*, Taylor and Francis, London.
66. Y. Waseda and K. Suzuki (1970). *Phys. Stat. Sol.*, **39**, 669.
67. J. E. Enderby and V. T. Nguyen (1975). *J. Phys.* C., **8**, L112.
68. J. Yvon (1966). *Correlations and Entropy in Classical Stat. Mechanics*, Pergamon, Oxford.
69. K. Tsuji, H. Endo, S. Minomura and K. Asaumi (1973). In S. Takeuchi (Ed.), *The Properties of Liquid Metals*, Taylor and Francis, London.
70. L. de Graaf and B. Mozer (1971). *J. Chem. Phys.*, **55**, 4967.

Chapter 14

Phase Transitions and Pretransition Phenomena in Liquid Crystals

S. CHANDRASEKHAR and N. V. MADHUSUDANA

Raman Research Institute, Bangalore 560006, India

14.1 Introduction

Liquid crystals represent states of matter that are intermediate between the crystalline solid and the amorphous liquid. They were so named by Lehmann[1] because they are fluid and at the same time anisotropic in their optical, electrical and magnetic properties. Friedel[2] preferred to call them 'mesomorphic states of matter' rather than 'liquid crystals', but both terms are employed synonymously in current usage. A necessary condition for the existence of a mesophase is that the molecule must be highly geometrically anisotropic, usually long and relatively narrow. Depending on the detailed molecular structure the system may exhibit one or more mesophases, transitions to which may be brought about by the effect of temperature (thermotropic mesomorphism) or by the influence of solvents (lyotropic mesomorphism). In this chapter, we shall be concerned only with the former kind of mesomorphism.

Thermotropic liquid crystals are classified broadly into three types: nematic, cholesteric and smectic. The nematic liquid crystal is characterized by a long range orientational order of the molecules but no long range translational order (Figure 14.1.1). The preferred molecular

Figure 14.1.1 Molecular arrangement in the nematic phase

direction (termed the 'director' and labelled as **n**, a dimensionless unit vector) usually varies from point to point in the medium, but a homogeneously aligned specimen is optically uniaxial and positive. The mesophase is non-ferroelectric, so that **n** and −**n** are fully equivalent. Recent X-ray studies[3] show that some nematics may develop a certain degree of short range smectic-like translational order; these are referred to as 'cybotactic' nematics.

The three types of deformations that can take place in the nematic liquid crystal are splay, twist and bend (Figure 14.1.2), each of these being associated with an elastic constant.[4] The nematic also flows quite freely, but its viscous properties are rather different from those of an ordinary liquid because of the coupling between the orientational motion of the director and the translational motion of the fluid. Assuming the medium to be incompressible, nonpolar and centrosymmetric, five independent viscosity coefficients are required to describe its dynamical behaviour.[5,6] Typically, the elastic constants are $\simeq 10^{-6}$ dyne and the viscosity coefficients $\simeq 10^{-1}$ poise.

The cholesteric mesophase is also a nematic type of liquid crystal except that it is composed of optically active molecules. As a consequence the structure acquires a spontaneous twist about an axis normal to the director (Figure 14.1.3). Thermodynamically the cholesteric is generally regarded as identical to the nematic, as the energy of twist forms a negligibly small part ($\sim 10^{-5}$) of that associated with the parallel alignment of the molecules.[7] The spiral structure of this mesophase is responsible for its unique optical properties, viz. a selective reflexion of circularly polarized light and a very high rotatory power.

Smectic liquid crystals have stratified structures, but different types of molecular arrangement are possible within each stratification. The simplest of these is smectic A in which the molecules are upright with their centres irregularly spaced in a liquid-like fashion (Figure 14.1.4*a*). Smectic B differs from A in that the molecular centres in each layer are hexagonal close packed. Smectic C is a tilted form of smectic A (Figure 14.1.4*b*). At least five

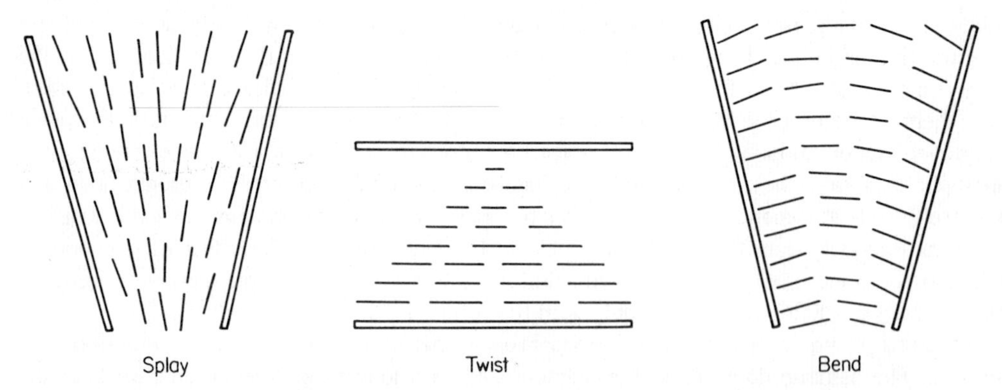

Figure 14.1.2 The three principal types of deformation in a nematic liquid crystal

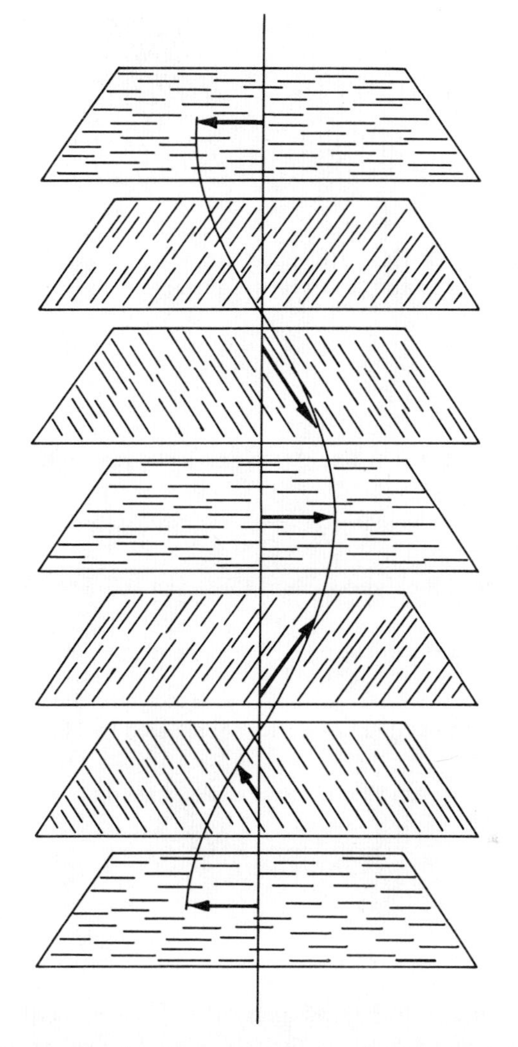

Figure 14.1.3 The cholesteric liquid crystal: schematic representation of the helical structure

other distinct smectic modifications have been identified,[8] but their structures are not yet known with any certainty and will not therefore be discussed further here.

The stratified structure of smectics imposes certain restrictions on the types of deformation that can take place in it. A compression of the layers involves very much more energy than a curvature elastic distortion in a nematic. Thus, for example, in smectic A the only distortion that is possible without affecting the inter-layer separation is splay; the twist and bend distortions are disallowed. The hydrodynamical behaviour of smectic A involves five viscosity coefficients and one 'permeation' coefficient.[9] The latter is related to fluid motion normal to the layers (which occurs by the permeation of the molecules from one layer to the next without the layers themselves moving) and accounts for the very high apparent viscosity for flow in this direction. For a general discussion of the hydrodynamics of the various smectic phases, reference may be made to a paper by Martin *et al.*[10]

Some examples of mesophase transitions and their latent heats in kilojoules/mole are given

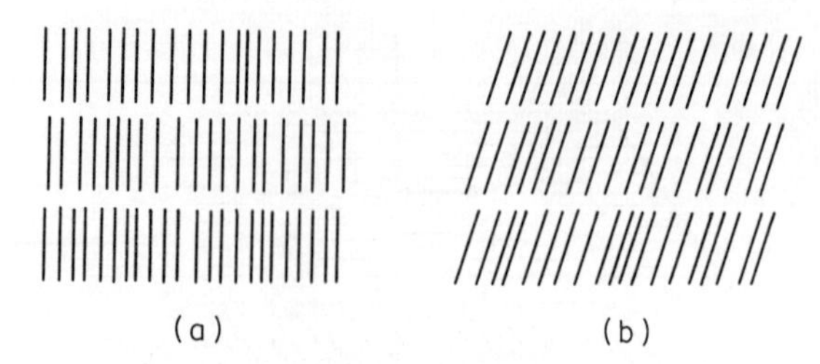

Figure 14.1.4 Molecular arrangement in the (a) smectic A and (b) smectic C phases

below:

4,4′-Di-methoxyazoxybenzene or p-Azoxyanisole (hereafter abbreviated to PAA)[11]

$$\text{Solid} \xleftrightarrow[29.57\ \text{kJ}]{118.2\,^\circ\text{C}} \text{nematic} \xleftrightarrow[0.57\ \text{kJ}]{135.3\,^\circ\text{C}} \text{isotropic}$$

p-Cyanobenzylidene p′-n-octyloxyaniline (hereafter abbreviated to CBOOA)[12]

$$\text{Solid} \xleftrightarrow[38.1\ \text{kJ}]{73.2\,^\circ\text{C}} \text{smectic A} \xleftrightarrow[\sim 0\ \text{kJ}]{82.8\,^\circ\text{C}} \text{nematic} \xleftrightarrow[1.1\ \text{kJ}]{107.5\,^\circ\text{C}} \text{isotropic}$$

Terephthal-bis- 4-n-butylaniline (hereafter abbreviated to TBBA)[13]

$$\text{Solid} \xleftrightarrow[18.17\ \text{kJ}]{113.0\,^\circ\text{C}} \text{smectic B} \xleftrightarrow[3.76\ \text{kJ}]{144.1\,^\circ\text{C}} \text{smectic C} \xleftrightarrow[\sim 0\ \text{kJ}]{172.5\,^\circ\text{C}} \text{smectic A}$$

$$\text{smectic A} \xleftrightarrow[0.29\ \text{kJ}]{199.6\,^\circ\text{C}} \text{nematic}$$

$$\text{isotropic} \xleftrightarrow[0.75\ \text{kJ}]{236.5\,^\circ\text{C}} \text{nematic}$$

A more detailed description of the structures and physical properties of the mesophases may be found in recent books and articles on the subject.[14] We shall be confining our attention to the specific problem of phase transitions and pretransition effects in thermotropic liquid crystal systems. In particular we shall be considering the nematic–isotropic (N–I), the smectic A–nematic (A–N), the smectic C–smectic A(C–A) and the smectic C–nematic (C–N) transitions.

14.2 The nematic–isotropic transition

14.2.1 The Maier–Saupe theory

For a molecule of arbitrary shape the long range orientational order parameter in the nematic phase may be expressed in the generalized form

$$s_{ij}^{\alpha\beta} = \tfrac{1}{2}\langle 3 i_\alpha j_\beta - \delta_{\alpha\beta}\delta_{ij}\rangle \tag{14.2.1}$$

where $\alpha,\beta = x,y,z$ refer to the space fixed axes; $i,j = x',y',z'$ refer to the principal axes of the molecule; i_α, j_β denote the projections of the unit vectors **i**, **j** along α, β; $\delta_{\alpha\beta}$, δ_{ij} are Kronecker deltas and the angular brackets denote a statistical average. $s_{ij}^{\alpha\beta}$ is symmetric in i,j and in α,β. It is also a traceless tensor with respect to either pair, that is,

$$s_{ij}^{\alpha\alpha} = s_{ii}^{\alpha\beta} = 0$$

where repeated tensor indices imply the usual summation convention.

Referred to the unique axis of the nematic medium and assuming the molecules themselves to be cylindrically symmetric rods with $i = j = 3$ to be the long axis, (14.2.1) reduces to the simple scalar form

$$s = \tfrac{1}{2}\langle 3\cos^2\theta - 1\rangle = \langle P_2(\cos\theta)\rangle \tag{14.2.2}$$

where θ is the angle which the molecular long axis makes with the unique axis of the medium, and P_2 $(\cos\theta)$ is the Legendre polynomial of the second order. s is 1 for perfectly parallel alignment and 0 for random orientation; in the nematic phase it has an intermediate value which decreases rapidly with temperature and drops abruptly to zero at the nematic–isotropic point T_{NI} in a weak first order transition. The order parameter can be determined by a variety of methods, e.g. from the diamagnetic anisotropy,[15] the optical anisotropy,[15–17] nuclear magnetic resonance spectra,[18] etc.

A microscopic approach that has proved to be useful in developing a theory of the long range orientational order and some of the related properties of the nematic phase is that due to Maier and Saupe.[19] It is based on the mean field approximation, very similar to that introduced by Weiss in ferromagnetism. The single particle potential is taken to be of the form

$$u_i = -AsP_2(\cos\theta_i) \tag{14.2.3}$$

where A is a function of volume. In their original presentation, Maier and Saupe regarded the dipole–dipole part of the anisotropic dispersion forces to be responsible for nematic stability and assumed $A \propto V^{-2}$, V being the molar volume. However, experimental data[20] on the pressure dependence of the order parameter of PAA indicate that $A \propto V^{-4}$. On the other hand, a recent argument due to Cotter[21] suggests that thermodynamic consistency of the mean field theory requires that $A \propto V^{-1}$. For the present purpose, we shall ignore the volume dependence of the potential as it will not affect the major conclusions of the theory. The free energy due to order is

$$F = N\left[\tfrac{1}{2}As(s+1) - k_{\mathrm{B}}T\ln\int\exp\left(\frac{3}{2}\frac{As}{k_{\mathrm{B}}T}\cos^2\theta_i\right)\mathrm{d}(\cos\theta_i)\right]. \tag{14.2.4}$$

The condition for the transition from the nematic to the isotropic phase is given by $F = 0$, which gives $A/k_{\mathrm{B}}T_{\mathrm{NI}} = 4.541$, or s at $T_{\mathrm{NI}} \simeq 0.43$. The theoretical plot of s versus T/T_{NI} is a universal curve for all substances. Experimentally, there are small but systematic deviations from this universal curve[7,17] (Figure 14.2.1) and the need for including higher order terms in the potential function has been emphasized.[22,23] The assumption that the molecule is a rigid rod is also an over-simplification. A realistic calculation has been made by Marcelja[24] of the contribution of the flexible end chain in the ordering process, which at once accounts for the alternation in the transition temperature, the order parameter and other related properties of the successive members of a homologous series.

A difficulty with the mean field approximation becomes apparent when we calculate the latent heat of transition ΔH. The theoretical value of ΔH turns out to be much too high, usually by a factor of 2 or 3. Large discrepancies are also found in the specific heat and isothermal compressibility.[22] This is not surprising as the mean field approach completely neglects the effect of near neighbour correlations. A naive way of accounting for these discrepancies is to assume that the nematic is composed of clusters of 2 or 3 perfectly aligned molecules and that the single particle potential applies to the cluster as a whole rather than to

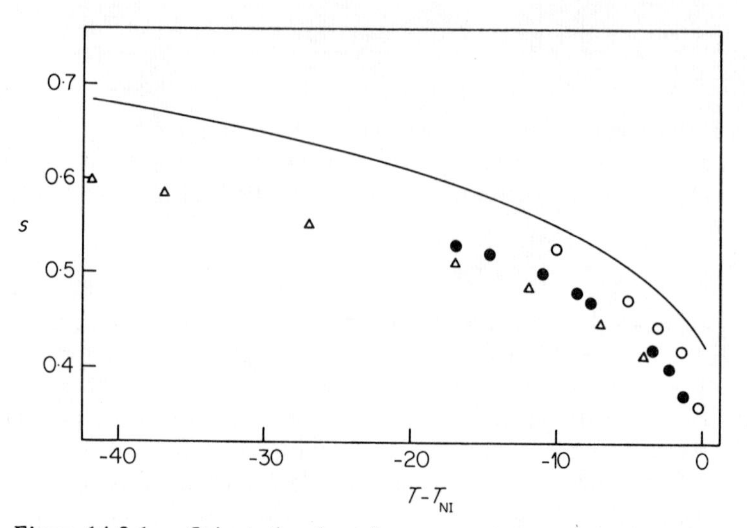

Figure 14.2.1. Orientational order parameter s plotted against temperature in the nematic phase of PAA; ○ from NMR measurements (McColl and Shih[20]); ● from diamagnetic anisotropy (Gasparoux *et al.*[15]); △ from refractive indices (Chandrasekhar and Madhusudana[17]). The curve represents the values predicted by the Maier–Saupe theory with $u_i \propto V^{-2}$

the individual molecules.[19,22] However, such a procedure cannot account for the remarkable short range order effects in the isotropic phase, which we shall be considering in the next section.

Another microscopic approach to the problem of the N–I transition is to regard the system as an assembly of hard rods with no forces between them other than the one preventing their interpenetration. While this approach is important in bringing out the role of molecular shape in the formation of the mesophase and may possibly yield interesting information on the molecular dynamics of such systems, it would be fair to say that the quantitative predictions of these calculations are as yet hardly applicable to real liquid crystals. A more satisfactory treatment would be to carry out the hard rod calculation with a superimposed single particle attractive potential of the Maier–Saupe form; however, such a theory has yet to be developed. For a critical summary of the various treatments of the hard rod fluid, reference may be made to a paper by Straley.[25] We do not propose to discuss this model any further.

14.2.2 Short-range order effects in the isotropic phase: the Landau–de Gennes model

It has long been known that the isotropic phase of a nematic liquid crystal shows certain anomalous properties that can be attributed to a nematic-like short range order. For example, the magnetic birefringence close to the transition may be as much as 100 times higher than in ordinary organic liquids.[26] Foex[26] noted many years ago that the behaviour is closely analogous to that of a ferromagnetic material above the Curie temperature. A phenomenological description of these effects has been proposed by de Gennes[27] on the basis of the Landau theory of phase transitions.[28] Consider an expansion of the free energy of the ordered system in the following form:

$$F = \tfrac{1}{2}A(T)s_{\alpha\beta}s_{\beta\alpha} - \tfrac{1}{3}B(T)s_{\alpha\beta}s_{\beta\gamma}s_{\gamma\alpha} + \tfrac{1}{2}L_1\partial_\alpha s_{\beta\gamma}\partial_\alpha s_{\beta\gamma} + \tfrac{1}{2}L_2\partial_\alpha s_{\alpha\gamma}\partial_\beta s_{\beta\gamma} - \tfrac{1}{2}H_\alpha H_\beta \chi_{\alpha\beta} \tag{14.2.5}$$

where $s_{\alpha\beta}$ is the order parameter given by (14.2.1) with $i = j = 3$ corresponding to the long molecular axis, the terms involving $\partial_\alpha = \partial/\partial x_\alpha$ describe the gradients in the order parameter, the last term represents the contribution due to a magnetic field ($\chi_{\alpha\beta}$ being the diamagnetic susceptibility tensor) and repeated tensor indices imply the usual summation convention. The term of order s^3 is not precluded by symmetry since $s_{\alpha\beta}$ and $-s_{\alpha\beta}$ represent two entirely different kinds of molecular arrangement. If $B > 0$, (14.2.5) leads to a first order phase transition.

For a second order transition, $B = 0$ and $A(T)$ may be taken to be of the form $a(T - T^*)$ where T^* is the transition temperature.[28] Though the N–I transition is first order, we may retain the same form of $A(T)$, with T^* now representing a hypothetical second order transition point slightly below T_{NI}. In principle, the free energy expression of this type may be used to evaluate the properties of the nematic phase, but as we have seen in Section 14.2.1, s is usually quite large below the transition ($\simeq 0.4$) so that very many more terms have to be included in the expansion in order to draw any valid conclusions. Consequently, the Landau theory is conveniently applied only to the weakly ordered isotropic phase. We shall discuss some of these applications.

Magnetic and electric birefringence To evaluate the magnetically induced order, we rewrite (14.2.5) in terms of the scalar order parameter s given by (14.2.2):

$$F = \frac{3}{4} A s^2 - \frac{H^2}{3} \chi_a s$$

neglecting higher order terms since s is very small ($\sim 10^{-5}$). Here $\chi_a = \chi_\parallel - \chi_\perp$ is the anisotropy of diamagnetic susceptibility for perfectly parallel alignment ($s = 1$). Using the condition $\partial F/\partial s = 0$ at once leads to

$$s_H = \frac{2\chi_a H^2}{9a(T - T^*)}. \tag{14.2.6}$$

Thus the magnetic birefringence, which is proportional to s_H, varies as $(T - T^*)^{-1}$ as the temperature approaches the transition point. This is borne out by experiments, with $T_{NI} - T^*$ usually of the order of 1 K[29] (Figure 14.2.2).

The case of electric birefringence is slightly more complicated. The orientational energy due to an electric field consists of two contributions: (*a*) that due to the anisotropy of low frequency polarizability α_a; and (*b*) that due to the permanent dipole moment μ. These are expressible as[30]

$$W_1 = -\frac{\alpha_a}{3} F h E^2 s$$

and

$$W_2 = -\frac{F^2 h^2 \mu^2 E^2}{6 k_B T} (3\cos^2\beta - 1) s$$

where $h = 3\epsilon/(2\epsilon + 1)$ is the cavity field factor, ϵ the average dielectric constant, $F = 1/(1 - \alpha f)$ the reaction field factor, α the average polarizability,

$$f = \frac{4\pi}{3} \frac{N}{V} \frac{(2\epsilon - 2)}{(2\epsilon + 1)}$$

N the Avogadro number, V the molar volume, and β is the angle that the permanent dipole makes with the long axis of the molecule. W_2 has been obtained by averaging over a

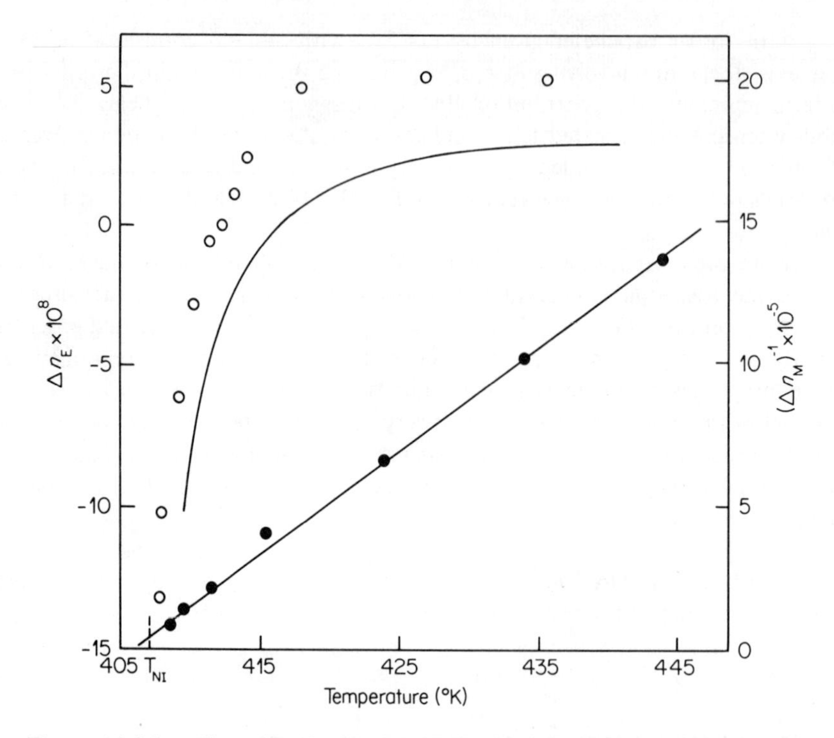

Figure 14.2.2 The electric birefringence (open circles, Tsvetkov and Ryumtsev[31]) and the reciprocal of the magnetic birefringence (full circles, Zadoc–Kahn[26]) in the isotropic phase of PAA plotted against temperature. The lines represent the theoretical variations (Madhusudana and Chandrasekhar[30])

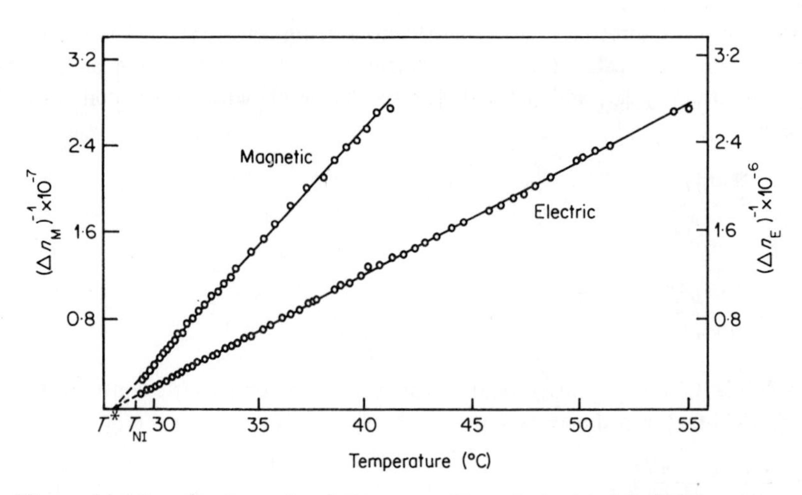

Figure 14.2.3 Reciprocals of the magnetic and electric birefringence in the isotropic phase of 4-hexyl-4′-cyanobiphenyl plotted against temperature. Both give the same value of T^* ($T^* = 28\ ^\circ$C, $T_{NI} - T^* = 1.1\ ^\circ$C) (From Ratna *et al.*[32])

distribution function involving only even powers of $\cos\theta$ in view of the assumed form of the free energy expression. In the present case, we get for the induced order parameter

$$s_E = \frac{2Fh^2E^2\,[\alpha_a - (F\mu^2/2k_BT)(1 - 3\cos^2\beta)]}{9a(T - T^*)} \,. \tag{14.2.7}$$

The electric birefringence, which is proportional to s_E, can be positive or negative depending on the angle β. Further, since the second term in [...] depends on temperature, there can even occur a reversal of sign of s_E with change of temperature. This has indeed been observed in PAA[31] (Figure 14.2.2). On the other hand, if the molecule has a strong dipole moment parallel to its length, the second term in [...] becomes positive and makes the dominant contribution, so that the variation of $1/s_E$ with temperature is linear over the entire temperature range[32] (Figure 14.2.3).

Light scattering In the absence of an external field, the mean value of the order parameter vanishes in the isotropic phase but its fluctuations about the zero value give rise to an anomalous scattering of light.[27] The coherence length ξ of the order parameter fluctuations, which may be interpreted physically as a measure of the average size of an ordered region, increases rapidly as the temperature approaches the transition point. Referring to equation (14.2.5) and setting $H = 0$, we find that in point of fact, two coherence lengths may be defined; for gradients along the local director axis

$$\xi_l^2(T) = \frac{L_1 + \tfrac{2}{3}L_2}{A(T)} \tag{14.2.8}$$

and for gradients perpendicular to the local director axis

$$\xi_t^2(T) = \frac{L_1 + L_2/6}{A(T)} \,. \tag{14.2.9}$$

Both ξ_l and ξ_t diverge as $T \to T^*$, a standard result in the theory of phase transitions. We now expand the order parameter fluctuations as a Fourier series:

$$s_{\alpha\beta}(\mathbf{r}) = \sum_q s_{\alpha\beta}(\mathbf{q}) \exp(i\mathbf{q}\cdot\mathbf{r}) \tag{14.2.10}$$

where $\mathbf{q}$ is the scattering wave vector. Substituting (14.2.10) in (14.2.5), applying the equipartition theorem and expressing the order parameter fluctuations in terms of the corresponding dielectric constant fluctuations $\delta\epsilon$, we obtain for the case in which the incident and scattered light are both polarized perpendicular to the scattering plane,[33]

$$\langle \delta\epsilon_{VV}^2(q) \rangle = \frac{4}{3}\left(\frac{2k_BT(\Delta\epsilon)^2}{9a(T - T^*)}\right)\left(1 - \xi_1^2q^2 - (\xi_2^2q^2/6)\right) \tag{14.2.11}$$

where $\Delta\epsilon$ is the anisotropy of the dielectric constant (for optical frequencies) for perfectly parallel alignment of the molecules, and $\xi_1 = L_1/A(T)$ and $\xi_2 = L_2/A(T)$. Similarly, for incident light polarized normal to the plane of scattering and the scattered light polarized parallel to it,

$$\langle \delta\epsilon_{VH}^2(q) \rangle = \left(\frac{2k_BT\,(\Delta\epsilon)^2}{9a(T - T^*)}\right)(1 - \xi_1^2q^2 - \tfrac{1}{2}\xi_2^2q^2\cos^2(\theta/2)) \tag{14.2.12}$$

where θ is the scattering angle. An analogous expression may be written for the HH geometry also. The intensity of scattering I is proportional to $\langle \delta\epsilon^2 \rangle$. Studies on the angular dependence of the scattering show that ξ is small compared with the wavelength of light, although not negligibly so. If we assume to a first approximation that $q\xi \ll 1$, we note that $I \propto (T - T^*)^{-1}$

and $I_{VV}/I_{VH} = 4/3$. Both these predictions have been confirmed experimentally.[33] Further, $I_{VH}(\theta) - I_{VH}(180 - \theta)$ yields ξ_1 (the ξ_2 term cancels out from this difference since $q = 4\pi n \sin(\theta/2)/\lambda$), and it is easily shown that $I_{VH} - I_{HH}$ yields ξ_2. Studies on p-methoxybenzylidene p-n-butylaniline (MBBA) have verified that the temperature dependence of the coherence lengths is in accord with the theory[33] and also that there is a slight anisotropy in the shape of the correlated region.[34]

$$\xi_l = 119(T - T^*)^{-1/2}\ \text{Å}$$

$$\xi_t = 104(T - T^*)^{-1/2}\ \text{Å}.$$

Pretransition effects in the dynamic properties In developing the hydrodynamic theory of truly isotropic liquids, one treats the velocity gradient tensor as a 'flux' and the corresponding viscous stress tensor $t_{\alpha\beta}$ as the conjugate 'force'. In the isotropic phase of a nematic, there is a fluctuating local order parameter, and therefore there is an additional flux given by

$$R_{\alpha\beta} = \frac{\delta s_{\alpha\beta}}{\delta t} \simeq \frac{\partial s_{\alpha\beta}}{\partial t} \tag{14.2.13}$$

the conjugate force being

$$\phi_{\alpha\beta} = -\frac{\partial F}{\partial s_{\alpha\beta}} = -A s_{\alpha\beta}. \tag{14.2.14}$$

If

$$d_{\alpha\beta} = \tfrac{1}{2}[(\partial v_\alpha/\partial x_\beta) + (\partial v_\beta/\partial x_\alpha)] \tag{14.2.15}$$

we have the following equations coupling the fluxes and forces:

$$\begin{bmatrix} t_{\alpha\beta} \\ \varphi_{\alpha\beta} \end{bmatrix} = \begin{bmatrix} \eta & \mu \\ \mu' & \nu \end{bmatrix} \begin{bmatrix} d_{\alpha\beta} \\ R_{\alpha\beta} \end{bmatrix} \tag{14.2.16}$$

where μ, η and ν are viscosity coefficients;[27] further, assuming Onsager's reciprocal relations, we have $\mu = \mu'$. We shall now discuss the application of these equations to two other properties which exhibit pretransition anomalies.

(a) Flow birefringence Consider a shear flow along x with a velocity grandient dv/dz. The flow induces a birefringence proportional to the velocity gradient with the principal axes of the index ellipsoid inclined at 45° to the x,z axes. In the steady state $R_{\alpha\beta} = 0$, $s_{xz} = \tfrac{1}{2}\mu(dv/dz)$ and hence

$$s_{xz} = -\frac{\mu}{2a(T - T^*)}\frac{dv}{dz}. \tag{14.2.17}$$

Now, the flow birefringence is proportional to s_{xz} and therefore shows a $(T - T^*)^{-1}$ dependence. This has been confirmed experimentally in the case of MBBA.[35]

(b) Spectrum of scattered light For small q (long wavelength fluctuations) one may neglect the velocity terms in (14.2.16) and write

$$\frac{\partial s_{\alpha\beta}}{\partial t} = -\Gamma s_{\alpha\beta} \tag{14.2.18}$$

where

$$\Gamma(T) = \frac{a(T - T^*)}{\nu}. \qquad (14.2.19)$$

$\Gamma(T)$ can be determined from the line-width of the scattered spectrum. However, since the viscosity coefficient ν is itself temperature dependent, the measurements have to be very precise.[36] Assuming that ν is proportional to the shear viscosity, it has been found that the experimental variation of $\Gamma(T)$ for MBBA is in agreement with (14.2.19).

Pretransition effects in the isotropic phase of cholesterics The isotropic phase of a cholesteric may also be expected to show pretransition anomalies in magnetic birefringence, light scattering, etc., since thermodynamically there is little difference between the nematic–isotropic and the cholesteric–isotropic transitions. However, there is an additional feature of the order parameter fluctuations in the cholesteric which is not present in the nematic, viz. a helical ordering arising from the chiral nature of the molecular interactions. The local correlations lack a centre of inversion which results in an enhancement of the optical activity in the isotropic phase, as was first shown by Cheng and Meyer.[37] The free energy in the isotropic phase may be written as

$$F_{\text{ch}} = F_{\text{nem}} + 2q_0 L' e_{\alpha\beta\gamma} s_{\alpha\mu} \partial_\gamma s_{\beta\mu} \qquad (14.2.20)$$

where F_{nem} is given by (14.2.5), $e_{\alpha\beta\gamma}$ is the Levi–Civita tensor; q_0 is a pseudo-scalar and L' a constant. The ratio q_0L'/L has the dimensions of twist. The detailed electromagnetic theory shows that the optical rotation arises from correlations of the form $\langle s^*_{x\alpha} s_{y\beta} \rangle$. The magnitude of this effect is just barely observable in most cholesterics since the anisotropy of molecular polarizability of these compounds is usually rather small. For this reason, Cheng and Meyer[37] specially chose for their studies p-ethoxybenzal-p′-(β-methyl butyl)aniline which has a very high anisotropy, and were able to observe a pronounced increase in the optical rotation near the transition. Their experimental results are shown in Figure 14.2.4. The natural optical activity of

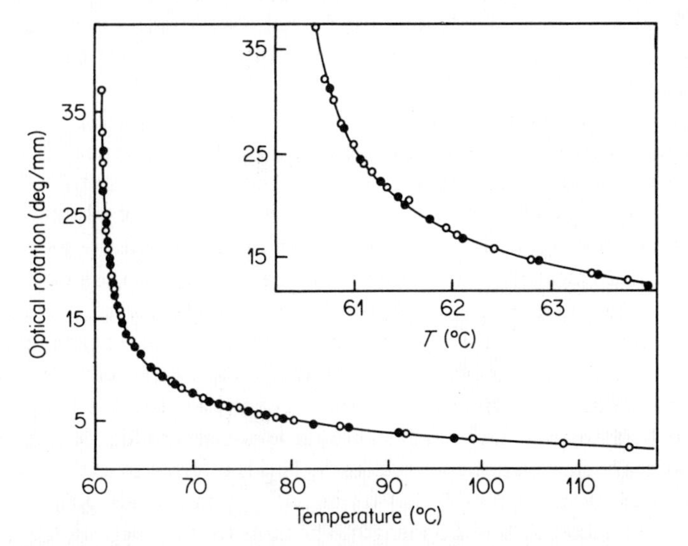

Figure 14.2.4 Anamalous optical rotation in the isotropic phase of a cholesteric liquid crystal. Open and closed circles are measurements on two different samples appropriately normalized. Cholesteric–isotropic transition temperature 60.57 °C. $\lambda = 0.6328\ \mu$m. (From Cheng and Meyer,[37] reproduced by permission of the American Institute of Physics)

the molecule (in the absence of correlations), determined by measuring the rotatory power of dilute solutions of varying concentrations and extrapolating to 100% concentration, was only about 1°/cm as compared with a total rotation of about 40°/cm close to the transition, proving that correlations play the predominant role. The rotation, which is directly related to the coherence length, varies as $(T - T^*)^{-1/2}$, in accordance with the predictions of the Landau–de Gennes model.

General remarks The pretransition anomalies in the isotropic phase of nematic (cholesteric) liquid crystals appear to agree with the classical Landau model. The magnetic birefringence, light scattering, etc., vary as $(T - T^*)^{-\gamma}$ and the coherence lengths as $(T - T^*)^{-\gamma/2}$, with $\gamma = 1$ in accordance with the mean field approximation. However, it is well known that the model does not work satisfactorily for all systems.[38,39] For example, it fails in the case of the ferromagnetic–paramagnetic phase transition for which $\gamma = 1.2$–1.4. A general approach that has been used with success is to write down the free energy in terms of certain 'homogeneous' thermodynamic functions. For instance, in the present case one may write

$$F = s^x f(\epsilon/s^y) \tag{14.2.21}$$

where $\epsilon = (T - T^*)/T^*$ is the reduced temperature and x and y are two exponents. One can then work out the critical exponents for various physical properties, e.g. the susceptibility, the specific heat, etc. The calculations lead to certain exponent relations or scaling laws, which are not generally satisfied by the classical exponents. Further, the Ornstein–Zernike relation for spatial correlation, viz.,

$$\langle s(0)s_1(R) \rangle \propto \frac{A}{R} \exp(-R/\xi) \quad \text{for } R \gg \xi \tag{14.2.22}$$

which is compatible with the classical theory, is no longer valid: it is also to be replaced by a homogeneous function. Exact calculations for model systems and more recently the general approach using the renormalization group theory have led to critical exponents which agree with experiments for a wide variety of systems.

It is of interest to examine the reason for the success of the classical theory in the present problem. The order parameter in the Landau expansion has been so defined that all fluctuations of wavelengths less than a certain length L are absorbed in it. The contribution of the longer wavelength fluctuations which have been ignored can be estimated assuming the Ornstein–Zernike relation, and if this contribution becomes significant then clearly the mean field theory breaks down. Detailed calculations[38,39] show that the breakdown is determined by a ratio of the form a/ξ_0, where a is the intermolecular distance and ξ_0 the correlation range at high temperature given by $\xi = \xi_0 \epsilon^{-\nu}$; this ratio has to be small as compared with unity for the mean field theory to be valid over a wide range of temperatures. Different criteria, depending essentially on the choice of L, have been worked out to calculate the temperature ϵ_c below which the mean field theory fails. For instance, in ferromagnets that have long range forces, choosing $L = \xi$ it has been shown that $\epsilon_c = (a/\xi_0)^6$. This is referred to as the Ginzburg criterion. It has been suggested by de Gennes[14] that for nematic liquid crystals, $a/\xi_0 \sim d/l \sim 1/5$, where d is the width of the molecule and l its length. The basis for this suggestion is presumably that at high temperatures the correlation ξ_0 would be such that the molecules tend to form spherical groups of diameter $\sim l$. This may be a possible explanation for the validity of the Landau theory in the case of the N–I transition, especially since the first order transition point T_{NI} is about 1 K above T^*.

14.2.3 Near neighbour correlations: the Bethe approximation

We shall now compare the phenomenological model just discussed and the molecular statistical theory of Maier and Saupe. According to the latter theory the free energy of the weakly ordered isotropic phase in the presence of an external magnetic field is

$$F = N\left[\tfrac{1}{2}As(s+1) - k_B T \ln \int_0^{\pi/2} \exp\left(\frac{3}{2k_B T}(As + \tfrac{1}{3}\chi_a H^2)\cos^2\theta\right)\sin\theta\, d\theta\right].$$

Expanding and integrating,

$$F = Nk_B T\left(\frac{As^2}{2k_B T^2}(T - T^*) - 0.0762\frac{A^3 s^3}{8k_B^3 T^3} + 0.0122\frac{A^4 s^4}{16k_B^4 T^4}\right) - \tfrac{1}{3}N\chi_a H^2 s$$

where $T^* = A/5k_B$. This expression is identical in form to the free energy expansion (14.2.5) of the Landau model, but it does not yield a satisfactory value of T^*. Since $(A/k_B T_{NI}) = 4.54$, $T^*/T_{NI} = 0.908$. For PAA, $T_{NI} = 408°K$ so that $T_{NI} - T^* \simeq 40°K$, whereas empirically $T_{NI} - T^* \simeq 1°K$.

Clearly, near neighbour correlations have to be taken into account in the molecular statistical approach to give a better description of the pretransition effects. A theory of this type has been developed recently[40-44] based on a method developed originally by Bethe[45] for treating order–disorder effects in binary alloys.

We suppose that every molecule is surrounded by z neighbours ($z \geqslant 3$) and that no two of the nearest neighbours are nearest neighbours of each other. Let the pair potential between the central molecule 0 and one of its neighbours j be $E(\theta_{0j})$, where θ_{0j} is a function of the usual spherical coordinates θ_0, ϕ_0; θ_j, ϕ_j, and let every outer shell molecule j be coupled with the remaining (external) molecules of the uniaxial medium through a mean field, i.e. with an interaction potential $V(\theta_j)$. The relative weight for a given configuration of a cluster of $z + 1$ molecules is then

$$\prod_{j=1}^{z} f(\theta_{0j})g(\theta_j) \tag{14.2.23}$$

where

$$f(\theta_{0j}) = \exp[-E(\theta_{0j})/k_B T] \tag{14.2.24}$$

$$g(\theta_j) = \exp[-V(\theta_j)/k_B T].$$

The relative probability of the central molecule assuming a certain orientation θ_0, ϕ_0 is

$$\int \cdots \int \prod_{j=1}^{z} f(\theta_{0j})g(\theta_j)\, d(\cos\theta_j)\, d\phi_j \tag{14.2.25}$$

while that for an outer shell molecule, say 1, to assume an orientation (θ_1, ϕ_1) is

$$\iint f(\theta_{01})g(\theta_1)\, d(\cos\theta_0)\, d\phi_0 \int \cdots \int \prod_{j=2}^{z} f(\theta_{0j})g(\theta_j)\, d(\cos\theta_j)\, d\phi_j\,. \tag{14.2.26}$$

Chang[46] postulated that these two probabilities must be identical if the orientations of the molecules 0 and 1 are the same, which leads to a consistency relation that must be satisfied for all values of (θ, ϕ).

This condition was expressed in a slightly different form by Krieger and James.[47] The relative probability that the central molecule 0 and one of its neighbours, say 1, are oriented along θ_0, ϕ_0 and θ_1, ϕ_1 respectively is

$$\psi(\theta_0, \phi_0; \theta_1, \phi_1) = f(\theta_{01}) g(\theta_1) \prod_{j=2}^{z} \int \ldots \int f(\theta_{0j}) g(\theta_j) \, d(\cos\theta_j) \, d\phi_j \, . \tag{14.2.27}$$

Krieger and James postulated that this probability should be the same irrespective of which molecule is regarded as the central one, that is,

$$\psi(\theta_0, \phi_0; \theta_1, \phi_1) = \psi(\theta_1, \phi_1; \theta_0, \phi_0)$$

so that

$$\frac{g(\theta_0)}{[\int \ldots \int f(\theta_{0j}) g(\theta_j) \, d(\cos\theta_j) \, d\varphi_j]^{z-1}} = \rho(\text{constant}) \tag{14.2.28}$$

which again has to be satisfied for all values by (θ_0, ϕ_0). It is easily seen that (15.2.28) is implied in Chang's consistency relation also.

In the first calculation using this model,[40,41] it was assumed that

$$E(\theta_{0j}) = -B^* P_2(\cos\theta_{0j}) \tag{14.2.29}$$

$$V(\theta_j) = -BP_2(\cos\theta_j). \tag{14.2.30}$$

Those pairs of values of $B^*/k_B T$ and $B/k_B T$ which give a constant ratio (14.2.28) for every $P_2(\cos\theta_0)$ represent solutions at that temperature. However, it turns out that the best choice of $B/k_B T$ for any given $B^*/k_B T$ still leads to discrepancies (2–3% near T_{NI} for $z = 8$) in fulfilling Chang's relation. On the other hand, there is a marked improvement in the accuracy by taking[43]

$$V(\theta_j) = -BP_2(\cos\theta_j) - CP_4(\cos\theta_j) \tag{14.2.31}$$

which is the form of the potential has been employed in recent extensions[22,23] of the Maier–Saupe theory. As before, the properties of the system can be derived in terms of the single parameter B^* and the solutions now satisfy Chang's relation to better than 0.1% at all angles near T_{NI} for $z = 8$.

Ypma and Vertogen[44] made a weaker assumption: using (14.2.29) and (14.2.30) they postulated that the *average* value of $P_2(\cos\theta)$ is the same for the central and outer shell molecules. With this assumption, the problem can be solved self-consistently, but then Chang's relation, which states that the orientational probability of the central and outer shell molecules should be the same at *every* angle, is very poorly satisfied. However, if we assume (14.2.31) and impose the additional condition that the average $P_4(\cos\theta)$ should also be the same for the central and outer shell molecules, the solutions fulfil Chang's relation to better than 0.1% near T_{NI} for $z = 8$, and the results become practically identical with those of the Krieger–James approximation.[43] It may be emphasized that thermodynamic equilibrium of the system requires that Chang's consistency relation be satisfied accurately. We shall therefore present only the results based on (14.2.31). The internal energy of the system is

$$U = -\tfrac{1}{2} N z B^* \langle P_2(\cos\theta_{0j}) \rangle \tag{14.2.32}$$

where $\langle P_2(\cos\theta_{0j}) \rangle$ is the statistical average over the distribution function for the entire cluster. From a plot of U versus $(1/T)$ at constant volume, which has a characteristic sigmoid shape, the first-order transition point T_{NI} can be located by the familiar equal areas of method (Figure 14.2.5). The value of $(T_{NI} - T^*)/T^*$ is now significantly better than the mean field value, but still rather high as compared with experiment. The agreement with the experimental

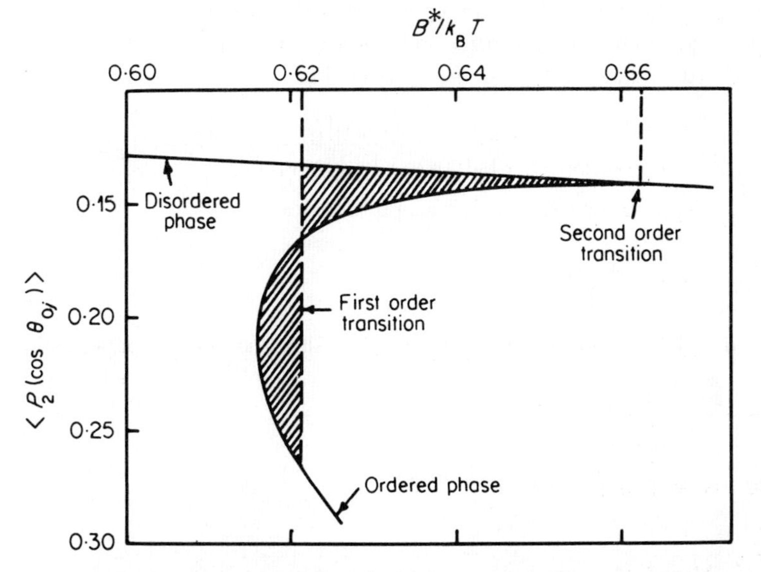

Figure 14.2.5 Plot of $\langle P_2(\cos\theta_{oj})\rangle$ $(=-2U/NzB^*)$ versus B^*/k_BT for $z = 8$. At the first order transition temperature the shaded areas are equal so that the Helmholtz free energy of the ordered and disordered phases are the same. At the second order transition temperature $\langle P_2(\cos\theta_{oj})\rangle = c_2 = 1/(z-1)$. (From Madhusudana *et al.*[43])

Table 14.2.1

	$\dfrac{T_{NI} - T^*}{T_{NI}}$
Bethe approximation: $z = 12$	0.0709
8	0.0620
4	0.0401
3	0.0305
Maier–Saupe theory: $z = \infty$	0.092
Experimental	0.003

value tends to improve with decreasing z, as was first noted by Ypma and Vertogen (see Table 14.2.1).

The isotropic phase We next consider the liquid phase of a nematic liquid crystal in which a weak orientational order ($\sim 10^{-4}$ or less) has been induced by an external field (magnetic or electric) acting along the z-axis of a space fixed coordinate system xyz. Let $W(\theta)$ be the orientational contribution to the potential energy of a molecule due to the field. The relative weight for a given configuration of the cluster of $(z+1)$ molecules is now

$$\prod_{j=1}^{z} f(\theta_{0j})g(\theta_j)h(\theta_0)h(\theta_j) \tag{14.2.33}$$

where $h(\theta_0) = \exp[-W(\theta_0)/k_BT]$, etc. Since we are dealing with a very weakly ordered system, the $P_4(\cos\theta)$ term may be neglected, and moreover the difference between the

Krieger–James and the Ypma–Vertogen approximations becomes insignificant. The exponentials involving $V(\theta_j)$ may be expanded and solutions of the consistency relation become straightforward. The induced order parameter defined as

$$s = \frac{\int \ldots \int P_2(\cos\theta_0)\psi(\theta_0,\phi_0;\theta_j,\phi_j)\,\mathrm{d}(\cos\theta_0)\,\mathrm{d}\phi_0\,\mathrm{d}(\cos\theta_j)\,\mathrm{d}\phi_j}{\int \ldots \int \psi(\theta_0,\phi_0;\theta_j,\phi_j)\,\mathrm{d}(\cos\theta_0)\,\mathrm{d}\phi_0(\cos\theta_j)\,\mathrm{d}\phi_j} \tag{14.2.34}$$

has the following form in the Krieger–James approximation:

$$s = \frac{c}{5k_\mathrm{B}T}\,\frac{1+c_2}{1-(z-1)c_2} \tag{14.2.35}$$

where c is given by $W(\theta_j) = -cP_2(\cos\theta)$, and c_2 is the short-range order parameter given by

$$c_2 = \frac{\int P_2(\cos\theta_{0j})f(\theta_{0j})\,\mathrm{d}(\cos\theta_{0j})}{\int f(\theta_{0j})\,\mathrm{d}(\cos\theta_{0j})}\,. \tag{14.2.36}$$

By expanding the integrand in (14.2.36), we get

$$s = \frac{c}{5k_\mathrm{B}T}\,\frac{1+c}{(T-T^*)} \tag{14.2.37}$$

where T^* is very slightly dependent on temperature. When $T = T^*$, we have by comparing (14.2.35) and (14.2.37),

$$c_2 = \frac{1}{(z-1)}\,.$$

That is, we have recovered the result of the Maier–Saupe theory, except that, as shown earlier, there is a considerable improvement in the value of $(T_{\mathrm{NI}} - T^*)$.

In the absence of an external field, the internal energy of the isotropic liquid is

$$U = -\tfrac{1}{2}NzB^*c_2 \tag{14.2.38}$$

and the specific heat due to near neighbour correlations is $(\partial U/\partial T)_v$.

Antiparallel correlations in strongly polar liquid crystals The vast majority of nematogens are polar compounds. With the advent of field effect nematic display devices, there is considerable interest in very highly polar materials of positive dielectric anisotropy, and a number of compounds have been synthesized with a strong dipole moment parallel to the long molecular axis. It is evident from simple energy considerations that neighbouring molecules in such systems would tend to be antiparallel. However, in the absence of long range translational order in the nematic fluid, we may rule out the possibility of any long range antiparallel correlation. The consequences of this type of short range antiparallel ordering have been discussed recently using the Bethe approximation.[42,43] We modify the pair potential (14.2.29) to

$$E(\theta_{0j}) = A^*P_1(\cos\theta_{0j}) - B^*P_2(\cos\theta_{0j}) \tag{14.2.39}$$

which favours an antiparallel orientation of the nearest neighbours, but let the potential energy of j in the mean field due the rest of the medium continue to be (14.2.31). Here, P_1 is the Legendre polynomial of the first order. Figure 14.2.6 gives the curves for the long range parameters s and $\langle P_4(\cos\theta)\rangle$ and the short range parameters $\langle P_1(\cos\theta_{0j})\rangle$ and $\langle P_2(\cos\theta_{0j})\rangle$.

Two important consequences of the theory are: (*a*) the magnetic and electric birefringence should both vary as $(T - T^*)^{-1}$ which, as we have seen earlier, is in agreement with observations

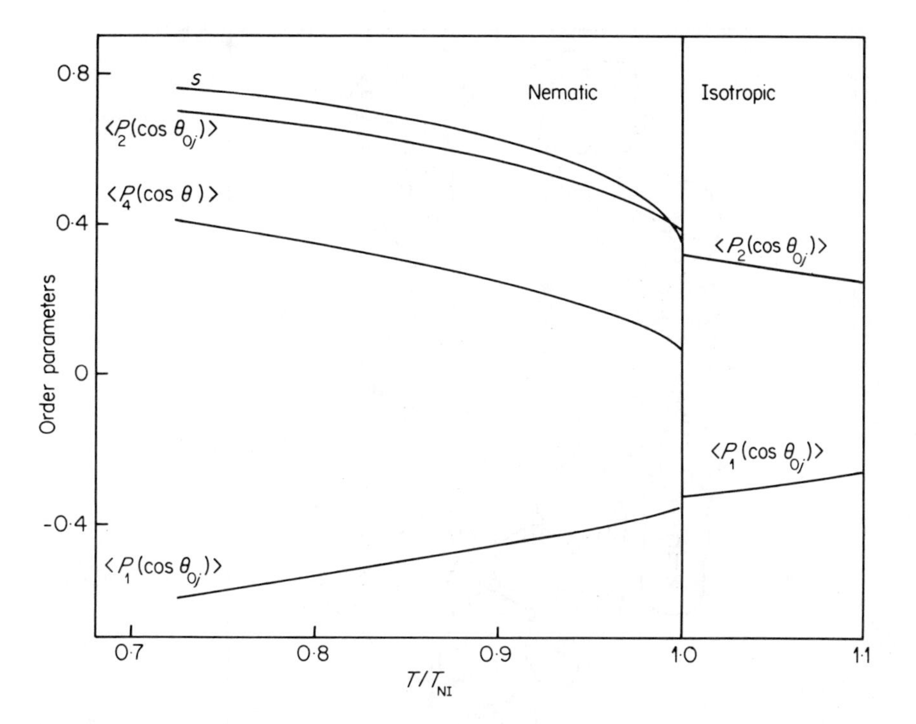

Figure 14.2.6 Short-range order parameters $\langle P_1(\cos\theta_{0j})\rangle$, $\langle P_2(\cos\theta_{0j})\rangle$ and the long-range order parameters s and $\langle P_4(\cos\theta\ )\rangle$ versus B^*/k_BT for $z\ = 4$. The negative value of $\langle P_1(\cos\theta_{0j})\rangle$ signifies antiparallel ordering. The curves are calculated for $A^*/B^* = 0.5$. (From Madhusudana *et al.*[43])

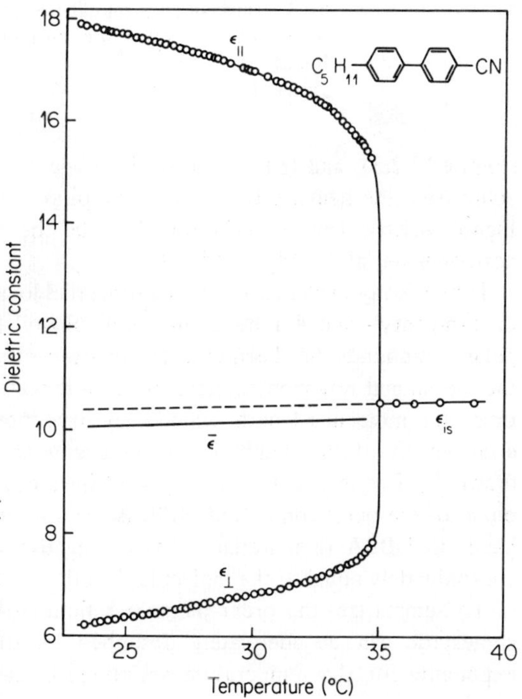

Figure 14.2.7 Principal dielectric constants $\epsilon_{\|}$ and $\epsilon_{\perp}$ in the nematic phase of 4-n-pentyl-4-cyanobiphenyl (5CB). $\bar{\epsilon} = \frac{1}{3}(\epsilon_{\|} + 2\epsilon_{\perp})$ is calculated from the measured values of $\epsilon_{\|}$ and $\epsilon_{\perp}$. The dashed line denotes the extrapolated value of ϵ_{is}, the dielectric constant in the isotropic phase. (From Ratna and Shashidhar[48])

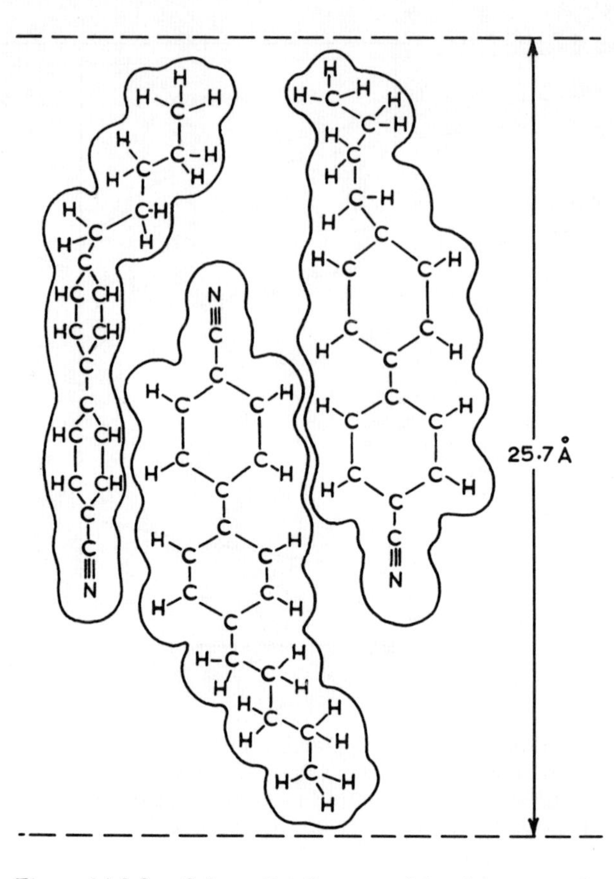

Figure 14.2.8 Schematic diagram of local structure in 5CB resulting in a repeat distance along the nematic axis of about 1.4 times the molecular length (proposed by Leadbetter *et al.*,[49] reproduced by permission of Journal de Physique, Paris)

(Figure 14.2.3); and (*b*) the mean dielectric constant $\bar{\epsilon}$ should increase by a few per cent on going from the nematic to the isotropic phase because of the diminution in $\langle P_1(\cos\theta_{0j})\rangle$ (see Figure 14.2.6). This is again found to be the case experimentally in a number of strongly positive materials[48] (Figure 14.2.7).

Direct X-ray evidence for such antiparallel local ordering in the nematic and isotropic phases of 4′-n-pentyl- and 4′-n-heptyl-4-cyanobiphenyl (5CB and 7CB), both of which are strongly polar compounds, has been reported very recently by Leadbetter *et al.*[49] They have found that the meridional reflexions correspond to a repeat distance along the preferred axis of about 1.4 times the molecular length, which they have interpreted as due to an overlapping head-to-tail arrangement of the neighbouring molecules (Figure 14.2.8). Similar conclusions have been drawn by Lydon and Coakley,[50] who have observed that the layer spacing in the smectic A phase of the octyl compound, 8CB, is again far in excess of the molecular length. On the other hand, in MBBA (a material of weak negative dielectric anisotropy) the repeat distance is approximately equal to the molecular length.

To summarize, the order parameter fluctuations in the isotropic phase of a nematic or a cholesteric can be adequately described by the Landau–de Gennes model, with classical exponents for the temperature variation of the properties in the pretransition region. The validity of the classical exponents most probably means that the range of interaction is much larger than the intermolecular distances. The Maier–Saupe type of mean field molecular theory gives a reasonably good value of the order parameter in the nematic phase, but leads to a large discrepancy in $(T_{NI} - T^*)$. This can be improved by taking explicit account of the near-neighbour interactions.

14.3 The smectic A–nematic transition

14.3.1 McMillan's molecular model for smectic A

The smectic A phase has one-dimensional translational order in addition to the usual orientational order common to all liquid crystals (Figure 14.1.4*a*). The director is normal to the layers and each layer is in effect a two-dimensional fluid. McMillan[51] has extended the Maier–Saupe theory to include an additional order parameter for characterizing the translational periodicity of such a layered structure. (A similar but somewhat more general treatment, based on the Kirkwood–Monroe[52] theory of melting, was developed independently by Kobayashi.[53]) The anisotropic part of the pair potential is conveniently taken in the form

$$V_{12}(r_{12}, \cos\theta_{12}) = -(V_0/Nr_0\pi^{3/2})[\exp\{-(r_{12}/r_0)^2\}]\,\frac{3\cos^2\theta_{12}-1}{2} \tag{14.3.1}$$

where the exponential term reflects the short range character of the interaction, r_{12} is the distance between the molecular centres and r_0 is of the order of the length of the rigid part of the molecules.

If the layer thickness is a, we may write the self-consistent single particle potential, retaining only the leading term in the Fourier expansion, as follows:

$$V_1(z, \cos\theta) = -V_0\,[s + \sigma\alpha\cos(2\pi z/a)]\,\tfrac{1}{2}(3\cos^2\theta - 1) \tag{14.3.2}$$

where

$$\alpha = 2\exp[-(\pi r_0/a)^2]\,. \tag{14.3.3}$$

This form of the potential ensures that the energy is a minimum when the molecule is in the smectic layer with its axis along z. s and σ are order parameters which we shall define presently.

The single particle distribution function is then

$$f_1(z, \cos\theta) = \exp[-V_1(z, \cos\theta)/k_BT] \tag{14.3.4}$$

and self consistency demands that

$$s = \left\langle \frac{3\cos^2\theta - 1}{2} \right\rangle$$

$$\sigma = \left\langle \cos(2\pi z/a)\left(\frac{3\cos^2\theta - 1}{2}\right)\right\rangle$$

where the angular brackets denote statistical averages over the distribution f_1. s is the usual orientational order parameter of the Maier–Saupe theory and σ is a new order parameter which is a measure of the amplitude of the density wave describing the layered structure. The last two equations can be solved numerically to obtain the following types of solutions:

(i) $\sigma = s = 0$ (isotropic phase)

(ii) $\sigma = 0,\quad s \neq 0$ (nematic phase)

(iii) $\sigma \neq 0,\quad s \neq 0$ (smectic phase).

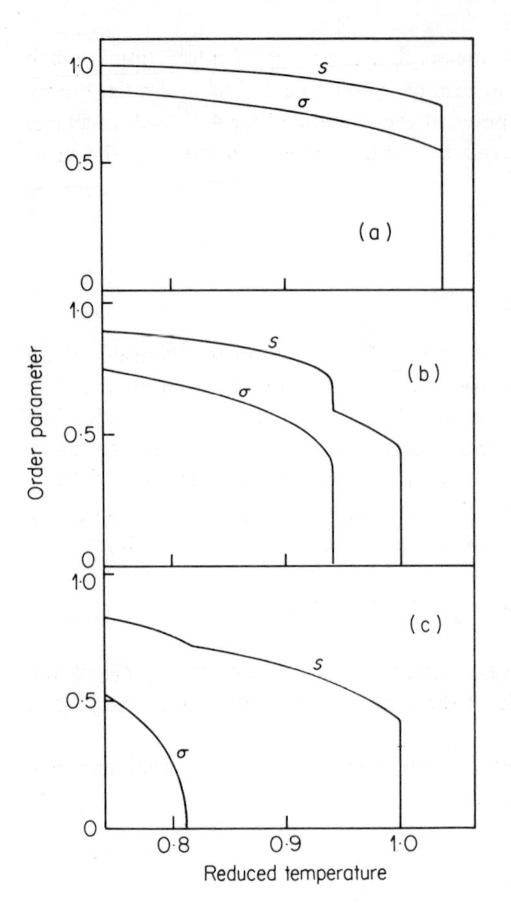

Figure 14.3.1 Order parameters s and σ plotted against reduced temperature $k_B T/0.2202 V_0$ predicted by McMillan's model for (a) $\alpha = 1.1$ showing first order smectic A-isotropic transition, (b) $\alpha = 0.85$ showing first order smectic A–nematic and nematic–isotropic transitions, and (c) $\alpha = 0.6$ showing second order smectic A–nematic transition and first order nematic–isotropic transition. (From McMillan,[51] reproduced by permission of the American Physical Society)

The free energy of the system can be calculated in the usual manner:

$$F = U - TS$$

where

$$U = -\tfrac{1}{2} N V_0 (s^2 + \alpha\sigma^2) \tag{14.3.5}$$

and

$$-TS = N V_0 (s^2 + \alpha\sigma^2) - N k_B T \ln\left[a^{-1} \int_0^a dz \int_0^1 d(\cos\theta) f(z, \cos\theta)\right]. \tag{14.3.6}$$

The two parameters characterizing the material are V_0, which determines the nematic–isotropic transition temperature, and α, a dimensionless interaction strength, which can vary between 0 and 2, and increases with increasing chain length of the alkyl tails of the molecule.

Curves of the order parameters for three representative values of α are presented in Figure 14.3.1. For $\alpha > 0.98$, the smectic A transforms directly into the isotropic phase, while for $\alpha < 0.98$ there is a smectic A–nematic (A–N) transition followed by a nematic–isotropic transition at a higher temperature. For $\alpha < 0.70$ the model predicts a second order A–N transition. The particular value of α (= 0.70) at which the line of first order A–N transition points goes over to a line of second order transition points corresponds to a tricritical point.

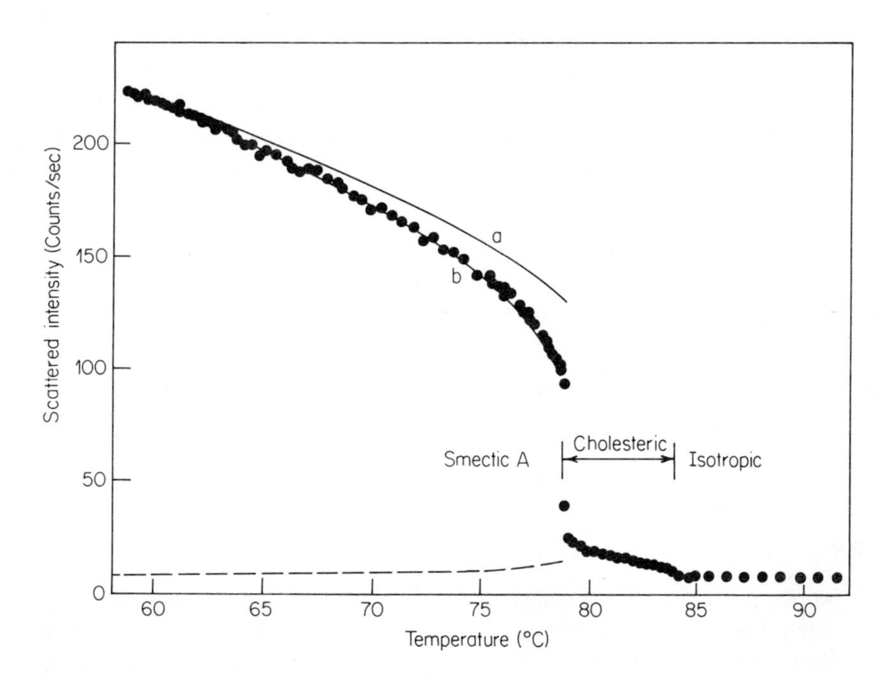

Figure 14.3.2 Measured intensity of X-ray scattering at the Bragg angle plotted against temperature for cholesteryl myristate. The dashed line is the calculated diffuse scattering and fluctuation scattering contribution. The full lines represent the theoretical curves for the total intensity due to Bragg, diffuse and fluctuation scattering derived from (a) the simpler model potential, and (b) the refined one. The theoretical intensity has been adjusted to be equal to the experimental value at the lowest temperature. (From McMillan,[54] reproduced by permission of the American Physical Society)

In a later paper McMillan[54] used the modified pair potential

$$V_{12}(r_{12}, \cos\theta_{12}) = -(V_0 N r_0^3 \pi^{3/2}) \exp[-(r_{12}/r_0)^2](\tfrac{3}{2}\cos^2\theta_{12} - \tfrac{1}{2} + \delta) \qquad (14.3.7)$$

where a partial decoupling has been introduced between translational and orientational ordering. There are now three model-potential parameters which are fixed by requiring the theory to fit T_{AN}, T_{NI} and S_{AN}. The results are essentially the same as those obtained with the simpler model but there are some quantitative improvements when compared with the experimental data.

A direct method of studying the translational order (or the amplitude of the density wave) is by measuring the intensity of the Bragg scattering from the smectic planes. McMillan's experimental results[54] on cholesteryl myristate are shown in Figure 14.3.2, and as can be seen there is excellent agreement with the refined model. The X-ray intensities reveal an appreciable pretransitional smectic-like behaviour in the cholesteric (nematic) phase. This aspect of the problem will be dealt with in a later section.

The orientational order parameter in the smectic and nematic phases, studied by magnetic resonance and other techniques, also follow the predicted type of behaviour as the length of the alkyl end-chain is increased. In particular, a continuous change of s at T_{AN}, as expected of a second order transition, has been found (within experimental limits) in n-p-ethoxybenzylidene-p-phenylazoaniline[55] and in CBOOA.[56]

14.3.2 Free energy of elastic deformation

In a nematic liquid crystal, three types of distortion are possible: splay, twist and bend (Figure 14.1.2). The elastic constants associated with these are k_{11}, d_{22} and k_{33} respectively, and the free energy per cm^3 of the deformed specimen relative to the undeformed one is given by[4]

$$F = \tfrac{1}{2}[k_{11}(\nabla \cdot \mathbf{n})^2 + k_{22}(\mathbf{n} \cdot \nabla \times \mathbf{n})^2 + k_{33}(\mathbf{n} \times \nabla \times \mathbf{n})^2]. \qquad (14.3.8)$$

In smectic A, twist and bend distortions are prohibited as will be evident from the following argument.[57] Assuming the layers to be incompressible, the integral

$$\frac{1}{a}\int_P^Q \mathbf{n} \cdot d\mathbf{r} \qquad (14.3.9)$$

represents the number of layers crossed on going from P to Q where a is the layer thickness. In a dislocation-free sample, this number should be independent of the path chosen so that

$$\nabla \times \mathbf{n} = 0. \qquad (14.3.10)$$

In other words, both twist and bend distortions are absent, leaving only the splay term in (14.3.8). A more complete description of smectic A needs to take into account the compressibility of the layers, though, of course, the elastic constant for compression may be expected to be quite large, almost comparable to that for a solid. The basic ideas of this model were put forward by de Gennes.[57] For small displacements u of the layers normal to their planes, the free energy in the presence of a magnetic field along z takes the form

$$F = \tfrac{1}{2}B\left(\frac{\partial u}{\partial z}\right)^2 + \tfrac{1}{2}\chi_a H^2\left[\left(\frac{\partial u}{\partial x}\right)^2 + \left(\frac{\partial u}{\partial y}\right)^2\right] + \tfrac{1}{2}k_{11}\left(\frac{\partial^2 u}{\partial x^2} + \frac{\partial^2 u}{\partial y^2}\right)^2$$

$$+ \tfrac{1}{2}K'\left(\frac{\partial^2 u}{\partial z^2}\right)^2 + \tfrac{1}{2}K''\frac{\partial^2 u}{\partial x^2}\left(\frac{\partial^2 u}{\partial x^2} + \frac{\partial^2 u}{\partial y^2}\right) \qquad (14.3.11)$$

where the first term is the elastic energy for the compression of the layers and χ_a the anisotropy of diamagnetic susceptibility. When $H = 0$, there will be no terms in $(\partial u/\partial x)^2$ or $(\partial u/\partial y)^2$ as a uniform rotation about y or x does not affect the free energy. The last two terms are usually negligible and may be omitted. Also, the physically reasonable assumption is made that twist and bend distortions are not allowed despite the fact that $\nabla \times \mathbf{n}$ does not strictly vanish when the layers are compressible.

We now write the free energy in terms of the Fourier components of u

$$u_K = \int u(\mathbf{r}) \exp(\mathrm{i}\mathbf{K} \cdot \mathbf{r})\, d\mathbf{r} \qquad (14.3.12)$$

$$F = \frac{1}{2}\sum_K |u_K|^2 [BK_z^2 + k_{11}(K_\perp^2 + \xi_H^{-2})K^2] \qquad (14.3.13)$$

where $K_\perp^2 = K_x^2 + K_y^2$ and $\xi_H = (k_{11}/\chi_a)^{1/2} H^{-1}$. Applying the equipartition theorem

$$\langle |u_K|^2 \rangle = \frac{k_B T}{BK_z^2 + k_{11}(K_\perp^2 + \xi_H^{-2})K_\perp^2} \qquad (14.3.14)$$

from which the mean square fluctuation

$$\langle u^2(\mathbf{r}) \rangle = \frac{k_B T}{4\pi (Bk_{11})^{1/2}} \log(\xi_H/a) \qquad (14.3.15)$$

where a is the layer spacing. As $H \to 0$, $\langle u^2 \rangle \to \infty$, implying that such a structure cannot be stable. However, we shall not discuss this basic question but consider only finite samples, which in any case are known to be quite stable and undergo sharp transitions to the nematic or the isotropic phase.

14.3.3 Phenomenological theory of the A–N transition

We now proceed to consider a Landau type of phenomenological description of the fluctuations attending the A–N transition. This approach to the problem is due to de Gennes[58] and to McMillan,[54] both of whom recognized the analogy with similar phenomena in superfluids.

We shall suppose for the present that the A–N transition is of second order, which as we have seen in Section 14.3.1 is a possibility predicted by McMillan's simple microscopic theory. We start with the density wave in the smectic phase

$$\rho(z) = \rho_0 [1 + 2^{-1/2} |\psi| \cos(q_s z - \phi)] \tag{14.3.16}$$

where ρ_0 is the mean density, $|\psi|$ the amplitude and $q_s = 2\pi/a$ the wave-vector of the density wave, a the interlayer spacing and ϕ a phase factor which gives the position of the layers. Thus the smectic order can be fully specified by the complex parameter

$$\psi = |\psi| \exp(i\phi). \tag{14.3.17}$$

Near the transition, the free energy may be expanded in powers of $|\psi|$ and its gradients. For a fixed orientation of the director,

$$F_s = \alpha |\psi|^2 + \frac{\beta}{2} |\psi|^4 + \frac{1}{2M_V} \left(\frac{\partial \psi}{\partial z}\right)^2 + \frac{1}{2M_T} \left[\left(\frac{\partial \psi}{\partial x}\right)^2 + \left(\frac{\partial \psi}{\partial y}\right)^2\right]. \tag{14.3.18}$$

From symmetry considerations, it is clear that only even powers of ψ may be included.

In the smectic phase ($T < T^*_{\mathrm{AN}}$), the amplitude of the density wave may be taken to be constant, so that only ϕ varies. The gradient terms of F therefore become

$$F_g = \frac{\psi_0^2}{2M_V} \left(\frac{\partial \varphi}{\partial z}\right)^2 + \frac{\psi_0^2}{2M_T} \left[\left(\frac{\partial \phi}{\partial x}\right)^2 + \left(\frac{\partial \phi}{\partial y}\right)^2\right]. \tag{14.3.19}$$

Comparing this with (14.3.11) it is at once clear that ϕ is related to the layer displacement u: $|\phi|^2 = q_s^2 |u|^2$ and $B = |\psi|^2 q_s^2 / M_V$. The terms $\partial\phi/\partial x$ and $\partial\phi/\partial y$ represent the tilt of the layers with respect to the director. If the director orientation is not fixed, it is the relative tilt between the layers and the director that should be considered and therefore (14.3.19) takes the generalized form

$$F = \alpha |\psi|^2 + \frac{\beta}{2} |\psi|^4 + \frac{1}{2M_V} \left(\frac{\partial \psi}{\partial z}\right)^2 + \frac{1}{2M_T} |\boldsymbol{\nabla}_\perp \psi - i q_s \delta \mathbf{n} \psi|^2 \tag{14.3.20}$$

where $\boldsymbol{\nabla}_\perp$ is the gradient operator in the plane of the layers. This equation is reminiscent of the Landau–Ginzburg expression[59] for the free energy of superconductors; $\mathbf{n}$ corresponds to the vector potential $\mathbf{A}$, $\boldsymbol{\nabla} \times \mathbf{A}$ being the local magnetic field.

The analogy may be extended further. By including the Frank elastic free energy terms in (14.3.19), one may define a characteristic length $\Lambda = (k/B)^{1/2}$. Making use of the condition $\partial F/\partial \psi_0 = 0$ and ignoring the difference between M_T and M_V,

$$\Lambda^2 = \frac{Mk\beta}{q_s^2 |\alpha|} \tag{14.3.21}$$

where k is an appropriate elastic constant. For a twist or bend, both of which involve $\boldsymbol{\nabla} \times \mathbf{n}$, Λ

may be interpreted as the depth to which the distortion penetrates into the smectic material. At temperatures much below T_{AN}, it is of the order of a the interlayer spacing. Thus Λ is equivalent to the *penetration depth* of the magnetic field in superconductors.

Above T_{AN}, we may ignore the term involving the fourth power in ψ and from the equipartition theorem obtain in the usual manner

$$\langle |\psi(K)|^2 \rangle = \alpha + (K_z^2/2M_V) + \{K_x^2 + K_y^2)/2M_T\}. \tag{14.3.22}$$

from which we may define the coherence lengths

$$\xi_\parallel^2 = \frac{1}{2M_V\alpha} \tag{14.3.23}$$

$$\xi_\perp^2 = \frac{1}{2M_T\alpha}. \tag{14.3.24}$$

Since $\alpha \to 0$ as $T \to T_{AN}^*$, $\langle |\psi(K)|^2 \rangle$, $\xi_\parallel$ and $\xi_\perp$ diverge. The variation of $\langle |\psi(K)|^2 \rangle$ can be seen directly in the intensity of the Bragg scattering (see Figure 14.3.2).

We have derived the theory for a second order transition, but it continues to be valid even if the transition is weakly first order, except that T_{AN}^* now represents a hypothetical second order transition temperature slightly below T_{AN}

Effect of including the director fluctuations on the A–N transition In liquid crystals which show a nearly continuous A–N transition, it is expected that the dimensionless ratio Λ/ξ (called the Landau–Ginzburg parameter in superconductors) is < 1 (type I behaviour). Halperin *et al.*[60] have shown that when the director fluctuations (which involve twist or bend) are included in the free energy expression, their contribution is likely to change the transition to first order. We shall give a brief outline of their theory.

We have already noted that $\langle u^2(r) \rangle$ given by (14.3.15) diverges for an unbounded sample. To avoid this complication in the theory, Halperin *et al.* effected a 'gauge transformation' which is based on the fact that a simultaneous local rotation of the layers and the director leaves the system invariant. The transformation is given by

$$\psi(\mathbf{r}) = \psi'(\mathbf{r}) \exp[-i\Phi(r)]$$

and

$$\delta\mathbf{n}(\mathbf{r}) = \delta\mathbf{n}'(\mathbf{r}) + \boldsymbol{\nabla}\chi(\mathbf{r}). \tag{14.3.25}$$

Substituting these in (14.3.20) and remembering that F has to remain invariant, we get the condition

$$\boldsymbol{\nabla}_\perp\Phi(\mathbf{r}) = -q_s\boldsymbol{\nabla}_\perp\chi(\mathbf{r}). \tag{14.3.26}$$

Expressing the fluctuation of director $\delta\mathbf{n}$ in terms of the longitudinal and transverse parts,

$$\delta\mathbf{n}(\mathbf{r}) = \delta\mathbf{n}_l(\mathbf{r}) + \delta\mathbf{n}_t(\mathbf{r}) \tag{14.3.27}$$

$(\boldsymbol{\nabla} \times \delta\mathbf{n}_l(\mathbf{r}) = 0$ and $\boldsymbol{\nabla} \cdot \delta\mathbf{n}_t(\mathbf{r}) = 0)$, we get

$$\delta\mathbf{n}'(\mathbf{r}) = \delta\mathbf{n}_t(\mathbf{r}) \quad \text{and} \quad \boldsymbol{\nabla}\chi(\mathbf{r}) = \delta\mathbf{n}_l(\mathbf{r}) \tag{14.3.28}$$

i.e. we can express the free energy in terms of $\psi'(\mathbf{r})$ and $\delta\mathbf{n}_t(\mathbf{r})$, or eliminate the longitudinal part of the director fluctuation. We write (in the smectic phase)

$$\psi'(\mathbf{r}) = |\psi'| \exp[\mathrm{i}\,\phi'(r)] \tag{14.3.29}$$

and neglect the fluctuations in the magnitude of ψ'. Rescaling the lengths along the z-direction such that $M_V = M_T = M$, say, we can write the free energy in the form

$$\alpha |\psi'|^2 + \frac{\beta}{2} |\psi'|^4 + \frac{1}{2M} |(\nabla - \mathrm{i}q_s \delta \mathbf{n}_t)\,\psi'|^2. \tag{14.3.30}$$

ϕ' and $\delta\mathbf{n}_t$ behave like independent variables (since $\nabla \cdot \delta\mathbf{n}_t = 0$) and the order parameter–order parameter correlation is given by

$$\langle \psi'^*(r)\psi'(0) \rangle = |\psi'|^2 \exp\{-\tfrac{1}{2} \langle [\phi'(r) - \phi(0)]^2 \rangle\}. \tag{14.3.31}$$

It turns out that the term in {...} tends to a finite value, i.e. $\langle \psi'^*(r)\psi'(0) \rangle$ does not diverge in the smectic phase, and hence $\psi'(r)$ behaves like a conventional order parameter. Using only $\delta\mathbf{n}_t(\mathbf{r})$, the nematic part of the free energy (14.3.8) can also be rewritten and thus, along with (14.3.30), it is again similar to the free energy functional of a superconductor.

We now analyse the effect of the director fluctuations $\delta\mathbf{n}_t$ on the nature of the A–N phase transition. The free energy relation is valid only for the long wavelength fluctuations of the order parameter ψ' and of the director fluctuations $\delta\mathbf{n}_t$, i.e. with a wave vector K less than a cut-off ζ. Assuming that ψ' has a given constant value, we can evaluate the expectation value of $\langle \delta n_t^2 \rangle$. Using the Fourier transform of the free energy expression, applying the equipartition theorem and transforming back to real space

$$\langle \delta n_t^2 \rangle_{\psi'} = \frac{k_B T^*_{AN}}{(2\pi)^3 k_{22}} \int_{|K|<\zeta} \frac{\mathrm{d}^3 K}{K^2 + K_s^2} \tag{14.3.32}$$

where we are considering a twist fluctuation and $K_s = 1/\Lambda$, Λ being the penetration depth given by (14.3.21). Near the phase transition point, ψ' is very small and hence $K_s \ll \zeta$. In this case, (14.3.32) can be easily integrated:

$$\langle \delta n_t^2 \rangle_{\psi'} = \frac{k_B T^*_{AN} \zeta}{\pi^2 k_{22}} - \frac{q_s k_B T^*_{AN}}{k_{22}(k_{22}M)^{1/2}} |\psi'| \tag{14.3.33}$$

If we insert (14.3.33) in (14.3.30), it is immediately obvious that the second term in the former gives rise to a negative $|\psi'|^3$ term in the free energy expression; thus the A–N transition should always be first order.

An estimate of the shift in the transition point due to this effect (relative to the second order transition point T^*_{AN}) shows that the first order transition lies outside the critical region as determined by the Ginzburg criterion discussed in Section 14.2.2. Therefore, subject to the assumption that most liquid crystals that show a nearly second order A–N transition are of type I, only the classical exponents are likely to be observed. In type II materials, the first order transition need not necessarily lie outside the critical region.

Chu and McMillan[61] have pointed out that owing to the renormalization of the elastic constants there will be a quenching of the director fluctuations, and in consequence there may be a weakening of the first order nature of the transition

Critical divergence of the elastic and viscous constants The growth of smectic-like short range order in the nematic phase near an A–N transition manifests itself strikingly in the temperature dependence of the elastic constants, as was first demonstrated by Gruler.[62] The twist and bend distortions, which are normally disallowed in the smectic phase, become increasingly difficult as the smectic-like regions increase in size and become longer lived, and

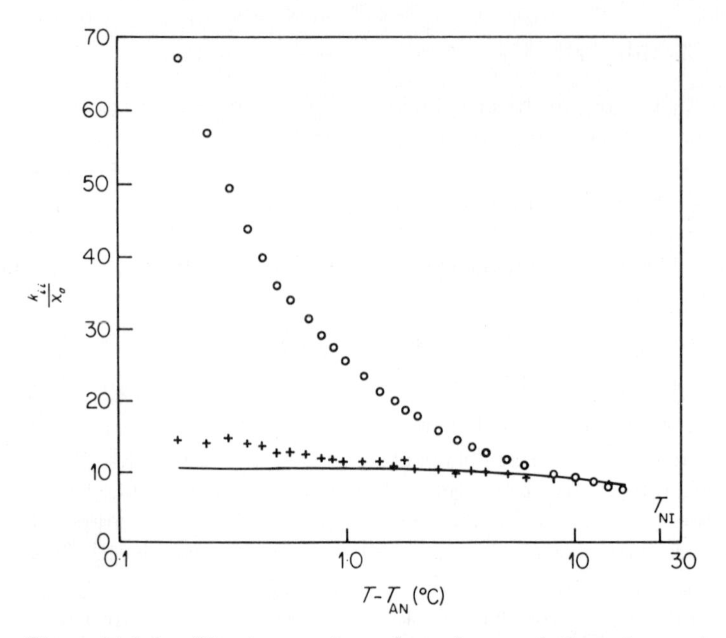

Figure 14.3.3 The temperature dependence of the splay and bend elastic constants, k_{11} (+) and k_{33} (○) respectively, in the nematic phase of CBOOA prior to the smectic A–nematic transition. The values are plotted as k_{ii}/χ_a, where χ_a is the anisotropy of the diamagnetic susceptibility of the nematic. The full line shows the order parameter s, normalized to fit k_{11}/χ_a at high temperature. The splay constant k_{11} deviates only very slightly from ordinary nematic behaviour while the bend constant k_{33} exhibits a critical increase near T_{AN} due to pretransition fluctuations. (From Cheung *et al.*,[63] reproduced by permission of the American Physical Society)

these two elastic constants rise more rapidly than expected from a simple Maier–Saupe type of theory. Figure 14.3.3 presents the data of Cheung *et al.*[63] for CBOOA; k_{33} diverges rapidly while k_{11} shows the normal temperature dependence.

Similarly, certain viscosity coefficients also exhibit a critical divergence. The origin of this effect may be explained physically in the following way. Consider, for example, the frictional torque associated with the twist viscosity coefficient λ_1. The formation of smectic clusters results in an extra torque due to flow normal to the smectic layers, and there is a net enhancement in the effective λ_1 (Figure 14.3.4).

We may estimate these effects approximately as follows. Let us consider first the elastic properties. We observe that the Frank free energy expression (14.3.8) should now include the contribution due to smectic short range order:

$$F = \tfrac{1}{2}k^0(\nabla n)^2 + F_s(\psi) \tag{14.3.34}$$

where k^0 is the usual nematic elastic constant in the absence of smectic-like order, and $F_s(\psi)$ when averaged over all ψ is of the form

$$F_s \sim \frac{-q_s^2}{M}\langle|\psi|^2\rangle(\xi\nabla n)^2 \tag{14.3.35}$$

where we have replaced δn by $\xi\nabla n$ and ignored the difference between M_V and M_T. For a

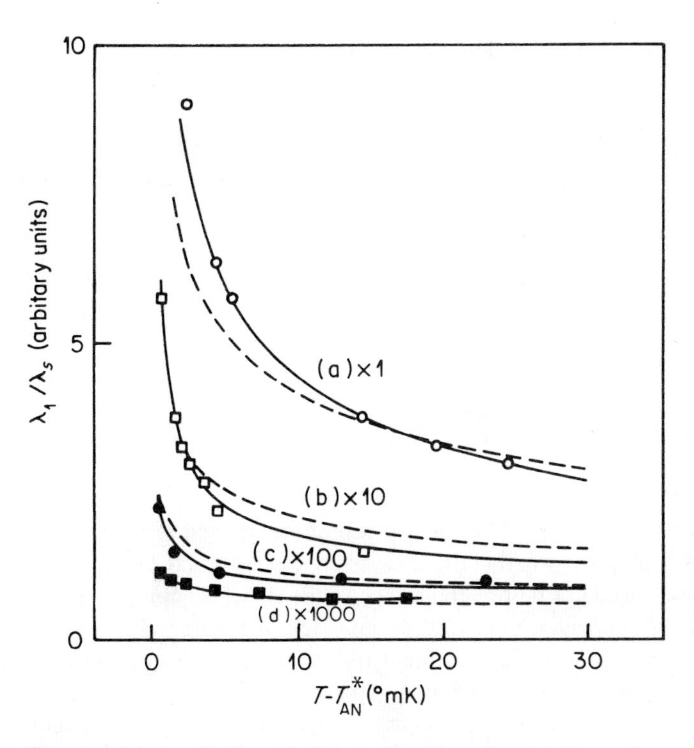

Figure 14.3.4 Ratio of the twist viscosity to the splay viscosity plotted against $T - T^*_{AN}$ for CBOOA: (a) open circles, 0–30 °mK; (b) open squares, 0–300 °mK; (c) filled circles, 0–3 °K; (d) filled squares, 0–30 °K. The solid line is the fit to the mean-field theory for the enhancement of λ_1, and the dashed line is the fit to the helium-analogy theory. (From Chu and McMillan,[61] reproduced by permission of the American Physical Society)

correlated region of volume ξ^3, it can easily be shown that in the mean field approximation

$$\langle |\psi|^2 \rangle \propto \frac{k_B T}{\alpha \xi^3} \tag{14.3.36}$$

Thus from (14.3.34), (14.3.35) and (14.3.36), the effective elastic constant for twist or bend will be $k^0 + \delta k$, where

$$\delta k \propto \frac{k_B T}{a^2} \xi .$$

Since the coherence length diverges rapidly near an A–N transition, the elastic constants for twist and bend should also show critical behaviour. In the mean field approximation,

$$\xi \propto (T - T^*_{AN})^{-1/2} .$$

However, invoking the analogy with superfluids, de Gennes[58] predicts

$$\xi \propto (T - T^*_{AN})^{-2/3} .$$

To estimate the enhancement of the twist viscosity, λ_1, we consider a slowly rotating magnetic field, having an angular frequency ω.[64,65] The director follows the field with a constant angle between **H** and **n**. The torque on the director is

$$\Gamma = \lambda_1 \omega. \tag{14.3.37}$$

Now, the layered arrangement of the smectic-like regions make an additional contribution to the torque, say Γ. Let the angle between the layer normal and the director be θ; then

$$\theta = \omega\tau_\psi \tag{14.3.38}$$

where τ_ψ is the relaxation time of $|\psi|$. We can derive $\tilde{\Gamma}$ from the elastic energy (14.3.35) with $\theta = \xi\nabla n$:

$$\tilde{\Gamma} = \frac{\partial F_s}{\partial \theta} = \frac{q_s^2}{M} \langle |\psi|^2 \rangle \tau_\psi \omega. \tag{14.3.39}$$

Using the value of $\langle |\psi|^2 \rangle$ from (14.3.36), the contribution to the excess viscosity is given by

$$\Delta\lambda_1 \propto \tau_\psi / \xi. \tag{14.3.40}$$

The temperature dependence of $\Delta\lambda_1$ depends on both τ_ψ and ξ and, as we shall see, the exact behaviour of $\tilde{\Gamma}$ depends on whether the classical or non-classical theory is valid.

We shall now consider a more detailed theory of these phenomena, applicable to both the elastic and viscous anomalies, following an approach of Jahnig and Brochard[66] who developed it in exact analogy with a similar treatment for the superconductor–normal transition due to Schmidt.[67] The slowing down of the order parameter fluctuations is calculated using the linear response theory and the fluctuation-dissipation theorem. We start by assuming the smectic and nematic contributions to the free energy (equations 14.3.20 and 14.3.8). Further, we assume both the local smectic order parameter as well as the local director to be time dependent, that is,

$$\psi = \psi(\mathbf{r}, t) \quad \text{and} \quad \delta\mathbf{n} = \delta\mathbf{n}(\mathbf{r}, t). \tag{14.3.41}$$

The static fluctuations of the order parameter are expressible as

$$\langle \psi^*(\mathbf{K})\psi(\mathbf{K}') \rangle = \langle |\psi(\mathbf{K})|^2 \rangle (2\pi)^3 \delta(\mathbf{K} - \mathbf{K}') \tag{14.3.42}$$

where $\langle |\psi(\mathbf{K})|^2 \rangle$ is given by (14.3.22).

The dynamical fluctuations are derived by assuming a simple relaxation for the order parameter. With

$$\psi(\mathbf{r}, t) = \int \frac{\mathrm{d}\mathbf{K}\,\mathrm{d}\Omega}{(2\pi)^4} \psi(\mathbf{K}, \Omega) \exp(-\mathrm{i}\Omega t + \mathrm{i}\mathbf{K}\cdot\mathbf{r}) \tag{14.3.43}$$

we have

$$[-\mathrm{i}\Omega + \Gamma(\mathbf{K})]\,\psi(\mathbf{K}, \Omega) = 0 \tag{14.3.44}$$

where the 'frequency factor' $\Gamma(\mathbf{K})$ is given by

$$\frac{1}{\Gamma(\mathbf{K})} = \frac{\tau_\psi}{(1 + K_\parallel^2\xi_\parallel^2 + K_\perp^2\xi_\perp^2)^m}. \tag{14.3.45}$$

(In the mean field theory this can be easily derived by assuming $\partial\psi(\mathbf{K})/\partial t \propto -\partial F/\partial\psi^*(\mathbf{K})$, in which case $m = 1$.) The frequency spectrum is then given by

$$\langle \psi^*(\mathbf{K}, \Omega)\psi(\mathbf{K}, \Omega') \rangle = 2\langle |\psi(\mathbf{K})|^2 \rangle \frac{\Gamma(\mathbf{K})}{\Omega^2 + \Gamma^2(\mathbf{K})} 2\pi\delta(\Omega - \Omega'). \tag{14.3.46}$$

We now introduce the molecular field **h** acting on the director by means of a variation of the free energy:

$$\delta F = -\int \mathrm{d}\mathbf{r}\, h^t(\mathbf{r}, t)\, \delta\mathbf{n}(\mathbf{r}, t) \tag{14.3.47}$$

where $\mathbf{h}^t = \mathbf{h} + \tilde{\mathbf{h}}$, $\tilde{\mathbf{h}}$, coming from the smectic-like fluctuations. Using (14.3.20),

$$\tilde{\mathbf{h}}(\mathbf{r}, t) = -\frac{\mathrm{i}q_s}{2M_T}\left[\psi^*(\mathbf{r}, t)\nabla_\perp\psi(\mathbf{r}, t) - \psi(\mathbf{r}, t)\nabla_\perp\psi^*(\mathbf{r}, t)\right]$$

$$-\frac{q_s^2}{M_T}\,\psi^*(\mathbf{r}, t)\psi(\mathbf{r}, t)\delta\mathbf{n}(\mathbf{r}, t) \tag{14.3.48}$$

If an external perturbation $\delta\mathbf{n}^{\mathrm{ext}}(\mathbf{r}', t')$ is applied to the medium, the thermal average of $\tilde{\mathbf{h}}$ can be evaluated using the linear response theory:

$$\langle \tilde{h}_\alpha(\mathbf{q}, \omega)\rangle = \tilde{s}_{\alpha\beta}(\mathbf{q}, \omega)\,\delta n_\beta^{\mathrm{ext}}(\mathbf{q}, \omega). \tag{14.3.49}$$

$\tilde{s}_{\alpha\beta}(\mathbf{q}, \omega)$ can be written as a sum of two quantities:

$$\tilde{s}_{\alpha\beta}(\mathbf{q}, \omega) = s_{\alpha\beta}^{\mathrm{s}} + s_{\alpha\beta}^{\mathrm{d}}(\mathbf{q}, \omega) \tag{14.3.50}$$

where the static part

$$s_{\alpha\beta}^{\mathrm{s}} = -\frac{q_s^2}{M_T}\,\langle |\psi|^2\rangle\,\delta_{\alpha\beta} \tag{14.3.51}$$

arises from the static average of $|\psi|^2$. The dynamical fluctuations of $|\psi|$ give rise to the second term in (14.3.50). Identifying $\delta\mathbf{n}$ with $\delta\mathbf{n}^{\mathrm{ext}}$ in (14.3.48), we can evaluate the imaginary part of $s_{\alpha\beta}^{\mathrm{d}}$ using the fluctuation-dissipation theorem:

$$s_{\alpha\beta}^{\mathrm{d}''}(\mathbf{q}, \omega) = \frac{\omega}{2k_{\mathrm{B}}T}\,\langle \tilde{h}_{1\alpha}(\mathbf{q}, \omega)\tilde{h}_{1\beta}(-\mathbf{q}, -\omega)\rangle \tag{14.3.52}$$

where the subscript 1 of $\tilde{\mathbf{h}}$ means that only the first term of (14.3.48) is contributing, and can be easily evaluated using (14.3.43). The real part of $s_{\alpha\beta}^{\mathrm{d}}$ can be obtained by using the dispersion relation:

$$s_{\alpha\beta}^{d'}(\mathbf{q}, \omega) = \int \frac{\mathrm{d}\omega'}{\pi}\,\frac{s_{\alpha\beta}^{d''}(\mathbf{q}, \omega')}{\omega' - \omega}. \tag{14.3.53}$$

Using the decoupling approximation[68] between the real and imaginary parts of the order parameter and the relation (14.3.42), Jahnig and Brochard derive the full response function

$$s_{\alpha\beta}^{d}(\mathbf{q}, \omega) = \frac{q_s^2}{k_{\mathrm{B}}TM_T^2}\int \frac{\mathrm{d}\mathbf{K}}{(2\pi)^3}\,\mathbf{K}_{\perp\alpha}\mathbf{K}_{\perp\beta}\,\langle |\psi(\mathbf{K} - \mathbf{q}/2)|^2\rangle$$

$$\times \langle |\psi(\mathbf{K} + \mathbf{q}/2)|^2\rangle\,\frac{\Gamma(\mathbf{K} - \mathbf{q}/2) + \Gamma(\mathbf{K} + \mathbf{q}/2)}{-\mathrm{i}\omega + \Gamma(\mathbf{K} - \mathbf{q}/2) + \Gamma(\mathbf{K} + \mathbf{q}/2)}. \tag{14.3.54}$$

The total nematic molecular field is obtained by taking the variation (equation 14.3.47) of the nematic free energy (equation 14.3.8) and including the dissipation due to the rotational motion of the director; it takes the following form:

$$h_x(\mathbf{q}, \omega) = \mathrm{i}\omega\lambda_1\delta n_x^{\mathrm{ext}}(\mathbf{q}, \omega) - (k_{11}q_x^2 + k_{22}q_y^2 + k_{33}q_z^2)\delta n_x^{\mathrm{ext}}(\mathbf{q}, \omega)$$

$$+ (k_{22} - k_{11})q_xq_y\delta n_y^{\mathrm{ext}}(\mathbf{q}, \omega) \tag{14.3.55}$$

with a similar expression for h_y.

We now derive the critical divergence of the elastic constants and viscosity coefficients using the response function given above. The elastic constants are obtained by evaluating the real part of the response function in the static limit. For this purpose, it is enough to treat the case $\alpha = \beta = x$. Using (14.3.22), the result is

$$s'_{xx}(\mathbf{q}, 0) = -\frac{k_B T q_s^2}{24\pi\xi_\parallel}\,(\xi_\perp^2 q_y^2 + \xi_\parallel^2 q_z^2). \tag{14.3.56}$$

We can now calculate $\langle \tilde{\mathbf{h}}_x(\mathbf{q}, 0) \rangle$ using equation (14.3.49). Comparing with (14.3.55), we see that the contribution of smectic-like fluctuations to the elastic constants are

$$\Delta k_{11} = 0 \tag{14.3.57}$$

$$\Delta k_{22} = \frac{k_B T q_s^2 \xi_\perp^2}{24\pi\xi_\parallel} \tag{14.3.58}$$

$$\Delta k_{33} = \frac{k_B T q_s^2}{24\pi}\,\xi_\parallel . \tag{14.3.59}$$

The above analysis for the response function holds in the hydrodynamic limit, i.e. $q\xi \ll 1$. The analysis can be carried out for a general case, and in the critical limit $q\xi \gg 1$, one gets

$$s'_{xx}(\mathbf{q}, 0) = -\frac{k_B T q_s^2}{16}\left(\frac{M_V}{M_T}\,q_y^2 + q_z^2\right)^{1/2} \tag{14.3.60}$$

which shows no critical temperature dependence. By comparison with (14.3.56), if we define

$$s'_{xx}(q_y, 0) = -\,\Delta k_{22}(q_y) q_y^2 \tag{14.3.61}$$

we get the wave-vector-dependent elastic constant:

$$\Delta k_{22}(q_y) = \frac{k_B T q_s^2}{16}\left(\frac{M_V}{M_T}\right)^{1/2}\frac{1}{q_y} \tag{14.3.62}$$

which is practically independent of temperature.

We now turn our attention to the twist viscosity coefficient. Using (14.3.55), (14.3.49) and (14.3.52) the fluctuation contribution to the twist viscosity coefficient,

$$\Delta\lambda_1(\mathbf{q}, \omega) = \frac{1}{\omega}\, s''_{xx}(\mathbf{q}, \omega) = \frac{1}{2k_B T}\,\langle \tilde{h}_{1x}(\mathbf{q}, \omega)\tilde{h}_{1x}(-\mathbf{q}, -\omega)\rangle \tag{14.3.63}$$

which is a Kubo formula for the twist viscosity. In the hydrodynamic limit, $q \to 0$, $\omega \to 0$ and $\Delta\lambda_1 = \Delta\lambda_1(0, 0)$. It can be derived using the imaginary part of the response function (14.3.54) along with the frequency factor given by equation (14.3.45). Assuming for convenience $m = 1$ in (14.3.45),

$$\Delta\lambda_1 = \frac{k_B T q_s^2}{16\pi}\,\frac{\tau_\psi}{\xi_\parallel}. \tag{14.3.64}$$

In the mean field approximation, $\tau_\psi \sim \xi^2$ and hence

$$\Delta\lambda_1 \sim \xi \sim (T - T^*_{AN})^{-0.5}. \tag{14.3.65}$$

On the other hand, as we shall see later (equation 14.3.84), if one uses dynamic scaling arguments,[69] and helium-like exponents, $\tau_\psi \sim \xi^{3/2}$ and $\Delta\lambda_1 \sim \xi^{1/2}$ or

$$\Delta\lambda_1 = c(T - T^*_{AN})^{-0.33}. \tag{14.3.66}$$

Further, it turns out that the coefficient c is extremely small. One has to go very close to T^*_{AN} to observe $\Delta\lambda_1$ which is of the order of 0.01 poise at $T^*_{AN} + 0.5°$.

The general case for arbitrary values of q can also be solved, and the result in the critical limit is

$$\Delta\lambda_1(\mathbf{q}, 0) \sim q^{-1/2}. \tag{14.3.67}$$

Hence, in the critical region, the spectrum of the director modes

$$\omega_s \sim \frac{\Delta k(\mathbf{q}, 0)}{\Delta\lambda_1(\mathbf{q}, 0)} q^2 \sim q^{3/2}. \tag{14.3.68}$$

For finite values of ω, but q tending to zero, we can use the approximation that $\tau_\psi = 1/\Gamma(\mathbf{K})$ is independent of the wave vector $\mathbf{q}$. Then the viscosity turns out as

$$\Delta\lambda_1(0, \omega) = \frac{\Delta\lambda_1(0, 0)}{1 + (\omega\tau_\psi)^2} \tag{14.3.69}$$

a result originally derived by Landau and Khalatnikov[70] for slowly relaxing processes.

Another quantity of interest is the damping time τ_n of the twist fluctuations of the nematic director,

$$\tau_n = \frac{\lambda 1}{k_{22}q^2}. \tag{14.3.70}$$

In the mean field theory, both $\Delta\lambda_1$ and Δk_{22} are proportional to ξ (see equations 14.3.58 and 14.3.65). Hence *close* to T^*_{AN}, τ_n should tend to a constant value. On the other hand, if the scaling laws are valid, $\Delta\lambda_1 \propto \xi^{1/2}$ (equation 14.3.66) and $\Delta k_{22} \propto \xi$ or $\tau_n \propto \xi^{-1/2} \propto (T - T_{AN})^{0.33}$, i.e. $\tau_n \approx 0$ as T^*_{AN} is approached.

The nematic liquid crystal is characterized by four other viscosity coefficients in addition to λ_1; however we will not discuss all of them. From the experimental point of view, only one of the Miesowicz coefficients, η_1, i.e. the viscosity measured with the director parallel to the flow direction but perpendicular to the velocity gradient, exhibits a critical enhancement.

The smectic A–cholesteric transition Lubensky[71] has argued that the smectic A–cholesteric transition should always be of first order, in analogy with the behaviour of a superconductor in an external magnetic field. In the cholesteric state the free energy due to twist is given by

$$F_{ch} = (k^0_{22}q^0_{ch})(\mathbf{n} \cdot \nabla \times \mathbf{n}) \tag{14.3.71}$$

where k^0_{22} and q^0_{ch} refer to the high temperature (i.e. unrenormalized) twist elastic constant and the twist angle per unit length in the medium. (In the unperturbed cholesteric state $\mathbf{n} \cdot \nabla \times \mathbf{n} = -q^0_{ch}$.) In the smectic phase, we can write $|\psi|^2 = -\alpha/\beta$ from (14.3.18), so that the free energy

$$F_s \simeq -\frac{\alpha'^2}{\beta}\left(\frac{T - T^*}{T^*}\right)^2$$

where we have assumed that $\alpha = \alpha'[(T - T^*)/T^*]$ according to the mean field theory. The medium undergoes a first order transition at a temperature $T^* + \Delta T$, where ΔT is determined by equating the two energies F_{ch} and F_s:

$$\frac{\Delta T}{T^*} = (\beta k_{22}^0)^{1/2} \frac{q_{ch}^0}{\alpha'} \tag{14.3.72}$$

which has been estimated by Lubensky to be $\sim 10^{-3}$.

Pretransition behaviour near a smectic A–cholesteric transition point In the cholesteric phase near a smectic A transition the relevant part of the free energy density, omitting a background term, is

$$F = \tfrac{1}{2} k_{22}^0 q_{ch}^2 - k_{22}^0 q_{ch}^0 q_{ch} + F_s(\psi) \tag{14.3.73}$$

where k_{22}^0 is the twist elastic constant of the medium in the absence of ψ-fluctuations, $q_{ch}^0 (=2\pi/P_0)$ is the corresponding twist per unit length. $F_s(\psi)$ can be written in the form $\frac{1}{2}\Delta k_{22} q_{ch}^2$ so that the actual value of the twist in the presence of the smectic-like short range order is given by

$$q_{ch} = \frac{q_{ch}^0 k_{22}^0}{k_{22}^0 + \Delta k_{22}} \tag{14.3.74}$$

Since Δk_{22} diverges as the temperature approaches the smectic A–cholesteric point, the pitch $P = 2\pi/q_{ch}$ also increases proportionately. The high temperature-sensitivity of the pitch has practical applications in thermography.[72] The material has to be so chosen that the pitch is of the order of the wavelength of visible light in the temperature range of interest. Small variations of pitch are shown up as changes in the colour of the Bragg scattered light and can be used for visual display of surface temperatures.

Acoustic wave propagation We write energy density in a general form to include the volume dilatation θ:[69,73]

$$F = \tfrac{1}{2} A_0 \theta^2 + \tfrac{1}{2} B_0 |\psi|^2 q_s^2 (\partial u/\partial z)^2 + C_0 |\psi|^2 q_s \theta (\partial u/\partial z) \tag{14.3.75}$$

where A_0, B_0 and C_0 are adiabatic coefficients of elasticity, and the last term describes the coupling between the two types of dilatations θ and $\partial u/\partial z$. Comparison of this expression with (14.3.20) shows that $B_0 \simeq 1/M_V$. Now, a force normal to the layers gives rise to a flow due to the 'permeation' of the molecules from layer to layer.[9] Neglecting this and other viscous effects, and using the conservation law

$$\operatorname{div} \mathbf{v} + \dot{\theta} = 0 \tag{14.3.76}$$

where $\mathbf{v}$ is the velocity of the particle, the propagating acoustic modes are given by the equation

$$\begin{aligned} x^2 - x[A_0 \sin^2\alpha + (A_0 + B_0 |\psi|^2 q_s^2 + C_0 |\psi|^2 q_s)\cos^2\alpha] \\ + \sin^2\alpha \cos^2\alpha |\psi|^2 q_s^2 (A_0 B_0 - |\psi|^2 C_0^2) = 0 \end{aligned} \tag{14.3.77}$$

where $x = \rho c^2$, ρ being the density, c the sound velocity and α the angle between the direction of propagation and the layer normal (z axis). Experimentally, C_0 is found to be much smaller than A_0 and B_0,[74] so that neglecting the coupling term we obtain

$$c_1 = \left(\frac{A_0 + B_0 |\psi|^2 q_s^2 \cos^4\alpha}{\rho} \right)^{1/2} \tag{14.3.78}$$

and

$$c_2 = \sin\alpha\cos\alpha\,|\,\psi\,|\,q_s\left(\frac{B_0}{\rho}\right)^{1/2}. \tag{14.3.79}$$

Thus there are two acoustic branches. The first represents the usual compressional wave whose velocity is practically independent of the direction of propagation (since usually $A_0 > B_0$). The other branch corresponds to changes in the layer spacing without appreciable changes in the density, and may be compared with the phonon branch in superfluids known as 'second sound'. The velocity of this mode is strongly orientation dependent. When the wave vector is parallel to the layers, the second sound becomes the highly damped 'undulation' mode (in which the layers undulate without any change in the layer spacing). Direct evidence for these two branches has been obtained from Brillouin scattering studies on monodomain smectic A samples.[75] From (14.3.79) we see that, close to T^*_{AN},

$$c_2 \sim |\,\psi\,|\,M^{-1/2} \sim \xi^{-1/2}. \tag{14.3.80}$$

Thus the second sound is a critical mode near the A–N transition. Its velocity goes to zero as $T \rightarrow T^*_{AN}$. When viscous and permeation effects are included, the critical damping of the two modes is given by[73]

$$D_1 = \rho^{-1}\Delta\eta\cos^4\alpha \tag{14.3.81}$$

$$D_2 = \rho^{-1}\Delta\eta\sin^2\alpha\cos^2\alpha + \frac{4k_{11}\Delta\lambda_1}{\mu^2}\sin^2\alpha + \nu_p B_0\,|\,\psi\,|^2 q_s^2\cos^2\alpha \tag{14.3.82}$$

where $\Delta\eta = (M_T/M_V)\Delta\lambda_1$, the second term in (14.3.82) arises from the damping of the undulation mode (k_{11} being the splay elastic constant and μ an effective viscosity) and the last term describes the permeation effects (ν_p being the permeation coefficient).

The relaxation frequency of the order parameter may be calculated using dynamic scaling. The characteristic frequency $\Omega_\psi(\mathbf{K}, T)$ is written as a homogeneous function of the form $K^z f(K\xi)$. In the hydrodynamic limit $K\xi \ll 1$,

$$\Omega_\psi(K, T) = c_2 K \sim \frac{K^{3/2}}{(K\xi)^{1/2}}. \tag{14.3.83}$$

If $K\xi \sim 1$, the distinction between the amplitude and phase disappears and

$$\tau_\psi^{-1}(T) = \frac{c_2}{\xi} = \left(\frac{|\,\psi\,|^2 q_s^2}{\rho M_V \xi^2}\right)^{1/2} \sim \frac{1}{\xi^{3/2}} \sim |\,T - T^*_{AN}\,|. \tag{14.3.84}$$

Interestingly, the same temperature dependence is predicted even by the mean field theory, for in this case

$$\frac{|\,\psi\,|^2}{M_V} \sim \frac{1}{\xi^2}$$

and

$$\tau_\psi^{-1} \sim \xi^{-2} \sim |\,T - T^*_{AN}\,|. \tag{14.3.85}$$

Similarly it can be shown that the permeation coefficient

$$\nu_p \sim \frac{1}{|\,T - T^*_{AN}\,|}. \tag{14.3.86}$$

Light scattering In the nematic phase, the fluctuations of the director give rise to a very intense scattering of light. The spectrum of the scattered light is determined by a slow mode which describes the relaxation of the orientational motion of the director. If n_1 is the component of the director fluctuation perpendicular to n_0 in the n_0q-plane (where n_0 is the unperturbed orientation of the director assumed to be along z and $q = (q_z^2 + q_\perp^2)^{1/2}$ the scattering vector) and n_2 is the component perpendicular to n_1, the frequencies of the n_1 and n_2 fluctuations are[76]

$$\omega_1 = \frac{i}{\eta_1}(k_{33}q_z^2 + k_{11}q_\perp^2)$$

and (14.3.87)

$$\omega_2 = \frac{i}{\eta_2}(k_{33}q_z^2 + k_{22}q_\perp^2)$$

where η_1, η_2 are the appropriate viscosity coefficients and k_{22}, k_{33} are the renormalized values of the elastic constants ($k_{33} = k_{33}^0 + \Delta k_{33}$, etc.). In the smectic phase, we have

$$\omega_1 = \frac{i}{\eta_1'}\left(k_{33}q_z^2 + k_{11}q_\perp^2 + \frac{|\psi|^2 q_s^2 q_z^2}{M_T q^2}\right)$$

and

$$\omega_2 = \frac{i}{\eta_2'}\left(k_{33}q_z^2 + k_{22}q_\perp^2 + \frac{|\psi|^2 q_s^2}{M_T}\right). \qquad (14.3.88)$$

In the smectic phase, as $q \to 0$, the director relaxes with a rate

$$\frac{1}{\tau_n} = \frac{|\psi|^2 q_s^2}{\eta' M_T}. \qquad (14.3.89)$$

If, in the smectic phase, the wave vector is in the plane of the layers ($q_z = 0$), we get the undulation mode which involves only the splay constant, and the light scattering is very strong as in a nematic. On the other hand, for an arbitrary **q**, the differential cross section per unit volume is of the form

$$\sigma_d \propto \frac{k_B T}{\{(|\psi|^2 q_s^2)/M_V\} + kq^2} \qquad (14.3.90)$$

where k is an effective elastic constant. It is clear that the light scattering should show critical behaviour. For example, as $q \to 0$, $\sigma_d \sim \xi$.

Experimental results Most of the experimental studies of the A–N transition have been made on CBOOA, but some results have also been reported on p-butoxybenzylidene-p′-octylaniline (BBOA) and more recently on n-p-octyloxycyanobiphenyl (8OCB). The data are summarized below.

CBOOA

Nature of the A–N transition

Technique	*Result*
Orientational order by ^{14}N nuclear magnetic resonance	Continuous[56]

Technique	*Result*
Calorimetry	(i) Transition entropy $<0.02\,R_0$[12] (ii) Very weakly first order[77] (iii) Transition entropy $\simeq 0.017\,R_0$. $C_p \propto (T - T^*_{AN})^{-\alpha}$, $\alpha = 0.14$–0.16[78]
Volumetric measurement	Very weakly first order[79]
Rayleigh scattered intensity (twist mode)	Continuous[61]

Elastic and viscosity constants

Technique	*Critical exponent*
k_{33}, Freedericksz transition	0.65[63] 0.5 for pure samples increasing to 1.0 for samples with 1% impurity.[80] The exponent 1.0 is attributed to a possible 2-dimensional behaviour[73]
k_{22}, Rayleigh scattering	0.66; (the critical regime $q\xi > 1$ was also studied in this experiment[81]
Penetration depth = $(k_{11}/B)^{1/2}$ from a study of undulation instability under dilatation	0.15[82]
k_{22}, Freedericksz transition	0.5[83]
k_{22}, Rayleigh scattering	0.5[61]
λ_1, rotating magnetic field	1.07; $\Delta\lambda_1 = 0.2$ poise at $T_{AN} + 0.5$[84]
λ_1, dynamics of Freedericksz transition	0.35; $\Delta\lambda_1 = 0.4$ poise at $T_{AN} + 0.5$[85]
λ_1, NMR relaxation time	0.4 ± 0.1[86]
Ultrasonic attenuation	helium-like[87]
τ_n, light beating spectroscopy	mean field[88]
τ_n, light beating spectroscopy	mean field[61]

BBOA

Weakly first order A–N transition	($\simeq$80 cals/mole)
k_{33}, Freedericksz transition	0.65: helium-like[89]
λ_1, dynamics of Freedericksz transition	helium-like rather than mean field. In this experiment, an unexplained anomalous pre-transition increase in k_{11} has also been detected[90]
λ_1, rotating magnetic field	0.6[84]
τ_n, light beating spectroscopy	mean field[88]

Technique	*Critical exponent*
Anisotropy of ultrasonic attenuation $\Delta\alpha$	A maximum in the anisotropy as a function of temperature is noted in the nematic phase. For $\nu \simeq 500$ MHz, $\Delta\alpha/\nu^2 \sim \Delta\eta \sim \Delta\lambda_1$, and the maximum is attributed to a Landau–Khalatnikov process (see 14.3.69) and a rather long τ_ψ.[91]

8OCB

This is a particularly stable compound exhibiting a very weak first order A–N transition ($\simeq$10 cals/mole). Light beating spectroscopy has been used to study both k_{22} and τ_n.[92] A stabilizing AC electric field was also used to vary the relaxation time. In the presence of the field, the relaxation time is given by

$$\tau_E = \frac{\lambda_1}{k_{22}q^2 + (\epsilon_a/4\pi)E^2}$$

where E is the rms field and ϵ_a the dielectric anisotropy. From $\tau(E)$ versus E, λ_1 could be evaluated independently of k_{22}. The results for k_{22}, τ_n and λ_1 favour the mean field theory.

The only compound for which the smectic A–cholesteric transition has been investigated quantitatively is cholesteryl nonanoate. Δk_{22}, determined from the divergence of the pitch,[93] as well as ultrasonic attenuation measurements[94] favour a helium-like behaviour.

The most sensitive of all these techniques is Rayleigh scattering (together with light beating spectroscopy) and the results for the A–N transition in all the three compounds investigated indicate a mean field behaviour. As discussed in Section 14.2.2, a possible reason for this is that a/ξ_0 is small. X-ray scattering studies in the nematic phases of some alkyl cyanobiphenyls (which are similar to 8OCB) reveal smectic-like correlations extending to some 4–5 layers;[49] thus a/ξ_0 may in fact be quite small ($\sim$1/5). A complicating feature in both CBOOA and 8OCB is that they are very strongly polar (with a dipole moment of about 4 debye along the long molecular axis) and there is strong evidence for antiparallel correlation between neighbours giving rise to 'double' molecular layers (see Section 14.2.3). Moreover, as dipolar forces are not short-ranged, the mean field theory would be favoured.

Yet another possibility has been suggested by Chu and McMillan.[61] It is well known that mixtures of ^{3}He and ^{4}He exhibit classical behaviour near the tricritical point. Thus, if the A–N transition in CBOOA (and in the other compounds) is located near the tricritical point, the mean field theory may indeed be expected to be valid.

14.4 The smectic C–smectic A transition

In smectic C, the molecules are disordered within the layers (as in smectic A) but inclined with respect to the layer normal (Figure 14.1.4*b*). The orientation of the director in such a structure is specified by two angles, the tilt angle ω and the azimuthal angle ϕ (Figure 14.4.1). The former is coupled with the layer thickness while the latter is not, and hence the ϕ-oscillations of the director can be quite large. If we define a unit vector **u** to represent the preferred orientation of the projection of the molecules on the basal (xy) plane, it is clear that **u** may be compared with the nematic director **n**, the only difference being that since **u** and −**u** correspond to tilts in opposite directions they are not equivalent. Thus the smectic C bears some interesting similarities to the nematic liquid crystal.[95] For example, as fluctuations of **u** are large, smectic C is quite turbid. It also exhibits optical textures which resemble the schlieren textures of a nematic.

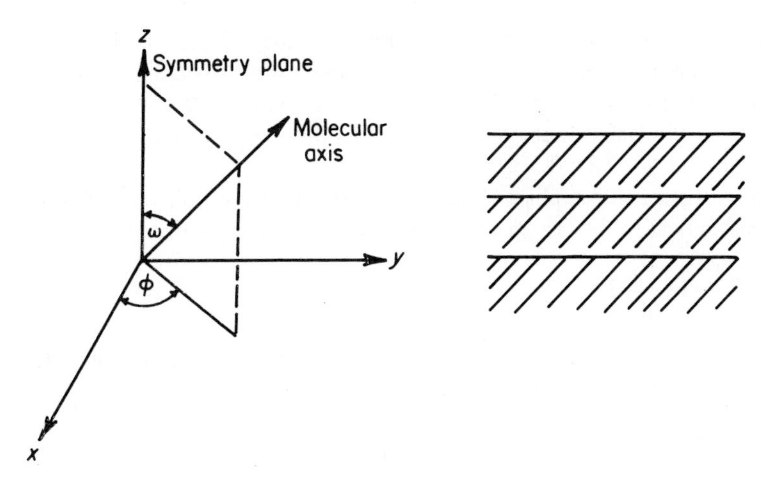

Figure 14.4.1 Diagram showing the tilt angle ω and the azimuthal angle ϕ in the smectic C structure

If the smectic C phase is followed by an A phase, the tilt angle ω decreases gradually and finally becomes zero at the C–A transition point (Figure 14.4.2). To discuss this transition, it is obvious that the order parameter requires two components, ω and ϕ, and may therefore be written as

$$\chi = \omega \exp(\mathrm{i}\phi). \tag{14.4.1}$$

This again brings out the analogy with superfluids.[96] The free energy may be written as before in the form

$$F = e(T)\,|\chi|^2 + \tfrac{1}{2}f(T)\,|\chi|^4 + \text{gradient terms}. \tag{14.4.2}$$

Using scaling arguments, de Gennes[96] has predicted that in the ordered phase, $T < T^*_{\mathrm{CA}}$,

$$\omega = |\chi| \sim (\Delta T)^{\beta} \tag{14.4.3}$$

where $\Delta T = T^*_{\mathrm{CA}} - T$, T^*_{CA} being the second order transition point, and $\beta \simeq 0.35$. In the mean field approximation, $\beta \simeq 0.5$.

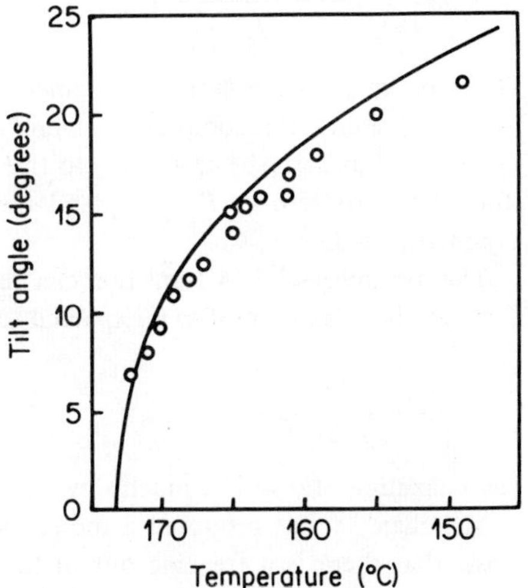

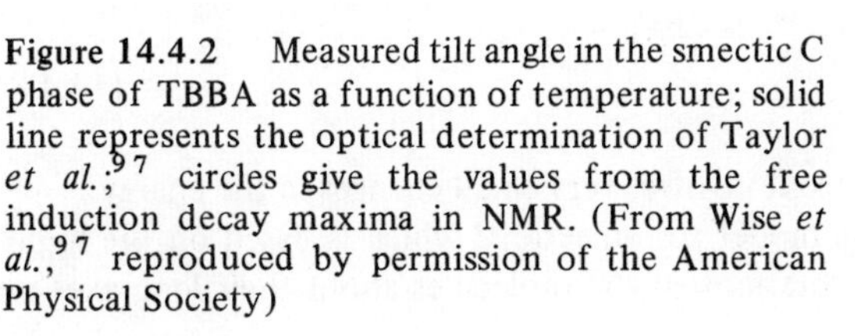

Figure 14.4.2 Measured tilt angle in the smectic C phase of TBBA as a function of temperature; solid line represents the optical determination of Taylor *et al.*;[97] circles give the values from the free induction decay maxima in NMR. (From Wise *et al.*,[97] reproduced by permission of the American Physical Society)

There have been two experimental determinations of β. From an NMR study of the free induction decay times measured as a function of the angle of rotation of an oriented sample with respect to the magnetic field, Wise *et al.*[97] have found $\beta = 0.40 \pm 0.04$ for TBBA. On the other hand, from the NMR doublet splitting of the probe molecule CH_2Cl_2 in TBBA, Luz and Meiboom[98] have reported $\beta = 0.5 \pm 0.1$.

Above T^*_{CA}, the smectic A phase should show some pretransition anomalies. For example an oblique magnetic field in the (yz) plane should induce a tilt angle given by

$$\omega \propto \frac{H_y H_z}{k_B T^*_{CA}} \left(\frac{T^*_{CA}}{\Delta T}\right)^{\gamma}. \tag{14.4.4}$$

There should also be some contribution to the light scattering due to fluctuations of ω, the intensity (in the hydrodynamic regime $q\xi \ll 1$) being proportional to

$$|\chi(q)|^2 \simeq \frac{k_B T^*_{CA}}{2e(T)} \sim (\Delta T)^{-\gamma}. \tag{14.4.5}$$

Ribotta *et al.*[99] have observed that a compressive stress applied normal to the layers can induce an A–C transition when the stress exceeds a threshold value. This is due to the coupling between the layer thickness and tilt angle. Assuming that a is the undistorted layer thickness in the A phase, a tilt of the molecules by an angle ω produces a change in the layer thickness which can now be taken to be $a(1 - \alpha\omega^2)$ where α is a coefficient depending on the shape of the molecule. Adding this contribution (see 14.3.11) to (14.4.2), the free energy of the strained medium may be written as

$$F = \frac{1}{2} B\left[\left(\frac{\partial u}{\partial z}\right)^2 + 2\alpha\left(\frac{\partial u}{\partial z}\right)\omega^2\right] + e\omega^2 + \frac{1}{2} f\omega^4 \tag{14.4.6}$$

where $\omega = |\chi|$ is taken as the order parameter. For a second order C–A transition $f > 0$, and in the A phase $e > 0$ and $\omega = 0$. Hence, a negative strain $(\partial u/\partial z)$ with a value larger than the threshold given by

$$\left(\frac{\partial u}{\partial z}\right)_{th} = -\frac{e}{\alpha B} \propto (T - T^*_{CA})^{\gamma} \tag{14.4.7}$$

induces a transition to the smectic C phase with a tilt

$$\omega = \left(-\frac{e + B\alpha(\partial u/\partial z)}{f}\right)^{1/2}.$$

The experiments were done on a homeotropically aligned sample of p-n-heptyloxybenzylidene-p-n-heptylaniline, the compression being applied by means of a piezoelectric ceramic driven by square-wave pulses. The transition to the smetic C phase was detected optically. The threshold strain tends to zero as $T \to T_{CA}$ and the data appear to be compatible with the mean field exponent, $\gamma = 1$.

The dynamics of C–A transition can be worked out as in the previous section. In the smectic C phase, the relaxation of ω is expected to follow a law of the type

$$\tau = \tau_0 \left(\frac{T_{CA}}{\Delta T}\right)^{\gamma}. \tag{14.4.8}$$

The relaxation of ϕ will be much slower, since it involves very small changes in the energy.

McMillan[100] has proposed a molecular model for smectic C which is based on the hypothesis that there is a freezing out of the rotations of the molecules about their long axes on passing from the A to the C phase. The molecule is supposed to have two antiparallel dipoles, located at the two ends, each having a transverse component μ_t and a longitudinal component μ_l. The model predicts a second order rotational transition, with all the dipoles within each layer becoming oriented in the lower-temperature phase. (The interlayer interactions are ignored.) The freezing of the rotation is in itself enough to make the medium biaxial. However, the presence of the vertical component μ_l of the dipole moment causes the molecules to tilt because of the torque exerted by the internal electric field. The tilt angle ω is given by

$$\omega \propto \frac{k_B T^*_{CA}}{k} \frac{\mu_l}{\mu_t} \eta \tag{14.4.9}$$

where k is an elastic constant characteristic of the A phase and

$$\eta = \langle \cos \phi \rangle \tag{14.4.10}$$

is a dipolar order parameter. The variation of ω near T^*_{CA} obeys the classical mean field theory, i.e. $\beta = 0.5$. It should be emphasized however that to date there is no experimental evidence of the freezing of the molecular rotations in the C phase.

Recently, Priest[101] has developed a theory of the C–A transition based on the second rank tensor orientational properties of the medium. The free energy expression includes, in addition to terms of the type (14.2.5) valid for nematics, terms which also involve tensors composed of components of intermolecular vectors. The formal theory leads to the result that the C–A transition can be of second order and, moreover, does not prohibit free rotation about the molecular long axis in the smectic C phase.

14.5 The smectic C–nematic transition

In principle, the C–N transition can be continuous. It is much more complicated than the A–N or the C–A transition since it involves layer formation as also a director tilt with reference to the layer normal. A pretransitional smectic C type of short range ordering occurs in the nematic phase which was first observed in X-ray studies[3] and designated by de Vries[3] as skew cybotactic groups. A Landau type of theory for these fluctuations has been proposed by de Gennes.[58] If $q_s = 2\pi/a$, a being the layer spacing, the smectic-like short range order builds up with the Fourier components of the density given by (see Figure 14.5.1)

$$K_z = q_1 = q_s \cos \omega$$

and

$$q_T = q_s \sin \omega \quad (q_T^2 = K_x^2 + K_y^2)\,. \tag{14.5.1}$$

The order parameter which is again a density wave is given by

$$\psi(\phi) = \rho(K_z = q_1, K_x = q_T \cos \phi, K_y = q_T \sin \phi) \tag{14.5.2}$$

and since ϕ can vary between 0 and 2π continuously, $\psi(\phi)$ is an order parameter with an infinite number of components, in principle. Taking the projection of q_s on the (xy) plane to be along the direction specified by the unit vector $\mathbf{u}(\cos \phi, \sin \phi, 0)$ the free energy may be written in the form (for the director oriented along z)

$$F = \sum_{\phi} \alpha \,|\, \psi_\phi |^2 + \frac{1}{2M_z}\left(\frac{\partial \psi_\phi}{\partial z}\right)^2 + \frac{1}{2M_u}\left(\frac{\partial \psi_\phi}{\partial u}\right)^2 + \frac{1}{2M_{uz}}\left(\frac{\partial \psi_\phi}{\partial z}\right)\left(\frac{\partial \psi_\phi}{\partial u}\right) + \frac{1}{2} g \left(\frac{\partial^2 \psi_\phi}{\partial v^2}\right)^2 \tag{14.5.3}$$

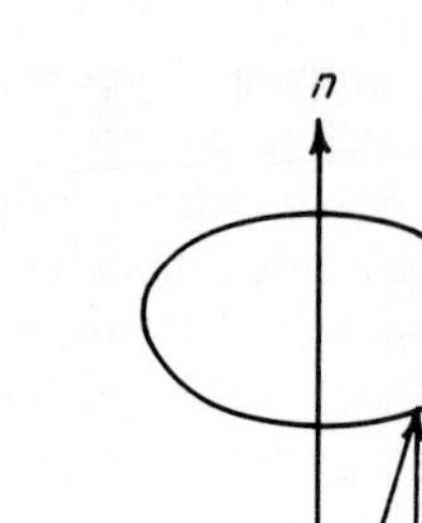
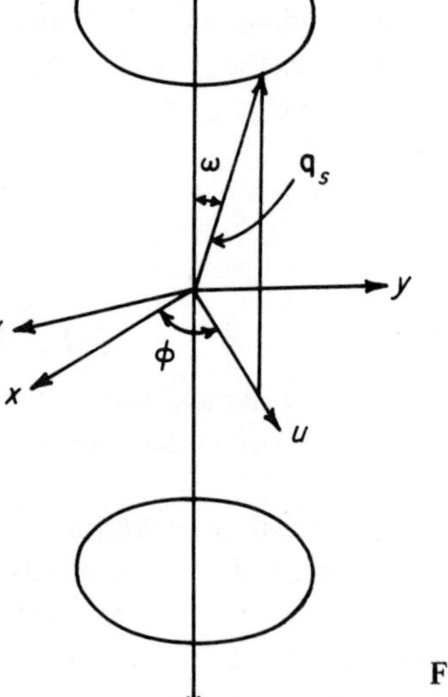

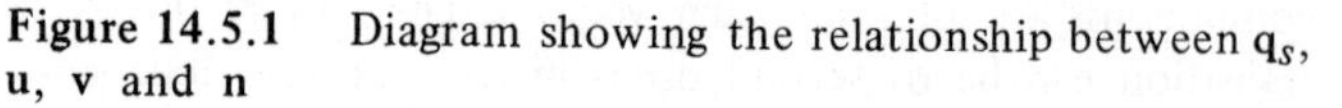

Figure 14.5.1 Diagram showing the relationship between $\mathbf{q}_s$, **u**, **v** and **n**

where the summation symbol over ϕ indicates the significance of ϕ as an index for components. A noteworthy feature of (14.5.3) is that contributions to the free energy from quantities of the order $(\nabla\psi_\phi)^2$ are confined to the vertical plane containing q_s and u. The term corresponding to gradients along v does not come in this order, since it merely represents a shift in ϕ. (Equivalently, the coefficient M in the v direction is assumed to have an infinite value.) The coefficients M_z, M_u and M_{zu} are expected to have orders of magnitude similar to the corresponding quantities in smectic A: they describe layer compressions which are involved in any fluctuations of q_s in the (zu) plane. The last term in (14.5.3) is analogous to the curvature term in (14.3.11) for smectic A. An important consequence of this form of the free energy functional is that the correlation lengths along z and u directions are of the usual form, viz. $\xi = (2M\alpha)^{-1/2}$, but that along v is reduced significantly. It is of the order of $(\xi a)^{1/2}$, where a is the layer thickness. $[(Mg)^{1/2}$ is comparable to the penetration depth Λ defined by (14.3.21).] Hence the volume of the correlated region turns out to be $\xi^2(\xi a)^{1/2}$ in this model and not ξ^3 as in smectic A.

To calculate the enhancement in the elastic constants of the nematic close to the C–N transition point, we have to introduce the director fluctuations as before. Taking δn_u and δn_v as the two independent components of the director fluctuation, the resulting fluctuations in q_s are:

$$\begin{aligned}\delta q_{su} &= q_1\,\delta n_u\\ \delta q_{sv} &= q_1\,\delta n_v\\ \delta q_{sz} &= q_T\,\delta n_u\,.\end{aligned}\tag{14.5.4}$$

Using these we can now work out the contribution to the elastic coefficients. Since the smectic C has tilted molecules in the layers, it is easy to see that, in general, even a splay deformation does not maintain the layer thickness. Hence all the three elastic constants are renormalized in the nematic phase near the C–N transition. Since the coherence volume is taken to be

$\xi^2(\xi a)^{1/2}$, we have

$$\langle |\psi^2| \rangle \simeq \frac{k_B T}{\alpha \xi^2 (\xi a)^{1/2}} \tag{14.5.5}$$

or

$$\Delta k \sim \frac{q_s^2 \langle |\psi|^2 \rangle \xi^2}{M} \sim \text{const. x } \xi^{3/2} \tag{14.5.6}$$

whereas $\Delta k > \xi$ near the A–N transition. Experimentally, k_{11} and k_{33} have been measured near the C–N transition but no quantitative estimates of the critical exponents have yet been made.[62]

Chu and McMillan[102] have recently constructed a unified Landau theory for the nematic, smectic A and the smectic C phases. Equations (14.3.8) and (14.3.20) give the contributions to the free energy functional from the nematic and the smectic A phases. The smectic C free energy is written down in terms of the dipolar order parameter η given by (14.4.10). It is analogous to (14.4.2), the gradient terms involving the curvature strains of η as in nematics, and being multiplied by $|\psi|^2$ to take into account the layering (η is supposed to exist only because of the layered arrangement). In contrast to de Gennes' formulation, the value of M in the v direction is now finite and leads to $\Delta k \propto \xi$ as in the case of the A–N transition. Also, the predicted structure factor in the nematic phase, which is proportional to $\langle |\psi_q|^2 \rangle$ fits much better with the experimental data for p-n-heptyloxy azoxybenzene (HAB)[103] than that derived from de Gennes' theory.

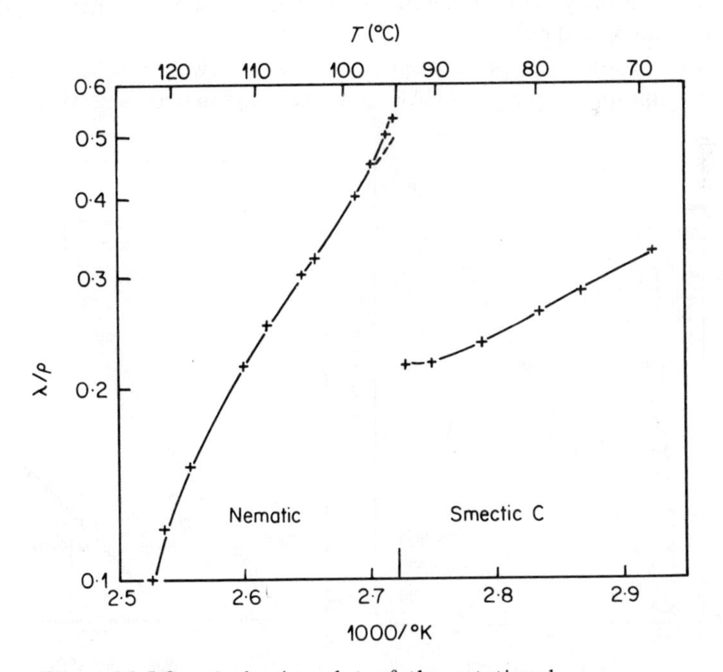

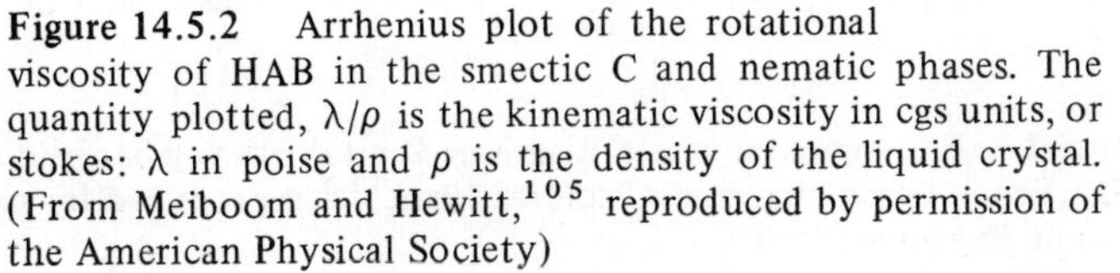
Figure 14.5.2 Arrhenius plot of the rotational viscosity of HAB in the smectic C and nematic phases. The quantity plotted, λ/ρ is the kinematic viscosity in cgs units, or stokes: λ in poise and ρ is the density of the liquid crystal. (From Meiboom and Hewitt,[105] reproduced by permission of the American Physical Society)

McMillan[104] has also presented a dynamical version of this theory. Some of the viscosity coefficients of the nematic phase (e.g. λ_1) become renormalized near the C–N transition (with $\Delta\lambda_1 \propto \xi$ in accordance with the mean field theory). However, λ_1 in the smectic C phase itself is not renormalized. Physically this is easily understood: once the long range layer order sets in, the twist viscosity of smectic C is low because fluid motion can take place parallel to the layers. This effect has been strikingly demonstrated in a recent experiment by Meiboom and Hewitt,[105] who measured the rotational viscosity of HAB by oscillating the sample in a magnetic field. Their results are shown in Figure 14.5.2.

14.6 Phase diagrams

14.6.1 Triple points

We conclude this chapter with a brief discussion of phase diagrams in liquid crystal systems.

Figure 14.6.1*a* presents the P–T diagram for p-n-ethoxybenzoic acid in which can be seen the solid–nematic–isotropic and the solid–smectic–nematic triple points.[106] This experiment was undertaken to verify a prediction of a theory of melting of molecular crystals,[107] namely, that mesomorphism can be induced by pressure in materials that are non-mesomorphic at atmospheric pressure. Studies were conducted on the first two members of the p-n-alkoxy-benzoic acid series. These two compounds, methoxy- and ethoxybenzoic acids, do not form liquid crystals at atmospheric pressure, whereas the third and higher homologues do. However, as the pressure is raised both the compounds exhibit mesophases, initially a nematic phase, and at higher pressures a smectic phase as well. The phase transitions were detected by differential thermal analysis, and the pressure induced mesophases identified by microscopic observations using a high pressure optical cell.

Figure 14.6.1*b* presents the phase diagram for HAB showing a solid–smectic C–nematic triple point. At atmospheric pressure, HAB shows two mesophases; smectic C and nematic. As

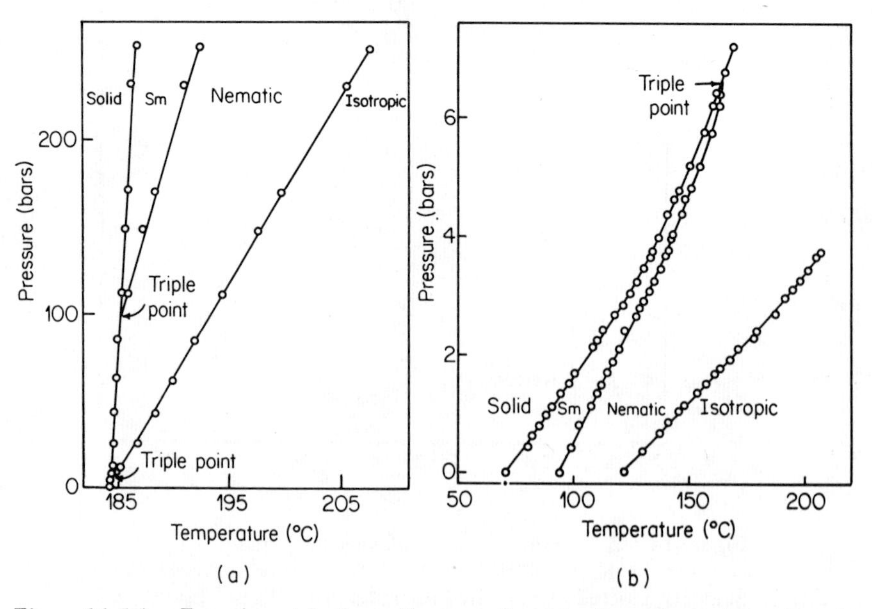

Figure 14.6.1 Experimental phase diagrams for (a) ethoxy-benzoic acid, and (b) HAB. (From Shashidhar and Chandrasekhar,[110] reproduced by permission of Journal de Physique, Paris)

the pressure is raised the temperature range of smectic C decreases, and above a certain pressure it disappears altogether. At the molecular level, a possible explanation for the suppression of the smectic C phase is that the tilt angle keeps increasing with pressure until the smectic layering collapses and only the nematic ordering remains. This question can be decided by a direct determination of the variation of the layer thickness with pressure.

14.6.2 Tricritical point

While discussing McMillan's molecular theory of smectic A, we noted that the order of the A–N transition depends on the value of a non-dimensional parameter α, which is related to the chain length of the molecule; for $0.70 \leqslant \alpha \leqslant 0.98$, the A–N transition is first order, while for $\alpha \leqslant 0.70$ it is second order. Thus $\alpha = 0.70$ represents a tricritical point.[108] In practice, α cannot be varied continuously and the location of the tricritical point by this method becomes difficult. The first observation of tricritical behaviour in liquid crystal systems was made by Keyes *et al.*[109] in their pressure studies on cholesteryl oleyl carbonate. From optical transmission measurements they concluded that the smectic A–cholesteric transition in this compound becomes second order (or very nearly so) at 2.66 kbar and 60.3 °C. This result has since

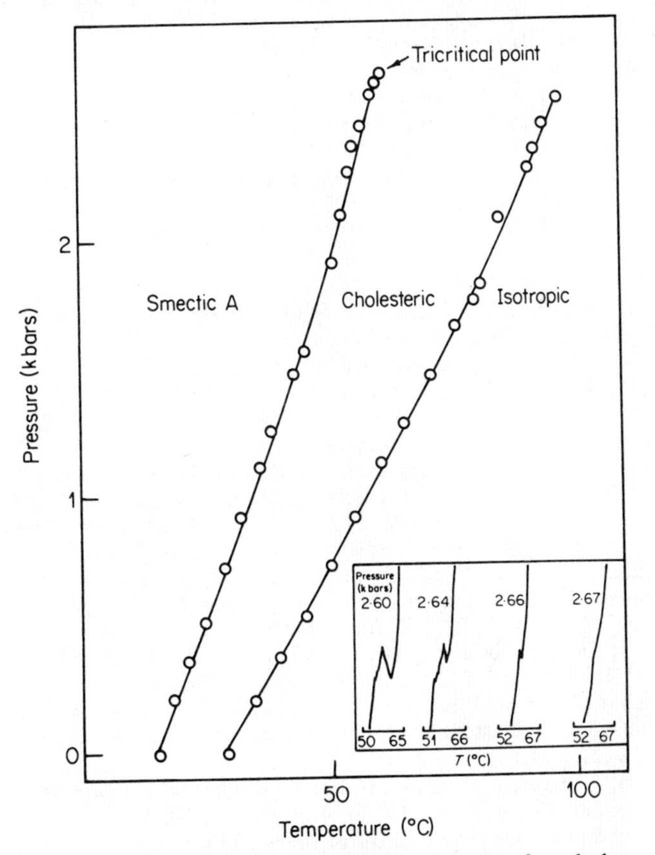

Figure 14.6.2 Experimental phase diagram for cholesteryl oleyl carbonate. Inset shows raw differential thermal analysis traces of the smectic A–cholesteric transition at different pressures in the vicinity of the tricritical point. (From Shashidhar and Chandrasekhar,[110] reproduced by permission of Journal de Physique, Paris)

been confirmed[110] using a more sensitive technique for detecting the phase transitions, viz. differential thermal analysis. The phase diagram is shown in Figure 14.6.2. More recently, a study of the orientational order parameter of p-cyano-benzylidene-p-nonylaniline as a function of pressure has shown that the discontinuity in the order parameter at T_{AN} decreases progressively with rise of pressure until it disappears altogether at the tricritical point.[111]

Tricritical behaviour is known to occur in $^3He-^4He$ mixtures. Alben[112] has suggested that the tricritical point in liquid crystal systems may similarly be found by making suitable mixtures. Calorimetric studies on mixtures of the second and fourth homologues of N–(alkoxy-benzylidene)-p-n-alkyl anilines appear to show such a behaviour.[113] The importance of trying to locate the tricritical point lies in the fact that the critical exponents near this point are expected to be different from those near an ordinary critical point.[112]

References

1. Lehmann, O. (1904). *Flussige Kristalle*, Engelmann, Leipzig.
2. Friedel, G. (1922). *Ann. Physique,* **18**, 273.
3. Chistyakov, I. G. and Chaikowsky, W. M. (1969). *Mol. Cryst. Liquid Cryst.,* **7**, 269.
   de Vries, A. (1970). *Mol. Cryst. Liquid Cryst.,* **10**, 219.
4. Frank, F. C. (1958). *Disc. Faraday Soc.,* **25**, 19.
5. Leslie, F. M. (1966). *Quart. J. Mech. Appl. Math.,* **19**, 357.
   Leslie, F. M. (1968). *Arch. Rational Mech. Anal.,* **28**, 265.
6. Parodi, O. (1970). *J. de Physique,* **31**, 581.
7. Saupe, A. (1968). *Angew Chem., Int. Edition,* **7**, 97.
8. Demus, D. and Demus, H. (1973). *Flussige Kristalle in Tabellen,* VEB Deutscher Verlag Für Grundstoffindustrie, Leipzig, 1973.
   de Vries, A. (1973). *Proc. Int. Liquid Crystals Conf.,* Bangalore, 1973, *Pramana Supplement* **1**, p. 93.
9. Helfrich, W. (1969). *Phys. Rev. Lett.,* **23**, 372.
10. Martin, P. C., Parodi, O. and Pershan, P. S. (1972). *Phys. Rev.,* **A26**, 2401.
11. Arnold, H. (1964). *Z. Phys. Chem.,* **226**, 146.
12. McMillan, W. L. (1973). *Phys. Rev.,* **A7**, 1419.
13. Taylor, T. R., Arora, S. L. and Fergason, J. L. (1970). *Phys. Rev., Lett.,* **25**, 722.
    Schnur, J. M., Sheridan, J. P. and Fontana, M. (1973). *Proc. Int. Liquid Crystals Conf.,* Bangalore, 1973. *Pramana Supplement,* **1**, p. 175.
14. Brown, G. H., Doane, J. W. and Neff, V. D. (1971). *A Review of the Structure and Physical Properties of Liquid Crystals,* Butterworths, London.
    de Gennes, P. G. (1974) *The Physics of Liquid Crystals,* Clarendon Press, Oxford.
    Stephen, M. J. and Straley, J. P. (1974). *Rev. of Mod. Phys.,* **46**, 617.
    Chandrasekhar, S. (1976). *Liquid Crystals,* Cambridge University Press, Cambridge.
    Chandrasekhar, S. (1976). Liquid Crystals. *Rep. Prog. Phys.,* **39**, 613.
15. Saupe, A. and Maier, W. (1961). *Z. Naturforsch.,* **16a**, 816.
    Gasparoux, H., Regaya, B. and Prost, J. (1971). *C.R. Acad. Sci.,* **272B**, 1168.
16. Chatelain, P. (1955). *Bull. Soc. franc. Miner. Crist.,* **78**, 262.
17. Chandrasekhar, S. and Madhusudana, N. V. (1969). *J. de Physique,* **30**, C4–24.
    Madhusudana, N. V., Shashidhar, R. and Chandrasekhar, S. (1972). *Mol. Cryst. Liquid Cryst.,* **16**, 21.
18. Spence, R. D., Gutowsky, H. S. and Holon, C. H. (1953). *J. Chem. Phys.,* **21**, 1891.
    Lippmann, H. and Weber, K. H. (1957). *Ann. Physik,* **20**, 265.
19. Maier, W. and Saupe, A. (1958). *Z. Naturforsch.,* **13a**, 564.
    Maier, W. and Saupe, A. (1959). *Z. Naturforsch.,* **14a**, 882.
    Maier, W. and Saupe, A. (1960). *Z. Naturforsch.,* **15a**, 287.
20. McColl, J. R. and Shih, C. S. (1972). *Phys. Rev. Lett.,* **29**, 85.
21. Cotter, M. A. Private communication.
22. Chandrasekhar, S. and Madhusudana, N. V. (1971). *Acta Cryst.,* **A27**, 303.
    Chandrasekhar, S., Madhusudana, N. V. and Shubha, K. (1971). *Faraday Soc. Symp.,* No. 5, 26.

23. Humphries, R. L., James, P. G. and Luckhurst, G. R. (1972). *J. Chem. Soc., Faraday Trans. II,* **68**, 1031.
24. Marcelja, S. (1974). *J. Chem. Phys.,* **60**, 3599.
25. Straley, J. P. (1973). *Mol. Cryst. Liquid Cryst.,* **22**, 333.
26. Zadoc-Kahn, J. (1936). *Ann. Physique,* Series II, **6**, 455.
    Foex, G. (1933). *Trans. Faraday Soc.,* **29**, 958.
27. de Gennes, P. G. (1971). *Mol. Cryst. Liquid Cryst.,* **12**, 193.
28. Landau, L. D. and Lifshitz, E. M. (1969). *Statistical Physics,* 2nd Ed. Pergamon, Oxford.
29. Stinson, T. W. and Litster, J. D. (1970). *Phys. Rev. Lett.,* **25**, 503.
30. Madhusudana, N. V. and Chandrasekhar, S. (1974). In J. F. Johnson and R. S. Porter (Eds), *Liquid Crystals and Ordered Fluids,* Vol. 2, Plenum, New York, p. 657.
31. Tsvetkov, V. N. and Ryumtsev, E. I. (1968). *Sov. Phys. Crystallogr.,* **13**, 225.
32. Ratna, B. R., Vijaya, M. S., Shashidhar, R. and Sadashiva, B. K. (1973). *Proc. Int. Liquid Crystals Conf.,* Bangalore. 1973. *Pramana Supplement,* **1**, p. 69.
33. Stinson, T. W., Litster, J. D. and Clark, N. A. (1972). *J. de Physique,* **33**, C1–69.
    Gulari, E. and Chu, B. (1975). *J. Chem. Phys.,* **62**, 798.
34. Courtens, E. and Koren, G. (1975). *Phys. Rev. Lett.,* **35**, 1711.
35. Martinoty, P., Candau, S. and Debeauvais, F. (1971). *Phys. Rev. Lett.,* **27**, 1123.
36. Stinson, T. W. and Litster, J. D. (1973). *Phys. Rev., Lett.,* **30**, 688.
37. Cheng, J. and Meyer, R. B. (1974). *Phys. Rev.,* **A9**, 2744.
38. Fisher, M. E. (1967). *Rep. Prog. Phys.,* **30**, 615.
39. Hohenberg, P. C. (1968). Lecture at *Conference on Fluctuations in Superconductors,* Asilonar, California, 1968. 40. Madhusudana, N. V. and Chandrasekhar, S. (1973).
40. Madhusudana, N. V. and Chandrasekhar, S. (1973). *Solid St. Commun.,* **13**, 377.
41. Madhusudana, N. V. and Chandrasekhar, S. (1973). *Pramana,* **1**, 12.
42. Madhusudana, N. V. and Chandrasekhar, S. (1973). *Proc. Int. Liquid Crystals Conf.,* Bangalore, 1973. *Pramana Supplement,* **1**, p. 57.
43. Madhusudana, N. V., Savithramma, K. L. and Chandrasekhar, S. (1976). *Pramana* (In press).
44. Ypma, J. G. J. and Vertogen, G. (1976). *Solid St. Commun.,* **18**, 475.
    Ypma, J. G. J. and Vertogen, G. To be published.
    The method is similar to that developed for magnetic systems by Weiss, P. R. (1948). *Phys. Rev.,* **74**, 1493.
45. Bethe, H. A. (1935). *Proc. Roy. Soc.,* **149**, 1.
46. Chang, T. S. (1937). *Proc. Camb. Phil. Soc.,* **33**, 524.
47. Krieger, T. J. and James, H. M. (1954). *J. Chem. Phys.,* **22**, 796.
48. Schadt, M. (1972). *J. Chem. Phys.,* **56**, 1494.
    Ratna, B. R. and Shashidhar, R. (1976). *Pramana,* **6**, 278.
49. Leadbetter, A. J., Richardson, R. M. and Colling, C. N. (1975). *J. de Physique,* **36**, C1–37.
50. Lydon, J. E. and Coakley, C. J. (1975). *J. de Physique,***36**, C1–45.
51. McMillan, W. L. (1971). *Phys. Rev.,* **A4**, 1238.
52. Kirkwood, J. G. and Monroe, E. (1941). *J. Chem. Phys.,* **9**, 514.
53. Kobayashi, K. (1971). *Mol. Cryst. Liquid Cryst.,* **13**, 137.
54. McMillan, W. L. (1972). *Phys. Rev.,* **A6**, 936.
55. Doane, J. W., Parker, R. S., Cvikl, B., Johnson, J. L. and Fishel, D. L. (1972). *Phys. Rev. Lett.,* **28**, 1694.
56. Cabane, B. and Clark, W. G. (1973). *Solid St. Commun.,* **13**, 129.
57. de Gennes, P. G. (1969). *J. de Physique,* **30**, C4–65.
58. de Gennes, P. G. (1972). *Solid St. Commun.,* **10**, 753.
    de Gennes, P. G. (1973). *Mol. Cryst. Liquid Cryst.,* **21**, 49.
59. See, e.g., Lynton, E. A. (1964). *Superconductivity,* 2nd Ed. Methuen, London.
60. Halperin, B. I., Lubensky, T. C. and Ma, S. K. (1974). *Phys. Rev. Lett.,* **32**, 292.
    Halperin, B. I. and Lubensky, T. C. Private communication.
61. Chu, K. C. and McMillan, W. L. (1975). *Phys. Rev.,* **A11**, 1059.
62. Gruler, H. (1973). *Z. Naturforsch.,* **28a**, 474.
63. Cheung, L., Meyer, R. B. and Gruler, H. (1973). *Phys. Rev. Lett.,* **31**, 349.
64. McMillan, W. L. (1974). *Phys. Rev.,* **A9**, 1720.
65. Jahnig, F. (1973). *Proc. Int. Liquid Crystals Conf.,* Bangalore, *Pramana Supplement,* **1**, p. 31.

66. Jahnig, F. and Brochard, F. (1974). *J. de Physique,* **35**, 301.
67. Schmidt, H. (1968). *Z. Phys.,* **216**, 336.
68. Ferrell, R. H. (1970). In J. I. Budnick and M. P. Kawatra (Eds), *Dynamical Aspects of Critical Phenomena,* Gordon & Breach, London.
69. Brochard, F. (1973). *J. de Physique,* **34**, 411.
70. Landau, L. and Khalatnikov, I. (1954). *Dokl. Acad. Nauk. SSSR,* **96**, 469.
71. Lubensky, T. C. (1975). *J. de Physique,* **36**, C1–151.
72. Fergason, J. L. (1964). *Scientific American,* **211**, 77.
73. Jahnig, F. (1976). (Private communication).
74. Miyano, K. and Ketterson, J. B. (1973). *Phys. Rev. Lett.,* **31**, 1047.
75. Liao, Y., Clark, N. A. and Pershan, P. S. (1973). *Phys. Rev. Lett.,* **30**, 639.
76. Orsay Liquid Crystals Group (1969). *J. Chem. Phys.,* **51**, 816.
77. Hardouin, F., Gasparoux, H. and Delhaes, P. (1975). *J. de Physique,* **36**, C1–127.
78. Djurek, D., Baturic-Rubcic and Franulovic, K. (1974). *Phys. Rev. Lett.,* **33**, 1126.
79. Torza, S. and Cladis, P. E. (1974). *Phys. Rev. Lett.,* **32**, 1406.
80. Cladis, P. E. (1974). *Phys. Lett.,* **48A**, 179.
81. Delaye, M., Ribotta, R. and Durand, G. (1973). *Phys. Rev. Lett.,* **31**, 443.
82. Ribotta, R. Private communication.
83. Madhusudana, N. V., Karat, P. P. and Chandrasekhar, S. (1973). *Proc. Int. Liquid Crystals Conf.,* Bangalore, 1973. *Pramana Supplement,* **1**, p. 225.
84. Gasparoux, H., Hardouin, F., Achard, M. F. and Sigaud, G. (1975). *J. de Physique,* **36**, C1–109.
85. Huang, C. C., Pindak, R. S., Flanders, P. J. and Ho, J. T. (1974). *Phys. Rev. Lett.,* **33**, 400.
86. Wise, R. A., Olah, A. and Doane, J. W. (1975). *J. de Physique,* **36**, C1–117.
87. Bacri, J. C. (1975). *J. de Physique,* **36**, C1–123.
88. Salin, D., Smith, I. W. and Durand, G. (1974). *J. de Physique,* **35**, L-165.
89. Leger, L. (1973). *Phys. Lett.,* **44A**, 535.
90. D'Humieres, D. and Leger, L. (1975). *J. de Physique,* **36**, C1-113.
91. Bacri, J. C. (1975). *J. de Physique,* **36**, L-177.
92. Delaye, M. (1976). *J. de Physique,* **37**, C3–99.
93. Pindak, R. S., Huang, C. C. and Ho, J. T. (1974). *Phys. Rev. Lett.,* **32**, 43.
94. Bacri, J. C. (1975). *J. de Physique,* **36**, L-259.
95. Saupe, A. (1969). *Mol. Cryst. Liquid Cryst.,* **7**, 59.
96. de Gennes, P. G. (1972). *Compt. Rend.,* **274B**, 758.
97. Wise, R. A., Smith, D. H. and Doane, J. W. (1973). *Phys. Rev.,* **A7**, 1366.
    Taylor, T. R., Fergason, J. L. and Arora, S. L. (1970). *Phys. Rev. Lett.,* **24**, 359.
98. Luz, Z. and Meiboom, S. (1973). *J. Chem. Phys.,* **59**, 275.
99. Ribotta, R., Meyer, R. B. and Durand, G. (1974). *J. de Physique,* **35**, L-161.
100. McMillan, W. L. (1973). *Phys. Rev.,* **8A**, 1921.
101. Priest, R. G. (1975). *J. de Physique,* **36**, 437.
102. Chu, K. C. and McMillan, W. L. Private communication.
103. McMillan, W. L. (1973). *Phys. Rev.,* **8A**, 328.
104. McMillan, W. L. Private communication.
105. Meiboom, S. and Hewitt, R. C. (1975). *Phys. Rev. Lett.,* **34**, 1146.
106. Chandrasekhar, S., Ramaseshan, S., Reshamwala, A. S., Sadashiva, B. K., Shashidhar, R. and Surendranath, V. (1973). *Proc. Int. Liquid Crystals Conf., Bangalore, Pramana Supplement,* **1**, 117.
107. Chandrasekhar, S., Shashidhar, R. and Tara, N. (1970). *Mol. Cryst. Liquid Cryst.,* **10**, 337.
108, Griffiths, R. B. (1970). *Phys. Rev. Lett.,* **24**, 715.
109. Keyes, P. H., Weston, H. T. and Daniels, W. B. (1973). *Phys. Rev. Lett.,* **31**, 628.
110. Shashidhar, R. and Chandrasekhar, S. (1975). *J. de Physique,* **36**, C1–49.
111. McKee, T. J. and McColl, J. R. (1975). *Phys. Rev. Lett.,* **34**, 1076.
112. Alben, R. (1973). *Solid St. Commun.,* **13**, 1783.
113. Johnson, D. L., Maze, C., Oppenheim, E. and Reynolds, R. (1975). *Phys. Rev. Lett.,* **34**, 1143.

Index